Advances in Experimental Medicine and Biology

Volume 1148

More information about this series at http://www.springer.com/series/5584

Nikolaos Labrou
Editor

Therapeutic Enzymes: Function and Clinical Implications

Springer

Editor
Nikolaos Labrou
Department of Biotechnology
Agricultural University of Athens
Athens, Greece

ISSN 0065-2598 ISSN 2214-8019 (electronic)
Advances in Experimental Medicine and Biology
ISBN 978-981-13-7708-2 ISBN 978-981-13-7709-9 (eBook)
https://doi.org/10.1007/978-981-13-7709-9

This Springer imprint is published by the registered company Springer Nature Singapore Pte Ltd.
The registered company address is: 152 Beach Road, #21-01/04 Gateway East, Singapore 189721, Singapore

Preface

Nowadays, the application of enzymes as pharmaceuticals is a growing field. Enzymes as pharmaceuticals have two fascinating features that distinguish them from all other types of drugs. First, enzymes bind their substrate with high affinity and specificity. Second, enzymes can convert their substrate to the desired products. These two features make enzymes valuable therapeutic tools, as, unlike common medicinal products which can temporarily solve the particular health problems, they address the underlying cause of health problem and the patient can achieve permanent relief. All these attributes make the enzymes potent drugs offering a platform and promising subcategory of modern biopharmaceuticals for the treatment of several severe diseases. Research and drug development efforts and the advancements in biotechnology over the past 20 years have greatly assisted the introduction of efficient and safe enzyme-based therapies for a range of both rare and common disorders. The introduction and regulatory approval of 20 different recombinant enzymes has enabled effective enzyme replacement therapy.

This volume aims to overview selected therapeutic enzymes, focusing in particular on more recently approved enzymes produced by recombinant DNA technology. It is impossible for a single book volume to cover all of the different aspects of this research area in which scientists have made significant progress. Thus, I have selected key examples covering a wide range of diverge scientific disciplines and state-of-the-art approaches, in order to provide the reader with a representative sample of the current status of the area. This volume highlights and provides an overview and recent progress on the three aspects of recombinant therapeutic enzymes and their clinical and pharmaceutical technology: (i) overview of the production process and biochemical characterization of therapeutic enzymes, (ii) focuses upon the engineering strategies and delivery methods of therapeutic enzymes, and (iii) clinical applications of selected therapeutic enzymes, including aspects on their mechanisms of action and information on safety and immunogenicity issues and various adverse events of the enzymes used for therapy.

This book is aimed at academics, researchers, and students undertaking advanced undergraduate/postgraduate programs in the biopharmaceutical/biotechnology area who wish to gain a comprehensive understanding of enzyme-based therapeutic molecules.

I sincerely hope that the readers will enjoy the information provided in this book and find its contents interesting and scientifically stimulating. I also hope that I have established a successful compilation of chapters within the exciting area of therapeutic enzymes. I wish to extend my sincere gratitude to all contributing authors for their enthusiasm and for the time they spent preparing the chapters for this book. Thanks also to those at Springer who have helped put this together so that it ultimately provides an invaluable resource to all those working in the field of therapeutic enzymes. I would especially like to thank my family for their understanding and patience during the editing and organization of the book chapters.

Athens, Greece Nikolaos Labrou

Contents

Production and Purification of Therapeutic Enzymes

1

M. Ângela Taipa, Pedro Fernandes, and Carla C. C. R. de Carvalho

Abstract

The use of therapeutic enzymes embraces currently a vast array of applications, abridging from diggestive disorders to cancer therapy, cardiovascular and lysosomal storage diseases. Enzyme drugs bind and act on their targets with great affinity and specificity, converting substrates to desired products in a reduced time frame with minimal side reactions. These characteristics have resulted in the development of a multitude of enzyme biopharmaceuticals for a wide range of human disorders.

The advances in genetic engineering and DNA recombination techniques facilitated the production of therapeutical human-like enzymes, using different cells as host organisms. The selection of hosts generally privileges those that secrete the enzyme into the culture medium, as this eases the purification process, and those that are able to express complex glycoproteins, with glycosylation patterns and other post-translational modifications close to human proteins. Moreover, engineering approaches such as pegylation, encapsulation in micro- and nanocarriers, and mutation of amino acid residues of the native enzyme molecule to yield variants with improved therapeutic activity, half-life and/or stability, have been also addressed. Engineered enzyme products have been designed to display enhanced delivery to target sites and reduced adverse side-effects (e.g., immunogenicity) upon continuous drug administration.

Irrespectively of the production method, the final formulation of therapeutic enzymes must display high purity and specificity, and they are often marketed as lyophilized pure preparations with biocompatible buffering salts and diluents to prepare the reconstituted aqueous solution before treatment.

Keywords

Therapeutic enzyme · Human therapy · Recombinant · Engineering · Production · Purification

M. Â. Taipa (✉) · C. C. C. R. de Carvalho
iBB-Institute for Biosciences and Bioengineering, Department of Bioengineering, Instituto Superior Técnico, Universidade de Lisboa, Lisbon, Portugal
e-mail: angela.taipa@tecnico.ulisboa.pt

P. Fernandes
iBB-Institute for Biosciences and Bioengineering, Department of Bioengineering, Instituto Superior Técnico, Universidade de Lisboa, Lisbon, Portugal

Faculty of Engineering, Universidade Lusófona de Humanidades e Tecnologias, Lisbon, Portugal

N. Labrou (ed.), *Therapeutic Enzymes: Function and Clinical Implications*, Advances in Experimental Medicine and Biology 1148, https://doi.org/10.1007/978-981-13-7709-9_1

Abbreviations

Ach	Acetylcholine
AGS	N-acetylgalactosamine 4-sulfatase
BSE	Bovine spongiform encephalomyelitis
CESD	Cholesteryl ester storage disease
CHO	Chinese hamster ovary
CocE	Cocaine esterase
CPD	Computational protein design
DE	Directed evolution
DNase	Deoxyribonuclease
EMEA	European Medicines Agency
FDA	Food and Drug Administration
hBche	Human butyrylcholinesterase
IC_{50}	Half-maximal inhibitory concentration
IMAC	Ion metal affinity chromatography
k_{cat}	Turnover number
K_M	Michaelis–Menten constant
L-ASNase	L-asparaginase
MPS	Mucopolysaccharidosis
PEG	Polyethylene glycol
PKU	Phenylketonuria
PMP	Plant-made pharmaceuticals
SCID	Severe combined immunodeficiency
SDM	Site-directed mutagenesis
T_m	Half-inactivation temperature
tPA	Tissue plasminogen activator
uPA	Urokinase-type plasminogen activator

1.1 Introduction and Historic Perpective

Therapeutic enzymes typically designates a set of biocatalysts that are used for the treatment of diseases and medical conditions. Several of these diseases are actually related to decreased activity or full enzyme deficiency. The overall concept of therapeutic enzymes centers around the highly specific nature and high catalytic activity of enzymes that allow the transformation of targeted substrate molecules with a minimal amount of catalyst in reduced time frame and with minimal, if any, side reactions that could have adverse effects (Dean et al. 2017; Kaur and Sekhon 2012).

The application of enzymes as therapeutic agents has its roots on their use as digestive aids (Kumar and Abdulhameed 2017). Yeasts have been traditionally used for this role, due to the hydrolytic activity of their amylases and proteases (Sizer 1972). Taka-Diastase, an amylase preparation, was marketed as a digestive aid throughout the world by Parke-Davis & Company in the late nineteenth century (Patil 2012) and by the 1930s the National Enzyme Company was founded to produce and market enzymes, e.g. proteases, to replace those that were destroyed during cooking and food processing (Kaur and Sekhon 2012; Petersen 2015). The onset of the twentieth century brought along further potential use for therapeutic enzymes, since pancreatic enzyme extracts, and particularly trypsin, were suggested as useful agents in the prevention of cancer (Gonzalez and Isaacs 1999). This led to the emergence of commercially available trypsin-based formulation for oral administration (Kumar and Abdulhameed 2017). This trend for enzyme therapy was relatively short-lived, yet it was revived in the 1960s/early 1970s, as it was suggested, and shortly after established, that enzyme replacement therapy could be used for the treatment of lysosomal storage diseases (Brady 2006; Li 2018). The approval in the USA of suitable legislation related to the treatment of rare diseases in 1983, resulted in significant financial incentives that prompted interest and investigation in this specific field, and concomitantly boosted research and development efforts dedicated to therapeutic enzymes (Li 2018).

Currently the use of therapeutic enzymes embraces a vast array of applications, abridging from diggestive disorders to cancer therapy, cardiovascular or lysosomal storage diseases (e.g. Gaucher disease). It is acknowledged as a growing market that is foreseen to increase at a compound annual growth rate (CAGR) of 9.4% during 2015–2020 (Allied Market Research 2018) and is expected to top USD 5.5 billion by 2020 (Market Research Engine 2018; Allied Market Research 2018). Enzymes approved for human therapy and their clinical indications, mechanisms of action and adverse effects have been extensively reviewed by different authors (Vellard 2003; Baldo 2015; Mane and Tale 2015; Kumar and Abdulhameed 2017; Yari et al. 2017; Kunamneni et al. 2018). Some of the most relevant applications of therapeutic enzymes are summarized in Table 1.1.

1.2 Therapeutic Enzymes: Sources and Enzyme Engineering

Most therapeutic enzymes were traditionally sourced from human or animal tissue, with limited availability and with complex/time-demanding requirements for extraction and isolation (Gurung et al. 2013; Shanley and Walsh 2005). Many of such enzymes were synthesized and secreted only in small quantities from human tissues and therefore in unsufficient quantities to meet clinical demand. Others were obtained from sources that, for a variety of reasons, were regarded as innapropriate. Examples include urokinase and ancrod, which were obtained from urine and snake venom, respectively. The introduction of recombinant DNA technology in the 1970s allowed heterologous protein expression and enabled the commercial production of a multitude of enzymes that were not previously available in the market (Gurung et al. 2013). This was particularly relevant for the dissemination of therapeutic enzymes. Recombinant production allows to circunvect the risk of transmission of diseases via infected source material while also surpassing low natural availability therefore benefiting both production levels and product safety. Most therapeutic enzymes currently marketed are thus recombinant (Yari et al. 2017; Baldo 2015). Microorganisms and plants are also valuable sources of therapeutic enzymes. Microbial enzymes are particularly appealing, as they are cheap to produce, can be used in lower dosages, have a wider pH range of activity than animal counterparts (Roxas 2008; Vellard 2003; Gurung et al. 2013; Kaur and Sekhon 2012). Nevertheless, they are likely to trigger the response of the immune system. A suitable source should provide an enzyme with high catalytic activity and high substrate affinity, thus screening efforts aim at finding enzymes with high turnover number and low Michaelis constant, so that the enzyme is extremely effective at low substrate and enzyme concentrations. Additionally, the enzyme source must be such that the presence of contaminants is avoided, to minimize the complexity of the production and purification methods (Gurung et al. 2013; Lutz et al. 2017; Dean et al. 2017).

Besides the risk of triggering immuneresponse, other major limitations often faced in the application of therapeutic enzymes are the low stability and lack of suitable delivery (Kumar and Abdulhameed 2017; Lutz et al. 2017). The risk of eliciting the response of the immune system can be overcome through covalent conjugation with polyethylene glycol (PEGylation) that has been shown to mask the immunogenicity of proteins, including therapeutic enzymes, and simultaneously increase enzyme stability (Fuhrmann and Leroux 2014; Mishra et al. 2016; Waugh 2011; Farhadi et al. 2018). Yet, although PEG is usually considered as non-immunogenic and non-antigenic, PEGylated compounds have been reported to prompt anti-PEG antibodies in animal studies, thus protein PEGylation must be carefully evaluated prior to delivery in humans (Baldo 2015). Other compounds, such as polysaccharides or polyoxazolines, are therefore being looked after as alternative to PEG for protein conjugation (Qi and Chilkoti 2015). PEGylation may also interfere negatively with the protein structural dynamics, more so as the level of modification increases, ultimately decreasing the catalytic activity of the enzyme (Rodríguez-Martínez et al. 2009). This shortcoming has been observed in commercial formulations of PEGylated asparaginase, where PEG is covalently bound to the enzyme randomly. By introducing cysteine residues at specific locations by site-directed mutagenesis, intramolecular thiol-maleimide PEG crosslinking of asparaginase subunits was performed with a minimal degree of modification, thus retaining catalytic activity, and simultaneously modulating immunogenicity and proteolysis. The resulting PEGylated asparaginase

Table 1.1 An overview of current uses of therapeutic enzymes (Mane and Tale 2015; Kaur and Sekhon 2012; Yari et al. 2017; Dean et al. 2017; Abderrezak 2018; Baldo 2015)

Therapeutic enzymes
Enzyme replacement therapy
Digestive enzymes: amylases, lactase (β-galactosidase), lipases, maltase (α-glucosidase), peptidases, proteases (chymotrypsin, pepsin trypsin) sucrase, (sacrosidase, invertase). Mixtures of amylase, lipase and protease (Creon, Pancreaze, Pertzye, Ultresa, Viokace and Zenpep): pancreatic insuficiency
N-acetylgalactosamine 4-sulfatase (galsulfase or Naglazyme, recombinant AGS): Maroteaux-Lamy syndrome or MPS VI
Deoxyribonuclease I, DNase (Dornase alfa or Pulmozyme; PEGylated derivative of DNase): cystic fibrosis
N-galactosamine-6-sulfate sulfatase (elosulfase alfa or Vimizim, recombinant): Morquio syndrome type A or MPS IV A
α-galactosidase (agalsidase beta or Fabrazyme, recombinant): Fabry disease
β-glucocerebrosidases, (taliglucerase alfa, velaglucerase alfa, imiglucerase, recombinants): Gaucher disease
α-glucosidase (alglucosidase alpha or Myozyme or Lumyzime, recombinant): Pompe disease
β-glucuronidase (Vestronidase alfa, recombinant): Sly syndrome or MPS VII
Iduronate 2-sulfatase (idursulfase or Elaprase, recombinant): Hunter syndrome or MPS II
α-L-iduronidase (laronidase or Aldurazyme, recombinant): Hurler syndrome or MPS I
Lysosomal acid lipase (sebelipase alfa or Kanuma, recombinant): Wolman Disease and CESD
Phenylalanine ammonia lyase, PAL (PEGvaliase or Palynziq, PEGylated derivative of PAL, recombinant): PKU
Tripeptidyl peptidase-1 (cerliponase alfa or Brineura): Batten disease
Cancer therapy
L-asparaginase (commercial formulations: from *Escherichia coli* with brand names Elspar, Kidrolase, Leunase and Spectrila; from *Erwinia chrysanthemi* with brand names Erwinase and Erwinaze; PEGasparaginase/Pegaspargase, a PEGylated derivative from *E. coli*, Oncaspar): acute lymphoblastic leukemia, acute myeloid leukemia, and non-Hodgkin's lymphoma
L-arginase, L-glutaminase, L-serine dehydratase, L-threonine deaminase, L-tryptophanase, L-tyroninase: antineoplastic
Arginine deiminase, PEGylated/PEG-ADI (Pegargiminase or Hepacid; Melanocid): hepatocellular carcinoma and melanoma
Carboxipeptidase G2 (glucarpidase or Voraxaze): cancer prodrug therapy. Reduces methotrexate toxicity by metabolization into inactive metabolites glutamate and 2,4-diamino-N-10-methylpteroic acid
Urate oxidase or uricase (Uricozyme or Fasturtec and Rasburicase or Elitek, a recombinant urate oxidase; Uricase peg20, a PEGylated form of the recombinant enzyme): Tackling hyperuricemia, a common side-effect of cancer chemotherapy, by degrading uric acid through oxidation. Uricase is also used in the treatment of gout
Topical enzymes
Collagenase, metalloproteinases that hydrolyze different types of collagen (Santyl, Xiaflex, Xiapex, collagenase from *Clostridium histolyticum*): treatment of skin ulcers, burns and wounds
Deoxyribonuclease and fibrinolysin, a bovine plasmin (both enzymes are combined in commercial formulation Elase): debridement of necrotic tissue
Ocriplasmin, also known as microplasmin, a truncated form of human plasmin: treatment of vitreomacular traction and decreasing macular edema
Hyaluronidase, traditionally known as spreading agent, degrades hyaluronic acid: aesthetic medicine (prevents complications from inappropriate injection of hyaluronic acid; promotes the diffusion of compounds injected subcutaneously)
Thrombolytics/fibrinolytics
Alteplase (Activase, Actilyse), streptokinase (Streptase), Urokinase (Kinlytic): lysis of blood clots, emergency treatment in the case of ischemic stroke, myocardial infarction, or a massive pulmonary embolism
Nattokinase: blood-clot dissolving role, prevention of cardiovascular disease
Miscellaneous
Butyrylcholinesterase and cocaine esterase: treatment of drug addiction; chondroitinases and hyaluronidase: regeneration of damaged nerve tissue; lisozyme: antibacterial; lactamase: penicillin allergy

CESD Cholesteryl Ester Storage Disease, *MPS* mucopolysaccharidosis, *SCID* severe combined immunodeficiency, *PKU* phenylketonuria

displayed a specific activity 1.3-fold higher than that of the native asparaginase, whereas the specific activity of the commercial PEGylated formulation was 0.7-fold higher than that of the native enzyme (Ramirez-Paz et al. 2018).

The large molecular size of therapeutic enzymes, namely those from microbial sources, has been also highlighted as a limiting factor for their wider dissemination, since it hampers proper distribution within the envisaged cells (Gurung et al. 2013; Kumar and Abdulhameed 2017; Kaur and Sekhon 2012). To enhance the delivery of therapeutic enzymes to the intended targets several strategies have been implemented, namely: glycosylation, e.g. with mannose-6-phosphate; production of recombinant enzymes as multifunctional fusion proteins that contain an affinity peptide in addition to the enzyme, to convey affinity or chimeras comprising the enzymatic module and an affinity peptide that enhances the attachment to the cell surface; covalent binding to monoclonal antibodies and more recently to nanocarriers (Kumar and Abdulhameed 2017; Oh 2015; Sharma and Bagshawe 2017; Andrady et al. 2011; Farhadi et al. 2018; Muro 2010). Other drawbacks often associated with therapeutic enzymes are the poor stability and low half-life, as they are sensitive to environmental conditions, e.g. temperature and pH leading to unfolding, and to the action of nucleases and proteases (Kumar and Abdulhameed 2017; Dean et al. 2017). Encapsulation in micro- and nanocarriers, such as liposomes, emulsion particles, hydrogels, inorganic (magnetic) nanoparticles, membrane vesicles, niosomes, viral capsids and red blood cells (Funaro et al. 2016; Dean et al. 2017; Farhadi et al. 2018; Abdel-Mageed et al. 2018), as well as PEGylation (Dozier and Distefano 2015) have been used to tackle the low stability of therapeutic enzymes (Dean et al. 2017). Alongside with these methodologies, the mutation of amino acid residues of the enzyme molecule to yield variants with improved therapeutic activity and/or stability has also been gradually introduced (Lutz et al. 2017; Tobin et al. 2014). These modifications can be introduced through different approaches, namely:

- Site-directed mutagenesis (SDM) that involves the deliberate and precise mutation of the amino acid sequence of the enzyme. It can include either single or combinatorial mutations and it is the classical approach for mutagenesis. It relies on previous studies addressing relationships between structure and function, which establish the role of a given amino acid in a specific position, and on biochemical principles to generate enzymes with improved/novel features as compared to the native enzyme (Tobin et al. 2014; Yang et al. 2017);
- Directed evolution (DE), where random mutagenesis or gene recombination is used to introduce/improve a specific enzyme feature, such as stability or activity. The mutants displaying the targeted feature are afterwards identified through screening. The random nature of this approach circumvents the need for detailed knowledge of structure-function mechanisms, but it requires a high throughput environment to ensure high odds of success in reasonable time frame and costs (Tobin et al. 2014; Xiao et al. 2015);
- Computational protein design (CPD), a molecular simulation approach that basically taps a wide array of potential amino acid sequences to select the most suitable (the one with the lowest energy) for a specific protein structure. The predicted structure is generated and evaluated for the desired feature, such as enzymatic activity or stability. The successful implementation of CPD strongly relies on understanding structure-function relationships and on the availability of suitable software (Sharabi et al. 2013; Samish 2017; Löffler et al. 2017). The outcome of mutants generated by computational design is typically validated by wet lab experiments that rely on site directed mutagenesis (Wijma and Janssen 2013).

Some recent examples of the application of these approaches to improve the performance of therapeutic enzymes are given in Table 1.2. The different approaches, involving either generation of mutants or functionalization/encapsulation of

Table 1.2 Examples of different engineering approaches for the generation of mutant therapeutic enzymes with improved properties

Methodology	Description	References
SDM	Operational stability enhancement of *Anabeana variabilis* phenylalanine ammonia lyase through Gln292 replacement with Cys. The concomitant introduction of a new disulfide bond improved the resistance against chemical unfolding, resulting in a 1.5-fold increase in the half-life, with no effect on the pH-activity profile. A minor negative impact in the kinetic parameters, k_{cat} and K_M, was reported when the mutant was compared with the native enzyme, still the former displayed a 1.5-fold increase in specific activity when compared to the later	Jaliani et al. (2013)
	Thermal stability enhancement of *Escherichia* sp. asparaginase II through Lys139 and Lys207 replacement with Ala. While improving thermal tolerance, these substitutions had no impact on the pH-activity profile as compared to the native enzyme. Still, the specific activity of mutants was slightly lower than that of the native form	Vidya et al. (2014)
	Thermal stability and activity enhancement of chondroitinase ABC I, which can be used to facilitate axon regrowh after spinal cord injury, through replacement of Glu140 with either Gly or Ala. The improved features of the mutants were assigned to their more compact structure and to increased k_{cat} together with reduced K_M values, when compared to the native enzyme	Nazari-Robati et al. (2013)
	Another successful attempt to improve the thermal stability and activity of chondroitinase ABC I, Asn806 and Gln810 replacement by Tyr, was shown to provide the best approach. Thus, k_{cat}/K_M and half-life of the mutant increased 1.25 and 1.4-fold when compared to the native enzyme. Again, the outcome of the mutation was ascribed to a more rigid structure, as an aromatic residue was introduced on the surface of the enzyme	Shahaboddin et al. (2017)
	Reduction of the glutaminase side activity of *Escherichia coli* asparaginase II towards its use as a drug for the treatment of acute lymphoblastic leukemia. The authors established that replacements of Asp248 had a major impact on glutamine turnover as compared to asparagine hydrolysis. The asparaginase activity of a selected mutant enzyme, where Asp248 was replaced with Asn, was favourable as compared to the native with, however, a decrease in stability. The authors suggested thus the need of further studies to evaluate the impact of the lower stability of the mutant in its half-life in vivo	Derst et al. (2000)
DE	Several rounds of mutation led to two mutants of urate oxidase, which degrades uric acid and can thus be used to treat tumor lysis syndrome. The catalytic efficiency of the mutants was 3.0 to 3.5-fold higher than that of the native enzyme. As compared to the later, the optimal reaction temperature of the mutants decreased from 40 to 35 °C, but the thermal stability increased, and the optimal reaction pH decreased from 10 to 9. Mutants obtained by SDM allowed to gain some insight on the modifications made but did not match the performance of the selected DE mutants	Li et al. (2017)
	Improvement of the pH activity profile of arginine deiminase from *Pseudomonas plecoglossicida*, which hydrolyses arginine to citrulline and ammonium, thus with application in the treatment of arginine-auxotrophic tumors. The optimal reaction pH of the triple mutant selected was shifted from 6.5 to 7.0 as compared to the native enzyme. Moreover, at pH 7.4, typical of human plasma, the k_{cat} of the mutant exceeded by four-fold that of the native enzyme, and it retained about 50% of its activity relative to its pH optimum, compared to about 10% for the wild type	Zhu et al. (2010)
	The outcome of three rounds of random mutation over arginine deiminase from *P. plecoglossicida* resulted in a mutant with a K_M of 0.17 mM, closer to the concentration of arginine in human plasma (~0.11 mM), than the K_M of the native enzyme (1.23 mM). The turnover of the mutant was of 26.4 s^{-1}, which compares favorably with that of the native enzyme (0.20 s^{-1}). Moreover, the IC50 of the mutant against two melanoma cell lines decreased from 4.5 and 8.2 μg/mL for the native to just 0.02 μg/mL for the mutant enzyme	Cheng et al. (2015a)
	Random mutation led to a mutant of L-asparaginase from *Erwinia chrysanthemi* with improved thermal stability, as highlighted by its T_m of 55.8 °C and its half-life of 159.7 h at 50 °C, 9.4 °C and 59-fold higher, respectively, than that of the native enzyme. The mutation, which stabilized the tertiary structure of the enzyme, resulted of Asp133 replacement with Val. Albeit far from the active site, the mutation had a negative impact on the catalytic efficiency	Kotzia and Labrou (2009)

(continued)

Table 1.2 (continued)

Methodology	Description	References
CPD	Mutants of human butyrylcholinesterase (hBche) against (−)-cocaine were designed. The aim was to significantly alter the affinity of native hBche that is about 1000-fold higher for the (+)-cocaine inactive enantiomer than to the biologically active (−)-cocaine, thus featuring a sound tool for anti-cocaine therapeutics. The design considered both structure of the enzyme and the catalytic mechanisms for hBche hydrolysis of both enantiomers. A quadruple mutant displayed a catalytic efficiency towards (−)-cocaine of 4.2×10^8 M/min, as compared to 9.1×10^5 M/min for the native enzyme	Zheng and Zhan (2008) and Pan et al. (2005)
CPD and SDM	Design of hBche mutants to adequately model the selectivity, to combine enhanced catalytic efficiency on (−)-cocaine with enhanced selectivity for the latter, rather than for the neurotransmitter acetylcholine (Ach), the natural substrate of hBche in the body. One quintuple mutant with relative catalytic efficiencies 1730-fold and 0.93-fold for (−)-cocaine and Ach, respectively, proved the most effective for protection of mice against cocaine toxicity. The study also highlighted the relevance of catalytic efficiency over k_{cat} as selective parameter for efficient mutants	Xue et al. (2011)
CPD and SDM	Generation of a quadruple mutant bacterial cocaine esterase (CocE), an enzyme acknowledged as the most efficient for hydrolyzing (−)-cocaine, with a roughly 3-fold higher catalytic efficiency as compared to the native enzyme. Moreover, the mutant had an in-vitro half-life at 37 °C of at least 100 days, largely exceeding that of 12 min for the native enzyme. When PEGylated, the mutant protected mice from a lethal dose of cocaine for about 3 days	Fang et al. (2014)
DE and SDM	A combination of two rounds of random mutation and SDM, where Val291 was replaced by Leu, yielded a mutant of arginine deiminase from *P. plecoglossicida* with a K_M of 0.35 mM, and a k_{cat} of 21.1 s^{-1}. The outcome corresponds to a 117-fold increase in turnover and 3.7-fold decrease in K_M. When compared to the native enzyme. The performance of the mutant was further highlighted as the IC_{50} for two melanoma cell lines decreased 100- to 200-fold	Cheng et al. (2015b)

IC_{50}: half-maximal inhibitory concentration; k_{cat}: turnover number; K_M: Michaelis–Menten constant; T_m: half-inactivation temperature; catalytic efficiency: k_{cat}/K_M

enzymes, can be merged to a faster and more efficient development of even more effective therapeutic enzymes (Tobin et al. 2014; Zhang et al. 2015; Fang et al. 2014).

1.3 Bioprocesses for the Production of Therapeutic Enzymes

The advances in genetic engineering and DNA recombination techniques facilitated the production of enzymes for different applications, with particular relevance to those for clinical use (Headon and Walsh 1994). Bioprocessing is an essential part of biotechnology industry, which is used for the production of biopharmaceuticals. Approximately 50% of all new medicines are classified as biopharmaceuticals. These drugs include various types of molecules such as peptide hormones, growth factors, cytokines, monoclonal antibodies and enzymes for human therapy (Aggarwal 2007). The demand in therapeutic enzymes is expected to increase over the coming years and thus bioprocesses for their production must meet the requirements of the global market (Singh et al. 2016; Tripathi 2016).

The commercial production of therapeutic enzymes has been achieved using different hosts, such as: mammalian cells, namely Chinese Hamster Ovary (CHO) cells, bacteria (e.g. *Bacillus* spp., *E. coli, E. chrysanthemi)*, yeasts (e.g. *Pichia pastoris*, *S. cerevisiae*) and molds (e.g. *Aspergillus sojae*) (Shanley and Walsh 2005; Yari et al. 2017; Vidya et al. 2017). At lab-scale many other organisms of the different kingdom have been used, in either submerged or solid state fermentations (Mane and Tale 2015; Sharifi-Sirchi and Jalali-Javaran 2016). For example, a wide range of microorganisms such as bacteria, fungi, yeast, algae and plants have proven to be proficient sources of L-asparaginase as reviewed by Verma et al. 2007. L-asparaginase (L-ASNase, also called L-asparagine amido- hydrolase with Enzyme Commission number 3.5.1.1) catalyses the conversion of L-asparagine in its free amino acid form to L-aspartate and ammonia. L-asparagine is an essential amino acid for the growth of lymphatic tumor cells, which require large ammounts of asparagine to keep their malignant growth. As opposed to normal cells, acute lymphoblatic leukemia cells are auxotrophic for asparagine (Dean et al. 2017; Verma et al. 2007). L-asparagine reduces asparagine blood concentration causing a selective inhibition of growth of sensitive malignant cells (Yari et al. 2017). For over 30 years, L-ASNase has been a mainstay of multidrug chemotherapeutic regimens extensively used for the treatment of malignancies of the lymphoid system, acute lymphoblastic leukemia, Hodgkin's lymphoma and melanosarcoma (Tundisi et al. 2017). A wide number of reports have also addressed the production and purification of recombinant L-asparaginase, using different gene sources and hosts (Verma et al. 2007). Commercial L-asparaginase (aproved by international regulatory entities as a biological therapeutic) is derived from *E. coli* and *E. chrysanthemi* cells (Table 1.1).

E. coli has been proven as an efficient expression host for the production of non-glycosylated proteins as it offers various advantages over yeast and other expression systems due to its well-understood genetics, cell biology, easy handling, and simple upstream process, which allows production of cost-effective large quantity of recombinant proteins (Tripathi 2016; Gupta and Shukla 2017). Major limitations of this expression bacterial system are incorrect folding/activity of expressed protein due to lack of post-translation modification machinery and the presence of endotoxins, which bears an inherent safety concern as patients medicated with *E. coli*-produced recombinant proteins may show an immune response to the administered drugs. Therefore, improved removal of these contaminants is a key issue through the purification process. Other disadvantages include loss of plasmid and antibiotic selection property, protein-mediated metabolic burden and stress, and poor secretion production medium (Tripathi 2016; Huang et al. 2012; Gupta and Shukla 2017).

Yeast systems are used when extracellular expression of recombinant proteins in large amounts is needed and when the product contains post-translational modifications (Tripathi 2016). The most used yeast expression systems are *S. cerevisiae* and *P. pastoris.* Both these hosts are capable of producing therapeutic proteins with proper folding and post-translational modifications close to human proteins. Advantages of yeast as host expression systems also include rapid growth in inexpensive culture media, high cell density cultivation and ease of genome modification (Tripathi 2016; Gupta and Shukla 2017). However, for the production of therapeutic glycoproteins intended for use in humans, yeasts have been less useful owing to their inability to modify proteins with human glycosylation patterns. Yeast *N*-glycosylation is of the high-mannose type, which confers a short half-life in vivo and thereby compromises the efficacy of most therapeutic glycoproteins (Wildt and Gerngross 2005). Nevertheless, *P. pastoris* cell lines have been utilized and optimized for the production of therapeutical enzymes such as orciplasmin or microplasmin, the first approved agent for vitreolysis therapy (Yari et al. 2017). This enzyme is a truncated form of plasmin that is expressed as an inactive zymogen in *P. pastoris* and then is activated by recombinant staphylokinase prior to formulation as a stable product (Noppen et al. 2014).

Yeast cells are also used as host system in the production of recombinant urate oxidase, an enzyme that controls efficiently hyperuricemia, a cause of gout disease (Richette and Bardin 2006) and a common side-effect of cancer chemotherapy (Yari et al. 2017). Urate oxidase converts uric acid into the soluble compound allantoin. The FDA-approved biotherapeutic (Rasburicase) is produced by genetically modified *S. cerevisiae* with cDNA originated from a strain of *Aspergillus flavus* (Leplatois et al. 1992; Yari et al. 2017).

Despite extensive research has been developed over the last decades, aiming at enginnering *E. coli* and yeast cell lines to overcome the limitations in recombinant protein production (Huang et al. 2012; Wildt and Gerngross 2005; Laurent et al. 2016; Laukens et al. 2015), mammalian cells are currently the most used for the production of recombinant therapeutic proteins. This is due to the fact that animal cells allow the expression of complex glycoproteins, with glycosylation patterns and other post-translational modifications close to human proteins (Gupta and Shukla 2017; Wells and Robinson 2017) although the enzymes might require modifications of the glycan structure to present therapeutic efficacy (Grabowski et al. 1995; Lalonde and Durocher 2017). The disadvantages of mammalian cell culture (compared, for example, with yeast-based expression systems) are that mammalian cells require more nutritious medium, stringent growth conditions and longer growth periods than yeast cells (Huang et al. 2010). Nevertheless, nowadays a great variety of commercially available serum-free and defined media allow the efficient growth of mammalian cell lines while minimizing variability between batches, simplifying downstream processing by decreasing the number of contaminants, and helping to comply with regulatory requirements (Agrawal and Bal 2012).

Within mammalian expression platforms, Chinese Hamster Ovary (CHO) cells are the most commonly used expression host system (Walsh 2014; Wells and Robinson 2017). CHO cells have been demonstrated as safe hosts for synthesis of biologics and a have a proven track record for producing proteins with glycoforms that are both compatible and bioactive in humans (Jayapal et al. 2007). One of the early concerns in recombinant protein production was that cultured mammalian cells were presumably derived from perturbation in oncogenes, and thus, can proliferate without the effects of senescence. However, CHO cells, as other mammalian cell lines, have proven safe, with the value of products generated considerably outweighing any associated risks (Jayapal et al. 2007).

The first FDA approved recombinant enzyme drug in 1987, Activase (alteplase, recombinant human tissue plasminogen activator, tPA) is produced in CHO cells (Jayapal et al. 2007). Likewise, Pulmozyme (deoxyribonuclease 1), Cerezyme (β-glucocerebrosidase), Benefix (Factor IX), ReFacto and Advate (Factor VIII), Tenecteplase (engineered tPA), Fabrazyme (α-galactosidase), Naglazyme (N-acetylgalactosamine-4-sulfatase, Aldurazyme (laronidase) and Myozyme (α-glucosidase) are examples of recombinant therapeutic enzymes produced in CHO cell lines (Jayapal et al. 2007). Other expression systems used for the production of recombinant therapeutic proteins include insect derived cell lines Sf9 and Sf21 from *Spodoptera frugiperda*, plants, transgenic animals and fungi (Gupta and Shukla 2017). In terms of biopharmaceuticals approved from 2004 to 2013, it has been reported that 56% were produced in mammalian cells, 24% in *E. coli*, 13% in *S. cerevisiae* and the remaining in other expression systems (Tripathi 2016).

The production of therapeutic proteins by transgenic animals has potential advantages such as the high production capacity associated with the use of larger animals such as sheep or goats, ease of collection of source material (by milking the animal) and the fact that producer animal numbers can be expanded by breeding programmes. The expression of both human tissue plasminogen activator (tPA) and urokinase-type plasminogen activator (uPA) has been studied in transgenic mice using a DNA construct consisting of the promoter and upstream regulatory sequence of the mouse whey acidic protein (the most abundant protein in mouse milk) fused to the nucleotide sequence coding for the human tPA and uPA. Active tPA and uPA could be recovered from mouse milk but expression levels obtained were low (Headon and Walsh 1994; Brandazza et al. 1991).

Taliglucerase-α, a carrot-cell expressed human β-glucocerebrosidase enzyme for intravenous treatment of Gaucher disease, was the first plant-cell expressed biotherapeutic approved for use in humans (Yari et al. 2017; Zimran et al. 2011). The recombinant protein is targeted to plant storage vacuoles, which generate N-glycans with exposed terminal mannose residues without an additional processing step (Oh 2015). The plant cell-based platform was easy to scale-up and provides long-term safety with regard to exposure to mammalian viral infections as well as cost efficacy compared to mammalian cell bioreactors (Zimran et al. 2011). Recently, expression of tPA in transgenic (Goojani et al. 2013) and transplastomic plants (Hidalgo et al. 2017; Abdoli-Nasab et al. 2016) was also evaluated.

The majority of enzymes approved for human therapy are recombinant proteins cloned in CHO cells. Mammalian cell culture generates complex broths with a variety of impurities. Typical impurities found in culture broths include host cells and cell debris, host cell protein and DNA released by the cells, aggregated proteins or cleaved proteins produced by the cells and medium additives (such as serum and protein used to support cell growth). Moreover, viruses (infecting cells) and bacterial pathogens or related contaminants such as endotoxins can also lead to serious problems with regard to the safety of the protein preparation. These impurities must therefore be removed to a final concentration below limits established by regulatory entities. Acceptable concentrations of impurities in a final protein/enzyme biopharmaceutical are host cell protein and serum <0.1–10 mg/L, DNA ≤10 ng/dose, virus pathogens $<10^{-6}$/dose, endotoxins ≤0.25 EU/mL and no cell debris (Zhu et al. 2017).

1.3.1 Development of Purification Strategies

Selection of the host organism for recombinant enzyme production commonly favours those that secrete the therapeutic enzyme into the culture medium, as this eases downstream processing (Shanley and Walsh 2005). If the enzyme of interest is produced intracellularly, it is necessary to disrupt the cells upon completion of fermentation and cell harvesting. This releases not only the target enzyme, but also the entire intracellular content of the cell, thereby rendering more complicated the subsequent purification procedure.

Purification is commonly considered the bottleneck in the manufacturing processes of biopharmaceuticals such as therapeutic enzymes. Besides the cell debris, host cell proteins, DNA, endotoxins, and viruses that need to be removed, product-related impurities such as protein moieties that are misfolded, aggregated, carry the wrong post-translational modifications, or are otherwise chemically degraded may complicate the purification, due to their similarity to the target molecule. As a rule, a significant percentage (50–80%) of the total manufacturing costs of biopharmaceutical molecules is incurred during downstream processing and critical challenges include improving process economics and efficiency, to reduce costs, and fulfilling increasingly demanding quality criteria for Food and Drug Administration (FDA) and European Medicines Agency (EMA) approval (Roque et al. 2004, 2007).

The level of downstream processing required for an enzyme in a bioprocess is dependent upon its intended application. While enzymes for industrial applicaton are used as relatively crude preparations those destined for clinical use are subjected to a far higher degree of downstream processing. Other than the very high purity required, there is no fundamental technical difference between the purification of enzymes or proteins for therapeutic use and the purification of proteins for other applications. However, all manufacturers of therapeutic products must comply with the principles and guidelines of cGMP (current Good Manufacturing Practices) in terms of facilities, equipment, personnel, quality systems, documentation/records and laboratory testing. Topics such as virus inactivation, endotoxin removal, process validation, use of materials derived from animal components are key considerations in the development of a purification process for a therapeutic protein (Hanke and Ottens 2014; Bonnerjea 2004).

Achieving the high levels of product purity required for use as an active pharmaceutical ingredient requires a complex cascade of unit operations (Hanke and Ottens 2014) that include primary recovery (cell separation), purification steps, virus removal and polishing. Many therapeutic enzymes are recombinant products produced in animal cell lines, particularly CHO cells. In this case, enzymes are secreted directly by the producing cells into the

Table 1.3 Name, properties and production organism of some FDA approved human-like therapeutic enzymes. Adapted from Wells and Robinson (2017), Dean et al. (2017), Jayapal et al. (2007), and Baldo (2015)

Generic and trade name	Enzyme properties	Production organism
Agalsidase β (Fabrazyme)	Recombinant α-D-galactosidase A; glycosylated, 2 subunits MW 100 kDa 398 amino acids	CHO cells
Alglucosidase α (Myozyme)	Recombinant human glycogen-specific acid α-glucosidase (GAA). Glycoprotein, MW 109 kDa, 883 amino acids	CHO cells
Alteplase (Activase)	Recombinant tPA, a serine protease; single glycosylated polypeptide chain, MW 59 kDa, 527 amino acids	CHO cells
Dornase α (Pulmozyme)	Recombinant human deoxyribonuclease I (rh DNASE I); 260 amino acids, identical to natural enzyme	CHO cells
Elosulfase α (Vimizim)	Recombinant N-acetylgalactosamine-6-sulfatase (rhGALNS); monomer 496 amino acids, MW 55.4 kDa, 2 glycosylation sites. Cys53 (in active site) modified to formylglycine	CHO cells
Imiglucerase (Cerezyme)	Recombinant monomeric glycoprotein human β-glucocerebrosidase; MW 60.4 kDa, 497 amino acids; His not Arg at position 495 of placental enzyme. Oligosaccharide chains modified to expose terminal mannose residues	CHO cells
L-Asparaginase (Elspar; Erwinase;	From *E. coli* (Elspar) From *Erwinia chrysanthemi* (Erwinaze) L-asparagine specific; 4 subunits; MW 35 kDa	*E.coli/E.chrysanthemi*
Laronidase (Aldurazyme)	Recombinant variant of human α-L iduronidase; MW 83 kDa, 628 amino acids; 6 glycosylation sites with at least 2 mannose; 6-phosphorylated	CHO cells
Orciplasmin (Jetrea)	Recombinant truncated human plasmin, MW ≈ 27.2 kDa 2 polypeptide chains of 19 and 230 aminoacids linked by two disulphide bridges.	*P. pastoris*
Pegasparagase (Oncaspar)	*E. coli* enzyme linked to monomethylpolyethylene glycol (PEG)	*E.coli*
Rasburicase (Elitek, Fasturtek)	Recombinant *Aspergillus*-derived urate oxidase expressed in yeast, tetrameric protein with identical subunits 301 amino acids, MW 34 kDa	*S. cerevisiae*
Taliglucerase-*α* (Elelyso)	Recombinant human lysosomal β-glucocerebrosidase; MW 60.8 kDa; differs from natural enzyme by 2 amino acids at N-terminal and up to 7 at C-terminal	Plant system (carrot cells)
Tenecteplase (TNKase; Metalyse)	Recombinant tPA of 527 amino acids with Asn for Thr at 103, Gln for Asn at 117, and tetra Ala at 296–299	CHO cells

culture medium. Extracellular production displays process advantages, by obviating the necessity of lysing (disrupting) cells in order to bring about the release of the protein (enzyme) of interest, thereby simplifying downstream processing. Whole cells can be removed by filtration or centrifugation of the extracellular medium. Therapeutic enzymes are invariably purified to homogeneity before final formulation, a process requiring multiple high-resolution chromatographic steps (Fig. 1.1).

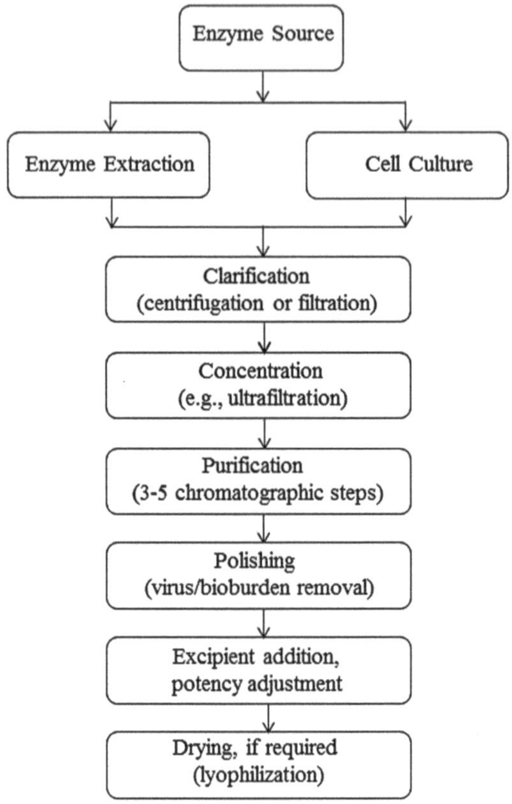

Fig. 1.1 Overview of a generalized manufacturing scheme for therapeutic enzymes

The development of the first chromatography steps is often the most difficult and time consuming aspect of developing a purification process, as a highly efficient and robust first step is critical to a successful bioprocess. This step is commonly an ion exchange chromatography that is designed/optimized to remove potentially deleterious impurities such as proteolytic enzymes released by the host organism in recombinant protein production (Bonnerjea 2004). Ion exchange resins have high loading capacities and can be operated at high flow rates, which is a fundamental issue in process scale-up. After this step, a sequence of other chromatographic separations is generally used to remove the remaining impurities and recover the target enzyme with a high purity level. These include hydrophobic interaction chromatography and gel filtration (or size exclusion) chromatography (Tripathi 2016).

Affinity chromatography allows high selectivity, high resolution and usually high capacity for the target enzyme while enabling the obtention of products with high purity in a single step. It requires a specific ligand, covalently bound to the chromatographic resin (Tripathi 2016). The ligand may recognize the target enzyme or an affinity tag fused to it. When possible and economically viable, this type of chromatography is frequently applied as the initial step of the purification process, followed by a chromatographic or polishing step to remove remaining impurities.

Before final formulation, polishing steps are essential to remove trace contaminants as bacterial pathogens, related endotoxins, host cell proteins and DNA, biological leachates from columns with immunogenic potential and adventitious pathogenic viruses into acceptable concentration levels. The polishing strategy is nowadays well established for the most common contaminants and is generally achieved by procedures for virus inactivation and bioburden removal by specific chromatographic processes and/or micro/nanofiltration of the final product preparation (Zhu et al. 2007; Gupta and Shukla 2017; Bonnerjea 2004). Therefore, downstream processes for CHO cell recombinant products have matured a stage where they can be purified to contain no more than picogram levels of contaminating CHO DNA per dose of product. Also, it has been found that many pathogenic viruses such as HIV, influenza, polio, herpes and measles do not replicate in CHO. Therefore, from a regulatory standpoint, CHO cells have withstood the test of time, and extensive testing and safety data eases approval of new CHO-based therapeutics (Jayapal et al. 2007).

At a laboratorial scale, different sources and bioprocesses have been explored to optimize the production of bioactive and stable therapeutic enzymes. Particularly, L-asparaginase (L-ASNase) has been widely studied in this context. The production and purification of L-ASNase (recombinant and non recombinant) has been recently reviewed by Verma et al. (2007). A wide range of microorganisms such as bacteria, fungi, yeast, Actinomycetes, algae and plants has proven to be proficient sources of this enzyme. Some organisms produce two forms of asparaginase, L-asaparaginase I and L-asparaginase II, which are two enzymes genetically and biochemically distinct. L asparaginase I is constitutive while L-aspariganse II is secreted in response to nitrogen source starvation. Commercially available L-ASNase are derived from *E. coli*, native and pegylated (PEG asparaginase) and from *E. chrysanthemy* (Tundisi et al. 2017).

Table 1.4 summarizes examples of purification strategies developed for the recovery of recombinant therapeutic enzymes in different expression systems at laboratorial scale studies. In most cases, the target enzyme was purified to near homogeneity but final activity yields were not always sufficiently high (i.e., >70–80%), which is a limitation for processes scale-up.

In some of the examples described, the gene coding for the enzyme is fused to another gene coding for an affinity tag to facilitate recovery by affinity chromatography. Affinity tags are robust tools that can be used to purify proteins, which makes them attractive candidates for implementation into platform processes. At a laboratory scale, tags can be applied to a wide variety of targets and typically yield high purity products, with high yield (>90%) after a single column chromatographic step. However, whereas many tagged proteins retain their structural integrity and biological activity, others do not (Waugh 2011). Due to the unpredictable changes that adding a tag may cause in a protein and its behaviour (Arnau et al. 2006), it is usually desirable to remove the tag at end of the process. This becomes more relevant when designing a process for the production of a recombinant protein that is intended for clinical use as some types of tags can create additional safety/regulatory issues (e.g., unpredictable immunogenic potential) when fused to human proteins/enzymes. Depending on the method used for tag removal, an additional separation step may be required to isolate the native, 'tagless' protein (Arnau et al. 2006). Therefore, tag removal remains a significant and cost-adding issue that must be resolved before affinity tags become widely used at manufacturing scale. One alternative is self-cleaving purification tags (e.g., inteins), which can provide the purity and versatility of more conventional tags but eliminate the need for proteolytic tag removal (Fong et al. 2010).

Despite the high number of commercially available enzyme therapeutics, retrieving information about the purification methodologies from scientific literature and patents is a difficult task. With regard to patents, several claims and embodiments are usually presented for several steps of the purification process, which difficult the assessment of those that are implemented in real industrial applications.

1.4 Large Scale Manufacturing of Therapeutic Enzymes

The numerous research studies regarding therapeutic enzymes, discussed in the previous sections, opened the possibility to treat a wide range of disorders. However, a therapy can only be implemented if the enzyme passes the

Table 1.4 Examples of purification strategies for the isolation and characterisation of recombinant therapeutic enzymes, in different expression systems

Enzyme	Producing organism	Purification Methodology	References
Human tissue plasminogen activator (tPA)	*E. coli* SF110 (Coexpression of DsbC[a])	Affinity chromatography (3 steps) on lysine-Sepharose, immobilized Erythrina caffra inhibitor and Zn-Sepharose resins	Qiu et al. (1998)
Human tissue plasminogen activator (tPA) –His_6	Transplastomic *Nicotiana tabacum L.* plant (chloroplasts)	Extraction of proteins from cloroplasts, dialysis, ion metal affinity chromatography (IMAC) on Ni-NTA agarose	Abdoli-Nasab et al. (2016) and Hidalgo et al. (2017)
Human microplasminogen (the inactive precursor of orciplasmin)	*P. pastoris*	Immuno affinity chromatograpy (with a monoclonal anti-orciplasmin antibody)	Noppen et al. (2014)
L-Asparaginase (isozyme II) from *E. coli* – His_6 (secreted)[b]	*E. coli* BLR (DE3)	Ion metal affinity chromatography (IMAC) on Ni-NTA agarose	Khushoo et al. (2004)
L-asparaginase from *Erwinia chrysanthemi*	*E. chrysanthemi*	Ion exchange chromatography on S-Sepharose FF	Goward et al. (1992)
L-asparaginase from *Erwinia chrysanthemi* 3937(*Er*l-ASNase)	*E. coli* BL21(DE3)pLysS	Ion exchange chromatography on S-Sepharose FF	Kotzia and Labrou (2009)
Carboxypeptidase G2 from *Pseudomonas* sp. strain RS16 and its individual domains – His_6[c]	*E. coli* (several strains)[d]	Ion metal affinity chromatography (IMAC) on Ni-NTA agarose	Jeyaharan et al. (2016)
Human acid β-glucosidase	Transgenic rice	Hydrophobic interaction chromatography on Octyl FF Sepharose, ion exchange chromatography on SP HR column, gel filtration on Sephacryl S-100 High Resolution	Patti et al. (2012)
Human α-L-iduronidase	CHO cells	Affinity chromatography on concanavalin A-Sepharose, heparin-Sepharose, gel filtration on Sephacryl S-200	Kakkis et al. (1994)
Human α-N-acetylglucosaminidase	CHO cells	Affinity chromatography (2 steps) on concanavalin A-Sepharose, perfusion chromatography on porous 20 HE, hydrophobic interaction chromatography (amino-octyl agarose)	Zhao and Neufeld (2000)
Urate oxidase from *Aspergillus flavus* (soluble)	*E. coli* BL21(DE3)	Ion exchange chromatography (DEAE SEpharose FF), hydrophobic interaction (Phenyl Sepharose FF) chromatography, Gel Filtration (Hiload 26/60 Superdex 75)	Li et al. (2006)
Urate oxidase from *Aspergillus flavus*	Not specified	Crystallization techniques[e]	Giffard et al. (2011)

(continued)

Table 1.4 (continued)

Enzyme	Producing organism	Purification Methodology	References
Urate oxidase from *Aspergillus flavus* – His_6	*P. pastoris* GS115	Ion metal affinity chromatography (IMAC) on Ni-NTA agarose	Fazel et al. (2014)
Urate oxidase from *Kluyveromyces marxianus* – cellulose binding affinity tag[f]	*E. coli* BL21(DE3)	Affinity chromatography (on cellulose)	Wang et al. (2006)

[a]Enzyme that catalyses disulfide bond isomerization in the periplasm, increasing the formation of active tPA
[b]Extracellular expression was obtained by fusing the gene coding for aspaginase II to an efficient pelB leader sequence and a N-terminal His_6 tag
[c]First high yield (250 mgL^{-1}) recombinant expression and purification of soluble and active carboxypeptidase two using the *E. coli* expression system
[d]When needed His Tag was removed in the end by incubation with tobacco etch virus (TEV) protease at 4 °C for 48 h
[e]Alternative to chromatography
[f]Urate oxidase was fused to a self-cleaving intein removed after purification

clinical trials and obtains the necessary regulatory approvals, and if an efficient manufacturing and purification process can be developed for commercial scale. The growing number of cancer cases and cardio vascular diseases is one of the most important driving forces for the growth of the therapeutic enzymes market (Persistence Market Research 2018) and increase in the production scale. Replacement therapy is also continuously demanding high quantities of therapeutic enzymes.

In most applications, oral administration is not recommended as enzymes may be digested by proteases. Formulations allowing the direct injection of the enzymes into the bloodstream are preferable but must comply with very high purity specifications. This makes crucial the purification steps from the bioreactor, where enzymes are produced, to product separation, and elimination of contaminants and unwanted compounds. Besides, to meet current good manufacturing practices, documented purity is critical along the purification process and, together with recovery yields, must be attained taking into consideration cost constraints (Willson and Ladisch 1990). The documented evidence assuring the production of a compound meeting the specifications published e.g. in FDA guidelines, allows process validation for large-scale production of a therapeutic enzyme.

Two important factors determine the overall manufacturing process: (i) the first is related to the production step, including the composition of the production medium, the type of producer (e.g. microorganism or animal cells), location of production (e.g. cytoplasm, periplasm, supernatant) and other production related aspects; (ii) the second relates to the desired compound, and is influenced by the form of product, type of impurities present, final purity and cost (Ho 1990). As previously mentioned, the extraction and purification of recombinant proteins may reach more than 50% of the total processing costs (Doran 2013; Roque et al. 2004).

1.4.1 Fermentation/Production Step

Biomass concentration and protein yields at the end of fermentation have a significant impact on enzyme cost. Ferreira et al. (2018) estimated 14 orders of magnitude between a case with low biomass and low recombinant enzyme content (1926 USD/kg) and the case with the highest biomass and recombinant enzyme productivities (135 USD/kg), with enzyme cost decreasing considerably with increasing volumetric productivity (Ferreira et al. 2018).

To minimize production cost and timeline, several upstream improvements have been developed, including enhanced engineered hosts, better cell lines, high-performing media free of animal-derived components, and improved bioreactor design (Chon and Zarbis-Papastoitsis 2011; Gupta and Shukla 2017). Recombinant DNA technology has been efficiently used to develop expression systems to produce enzymes from the original organism using industrial microorganisms such as *E. coli* and *Bacillus subtilis*. *E. coli* provides the least expensive, fastest and easiest expression system but this bacterium cannot express very large proteins (>30 kD), S-S rich proteins, or proteins requiring post-translational modifications (Demain and Vaishnav 2009).

In *E. coli*-based bioprocesses, nutrient composition of the cultivation medium and fermentation parameters, such as temperature, pH and dissolved oxygen, largely affect cell density and production yields. When high cell densities are reached, nutrient depletion, especially of dissolved oxygen, and acetate accumulation occur, which affects the production of recombinant proteins. During aerobic growth on glucose, *E. coli* converts 10–30% of the carbon flux from glucose into acetate (Tripathi et al. 2009). Excessive production of acetate may be prevented by controlling glucose concentration, lowering the growth temperature and using a low-acetate-producing *E. coli* strain such as BL21 (Phue and Shiloach 2004; Ferreira et al. 2018). However, in industrial fermenters with glucose feeding, cells at the top of the cultivation medium are challenged by high glucose concentrations whist those at the bottom may suffer from glucose starvation, resulting in heterogeneous acetate profiles, decreased plasmid stability, and lower protein production rates inside the vessels (Tripathi et al. 2009).

Soluble recombinant proteins produced in the cytoplasm may be recovered after cell lysis. However, they will be susceptible to proteolysis, before and after cell lysis. Several strategies have been tested to prevent proteolysis of the produced proteins, including: the use of protease-defective *E. coli* strains, yeast or mammalian cells; lowering the incubation and cell processing temperatures; changing the induction conditions; decreasing the processing time; and, addition of protease inhibitors (Ryan and Henehan 2013).

Bacillus strains such as *B. subtilis*, *B. licheniformis*, *B. megaterium*, and *B. brevis*, were proven as preferred homologous expression systems for the production of e.g. proteases and amylases since they are easy to manipulate and metabolically robust. Furthermore, they are genetically well characterised systems, regarded as "generally recognised as safe" (GRAS), grow well and have high secretion with no involvement of intracellular inclusion bodies, resulting in efficient and cost effective recovery systems (Demain and Vaishnav 2009; Westers et al. 2004; Hohmann et al. 2016). Nijland and Kuipers (2008) reviewed the numerous patents that have been published to optimize protein secretion by *B. subtilis*, including the use of strong promoters, overexpression of targeting chaperones, deletion of proteases, and changes in charge of the cell wall or of the secreted protein (Nijland and Kuipers 2008).

It has also been shown that *Lactococcus lactis* may be used at industrial scale to produce heterologous proteins such as lysostaphin using the nisin-controlled gene expression system (Mierau et al. 2005). Lysostaphin is an antibacterial protein effective against *Staphylococcus aureus*. To guarantee that the product was Bovine Spongiform Encephalomyelitis (BSE) agent-free, a new media containing only plant-based peptone, yeast extract, and certified BSE free lactose was developed. Scale-up to 3000 L fermenters resulted in the production of 100 mg/L lysostaphin.

Yeasts present several advantages as cloning hosts, when compared to bacteria, since they grow rapidly in simple media to high cell densities, may produce proteins larger than 50 kD, and may secrete heterologous proteins into the extracellular cultivation media (Demain and Vaishnav 2009; Adrio and Demain 2014). The two most used yeasts as expression systems are *S. cerevisiae* and *P. pastoris*, but lately other yeasts such as *Komagataella* sp., *Kluyveromyces lactis*, and *Yarrowia lipolytica* have emerged as advantageous hosts (Gomes et al. 2018).

To improve recombinant protein yields in manufacturing processes using yeasts as hosts, several cultivation strategies have been applied, including control of dissolved oxygen concentration, temperature, pH and osmolarity (Gomes et al. 2018; Cereghino et al. 2002). Wang et al. (2006) evaluated different procedures for harvesting a therapeutic protein from high cell density fermentation broth of *P. pastoris*: centrifugation followed by depth filtration (Option1), centrifugation followed by filter-aid enhanced depth filtration (Option2), and microfiltration (Option3) (Wang et al. 2006). The authors studied the scale-up to 3000 L fermentation broth, maintaining the filter or membrane loading constant, and using Prostak membranes or hollow fibber cartridges in series in microfiltration to reduce total pumping requirements. All three approaches allowed product recoveries above 80%, harvest process time below 15 h, and clarification lower than 6 nephelometric turbidity units. However, Options 1 and 2 require high capital investment, but low investment in consumables and reuse of equipment, contrary to Option3.

Mammalian cells, in particular Chinese hamster ovary (CHO) cells, are used for large scale manufacturing of many FDA approved therapeutic enzymes. Several biopharmaceutical companies have invested in single-use bioreactors for the cultivation of mammalian cells, with the major advantages being a reduced capital investment in facilities and equipment, and the time saved in necessary operations in stainless steel bioreactors such as sterilization and cleaning, i.e., a faster campaign turnaround time is possible (Agrawal and Bal 2012; Langer 2017). Cellular growth and metabolism of several CHO and NS0 cell lines in disposable bioreactors of 50, 250, and 1000 L was similar to that obtained in traditional stainless steel and glass tanks while the set-up and disposal times were much shorter (Smelko et al. 2011). Approximately 7.5 kg of a therapeutic antibody could be harvested from the 1100 L disposable bioreactor with similar quality to the product produced in traditional bioreactors by applying a high cell mass fed-batch strategy.

Perfusion systems allow the feeding of new media to the bioreactor in continuous mode, resulting in increased cell density (ca. $30–100 \times 10^6$ cells/mL) when compared to conventional batch and fed-batch processes (ca. $5–25 \times 10^6$ cells/mL), although cell-retention systems may increase the complexity of operation (Lim et al. 2011). Additionally, the centrifugation and depth filtration steps, necessary in fed-batch systems in traditional bioreactors to remove the cells, are not required because the cells are retained in perfusion systems.

The application of plants for the production of recombinant proteins could present several advantages and flexibility when compared with microbial and mammalian production systems, in terms of e.g. costs, scalability and safety (Hellwig et al. 2004; Merlin et al. 2014; Sabalza et al. 2014). While plant suspension cells and algae may be used for low-volume processes, field-grown commodity crops may be used to produce metric tons of recombinant protein at competitive costs (Merlin et al. 2014). Besides, a unique advantage of plant-based systems for the production of pharmaceuticals for oral administration is the "bioencapsulation" of the product by edible plants (Daniell et al. 2009). However, several aspects have to be overcome before plant systems may be used extensively, such as: lack of control over growth conditions; batch-to-batch product consistency; a high level of containment; maintenance of protein stability in plant tissues; and, storage of plant material containing the desired pharmaceutical enzymes as some plants used as hosts have limited shelf live in fresh state (Streatfield 2007; Hellwig et al. 2004; Daniell et al. 2009). To date only few plant-made pharmaceuticals (PMP) have been approved by regulatory agencies for commercial sale, including a recombinant glucocerebrosidase to be used as enzyme replacement therapy for the long-term treatment of adult patients with Gaucher disease (Protalix Biotherapeutics, Israel).

1.4.2 Product Isolation and Purification

There is no general protocol to develop a large-scale downstream process for the purification of therapeutic enzymes, although the purification process usually involves the following steps:

removal of insoluble material such as biomass and cell debris; concentration; fractioning and isolation of the product; and purification.

Following enzyme production at large scale in fermenters using microbial cells, continuous centrifugation or filtration is usually the first step for separation of cell biomass from the cultivation broth. A concentration step may be carried by ultrafiltration, which will further purify the enzyme containing solution by removing ions and proteins with lower molecular weight than the desired enzyme. The increased enzyme concentration will favour chromatographic purification steps by decreasing the amount of support needed. Other intermediate purification steps may include salt precipitation followed by dialysis and two-phase liquid extraction.

The purification of L-asparaginase represents up to 80% of the total production costs since the highly purified solution enables reduced risk of toxicity and allergies (Tundisi et al. 2017). Aqueous two-phase systems (ATPS) were found useful to overcome long processing times, and combined with precipitation and/or fractioning steps are a viable process for protein purification (Tundisi et al. 2017). Besides, since clarification, concentration and purification are done in just one step, and the scale-up is possible by using equipment found traditionally in the industry for liquid-liquid extraction, total costs decrease significantly.

Varga et al., patented a new method to recover polymyxin B, which is a complex of very similar polypeptide antibiotics, produced by *Bacillus* strains (US 7951913 B2 (Varga et al. 2011). The invention describes the purification of polymyxin B as follows: the filtrate of the fermentation broth has the pH adjusted to 2.5–3.5 and is purified using polystyrene non-ionogenic synthetic adsorbents, with the compound being eluted by an aqueous solution of an organic solvent (e.g. methanol, ethanol, propanol, acetone); polymyxin B in the water-organic eluate is precipitated by the use of an alkaline agent such as NaOH or KOH at pH 9–11; the precipitated product is filtered, washed with deionized water at temperature from 30 °C to 80 °C, and neutralized with an acid solution of e.g. HCl; the resulting product solution may be discoloured by activated carbon, and the pure solution of polymyxin B is dried to obtain crystals of the product.

Despite increasing competition from non-chromatographic techniques (such aqueous two-phase extraction and precipitation), and pressure to reduce costs and increase throughput at a manufacturing scale, packed-bed chromatography is still the dominant technique in biopharmaceutical proteins purification. This prevalence is mainly due to the high-resolution purification that can be achieved even for similar components. Advances in resin chemistry have alleviated some of the concerns that packed-bed chromatography could not handle the throughput and production needs in coming years. If not during early process stages, then during product polishing where very high purities are required for therapeutics, it seems unlikely that chromatography will lose its place in biopharmaceutical manufacturing in the near future (Hanke and Ottens 2014).

Lysostaphin, a metalloendopeptidase, produced by *L. lactis* was recovered from a 3000 L batch after concentration and washing of the cells, followed by cell disruption, removal of cell debris, and purification by chromatography techniques (Mierau et al. 2005). The fermentation broth was initially concentrated ca. 20-fold by tangential flow filtration and 200% diafiltrated to remove medium components. After continuous homogenization at 1400 bar at a rate of 80 L/h of the retentate, cell debris were removed by another tangential flow filtration. Lysostaphin was purified by cation-exchange chromatography using SP-Sepharose FF as support. At the end of the downstream process, ca. 40% of the originally produced lysostaphin could be recovered.

1.4.3 Enzyme Formulations

Irrespectively of the production method, the final formulation of therapeutic enzymes must display high purity and specificity, and they are often marketed as a lyophilized pure preparations with biocompatible buffering salts and diluents to prepare the reconstituted aqueous solution (Gurung et al. 2013).

Intravenous, intramuscular and subcutaneous injections are the most common ways to deliver biopharmaceuticals but oral and pulmonary delivery could provide also good alternatives (Mitragotri et al. 2014). However, recombinant proteins are susceptible to degradation (e.g. by proteases during digestion) and present a considerable molecular size.

To prevent enzymatic degradation in the stomach five of the six pancrelipase products approved by FDA for the treatment of exocrine pancreatic insufficiency are formulated as enteric-coated capsules to delay drug release until they reach the lower digestive tract (NDHHS 2012). The other pancrealipase that is not enteric-coated must be administered with a proton pump inhibitor to reduce the gastric pH and thus protein degradation.

Glutenases, which may be used for the treatment of Celiac Sprue and/or dermatitis herpetiformis in gluten intolerant patients, may be administered as: (i) oral preparations of enzymes isolated and purified from wild strains, such as *Penicillum citrinum* or *Lactobacillus helveticus*, or recombinant strains; (ii) engineered polymer microspheres containing the former enzymes; or (iii) a microorganism capable of producing glutenase (WO 2005/107786 A1; Shan et al. 2005).

The application of enzymes through inhalation is effective e.g. in patients with cystic fibrosis. Dornase alfa (Pulmozyme®), is a highly purified solution of recombinant human deoxyribonuclease I, an enzyme that selectively cleaves DNA. The enzyme is produced by CHO cells and is purified by tangential flow filtration and column chromatography (report 1993–2015).

Administration of insulin via an inhaler was not the expected success, with Pfizer and Sanofi presenting disappointing sales probably due to cost and lack of trust from physicians who questioned the control of the dose received by the patient (Oleck et al. 2016).

Acknowledgements CCCR de Carvalho acknowledges Fundação para a Ciência e a Tecnologia, I.P. (FCT), Portugal, for financial support under program "FCT Investigator 2013" (IF/01203/2013/CP1163/CT0002).

References

Abdel-Mageed HM, Fahmy AS, Shaker DS, Mohamed SA (2018) Development of novel delivery system for nanoencapsulation of catalase: formulation, characterization, and in vivo evaluation using oxidative skin injury model. Artif Cells Nanomed Biotechnol 46:1–10. https://doi.org/10.1080/21691401.2018.1425213

Abderrezak K (2018) Therapeutic enzymes used for the treatment of non-deficiency diseases. In: Shashi Lata B, Pankaj Kumar C (eds) Research advancements in pharmaceutical, nutritional, and industrial enzymology. IGI Global, Hershey, pp 46–70. https://doi.org/10.4018/978-1-5225-5237-6.ch003

Abdoli-Nasab M, Javaran MJ, Cusido RM, Palazon J (2016) Purification of recombinant tissue plasminogen activator (rtPA) protein from transplastomic tobacco plants. Plant Physiol Biochem 108:139–144. https://doi.org/10.1016/j.plaphy.2016.06.029

Adrio JL, Demain AL (2014) Microbial enzymes: tools for biotechnological processes. Biomol Ther 4(1):117–139. https://doi.org/10.3390/biom4010117

Aggarwal S (2007) What's fueling the biotech engine? Nat Biotechnol 25:1097–1104. https://doi.org/10.1038/nbt1007-1097

Agrawal V, Bal M (2012) Strategies for rapid production of therapeutic proteins in mammalian cells. Bioprocess Int 10(4):32–48

Allied_Market_Research (2018) Specialty enzymes market is expected to reach $947.5 million, globally, by 2020. https://www.alliedmarketresearch.com/press-release/specialty-enzymes.html. Accessed 17 Aug 2018

Andrady C, Sharma SK, Chester KA (2011) Antibody–enzyme fusion proteins for cancer therapy. Immunotherapy 3(2):193–211. https://doi.org/10.2217/imt.10.90

Arnau J, Lauritzen C, Petersen GE, Pedersen J (2006) Current strategies for the use of affinity tags and tag removal for the purification of recombinant proteins. Protein Expr Purif 48(1):1–13. https://doi.org/10.1016/j.pep.2005.12.002

Baldo BA (2015) Enzymes approved for human therapy: indications, mechanisms and adverse effects. BioDrugs 29(1):31–55. https://doi.org/10.1007/s40259-015-0116-7

Bonnerjea J (2004) Purification of therapeutic proteins. In: Cutler P (ed) Protein purification protocols. Humana Press, Totowa, pp 455–462. https://doi.org/10.1385/1-59259-655-x:455

Brady RO (2006) Enzyme replacement for lysosomal diseases. Annu Rev Med 57(1):283–296. https://doi.org/10.1146/annurev.med.57.110104.115650

Brandazza A, Lee E, Ferrera M, Tillman U, Sarmientos P, Westphal H (1991) Use of the urokinase-type plasminogen activator gene as a general tool to monitor expression in transgenic animals: study of the tissue-specificity of the murine whey acidic protein (WAP) expression signals. J Biotechnol 20(2):201–212. https://doi.org/10.1016/0168-1656(91)90228-N

Cereghino GPL, Cereghino JL, Ilgen C, Cregg JM (2002) Production of recombinant proteins in fermenter cultures of the yeast *Pichia pastoris*. Curr Opin Biotechnol 13(4):329–332. https://doi.org/10.1016/S0958-1669(02)00330-0

Cheng F, Kardashliev T, Pitzler C, Shehzad A, Lue H, Bernhagen J, Zhu L, Schwaneberg U (2015a) A competitive flow cytometry screening system for directed evolution of therapeutic enzyme. ACS Synth Biol 4(7):768–775. https://doi.org/10.1021/sb500343g

Cheng F, Zhu L, Lue H, Bernhagen J, Schwaneberg U (2015b) Directed arginine deiminase evolution for efficient inhibition of arginine-auxotrophic melanomas. Appl Microbiol Biotechnol 99(3):1237

Chon JH, Zarbis-Papastoitsis G (2011) New Biotechnol 28(5):458–463. https://doi.org/10.1016/j.nbt.2011.03.015

Daniell H, Singh ND, Mason H, Streatfield SJ (2009) Plant-made vaccine antigens and biopharmaceuticals. Trends Plant Sci 14(12):669–679. https://doi.org/10.1016/j.tplants.2009.09.009

Dean SN, Turner KB, Medintz IL, Walper SA (2017) Targeting and delivery of therapeutic enzymes. Ther Deliv 8(7):577–595. https://doi.org/10.4155/tde-2017-0020

Demain AL, Vaishnav P (2009) Production of recombinant proteins by microbes and higher organisms. Biotechnol Adv 27(3):297–306. https://doi.org/10.1016/j.biotechadv.2009.01.008

Derst C, Henseling J, Röhm KH (2000) Engineering the substrate specificity of *Escherichia coli* asparaginase. II. Selective reduction of glutaminase activity by amino acid replacements at position 248. Protein Sci 9(10):2009–2017

Doran PM (2013) Chapter 11: Unit operations. In: Doran PM (ed) Bioprocess engineering principles, 2nd edn. Academic, London, pp 445–595. https://doi.org/10.1016/B978-0-12-220851-5.00011-3

Dozier KJ, Distefano DM (2015) Site-specific PEGylation of therapeutic proteins. Int J Mol Sci 16(10):25831–25864. https://doi.org/10.3390/ijms161025831

Fang L, Chow KM, Hou S, Xue L, Chen X, Rodgers DW, Zheng F, Zhan C-G (2014) Rational design, preparation, and characterization of a therapeutic enzyme mutant with improved stability and function for cocaine detoxification. ACS Chem Biol 9(8):1764–1772. https://doi.org/10.1021/cb500257s

Farhadi SA, Bracho-Sanchez E, Freeman SL, Keselowsky BG, Hudalla GA (2018) Enzymes as immunotherapeutics. Bioconjug Chem 29(3):649–656. https://doi.org/10.1021/acs.bioconjchem.7b00719

Fazel R, Zarei N, Ghaemi N, Namvaran MM, Enayati S, Mirabzadeh Ardakani E, Azizi M, Khalaj V (2014) Cloning and expression of *Aspergillus flavus* urate oxidase in *Pichia pastoris*. Springerplus 3(1):395. https://doi.org/10.1186/2193-1801-3-395

Ferreira RG, Azzoni AR, Freitas S (2018) Techno-economic analysis of the industrial production of a low-cost enzyme using *E. coli*: the case of recombinant β-glucosidase. Biotechnol Biofuels 11(1):81. https://doi.org/10.1186/s13068-018-1077-0

Fong BA, Wu W-Y, Wood DW (2010) The potential role of self-cleaving purification tags in commercial-scale processes. Trends Biotechnol 28(5):272–279. https://doi.org/10.1016/j.tibtech.2010.02.003

Fuhrmann G, Leroux J-C (2014) Improving the stability and activity of oral therapeutic enzymes-recent advances and perspectives. Pharm Res 31(5):1099–1105. https://doi.org/10.1007/s11095-013-1233-y

Funaro MG, Nemani KV, Chen Z, Bhujwalla ZM, Griswold KE, Gimi B (2016) Effect of alginate microencapsulation on the catalytic efficiency and in vitro enzyme-prodrug therapeutic efficacy of cytosine deaminase and of recombinant *E. coli* expressing cytosine deaminase. J Microencapsul 33(1):64–70. https://doi.org/10.3109/02652048.2015.1115902

Giffard M, Ferté N, Ragot F, El Hajji M, Castro B, Bonneté F (2011) Urate oxidase purification by salting-in crystallization: towards an alternative to chromatography. PLoS One 6(5):e19013. https://doi.org/10.1371/journal.pone.0019013

Gomes AMV, Carmo TS, Carvalho LS, Bahia FM, Parachin NS (2018) Comparison of yeasts as hosts for recombinant protein production. Microorganisms 6(2):38. https://doi.org/10.3390/microorganisms6020038

Gonzalez NJ, Isaacs LL (1999) Evaluation of pancreatic proteolytic enzyme treatment of adenocarcinoma of the pancreas, with nutrition and detoxification support. Nutr Cancer 33(2):117–124. https://doi.org/10.1207/s15327914nc330201

Goojani HG, Javaran MJ, Nasiri J, Goojani EG, Alizadeh H (2013) Expression and large-scale production of human tissue plasminogen activator (t-PA) in transgenic tobacco plants using different signal peptides. Appl Biochem Biotechnol 169(6):1940–1951. https://doi.org/10.1007/s12010-013-0115-4

Goward CR, Stevens GB, Tattersall R, Atkinson T (1992) Rapid large-scale preparation of recombinant *Erwinia chrysanthemi* L-asparaginase. Bioseparation 2(6):335–341

Grabowski GA, Barton NW, Pastores G et al (1995) Enzyme therapy in type 1 gaucher disease: comparative efficacy of mannose-terminated glucocerebrosidase from natural and recombinant sources. Ann Intern Med 122(1):33–39. https://doi.org/10.7326/0003-4819-122-1-199501010-00005

Gupta SK, Shukla P (2017) Sophisticated cloning, fermentation, and purification technologies for an enhanced therapeutic protein production: a review. Front Pharmacol 8:419. https://doi.org/10.3389/fphar.2017.00419

Gurung N, Ray S, Bose S, Rai V (2013) A broader view: microbial enzymes and their relevance in industries, medicine, and beyond. Biomed Res Int 2013:329121. https://doi.org/10.1155/2013/329121

Hanke AT, Ottens M (2014) Purifying biopharmaceuticals: knowledge-based chromatographic process development. Trends Biotechnol 32(4):210–220. https://doi.org/10.1016/j.tibtech.2014.02.001

Headon DR, Walsh G (1994) The industrial production of enzymes. Biotechnol Adv 12(4):635–646. https://doi.org/10.1016/0734-9750(94)90004-3

Hellwig S, Drossard J, Twyman RM, Fischer R (2004) Plant cell cultures for the production of recombinant proteins. Nat Biotechnol 22:1415–1422. https://doi.org/10.1038/nbt1027

Hidalgo D, Abdoli-Nasab M, Jalali-Javaran M, Bru-Martínez R, Cusidó RM, Corchete P, Palazon J (2017) Biotechnological production of recombinant tissue plasminogen activator protein (reteplase) from transplastomic tobacco cell cultures. Plant Physiol Biochem 118:130–137. https://doi.org/10.1016/j.plaphy.2017.06.013

Ho SV (1990) Strategies for large-scale protein purification. In: Protein purification. vol 427. ACS symposium series, vol 427. American Chemical Society, pp 14–34. https://doi.org/10.1021/bk-1990-0427.ch002

Hohmann H, van Dijl JM, Krishnappa L, Prágai Z (2016) Host organisms: *Bacillus subtilis*. In: Wittmann C, Liao JC (eds) Industrial biotechnology, vol 1. Wiley, Weinheim. https://doi.org/10.1002/9783527807796.ch7

Huang C Jr, Lin H, Yang X (2012) Industrial production of recombinant therapeutics in *Escherichia coli* and its recent advancements. J Ind Microbiol Biotechnol 39(3):383–399. https://doi.org/10.1007/s10295-011-1082-9

Huang Y-M, Hu W, Rustandi E, Chang K, Yusuf-Makagiansar H, Ryll T (2010) Maximizing productivity of CHO cell-based fed-batch culture using chemically defined media conditions and typical manufacturing equipment. Biotechnol Prog 26(5):1400–1410. https://doi.org/10.1002/btpr.436

Jaliani HZ, Farajnia S, Mohammadi SA, Barzegar A, Talebi S (2013) Engineering and kinetic stabilization of the therapeutic enzyme *Anabeana variabilis* phenylalanine ammonia lyase. Appl Biochem Biotechnol 171(7):1805–1818. https://doi.org/10.1007/s12010-013-0450-5

Jayapal KP, Wlaschin KF, Hu WS, Yap MGS (2007) Recombinant protein therapeutics from CHO cells – 20 years and counting. Chem Eng Prog 103:40–47

Jeyaharan D, Aston P, Garcia-Perez A, Schouten J, Davis P, Dixon AM (2016) Soluble expression, purification and functional characterisation of carboxypeptidase G2 and its individual domains. Protein Expr Purif 127:44–52. https://doi.org/10.1016/j.pep.2016.06.015

Kakkis ED, Matynia A, Jonas AJ, Neufeld EF (1994) Overexpression of the human lysosomal enzyme α-L-iduronidase in Chinese hamster ovary cells. Protein Expr Purif 5(3):225–232. https://doi.org/10.1006/prep.1994.1035

Kaur R, Sekhon BS (2012) Enzymes as drugs: an overview. J Pharm Educ Res 3(2):29–41

Khushoo A, Pal Y, Singh BN, Mukherjee KJ (2004) Extracellular expression and single step purification of recombinant *Escherichia coli* L-asparaginase II. Protein Expr Purif 38(1):29–36. https://doi.org/10.1016/j.pep.2004.07.009

Kotzia GA, Labrou NE (2009) Engineering thermal stability of L-asparaginase by in vitro directed evolution. FEBS J 276(6):1750–1761. https://doi.org/10.1111/j.1742-4658.2009.06910.x

Kumar SS, Abdulhameed S (2017) Therapeutic enzymes. In: Sugathan S, Pradeep NS, Abdulhameed S (eds) Bioresources and bioprocess in biotechnology: volume 2 : Exploring potential biomolecules. Springer, Singapore, pp 45–73. https://doi.org/10.1007/978-981-10-4284-3_2

Kunamneni A, Ogaugwu C, Goli D (2018) Chapter 15: Enzymes as therapeutic agents. In: Nunes C, Kumar V (eds) Enzymes in human and animal. Nutrition Academic Press, Oxford, pp 301–312. https://doi.org/10.1016/b978-0-12-805419-2.00015-0

Lalonde M-E, Durocher Y (2017) Therapeutic glycoprotein production in mammalian cells. J Biotechnol 251:128–140. https://doi.org/10.1016/j.jbiotec.2017.04.028

Langer ES (2017) Trends and growth in single-use system (SUS) adoption. Am Pharm Rev 343797

Laukens B, Visscher CD, Callewaert N (2015) Engineering yeast for producing human glycoproteins: where are we now? Future Microbiol 10(1):21–34. https://doi.org/10.2217/fmb.14.104

Laurent JM, Young JH, Kachroo AH, Marcotte EM (2016) Efforts to make and apply humanized yeast. Brief Funct Genomics 15(2):155–163. https://doi.org/10.1093/bfgp/elv041

Leplatois P, Le Douarin B, Loison G (1992) High-level production of a peroxisomal enzyme: *Aspergillus flavus* uricase accumulates intracellularly and is active in *Saccharomyces cerevisiae*. Gene 122(1):139–145. https://doi.org/10.1016/0378-1119(92)90041-M

Li M (2018) Enzyme replacement therapy: a review and its role in treating lysosomal storage diseases. Pediatr Ann 47(5):e191–e197. https://doi.org/10.3928/19382359-20180424-01

Li J, Chen Z, Hou L, Fan H, Weng S, Ce X, Ren J, Li B, Chen W (2006) High-level expression, purification, and characterization of non-tagged *Aspergillus flavus* urate oxidase in *Escherichia coli*. Protein Expr Purif 49(1):55–59. https://doi.org/10.1016/j.pep.2006.02.003

Li W, Xu S, Zhang B, Zhu Y, Hua Y, Kong X, Sun L, Hong J (2017) Directed evolution to improve the catalytic efficiency of urate oxidase from *Bacillus subtilis*. PLoS One 12(5):e0177877. https://doi.org/10.1371/journal.pone.0177877

Lim J, Sinclair A, Shevitz J, Carter JB (2011) An economic comparison of three cell culture techniques: fed-batch, concentrated fed-batch, and concentrated perfusion. BioPharm Int 24(2):54–60

Löffler P, Schmitz S, Hupfeld E, Sterner R, Merkl R (2017) Rosetta:MSF: a modular framework for multistate computational protein design. PLoS Comput Biol 13(6):e1005600. https://doi.org/10.1371/journal.pcbi.1005600

Lutz S, Williams E, Muthu P (2017) Engineering therapeutic enzymes. In: Alcalde M (ed) Directed enzyme evolution: advances and applications. Springer, Cham, pp 17–67. https://doi.org/10.1007/978-3-319-50413-1_2

Mane P, Tale V (2015) Overview of microbial therapeutic enzymes. Int J Curr Microbiol Appl Sci 4(4):17–26

Market Research Engine (2018) Specialty enzymes market to touch US$ 5.5 Billion by 2022. Herald Keeper. http://heraldkeeper.com/market/specialty-enzymes-market-touch-us-5-5-billion-2022-2-96559.html. Accessed 17 Aug 2018

Merlin M, Gecchele E, Capaldi S, Pezzotti M, Avesani L (2014) Comparative evaluation of recombinant protein production in different biofactories: the green perspective. Biomed Res Int 2014:136419–136419. https://doi.org/10.1155/2014/136419

Mierau I, Leij P, van Swam I, Blommestein B, Floris E, Mond J, Smid EJ (2005) Industrial-scale production and purification of a heterologous protein in *Lactococcus lactis* using the nisin-controlled gene expression system NICE: the case of lysostaphin. Microb Cell Factories 4(1):15. https://doi.org/10.1186/1475-2859-4-15

Mishra P, Nayak B, Dey RK (2016) PEGylation in anti-cancer therapy: an overview. Asian J Pharm Sci 11(3):337–348. https://doi.org/10.1016/j.ajps.2015.08.011

Mitragotri S, Burke PA, Langer R (2014) Overcoming the challenges in administering biopharmaceuticals: formulation and delivery strategies. Nat Rev Drug Discov 13(9):655–672. https://doi.org/10.1038/nrd4363

Muro S (2010) New biotechnological and nanomedicine strategies for treatment of lysosomal storage disorders. Wiley Interdiscip Rev Nanomed Nanobiotechnol 2(2):189–204. https://doi.org/10.1002/wnan.73

Nazari-Robati M, Khajeh K, Aminian M, Mollania N, Golestani A (2013) Enhancement of thermal stability of chondroitinase ABC I by site-directed mutagenesis: an insight from Ramachandran plot. BBA-Proteins Proteomics 1834(2):479–486. https://doi.org/10.1016/j.bbapap.2012.11.002

NDHHS NDoHaHS (2012) Therapeutic class overview pancreatic enzymes. https://www.medicaid.nv.gov/Downloads/provider/Pancreatic%20Enzymes.pdf

Nijland R, Kuipers O (2008) Optimization of protein secretion by *Bacillus subtilis*. Recent Pat Biotechnol 2(2):79–87. https://doi.org/10.2174/187220808784619694

Noppen B, Fonteyn L, Aerts F, De Vriese A, De Maeyer M, Le Floch F, Barbeaux P, Zwaal R, Vanhove M (2014) Autolytic degradation of ocriplasmin: a complex mechanism unraveled by mutational analysis. Protein Eng Des Sel 27(7):215–223. https://doi.org/10.1093/protein/gzu015

Oh D-B (2015) Glyco-engineering strategies for the development of therapeutic enzymes with improved efficacy for the treatment of lysosomal storage diseases. BMB Rep 48(8):438–444. https://doi.org/10.5483/BMBRep.2015.48.8.101

Oleck J, Kassam S, Goldman JD (2016) Diabetes Spectr 29(3):180–184. https://doi.org/10.2337/diaspect.29.3.180

Pan Y, Gao D, Yang W, Cho H, Yang G, Tai H-H, Zhan C-G (2005) Computational redesign of human butyrylcholinesterase for anticocaine medication. Proc Natl Acad Sci U S A 102(46):16656–16661. https://doi.org/10.1073/pnas.0507332102

Patil PN (2012) Discoveries in pharmacological sciences. Discoveries in pharmacological sciences. World Scientific Publishing Co., Singapore. https://doi.org/10.1142/8174

Patti T, Bembi B, Cristin P, Mazzarol F, Secco E, Pappalardo C, Musetti R, Martinuzzi M, Versolatto S, Cariati R, Dardis A, Marchetti S (2012) Endosperm-specific expression of human acid beta-glucosidase in a waxy rice. Rice 5(1):34. https://doi.org/10.1186/1939-8433-5-34

Persistence Market Research (2018) Therapeutic enzymes market: global industry trend analysis 2012 to 2017 and forecast 2017–2025. https://www.persistencemarketresearch.com/market-research/therapeutic-enzymes-market.asp. Accessed 14 Sept 2018

Petersen C (2015) History of health. Manuf Today :90–92

Phue J-N, Shiloach J (2004) Transcription levels of key metabolic genes are the cause for different glucose utilization pathways in *E. coli* B (BL21) and *E. coli* K (JM109). J Biotechnol 109(1):21–30. https://doi.org/10.1016/j.jbiotec.2003.10.038

Qi Y, Chilkoti A (2015) Protein-polymer conjugation – moving beyond PEGylation. Curr Opin Chem Biol 28:181–193. https://doi.org/10.1016/j.cbpa.2015.08.009

Qiu J, Swartz JR, Georgiou G (1998) Expression of active human tissue-type plasminogen activator in *Escherichia coli*. Appl Environ Microbiol 64(12):4891–4896

Ramirez-Paz J, Saxena M, Delinois LJ, Joaquín-Ovalle FM, Lin S, Chen Z, Rojas-Nieves VA, Griebenow K (2018) Thiol-maleimide poly(ethylene glycol) crosslinking of L-asparaginase subunits at recombinant cysteine residues introduced by mutagenesis. PLoS One 13(7):e0197643. https://doi.org/10.1371/journal.pone.0197643

Richette P, Bardin T (2006) Successful treatment with rasburicase of a tophaceous gout in a patient allergic to allopurinol. Nat Clin Pract Rheumatol 2(6):338–342.; quiz 343. https://doi.org/10.1038/ncprheum0214

Rodríguez-Martínez JA, Rivera-Rivera I, Solá RJ, Griebenow K (2009) Enzymatic activity and thermal stability of PEG-α-chymotrypsin conjugates. Biotechnol Lett 31(6):883–887. https://doi.org/10.1007/s10529-009-9947-y

Roque ACA, Lowe CR, Taipa MÂ (2004) Antibodies and genetically engineered related molecules: production and purification. Biotechnol Prog 20(3):639–654. https://doi.org/10.1021/bp030070k

Roque ACA, Silva CSO, Taipa MÂ (2007) Affinity-based methodologies and ligands for antibody purification: advances and perspectives. J Chromatogr A 1160(1):44–55. https://doi.org/10.1016/j.chroma.2007.05.109

Roxas M (2008) The role of enzyme supplementation in digestive disorders. Altern Med Rev 13(4):307–314

Ryan BJ, Henehan GT (2013) Overview of approaches to preventing and avoiding proteolysis during expression and purification of proteins. Curr Protoc Protein Sci 71(1):5.25.21–25.25.27. https://doi.org/10.1002/0471140864.ps0525s71

Sabalza M, Christou P, Capell T (2014) Recombinant plant-derived pharmaceutical proteins: current technical and economic bottlenecks. Biotechnol Lett 36(12):2367–2379. https://doi.org/10.1007/s10529-014-1621-3

Samish I (2017) Achievements and challenges in computational protein design. In: Samish I (ed) Computational protein design. Springer, New York, pp 21–94. https://doi.org/10.1007/978-1-4939-6637-0_2

Shahaboddin ME, Khajeh K, Maleki M, Golestani A (2017) Improvement of activity and stability of Chondroitinase ABC I by introducing an aromatic cluster at the surface of protein. Enzym Microb Technol 105:38–44. https://doi.org/10.1016/j.enzmictec.2017.06.004

Shan L, Bethune M, Khosla C, Gass J (2005) Therapeutic enzyme formulations and uses thereof WO/2005/107786

Shanley N, Walsh G (2005) Applied enzymology, an overview. In: McGrath BM, Walsh G (eds) Directory of therapeutic enzymes. CRC Press, Boca Raton, pp 1–16

Sharabi O, Erijman A, Shifman JM (2013) Chapter 3: Computational methods for controlling binding specificity. In: Keating AE (ed) Methods in enzymology, vol 523. Academic, Burlington, pp 41–59. https://doi.org/10.1016/B978-0-12-394292-0.00003-5

Sharifi-Sirchi GR, Jalali-Javaran M (2016) Selecting appropriate hosts for recombinant proteins production: review article. HMJ 20(3):214–222

Sharma SK, Bagshawe KD (2017) Antibody directed enzyme prodrug therapy (ADEPT): trials and tribulations. Adv Drug Deliv Rev 118:2–7. https://doi.org/10.1016/j.addr.2017.09.009

Singh R, Kumar M, Mittal A, Mehta PK (2016) Microbial enzymes: industrial progress in 21st century. 3 Biotech 6(2):174–174. https://doi.org/10.1007/s13205-016-0485-8

Sizer IW (1972) Medical applications of microbial enzymes. In: Perlman D (ed) Advances in applied microbiology, vol 15. Academic, New York, pp 1–11. https://doi.org/10.1016/S0065-2164(08)70087-X

Smelko JP, Wiltberger KR, Hickman EF, Morris BJ, Blackburn TJ, Ryll T (2011) Performance of high intensity fed-batch mammalian cell cultures in disposable bioreactor systems. Biotechnol Prog 27(5):1358–1364. https://doi.org/10.1002/btpr.634

Streatfield SJ (2007) Approaches to achieve high-level heterologous protein production in plants. Plant Biotechnol J 5(1):2–15. https://doi.org/10.1111/j.1467-7652.2006.00216.x

Tobin PH, Richards DH, Callender RA, Wilson CJ (2014) Protein engineering: a new frontier for biological therapeutics. Curr Drug Metab 15(7):743–756

Tripathi NK (2016) Production and purification of recombinant proteins from *Escherichia coli*. ChemBioEng Rev 3(3):116–133. https://doi.org/10.1002/cben.201600002

Tripathi N, Kannusamy S, Jana A, Rao PVL (2009) High yield production of heterologous proteins with *Escherichia coli*. Def Sci J 59(2):137–146. https://doi.org/10.14429/dsj.59.1501

Tundisi LL, Coêlho DF, Zanchetta B, Moriel P, Pessoa A, Tambourgi EB, Silveira E, Mazzola PG (2017) L-Asparaginase purification. Sep Purif Rev 46(1):35–43. https://doi.org/10.1080/15422119.2016.1184167

Varga I, Bobalova M, Michalkova E, Jakubcova M (2011) Method of polymyxin B recovery from fermentation broth. USA Patent US7951913B2, 2011-05-31

Vellard M (2003) The enzyme as drug: application of enzymes as pharmaceuticals. Curr Opin Biotechnol 14(4):444–450. https://doi.org/10.1016/S0958-1669(03)00092-2

Verma N, Kumar K, Kaur G, Anand S (2007) L-asparaginase: a promising chemotherapeutic agent. Crit Rev Biotechnol 27(1):45–62. https://doi.org/10.1080/07388550601173926

Vidya J, Ushasree MV, Pandey A (2014) Effect of surface charge alteration on stability of L-asparaginase II from *Escherichia* sp. Enzym Microb Technol 56:15–19. https://doi.org/10.1016/j.enzmictec.2013.12.012

Vidya J, Sajitha S, Ushasree MV, Binod P, Pandey A (2017) 12 – Therapeutic enzymes: l-Asparaginases. In: Pandey A, Negi S, Soccol CR (eds) Current developments in biotechnology and bioengineering. Elsevier, Amsterdam, pp 249–265. https://doi.org/10.1016/B978-0-444-63662-1.00012-9

Walsh G (2014) Biopharmaceutical benchmarks 2014. Nat Biotechnol 32:992. https://doi.org/10.1038/nbt.3040. https://www.nature.com/articles/nbt.3040#supplementary-information

Wang A, Lewus R, Rathore AS (2006) Comparison of different options for harvest of a therapeutic protein product from high cell density yeast fermentation broth. Biotechnol Bioeng 94(1):91–104. https://doi.org/10.1002/bit.20816

Waugh DS (2011) An overview of enzymatic reagents for the removal of affinity tags. Protein Expr Purif 80(2):283–293. https://doi.org/10.1016/j.pep.2011.08.005

Wells E, Robinson AS (2017) Cellular engineering for therapeutic protein production: product quality, host modification, and process improvement. Biotechnol J 12(1):1600105. https://doi.org/10.1002/biot.201600105

Westers L, Westers H, Quax WJ (2004) *Bacillus subtilis* as cell factory for pharmaceutical proteins: a biotechnological approach to optimize the host organism. BBA-Mol Cell Res 1694(1):299–310. https://doi.org/10.1016/j.bbamcr.2004.02.011

Wijma HJ, Janssen DB (2013) Computational design gains momentum in enzyme catalysis engineering. FEBS J 280(13):2948–2960. https://doi.org/10.1111/febs.12324

Wildt S, Gerngross TU (2005) The humanization of *N*-glycosylation pathways in yeast. Nat Rev Microbiol 3:119–128. https://doi.org/10.1038/nrmicro1087

Willson RC, Ladisch MR (1990) Large-scale protein purification. In: Protein purification. vol 427. ACS Symposium Series, vol 427. American Chemical Society, pp 1–13. https://doi.org/10.1021/bk-1990-0427.ch001

Xiao H, Bao Z, Zhao H (2015) High throughput screening and selection methods for directed enzyme evolution. Ind Eng Chem Res 54(16):4011–4020. https://doi.org/10.1021/ie503060a

Xue L, Ko M-C, Tong M, Yang W, Hou S, Fang L, Liu J, Zheng F, Woods JH, Tai H-H, Zhan C-G (2011) Design, preparation, and characterization of high-activity mutants of human butyrylcholinesterase specific for detoxification of cocaine. Mol Pharmacol 79(2):290–297. https://doi.org/10.1124/mol.110.068494

Yang H, Li J, Du G, Liu L (2017) Chapter 6: Microbial production and molecular engineering of industrial enzymes: challenges and strategies. In: Brahmachari G (ed) Biotechnology of microbial enzymes. Academic, Amsterdam, pp 151–165. https://doi.org/10.1016/B978-0-12-803725-6.00006-6

Yari M, Ghoshoon MB, Vakili B, Ghasemi Y (2017) Therapeutic enzymes: applications and approaches to pharmacological improvement. Curr Pharm Biotechnol 18(7):531–540. https://doi.org/10.2174/1389201018666170808150742

Zhang L, Liu M, Jamil S, Han R, Xu G, Ni Y (2015) PEGylation and pharmacological characterization of a potential anti-tumor drug, an engineered arginine deiminase originated from *Pseudomonas plecoglossicida*. Cancer Lett 357(1):346–354. https://doi.org/10.1016/j.canlet.2014.11.042

Zhao K-W, Neufeld EF (2000) Purification and characterization of recombinant human α-*N*-acetylglucosaminidase secreted by Chinese hamster ovary cells. Protein Expr Purif 19(1):202–211. https://doi.org/10.1006/prep.2000.1230

Zheng F, Zhan C-G (2008) Rational design of an enzyme mutant for anti-cocaine therapeutics. J Comput Aided Mol Des 22(9):661–671. https://doi.org/10.1007/s10822-007-9144-9

Zhu MM, Mollet M, Hubert RS (2007) Industrial production of therapeutic proteins: cell lines, cell culture, and purification. In: Kent JA (ed) Kent and Riegel's handbook of industrial chemistry and biotechnology. Springer, Boston, pp 1421–1448. https://doi.org/10.1007/978-0-387-27843-8_32

Zhu L, Tee KL, Roccatano D, Sonmez B, Ni Y, Sun Z-H, Schwaneberg U (2010) Directed evolution of an antitumor drug (arginine deiminase PpADI) for increased activity at physiological pH. Chembiochem 11(5):691–697. https://doi.org/10.1002/cbic.200900717

Zhu MM, Mollet M, Hubert RS, Kyung YS, Zhang GG (2017) Industrial production of therapeutic proteins: cell lines, cell culture, and purification. In: Kent JA, Bommaraju TV, Barnicki SD (eds) Handbook of industrial chemistry and biotechnology. Springer, Cham, pp 1639–1669. https://doi.org/10.1007/978-3-319-52287-6_29

Zimran A, Brill-Almon E, Chertkoff R, Petakov M, Blanco-Favela F, Terreros Muñoz E, Solorio-Meza SE, Amato D, Duran G, Giona F, Heitner R, Rosenbaum H, Giraldo P, Mehta A, Park G, Phillips M, Elstein D, Altarescu G, Szleifer M, Hashmueli S, Aviezer D (2011) Pivotal trial with plant-cell–expressed recombinant glucocerebrosidase, taliglucerase alfa, a novel enzyme replacement therapy for Gaucher disease. Blood 118(22):5767–5773. https://doi.org/10.1182/blood-2011-07-366955

2 Production of Therapeutic Enzymes by Lentivirus Transgenesis

María Celeste Rodríguez, Natalia Ceaglio, Sebastián Antuña, María Belén Tardivo, Marina Etcheverrigaray, and Claudio Prieto

Abstract

Since ERT for several LSDs treatment has emerged at the beginning of the 1980s with Orphan Drug approval, patients' expectancy and life quality have been improved. Most LSDs treatment are based on the replaced of mutated or deficient protein with the natural or recombinant protein.

One of the main ERT drawback is the high drug prices. Therefore, different strategies trying to optimize the global ERT biotherapeutic production have been proposed. LVs, a gene delivery tool, can be proposed as an alternative method to generate stable cell lines in manufacturing of recombinant proteins. Since LVs have been used in human gene therapy, clinical trials, safety testing assays and procedures have been developed. Moreover, one of the main advantages of LVs strategy to obtain manufacturing cell line is the short period required as well as the high protein levels achieved.

In this chapter, we will focus on LVs as a recombinant protein production platform and we will present a case study that employs LVs to express in a manufacturing cell line, alpha-Galactosidase A (rhαGAL), which is used as ERT for Fabry disease treatment.

Keywords

ERT · LSDs · Fabry disease · Lentiviral Vectors (LVs) · rhαGAL

M. C. Rodríguez · N. Ceaglio · M. Etcheverrigaray
Cell Culture Laboratory, UNL, CONICET, FBCB, Santa Fe, Argentina

S. Antuña · M. B. Tardivo
Zelltek S.A., Santa Fe, Argentina

C. Prieto (✉)
Cell Culture Laboratory, UNL, FBCB, Santa Fe, Argentina
e-mail: cprieto@fbcb.unl.edu.ar

Abbreviations

E. coli	*Escherichia coli*
ER	endothelial reticulum
PEG	polyethylene glycol
FDA	Food and Drug Administration
LSD	lysosomal storage disease
MPS	mucopolysaccaridosis
ERT	enzyme replacement therapy
LVs	lentiviral vectors
CHO	Chinese Hamster Ovary
Neu5Gc	*N*-glycolylneuraminic acid
BHK-21	baby hamster kidney cells
HEK293	Human Embryonic Kidney 293

N. Labrou (ed.), *Therapeutic Enzymes: Function and Clinical Implications*, Advances in Experimental Medicine and Biology 1148, https://doi.org/10.1007/978-981-13-7709-9_2

EMA	European Medicines Agency
CAP	CEVEC´s Amniocyte Production
ICH	International Conference of Harmonization
TGE	transient gene expression
DHFR	dihydrofolate reductase
GS	glutamine synthetase
MTX	methotrexate
MSX	methionine sulfoximine
HPRT	hypoxanthine phosphoribosyl transferase
RMCE	recombinase-mediated cassette exchange
FRT	flippase recognition target sites
ZFNs	zinc finger nucleases
TALENs	transcription activator-like effector nucleases
NHEJ	non-homologous end-joining
HDR	homology-directed repair
CRISPR	clustered regularly interspaced short palindromic repeats
SpCas9	*Streptococcus pyogenes* Cas9 endonuclease
gRNA	guide RNA
trcRNA	trans-acting antisense RNA
PAM	protospacer adjacent motif
S/MARs	matrix attachment regions
UCOEs	Ubiquitous Chromatin Opening Elements
PEI	Polyethylenimine
RT	reverse transcriptase
IN	integrase
TG	transgene
VSV	Vesicular Stomatitis Virus
LTR	long terminal repeats
SIN	self-inactivating
RRE	rev response elements
cPPT	Central Polypurine tract
WPRE	post-transcriptional regulatory element of the woodchuck
qPCR	real time polymerase chain reaction
MOI	multiplicity of infection
CAR	chimeric antigen receptor
RCL	replication-competent lentivirus
ELISA	enzyme-linked immunosorbent assay
PERT	product-enhanced reverse transcriptase
LOD	limit of detection
M6P	phosphate-6-O-mannose
GlcNAc	*N*-acetylglucosamine
M6PR	M6P receptor
C6S	chondroitin-6-sulfate
KS	keratan sulfate
GALNS	*N*-acetylgalactosamine-6-sulfate sulfatase
LAL	Lysosomal acid lipase
Gb3	globotriaosylceramide
rhαGAL	recombinant human alpha galactosidase A
LP	lentiviral particle
IEX	ionic exchange
HIC	Hydrophobic interaction chromatography
RP-HPLC	reversed phase high performance liquid chromatography
IEF	isoelectrofocusing
HPAEC-PAD	high-pH anion-exchange chromatography with pulsed amperometric detection
WAX	weak anion exchange
Man	Mannose
Gal	Galactose
GlcNAc	*N*-acetylglucosamine
Fuc	Fucose
4MU-α-Gal	4-Methylumbelliferyl α-D-galactopyranoside

2.1 Introduction

Recombinant protein production has emerged in the 1980s and, since then, have gained importance. Indeed, different strategies have been developed to improve the protein production levels, and thus, reducing the cost. Nowadays, there are extensive expression systems available for large-scale recombinant protein production: *E. coli*, baculovirus-mediated insect cell expression, yeast, as well as several mammalian-based systems can be mentioned as examples. Platforms like *E. coli*, yeast and insect cells are still the preferred systems for many protein expression groups; however, to obtain optimum yields and activities, mammalian proteins often require to be expressed in animal cells systems (Bandaranayake and Almo 2014). In addition, an adequate protein glycosylation profile is always desirable, determining proper biological functions of many proteins. For instance, many cell surface proteins are glycosylated and will not assemble correctly into membranes or function properly if lacking their sugar residues (Clark and Pazdernik 2016).

Bacterial cells are by far the cheapest and fastest platform for recombinant protein expression. In addition, they are able to glycosylate some proteins through the attachment of glycosyl groups to side-chain oxygens by a process known as O-glycosylation. However, they do not *N*-glycosylate and, in many cases, they do not allow correct folding of mammalian proteins. Other recombinant proteins are produced in yeast, which can also divide rapidly and generate high yields. However, these cells produce glycoproteins with high-mannose glycan structures, which may be immunogenic if they are used as biotherapeutics. Examples of approved therapeutics from yeast expression systems are ocriplasmin (JETREA®) and catridecagog (TRETTEN®). Plants and baculovirus are able to produce recombinant proteins with complex glycans but, like yeast, these could be immunogenic. In addition, both plant and baculovirus system glycoproteins lack sialic acid residues, a monosaccharide playing an essential role in prolonging protein's serum half-life. Currently, the only therapeutics approved from insect cell expression systems are the human papillomavirus vaccine (CERVARIX®), the prostate cancer immunotherapy vaccine (PROVENGE®) and the flu vaccine (FLUBLOK®) (Dumont et al. 2016; Cox 2012;). Regarding plant cells, several attempts have been made to improve glycosylation profile. For instance, Pegunigalsidase alfa (PRX-102, Protalix Biotherapeutics) was expressed in Tobacco BY2 cell culture and targeted to the ER adding of a C-terminal KDEL signal peptide. Consequently, as expected of an ER retained protein, it exhibited mainly high mannose structures, in addition to paucimannose structures. However, in the absence of sialic acid residues, the long circulatory half-life of PRX-102 was achieved by chemical modification with a homo-bifunctional PEG molecule (Kizhner et al. 2015).

On the contrary, mammalian cells allow proper glycosylation and folding of proteins, but their culture is more expensive, growth rate is slower and recombinant protein yields tend to be lower. Therefore, determining the system to express recombinant proteins depends on the protein and its particular properties.

About 60–70% of the FDA-approved recombinant proteins for therapeutic use are obtained employing mammalian cells; around 20% using *E. coli* and the remainder using baculovirus, yeast and transgenic animals as expression systems (Clark and Pazdernik 2016). Among therapeutic proteins, enzymes represent a small and profitable market (Kumar and Abdulhameed 2017). Enzymes are biological catalysts that promote the majority of reactions occurring in biological systems. They are crucial pillars to maintain life, as without them these reactions would not take place in a proper period of time (Porter et al. 2016). The use of enzymes as biotherapeutics was in practice long ago, employing crude preparations as digestive aids for gastrointestinal diseases. Their potential as a therapeutic tool was first described during the development of an effective anthrax treatment using crude secretions from *Bacillus pyocyaneus*was, suggesting that enzymes are responsible for the action. Today, enzymes have clinical applications as cardiovascular agents, oncolytics, digestive aids, in the replacement therapy of lysosomal storage disorders and wound debridement agents, just to mention a few. Enzymes' specificity represents an advantage in comparison with non-enzymatic drugs (Kumar and Abdulhameed 2017).

Development of enzyme therapeutics against rare diseases such as LSDs was initially supported by the Orphan Drug Act approved in 1983 in the USA. This legislation offer incentives such as a 7-year marketing exclusivity, and availability of grant funding (~$400,000 per year for 4 years) for rare diseases, defined as those affecting less than 200,000 individuals in the USA. Since 2000, in the European Union, orphan designation, defined by an incidence of not more than five in 10,000 individuals, was available, conferring a 10-year market exclusivity, tax benefits and application fees waivers. Indeed, similar orphan drug legislations were approved in several countries such as Australia (1988), Japan (1993) and Singapore (1997) (Solomon and Muro 2017).

Alglucerase (Ceredase®) was the first enzyme for replacement therapy approved by the FDA in 1991 against a LSD named Gaucher disease. The LSDs are a group of 50–60 distinct metabolic disorders exhibiting an abnormal, progressive store of non-degraded substances in altered cells, which can be debilitating, even fatal (Kumar and Abdulhameed 2017; Solomon and Muro 2017). All LSDs are monogenetic diseases inherited as autosomal recessive manner, except Fabry disease and MPS type II, which are both inherited as X-linked recessive condition, and Danon disease inherited in an X-linked dominant pattern (Solomon and Muro 2017).

In general, the LSD management can be grouped into two types of treatments: palliative approaches and disease specific treatments. The former is used to treat symptoms, whereas the latter is used to correct or ameliorate the biochemical defect. Among other approaches such as gene therapy, small molecules therapy and organ/cell transplantation, ERT represents an adequate strategy for eight out of the 50–60 LSDs. Although the average cost of ERT ranges from $50,000–200,000 per year, it is the most effective therapy to this date even when other shortcomings like size, antigenicity and stability are involved. Consequently, ERT improvements are being continuously investigated trying to increase therapeutic efficacy, reduce dosage and adverse effects, and ultimately increase ERT feasibility (Kumar and Abdulhameed 2017; Solomon and Muro 2017).

In this chapter, we will focus on this final issue, providing an overview of the different strategies trying to optimize the active recombinant enzymes' production processes. Besides, we will focus our discussion on a specific gene delivery tool that enhances protein expression based on LVs. Finally, we will present a case study which employs LVs to express alpha-Galactosidase A, used as ERT for Fabry disease, in a manufacturing cell line.

2.2 Production of Therapeutic Enzymes in Animal Cell Cultures

Protein therapeutics represent the largest group of new products in development by the biopharmaceutical industry (Dumont et al. 2016). The manufacturing and production of these drugs imply many steps. Indeed, protein therapeutics exhibit complex secondary, tertiary and quaternary structures that must be maintained, and many require posttranslational modifications. These probably are the main reasons why protein therapeutics cannot be completely synthesized by chemical processes and have to be manufactured in living cells (Lagassé et al. 2017).

As we have mentioned above, biotherapeutic proteins are produced in a wide variety of platforms, including non-mammalian hosts (bacteria, yeast, plant and insect) and mammalian expression systems (including human cell lines). In addition, the most appropriate expression platform depends on the particular protein to be expressed. Consequently, the choice of the cell line, species origin, and culture conditions all affect the final properties of the proteins that will reach the market. Thus, it is desirable to obtain a high-quality final product maintaining not only its safety attributes but also its efficacy (Lagassé et al. 2017). Regarding this, the mammalian cells are the preferred expression system due to their ability to produce complex proteins with a proper glycosylation profile similar to those produced in humans. Moreover, contrary to other expression platforms such as bacteria, most proteins can be secreted rather than requiring cell lysing facilitating and reducing the cost of the downstream processing.

The workhorse of mammalian protein production, especially at industrial scale, is the CHO cell line, isolated by Theodore Puck in the late 1950s. The wide spread of CHO cells as recombinant protein production platforms is due to its ability to achieve high densities in suspension culture mode and the facility to adapt them to grow in serum and protein free conditions (Hacker and Balasubramanian 2016; Bandaranayake and Almo 2014; Butler and Spearman 2014). Besides, several gene amplification systems have been established with CHO cells, allowing a high gene copy number, a subsequent high protein yield (up to the gram per liter range for some proteins), and a high protein productivity (Dumont et al. 2016). Furthermore, as this cell line has been used for more than 50 years, there is a well-established process accepted by organizations, regulatory agencies and suppliers, simplifying and reducing the costs of the entire protein production process. The first approved biotherapeutic using a CHO platform was tissue plasminogen activator, produced in 1986 by Genentech Inc. (Khan 2013). Today, the overall value of products originating from these cells exceeds 50 billons dollars annually (Wurm 2013).

However, one drawback in biotherapeutics production using the CHO platform is the fact that it produces two types of monosaccharides present in many mammals, though not in humans, named galactose-α1,3-galactose and Neu5Gc. Because glycans containing these two types of residues have been described as antigenic, non-human cell lines must be screened during the production step to identify clones with acceptable glycan profiles (Higgins 2010).

Other non-human cell lines used for the production of biotherapeutics proteins include suspension BHK-21, used in the production of coagulation factors such as factor VIII (Wurm 2004), and murine myeloma cell lines (NS0 and Sp2/0), employed for the production of mAbs, for instance, palivizumab and ofatumumab (Dumont et al. 2016).

Developed almost two decades after the CHO platform, the HEK293 cell line was the first human cell line to be transformed using adenovirus DNA fragments of the Ad5 serotype. Since then, the HEK293 system has become one of the preferred used human cell lines for protein production (Bandaranayake and Almo 2014). HEK293 cells are easily grown in suspension serum-free culture, reproduced rapidly, adaptable to several of transfection methods, and highly efficient at protein production (Dumont et al. 2016). To improve transient gene expression, two additional cell lines were developed, the HEK293-T line expressing the SV40 large T-antigen and the 293-E line expressing the Epstein-Barr virus EBNA1 protein. Both of them support the episomal replication of plasmids containing either the SV40 origin of replication or the EBV oriP, respectively (Bandaranayake and Almo 2014). Using Fc fusion technology, recombinant factor VIII Fc (rFVIIIFc, Eloctate, Bioverativ) and recombinant factor IX Fc (rFIXFc, Alprolix, Bioverativ) were developed for the treatment of hemophilia A and B respectively, fusing a single molecule of rFVIII or rFIX to the Fc domain of IgG1. Both drugs are examples of bioterapheutics produced by recombinant DNA technology in HEK-293 cells (Dumont et al. 2016).

Another human cell line, called HT-1080, was obtained from a fibrosarcoma with an epithelial-like phenotype and has been used by Shire Human Genetics Therapies Inc. to produce four commercial therapeutic proteins, as outlined below. (i) Epoetin delta (DYNEPO®), indicated for the treatment of symptomatic anemia related with chronic renal failure in adult patients and approved by EMA in 2002. Indeed, since 2008 this drug was discontinued by the manufacturer for commercial reasons; (ii) Iduronate-2-sulfatase (idursulfase, ELAPRASE®), licensed by EMA in 2007 and FDA in 2006 for the treatment of Hunter disease; (iii) Agalsidase alfa (Replagal), authorized by EMA in 2001, but not by FDA, for the treatment of Fabry disease. An alternative commercial product produced using CHO cells (Fabrazyme, Genzyme) was approved by both regulatory agencies EMA and FDA; (iv) Velaglucerasealfa (VPRIV®), approved in 2010 by both EMA and FDA for the treatment of type 1 Gaucher disease. Production of these proteins by Shire Human Genetics Therapeutics is based on the gene-activation technology, involving the introduction of a DNA promoter into the upstream of an endogenous gene in the HT-1080 cell line. However, up to now, it has not been reported any information regarding the characteristics of this cell line used in the bioprocess such as growth rate or adherence dependency (Dumont et al. 2016; Swiech et al. 2012). In addition, three additional human cell lines have been utilized to produce biotherapeutics. The PER.C6 (Crucell) cell line was originally generated from human retinoblastoma cells immortalized by transfection with the adenovirus E1 gene to produce human adenovirus vectors in vaccine development and gene therapy (Bandaranayake and Almo 2014; Dumont et al. 2016). However, this cell line also has some features making it ideal for recombinant protein production such as its ability to grow in suspension culture mode in serum free medium reaching high cell densities. Currently, a combination of two monoclonal antibodies used against rabies virus, named CL184 (Foravirumab, Sanofi/Crucell), is produced in PER.C6 cells and is being tested in clinical phase 1/2 (Lalonde and Durocher 2017). Like PER.C6 cell line, CAP cell line was originally developed for adenovirus production by transforming primary human cells (amniocytes) through adenovirus type five E1 gene. Nevertheless, this cell line has been optimized to grow in suspension mode and can be very efficiently transfected with commercially available reagents. Moreover, a master cell bank of CAP cells has been certified according to ICH guidelines and European Pharmacopeia (Swiech et al. 2012). During 2017, CEVEC has been granted a patent by the Australian Patent Office for its CAP®Go expression technology, a platform allowing individual glyco-optimized cell lines for difficult-to-express protein production such as C1 esterase Inhibitor. Also, CEVEC has filed similar patent applications in other important markets around the world. Finally, because one of the main disadvantages of using HEK293 for commercial protein production is that these cells form aggregates in bioreactors, Cho and Chen (Cho et al. 2002) fused HEK293 and a human B-cell line using PEG, giving rise to HKB-11 cell line. The latter has been used to produce a recombinant factor VIII protein and tissue factor (Dumont et al. 2016).

Despite human cell lines have emerged as the most promising alternative to substitute the CHO platform, there is no universal production system. Indeed, the expression of a recombinant protein must be optimized according to the protein itself, regulatory and cost implications. In addition, as we will explain below, gene expression in mammalian cells needs a suitable vector that should act as a vehicle to transport the TG into the required cell line and an adequate strategy to improve both protein quality and quantity.

2.2.1 Strategies to Achieve High Level Expression of Therapeutic Enzymes

In biopharmaceutical companies, the production of glycoproteins is achieved by either transient or stable gene expression (Lalonde and Durocher 2017). Generally, generating stable cell lines expressing recombinant proteins takes from 6 to 12 months. On the contrary, TGE, defined as the expression of a gene of interest during a short period without being stable integrated into the host genome, takes only several weeks to set up a production system. Moreover, skipping the lengthy selection process involved in the plasmid integration within the host genome, the TGE upstream process is much simpler than the stable gene expression, often generating adequate amount of proteins (Lalonde and Durocher 2017; Ding et al. 2017; Hacker and Balasubramanian 2016). However, TGE requires more than 1 mg of plasmid DNA per liter of transfection, representing a drawback for projects requiring large culture volumes (Hacker and Balasubramanian 2016; Bandaranayake and Almo 2014). In addition, TGE has not been accepted as manufacturing therapeutic drugs, except for vaccine products and viral vectors used in gene therapy (Lalonde and Durocher 2017; Ding et al. 2017). The production process reproducibility and product quality consistency using TGE have been the main concerns.

Because using the stable gene expression system guarantees the process robustness and large protein amounts, it is still being the most widely used protocol. Other advantages of this expression system are the greater scalability and the renewable source of the recombinant cell line from a frozen cell bank (Hacker and Balasubramanian 2016).

Currently, different protein-engineering platforms have been proposed aiming to increase the circulating half-life, targeting and functionality of therapeutic proteins as well as to improve the production yield (Lagassé et al. 2017). Regarding stable expression, a gene marker is usually integrated in the expression plasmid along with the cDNA encoding the TG, conferring a selective advantage to the host cells that integrated the plasmid into their genome. The DHFR and the GS genes are the most frequently used selection markers. The first encodes the DHFR enzyme required for the synthesis of the essential cofactor folic acid. Mammalian cells lacking the DHFR DNA sequence (CHO-DXB11 and CHO-DG44) have allowed the genes amplification using an antagonist of DHFR, the chemical drug MTX. These cells require the presence of glycine, hypoxanthine and thymidine to survive. The selection of recombinant cell lines using stepwise increases the MTX concentration in the culture medium (without hypoxanthine and thymidine), resulting in amplified copies of the transfected DHFR genes together with the gene of interest, thus increasing the protein productivity levels (Lalonde and Durocher 2017; Clark and Pazdernik 2016; Bandaranayake and Almo 2014; Wurm 2013). Similarly, the GS amplification system, originally developed by Celltech (Lonza Biologics), has been successfully used with GS-negative NS0 and Sp2/0 cells, as well as with GS-expressing CHO cells. The CHO cells, containing an endogenous GS gene, need to be treated with MSX plus the removal of glutamine in the culture media to inhibit the endogenous GS activity, achieving enough selection pressure. In addition, a CHO GS knock out cell line has been developed, thus increasing the stringency of this amplification method. Other selection systems have been developed, such as the OSCAR™ system from the University of Edinburgh, that uses minigene vectors encoding the HPRT, essential for purine synthesis. However, in the industry, only the DHFR and the GS systems have been used for high-scale production (Lalonde and Durocher 2017).

Copy number of the TG as well as the integration site(s) within the host cell genome are some of the key factors to be considered. Recently, several tools have been developed for targeted site gene integration. As integration into a host chromosome is mainly a random event, usually the TG of interest is not expressed, as only a small percentage of the cell's genome (~0.1%) is actively transcripted. Thus, finding, identifying and targeting the TG of interest into transcriptionally active genome sites represents one of the main challenge to guarantee high-producing cell lines. In the biopharmaceutical industry, the RMCE strategy is one of the technologies currently used. Briefly, this tool involves two steps. First, the development of a master cell line that contains a single copy of a reporter gene at an active and stable genome region (hotspot), flanked by two heterospecific integration target sequences, such as FRT or loxP (*locus of crossover in phage P1*) sites, for the Cre system. These site-specific recombination systems are common in bacteria and lower eukaryotes and act in a reversible manner comprising the different recombinase alternatives (integration, excision/resolution and inversion) depending on the two recombinase target sites (Turan et al. 2013). Second, flanked by matching heterospecific recombinase target sites, the TG of interest can be introduced into this cell line in the presence of the appropriate recombinase (Flp or Cre), leading to the exchange of the reporter gene with the TG of interest. The main advantage of this system is that the recombinant cell lines have a single transgene copy at a hotspot genome region. However, despite this system promises to improve the cell line generation process, speed and efficiency, the actual improvement reported up to know in protein yields has only been moderate (Hacker and Balasubramanian 2016; Bandaranayake and Almo 2014; Wirth et al. 2007). In addition, as the first RMCE step implies the development of a cell line or clone containing a unique integration site within the host genome, multiple screening rounds must be done which are comparable to or even more than those needed for the classical random integration protocols.

Other genome engineering methods have been employed, based on programmable nucleases, such as ZFNs and TALENs. These nucleases induce DNA double-strand breaks at a specific location into the host genome, facilitating the integration of the cassette by NHEJ or HDR. Both ZFN and TALENs technologies require the synthesis of a DNA-binding domain that recognize a particular sequence (the targeted sequence for cleavage) combined to a nuclease effector domain (typically FokI). These nucleases are thus dependent on the ability of developing a good DNA recognition motif, with high specificity and affinity. Taken together, ZFNs and TALENs are robust genome engineering systems, but have some drawbacks such as limited multiplex genome targeting (Lalonde and Durocher 2017; Moreno and Mali 2017).

An alternative to these genome-editing nucleases is the recently developed CRISPR/Cas9 technology. In bacteria, the CRISPR system provides acquired immunity against foreign DNA via RNA-guided DNA cleavage. Three types of CRISPR/Cas systems exist, being the most widely used the type II using the SpCas9. Engineered CRISPR system contains two components: a gRNA and a CRISPR-associated endonuclease (Cas9). The gRNA is a short synthetic RNA comprising two functional regions: a variable spacer region which guides target *loci* recognition (crRNA), and a constant scaffold region which forms hairpin loops facilitating the binding to the Cas9 (trcRNA). In general, the variable spacer region is a short ~20 bp sequence, complementary to the target *loci*, and must be flanked at the 3′ end by a conserved PAM sequence, for instance, the NGG in SpCas9. The PAM sequence is necessary for CRISPR-Cas targeting (Moreno and Mali 2017; Gupta and Shukla 2017; Merkle et al. 2015; Gaj et al. 2013).This technique has already been validated in CHO cells and reduces the production variability between clones (Ronda et al. 2014).

However, ZFN, TALENs and CRISPR/Cas9 technologies are mostly used for gene-specific knockout. The ZFN knockout approach was particularly successful for specific deletion of the GS and the DHFR genes in CHO cells, thus improving the selection stringency of the generated cell lines. These tools could be very useful for site-specific integration, yet one of the key challenges is to identify good hotspots in the host genome. In addition, regarding CRISPR/Cas technology, Cas9 nuclease has been reported to cut at imperfectly matched sequences elsewhere in the genome causing off-target insertions and deletions via the NHEJ process. Moreover, such specific integration may not be adequate in all the approaches as some therapeutic proteins may require a particular level of expression to fold properly or to acquire adequate quality attributes (e.g. glycosylation, proteolytic processing, etc.) (Lalonde and Durocher 2017; Merkle et al. 2015).

The influence of all proximal and remote cis-acting chromosomal elements which modulate binding of transcription factors and thereby, the position-effect can be controlled by using cis-acting epigenetic regulatory elements (Spencer et al. 2015). These epigenetic regulatory elements are DNA sequences that can modulate the genome chromatin landscape and are essential for the proper gene expression regulation. Indeed, these elements are useful tools to overcome the position-effect and the associated loss in protein yield. One of the most well-known epigenetic elements are the S/MARs and UCOEs. The S/MARs are DNA sequences serving as attachment points to the nuclear matrix, maintaining a transcriptionally active chromatin structure. Also, these sequences enhance DNA demethylation, making genes more transcriptional active. However, in general S/MAR's activity depends on the type of sequence and vector configuration. UCOEs (Merck Millipore) are methylation-free CpG-rich sequences maintaining chromatin in an open conformation. They can also prevent DNA methylation of the CMV constitutive promoter sequence, that is usually susceptible to gene silencing in many cell lines including CHO cells (Lalonde and Durocher 2017; Hacker and Balasubramanian 2016; Bandaranayake and Almo 2014; Harraghy et al. 2015).

Another point to be considered is the gene delivery system. The classic calcium phosphate coprecipitation method as well as PEI, a cationic polymer, are useful plasmid DNA-transfection systems. These reagents are economical, efficient, applicable at large scale and present a high gene transfer efficiency (Corchero et al. 2011). Other routinely used non-viral gene transfer methods are electroporation, biolistic and those which use cation liposomes, such as lipofectamine (Invitrogen). Another genetic tool that can be used to introduce a foreign DNA into a cell host genome are the transposons. Transposons were discovered in maize and later found in all the genomes but most of them proved to be non-functional. These elements are DNA sequences able to move within the genome (class II transposons). Several naturally-occurring transposons, including Piggy-Bac, Sleeping Beauty, the Tc1/mariner element Mos1, and the medaka fish Tol2 element, have been genetically modified to properly adapt in mammalian cells. The essential components of these class II transposons are the transposase gene as well as the two terminal repeats that form the extremities of the transposon. The latter are specifically recognized by the transposase directing the cleavage and ligation reactions necessary for the transposon movement, from one DNA sequence to another. When used as gene delivery vehicles in mammalian cells, the transposons are usually employed as dual vector systems: a helper vector bearing the transposase gene under the control of a strong constitutive promoter and a donor vector carrying the terminal repeats flanking one or more TG of interest and a selection marker along with their respective promoters. The transposases improved the mammalian cells use, enhance transposition in or near actively transcribed genes, thus increasing the productivity levels and the gene stability (Hacker and Balasubramanian 2016). Finally, another widespread used delivery system is the viral vector system. Viral vectors have naturally evolved to transduce mammalian cells efficiently, and, consequently, they have become one of the preferred gene delivery tool. They are more efficient than non-viral strategies and can be broadly classified into two major classes, those which integrate into the host chromatin, such as retroviruses and lentiviruses, and those that persist episomally in the host cell, such as alphavirus or adenoviruses. The alphaviruses and adenoviruses are the most commonly used systems for transient protein production whereas the lentiviruses are the most common tools for generating stable protein expression.

In the next section, we will focus on the LVs as gene delivery method to generate stable cell lines in manufacturing of recombinant proteins. They are an excellent tool enabling the transduction of a broad range of cell types and, as transposons, LVs tend to integrate in actively transcribed regions.

2.2.2 Lentivirus Transgenesis Versus Other Transgenesis Methods in Animal Cell Culture

Lentivirus are members of the viral *Retroviridae* (retroviruses) family, which includes HIV-1, HIV-2, simian immunodeficiency virus and feline immunodeficiency virus. They are characterized by the use of their viral RT and IN for stable insertion of viral genomic sequence into the host genome. Unlike other retrovirus, lentivirus can replicate in non-dividing cells causing slowly progressive disease, such as immunodeficiency, anemia and encephalitis in their specific hostage (Sakuma et al. 2012).

LVs are replication-defective derivatives of retroviruses that can be used for gene delivery via a process called transduction. Currently, those derived from HIV type 1 are the most used vectors. The evolving LVs backbone and the virus ability to deliver recombinant DNA sequences into the host genome have turned them into a useful tool, being clinical gene therapy and recombinant protein production the main applications (Picanço-Castro et al. 2012).

Because of the HIV-1 human pathogenesis, different generations of LV vector systems have been developed, of which the third-generation is the most widely used. It is a four-plasmid system, consisting of three helper plasmids and one transfer plasmid vector containing the TG of interest. The native HIV envelope glycoprotein has been replaced by the VSV-G-protein (Merten et al. 2016; Sakuma et al. 2012). These VSV-G-pseudotyped retroviral vectors have two primary advantages over unmodified vectors. First, VSV-G is more stable than retroviral or lentiviral envelopes, allowing pseudotyped viruses to be concentrated by ultracentrifugation obtaining higher titers. Second, it is known that VSV-G binds the ubiquitous membrane component phosphatidyl serine, allowing these vectors to transduce a markedly wider set of cells.

In addition, to increase safety, in third-generation LVs systems, different lentivirus elements have been omitted or replaced; for instance, the LTR elements have been replaced by heterologous promoters and polyadenylation sequences. Thus, the Tat dependent U3 sequence localized in the 5′-LTR LV sequence has been removed and replaced by strong constitutive promoters like CMV. Also, SIN-LV vectors have been developed by introducing a deletion into the U3 element of the 3′-LTR. Furthermore, as the optimal vector production needs the interaction between Rev. and RRE facilitating the nuclear export of both gag-pol and transfer vector mRNA, two considerations must be taken into account. First, the Rev sequence was provided from a separate plasmid. Second, the virus RRE elements have been replaced by heterologous sequences (Merten et al. 2016; Sakuma et al. 2012).

To improve the transfer vector sequence some *cis*-acting motifs have been incorporated. For instance, the LV cPPT increased the percentage of transduce cells. In addition, some post-transcriptional elements have demonstrated the improvement in TG expression such as the WPRE.

Thus, the four plasmids used to generate third-generation vectors are: (i) a packaging construct containing only *gag* and *pol (reverse transcriptase)* genes, which are responsible for coding the core viral proteins and the enzymes involved in viral replication, respectively; (ii) a plasmid expressing Rev; (iii) the VSV-G plasmid; and (iv) a transgene plasmid. This vector system has only three of the nine genes of HIV, increasing its biosafety. Since the vector elements are split into four plasmids, at least three recombination events are required to generate a replication-competent HIV-1-like virus (Merten et al. 2016; Sakuma et al. 2012; Plewa 2010).

Most of the current LVs vector production methods comprise a first step involving cotransfection of HEK293T or HEK293 cells (Fig. 2.1). As we have mentioned before, the HEK293T are widely preferred as they enable a more efficient vector production. Thereafter, the LVs are budded off into the cell-conditioned medium from which they can be collected and concentrated or purified in additional steps. Finally, those LVs are used to transduce a target cell line that will express the recombinant protein of interest (Fig. 2.1).

At small scale, LVs production is performed using adherent cells growing in Petri dishes. At optimal confluence, cells are transfected using the classical calcium-phosphate coprecipitation protocol or cationic lipids, such as PEI or lipofectamine. Since attachment of HEK293T cells to the surface is not very strong, vector production is more difficult to perform in roller flasks, and transfection conditions in these devices must be optimized. Larger scale production implies an increase in the culture surface, working with parallel cultures. On the contrary, performing the cotransfection step in suspension conditions is easier to scale-up. However, the calcium-phosphate coprecipitation method is less effective to transfect suspension cells, so other transfection systems like cationic polymers are used more often. Afterwards, either in adherent or suspension production, culture supernatants containing LVs should be first clarified by centrifugation and then subjected to an ultracentrifugation step at about 70,000 g (Merten et al. 2016; Gaillet et al. 2010; Plewa 2010). These purification steps contribute to eliminate all cell-derived elements, retaining the vector potency (Picanço-Castro et al. 2012).

Before LVs infection of target cells, titration of produced LVs is critical to evaluate transduction efficiency. Various methods have been developed, such as those involving the quantification of viral components in culture supernatants (physical particle numbers, p24 capsid protein concentration, RT activity or

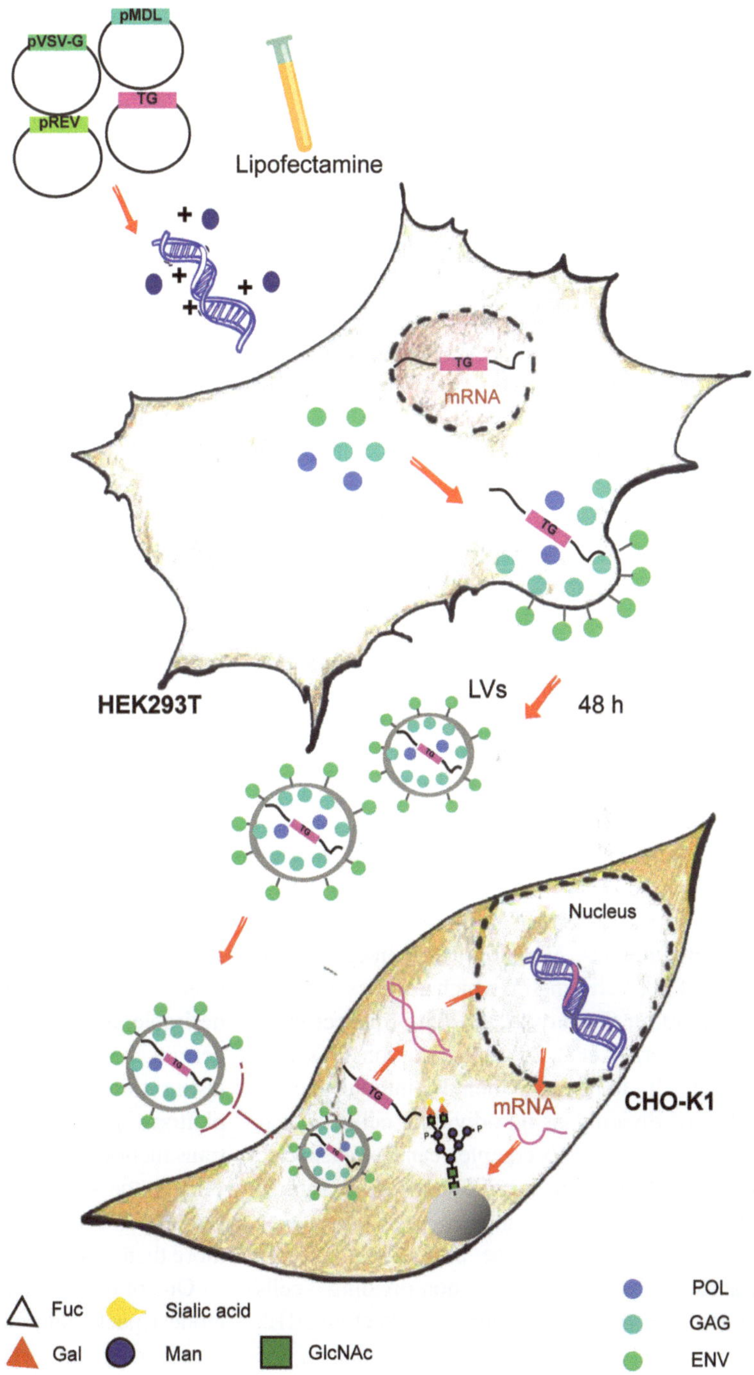

Fig. 2.1 The LVs production and cell transfection. The first step comprises the cotransfection of HEK293T cell with the four plasmid system using lipofectamine: (i) a packaging construct containing only *gag* and *pol* genes (pMDL); (ii) a plasmid expressing Rev. (pREV); (iii) the VSV-G plasmid (p-VSV-G); and (iv) a transgene plasmid (TG). After 48 h of the cotransfection event, the cell culture supernatant contains recombinant LVs that can be used to transduce the target cells (CHO-K1). Thereafter, the CHO-K1 cells overexpress the protein of interest

genome copy numbers). The main drawback of these methodologies is the equal detection of infective and non-infective vectors. Other strategy comprises the detection of proviral copy numbers in infected cells by qPCR, although it not always correlates with the expression level achieved in the target cells. Finally, to determine the percentage of transduced cells by transgene expression is the most reliable method to evaluate LVs titers. However, this last method not always can be applied because a reliable marker gene, as green fluorescent protein, is necessary. Both in suspension and adherent cultures, LVs titers achieved in culture supernatants are in the range of 10^6–10^8 TU.mL^{-1} (Merten et al. 2016; Sakuma et al. 2012).

Lentivirus vectors have been applied to the generation of highly productive CHO, BHK, and HEK293 cell lines using either inducible or constitutive promoters (Rodríguez et al. 2017; Shestopal et al. 2017; Mao et al. 2015; Mufarrege et al. 2014; Baranyi et al. 2010; Johnston et al. 2012; Prieto et al. 2011; Spencer et al. 2011; Gaillet et al. 2010; Oberbek et al. 2011; Plewa 2010). Besides, LVs can be used to express recombinant proteins in primary cultures, as hematopoietic stem cells (Cappellino et al. 2017), neurons or T-lymphocytes (Moreno and Mali 2017; Gaillet et al. 2010).

In comparison with non-viral delivery methods, viral vectors improve the gene transfer, because the viruses from which they derived have specific entry mechanisms into the infect cells. Still better, LVs convert their RNA genomes into DNA and integrate permanently into the host cells genome. Thus, this stable genetic modification allows the generation of stable cell lines for recombinant protein production in shorter timelines. Moreover, they integrate into areas of open chromatin, which are sites of active transcription (Baranyi et al. 2010; Spencer et al. 2011; Oberbek et al. 2011). In addition, LVs have the capacity to integrate their TG into the nuclear DNA of dividing or non-dividing cells. This is possible because LVs complex can be transported into the nucleus in an ATP-dependent manner. Hence, LVs have advantages over other gene delivery systems because they can efficiently transduce dividing and non-dividing cells (Baranyi et al. 2010; Picanço-Castro et al. 2012; Plewa 2010). Besides, the copy number of the integrated viral vector can be controlled by varying the MOI. To achieve a high copy number, host cells are transduced at a high MOI so that the expression starts out at relatively high levels (Baranyi et al. 2010). Moreover, multiple rounds of serial transduction can be done to increase protein production levels (Rodríguez et al. 2017; Spencer et al. 2011). Taking all this together, LVs have distinguish characteristics in comparison with classical methods employing plasmid vectors, which are suboptimal, random and single site integration (Spencer et al. 2011).

Transient transfection followed by drug selection has been the traditional method for DNA delivery and integration to obtain producer cell lines. While successful, this is often a prolonged (4–6 months) process. In addition, there is no guarantee that the recombinant cell line will maintain a high expression level through many cell generations without selection (Bandaranayake and Almo 2014; Gaillet et al. 2010). Besides, as we have previously mentioned, TGE of HEK293 and CHO cells have been also applied to produce recombinant proteins and it is currently utilized like the main method for the fast synthesis of proteins. This method is very efficient with HEK293, but in CHO cells, the yields are generally much lower due to the difficulty to transfect these cells (generally under 40%) (Gaillet et al. 2010). On the contrary, LVs platform eliminates the need for using drugs such as MTX or MSX and, as this procedure allows starting with a highly expressing pool, it may reduce the number of clones that need to be screened. Furthermore, it is known that the ability to adapt host cells to grow in suspension is highly desirable, as it allows volumetric scalability and permits the use of protein- and serum-free media. Using third-generation LVs as an expression platform gives the possibility to perform cell transduction directly in suspension conditions. Finally, the LVs system allows the generation of stable cell lines in a short period of time, in no more than 4 weeks (Spencer et al. 2011).

One of the drawbacks of the LVs system is the modest packaging size of the lentivirus capsid, which may limit the maximum size of the recombinant protein being expressed. However, several reports have demonstrated the use of LVs in manufacturing cell lines allowing the insertion of cDNA sequence around 6 kilobases in length, such as the coding FVIII sequence (Shestopal et al. 2017; Johnston et al. 2012; Spencer et al.

2011). Another issue to be considered is that LVs derived from HIV-1 should be manipulated in a biosafety level two containment.

Since LVs have been used in human gene therapy and clinical trials, safety testing assays and procedures have been developed. Moreover, LVs assembly processes involved in good manufacture practices which could be easily adapted in manufacturing cell lines for protein therapeutics. In the next section, we will explain the main aspects concerning the safety of this method.

2.2.3 Quality Control in Manufacturing Cell Lines Using LVs as Transgenesis System

There are safety concerns that LVs might be contaminated with replicating virus or viral proteins that could be pathogenic as they derive from HIV-1. A replicating virus could theoretically arise because of recombination events between the vector genome containing cis-acting viral sequence elements and the genes encoding the viral proteins during the packaging process. In addition, the possibility of recombination between packaging plasmids and human endogenous retroviral sequences must also be considered (Cornetta et al. 2011). Despite this drawback, LVs are also in development for human gene therapy and have been used in the treatment of hereditary disorders, for instance, Wiskott-Aldrich syndrome and severe combined immunodeficiency disorders (Skrdlant et al. 2017). During 2017 the FDA approved the first gene therapy using LVs, CD19 CAR T cells for the therapy of relapsed leukemia (CTL019, Novartis) (Cornetta et al. 2017; Plewa 2010).

Some aspects must be considered if LVs are used to engineer production cell lines. It is well known that standard manufacturing cell lines protocols avoid the use of animal-derived products as much as possible. However, in those systems in which adherent HEK293 packaging cells are used, further tests must be applied to assess an animal free pathogen in the manufacturing process (Merten et al. 2016; Plewa 2010).

Importantly, the assays that have been developed for the detection of RCL in human gene therapy using LVs could be easily adapted to biopharmaceutical manufacturing processes. The HIV-1 genes and recombination products could be detected using qPCR assay. Viral proteins detection could be done by ELISA, western blot or PERT assay. However, the most critical test is the detection of RCL. In the USA, a LVs lot must be screened for RCL. The FDA requires a percentage of the cell product be screened for RCL, that is equivalent to 1×10^8 cells (Cornetta et al. 2017). Basically, in this test, viral vector particles or transduced cells are exposed to a human T cell line C8166-45, that is very permissive for HIV replication, in an extended period (3 weeks approximately). Thus, the transduced cells and replicative virus that might be present are amplified several times. Following multiple passages (a minimum of five), an indicator line is infected and one of the previously mentioned quantification methods is applied. However, this technique has some limitations. The identification of a positive control virus that represents the probable characteristics of an RCL is difficult to develop because of the LVs safety features. Currently, an attenuated HIV strain pseudotyped with the VSV-G, R8.71, is the best approximation of RCL (Plewa 2010; Escarpe et al. 2003). The detection is also complicated because many components of RCL will be similar to those of a vector particle (capsid, IN and RT). Moreover, if LVs are generated using transient transfection methods, those present in the supernatants will be contaminated with the packaging plasmid DNA that contains the same viral sequences likely to be present in RCL (Cornetta et al. 2017; Cornetta et al. 2011; Sastry et al. 2003). Another issue regarding measurement of HIV proteins and RCL is the LOD. For instance, the LOD of the p24 ELISA that detects the p24-gag protein is 3 $pg.mL^{-1}$ (or about 3.6×10^4 viral $particles.mL^{-1}$) or the LOD of a qPCR assay detecting the VSV-G is around ten copies. However, a negative result cannot be an absolute conclusion that a sample lacks viral protein or particles because there are levels that might not be detectable. What is more, the potential growth rate of RCL could be attenuated when compared with that of the wild-type lentivirus due to the absence of accessory genes. Therefore, the biologic assay must be done during a prolonged period to amplify any slow-growing virus (Cornetta et al. 2017). One of the last RCL assays that has been developed recently demonstrated redundancy, using two orthogonal detection methods (psi-gag PCR and PERT), to minimize false-positive or false-negative results due to technical mistakes (Cornetta et al. 2017).

Despite its limitations, RCL assay is quite robust and should be sufficient for testing during manufacturing process. In addition, until now, no RCL particle has been detected (Cornetta et al. 2017). The ability to split the vector components into four plasmids, the use of SIN LTRs and the retention of Rev. dependence, all contribute to the safety profile of LVs. The HIV-1 virus also depends on several accessory genes and regulatory sequences that are deleted in LVs system, limiting the potential growth of any RCL. Furthermore, LVs are generally produced by transient transfection, reducing the time for feasible recombination events (Cornetta et al. 2017).

2.2.4 Production of Therapeutic Enzymes for Lysosomal ERT

Lysosomes are catabolic organelles containing more than 60 hydrolytic enzymes that break down and recycle a range of complex substrates (Oh 2015; Parenti et al. 2015). They are acidic, membrane-bound cytoplasmatic organelles ubiquitously located through all the body cells, occupying approximately five percentage of the intracellular volume. In addition, lysosomes are involved with numerous cellular homeostatic processes, not only in catabolic ones but also in other cellular functions such as sensing, signaling, vesicle trafficking and cellular growth (Solomon and Muro 2017; Parenti et al. 2015). Substrates are transported to lysosomes through different routes. Specialized endocytic mechanisms (phagocytosis, micropinocytosis, clathrin-mediated endocytosis, caveolin-mediated endocytosis and clathrin- or caveolin-independent endocytosis) are preferentially exploited according to the cargo nature. Intracellular materials are mainly transported to the lysosomes through autophagy (Parenti et al. 2015).

Lysosomal enzymes maturation involves a series of common process within ER and the Golgi apparatus. Firstly, the nascent peptide is directed to the ER guided by its signal peptide in which the protein undergoes two mainly modifications, (i) the *N*-glycosylation of specific asparagines in the consensus motif (*N*-X-S/T, where X is any amino acid except proline), consisting in the transfer of an oligosaccharide formed by two glucoses, nine mannoses and two *N*-acetylgalactosamine residues; (ii) the Cys-to-Formylglycine conversion for sulfatases. In the case of the *N*-glycosylation, three glucoses and one mannose residues are removed before the lysosomal enzymes leave the ER (Espejo-Mojica et al. 2015). Thereafter, they move to the Golgi apparatus where high-mannose type glycans of lysosomal enzymes are modified, undergoing further modifications that produce M6P in a two-step reaction. Firstly, in the *cis*-Golgi complex, the enzyme GlcNAc-1-phosphotransferase recognizes lysosomal enzymes and then transfers GlcNAc-1-phosphate from UDP-GlcNAc to the six-carbon hydroxyl groups of the GlcNAc-1-phosphate-6-*O*-mannose group. Secondly, in the *trans*-Golgi complex, the GlcNAc-1-phosphodiester α-*N*-acetylglucosaminidase removes the outer GlcNAc and leaves a phosphate group linked to the M6P glycan (Espejo-Mojica et al. 2015; Oh 2015; Braulke and Bonifacino 2009). Thereafter, the lysosomal enzymes leave Golgi in clathrin-coated vesicles fused to the early endosome, where, due to lysosome's acidic pH, the enzymes are released from the M6PR. The released enzymes go through a proteolytic process that occurs mainly at the *N*-terminus. Most of the mature enzymes are targeted to the lysosome, but between 2% and 20% are secreted and they can be recaptured by M6PR located in the cell plasmatic membrane, representing the bases of the ERT, gene and cellular therapies (Espejo-Mojica et al. 2015).

As lysosomes are ubiquitous organelles involved in numerous homeostatic processes, any alteration can modify the normal cell functioning in several organs, leading to multi-systemic diseases. Lysosomal dysfunctions have been described for several neurodegenerative diseases such as Alzheimer's, Parkinson's and Huntington's disease, yet the most studied class of diseases are the LSDs, directly caused by abnormal lysosomal function (Solomon and Muro 2017).

For each LSD, single or multiple gene mutations have been identified in the causing of the disease. Many of the mutations involve a single base pair change altering one amino acid in the encoded protein. They reduce either the enzyme biological activity or impair translocation to the lysosome, causing premature protein degradation. The LSDs genetic causes are known, but in most cases because of their complexity, they are confused with other conditions having the same clinical symptoms (Solomon and Muro 2017; Xu et al. 2016).

The clinical consequences of substrate storage in multiple organs and systems imply visceral, ocular, hematological, skeletal, and neurological manifestations. Symptoms may occur at different times in life, in some cases starting in utero, during the newborn period, or becoming evident in late adulthood. In general, LSDs are chronic and progressive, and often responsible for physical and neurological disabilities, affecting patients' health and life expectancy (Parenti et al. 2015).

Current approved treatments for LSDs are mainly based on the replacement of the mutated protein with natural or recombinant protein to restore its wild-type function, employing the interaction between M6P and M6PR presented in the plasmatic membrane. To date, there have been successful drug approvals for the LSDs. As the Orphan Drug Act stimulates investment into therapeutic development for rare diseases through financial incentives, 17 drugs for LSDs received FDA approval between 1983 and 2015, but only covering eight different LSDs: Gaucher disease, MPS I (Hurler disease), MPS II (Hunter syndrome), MPS IVA (Morquio A syndrome), MPS VI, Pompe disease, lysosomal acid lipase deficiency and Fabry disease. Meanwhile, in the European Union, 15 products for LSDs were authorized for treatments between 1995 and 2015, most of them being the same products approved by the FDA (Xu et al. 2016).

Initially, the deficient enzymes were purified from human tissues, but nowadays they are produced by recombinant DNA technology. Mammal-derived cells have been traditionally used as platforms to produce lysosomal enzymes (Solomon and Muro 2017). The CHO cells are the most widely used system to produce ERT recombinant enzymes. Other mammalian-based systems include the HT-1080 cells.

In 1991 the FDA approved the first ERT drug (Ceredase®, alglucerase, Genzyme), the placental-derived macrophage-targeted glucocerebrosidase, for patients with Gaucher disease type 1. ERT helps the body break down glucocerebroside, which accumulates in patient's monocytes-macrophages system due to the lack of β-glucocerebrosidase (GBA, E.C. 3.2.1.45). Gaucher disease is classified into three types, according to the presence of neurological deterioration, age of appearance and disease progression (Bennett and Mohan 2013). Management of ERT type I disease (OMIM 230800), with no neuronal involvement, has been effective, but it has not been that effective for the treatment of type II (OMIM 230900) and III (OMIM 231000) diseases, that have neuronal implications (Xu et al. 2016). Imiglucerase (Cerezyme, Genzyme), an analog of glucocerebrosidase produced by DNA technology using CHO cells, was approved by the FDA in 1994 replacing alglucerase. In 2010, Velaglucerase alfa (VPRIV, Shire Human Genetics Therapies Inc.), an analog of recombinant glucocerebrosidase produced in human HT-1080 cells, became the third ERT approved by the FDA. Velaglucerase alfa was manipulated to contain *N*-glycans with terminal mannose residues by kifunensine treatment of the culture medium, obtaining high mannose-type glycans (Oh 2015). In May 2012, the FDA approved Taliglucerase alfa (ELELYSO, Pfizer Inc., or, outside the USA, Protalix BioTherapeutics), which is produced by genetically modified carrot cells. A comparison study of the three enzymes showed similar results in in vitro enzyme activity, ex vivo cellular uptake, and in vivo tissue distribution assays. However, in another study using monocytes isolated from a Gaucher patient, Taliglucerase alfa had a lower uptake efficiency than the others. The prolonged use of Taliglucerase alfa containing the plant-specific glycan structures (β-(1,2)-xylose and core α-(1,3)-fucose) may induce an immune reaction. In addition, Velaglucerase alfa has a concern related to its longer mannose chains, which was shown to bind, in serum, more efficiently to an undesirable mannose binding lectin (Oh 2015). As a second alternative to these therapies, a substrate-reducing agent, a small molecule that reduce the accumulation of substrates by inhibiting their biosynthesis, miglustat (Actelion Pharmaceuticals), has been approved for clinical use to treat type 1 Gaucher disease and Niemann–Pick disease type C, another LSD. A novel substrate inhibitor, eliglustat tartrate (Cerdelga™, Genzyme Corp.), has been introduced recently and evaluated in a phase II clinical trial for treating Gaucher disease, and Ibiglustat (GZ/SAR402671) is in clinical trials not only for Gaucher, but also for Fabry and Parkinson's diseases (Solomon and Muro 2017; Parenti et al. 2015).

Hurler syndrome (MPS type I, OMIM 607014) is an autosomal recessive disease caused by mutations in the gene encoding α-L-iduronidase (E.C. 3.2.1.76). It presents three clinical subtypes: Hurler syndrome, Hurler-Scheie syndrome and Scheie syndrome, being Hurler syndrome the most severe and Scheie syndrome the least one (Jameson et al. 2013). Laronidase (Aldurazyme™, manufactured by BioMarin and commercialized by Genzyme Corp), produced in CHO cells, has been approved by the FDA and EMA in 2003 and is administrated as intravenous infusions of 0.58 $mg.kg^{-1}$ per week.

Hunter syndrome (MPS II, OMIM 309900) is an X-linked lysosomal storage disease produced by the deficiency of iduronate-2-sulfatase (E.C. 3.1.6.13), leading a lysosomal accumulation of heparan sulfate and dermatan sulfate, leading to serious complications such as developmental delays and mental impairment. Currently, two recombinant enzymes are available for the treatment of Hunter disease: idursulfase (Elaprase®, Shire Human Genetic Therapies), approved by both FDA and EMA, and idursulfase-beta (Hunterase®, Green Cross Corp., Yongin, Korea) approved by the Korea Food and Drug Administration (KFDA) in 2012 (Sestito et al. 2015; Chung et al. 2014; Sohn et al. 2013a, b). Elaprase® is produced in the HT-1080 cells, whereas Hunterase® in CHO cells. A biochemical comparison of both commercial enzymes showed that idursulfase beta exhibited significantly higher specific enzyme activity than idursulfase because of the higher formylglycine content (Chung et al. 2014).

Mucopolysaccharidosis IVA (MPS IVA; Morquio A syndrome, OMIM 253000) is an autosomal recessive LSD caused by deficiency of GALNS (E.C. 3.1.6.4). Absence of this enzyme leads to a progressive accumulation of the glycosaminoglycans: C6S and KS and subsequent cellular pathology, especially in connective tissues rich in KS, like cartilage, cornea and heart valve (Tomatsu et al. 2015; Dvorak-Ewell et al. 2010). Recombinant enzyme GALNS for ERT has been produced in CHO cells (Elosulfasealfa, Vimzim®; BioMarin Europe Ltd). This drug has achieved both FDA and EMA approval during 2014 and is administered as a 2 $mg.kg^{-1}$ infusion per week.

Mucopolysaccharidosis type VI (MPS VI, Maroteaux-Lamy syndrome, OMIM 253200) is another LSD caused by functional absence of the enzyme *N*-acetylgalactosamine 4-sulfatase (arylsulfatase B; E.C. 3.1.6.12). Deficiency of this enzyme results in the accumulation of dermatan-sulfated glycosaminoglycans within lysosomes of various organs. The progressive accumulation of these compounds results in multi-organ system dysfunction such as joint contractures, short stature, dysostosis multiplex, decreased pulmonary function, cardiac abnormalities and, ultimately, shortened life span. Galsuflase (Naglazyme™, BioMarin), produced in CHO cells, received the marketing authorization from the FDA in 2005, and in January 2006, from EMA (Solomon and Muro 2017; Braunlin et al. 2013).

Pompe disease (OMIM 232300) is another LSD caused by a deficiency in lysosomal acid α-glucosidase (GAA, E.C. 3.2.1.20) involved in the breakdown of glycogen. The disease presents two classical phenotypes, infantile and late-onset forms. Individuals with the infantile form have severe cardio-respiratory disease and hypotonia and frequently die at 1 year of age. The late-onset form is characterized by a variable progression and clinical manifestations, including elevation in creatine kinase, muscle weakness, and acute respiratory alterations. Before 2006, therapy for Pompe disease was only palliative (Oh 2015; Vernon 2015). For this disease, two ERT products have been approved: (i) recombinant α-glucosidase (Myozyme™, Genzyme Corp) produced in CHO cells in a 160 l bioreactor has been approved in Europe, USA, Canada and Japan in 2006 for the treatment of infantile-onset disease; (ii) the same enzyme, produced at a larger scale in a 4000 l bioreactor (Lumizyme™), has been approved for the treatment of the late-onset disease in 2010. In addition, during 2014, Lumizyme™ was approved for the treatment for the child-onset disease (Solomon and Muro 2017).

Lysosomal acid lipase (LAL, E.C. 3.1.1.13) deficiency (OMIM 278000) is an autosomal recessive and progressive metabolic liver disease caused by mutations in the LAL gene. This enzyme is responsible for hydrolysing cholesteryl esters and triglycerides within low-density lipoprotein cholesterol particles into free cholesterol and free fatty acids leading to their accumulation in the liver, blood vessel walls and other organs. The LAL deficiency has two distinct phenotypes: (i) early onset, resulting of an absent (or nearly almost absent) LAL activity affecting the infants; and (ii) late-onset, which is the result of a partial loss of the enzyme activity, presented in children and adults (Frampton 2016). The recombinant form of LAL (Sebelipase α, Kanuma®, Alexion Pharmaceuticals) has been produced in transgenic hen eggs and approved by EMA and FDA for the treatment of both LAL deficiency variants during 2015 (Solomon and Muro 2017).

Fabry disease (OMIM 301500) is an X-linked recessive gene disorder caused by α-Galactosidase A (αGAL, EC 3.2.1.22) deficiency, leading to a systemic lysosomal accumulation of neutral glycosphingolipids, predominantly Gb3. Currently, two distinct recombinant protein replacement drugs are used for the treatment of Fabry patients: agalsidase beta (Fabrazyme®, Genzyme-Sanofi), and agalsidase alfa (Replagal®; Shire Human Genetic Therapies). Different reports have shown that the two recombinant enzymes exhibit identical biochemical properties and are comparable with each other in terms of amino acid composition, specific activity, stability and uptake by cultured fibroblasts, with only minor differences in glycosylation composition and mannose-6-phosphate receptor-mediated cellular uptake (Sakuraba et al. 2006; Lee et al. 2003). Agalsidase alfa is produced in HT-1080 cell line and is administered by intravenous infusion at 0.2 mg.kg^{-1} every week; whereas agalsidase beta is produced in CHO-K1 cells and is administered by intravenous infusion at a dose of 1.0 mg.kg^{-1}. Both agalsidase alfa and beta have been approved during 2001 by EMA, but only algasidase beta has been granted the FDA approval during 2003 (Desnick 2004). In addition, another recombinant version, Pegunigalsidase alfa (PRX-102, Protalix, Biotherapeutics) is in advanced stage of development (phase III clinical trial). This therapeutic is produced in a plant cell-based protein expression system and is chemically modified with a homo-bifunctional PEG, resulting in a PEGylated variant (Kizhner et al. 2015).

Despite the different ERT alternatives developed until now, this technology has some limitations, including patient immune reactions which reduce the clinical efficacy, low efficiency of lysosomal targeting and difficult delivery to the central nervous system. Moreover, ERT cost ranges from $50,000–200,000 per year per patient, as dosages depend on patients' weight. However, small increases in enzyme activity (as little as 10%) may be sufficient to produce clinical benefits and phenotypic correction of LSDs (Solomon and Muro 2017; Oh 2015; Parenti et al. 2015). In addition, several strategies have been continuously investigated aiming to increase ERT efficacy, reduce dosage, adverse effects and the cost of the global production process of the recombinant enzyme variant.

In the next section we present a case study showing the production of rhαGAL, for which a reduction of the timeline and the global process cost were achieved using LVs to deliver the transgene of rhαGAL and drive its expression in CHO suspension manufacturing cell lines.

2.3 A Case Study: Production of Active Human α-Galactosidase A in Suspension CHO-K1 Cells

Drugs for rare diseases are of important public health concern and a challenge for the medical community since they are often difficult to discover, develop and market. However, access to orphan drugs continues to be limited by high prices (Kumar Kakkar and Dahiya 2014).

Positive effects of rhαGAL enzyme replacement therapy has been demonstrated on clearing Gb3 from plasma and target organs, reversing the pathogenesis of the disease. However, this approach presents some disadvantages related to: (i) limited efficacy in patients with an advanced stage; (ii) short enzyme's half-life mainly due to the capture of the enzyme by the liver; and (iii) high cost of the treatment, as this disease is chronic and needs repeated doses throughout life (Cabrera et al. 2016). Currently, different strategies have been applied to optimize the global production process of rhαGAL. A selection of best-performing clones is one of the major time-consuming tasks and it further increases the cost of development and production. With the proper biochemical characteristics described until now, the highest enzyme expressing clone AGA5.3, obtained by MTX amplification, secretes rhαGAL levels of 7.5 $pg.cell^{-1}.d^{-1}$ with specific 1.3×10^4 $U.mg^{-1}$ enzymatic activity (Ioannou et al. 1992). Sohn et al. (2013a, b) reported the development of CHO clones with 30-fold higher productivities of rhαGAL (more than 150 $mg.L^{-1}.d^{-1}$) than the one achieved by AGA5.3 clone in a hollow fiber bioreactor (5 $mg.L^{-1}.d^{-1}$). However, the resulting enzyme (I303) appeared to experience incomplete sialylation due to the gap between the high level of protein production and the low capacity of the glycosylation process taking place during secretion. Despite obtaining complete sialic acid capping after carrying out an in vitro enzyme reaction, this extra step makes the production process time-consuming, laborious and more expensive. The TGE has also been proposed as an alternative strategy to obtain rhαGAL in suspension human HEK293 cells (Corchero et al. 2011). Likewise, with the advances in plant technology, transgenic plants have also been explored as source for rhαGAL production. For instance, in PRX-102, rhαGAL has been chemically modified with a homo-bifunctional PEG cross-linker attached to the two rhαGAL's subunits, improving serum half-life as well as plasma and lysosomal stability. However, PEGylation is not a simple process, as some significant parameters must be taken into account, including the molecular weight and the PEG architecture, the PEG number attached sites and their possible location. Moreover, PEG itself does carry some potential risks, such as the antibody formation against PEG (anti-PEG), hypersensitivity to PEG and vacuolation (Zhang et al. 2014). Other approaches focus on correcting the trafficking defects of enzymes to lysosomes using molecular chaperones assisting enzymes to fold correctly. Galafold™ (Migalastat; Amicus Therapeutics) has been clinically approved for Fabry disease in Europe. Nonetheless, this type of approach is only valid for patients with certain mutations characterized by an unstable protein phenotype.

Under these circumstances, we herein report the production of rhαGAL in suspension CHO-K1 cells, using third-generation LVs as an optimized expression platform. Indeed, we have replaced the amplification strategy based in DHFR/MTX selection by successive lentiviral transductions of suspension cells, obtaining productivities up to 59 $pg.cell^{-1}.day^{-1}$.

2.3.1 The rhαGAL Cell Line Development, Cloning and Bioreactor Production

The LVs were produced by simultaneous co-transfection of HEK293T cells with the four plasmids system, using the lipid reagent

Lipofectamine 2000 (Invitrogen™, USA). Thereafter, the culture supernatant containing the LVs was harvested after 48 h, clarified by low speed centrifugation and frozen at −70 °C prior to use. The titer was calculated with a HIV-1 p24 ELISA kit (QuickTiter™Lentivirus Titer Kit, Cell Biolabs Inc., USA), obtaining a value of 1.9×10^8 TU.mL^{-1}.

Suspension adapted CHO-K1 cells were serially transduced (three times) at a MOI of ~35 LP per cell. Performing repeated transductions can increase the copy number of the transgene and, thus, the expression of the protein of interest. This procedure eliminates the need of using drugs such as MTX or MSX and, as this protocol allows starting with a highly expressing pool, it may reduce the number of screened clones. Indeed, it is known that the ability to adapt host cells to grow in suspension is highly desirable as it allows volumetric scalability and permits the use of protein- and serum-free media (Van der Valk et al. 2010).

From the population of cells that underwent three rounds of transductions, clonal isolates were obtained by limiting dilution. Fifteen clones, selected for further analysis based on the rhαGAL high level expression, displayed expression rates from 3.5 to 59.4 pg.cell^{-1}.d^{-1} and specific enzyme activities from 3.0×10^3 to 2.4×10^4 U.mg^{-1}. This study demonstrated, once more, the potential of LVs as a gene delivery tool. Our clones achieved up to eightfold higher productivities and up to twofold higher specific enzymatic activities in culture supernatants compared to the AGA5.3 clone, which is the platform for the production of algasidase beta, Fabrazyme®.

Two of the highest producing clones (P3G2 and P3E8) were selected for metabolism, cell growth analysis, and subsequent scale up. The transgene copy number of the selected clones were determined by qPCR, indicating the existence of 2.2 ± 0.1 copies of the transgene and 5.6 ± 0.1 copies of the transgene for P3G2 and P3E8, respectively, in comparison with the 43 ± 2 copies of the TD3 cell population from which they derived. Importantly, no copies of LVs-derived components were identified in 1.0×10^4 copies of the transgene in the rhαGAL producer cell line, assessed by a qPCR assay.

For the rhαGAL production, the P3G2 clone was cultured in serum-free medium in a one-liter bioreactor (Biostat Q Plus, Sartorius), in perfusion mode, during 14 days, reaching 1.8×10^7 cell.mL^{-1} cell densities and up to 95% cell viabilities. Perfusion rate varied between 0.5 and 1.0 reactor volumes per day (Fig. 2.2a). Concentrations up to 35 μg mL^{-1} of rhαGAL were attained in ~10 L of harvest. In addition, the P3E8 clone was cultured in the same bioreactor system during 14 days, reaching up to 1×10^7 cell.mL^{-1} cell densities, and up to 90% cell viabilities. The temperature systems varies from 37 to 32 °C, obtaining nearly 60 μg mL^{-1} rhαGAL concentrations, in ~10 L harvest (Fig. 2.2b).

2.3.2 The rhαGAL Purification

The rhαGAL present in clarified culture supernatant of both clones was purified using two sequential steps. Briefly, the bioreactor harvest was loaded onto a weak IEX resin (DEAE Sepharose Fast Flow, GE Healthcare), and equilibrated with 10 mM phosphate buffer, pH 6.0. No enzyme leakage was observed during the loading or washing steps of this first chromatography. The enzyme was eluted using an isocratic 0.18 M NaCl gradient in the same buffer. Next, a second purification step was performed, based in a HIC (Butyl Sepharose 4 Fast Flow, GE Helthcare). Most of the enzyme contaminants were removed during the washing step, while the elution condition (0.75 M NaCl, 10 mM phosphate buffer, pH 7.5) allowed recovering the enzyme with a 60% yield. This step was essential to remove most protein contaminants, as revealed by SDS-PAGE (Fig. 2.3a, b). As a result, after two purification steps, the active enzyme was recovered with high purity and 60% overall yield (Tables 2.1 and 2.2). A purity of 100% was obtained considering the rhαGAL amount and total protein content determined by ELISA and Bradford assay, respectively. This apparent 100% purity value can be explained as resulting from experimental errors.

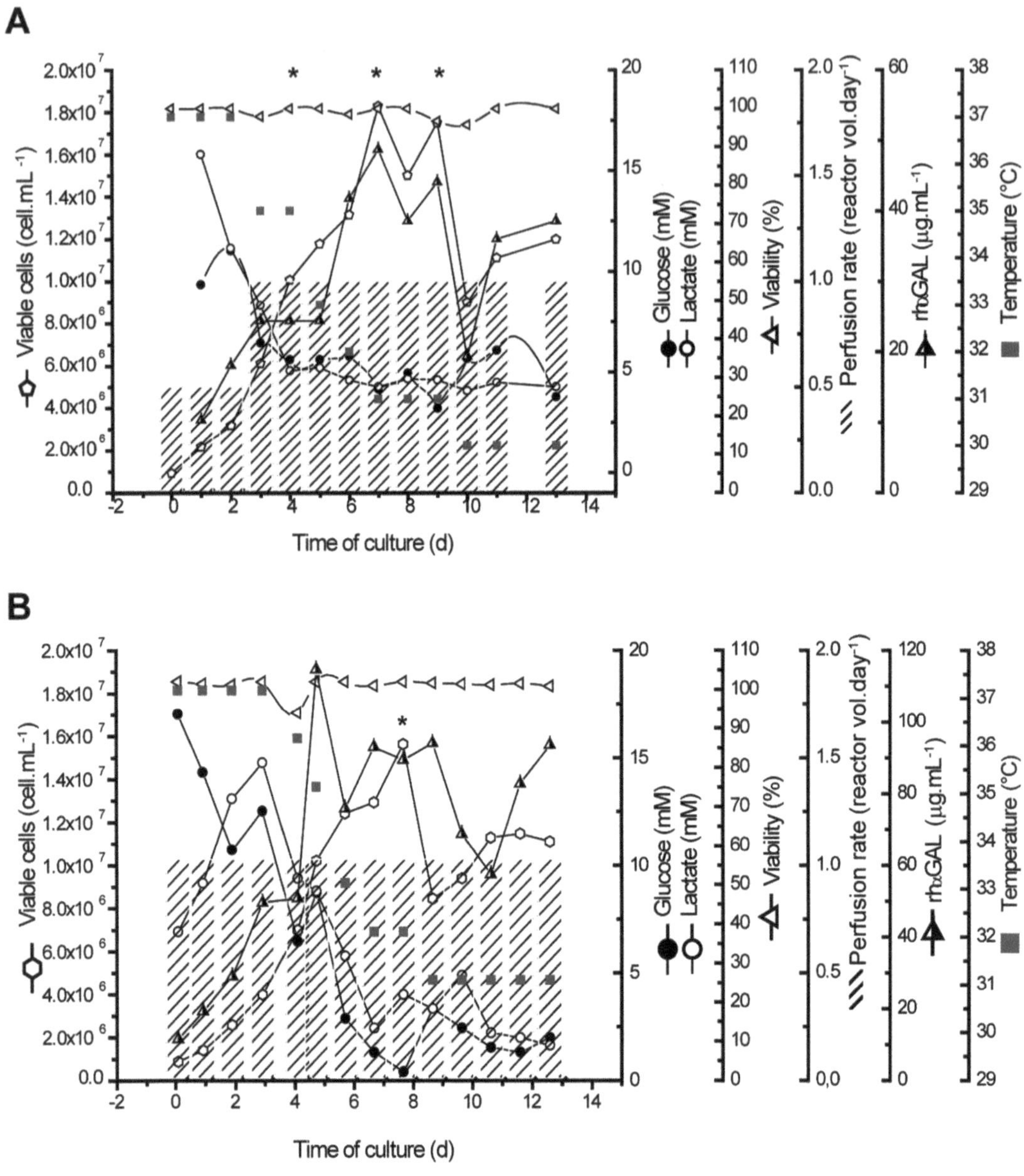

Fig. 2.2 Culture of P3G2 (**a**) and P3E8 (**b**) in 1 L bioreactor in perfusion mode. Profiles of cell density, cell viability, glucose, lactate, rhαGAL concentrations, perfusion rate and temperature are shown. Asterisk character indicates cell bleeding

In addition, the purity of rhαGAL variants was assessed by SDS-PAGE and compared to Fabrazyme®. A nearly 98% purity was obtained for both purified molecules. However, an extra polishing step should be applied to reach a higher than 99.00% purity.

As an additional characterization step of the purified proteins in comparison with Fabrazyme®, a C4 RP-HPLC was made, identifying three distinct protein peaks. As Lee et al. (2003) reported, these different peaks could be attributed to C-terminal heterogeneity with truncated species, lacking either two (peak 1) or one (peak 2) C-terminal residues present in each preparation as well as the full-length protein (peak 3). The relative amount of protein form varied between

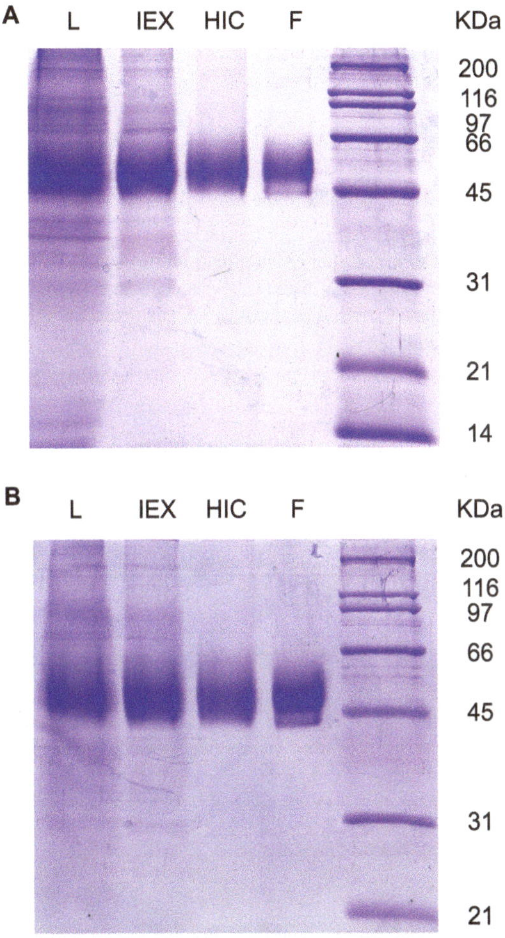

Fig. 2.3 **SDS-PAGE of fractions corresponding to different stages of the purification process of rhαGAL-P3G2 (a) and rhαGAL-P3E8 (b).** Lane 1, clarified crude supernatant (L, Load); lane 2, eluate from first chromatographic step (IEX); lane 3, eluate from second chromatographic step (HIC); lane 4, Fabrazyme®, reference molecule (F); lane 5, molecular mass standards

the three preparations; in Fabrazyme® the predominant specie was peak 3, whereas in the rhαGAL deriving from P3G2 clone (rhαGAL-P3G2), the peak 2 was the more abundant one. The rhαGAL deriving from P3E8 clone (rhαGAL-P3E8) exhibited only two peaks, being also the peak 2 the predominant one (Fig. 2.4).

Importantly, the rhαGAL-P3G2 specific activity was within the $1.6–4.8 \times 10^6$ U.mg $^{-1}$range which, in our work, was determined for Fabrazyme®. Such activity was also included in the values reported in literature (Selden et al. 2000). Interestingly, the specific activity of the rhαGAL-P3E8 was higher than that determined for Fabrazyme®, which could be attributed to different protein profile assessed by the RP-HPLC assay.

2.3.3 Physicochemical and Biochemical Characterization

Coomasie Blue staining was performed on SDS-PAGE and IEF to analyze the apparent molecular mass and isoform distribution of the purified rhαGAL variants in comparison with Fabrazyme®. The apparent molecular masses of the three molecules was nearly the same: a monomer band of about 50 kDa was observed (Fig. 2.3a, b). Furthermore, both purified products exhibited complex isoelectric focusing patterns similar to Fabrazyme®, with a wide diversity of low isoelectric point glycoforms (Fig. 2.5a). Although the three molecules shared a considerable number of isoforms, both the rhαGAL-P3G2 and the rhαGAL-P3E8 showed a higher proportion of glycoforms concentrated in the more acidic zone of the pH range compared to Fabrazyme®.

The in vitro stability of both purified rhαGAL variants and Fabrazyme® following exposure to human plasma was assessed by analysis of their residual enzymatic activity. The rhαGAL-P3G2 and Fabrazyme® molecules showed similar profiles ($p > .05$): as soon as they were diluted in plasma their activity began to fall, retaining ~29% of the initial value after 45 min of incubation. However, the rhαGAL-P3E8 retained up to 60% of its original activity after 45 min of incubation (Fig. 2.5b).

During purification and subsequent storage, many processes may occur, probably interfering with the enzyme biological activity. To assess the proper folding and enzymatic properties of both purified rhαGAL variants, we compared their kinetic parameters with those of Fabrazyme®. Michaelis-Menten analysis demonstrated that both rhαGAL molecules were capable of hydrolyzing the synthetic substrate (4MU-α-Gal) at a comparable rate to Fabrazyme® without

Table 2.1 Summary of purification rhαGAL-P3G2

	V^a (mL)	Activity ($U.mL^{-1}$)	Total protein ($mg.mL^{-1}$)	SA^b ($U.mg^{-1}$)	R^c (%)	Enrichment (fold)	Purity (%)
Cell harvest	75.0	4.3×10^4	1.1	3.9×10^4	100	1	2.0
IEX eluate	9.0	4.2×10^5	0.8	5.3×10^5	117	13.5	23.1
HIC eluate	1.0	1.9×10^6	0.8	2.4×10^6	59	61.5	100

All purification steps were performed at 4 °C. Enzyme activity was determined after combining fractions using 4MU-α-Gal as a substrate. Total protein was determined using modified Bradford assay. Recovery was estimated from total enzyme activity. Enrichment was calculated as a relation between specific activities. Purity (%) was determined as a relation between enzyme concentration and total protein concentration ($mg.ml^{-1}$)

[a]*V* Volume

[b]*SA* Specific Activity

[c]*R* Recovery

Table 2.2 Summary of purification rhαGAL-P3E8

	V^a (ml)	Activity ($U.mL^{-1}$)	Total protein ($mg.mL^{-1}$)	SA^b ($U.mg^{-1}$)	R^c (%)	Enrichment (fold)	Purity (%)
Cell harvest	250	2.6×10^4	0.22	1.2×10^5	100	1	16.0
IEX eluate	15	3.8×10^5	0.64	6.0×10^5	87.3	5	86.1
HIC eluate	0.4	1.0×10^7	1.33	7.6×10^6	61.1	63	100

All purification steps were performed at 4 °C. Enzyme activity was determined after combining fractions using 4MU-α-Gal as a substrate. Total protein was determined using modified Bradford assay. Recovery was estimated from total enzyme activity. Enrichment was calculated as a relation between specific activities. Purity (%) was determined as a relation between enzyme concentration and total protein concentration ($mg.ml^{-1}$)

[a]*V* Volume

[b]*SA* Specific Activity

[c]*R* Recovery

significant differences between their kinetic parameters (the K*m* value was 1.5 ± 0.6, 1.3 ± 0.1 and 1.1 ± 0.3 mM for Fabrazyme®, rhαGAL-P3G2 and rhαGAL-P3G8, respectively. The K_{cat} value was 1200 ± 300, 1400 ± 400 and 1000 ± 600 min^{-1} for Fabrazyme®, rhαGAL-P3G2 and rhαGAL-P3G8, respectively). Additionally, no apparent differences were detected in their structural conformation, as evaluated by the fluorescence spectrometry (data not shown).

Type and amount of monosaccharides, sialic acid and M6P present in the purified rhαGAL variants and Fabrazyme® glycans were determined by acid hydrolysis of the samples followed by HPAEC-PAD using a DIONEX ICS-5000 system equipped with a CarboPac™ PA20 column (Thermo Fisher Scientific Dionex). Moreover, to analyze the relative amount of the charged *N*-glycans of the enzymes, WAX chromatography was performed using an ASAHIPAK ES-502N7C column (100 × 7.5 mm; SHODEX, Japan) connected to a HPLC system equipped with a fluorescence detector module. Previously, the *N*-glycans were enzymatically released using PNGase F, followed by the 2-AB glycan derivatization reaction.

Regarding monosaccharide composition, the three molecules presented similar Man, Gal and GlcNAc content. The rhαGAL-P3G2 exhibited a 75% higher Fuc amount than Fabrazyme®, whereas the rhαGAL-P3E8 variant presented nearly the same amount as Fabrazyme® (Fig. 2.6).

For successful ERT for Fabry disease, the importance of the M6P and sialic acid content in the rhαGAL products has been reported (Sohn

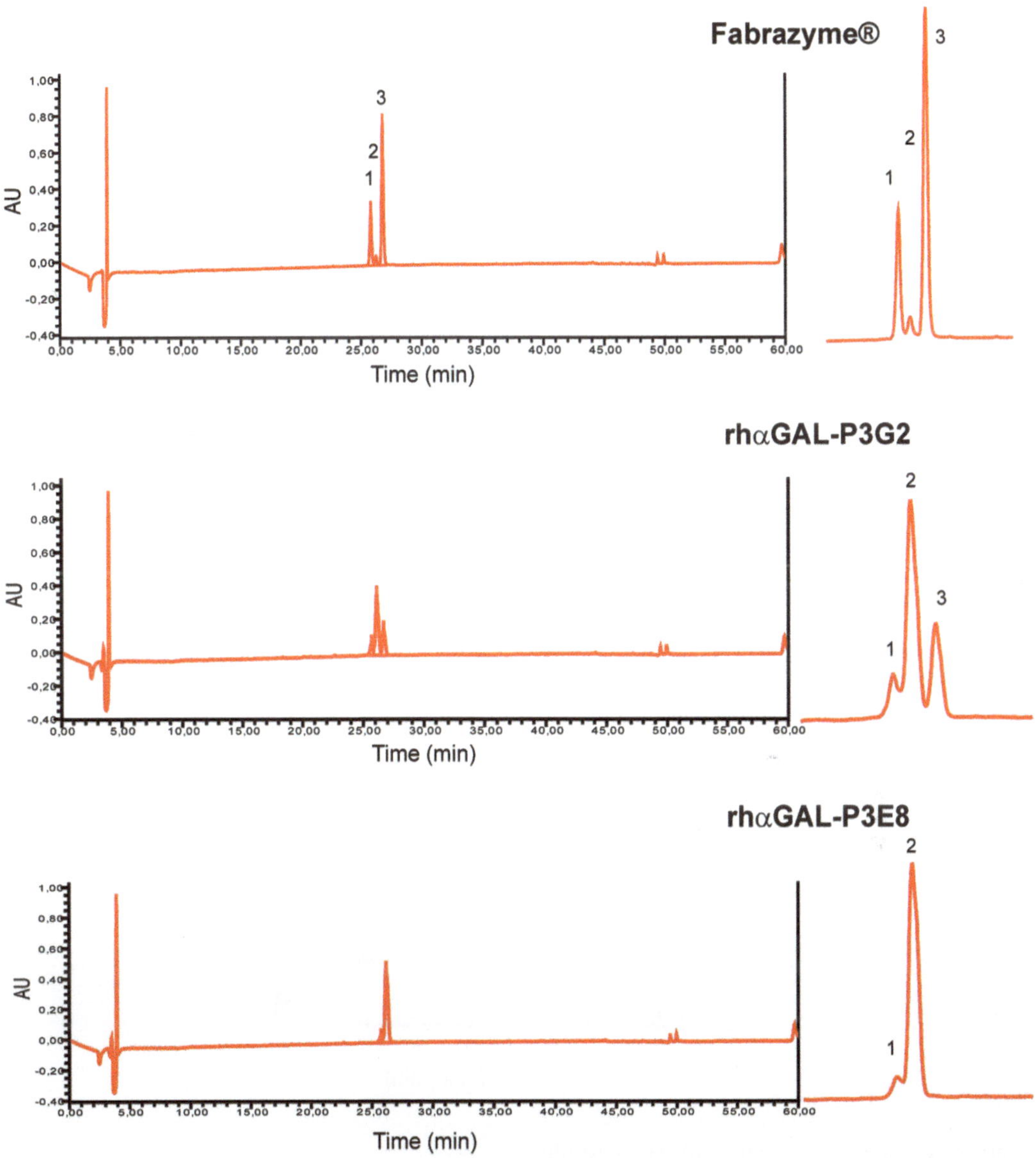

Fig. 2.4 C4 RP-HPLC analysis of Fabrazyme®, rhαGAL-P3G2 and rhαGAL-P3E8, respectively

et al. 2013a, b). Analysis of M6P glycans is important to determine the efficiency of lysosomal targeting and, therefore, therapeutic efficacy (Kang et al. 2016). Sialic acid capping is essential to mask terminal galactose, which reduces uptake by the asialoglycoprotein receptor in the liver and, thus results in proper tissue distribution with longer half-life in the body. Moreover, charge and size of proteins are important for glomerular filtration: since the glomerular filter is negatively charged, the anionic molecules are repelled and not filtered (Mahmood and Green 2005). The M6P content for both rhαGAL variants was the same as that determined for Fabrazyme®, whereas the Neu5Ac amount of rhαGAL-P3G2 and rhαGAL-P3E8 molecules was about 50% higher than the corresponding one for Fabrazyme®. Also, both rhαGAL purified variants exhibited a 33% and a 54% reduced content of the immunogenic type of sialic acid (Neu5Gc)

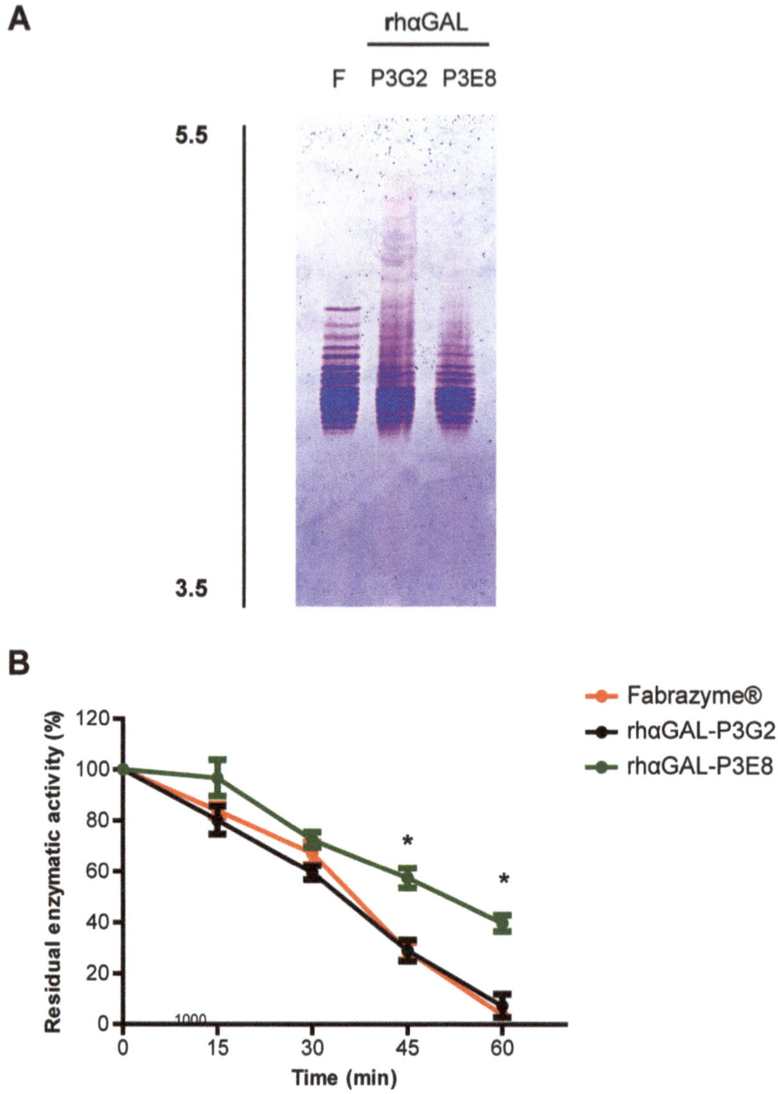

Fig. 2.5 (**a**) IEF profiles of Fabrazyme® (F), rhαGAL-P3G2 and rhαGAL-P3E8, respectively. The line indicates the pH range employed. (**b**) In vitro stability of Fabrazyme®, rhαGAL-P3G2 and rhαGAL-P3E8, respectively, following exposure to human plasma. Stability was measured as the remaining activity of the enzymes analyzed by the hydrolysis of the synthetic substrate 4MU-α-Gal after incubation in human plasma. The results are expressed as the percentage of the initial enzymatic activity. Results are an average of six measurements ± SD

compared to Fabrazyme® (Fig. 2.6). These results were comparable to the ones obtained by WAX-HPLC, which indicated the presence of more negatively charged structures and less neutral structures for the rhαGAL variants (Fig. 2.7).

Taking into account the higher sialic acid content, the longer the rhαGAL serum half-life, therefore, the results mentioned above may have a significant impact on the protein therapeutic efficacy by reducing both, the drug amount and its administration frequency (Varki 2008; Byrne et al. 2007). However, these results should be verified by in vivo assays in the future.

2.3.4 Cellular Uptake and Localization of the rhαGAL Variants

The uptake of the rhαGAL variants into primary skin fibroblasts of Fabry patients (NIGMS Coriell Institute), lacking endogenous α-Gal, was evaluated by fluorescent microscopy. The subcellular localization of the recombinant enzyme was determined by an immunofluorescent assay. Both rhαGAL variants and Fabrazyme® showed a characteristic granular pattern that would indicate a lysosomal

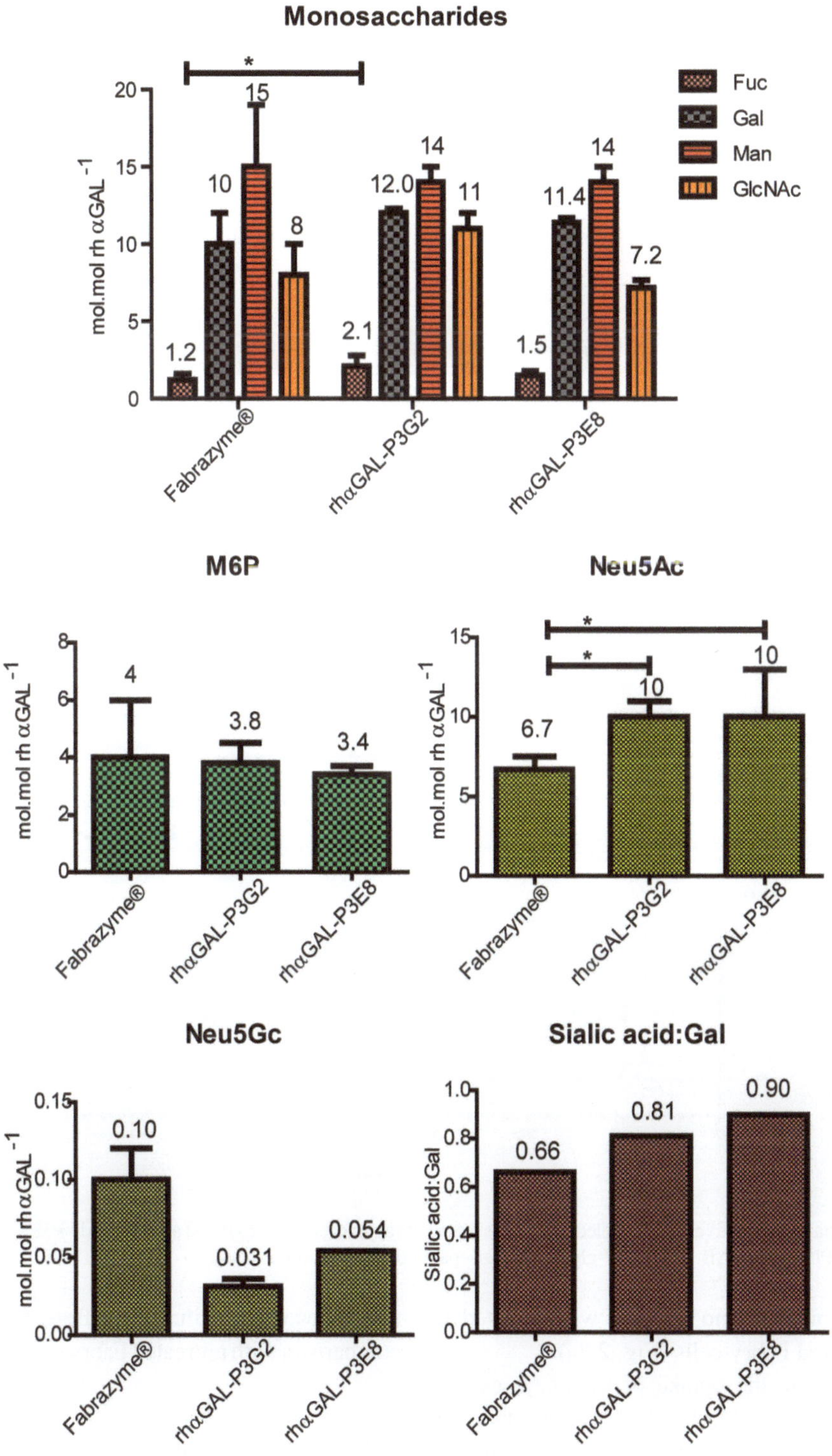

Fig. 2.6 Monosaccharide composition of Fabrazyme®, rhαGAL-P3G2 and rhαGAL-P3E8, respectively

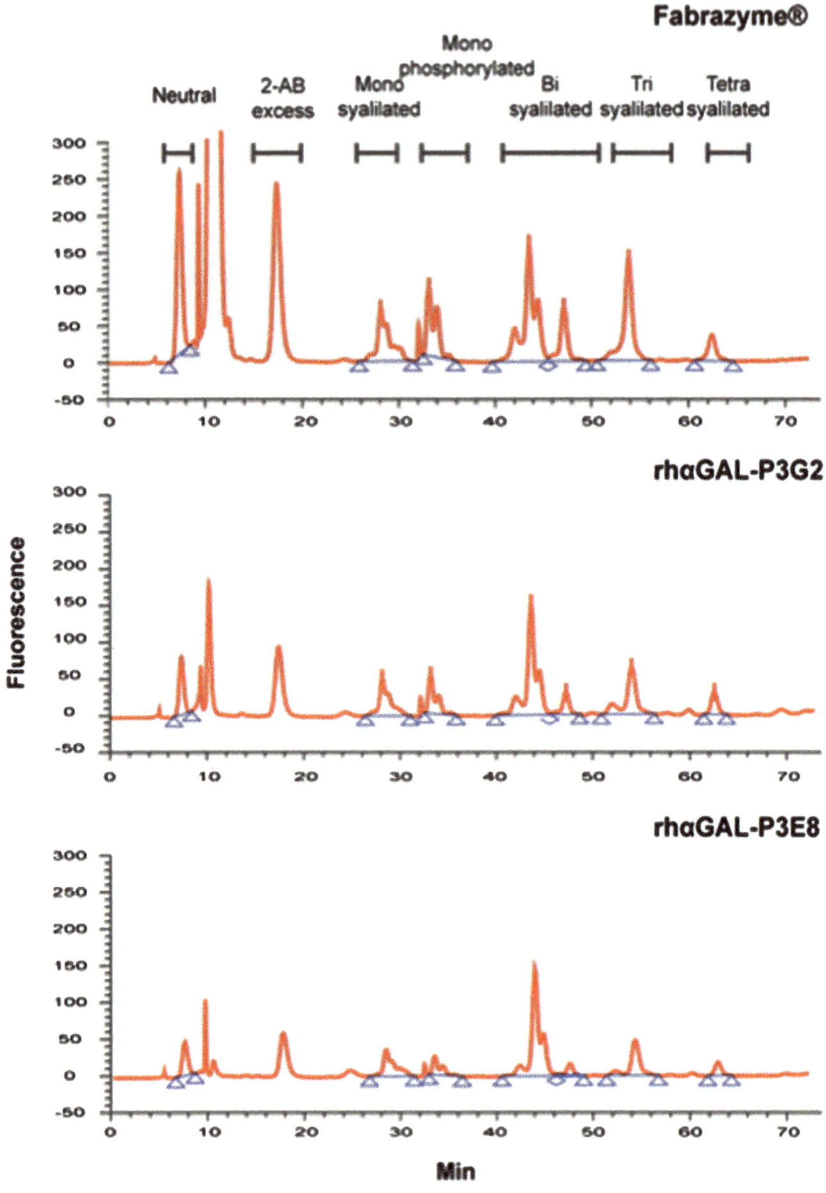

Fig. 2.7 Analysis of native 2-AB labeled *N*-glycans from Fabrazyme®, rhαGAL-P3G2 and rhαGAL-P3E8, respectively, by WAX-HPLC. The 2AB-labelled *N*-glycans were separated based on their charge

localization. Also, no enzyme was detected in the untreated Fabry cells (Fig. 2.8a).

In addition, the uptake of the enzymes was measured in fibroblast cell homogenates after 24 h of incubation. The total protein content was determined using Bradford assay, and the enzymatic activity was assayed using the artificial 4MU-α-Gal substrate. The uptake of three molecules was significantly different ($p < .05$) in comparison with untreated Fabry cells (Fig. 2.8b).

2.4 Conclusion

Currently, mammalian expression systems, mainly CHO cell line, are extensively used for biotherapeutics production. The wide spread suc-

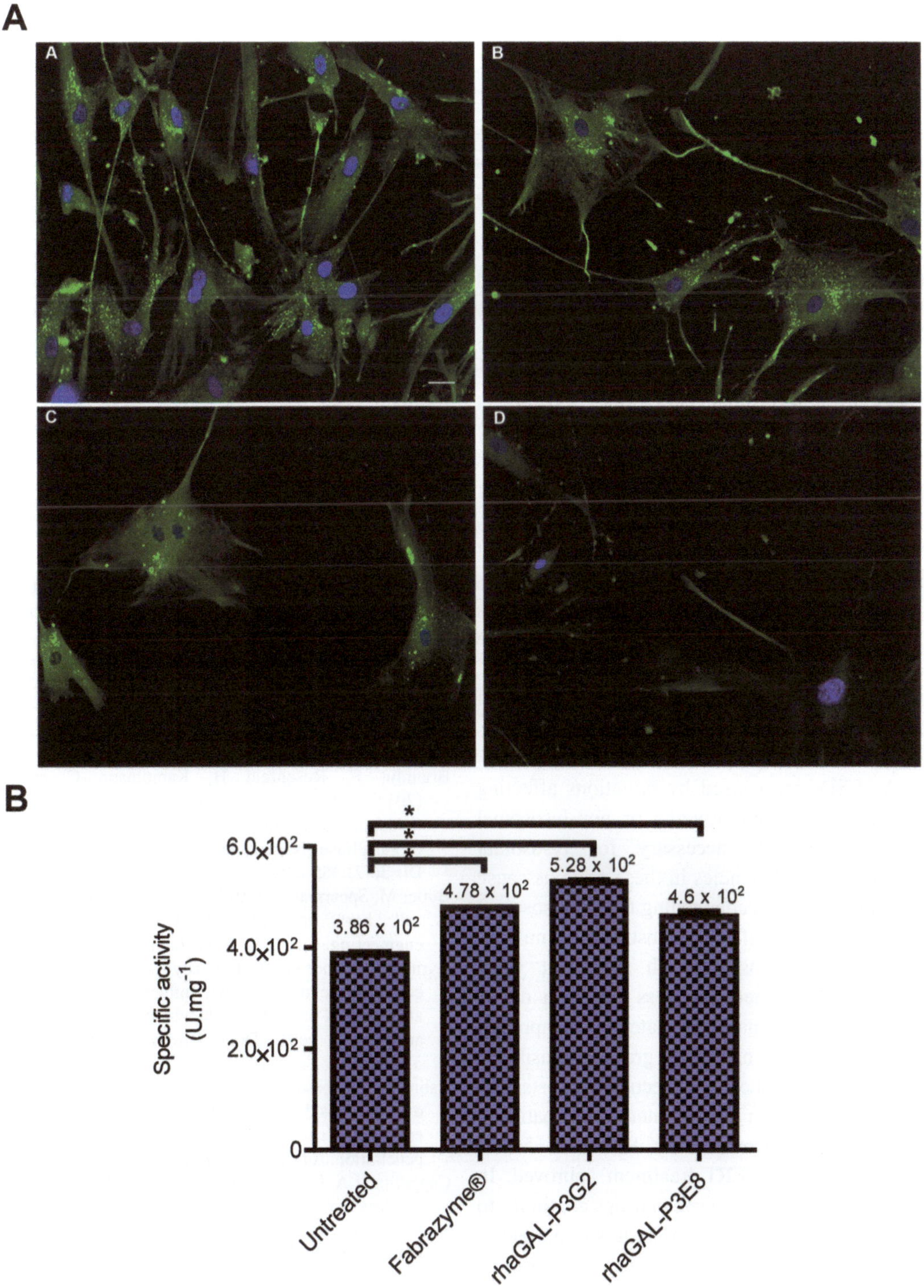

Fig. 2.8 (**a**) Fluorescence microscopy to study the subcellular localization of rhαGAL. First, Fabry fibroblast were incubated with Fabrazyme® (A), rhαGAL-P3G2 (B), rhαGAL-P3E8 (C) or without any protein (D), then fixed, incubated with an anti-rhαGAL antibody and incubated with a second antibody *Alexa Fluor488 goat anti-rabbit igG H and L* (ab 150,081). Nuclei were stained with DAPI. Bar scale: 19 μm. (**b**) Specific activity of rhαGAL variants in fibroblast cell homogenates after 24 h of incubation. The total protein content was determined using Bradford assay and the enzymatic activity was assayed using the artificial 4MU-α-Gal substrate. (* $p < .05$)

cess of the CHO platform can be attributed due to its adaptability to grow at high densities in suspension cultures, ease of adaptation to serum-free conditions and the addition of human-like glycosylation. To obtain higher protein expression levels, several strategies have been applied, such as those employing epigenetic regulatory elements or those using the revolutionary artificially engineered scissors as the CRISPR/Cas technology. In addition, when the production of a stable cell line is required, the selection of a proper delivery system is crucial. As LVs can reduce process timelines, increase production levels and enable use of difficult-to-transfect cell lines, they are an attractive delivery tool system. As they derive from HIV-1, there are safety concerns that LVs might be contaminated with replicating virus or viral proteins that could be pathogenic. However, several sensitive assays have been developed that guarantee any RCL absence.

Among therapeutic proteins, enzymes represent a small and profitable market. With the approval of the Orphan Drug Act in 1983 in USA, the development of enzyme therapeutics against rare diseases such as lysosomal disorders has gained notable importance.

As LSDs are caused by mutations affecting essential lysosomal proteins or non-lysosomal proteins that are necessary for lysosomal function, the deficiencies in these proteins cause a pathogenetic cascade leading to intralysosomal accumulation of different substrates in multiple tissues and organs. Though in recent years remarkable treatment progress has been made, these disorders remain associated with important neglected medical needs and great responsibility in terms of public health and economic costs. The ERT has offered an improvement in the patient's life quality. However, only 8 out of 50–60 disorders have an ERT treatment approved. In addition, the access to orphan drugs continues to be limited by high prices. Also, it is important to ameliorate the enzyme half-life and lysosomal targeting. Consequently, an interdisciplinary effort could be necessary to address these issues from different angles.

In this chapter, we have presented a strategy trying to reduce the timeline of the overall production process of a promising therapeutic alternative for Fabry disease. In this way, we have not only obtained clones, demonstrating a suitable balance between the enzyme high productivity and quality, but we have also enhanced the gene delivery efficiency and reduced the timeline achievement to produce a cell line. Moreover, the purified variants exhibit improved properties regarding glycosylation, in particular, higher sialic acid content and a reduced percentage of antigenic Neu5Gc glycan, attributes which may derive in an enhanced rhαGAL in vivo efficacy.

References

Bandaranayake AD, Almo SC (2014) Recent advances in mammalian protein production. FEBS Lett 588(2):253–260

Baranyi L, Roy A, Embree HD, Dropulic B (2010) Lentiviral vector-mediated genetic modification of cell substrates for the manufacture of proteins and other biologics. PDA J Pharm Sci Technol 64:379–385

Bennett LL, Mohan D (2013) Gaucher disease and its treatment options. Ann Pharmacother 47(9):1182–93

Braulke T, Bonifacino JS (2009) Sorting of lysosomal proteins. BBA – Mol Cell Res 1793:605–614

Braunlin E, Rosenfeld H, Kampmann C et al (2013) Enzyme replacement therapy for mucopolysaccharidosis VI: long-term cardiac effects of galsulfase (Naglazyme®) therapy. J Inherit Metab Dis 36(2):385–394

Butler M, Spearman M (2014) The choice of mammalian cell host and possibilities for glycosylation engineering. Curr Opin Biotechnol 30:107–112

Byrne B, Donohoe GG, O'Kennedy R (2007) Sialic acids: carbohydrate moieties that influence the biological and physical properties of biopharmaceutical proteins and living cells. Drug Discov Today 12(7–8): 319–326

Cabrera I, Abasolo I, Corchero JL et al (2016) α -galactosidase-A-loaded Nanoliposomes with enhanced enzymatic activity and intracellular penetration. Adv Healthc Mater 5:829–840

Cappellino LA, Kratje RB, Etcheverrigaray M et al (2017) Strategy for erythroid differentiation in ex vivo cultures: lentiviral genetic modification of human hematopoietic stem cells with erythropoietin gene. J Biosci Bioeng 124(5):591–598

Cho MS, Yee H, Chan S (2002) Establishment of a human somatic hybrid cell line for recombinant protein production. J Biomed Sci 9:631–638

Chung YK, Sohn YB, Sohn JM et al (2014) A biochemical and physicochemical comparison of two recombinant enzymes used for enzyme replacement therapies of hunter syndrome. Glycoconj J 31(4):309–315

Clark DP, Pazdernik NJ (2016) Chapter 10- recombinant proteins. In: Clark DP, Pazdernik NJ (ed) Biotechnology, 2nd edn. Academic Cell, p 355–363

Corchero JL, Mendoza R, Lorenzo J et al (2011) Integrated approach to produce a recombinant, his-tagged human α-galactosidase A in mammalian cells. Biotechnol Prog 27(5):1206–1217

Cornetta K, Yao J, Jasti A et al (2011) Replication-competent lentivirus analysis of clinical grade vector products. Mol Ther 19(3):557–566

Cornetta K, Duffy L, Turtle CJ et al (2017) Absence of replication-competent lentivirus in the clinic: analysis of infused T cell products. Mol Ther 25(12):1–9

Cox MMJ (2012) Recombinant protein vaccines produced in insect cells. Vaccine 30(10):1759–1766

Desnick RJ (2004) Enzyme replacement therapy for Fabry disease: lessons from two alpha-galactosidase A orphan products and one FDA approval. Expert Opin Biol Ther 4(7):1167–1176

Ding K, Han L, Zong H et al (2017) Production process reproducibility and product quality consistency of transient gene expression in HEK293 cells with anti-PD1 antibody as the model protein. Appl Microbiol Biotechnol 101(5):1889–1898

Dumont J, Euwart D, Mei B et al (2016) Human cell lines for biopharmaceutical manufacturing: history, status, and future perspectives. Crit Rev Biotechnol 36(6):1110–1122

Dvorak-Ewell M, Wendt D, Hague C et al (2010) Enzyme replacement in a human model of mucopolysaccharidosis IVA in vitro and its biodistribution in the cartilage of wild type mice. PLoS One 5(8):1–11

Escarpe P, Zayek N, Chin P et al (2003) Development of a sensitive assay for detection of replication-competent recombinant lentivirus in large-scale HIV-based vector preparations. Mol Ther 8(2):332–341

Espejo-Mojica ÁJ, Alméciga-Díaz CJ, Rodríguez A et al (2015) Human recombinant lysosomal enzymes produced in microorganisms. Mol Genet Metab 116(1–2):13–23

Frampton JE (2016) Sebelipase alfa: a review in lysosomal acid lipase deficiency. Am J Cardiovasc Drugs 16(6):461–468

Gaillet B, Gilbert R, Broussau S et al (2010) High-level recombinant protein production in CHO cells using lentiviral vectors and the cumate gene-switch. Biotechnol Bioeng 106(2):203–215

Gaj T, Gersbach CA, Barbas CF (2013) ZFN, TALEN, and CRISPR/Cas-based methods for genome engineering. Trends Biotechnol 31(7):397–405

Gupta SK, Shukla P (2017) Gene editing for cell engineering: trends and applications. Crit Rev Biotechnol 37(5):672–684

Hacker DL, Balasubramanian S (2016) Recombinant protein production from stable mammalian cell lines and pools. Curr Opin Struct Biol 38:129–136

Harraghy N, Calabrese D, Fisch I et al (2015) Epigenetic regulatory elements: recent advances in understanding their mode of action and use for recombinant protein production in mammalian cells. Biotechnol J 10(7):967–978

Higgins E (2010) Carbohydrate analysis throughout the development of a protein therapeutic. Glycoconj J 27(2):211–225

Ioannou YA, Bishop DF, Desnick RJ (1992) Overexpression of human a-galactosidase a results in its intracellular aggregation, crystallization in lysosomes, and selective secretion. J Cell Biol 119:1137–50

Jameson E, Jones S, Remmington T (2013) Enzyme replacement therapywith laronidase (Aldurazyme®) for treating mucopolysaccharidosis type I. Cochrane Database Syst Rev 9(4):CD009354

Johnston JM, Denning G, Doering CB et al (2012) Generation of an optimized lentiviral vector encoding a high- expression factor VIII transgene for gene therapy of hemophilia A. Gene Ther 20(6):607–61576

Kang J-Y, Kwon O, Gil JY et al (2016) Comparison of fluorescent tags for analysis of mannose-6-phosphate glycans. Anal Biochem 501:1–3

Khan KH (2013) Gene expression in mammalian cells and its applications. Adv Pharm Bull 3(2):257–263

Kizhner T, Azulay Y, Hainrichson M et al (2015) Characterization of a chemically modified plant cell culture expressed human α-galactosidase-A enzyme for treatment of Fabry disease. Mol Genet Metab 114:259–267

Kumar Kakkar A, Dahiya N (2014) From Blockbusters to Niche Busters. Drug Dev Res 75:231–234

Kumar S, Abdulhameed S (2017) Therapeutic enzymes. Bioresour Bioprocess Biotechnol 2:45–73

Lagassé HAD, Alexaki A, Simhadri VL et al (2017) Recent advances in (therapeutic protein) drug development [version 1; referees: 2 approved]. F1000Res. 2017 6(113):1–7

Lalonde ME, Durocher Y (2017) Therapeutic glycoprotein production in mammalian cells. J Biotechnol 251:128–140

Lee K, Jin X, Zhang K et al (2003) A biochemical and pharmacological comparison of enzyme replacement therapies for the glycolipid storage disorder Fabry disease. Glycobiology 13(4):305–313

Mahmood I, Green MD (2005) Pharmacokinetic and pharmacodynamic considerations in the development of therapeutic proteins. Clin Pharmacokinet 44(4):331–347

Mao Y, Yan R, Li A et al (2015) Lentiviral vectors mediate long-term and high efficiency transgene expression in HEK 293T cells. Int J Med Sci 12(5):407–415

Merkle FT, Neuhausser WM, Santos D et al (2015) Efficient CRISPR-Cas9-mediated generation of knockin human pluripotent stem cells lacking undesired mutations at the targeted locus. Cell Rep 11(6):875–883

Merten O-W, Hebben M, Bovolenta C (2016) Production of lentiviral vectors. Mol Ther Methods Clin Dev 3:1–14

Moreno AM, Mali P (2017) Therapeutic genome engineering via CRISPR-Cas systems. Syst Biol Med 9:e1380
Mufarrege EF, Antuña S, Etcheverrigaray M et al (2014) Development of lentiviral vectors for transient and stable protein overexpression in mammalian cells. A new strategy for recombinant human FVIII (rhFVIII) production. Protein Expr Purif 95:50–56
Oberbek A, Matasci M, Hacker DL (2011) Generation of stable, high-producing cho cell lines by lentiviral vector-mediated gene transfer in serum-free suspension culture. Biotechnol Bioeng 108(3):600–610
Oh DB (2015) Glyco-engineering strategies for the development of therapeutic enzymes with improved efficacy for the treatment of lysosomal storage diseases. BMB Rep 48(8):438–444
Parenti G, Andria G, Ballabio A (2015) Lysosomal storage diseases: from pathophysiology to therapy. Annu Rev Med 66(1):471–486
Picanço-Castro V, de Sousa Russo-Carbolante EM, Tadeu Covas D (2012) Advances in lentiviral vectors: a patent review. Recent Pat DNA Gene Seq 6(2):82–90
Plewa C (2010) Application of lentiviral vectors for development of production cell lines and safety testing of lentiviral-derived cells or products PDA. J Pharm Sci Technol 64:386–391
Porter JL, Rusli RA, Ollis DL (2016) Directed evolution of enzymes for industrial biocatalysis. Chembiochem 17(3):197–203
Prieto C, Fontana D, Etcheverrigaray M et al (2011) A strategy to obtain recombinant cell lines with high expression levels. Lentiviral vector-mediated transgenesis. BMC Proc 5(Suppl 8):7–8
Rodríguez MC, Ceaglio N, Antuña S et al (2017) High yield process for the production of active human a -galactosidase a in CHO-K1 cells through lentivirus transgenesis. Biotechnol Prog 33(5):1334–1345
Ronda C, Pedersen LE, Hansen HG et al (2014) Accelerating genome editing in CHO cells using CRISPR Cas9 and CRISPy, a web-based target finding tool. Biotechnol Bioeng 111(8):1604–1616
Sakuma T, Barry MA, Ikeda Y (2012) Lentiviral vectors: basic to translational. Biochem J 443:603–618
Sakuraba H, Murata-Ohsawa M, Kawashima I et al (2006) Comparison of the effects of agalsidase alfa and agalsidase beta on cultured human Fabry fibroblasts and Fabry mice. J Hum Genet 51:180–188
Sastry L, Xu Y, Johnson T et al (2003) Certification assays for HIV-1-based vectors: frequent passage of gag sequences without evidence of replication-competent viruses. Mol Ther 8(5):830–839
Selden RF, Borowski M, Kinoshita CM et al (2000) Medical preparations for the treatment of alpha-galactosidase a deficiency. WO 00/53730 A3 14 Sept 2000
Sestito S, Grisolia M, Concolino D (2015) Profile of idursulfase for the treatment of Hunter syndrome. Res Reports Endocr Disord 5:79–90
Shestopal SA, Hao J-J, Karnaukhova E et al (2017) Expression and characterization of a codon-optimized blood coagulation factor VIII. J Thromb Haemost 15(4):709–720
Skrdlant LM, Armstrong RJ, Keidaisch BM et al (2017) Detection of replication competent lentivirus using a qPCR assay for VSV-G. Mol Ther Methods Clin Dev 8:1–7
Sohn YB, Cho SY, Park SW et al (2013a) Phase I/II clinical trial of enzyme replacement therapy with idursulfase beta in patients with mucopolysaccharidosis II (Hunter syndrome). Orphanet J Rare Dis 8(1):2–9
Sohn Y, Lee JM, Park H et al (2013b) Enhanced sialylation and in vivo efficacy of recombinant human α-galactosidase through in vitro glycosylation. BMB Rep 46(3):157–162
Solomon M, Muro S (2017) Lysosomal enzyme replacement therapies: historical development, clinical outcomes, and future perspectives Melani. Adv Drug Deliv Rev 118:109–134
Spencer HT, Denning G, Gautney RE et al (2011) Lentiviral vector platform for production of bioengineered recombinant coagulation factor VIII. Mol Ther 19(2):302–309
Spencer S, Gugliotta A, Koenitzer J et al (2015) Stability of single copy transgene expression in CHOK1 cells is affected by histone modifications but not by DNA methylation. J Biotechnol 195(1):15–29
Swiech K, Picanço-Castro V, Covas DT (2012) Human cells: new platform for recombinant therapeutic protein production. Protein Expr Purif 84(1):147–153
Tomatsu S, Sawamoto K, Shimada T et al (2015) Enzyme replacement therapy for treating mucopolysaccharidosis type IVA (Morquio A syndrome): effect and limitations. Expert Opin Orphan Drugs 3(11):1279–1290
Turan S, Zehe C, Kuehle J et al (2013) Recombinase-mediated cassette exchange (RMCE) – a rapidly-expanding toolbox for targeted genomic modifications. Gene 515(1):1–27
Van der Valk J, Brunner D, De Smet K et al (2010) Optimization of chemically defined cell culture media – replacing fetal bovine serum in mammalian in vitro methods. Toxicol In Vitro 24(4):1053–1063
Varki A (2008) Sialic acids in human health and disease. Trends Mol Med 14(8):351–360
Vernon HJ (2015) Inborn errors of metabolism: advances in diagnosis and therapy. JAMA Pediatr 169(8):778–782
Wirth D, Gama-Norton L, Riemer P et al (2007) Road to precision: recombinase-based targeting technologies for genome engineering. Curr Opin Biotechnol 18:411–419
Wurm FM (2004) Production of recombinant protein therapeutics in cultivated mammalian cells. Nat Biotechnol 22(11):1393–1398
Wurm FM (2013) CHO Quasispecies – implications for manufacturing processes. Processes 1(3):296–311
Xu M, Motabar O, Ferrer M et al (2016) Disease models for the development of therapies for lysosomal storage diseases. Ann N Y Acad Sci 1371(1):15–29
Zhang F, Liu M, Wan H (2014) Discussion about several potential drawbacks of PEGylated therapeutic proteins. Biol Pharm Bull 37(3):335–339

Quality Control and Downstream Processing of Therapeutic Enzymes

3

David Gervais

Abstract

Therapeutic enzymes are a commercially minor but clinically important area of biopharmaceuticals. An array of therapeutic enzymes has been developed for a variety of human diseases, including leukaemia and enzyme-deficiency diseases such as Gaucher's disease. Production and testing of therapeutic enzymes is strictly governed by regulatory bodies in each country around the world, and batch-to-batch consistency is crucially important. Manufacture of a batch starts with the fermentation or cell culture stage. After expression of the therapeutic enzyme in a cell culture bioreactor, robust and reproducible protein purification, or downstream processing (DSP) of the target product, is critical to ensuring safe delivery of these medicines. Modern processing technology, including the use of disposable processing equipment, has greatly improved the DSP development pathway in terms of robustness and speed to clinic. Once purified, the drug substance undergoes rigorous quality control (QC) testing according to current regulatory guidance, to enable release to the clinic and patient. QC testing is conducted to ensure the safety, purity, identity, potency and strength of the medicinal product, requiring multiple analytical methods that are rigorously validated and monitored for robust performance. Several case studies, including L-asparaginase and asfotase alfa, are discussed to illustrate the methods described herein.

Keywords

Therapeutic enzymes · Downstream processing · Quality control · Enzyme characterisation · Enzyme manufacturing

Abbreviations

2D-GE	Two-dimensional gel electrophoresis
ABG	Acid β-glucosidase
ADA	Anti-drug antibodies
ADC	Antibody-drug conjugates
AUC	Analytical ultracentrifugation
CFU	Colony-forming units
CGE	Capillary gel electrophoresis
CHO	Chinese hamster ovary
cIEF	Capillary isoelectric focussing
CM	Carboxymethyl
CNS	Central nervous system
DEAE	Diethyl amino ethyl
DNA	Deoxyribonucleic acid
DOE	Design of experiments

D. Gervais (✉)
Porton Biopharma Limited, Salisbury, Wiltshire, UK
e-mail: dave.gervais@portonbiopharma.com

N. Labrou (ed.), *Therapeutic Enzymes: Function and Clinical Implications*, Advances in Experimental Medicine and Biology 1148, https://doi.org/10.1007/978-981-13-7709-9_3

DP	Drug product
DS	Drug substance
DSP	Downstream processing
ELISA	Enzyme-linked immunosorbent assay
EMA	European medicines agency
ERT	Enzyme replacement therapy
EU	Endotoxin units
F_c	Constant domain (antibody)
FDA	United States Food and Drug Administration
GD	Gaucher's disease
GMP	Good manufacturing practice
HCP	Host cell proteins
HER2	Human epidermal growth factor receptor 2
HIC	Hydrophobic interaction chromatography
HPLC	High-pressure liquid chromatography
HPP	Hypophosphatasia
HVAC	Heating, ventilation and air conditioning
ICH	International Council on Harmonisation
IEX	Ion-exchange
IEX-HPLC	Ion-exchange high-pressure liquid chromatography
IgG	Immunoglobulin
IMAC	Immobilised-metal affinity chromatography
ITC	Isothermal calorimetry
IU	International unit (of enzymatic activity)
k_{cat}	Enzyme catalytic constant
K_M	Michaelis constant
LAL	Limulus amoebocyte lysate
LAL-D	Lysosomal acid lipase deficiency
LC-MS	Liquid chromatography coupled mass spectrometry
LOD	Limit of detection
mAbs	Monoclonal antibodies
MALDI	Matrix-assisted laser desorption/ionisation
MF	Microfiltration
MS	Mass spectrometry
MS/MS	Tandem mass spectrometry
NAD	Nicotinamide adenine dinucleotide
NADP	Nicotinamide adenine dinucleotide phosphate
NCPPB	National Collection of Plant Pathogenic Bacteria (UK)
PCR	Polymerase chain reaction
PEG	Poly(ethylene glycol)
pI	Isoelectric point
PMDA	Japanese Pharmaceuticals and Medical Devices Agency
Q	Quaternary ammonium
QA	Quality assurance
QbD	Quality by design
QC	Quality control
QMS	Quality management system
qPCR reaction	Quantitative polymerase chain
R&D	Research and development
RP-HPLC	Reversed-phase high-pressure liquid chromatography
RT-PCR	Real-time polymerase chain reaction
S	Sulphopropyl
SDS-PAGE	Sodium dodecyl sulphate polyacrylamide gel electrophoresis
SEC	Size-exclusion chromatography
SOP	Standard operating procedures
SPR	Surface plasmon resonance
TFF	Tangential-flow filtration
TFF-MF	Tangential-flow filtration – microfiltration
TNALP	Tissue-nonspecific alkaline phosphatase
U	Units (of enzyme activity)
US	United States
UV	Ultraviolet
V_{max}	Maximum enzymatic reaction velocity

3.1 Introduction to Production of Therapeutic Enzymes

Therapeutic enzymes represent a small fraction ($6.6 billion in worldwide sales in 2015) of the overall global recombinant therapeutic protein market worldwide ($175 billion in 2015), but are nonetheless important economically and clinically (Dewan and Sullivan 2016). Therapeutic enzymes have been developed over the past few decades to treat a vast array of human diseases, from oncology (i.e., leukaemia) to rare genetic enzyme deficiency disorders (i.e., Gaucher's disease). These enzymes occupy a distinct part of the overall recombinant protein therapeutic product area, together with cytokines (e.g., interferon-β-1b for multiple sclerosis), monoclonal antibodies (e.g., anti-HER2 antibody for treatment of breast cancer), hormones (e.g., insulin) and vaccines (e.g., recombinant hepatitis B vaccine).

Critical to the production of these important products is the purification of the enzyme, and demonstrating that it is of adequate purity for human administration. Such proof is provided by Quality Control (or QC), which is mandated by regulatory Good Manufacturing Practices (GMP) worldwide and an important part of ensuring that the patient receives medicinal products that are safe to use. The purpose of this chapter is to describe examples of some key therapeutic enzyme products, introduce and explain the nature of the downstream processes used to purify them, and provide insight into the QC techniques used to evaluate of them.

3.1.1 Types of Therapeutic Enzymes

A staggering array of therapeutic enzymes is currently approved by regulatory agencies for clinical use in humans. Whilst it is not possible to list all of the enzymes here, a selected list of commercially- and therapeutically-important enzymes is shown in Table 3.1.

As can be observed from Table 3.1, therapeutic enzymes are produced from a variety of cellular expression platforms to treat an array of human diseases. Some of the technologies used to produce these products include bacterial cells (e.g. *Erwinia chrysanthemi, Escherichia coli*), mammalian cells (e.g. CHO, human cell lines), and more recently, plant cells (Aviezer et al. 2009). A particularly novel example is the recently-approved treatment for lysosomal acid lipase deficiency (LAL-D, also referred to as Wolman disease), sebelipase alfa or Kanuma from Alexion Pharmaceuticals (Su et al. 2016). Individuals with LAL-D, a genetic disorder, do not produce enough lysosomal acid lipase, which is used by the body to break down fatty acids and triglycerides. Sebelipase alfa is used as an enzyme replacement therapy for these individuals. Sebelipase alfa is notable because it is a recombinant enzyme manufactured from the eggs of transgenic chickens, the first biological drug to be developed and manufactured in this way.

Regardless of the method of protein expression, there are several common themes to therapeutic enzyme production. Like all biological products produced for clinical use, enzymes must undergo a rigorous and robust purification procedure in order to become appropriate for human administration. An overview of routes to therapeutic enzyme manufacture is provided in Sect. 3.1.2.

3.1.2 Production of Therapeutic Enzymes

Production of therapeutic enzymes (Fig. 3.1) in common with all biopharmaceutical manufacture is a complex and time-consuming effort that requires multiple departments within an organisation to work together, including Manufacturing, Engineering, Quality Assurance (QA) and Quality Control (QC). The most critical aspect of production of human clinical enzymes is adherence to Good Manufacturing Practices (GMP), designed by regulatory agencies around the world (*i.e.,* the US Food and Drug Administration or FDA, and the Japanese Pharmaceuticals and Medical Devices Agency or PMDA) to provide assurance that products are safe and effective for use. Although each country in effect has its own interpretation of GMP, in recent years the International Council on Harmonisation (ICH) has worked to provide alignment between the regulatory bodies on the scientific and technical aspects of pharmaceutical production. A good

Table 3.1 Selected therapeutic enzymes of industrial importance

Enzyme name (drug name)	Manufacturer	Disease indication	Expression platform	References
L-asparaginase *Erwinia chrysanthemi* (Erwinase)	Porton biopharma	Acute lymphoblastic Leukaemia	*Erwinia chrysanthemi* NCPPB 1066	Salzer et al. (2014)
PEGylated L-asparaginase *Escherichia coli* (Oncaspar)	Shire Pharmaceuticals	Acute lymphoblastic leukaemia	*Escherichia coli*	Graham (2003)
Carboxypeptidase (Voraxase)	Protherics	Methotrexate clearance in patients with impaired renal function	*Escherichia coli*	Patterson and Lee (2010)
β-glucocerebrosidase (Elelyso)	Pfizer	Gaucher's disease	Plant cells (carrots)	Mor (2015)
Imiglucerase (Cerezyme)	Genzyme	Gaucher's disease	Chinese hamster ovary (CHO)	Edmunds (2005)
Elosulfase alfa (Vimizim)	BioMarin	Mucopolysaccharide processing disorder (Marquio A)	Chinese hamster ovary (CHO)	Sanford and Lo (2014)
Lysosomal acid lipase (Kanuma)	Alexion	Lysosomal acid lipase deficiency	Eggs from transgenic chickens (*Gallus gallus*)	Su et al. (2016)
Dornase alfa (Pulmozyme)	Genentech	Cystic fibrosis	Chinese hamster ovary (CHO)	Wagener and Kupfer (2012)
Human α-L-iduronidase (Aldurazyme)	Genzyme	Hurler and Hurler-Scheie syndromes	Chinese hamster ovary (CHO)	Kakkis (2005)
Human α-galactosidase A (Fabrazyme)	Genzyme	Fabry disease	Chinese hamster ovary (CHO)	Lee et al. (2003)
Human iduronate-2-sulfatase (Elaprase)	Shire	Hunter's syndrome	Human cell line HT-1080	Heartlein and Kimura (2014)
Human N-acetylgalactosamine-4-sulfatase (Naglazyme)	BioMarin	Maroteaux-Lamy syndrome	Chinese hamster ovary (CHO)	Harmatz et al. (2014)

description of pharmaceutical GMP is provided by Tobin and Walsh (2008). However, GMP is constantly evolving as a result of advances in technology and capabilities. Thus, it is important for potential therapeutic manufacturers to stay current with GMP and consult individual guidance documents from each agency as well as the ICH, which are freely available *via* the internet.

In therapeutic enzyme production, QC is critical, in order to ensure the identity, strength, potency, purity and safety of the drug. Thus, QC operates across the entire biopharmaceutical process workflow (Fig. 3.1), through release of process intermediates (e.g., cell paste), the final bulk product or Drug Substance (DS), and the final dosage form or Drug Product (DP). Release of the DS for formulation and filling and the DP to the clinic depends on meeting specifications during QC testing, and oversight and approval from QA. QC also perform the necessary in-process controls, for example to determine the protein content of a process stream prior to load onto a chromatography column.

In this chapter, the methods and technologies for purification, or Downstream Processing, of therapeutic proteins (Sect. 3.3.2) are reviewed together with the QC analytical methods (Sect. 3.3.3). Several examples from industry are provided (Sect. 3.4), to illustrate some of the various permutations and possibilities in this vast field.

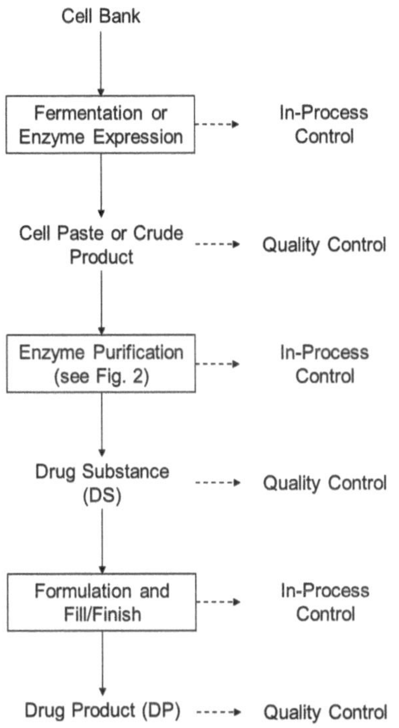

Fig. 3.1 Overview of therapeutic enzyme production workflow

3.2 Downstream Processing of Therapeutic Enzymes

3.2.1 Downstream Processing

The term Downstream Processing (or Downstream Process, DSP) with respect to biological products refers to the process of purification of the biologically active molecule(s) that comprise that product into a sufficiently pure form, such that it is possible to administer to humans in a clinical setting. Typically, DSP consists of a number of linked manufacturing steps, each of which contributes to an improvement in the purity of the target molecule. Some types of impurities that must be removed include expression-cell or host-cell non-product proteins, nucleic acids, lipids and phospholipids. Any materials introduced during DSP, such as protease inhibitors or antifoaming agents, must also be removed in the purification train. In a therapeutic or biopharmaceutical setting, the design of this process is critical, as in addition to being economically viable, DSP must also be performed in such a way to achieve the desired purity while conserving the biological activity and efficacy of the product.

In the context of DSP for therapeutic enzymes, the critical factor to preserve is enzymatic activity. An enzyme's activity is necessarily and intrinsically linked to the three-dimensional structure of the enzyme and the molecular architecture within the active site, which admits the substrate and performs the chemical reaction. Enzymes, as proteins, have four potential levels of structure: primary (sequence), secondary (i.e., α-helices and β-sheet structures), tertiary (i.e., whole-protein folding) and potentially, quaternary (i.e., multi-protein complexes). Preservation of these levels must be accomplished during DSP, so processing steps must be sufficiently gentle to achieve this. Some of the measures typically taken in this regard include processing in cold or refrigerated units, restricting reagents and buffers to those which promote or preserve structure (e.g., avoiding most organic solvents), selecting appropriate ranges of pH, use of biocompatible product-contact surfaces and purification units, and, if necessary, the use of protease inhibitors to prevent degradation.

In recent years, the biopharmaceutical industry, in collaboration with academic institutions and regulatory bodies, has made great advances in streamlining and improving the quality of DSP for therapeutic enzymes and biologics. The introduction of disposable technology (Allison and Richards 2014; Gao and Allison 2016) for small-volume, high-value products such as therapeutic enzymes has helped manufacturers achieve easier compliance with regulatory guidance for cleaning and validation. Similarly, the introduction of quality-by-design (QbD) principles (ICH Q8(R2) 2009; Yu et al. 2014) into biopharmaceutical R&D has helped developers and manufacturers produce DSP strategies that are both robust and scaleable.

The output of DSP is the Drug Substance (DS), normally stored as a bulk product that is later processed through to the final formulated presentation (vials, pre-filled syringes, or other dosage forms) in a separate, formulation process.

The DS is typically stored under controlled conditions at a selected temperature (e.g., 2–8 °C or −20 °C) appropriate for the product. During storage, the DS must be stable over time, and rigorous studies must be conducted to demonstrate stability during development and after product licensure. The stability studies are typically conducted in concert with the QC unit, who conduct the analytical testing. The stability acceptance criteria will be established based on the data collected during development in order to ensure that the product efficacy is maintained over the intended shelf life.

In this section, some of the techniques and strategies for therapeutic enzyme DSP are described and illustrated. In addition, some promising new procedures are outlined and potential future directions in the field are proposed.

3.2.2 Unit Operations Used in Downstream Processing

Often, the most extensive part of the overall manufacturing process is the DSP section, as several steps are usually required to obtain enzyme of the required purity. Nonetheless, there is a typical workflow or pattern to therapeutic enzyme DSP, consisting of cell lysis, crude product extraction, product polishing, and final product filtering (Fig. 3.2). These stages are described in further detail below. It is important to note that, although this DSP workflow also applies to non-enzymatic proteins, certain aspects are unique to therapeutic enzymes, and these will be highlighted below.

The first phase of DSP is the separation of the enzyme from the cellular material used during protein expression in the bioreactor or fermenter. As mentioned above, many different expression platforms have been used historically. However, regardless of the expression system, in the majority of cases the initial phase of the DSP follows a common theme. In cases where the target enzyme is a soluble protein secreted into the growth medium, such as is the case for many mammalian and insect-cell applications, simple physical separation of the cells and medium is sufficient.

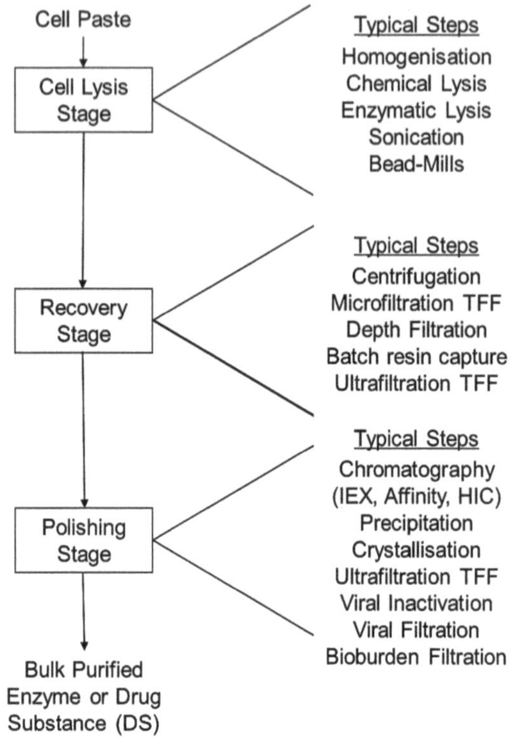

Fig. 3.2 Typical workflow for therapeutic protein DSP.

Secreted enzyme production may also be linked with a perfusion process (Voisard et al. 2003) in which a continuous, cell-free stream of crude product is obtained from the bioreactor.

For intracellular products, such as many enzymes expressed in bacterial and yeast cells, the host cells typically require a lysis step to release the target enzyme. Most commonly, manufacturers employ high-pressure homogenisation or microfluidisation (Middelberg 1995) to release the target protein. In homogenisation, the cell suspension is pumped through a high-pressure orifice or cell and the resulting pressure drop causes the cells to lyse. Pressure drops routinely used during homogenisation may exceed 15,000–20,000 psi. The outputs of homogenisation include the target protein and other cellular contents as well as insoluble cellular debris, which must be removed (e.g. by centrifugation). One advantage of homogenisation as a lysis step is the relatively short residence time in the equipment, which minimises heat input to the process cell fluids, and therefore minimises proteolytic activity which may arise from unwanted enzymes in the host cell. Post-homogenisation cooling, often provided by an outlet heat exchanger, may be critical to minimise such effects.

The choice of lytic method may also be critical for the function of the remaining DSP (Balasundaram et al. 2009) and may impact yields downstream. Depending on the pI of the target enzyme, the pH of the growth medium or lysis buffer, and the characteristics of the cellular debris, significant losses may be encountered due to product-cell debris adsorption and subsequent removal during centrifugation (Bierau et al. 2001; Quirk and Woodrow 1984). Care must therefore be taken in the design of lysis steps of the DSP, to minimise these losses. Apart from homogenisation, other techniques for cell lysis include enzymatic lysis with reagents such as lysozyme (Salazar and Asenjo 2007), chemical lysis including alkaline treatments (Wade 1972), sonication, and bead-mills, but homogenisation remains the most easily scaleable and industrially-achievable technique.

After lysis is complete, the next phase of DSP is sometimes referred to as extraction, or the production of a crude enzyme intermediate preparation. Immediately after the lysis step, the unwanted insoluble components (i.e., cellular debris) must first be removed from the soluble target protein and other impurities. Typically, centrifugation is employed to achieve this essentially solid-liquid separation, but other techniques such as tangential-flow microfiltration (TFF-MF) may be used. In centrifugation, either batch-mode or continuous (*i.e.*, disc-stack) equipment may be employed, depending on the scale of the DSP. Depth filtration is another technology that is sometimes used for clarification. The extent of the clarification achieved may be critical to the later polishing steps, with poorly-clarified lysates requiring further filtration steps downstream, as found for alcohol dehydrogenase production in yeast (Bracewell et al. 2008). In TFF-MF, membrane fouling may hamper the capacity of the operation to perform the separation efficiently, so researchers have investigated backpulsing the membranes with air pressure periodically to solve this problem (Kuberkar and Davis 2001; Levesley and Hoare 1999; Parnham and Davis 1996). The type and area of the MF membrane may also be a critical factor in development, to minimise effects of fouling and potential yield losses.

The crude enzyme isolate must then be forward-processed through polishing steps to achieve the purity required for human therapeutic use (the specifics of which are outlined further in Sect. 3.3.3). The workhorse unit operation of the polishing phase, and indeed the entire DSP for most therapeutic enzymes, is column chromatography. Column chromatography is an adsorptive separation technique (Carta and Jungbauer 2010) carried out using a stationary phase (a porous media) packed in a tube (a column). The separation in chromatography occurs when certain components (e.g. impurities or the target protein) adsorb to the stationary phase while others are allowed to flow through the column. Adsorbed components may be eluted or removed from the column by changing the composition of the liquid phase. Thus, critical to the design of chromatographic separations are several factors including the type of stationary phase (discussed further below), the liquid composition and pH, the composition of the crude protein feed to the column, the temperature, and the flow rate through the column. Further important factors in column chromatography step development are the kinetics of adsorption between the enzyme and the stationary phase, and the degree of dispersion or peak spreading that may occur during operation (Carta and Jungbauer 2010).

First and foremost in therapeutic enzyme chromatography is the choice of stationary phase, sometimes referred to as resin. The stationary phase is also referred to as the 'packing' in the column while the liquid phase is the fluid that flows through the column. Typical stationary phases are often spherical and have pore structures to allow proteins to interact with a larger liquid-solid interfacial area compared with the outside surface area of each spherical bead. The types of available stationary phases for protein and enzyme purification (Table 3.2) are designed in order to exploit one or more of the properties of proteins, such as charge, hydrophobicity, and

Table 3.2 Types of chromatography matrices for therapeutic enzyme production

Chromatography ligand type	Mechanism of separation	References
Cation-exchange (i.e., S or sulphopropyl and CM or carboxymethyl)	Enzymes and proteins in the process stream bind to the resin if they have a net positive charge at the processing pH	Janson (2012) and Scopes (2013)
	Proteins are eluted by an increase in pH above the protein pI, or increase in salt concentration	
Anion-exchange (e.g., Q or quaternary ammonium and DEAE or diethylaminoethyl)	Enzymes and proteins in the process stream bind if they have a net negative charge at the processing pH	Janson (2012) and Scopes (2013)
	Proteins are eluted by a decrease in pH below the protein pI, or increase in salt concentration	
Hydrophobic-interaction (i.e., phenyl, butyl)	Hydrophobic surface patches on the enzyme or protein mediate binding to the resin ligand	Cummins and O'Connor (2011) and Kennedy (1995)
	Various salts are used to promote protein-resin binding, such as ammonium sulphate	
	Proteins are eluted by changing the salt concentration, pH, or both	
Affinity	A molecule with high specificity towards the target enzyme, such as the substrate or substrate analogue, is immobilised on a chromatographic support	Cuatrecasas et al, (1968) and Hage and Cazes (2005)
	The target binds while most or all other proteins are separated in the flow-through	
	The target is eluted by introducing a high concentration of the substrate or other molecule into the mobile phase	
Dye-ligand	A form of affinity chromatography using certain dye molecules as the ligand	Clonis et al. (2000)
	The dye molecules serve as a mimic of the enzyme substrate	
	Useful for purification of certain enzymes, i.e., kinases	
	Residual amounts of bound dye may leach from the column over time; therapeutic products would require QC testing for absence of free dye	
IMAC	A form of affinity chromatography immobilised metal cations, usually Ni+2, as the ligand	Kågedal (2011) and Porath (1992)
	A purification tag (usually His_6) expressed on the enzyme at the N- or C- terminus, binds to the metal cation	
	The bound protein may be eluted by adding a histidine mimic, such as imidazole, to the mobile phase	
Hydroxylapatite	Hydroxylapatite is a mineral with the composition $Ca_{10}(PO_4)_6(OH)_2$	Doonan (1996)
	As a chromatography matrix it may be difficult to pack into columns but offers multi-mode (i.e., cationic and anionic) binding properties	
Gel-filtration	Gel-filtration is a technique used to separate proteins and enzymes by molecular size and/or shape	Hagel (2001)
	The chromatography medium has no ligand but a defined pore structure that admits some proteins and not others; smaller proteins spend more time in the pore structure before eluting	
	Not often used in industrial or therapeutic settings due to the lengths of columns that may be required	

molecular size. The resins typically consist of a structural inert support (i.e., agarose) and a chemical ligand that has interacts with a particular property or characteristic of proteins. For example, ion-exchange resins adsorb proteins based on the protein's surface charge, which is in part dependent on the pH and ionic strength of the chromatography buffers (or liquid phase). Typical ion-exchangers include those with positively-charged ligands (i.e., quaternary ammonium or Q, and diethylaminoethyl or DEAE) or negatively-charged ligands (i.e., sulphopropyl or S, and carboxymethyl or CM). Elution is achieved by shifts in either the salt concentration (typically, sodium chloride) or pH of the liquid phase.

Although ion-exchange chromatography is probably the most common form in therapeutic enzyme purification, many other types of protein chromatography may also be employed (Table 3.2). These include chromatography resins with a hydrophobic ligand, referred to as Hydrophobic-Interaction Chromatography (HIC) matrices (Cummins and O'Connor 2011). HIC resins adsorb proteins based on solvent-accessible non-polar amino acid residues such as valine, leucine, phenylalanine, and tryptophan. HIC adsorption is promoted by relatively high concentrations of certain salts, such as ammonium sulphate, and elution of proteins may be achieved by lowering the ammonium sulphate concentration in the liquid phase. Often, a gradient of slowly decreasing ammonium sulphate or other salt may selectively elute proteins from the resin in the order of increasing strength of adsorption (i.e., the most strongly hydrophobically bound proteins would elute last).

A further type of chromatography that is often used in enzyme purification is affinity chromatography (Hage and Cazes 2005), which can often achieve high purification factors in one step. In affinity chromatography, the ligand is a molecule with high specificity to the target enzyme. Typically, this is the enzyme's substrate, an analogue of the substrate, or inhibitor compound. During an affinity chromatography process, the target enzyme binds to the affinity ligand strongly, and impurities and non-target proteins pass through the column. The high-specificity interaction between enzyme and substrate makes affinity chromatography particularly useful in therapeutic enzyme purification. In order to elute the bound enzyme, a change must be made to the mobile phase which may often be introduction of the substrate (or substrate analogue) as a solute in the mobile phase. This provides competition between enzyme binding and enzyme desorption, and at a sufficiently high concentration of substrate the enzyme will desorb from the resin.

One example of affinity chromatography was studied for the therapeutic enzyme L-asparaginase (Lee et al. 1989; Gervais and Foote 2014), in which L-asparagine was covalently linked to the chromatographic resin and telution of bound asparaginase mediated with free L-asparagine in the elution buffer. Another good example of enzyme affinity chromatography is dye-ligand chromatography, where the bound substrate mimetic is a dye molecule such as Cibacron blue F_3G-A (McGettrick and Worrall 2004). Some applications of dye affinity matrices and some issues relating to their use in pharmaceutical applications were reviewed by Labrou (2003).

A particular variant of affinity chromatography is immobilised-metal affinity chromatography (IMAC), in which the binding is between histidine-rich sequences (usually a tag of six histidine residues fused to either the N- or C- termini of the enzyme) and an immobilised metal cation (most often, Ni^{2+}) (Kågedal 2011). Elution of these types of resins is achieved by introduction of a relatively high concentration of free histidine or imidazole (a histidine mimetic) in the mobile phase. IMAC is used in many protein and enzyme purification labs worldwide in order to provide research-grade materials for laboratory in vitro studies. One issue with the use of IMAC for therapeutic protein production, however, is the potential for leaching of toxic heavy metals (such as Ni^{2+}) into the product stream, which may be difficult to remove later in the DSP and may inhibit enzyme activity. Any compounds added or leached (e.g., Ni^{2+} ions during IMAC, or leached dyes from affinity media) to the product during manufacture must be cleared by the subsequent process steps. Ideally, the clearance mechanism should be validated over the course of several batches to show the robustness of the process, as well as to avoid routine testing of the compound during DS release in QC. Furthermore, the presence of the purification tag (e.g. an N- or C- terminal His_6 sequence) may affect enzyme activity or the in vivo performance of the drug, and may need to be cleaved or otherwise removed. Such materials used during therapeutic enzyme DSP need careful consideration early in the drug development cycle, as including them out of convenience early on may result in the need for re-running expensive clinical studies subsequent to their removal.

In addition to chromatography, other techniques are commonly used during therapeutic enzyme DSP. Ultrafiltration, in a tangential-flow filtration (TFF) arrangement, may be used to achieve partial removal of small-molecule contaminants, as well as exchange of buffer from one type to another (van Reis and Zydney 2001). Solvent-mediated crystallisation, while not typically used to purify therapeutic proteins, has been used in the production of *E. coli* L-asparaginase (Rauenbusch et al. 1970). Ammonium sulphate precipitation may also be successfully used to purify enzymes such as L-asparaginase without negatively affecting the catalytic activity (Santos et al. 2018).

3.3 Quality Control of Therapeutic Enzymes

3.3.1 Introduction to Quality Control

Quality is critical for batches of therapeutic enzymes for human administration. Unlike many other consumer products such as foodstuffs, it is usually not possible for the end user (i.e., the patient) to discern whether a pharmaceutical product is damaged, tainted or spoiled. Quality Control (QC) plays a critically important role in ensuring that pharmaceutical products, such as therapeutic enzymes, are fit-for-purpose.

In a typical biopharmaceutical manufacturing setting, QC is provided by a dedicated department or unit, with highly-trained personnel. Clear, accurate recording of the work performed in QC is a critical part of the operation. Methods utilised for each therapeutic enzyme product are pre-specified and agreed with regulatory bodies. It is not usually permissible or possible to change a method or make changes within a method spontaneously subsequent to product licensure. Any changes require extensive development work and robustness studies to demonstrate equivalence prior to seeking regulatory approval for the method change. Likewise, agreed and appropriate specification ranges (e.g. protein content must be within the range 10–20 mg/mL) will need to be set for the final product in order for release to the clinic or market. The QC methods also require validation, described in more detail below, to ensure that they provide reproducible and accurate results.

Like the DSP of therapeutic enzymes, as described above, QC is conducted under the rules and regulations of GMP. This means, in addition to the factors previously described, that QC staff must be trained and qualified to run the methods, that the methods are run according to written and approved procedures (typically referred to as Standard Operating Procedures or SOPs), that methods performed are documented appropriately, and that there are controls in place to prevent mix-ups or confusion between tests on different products or process stages. Any deviations to practice, procedure or results must be investigated and the investigation documented, with improvements to practice implemented if necessary. The chain-of-custody for analytical test samples must be documented and samples must be clearly identified and traceable. The QC unit must operate under the control of an overarching Quality Management System (QMS) designed to ensure that procedures and documents are recorded and kept up-to-date in an organised way.

3.3.2 Types of Quality Control Methods for Therapeutic Enzymes

The primary purpose of QC testing, as described above, is to ensure the safety and efficacy of the therapeutic enzyme product. Methods used to analyse and release product in a QC setting are selected based on the capacity to answer these fundamental questions. However, it is possible to further subdivide the analytical methods up into more specific categories according to the characteristics of each method. One way of categorising QC methods, defined by the US FDA, arranges them according to fit in one of the following groups: Identity, Purity, Potency, Strength and Safety. These categories are discussed below along with a selection of the types of methods a typical QC laboratory may choose to employ.

Many QC methods rely on a comparison of the test material with a reference standard. Each time a new batch is manufactured, it will be measured using the QC release methods against the reference standard, and the results compared. There will be defined acceptance criteria for each QC method which must be met in order to release the batch. The reference standard is typically made from a standard manufacturing batch of the therapeutic enzyme drug product, and characterised using a selection of methods more extensive than those used for routine product release. For example, aggregate profile of each batch made during routine manufacture may be analysed using size-exclusion chromatography (SEC), but in addition to this, upon the introduction of a new batch of reference standard, additional orthogonal methods such as analytical ultracentrifugation (AUC) may be employed. This is done in order to provide a high degree of assurance that the reference material is appropriate for the routine release of drug substance and drug product. The reference standard must be stored under controlled conditions according to GMP procedures, and be monitored for stability under protocol at defined time intervals.

3.3.2.1 Identity Testing

The purpose of an identity test is to establish that the product batch contains the correct therapeutic enzyme. Therefore, an identity test needs to provide definitive proof that the enzyme, or the active ingredient of the drug, is the main component or, at least is present in the batch. Results for identity tests are not normally numeric in nature; they will usually be a 'pass' or 'fail' result according to an agreed specification suitable for the test and validated before any product is released to the clinic or market.

A good example of an identity test for a therapeutic enzyme product would be a peptide-mapping method, either with or without mass spectrometry (Wei et al. 2005). In peptide mapping, the enzyme is degraded into peptides using a proteolytic enzyme (*i.e.*, trypsin) under controlled conditions. When the enzyme amino acid sequence is known, the proteolytic cut sites (and therefore, the peptides created) can be predicted. The resulting peptide digest is then analysed using Reversed-Phase High-Pressure Liquid Chromatography (RP-HPLC) or equivalent, to separate and quantify the peptides. Each digested enzyme or protein gives a specific peptide signature. In developing peptide mapping methods for a particular therapeutic enzyme, RP-HPLC coupled with mass spectrometry (MS) is very useful in order to identify the various peptides. Once known, a given peptide's HPLC retention time (or a collection of digested peptide elution times) may be very accurately predicted. In a QC setting, the unknown (or test batch) peptide map would be compared against a known reference standard of the same product, with a threshold set (e.g.., >95% peptide similarity) for acceptance of the result as a positive identity test. An example of an ultraviolet (UV) absorbance peptide mapping chromatogram for a therapeutic enzyme, showing the peaks corresponding to various digested peptides, is depicted in Fig. 3.3.

Other analytical techniques may be equally valid as identity tests for therapeutic enzymes. For example, peptide mapping with LC-MS or MS/MS would provide further detail for an identity test, but the MS component may not be robust enough in a regulated, GMP environment like

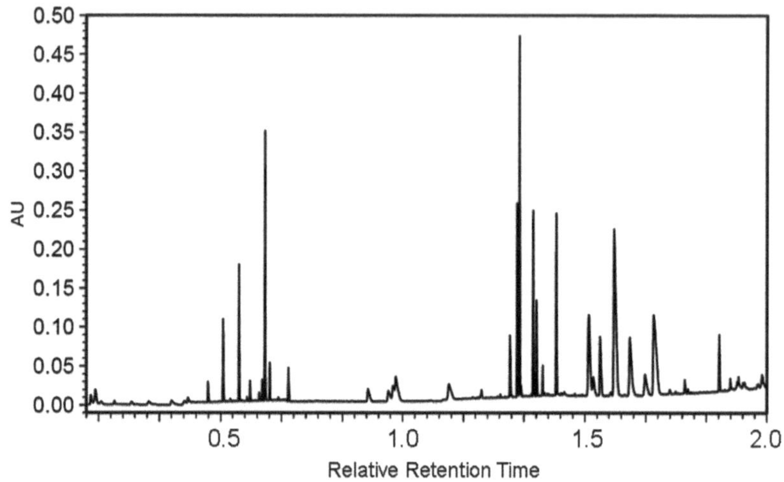

Fig. 3.3 Example of peptide mapping UV Chromatogram

QC. An N-terminal sequence (or Edman degradation) against a reference standard for a defined number of residues could provide a proof of identity and also demonstrate protein integrity. Equally, a western blot of both the test item and reference standard, using an antibody specific to the therapeutic enzyme, would be a plausible identity test strategy.

3.3.2.2 Purity Testing

A biopharmaceutical product must be free of contaminating substances, whether they are product-related or derived from the manufacturing process. The presence of substantial amounts of impurities may impact the performance of the drug in vivo, or cause unwanted and potentially serious side effects. Purity analysis is, therefore a critical QC activity during testing of the therapeutic enzyme product. The type of test employed for a purity measurement necessarily depends on the target analyte. An overview of selected impurities and potential QC purity techniques is shown in Table 3.3.

Impurities may be product-related or process-related. Product-related impurities include those arising from the target enzyme, such as degradation products. Clipped or truncated forms of the enzyme are a good example of a product-related impurity. Process-related impurities are substances that may be used or present during manufacture but would normally be expected to be cleared or removed from the process stream to a

Table 3.3 Selected impurity types and associated measurement techniques

Impurity type	Techniques used for purity assessment in QC	References
Protein impurity	SDS-PAGE, CGE, RP-HPLC	van Tricht et al. (2015)
Residual host cell protein	ELISA, LC-MS/MS	Wang et al. (2009)
Residual host cell DNA	Quantitative PCR, others	Lee et al. (2010) and Hu et al. (2014)
Antifoam, fermentation media components, protease inhibitors, etc.	RP-HPLC	DiPaolo et al. (1999)
Aggregated enzyme	SEC-HPLC, AUC	Fekete et al. (2014) and Hong et al. (2012)
Truncated or clipped product forms	SDS-PAGE, CGE	Rustandi and Wang 2011
Methionine oxidised variants	RP-HPLC, LC-MS/MS	Torosantucci et al. 2014 and Houde et al. (2006)
Deamidated or other charge variants	cIEF, IEX-HPLC, CZE	Gervais, (2016) and Yao et al. (2018)
Glycosylation variants	RP-HPLC	Shang et al. (2014)

certain, acceptable level. Examples of process-related impurities may include other host cell proteins (HCP) (i.e., non-target enzyme) from the cellular expression platform, antifoaming agents, antibiotics, residual media components from fermentation, and leachates from chromatography media (i.e., ligands from affinity columns). Each impurity, or potential impurity, that could be present in the final product should be covered by a purity method. In some cases, it may be possible to remove the need to test on a routine basis by validating, or risk-assessing the removal of certain components or impurities during process validation. Process validation for impurity removal is not covered in this chapter, but a good example is provided by Shukla and co-workers for HCP (Shukla et al. 2008).

Enzyme variants with post-translational modifications, such as methionine oxidation, asparagine or glutamine deamidation, and changes to glycosylation patterns, may also be considered as product-related impurities. However, the exact definition of these forms depends highly on the particular therapeutic enzyme product and will vary on a case-by-case basis. In some cases, post-translationally modified forms of the target enzyme may be considered as part of the product, if certain criteria are met. One of these criteria is that on a batch-to-batch basis, the modified forms would need to be present in a consistent ratio relative to the unmodified product. Further, during pre-clinical and/or clinical testing the modified forms may be demonstrated to have the same efficacy, safety and potency as the unmodified enzyme. Some enzyme products may only be acceptable as one chemical form with uniform post-translational modification profile, and for others, a collection of forms or variants may be considered together as the product. For example, deamidated forms of *Erwinia chrysanthemi* L-asparaginase were found in vitro to have the same activity and characteristics as the wild-type enzyme (Gervais and Foote 2014), supporting the hypothesis that these forms, if present, could be considered as part of the product. For the purposes of this chapter, we consider only the case where post-translationally modified forms of the target enzyme are product-related impurities.

The workhorse of biopharmaceutical purity testing is high-performance liquid chromatography (HPLC). In a QC environment, HPLC has the added benefits of similarity of equipment and procedures across a wide range of functions and column chemistries, making it ideal for this high-pressure environment. HPLC is also straightforward to validate and maintain compliance with regulatory guidance (Epshtein 2004). HPLC is used in a variety of forms for protein analysis. Differences in enzyme charge, as well as charge variants, may be analysed using ion-exchange (IEX) HPLC (Fekete et al. 2015). RP-HPLC, described earlier in the context of peptide-mapping identity methods, may also be used for intact protein analysis to quantify oxidised (i.e., methionine or tryptophan oxidised) forms and small amounts of non-target protein impurities (Yang et al. 2007). RP-HPLC may also be used to measure and analyse the presence and nature of glycosylation moieties in the enzyme (Shang et al. 2014). Away from product-related species, HPLC may be used to accurately measure residual process-related impurities, such as antifoam added during fermentation, to ensure that such materials are reduced to below a threshold value during processing.

Apart from HPLC, a variety of methods may be employed for biopharmaceutical purity testing. SDS-PAGE, a relatively low-resolution technique, is still used to determine the in-process level of non-target protein impurities (Gervais et al. 2012). However, the more quantitative capillary gel electrophoresis (CGE), is now replacing traditional SDS-PAGE in the modern QC laboratory (van Tricht et al. 2015). Other capillary electrophoresis techniques, such as capillary isoelectric focussing (cIEF) for protein charge variants (analogous to IEX-HPLC) (Rodriguez-Diaz et al. 1997), and capillary zone electrophoresis (Dolnik 1997), are also gaining traction in the biopharmaceutical QC laboratory for impurity measurements.

Quantitation of protein-protein aggregates is another critical area for biopharmaceutical QC testing. Such aggregates are usually present in pharmaceutical preparations of proteins and enzymes, and a small quantity is usually tolerable.

However, increased amounts of aggregate may enhance or accelerate the chance of undesired immunological reactions in the patient (Ratanji et al. 2014), or cause formation of anti-drug antibodies (ADAs) which may negate the efficacy of the drug. Size-exclusion chromatography (SEC or SEC-HPLC) is typically and routinely used to assess the content of protein-protein aggregation, as it is relatively simple to use, normally uses the same laboratory equipment for other HPLC techniques, and is reproducible (Fekete et al. 2014). Ideally there should be no interaction of the analyte protein with the resin matrix surface, as the differential migration of proteins through the column is created solely by the relationship between the size of each protein species and the pore structure of the column matrix. Thus, the choice of column matrix for SEC-HPLC may be quite important, as certain proteins (particularly those with a high isoelectric point) may partially bind to certain chromatography resins with a negative overall surface charge (i.e., uncoated silica), creating artefacts in analysis (Gervais et al. 2017).

The presence of residual DNA in the biopharmaceutical product also forms an important part of QC release testing. The World Health Organisation and US FDA have issued regulatory guidance on the level of residual DNA (<10 ng/dose) that is acceptable in biologic products (Yang 2013). The reason for this tight specification for residual DNA is based on the theoretical risk of certain nucleic acid sequences (e.g., oncogenes) posing a health risk to the patient receiving the therapeutic enzyme. Therefore, the DSP for a therapeutic enzyme must be able to achieve these levels, and analytical methods must have a limit of detection (LOD) appropriate to measure these levels in the final product. The most common DNA testing method used in a QC setting is the real-time polymerase chain reaction (RT-PCR) or quantitative PCR (qPCR), but other methods are also used (Wang et al. 2012).

Similarly to residual DNA, residual levels of host-cell proteins (HCP), not detectable by SDS-PAGE or other common techniques, must also be determined and demonstrated to pass an appropriate specification (Wang et al. 2009). Typical levels of HCP present in finished biologic drugs are of the order of ng HCP per mg of therapeutic protein, and higher-than-intended levels of HCP in the enzyme DP may possibly be toxic or cause undesired immunological reactions in the patient. Sensitive measurement techniques are therefore used in QC for HCP detection and quantitation (Tscheliessnig et al. 2013), and most commonly, the enzyme-linked immunosorbent assay or ELISA (Zhu-Shimoni et al. 2014). In an ELISA method, the antibodies used are critical to the success of the assay. Therefore, it is important during development of an HCP ELISA to demonstrate that the polyclonal antibodies used for the assay provide appropriate coverage of the expression organism's proteome. The use of two-dimensional gel electrophoresis (2D-GE) (Jin et al. 2010) combined with western blotting for assessment of antibody coverage in HCP ELISA is becoming the regulatory expectation. Among the non-ELISA approaches to HCP measurement, most notable is mass spectrometry (Schenauer et al. 2012).

3.3.2.3 Potency Testing

Potency, or strength testing, is critical to the safe delivery of a medicinal product. Obviously, the clinician overseeing or administering the product has to ensure that the correct amount of active pharmaceutical ingredient is delivered to the patient. An overdose can be toxic or fatal, and a lower-than-intended dose may cause a lack of efficacy of the drug. Potency may be measured by a number of different means for pharmaceutical products, and normally the unit of dose (i.e. milligrams, etc.) would be established during testing of the product in clinical trials.

For therapeutic enzymes, the usual measurement of potency is enzyme activity, which is typically defined in terms of units. One unit of enzyme is described as the amount of enzyme required to catalyse the reaction of a certain amount of substrate in a given time. For example, one unit of L-asparaginase is defined as the amount of enzyme required to catalyse the hydrolysis of 1 μmol of L-asparagine in 1 min (Broome 1968). The temperature used for measurement of the enzymatic reaction is normally specified in the definition of a unit of activity, and it is usual to refer to units as 'international units' or IU when performing enzyme activity analysis under controlled conditions.

The design of an activity method for a therapeutic enzyme is highly dependent on the reaction that the enzyme catalyses. Thus, enzyme activity methods vary on a case-by-case basis. For some therapeutic enzymes such as glucocerebrosidase, a fluorescent analogue of the normal substrate may be used for the in vitro activity measurement (Shaaltiel et al. 2007). Other enzymes may create a product from the enzymatic reaction that may be measured. For example many dehydrogenases utilise cofactors such as NAD or NADP, whose reduced forms can be followed spectrophotometrically. Measurement of L-asparaginase activity is typically based on the Nessler reaction and the ability to measure liberated NH_3 from the hydrolysed L-asparagine. Yao and co-workers described a robustness study on L-asparaginase activity measurement using a design of experiments (DOE) approach (Yao et al. 2016). In most cases, the activity of the test batch of therapeutic enzyme would be measured against a known reference standard, as described above.

In addition to enzyme activity, the specific activity of the enzyme is often determined in QC. Specific activity is defined as the enzyme activity per unit mass of protein; typical units for specific activity would be enzyme units per mg protein (U/mg). The specific activity provides a crude measurement of enzyme purity; contaminating proteins would lead to a lower specific activity value in crude product intermediates. In order to measure specific activity, it is necessary to measure the protein content of a sample. Standard spectrophotometric or colorimetric protein assays, such as the methods of Lowry, the BCA method, or others, would be employed for these (Noble et al. 2007). Whether specific activity, enzyme activity, or both are used to assess potency, the accuracy of the method(s) used is very important and should be stringently assessed and qualified during development and method validation.

3.3.2.4 Safety Testing in QC

First and foremost, therapeutic enzymes and biopharmaceuticals must be safe and do no harm to the patient. As part of the assurance of product safety, the absence of adventitious substances (i.e., endotoxins) and organisms (i.e., bacteria) must be established during QC testing. Generally, methods for these analytes are specified in compendial registers (e.g., the US Pharmacoepia, www.usp.org) and accepted as standard by regulatory agencies. A summary is provided, below, of the major types of safety testing that may be required in a therapeutic enzyme product.

The term bioburden refers to the amount or quantity of microorganisms (including bacteria, fungi and yeasts) in a non-sterile sample. Biologics manufacture typically occurs in two states: aseptic or pre-sterile filtration. Aseptic processing areas are costly to build and maintain, so generally manufacturers only use aseptic processing for the latter stages of a process, with a sterile filtration step carried out prior to steps in aseptic areas. In the non-sterile areas of the manufacturing plant, it is accepted that a degree of bioburden will be present, but this must be under control. The absence of certain types of microbes or known pathogens (lists of which are available from various regulatory agencies) must still be demonstrated in non-sterile product streams, and rigorous investigations and preventative actions performed if any are detected.

For non-sterile processing steps and manufacturing areas, bioburden testing (Hentz 2013) is routinely used to monitor levels of microorganisms in the product intermediate streams, and may be very important to understand root causes of contamination should a failure occur downstream. Bioburden testing is also used to assess the quality of the facility water systems, of the air handling system (HVAC) and particulate matter in the manufacturing suite, of the gowning procedures and hygiene of personnel, and other aspects. In bioburden testing, a sample of the product or water is taken and grown on sterile media in a controlled QC environment. Results are typically reported in terms of colony-forming units (CFU) per volume of the original sample.

Sterility testing is essentially the same as bioburden testing, but with more stringent procedures and specifications (i.e., the sample is only considered sterile if no CFU are found).

Endotoxins are lipopolysaccharide-based molecules derived from the outer membrane of gram-negative bacteria. They are highly immunogenic and pyrogenic in humans and animals and administration of endotoxin along with a biopharmaceutical product may result in fever, sepsis, or death. Many cellular expression platforms are based on gram-negative organisms (e.g. *E. coli*), where endotoxin is a particular concern; however, endotoxin may be present in any biopharmaceutical preparation due to contamination of a product intermediate or final product with bacteria. Endotoxin, therefore, must be rigorously controlled and tested for in a QC environment (Dullah and Ongkudon 2017). There are compendial methods (from the US Pharmacoepia, European Pharmacoepia, and others) for endotoxin measurement. Most of these involve a method known as the Limulus Amoebocyte Lysate (LAL) test, which can detect very low levels (*ca.* 0.01 endotoxin units (EU) per mL) but other methods also exist. It is worth mentioning that, particularly for those products manufactured using gram-negative bacteria, that DSP steps must be capable of endotoxin clearance to regulatory-acceptable levels. In these cases, endotoxin testing may be an important part of process design during the development phase. As endotoxin typically carries a negative overall charge, anion-exchange chromatography steps during DSP are one good strategy to achieve endotoxin removal.

For those products produced using mammalian cell lines or cell cultures, viral safety is another important factor. Cell lines may contain or become infected with adventitious viruses, which could theoretically end up in the final product. While QC viral testing of finished products is possible (Darling et al. 1998), it may be complex and time-consuming, so manufacturers tend to design DSP steps for viral clearance and validate the removal of some or all viruses from the product (Darling 2002). Some steps that are used for viral clearance during DSP include viral inactivation (using low-pH, sodium caprylate, and others) (Farshid et al. 2005), ion-exchange chromatography, and viral filtration steps (Burnouf et al. 2005). One strategy for virus removal validation involves taking samples from a manufacturing run, spiking it with a panel of viruses in a controlled laboratory, and conducting scale-down DSP models to show removal (Valera et al. 2003).

3.3.2.5 Analytical Characterisation

Certain analytical techniques may be employed during biopharmaceutical production and development, but not used as part of the product specification to release material to the clinic. These types of tests, which would be outside the validation or GMP envelope of the quality or QC system, are referred to as characterisation methods. Characterisation methods may be difficult to validate in a GMP environment due to inherent variability of the method, whilst still providing useful information for the organisation (Razinkov et al. 2013). The results of such methods would typically be classified as 'for information only' and, unlike a release test, would not be compared against a specification range. The results of characterisation methods may be useful to the manufacturer to track process performance or provide further information to a technical support group. Such methods must be clearly indicated and communicated as 'not for release of product', and may be performed either by a QC unit, or more typically by a development or technical support unit. Typical characterisation techniques include biophysical methods that were used during product development but not included on the release specification, such as surface plasmon resonance (SPR) or isothermal titration calorimetry (ITC) which are continued after licensure for monitoring purposes (Holdgate et al. 2013). For some products, the determination of enzymatic constants, such as the Michaelis constant (K_M) or catalytic constant (k_{cat}), may be classed as characterisation methods, but for others these measurements will form part of the release specification. Other examples of characterisation methods may include structural methods designed to understand the presence of disulphide bonds within the enzyme (Zhang et al. 2011), if not detected using other means described above such as SDS-PAGE or SDS-CGE.

3.3.2.6 Stability Testing

Therapeutic enzyme products must be stable upon storage, both as a bulk DS prior to formulation and as a final product in the vial. This is critical to ensure that the patient receives the drug as it was intended and not in a degraded or altered state. Although some degree of change in product quality over time is inevitable for protein therapeutics, any allowable change must not affect the overall efficacy or utility of the product in the clinic. The company or manufacturer must generate the necessary data to demonstrate that the product is stable over a specified time period under controlled storage conditions. Therefore, stability testing is an important part of biopharmaceutical development and production.

Manufacturers may choose to place each product batch on stability testing or just a selected number of batches, depending on the standard procedures and protocols used by the organisation. Typically, the release methods used for the therapeutic enzyme, or a subset thereof, (in the categories described above, purity, identity, safety) are used for stability testing. Stability testing is normally conducted by the QC unit of the manufacturing organisation. Certain methods are classed as stability-indicating (i.e., determined during stress testing during development) and would be included in the stability specification for the product. Acceptance criteria for a stability-indicating method may be wider than those for product release at the time of manufacture, as some change may be expected in the method over time. For sterile products, it is usually necessary to include periodic sterility tests at certain time intervals during stability testing. Stability testing may be conducted in so-called real time conditions (i.e. at the intended storage temperature and humidity) or at accelerated conditions (i.e., at higher-than intended storage temperature, humidity or both). Theoretically, accelerated conditions are used to demonstrate any potential changes sooner than real-time testing, but data from accelerated conditions must be interpreted with caution, as the changes may not truly reflect what would occur in real-time storage.

3.3.3 Method Lifecycle Management

As for all processes, the lifecycle of analytical QC methods must be managed. The concept of lifecycle management of processes encompasses all phases of the process over time, from initial design and concept to development, implementation, monitoring and finally, retirement. Lifecycle management is of increasing interest in the industry, and particularly in the context of analytical methods (Parr and Schmidt 2018). Results from QC methods may drift over time due to extraneous factors such as changes in laboratory staff, changes in equipment, reagents, or consumables, environmental issues and other factors.

Particular phases of lifecycle management that are of note in the context of QC analytical methods include the development and monitoring phases. As described above, QbD is a set of techniques that may be used to improve the robustness of processes in the pharmaceutical industry. These approaches may also be applied to analytical methods (Hubert et al. 2015) in order to generate the most robust method possible. While method validation is important (Taverniers et al. 2004), the robustness of a method is also critical as unexplained or spurious results in a QC context may result in problems with regulatory bodies or release of product. Thus, continuous method monitoring must be performed by the QC unit to ensure that performance of the method is consistent over time.

3.4 Examples of Therapeutic Enzyme Bioprocessing and QC

In order to illustrate the principles and strategies discussed above, a number of examples, describing licensed enzyme products, are presented in this section.

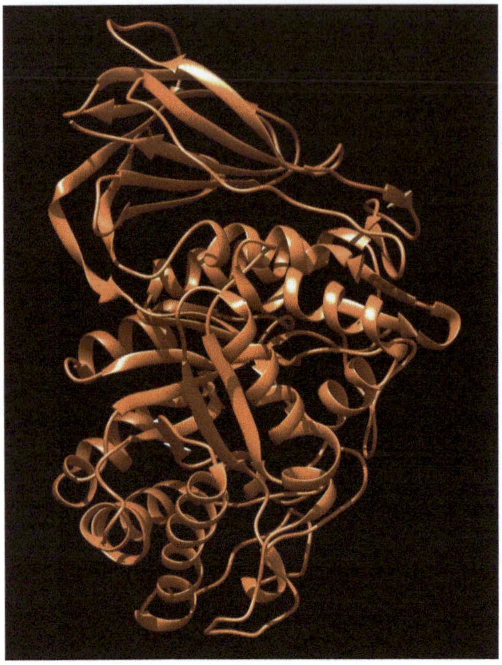

Fig. 3.4 Crystal structure of human ABG. The structure was rendered using Chimera 1.9 (Pettersen et al. 2004) and PDB entry 1OGS (Dvir et al. 2003)

3.4.1 Taliglucerase Alfa or Glucocerebrosidase

Enzyme-replacement therapy (ERT) is used to treat mostly rare genetic and metabolic disorders that cause a lack of a certain enzyme in the patient. Gaucher's Disease (GD) is a genetic condition also referred to as lysosomal storage disorder, caused by a lack of the enzyme acid β-glucosidase (ABG) in humans (Grabowski et al. 2014). Patients with GD have an accumulation of ABG substrates in a variety of tissues, which leads to symptoms and problems primarily with bone and bone marrow and the central nervous system (CNS). As a 56 kDa (unglycosylated molecular weight) glycoprotein with post-translationally added glycosylation at various sites, ABG cannot be expressed in bacteria and must be manufactured using cell culture. The structure (Fig. 3.4) of ABG was reported in the literature (Dvir et al. 2003) with the active site and catalytic mechanism understood, and the glycosylation sites of ABG were also elucidated (Hermans et al. 1993).

Several companies have developed and manufactured ABG for clinical ERT in humans. The glycosylation of recombinant ABG is critical for the efficacy of the drug (Grabowski et al. 2014), with terminal mannose glycan moieties necessary to mediate uptake of the enzyme by tissue macrophages. Earlier, the companies Genzyme (Boston, MA, USA) and Shire (Dublin, Ireland) had developed versions (Imiglucerase and Velaglucerase alfa, respectively) of ABG with the correct mannose-terminated glycosylation. These versions were produced in mammalian cell lines, and required post-expression alteration of the glycans with exoglycosidase enzymes to expose the important mannose residues at the end of each glycan chain (Tekoah et al. 2013).

A new version of ABG, taliglucerase alfa, was approved recently for human clinical use. Taliglucerase alfa is marketed as Elelyso® and produced by Protalix (Israel) and Pfizer (United States). Taliglucerase alfa is notable as the first approved biologic product to be produced in plant cells. The enzyme is expressed in a bioreactor using *Daucus carota* (carrot root) cells in a suspension culture, with a plasmid encoding the taliglucerase alfa gene and a kanamycin selection marker (Grabowski et al. 2014). Expression in plant cells shares some advantages in terms of design of growth media and costs with bacterial expression system, but unlike bacteria, plant cells enable N-linked glycosylation of the expressed protein. In the case of Elelyso, the carrot cell expression system achieves removal of the end residues from the N-linked glycosylation sites, exposing the mannose glycan residues critical for ABG function without the use of exogenous exoglycosidase treatment during DSP.

While many cell-culture based proteins and enzymes are secreted into the growth medium, this was not an option for taliglucerase alfa. The reason for this is the need for mannose-terminated glycans to enable uptake in vivo by the target (GD) cells. Secreted proteins have different glycan termination residues, which would disable uptake of the ABG by the receptor cells. Thus, a vacuole-based accumulation within the host (carrot) cells was engineered into the taliglucerase alfa expression system, as the vacuole-targeted expression resulted in high terminal mannose content on the glycan chains (Tekoah et al. 2015).

DSP for taliglucerase alfa was designed around the intracellular accumulation of the product (Tekoah et al. 2015). The recovery of the product from the host plant cells is a critical part of the DSP, and development of the taliglucerase alfa DSP had to take into account the presence of the relatively difficult-to-break-down plant cell wall. Several strategies for extraction and recovery of proteins from plant cell expression systems, including freeze-thaw cycles, detergents, or grinding strategies, have been developed recently (Hassan et al. 2008). After extraction of taliglucerase alfa from the carrot cells using the detergent Triton X-100 (Shaaltiel et al. 2007), the waste cell mass is removed and the product purified using column chromatography steps including cation-exchange and hydrophobic interaction chromatography (Shaaltiel et al. 2007; Tekoah et al. 2015).

From a QC point of view, the most critical features of taliglucerase alfa are the enzyme activity and the glycosylation patterns, including the terminal mannose content of the enzyme. The glycosylation profile of taliglucerase alfa was characterised extensively using MALDI MS (Shaaltiel et al. 2007). Glycan sequencing was carried out by exoglycosidase treatment to remove the glycan structures and perform sequential digestion before reversed-phase HPLC analysis. Batch-to-batch consistency of the glycosylation profile of taliglucerase alfa has been demonstrated to be highly reproducible (Tekoah et al. 2015). In addition to analysis of glycosylation, the enzymatic activity of taliglucerase alfa is studied using a fluorescent substrate analogue, with activities and enzymatic constants (i.e., V_{max} and K_M) comparing very well with competitor products such as imiglucerase.

Although special considerations were required for the production of taliglucerase alfa due to the plant expression systems, it is clear that the developed process results in a highly reproducible, efficacious product. As a case study, taliglucerase alfa is a trailblazing product that paves the way for future plant-expressed products to be developed. The possibility of further such products seems likely, with purification of proteins from plant cells attracting increasing interest in the biopharmaceutical field (Wilken and Nikolov 2012).

3.4.2 L-Asparaginase *Erwinia chrysanthemi*

L-asparaginase is used to treat Acute lymphoblastic leukaemia (ALL), the main type of cancer affecting children. L-asparaginase for human clinical use is derived from one of two sources: *Escherichia coli* or *Erwinia chrysanthemi*. Since the late 1960s, *Erwinia chrysanthemi* L-asparaginase (formerly referred to as *Erwinia carotovora* L-asparaginase, marketed as Erwinase® or Erwinaze®) has been produced from bacterial strain NCPPB 1066 (Buck et al. 1971; Minton et al. 1986) at Porton Down, UK. Erwinase® is currently manufactured by Porton Biopharma (Porton Down, UK), and sold worldwide by Jazz Pharmaceuticals (Dublin, Ireland). In the past few years, Erwinase® has been the subject of an extensive validation (Gervais et al. 2013) and product characterisation exercise. *Erwinia* L-asparaginase is a homotetramer consisting of four identical 35 kDa subunits. It is not a recombinant product; it is made from the native organism under controlled fermentation conditions. The crystal structure of *Erwinia* L-asparaginase has been solved (Aghaiypour et al. 2001) and shows the tetrameric nature of the enzyme (Fig. 3.5).

The Erwinase® manufacturing process (Fig. 3.6) involves growth of the host organism using fermentation, cell lysis and extraction of the crude enzyme preparation, DSP purification of the enzyme, and formulation and lyophilisation to make the final product. Between processing stages, the intermediates of Erwinase® are stored frozen at −20 °C, in order to facilitate flexibility in the manufacturing schedule. The DSP (Fig. 3.6b) involves a multi-step process which includes both anion- and cation-exchange chromatography, ammonium sulphate precipitation, crystallisation and dialysis to manufacture DS. The capacity of the DSP to remove protein

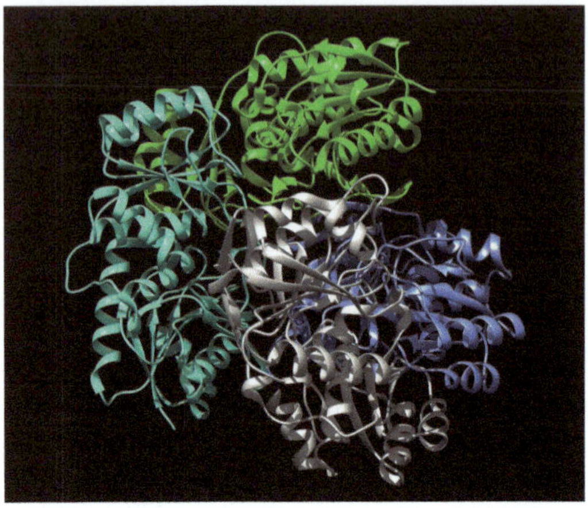

Fig. 3.5 Crystal Structure of *Erwinia chrysanthemi* L-asparaginase. Each identical subunit in the tetrameric structure is depicted using a separate colour. The structure was rendered using Chimera 1.9 (Pettersen et al. 2004) and PDB entry 1HG1 (Aghaiypour et al. 2001)

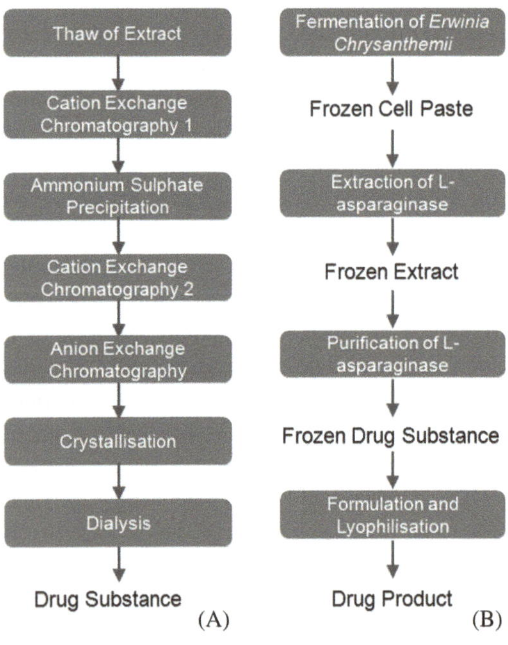

Fig. 3.6 Production Process for *Erwinia chrysanthemi* L-asparaginase (Erwinase®). (**a**) Overall Process. (**b**) DSP. (Reprinted with permission from Gervais et al. 2013)

contaminants may be observed by comparing the reducing SDS-PAGE impurity profile of a cation-exchange intermediate with the final DS (Fig. 3.7). The Erwinase® DSP successfully underwent process validation in recent years, confirming the process robustness for routine biopharmaceutical production. In addition, supportive laboratory studies (Gervais and Foote 2014) have been carried out to understand the impact of certain post-translational changes, such as deamidation, on the activity of the enzyme. These studies have proved invaluable in interactions with various regulatory agencies.

The set of QC methods used in the production of Erwinase® has evolved over the long history of the product. For example, older purity-indicating methods, such as cellulose-acetate membrane electrophoresis, were replaced with newer, more robust techniques (i.e., SDS-PAGE using commercial pre-cast gels and buffers) that could be validated in a QC environment (Fig. 3.7). In addition, new methods were introduced, such as HPLC-based methods, to provide high-resolution assessment of batch purity. SEC-HPLC methods were developed to fully quantitate the aggregate profile of Erwinase®, which includes both dimeric species and higher-order (i.e. trimeric and above) aggregates (Gervais et al. 2017). Ion-exchange HPLC is used to quantitate charge variants, which in Erwinase® were shown to include both deamidated species and slight conformational variants (Gervais et al. 2015). Erwinase is QC released against the results from RP-HPLC for the measurement of methionine oxidation, and a peptide mapping method is used for an identity test. The release tests currently used for DS production were briefly discussed previously (Gervais et al. 2013).

In summary, Erwinase® is an excellent example of a therapeutic enzyme used in a different clinical context, oncology. Compared to the first-line *E. coli* therapy, Erwinase® provides a different immunologic profile and is therefore an important drug in ALL protocols.

3.4.3 Asfotase Alfa

Asfotase alfa is an enzyme fusion protein developed to treat hypophosphatasia (HPP) in infants. HPP is caused by a genetic mutation that results in a lack of the enzyme tissue-nonspecific alkaline phosphatase (TNALP). Bone growth and

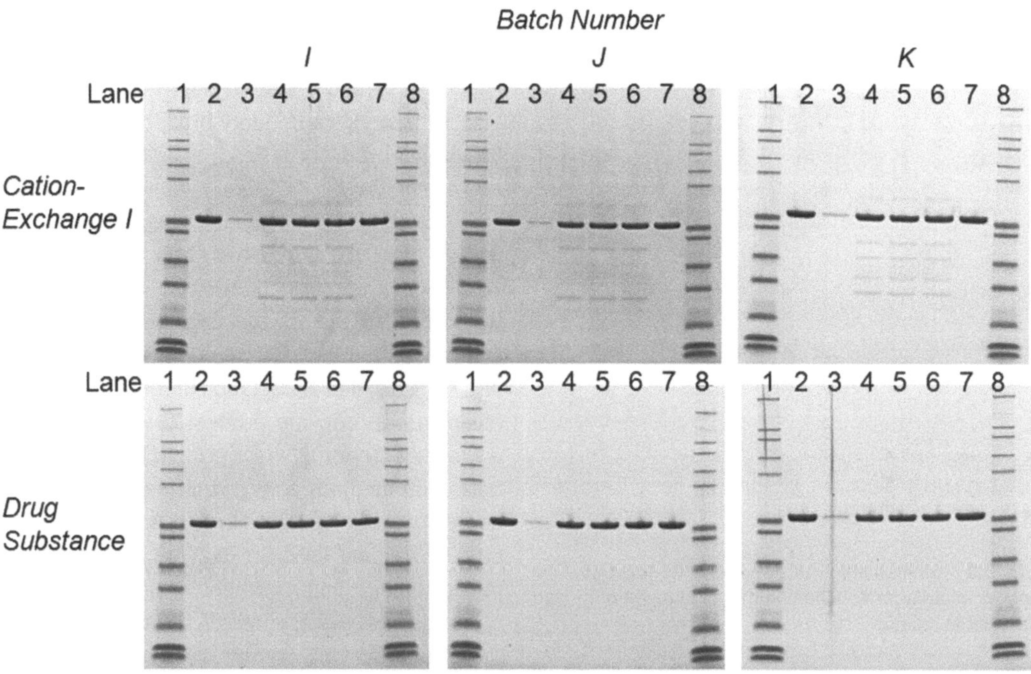

Fig. 3.7 SDS-PAGE QC Analysis of Erwinase® at the Intermediate and DS (Drug Substance) Stages. (Reprinted with permission from Gervais et al. 2013)

mineralisation are critically dependent on the breakdown of an inhibitor, pyrophosphate. In healthy individuals, TNALP is involved in the breakdown of pyrophosphate, so babies with the genetic deficiency accumulate pyrophosphate, resulting in rickets and problems in bone formation. Prior to the introduction of Asfotase alfa, the mutation was often fatal, as there was no other treatment (Millàn and Whyte 2016). The rare nature of HPP led to Asfotase alfa being granted orphan drug status in the US, Japan and the EU.

Asfotase alfa is marketed by Alexion Pharmaceuticals (New Haven, USA) under the name Strensiq®. As a fusion protein, Asfotase alfa is distinct from many other therapeutic enzymes and from those discussed in Sects. 3.4.1 and 3.4.2 above. Asfotase alfa is a dimeric fusion glycoprotein, consisting of two identical 726-residue polypeptide chains. The protein has six N-linked glycosylation moieties on each protein monomer, and has two inter-chain disulphide bridges holding the two monomers together in dimeric form. Starting with the N-terminus, the fusion protein consists of a soluble form of human TNALP followed by the constant (F_c) region of human IgG1, with a c-terminal deca-aspartate (Asp_{10}) sequence. The structure of Asfotase alfa is depicted in cartoon form in Fig. 3.8. The C-terminal sequence of ten aspartic acid residues may enable Asfotase alfa to bind to bone tissue to facilitate biological activity (EMA Strensiq Public Assessment Report 2015). Asfotase alfa is expressed in CHO cell culture, with the protein secreted into the growth medium.

The DSP for Asfotase alfa is distinct from other therapeutic enzyme products, due to the presence of the IgG1 F_c domain. This domain allows affinity purification of the enzyme fusion directly from the CHO cell culture supernatant. A Protein A column (Millàn et al. 2008) facilitates binding of the F_c region, and elution is achieved by modulation of pH. Although the Protein A step is likely to be the workhorse of the Asfotase alfa purification process, a series of further chromatography steps are used along with a viral inactivation step, virus filtration step, and sterile filtration. Asfotase alfa is stored as a DS at an elevated

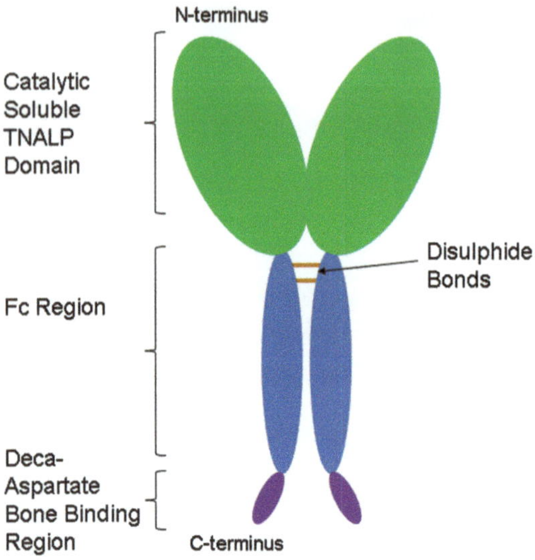

Fig. 3.8 Representation of Asfotase alfa Structure. The mature protein forms a homodimer held together by two disulphide bonds

protein concentration of approximately 100 mg/mL. (EMA Strensiq Public Assessment Report 2015).

The approach to QC and release of the DS for Asfotase alfa reflects the complexity of the molecule. Enzyme activity is measured in vitro using several different methods, including the activity of soluble enzyme against an inorganic phosphate substrate, a bound (to hydroxylapatite *via* the deca-aspartate region) enzymatic activity assay, and the determination of kinetic constants, such as K_M and k_{cat}. Purity assessment follows a standard protocol for therapeutic enzymes. Chromatographic methods such as SEC-HPLC for aggregate quantitation and ion-exchange HPLC for charge variant quantitation are used, and residual HCP and DNA are measured (EMA Strensiq Public Assessment Report 2015). The stability of the DS was also rigorously demonstrated.

Compared to more standard therapeutic enzymes such as those discussed in Sects. 3.4.1 and 3.4.2, Asfotase alfa represents a different class of product. The fusion of the catalytic enzyme domain with a domain of human IgG facilitates easier purification, using a Protein A-based protocol similar to the platform approach widely used in the monoclonal antibody area. It is anticipated that this type of fusion-protein approach, greatly simplifying the molecule's DSP, will be taken up more broadly in the future for similar products.

3.4.4 Future Outlook: New Products

Innovation in therapeutic enzyme development and production has consistently evolved over the past decades, and is likely to continue to do so. In recent times, enzyme product development has taken cues from the wider protein biotherapeutic area, with methods common to monoclonal antibodies (mAbs) (such as Protein A chromatography) now being used for enzyme fusion protein purification. Improvements in removing unwanted immunogenicity, through improvements in achieving correct patterns of glycosylation within the production host cell, will also continue to be a focus area going forward.

Antibody-drug conjugates (ADCs) are an evolution of mAbs, where cytotoxic small-molecule drugs are covalently bound to antibodies for targeted delivery of potent drugs to the intended cell in vivo. Although certain conjugates (e.g., polyethylene glycol or PEG) of enzymes are common in the industry, it is possible that the enzyme-drug conjugates will also be developed. Such enzyme-drug conjugates may have the advantage of increasing the half-life of the cytotoxic compound. Another possibility is targeted enzyme delivery using antibody-enzyme conjugates.

An important and emerging area in the biopharmaceutical industry is the rise of personalised medicine and cellular therapies. The cost of these therapies is currently prohibitive for many routine applications, but in future it is possible that enzyme delivery to the patient will occur through such cell therapies. Gene-edited autologous cells could be isolated from tissues, engineered to express the enzyme of interest, and re-introduced into the same patient. Although such concepts may still be years or decades away, they may eventually become cost-effective and replace established standards of care in the therapeutic enzyme arena, such as ERT. Biopharmaceutical companies will therefore have to evolve and respond to these challenges over time to ensure that patients and shareholders alike are to benefit.

3.5 Concluding Remarks

Therapeutic enzymes, whilst occupying a relatively small overall part of the global biologics area, are important both clinically and commercially. A vast array of therapeutic enzymes, for the treatment of a variety of human diseases, are currently approved and on the market. In some cases, enzyme therapeutics and ERT are the only treatment for these diseases. The production methods, including protein cellular expression platforms, continue to evolve and are becoming more complex over time, so DSP and QC methods must also develop and respond to these challenges.

Enzyme products are distinct from others in the sector, including mAbs. In the design of DSP strategies for therapeutic enzymes, care must be taken so that the enzymatic activity, the critical feature of the enzyme, is preserved. QC methods must be appropriately designed and maintained over the lifetime of the product to effectively enable release and stability assessment of each product according to current regulatory guidelines. Further particular challenges around the development of enzyme products include measurement of the catalytic kinetics of the molecule and avoiding inactivation of the enzyme. Biopharmaceutical companies must tailor their procedures and strategies to account for these differences during manufacturing, development and QC.

References

Aghaiypour K, Wlodawer A, Lubkowski J (2001) Structural basis for the activity and substrate specificity of *Erwinia chrysanthemi* L-asparaginase. Biochemistry 40(19):5655–5664

Allison N, Richards J (2014) Current status and future trends for disposable technology in the biopharmaceutical industry. J Chem Technol Biotechnol 89(9):1283–1287

Aviezer D, Brill-Almon E, Shaaltiel Y et al (2009) A plant-derived recombinant human glucocerebrosidase enzyme – a preclinical and phase I investigation. PLoS One 4(3):e4792

Balasundaram B, Harrison S, Bracewell DG (2009) Advances in product release strategies and impact on bioprocess design. Trends Biotechnol 27(8):477–485

Bierau H, Hinton RJ, Lyddiatt A (2001) Direct process integration of cell disruption and fluidised bed adsorption in the recovery of labile microbial enzymes. Bioseparation 10(1):73–85

Bracewell DG, Boychyn M, Baldascini H et al (2008) Impact of clarification strategy on chromatographic separations: pre-processing of cell homogenates. Biotechnol Bioeng 100(5):941–949

Broome JD (1968) Studies on the mechanism of tumor inhibition by L-asparaginase: effects of the enzyme on asparagine levels in the blood, normal tissues, and 6C3HED lymphomas of mice: differences in asparagine formation and utilization in asparaginase-sensitive and-resistant lymphoma cells. J Exp Med 127(6):1055–1072

Buck PW, Elsworth R, Miller GA et al (1971) The batch production of L-asparaginase from *Erwinia carotovora*. J Gen Microbiol 65:i

Burnouf T, Radosevich M, Goubran HA et al (2005) Place of nanofiltration for assuring viral safety of biologicals. Cur Nanosci 1(3):189–201

Carta G, Jungbauer A (2010) Protein chromatography: process development and scale-up. Wiley, Hoboken

Clonis YD, Labrou NE, Kotsira VP et al (2000) Biomimetic dyes as affinity chromatography tools in enzyme purification. J Chromatogr A 891(1):33–44

Cuatrecasas P, Wilchek M, Anfinsen CB (1968) Selective enzyme purification by affinity chromatography. Proc Natl Acad Sci U S A 61(2):636–643

Cummins PM, O'Connor BF (2011) Hydrophobic interaction chromatography. In: Walls D, Loughran S (eds) Protein chromatography. Humana Press, New York, pp 431–437

Darling A (2002) Validation of biopharmaceutical purification processes for virus clearance evaluation. Mol Biotechnol 21(1):57–83

Darling AJ, Boose JA, Spaltro J (1998) Virus assay methods: accuracy and validation. Biologicals 26(2):105–110

Dewan SS, Sullivan LL (2016) Protein therapeutics market – technology advances spur market growth of protein therapies. Drug Dev Deliv 16(9):44–47

DiPaolo B, Pennetti A, Nugent L et al (1999) Monitoring impurities in biopharmaceuticals produced by recombinant technology. Pharm Sci Technol Today 2(2):70–82

Dolník V (1997) Capillary zone electrophoresis of proteins. Electrophoresis 18(12–13):2353–2361

Doonan S (1996) Chromatography on hydroxyapatite. In: Doonan S (ed) Protein purification protocols. Humana Press, New York, pp 211–215

Dullah EC, Ongkudon CM (2017) Current trends in endotoxin detection and analysis of endotoxin–protein interactions. Crit Rev Biotechnol 37(2):251–261
Dvir H, Harel M, McCarthy AA et al (2003) X-ray structure of human acid-β-glucosidase, the defective enzyme in Gaucher disease. EMBO Rep 4(7):704–709
Edmunds T (2005) β-glucocerebrosidase, ceredase and cerezyme. Dir Ther Enzymes 12:117–133
Epshtein NA (2004) Validation of HPLC techniques for pharmaceutical analysis. Pharm Chem J 38(4):212–228
European Medicines Agency (2015) Public assessment report for Strensiq. http://www.ema.europa.eu/docs/en_GB/document_library/EPAR_-_Public_assessment_report/human/003794/WC500194340.pdf. Accessed 7 Mar 2018
European Medicines Agency Public Assessment Report for Strensiq 25 June (2015). http://www.ema.europa.eu/docs/en_GB/document_library/EPAR_-_Public_assessment_report/human/003794/WC500194340.pdf. Accessed 7 Mar 2018
Farshid M, Taffs RE, Scott D et al (2005) The clearance of viruses and transmissible spongiform encephalopathy agents from biologicals. Cur Opin Biotechnol 16(5):561–567
Fekete S, Beck A, Veuthey JL et al (2014) Theory and practice of size exclusion chromatography for the analysis of protein aggregates. J Pharm Biomed Anal 101:161–173
Fekete S, Beck A, Veuthey JL et al (2015) Ion-exchange chromatography for the characterization of biopharmaceuticals. J Pharm Biomed Anal 113:43–55
Gao Y, Allison N (2016) Extractables and leachables issues with the application of single use technology in the biopharmaceutical industry. J Chem Technol Biotechnol 91(2):289–295
Gervais D (2016) Protein deamidation in biopharmaceutical manufacture: understanding, control and impact. J Chem Technol Biotechnol 91(3):569–575
Gervais D, Foote N (2014) Recombinant deamidated mutants of *Erwinia chrysanthemi* L-asparaginase have similar or increased activity compared to wild-type enzyme. Mol Biotechnol 56(10):865–877
Gervais D, Allison N, Jennings A et al (2013) Validation of a 30-year-old process for the manufacture of L-asparaginase from *Erwinia chrysanthemi*. Bioprocess Biosyst Eng 36(4):453–460
Gervais D, King D, Kanda P et al (2015) Structural characterisation of non-deamidated acidic variants of *Erwinia chrysanthemi* L-asparaginase using small-angle X-ray scattering and ion-mobility mass spectrometry. Pharm Res 32(11):3636–3648
Gervais D, Downer A, King D et al (2017) Robust quantitation of basic-protein higher-order aggregates using size-exclusion chromatography. J Pharm Biomed Anal 139:215–220
Grabowski GA, Golembo M, Shaaltiel Y (2014) Taliglucerase alfa: an enzyme replacement therapy using plant cell expression technology. Mol Genet Metab 112(1):1–8
Graham ML (2003) Pegaspargase: a review of clinical studies. Adv Drug Del Rev 55(10):1293–1302
Hage DS, Cazes J (eds) (2005) Handbook of affinity chromatography. CRC Press, Boca Raton
Hagel L (2001) Gel-filtration chromatography. Cur Prot Mol Biol 44(1):10.9.1–10.9.2
Harmatz PR, Garcia P, Guffon N et al (2014) Galsulfase (Naglazyme®) therapy in infants with mucopolysaccharidosis VI. J Inherit Metab Dis 37(2):277–287
Hassan S, Van Dolleweerd CJ, Ioakeimidis F et al (2008) Considerations for extraction of monoclonal antibodies targeted to different subcellular compartments in transgenic tobacco plants. Plant Biotechnol J 6:733–748
Heartlein M, Kimura A (2014) Discovery and clinical development of idursulfase (Elaprase®) for the treatment of mucopolysaccharidosis II (Hunter syndrome). Orphan Drugs Rare Dis 38:164
Hentz NG (2013) Pharmaceutical bioburden testing. In: Flickinger MC (ed) Encyclopedia of industrial biotechnology. Wiley, New York, pp 757–774. https://doi.org/10.1002/9780470054581.eib655
Hermans MM, Wisselaar HA, Kroos MA et al (1993) Human lysosomal alpha-glucosidase: functional characterization of the glycosylation sites. Biochem J 289(3):681
Holdgate G, Geschwindner S, Breeze A et al (2013) Biophysical methods in drug discovery from small molecule to pharmaceutical. In: Williams M, Daviter T (eds) Protein-ligand interactions. Methods in molecular biology (methods and protocols) 1008. Humana Press, Totowa
Hong P, Koza S, Bouvier ES (2012) A review, size-exclusion chromatography for the analysis of protein biotherapeutics and their aggregates. J Liq Chromatogr Rel Technol 35(20):2923–2950
Houde D, Kauppinen P, Mhatre R et al (2006) Determination of protein oxidation by mass spectrometry and method transfer to quality control. J Chromatogr A 1123(2):189–198
Hu B, Sellers J, Kupec J et al (2014) Optimization and validation of DNA extraction and real-time PCR assay for the quantitative measurement of residual host cell DNA in biopharmaceutical products. J Pharm Biomed Anal 88:92–95
Hubert C, Houari S, Rozet E et al (2015) Towards a full integration of optimization and validation phases: an analytical-quality-by-design approach. J Chromatogr A 1395:88–98
International Conference on Harmonisation (2009) Q8(R2), Pharmaceutical Development
Janson JC (ed) (2012) Protein purification: principles, high resolution methods, and applications. Wiley, New York
Jin M, Szapiel N, Zhang J et al (2010) Profiling of host cell proteins by two-dimensional difference gel electrophoresis (2D-DIGE): implications for downstream process development. Biotechnol Bioeng 105(2):306–316

Kågedal L (2011) Immobilized metal ion affinity chromatography. In: Janson J (ed) Protein purification. Wiley-VCH, New York, pp 183–201

Kakkis E (2005) α-L-Iduronidase: the development of Aldurazyme (Laronidase). Dir Ther Enzymes 12:239–260

Kennedy RM (1995) Hydrophobic-interaction chromatography. Cur Prot Protein Sci 00(1):8.4.1-8.4.21

Kuberkar VT, Davis RH (2001) Microfiltration of protein-cell mixtures with crossflushing or backflushing. J Membr Sci 183(1):1–4

Labrou NE (2003 Jun 25) Design and selection of ligands for affinity chromatography. J Chromatogr B 790(1–2):67–78.

Lee SM, Wroble MH, Ross JT (1989) L-asparaginase from *Erwinia carotovora*. Appl Biochem Biotechnol 22(1):1–1

Lee K, Jin X, Zhang K et al (2003) A biochemical and pharmacological comparison of enzyme replacement therapies for the glycolipid storage disorder Fabry disease. Glycobiology 13(4):305–313

Lee JH, Shin JS, Bae JE (2010) Quantitative detection of residual *E. coli* host cell DNA by real-time PCR. J Microbiol Biotechnol 20(10):1463–1470

Levesley JA, Hoare M (1999) The effect of high frequency backflushing on the microfiltration of yeast homogenate suspensions for the recovery of soluble proteins. J Membr Sci 158(1):29–39

McGettrick AF, Worrall DM (2004) Dye-ligand affinity chromatography. In: Cutler P (ed) Protein purification protocols. Humana Press, Totowa, pp 151–157

Middelberg AP (1995) Process-scale disruption of microorganisms. Biotechnol Adv 13(3):491–551

Millán JL, Whyte MP (2016) Alkaline phosphatase and hypophosphatasia. Calcif Tissue Int 98(4):398–416

Millán JL, Narisawa S, Lemire I et al (2008) Enzyme replacement therapy for murine hypophosphatasia. J Bone Miner Res 23(6):777–787

Minton NP, Bullman HM, Scawen MD et al (1986) Nucleotide sequence of the *Erwinia chrysanthemi* NCPPB 1066 L-asparaginase gene. Gene 46(1):25–35

Mor TS (2015) Molecular pharming's foot in the FDA's door: Protalix's trailblazing story. Biotechnol Lett 37(11):2147–2150

Noble JE, Knight AE, Reason AJ et al (2007) A comparison of protein quantitation assays for biopharmaceutical applications. Mol Biotechnol 37(2):99–111

Parnham CS, Davis RH (1996) Protein recovery from bacterial cell debris using crossflow microfiltration with backpulsing. J Membr Sci 118(2):259–268

Parr MK, Schmidt AH (2018) Life cycle management of analytical methods. J Pharm Biomed Anal 147:506–517

Patterson DM, Lee SM (2010) Glucarpidase following high-dose methotrexate: update on development. Expert Opin Biol Ther 10(1):105–111

Pettersen EF, Goddard TD, Huang CC et al (2004) UCSF Chimera—a visualization system for exploratory research and analysis. J Comput Chem 25(13):1605–1612

Porath J (1992) Immobilized metal ion affinity chromatography. Protein Expr Purif 3(4):263–281

Quirk AV, Woodrow JR (1984) Investigation of the parameters affecting the separation of bacterial enzymes from cell debris by tangential flow filtration. Enz Microb Technol 6(5):201–206

Rauenbusch E, Bauer K, Kaufmann W et al (1970) Isolation and crystallisation of L-asparaginase from *E. coli*. In: Grundmann E, Oettgen HF (eds) Experimental and clinical effects of L-Asparaginase. Springer, Berlin, pp 31–38

Razinkov VI, Treuheit MJ, Becker GW (2013) Methods of high throughput biophysical characterization in biopharmaceutical development. Cur Drug Disc Technol 10(1):59–70

Rodriguez-Diaz R, Wehr T, Zhu M (1997) Capillary isoelectric focusing. Electrophoresis 18(12–13):2134–2144

Rustandi RR, Wang Y (2011) Use of CE-SDS gel for characterization of monoclonal antibody hinge region clipping due to copper and high pH stress. Electrophoresis 32(21):3078–3084

Salazar O, Asenjo JA (2007) Enzymatic lysis of microbial cells. Biotechnol Lett 29(7):985–994

Salzer WL, Asselin BL, Plourde PV et al (2014) Development of asparaginase *Erwinia chrysanthemi* for the treatment of acute lymphoblastic leukemia. Ann N Y Acad Sci 1329(1):81–92

Sanford M, Lo JH (2014) Elosulfase alfa: first global approval. Drugs 74(6):713–718

Santos JH, Flores-Santos JC, Meneguetti GP et al (2018) In situ purification of periplasmatic L-asparaginase by aqueous two phase systems with ionic liquids (ILs) as adjuvants. J Chem Technol Biotechnol 93(7):1871–1880

Schenauer MR, Flynn GC, Goetze AM (2012) Identification and quantification of host cell protein impurities in biotherapeutics using mass spectrometry. Anal Biochem 428(2):150–157

Scopes RK (2013) Protein purification: principles and practice. Springer, Berlin

Shaaltiel Y, Bartfeld D, Hashmueli S et al (2007) Production of glucocerebrosidase with terminal mannose glycans for enzyme replacement therapy of Gaucher's disease using a plant cell system. Plant Biotechnol J 5(5):579–590

Shang TQ, Saati A, Toler KN et al (2014) Development and application of a robust N-glycan profiling method for heightened characterization of monoclonal antibodies and related glycoproteins. J Pharm Sci 103(7):1967–1978

Shukla AA, Jiang C, Ma J et al (2008) Demonstration of robust host cell protein clearance in biopharmaceutical downstream processes. Biotechnol Prog 24(3):615–622

Su K, Donaldson E, Sharma R (2016) Novel treatment options for lysosomal acid lipase deficiency: critical appraisal of sebelipase alfa. Appl Clin Genet 9:157

Taverniers I, De Loose M, Van Bockstaele E (2004) Trends in quality in the analytical laboratory. II. Analytical method validation and quality assurance. Trends Anal Chem 23(8):535–552

Tekoah Y, Tzaban S, Kizhner T et al (2013) Glycosylation and functionality of recombinant β-glucocerebrosidase from various production systems. Biosci Rep 33(5):e00071

Tekoah Y, Shulman A, Kizhner T et al (2015) Large-scale production of pharmaceutical proteins in plant cell culture—the protalix experience. Plant Biotechnol J 13(8):1199–1208

Tobin JJ, Walsh G (2008) Good manufacturing practice. In: Tobin JJ, Walsh G (eds) Medical product regulatory affairs: pharmaceuticals, diagnostics, medical devices. Wiley, New York, pp 209–235

Torosantucci R, Schöneich C, Jiskoot W (2014) Oxidation of therapeutic proteins and peptides: structural and biological consequences. Pharm Res 31(3):541–553

Tscheliessnig AL, Konrath J, Bates R et al (2013) Host cell protein analysis in therapeutic protein bioprocessing–methods and applications. Biotechnol J 8(6):655–670

Valera CR, Chen JW, Xu Y (2003) Application of multivirus spike approach for viral clearance evaluation. Biotechnol Bioeng 84(6):714–722

Van Reis R, Zydney A (2001) Membrane separations in biotechnology. Curr Opin Biotechnol 12(2):208–211

van Tricht E, Geurink L, Pajic B et al (2015) New capillary gel electrophoresis method for fast and accurate identification and quantification of multiple viral proteins in influenza vaccines. Talanta 144:1030–1035

Voisard D, Meuwly F, Ruffieux PA et al (2003) Potential of cell retention techniques for large-scale high-density perfusion culture of suspended mammalian cells. Biotechnol Bioeng 82(7):751–765

Wade HE (1972) Extraction of asparaginase from bacterial culture. US Patent 3,660,238. 2 May 1972

Wagener JS, Kupfer O (2012) Dornase alfa (Pulmozyme). Curr Opin Pulm Med 18(6):609–614

Wang X, Hunter AK, Mozier NM (2009) Host cell proteins in biologics development: identification, quantitation and risk assessment. Biotechnol Bioeng 103(3):446–458

Wang X, Morgan DM, Wang G et al (2012) Residual DNA analysis in biologics development: review of measurement and quantitation technologies and future directions. Biotechnol Bioeng 109(2):307–317

Wei Z, Tous G, Yim A et al (2005) Validation of peptide mapping with electrospray mass spectrometry for recombinant proteins of biopharmaceutical interest and its applications as an identity test and a characterization tool. Dev Biol 122:29–47

Wilken LR, Nikolov ZL (2012) Recovery and purification of plant-made recombinant proteins. Biotechnol Adv 30(2):419–433

World Health Organisation (2013) Guidelines on the quality, safety and efficacy of biotherapeutic protein products prepared by recombinant DNA technology. Geneva

Yang H (2013) Establishing acceptable limits of residual DNA. PDA J Pharm Sci Technol 67(2):155–163

Yang J, Wang S, Liu J et al (2007) Determination of tryptophan oxidation of monoclonal antibody by reversed phase high performance liquid chromatography. J Chromatogr A 1156(1–2):174–182

Yao H, Vancoillie J, D'Hondt M et al (2016) An analytical quality by design (QbD) approach for a l-asparaginase activity method. J Pharm Biomed Anal 117:232–239

Yao H, Vandenbossche J, Sänger-van de Griend CE et al (2018) Development of a capillary zone electrophoresis method to quantify *E. coli* l-asparaginase and its acidic variants. Talanta 182:83–91

Yu LX, Amidon G, Khan MA et al (2014) Understanding pharmaceutical quality by design. AAPS J 16(4):771–783

Zhang L, Chou CP, Moo-Young M (2011) Disulfide bond formation and its impact on the biological activity and stability of recombinant therapeutic proteins produced by *Escherichia coli* expression system. Biotechnol Adv 29(6):923–929

Zhu-Shimoni J, Yu C, Nishihara J et al (2014) Host cell protein testing by ELISAs and the use of orthogonal methods. Biotechnol Bioeng 111(12):2367–2379

X-Ray Crystallography in Structure-Function Characterization of Therapeutic Enzymes

4

Anastassios C. Papageorgiou

Abstract

Enzymes are key biological macromolecules that support life by accelerating the conversion of target molecules to desired products in many biochemical reactions. Enzymes are characterized by high affinity, specificity and great catalytic efficiency. Owing to their unique characteristics, enzymes have attracted significant attention for use in therapeutic settings as a distinct class of drugs different from other types of medicines. Enzyme-based therapies are currently in use for the treatment of a wide range of diseases, including leukemia, metabolic disorders, inflammation and cardiovascular disease. However, several challenges, such as immunogenicity and stability, remain. X-ray crystallography has provided key structural insights into the understanding of the molecular basis of diseases and development of enzyme-based therapies. Here, the role of X-ray crystallography in the development of therapeutic enzymes is examined and several examples are provided.

Keywords

Enzyme therapy · Enzyme deficiency · Enzyme stability · Chaperones · Structure stabilizers

A. C. Papageorgiou (✉)
Turku Centre for Biotechnology, University of Turku and Åbo Akademi University, Turku, Finland
e-mail: tassos.papageorgiou@btk.fi

Abbreviations

ADI	Arginine deiminase
ASNase	Asparaginase
CHO	Chinese hamster ovary
DGJ	1-deoxygalactonojirimycin
DNJ	1-deoxynojirimycin
EcA	*E. coli* ASNase
ErA	*Erwinia chrysanthemi* L-asparaginase
ERT	Enzyme replacement therapy
GAA	Acid alpha-glucosidase
GAG	Glycosaminoglycan
GALC	β-galactocerebrosidase
GALNS	N-acetylgalactosamine-6-sulfate sulfatase
GCase	Acid beta-glucosidase
GD	Gaucher disease
GUSB	β-glucuronidase
IDS	Iduronate-2-sulfatase
IDUA	α-L-iduronidase
LK	Lumbrokinase
MPS	Mucopolysaccharidosis
NAC	N-acetylcysteine
NEP	Neprilysin
PAI-1	Plasminogen activator inhibitor-1
PAL	Phenylalanine ammonia lyase
PCT	Pharmacological chaperone therapy
PDB	Protein data bank
PEG	Polyethylene glycol
PKU	Phenylketonuria

N. Labrou (ed.), *Therapeutic Enzymes: Function and Clinical Implications*, Advances in Experimental Medicine and Biology 1148, https://doi.org/10.1007/978-981-13-7709-9_4

SGSH	N-sulfoglucosamine sulfohydrolase
tPA	Tissue plasminogen activator
uPA	Urokinase-type plasminogen activator

design of new therapeutic strategies. In this review, the role of X-ray crystallography in the understanding and development of therapeutic enzymes is discussed.

4.1 Introduction

Enzymes are essential biological macromolecules that support life by catalyzing many biological processes. They are characterized by unique properties, such as specificity, high affinity and catalytic efficiency, which make them different from other biological macromolecules. Owing to their properties, enzymes have attracted considerable interest for use as novel pharmaceutical agents in a variety of diseases, including cancers, autoimmunity, inflammation, bacterial infections, and metabolic disorders.

The concept of therapeutic enzymes was introduced in the 1960s by Christian de Duve (Desnick and Schuchman 2012) as part of replacement therapies for genetic deficiencies. The first genetically-engineered enzyme drug, a recombinant tissue plasminogen activator known as alteplase, was approved by the Food and Drug Administration in the U.S.A in 1987 (Activase®). Later, many cardiovascular enzymes were developed and approved by the USFDA. In 1990, Adagen, a form of bovine adenosine deaminase (ADA) conjugated with polyethylene glycol (PEG) was approved for the treatment of a type of severe combined immunodeficiency disease (SCID), which is caused by the chronic deficiency of ADA and affects lymphocyte maturation. Enzymes are nowadays widely used for different therapeutic purposes from digestion and celiac disease to metabolic diseases, cancer, cardiovascular diseases either separately or in combination with other therapies. Enzymes are also employed in diagnosis, biochemical investigations and monitoring of many alarming diseases. Detailed knowledge of the three-dimensional structure of disease-related enzymes and their interactions with substrates and ligands has provided critical information for structure-based improvements of enzymes and

4.2 Enzyme Therapy in Metabolic Disorders

Defects in enzymes involved in metabolic pathways have been implicated in a number of pathological conditions. In these cases, various strategies to compensate for the loss of the enzymatic activity have been pursued, including the use of enzyme replacement (ER) or pharmacological chaperones (PC) as structure stabilizers (Mohamed et al. 2017).

4.2.1 Phenylketonuria

Phenylketonuria (PKU; OMIM 261600) is an inherited metabolic disorder characterized by accumulation of phenylalanine (L-Phe) in the body as the result of the loss or impairment of phenylalanine hydroxylase (EC 1.14.16.1), an enzyme that metabolizes L-Phe. Excess of L-Phe can damage tissues, resulting in intellectual disability, delayed development, behavioral problems and mental disorders. An estimated of 50,000 people under the age of 40 are believed to be affected in the developed world. Conversion of L-Phe into harmless metabolite trans-cinnamic acid (tCA) and ammonia is catalyzed by phenylalanine ammonia lyase (PAL; EC 4.3.1.5). tCA is further metabolized and excreted in the urine as hippurate. Since the reaction catalyzed by PAL produces non-toxic compounds (at the generated levels), PAL has been proposed as a therapeutic agent for PKU (Wang et al. 2005b).

Use of *Anabaena variabilis* and *Rhodosporidium Toruloides* PALs in a mouse model has given promising results in treating phenylketonuria (Gamez et al. 2005). Owing, however, to problems with protease susceptibility, thermal stability, and aggregation, the structure of PAL from *R. Toruloides* (Calabrese et al. 2004) (PDB id 1t6p) was optimized with pegylation. Various PAL/PEG molar ratios of linear 20 kDa PEG were prepared and tested for in vitro and in vivo effectiveness and the optimum ratios were determined. Furthermore, a preparation of PEGylated *A. variabilis* PAL (*Av*PAL) from BioMarin Pharmaceutical (with the trade name pegvaliase) has been used in human clinical trials with results from single-dose injections being well tolerated by adult patients with PKU, although all participants developed an immunogenic response to PEG. Pegvaliase has been given an orphan drug status in the United States and the European Union and recently entered phase 3 clinical trials with initial results already available (Thomas et al. 2018).

The structures of two PALs from the cyanobacteria *A. variabilis* ATCC 29413 (PDB id: 2nyn) and *Nostoc punctiforme* ATCC 29133 (PDB id: 2nyf), respectively, were reported in 2007 (Moffitt et al. 2007). Compared to plant and yeast PALs, these two cyanobacterial PALs are approximately 20% smaller in size but exhibit similar substrate selectivity and kinetics properties. Mutation of two surface cysteine residues (C503 and C565) to serines (Wang et al. 2008) resulted in reduced aggregation susceptibility of the *Av*PAL (Fig. 4.1). Currently, this double mutant after PEGylation (r*Av*PAL-PEG) is in Phase 3 clinical trials for assessment of its efficacy in PKU patients (Bell et al. 2017).

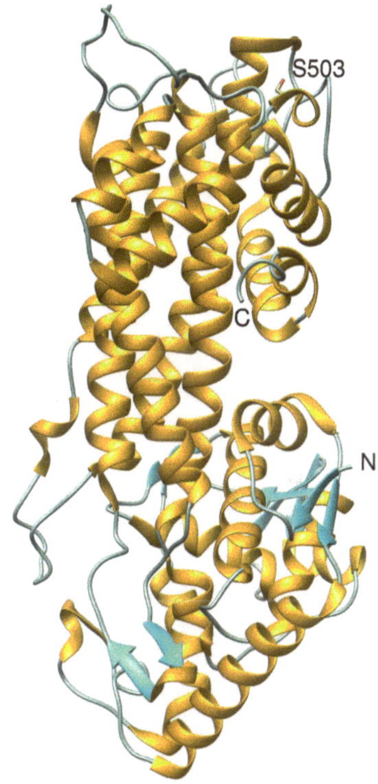

Fig. 4.1 Ribbon diagram of *Av*PAL C503S/C565S double mutant (PDB id 3czo). Residue 565 is at the flexible end of the C-terminal and not located in the structure. All figures, including this one, of protein structures were created with CHIMERA (Pettersen et al. 2004)

4.2.2 Morquio Syndrome

Known also as mucopolysaccharidosis type IV, Morquio syndrome belongs to a group of inherited metabolic disorders collectively termed mucopolysaccharidoses (MPSs). The cause of MPS IV is a deficiency of the lysosomal enzyme N-acetylgalactosamine-6-sulfate sulfatase (GALNS), which is required for the degradation of mucopolysaccharides or glycosaminoglycans (GAGs). GALNS deficiency leads to specific accumulation of two GAGs: keratan sulfate (KS) and chondroitin-6-sulfate (C6S). The majority of disease-causing mutations result in disruption of the hydrophobic core of the protein. Elosulfase alfa, a synthetic version of this enzyme (trade name Vimizim) produced in Chinese hamster ovary (CHO) cell line, is identical to the naturally occurring human enzyme in amino acid sequence and N-linked glycosylation sites. The enzyme is usually injected into patients. Side-effects may include neutralizing antibodies to elosulfase alfa. In addition, hypersensitivity reactions or immunoglobulin E (IgE)-mediated anaphylaxis may occur that require withdrawal of treatment. In an effort to improve delivery of recombinant GALNS to the bones of MPS IV A patients, one proposed variant of GALNS contains six additional glutamate residues at the N terminus (Tomatsu et al. 2010). The structure of GALNS (Fig. 4.2) has been solved at 2.2 Å resolution (Rivera-Colón

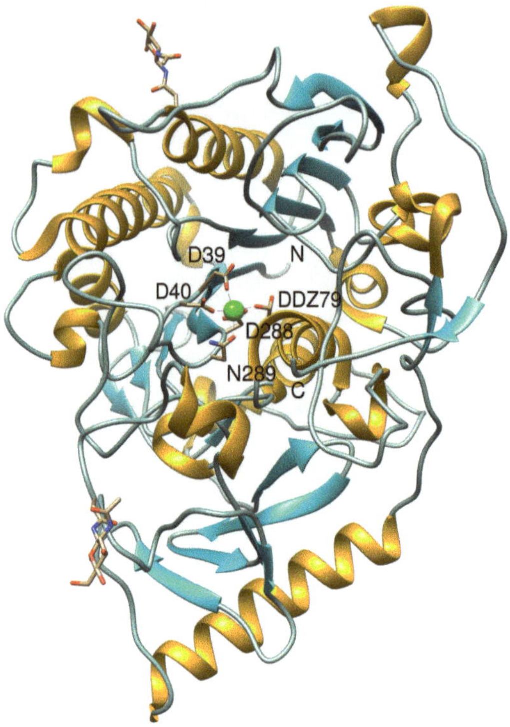

Fig. 4.2 Ribbon diagram of GALNS (PDB id 4fdi). The bound Ca^{2+} is depicted as a green sphere. The modified Cys79 (DDZ; 3,3 dihydroxy-L-alanine) that acts as the catalytic nucleophile is shown. Glycans are shown with coral colour for the carbon atoms

et al. 2012). From a total of 120 disease-causing missense mutations found in MPS IV A patients, only 6 were found to affect the active site. The rest are buried residues in the core of the protein (65%) and 32 mutations (27%) concern surface residues. The active site comprises a large convoluted pocket suggesting that chaperones or small molecules may be used in drug design efforts as stabilizers of the structure.

4.2.3 Sanfilippo A Syndrome

Mucopolysaccharidosis type IIIA is caused by deficiency in the enzyme N-sulfoglucosamine sulfohydrolase (SGSH; sulfamidase). The structure of the recombinant enzyme expressed in human embryonic kidney (HEK) cells has been determined to 2.0 Å resolution (Sidhu et al. 2014). It consists of a central β-sheet surrounded by two domains, a large N-terminal domain (domain 1) and a smaller C-terminal domain (domain 2), as is typical for the sulfatase fold (Fig. 4.3). The active site is found in a narrow cleft in domain 1. A highly conserved lysine in O-sulfatases is replaced in SGSH by an arginine (R282) that is positioned to bind the N-linked sulfate substrate. About 80 mutations that have been described in patients were mapped onto the structure and about a quarter of them were found located on exposed regions of the structure. A low flexibility detected in the SGSH structure have led to suggestions for use of small molecules with allosteric or stabilizing effects for mutations located at the surface and away from the active site and molecular chaperones that bind to the active site to reconstitute in missense mutations.

4.2.4 Pompe Disease

Pompe disease (OMIM 232300, ORPHA365) is an autosomal recessive disorder of glycogen metabolism caused by a deficiency of the lysosomal enzyme acid alpha-glucosidase (GAA) that leads to lysosomal accumulation of glycogen in all tissues, particularly in cardiac and skeletal muscle (Kishnani and Beckemeyer 2014). The disease begins within a few months of birth and infants with this disorder typically experience muscle weakness (myopathy), poor muscle tone (hypotonia), an enlarged liver (hepatomegaly), and heart defects. There are two categories of Pompe's disease: infantile-onset Pompe disease (IOPD) with symptoms that start within 2 months after birth often with cardiomyopathy. Death occurs before the 1st year of age if it remains untreated. Late-onset Pompe disease (LOPD) appears in the later stages of life, from 1 year to adulthood and progresses slowly, leading eventually to respiratory failure.

Recombinant human GAA (rhGAA) produced in CHO cells or in transgenic rabbit milk (Van Hove et al. 1996; Bijvoet et al. 1999) was approved in 2006 for ER therapy (ERT) to treat Pompe disease and has proven to be beneficial for patient's survival and to stabilize the disease course. ERT is currently the only approved treat-

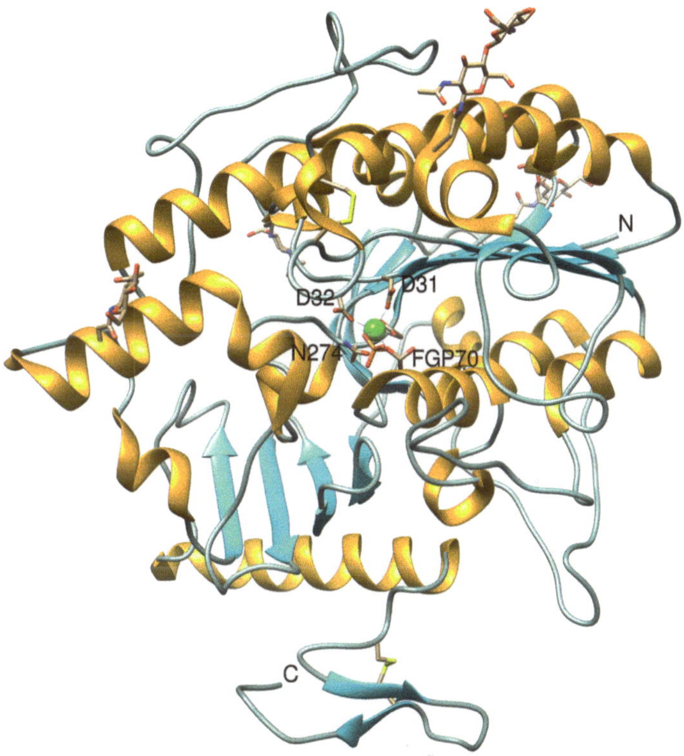

Fig. 4.3 Ribbon diagram of SGSH (PDB id 4mhx). The active site with the phosphorylated formylglycine 70 (FGP70) and the bound Ca^{2+} is shown

ment for Pompe disease. However, despite the clinical benefits, patient's response varies considerably, and the efficiency of the treatment is limited by insufficient targeting and uptake in muscle tissues, immunogenic reactions and build-up of autophagic compartments in myocytes. Hence, alternative therapeutic strategies based on different approaches and rationale are currently in need. Myozyme® (alglucosidase alfa) was approved in 2006 for ERT of IOPD and Lumizyme® was approved in 2010 for ERT of LOPD by FDA.

PC therapy (PCT) using small-molecule "chaperones" that prevent enzyme degradation by the endoplasmic reticulum-associated degradation system has been proposed also in Pompe disease, either as alternative therapy or complementary to ERT. Studies have shown an increase in GAA activity, suggesting increased stability in the presence of 'chaperones' (Parenti et al. 2014). The most extensively studied GAA chaperone molecules are two glucose analogues, the imino sugars 1-deoxynojirimycin (DNJ, AT2220) and N-butyldeoxynojirimycin (NB-DNJ, miglustat). Until recently, the enzyme in its commercial form (Myozyme R) could not be crystallized. However, in situ proteolysis carried out by α-chymotrypsin to remove putative flexible loops as suggested by disorder predictors resulted in an enzyme form with a ~5 kDa lower mass that crystallized readily. Four protease-labile regions were cleaved off (Q57-T80, G116-M122, R199-A204 and A782-R794), without affecting the catalytic activity of the resultant GAA form. The cleaved form resulted in high resolution crystal structures of recombinant human GAA in free state and in complex with active site inhibitors (Roig-Zamboni et al. 2017). The structure consists of an N-terminal trefoil type-P domain, followed by a β-sheet N-terminal domain, a catalytic $(\beta/\alpha)_8$-barrell domain similar to the family 31 of glycoside hydrolases (GH31), two β-sheet inserts (I and II), and proximal and distal β-sheet domains at the C-terminal. The catalytic nucleophile and acid/base can be assigned to D518 and D616 residues, respectively.

The crystal structure of the rhGAA-DNJ revealed no major conformational changes upon

DNJ binding in the rhGAA active site apart from a 20° tilt in the side-chain of W376. In contrast, binding of derivative N-hydroxyethyl-deoxynojirimycin (NHE-DNJ) showed a stabilizing interaction of the hydroxyethyl substituent of NHE-DNJ with W481, located on the tip of insert I. N-acetylcysteine (NAC) has been shown to act as allosteric pharmacological chaperone, enhancing the stability of mutated endogenous GAA and rhGAA used for ERT, without affecting the catalytic activity. Importantly, NAC is specific towards GAA and shows no stabilizing effect on a GH31 homologue from rice. To understand the stabilising effect of NAC, the crystal structure of rhGAA in complex with NAC was determined by crystal soaking experiments. The 1.83 Å resolution structure of the complex revealed two NAC-binding sites at 30 Å and 25 Å away, respectively, from the active site (Fig. 4.4).

Comparison of residues surrounding the NAC-binding sites with respect to that of unbound rhGAA showed that their atomic thermal displacement parameters are lower, suggesting an overall inter-domain stabilizing function driven by NAC. Moreover, NAC may also exhibit an anti-oxidative effect as residue C938 is oxidized to the sulfenic acid form in all the crystal structures, except for the structure of the rhGAA-NAC complex.

4.2.5 Maroteaux-Lamy Syndrome

MPS type VI (OMIM 253200) is a recessive autosomal lysosomal disease, leading to an accumulation of glycosaminoglycans within lysosomes owing to partial or complete lack of activity in N-Acetylgalactosamine-4-sulfatase (arylsulfatase B). MPS VI patients experience a progressive, chronic and multisystemic disease that causes not only significant morbidity but also early mortality. Clinical features of MPS VI are similar to the other MPS disorders: skeletal dysostosis, coarse face, corneal opacification, visceromegaly, upper airway obstruction and valvular heart disease. Intellectual development is preserved in this disease.

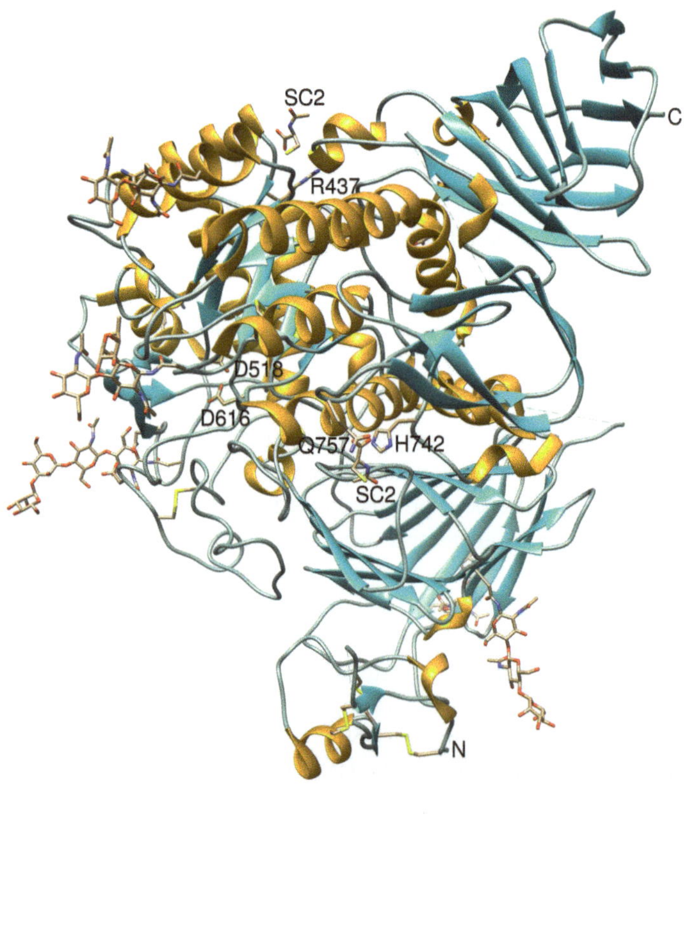

Fig. 4.4 Ribbon diagram of rhGAA in complex with NAC (SC2 in PDB naming system; PDB id 5nn4). The two SC2 binding sites are shown. Catalytic residues D518 and D616 are also depicted

Specific treatments for MPS VI include hematopoietic stem cell transplantation (HSCT) and ERT. A recombinant form of human N-acetylgalactosamine-4-sulfatase, known as Naglazyme (galsulfase), produced in CHO has recently been approved as an ERT in the US (2005) and the EU (2006) for use in MPS VI patients (Lourenço and Giugliani 2014). Galsulfase acts by cleaving the sulfate moieties of the GAG dermatan sulfate, which are abnormally stored in the lysosomes of cells in patients with mucopolysaccharidosis VI. Galsulfase is a glycoprotein with a single-chain protein of 56 kDa, comprising 495 amino acids and containing six asparagine-linked glycosylation sites, four of which carry a bis-mannose-6-phosphate mannose oligosaccharide for specific cellular recognition.

The structure of N-acetylgalactosamine-4-sulfatase has been determined at 2.5 Å resolution and consists of two domains (Bond et al. 1997). The larger domain has an α/β topology with similarities to the alkaline phosphatase family and contains the active site (Fig. 4.5). A calcium-binding site coincides with the zinc-binding site found in alkaline phosphatases.

4.2.6 Gaucher Disease

Gaucher disease (GD) is the most common autosomal recessive lysosomal storage disease (LSD). It is caused by deficiency or absence of the activity of the enzyme acid β-glucosidase (GCase; E.C.3.2.1.45), known also as glucocerebrosidase or glucosylceramidase, leading to accumulation of glucocerebroside (glucosylceramide; GLC) in tissue macrophages. The disease is characterized by hepatomegaly, splenomegaly, anemia, skeletal complications, and neurological complications.

GD is classified into three types: type 1 GD (GD1) is chronic and non-neuronopathic, accounting for 95% of GDs, and types 2 and 3 (GD2, GD3) cause nerve cell destruction. GD2 occurs in the first few months of life and rapidly progresses rapidly until death during infancy.

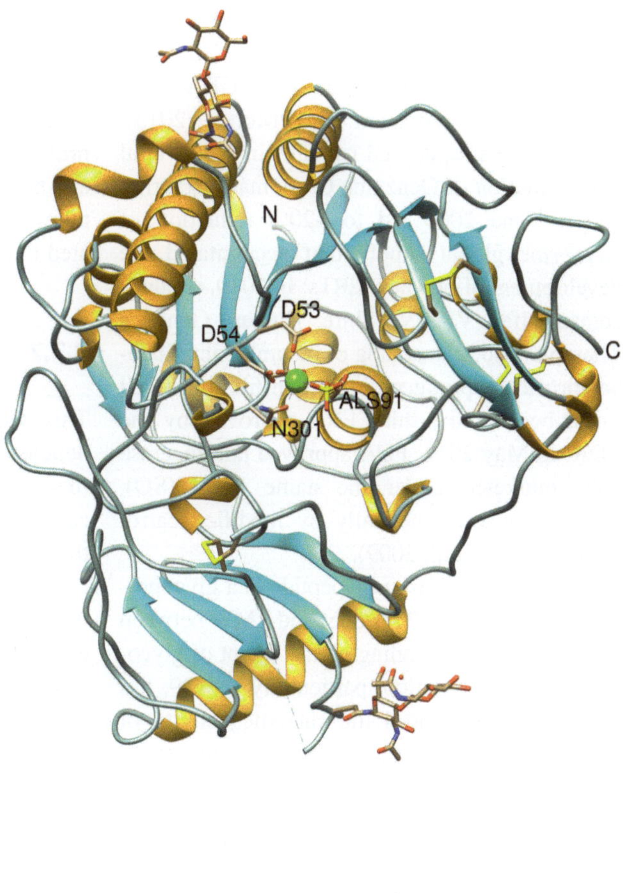

Fig. 4.5 Ribbon diagram of human N-acetylgalactosamine-4-sulfatase crystal structure (PDB id 1fsu). The active site Cys91 has been found modified in the structure as oxo-alanine (ALS). Bound Ca^{2+} ion is shown as a green sphere

GD3 patients survive infancy but they suffer from neurological features for the rest of their lives. Whereas type I is currently treatable with enzyme replacement therapy, there is no equivalent treatment for types II and III since the recombinant enzyme does not pass the blood-brain barrier. For patients with the acute neuronopathic GD2 form, ERT is inefficient.

Over 300 mutations have been identified, including nonsense ones, deletions and insertions, and complex alleles. Two important disease-causing missense mutations are N370S and L444P, which account for the majority of mutations found in the Ashkenazi Jewish population. N370S causes mild GD1 whereas L444P results in severe neurological disorders, characteristic of GD2 and GD3. In recent years it has emerged that GD may increase the risk for Parkinson disease (Osellame and Duchen 2013). The N370S mutation is the most prevalent mutation in GD patients with early-onset parkinsonism (Ran et al. 2016).

The development of two effective ERTs (imiglucerase [Cerezyme] and velaglucerase alfa [VPRIV]) for GD and, recently, the discovery of a new plant-derived ERT (taliglucerase alfa [ELELYSO]) have made treatment of GD1 possible by supplementing the defective GBA1 with active enzyme (Bennett and Mohan 2013). Cerezyme was approved by FDA in 1994. Viral contamination of Genzyme Corp. manufacturing site in June 2009 led to ~20% reduction of Cerezyme global supply that necessitated the development of two new ERTs: In 2010, velaglucerase alfa (VPRIV, Shire Human Genetics Therapies Inc.), an analog of recombinant glucocerebrosidase produced in human fibroblast cell lines, became the third ERT approved by the FDA. In May 2012, FDA approved taliglucerase alfa (marketed under the name ELELYSO), which is made genetically by modified carrot cells (Aviezer et al. 2009).

The mature GCase polypeptide is a glycoprotein consisting of 497 residues and a MW between 59 and 69 kDa depending on the extent of glycosylation. Glycosylation, particularly at N19, has been found to control the catalytic activity of GCase, (Pol-Fachin et al. 2016). The structure of the partially deglycosylated form was determined in 2003 (PDB id 1ogs) (Dvir et al. 2003). The structure of the fully glycosylated Cerezyme was determined in 2006 (Fig. 4.6) and found to be virtually identical to that of the partially deglycosylated enzyme (Brumshtein et al. 2006).

The overall architecture comprises three discontinuous domains: I, II and III. Domain I interacts tightly with domain III whereas domains II seem to be connected to domain III by a flexible hinge. Domain I (residues 1–27 and 383–414) consists of one main three-stranded, anti-parallel β-sheet that is flanked by a perpendicular amino-terminal strand and a loop. Domain II (residues 30–75 and 431–497) resembles an immunoglobulin (Ig) fold consisting of two closely associated β-sheets that form an independent domain. Domain III (residues 76–381 and 416–430) adopts a (β/α) eightfold. TIM barrel which contains the catalytic site, consistent with homology to GH-A clan members. Two glutamate residues (E235 and E340) have been identified as the acid/base catalyst and the nucleophile, respectively. Seven aromatic side chains (F128, W179, Y244, F246, Y313, W381 and F397) line one side of the active-site pocket and, as in other β-glycosidases, they may be involved in substrate recognition. The entrance to the active site is lined up with several hydrophobic residues that may facilitate protein interaction with the lysosomal membrane and help the protein to reach areas where activity takes place. Asparagine 19 was found glycosylated in the structures as expected.

4.2.7 Fabry Disease

This is an X-linked inherited disease caused by a deficiency of the enzyme alpha-galactosidase A (α-Gal A) that results in a buildup of a type of fat called globotriaosylceramide (Gb3, or GL-3) in the body. Current treatment of Fabry disease includes ER and PC therapies. Fabrazyme, the first enzyme for use in ER treatment of Fabry disease was approved in 2003. Migalastat (1-deoxygalactonojirimycin; DGJ) is currently in Phase II clinical trials for oral administration in PC treatment. DGJ binds at the active site of the

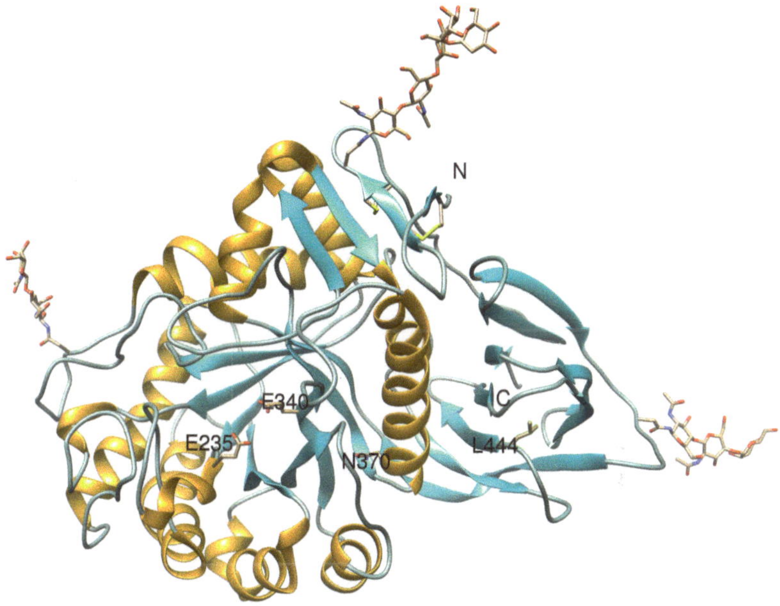

Fig. 4.6 Ribbon diagram of GCase crystal structure (PDB id 2j25, B-subunit). Catalytic residues E340 and E235 are shown. Residues N370 and L444 which are found frequently mutated in GD patients are also depicted in ball-and-stick

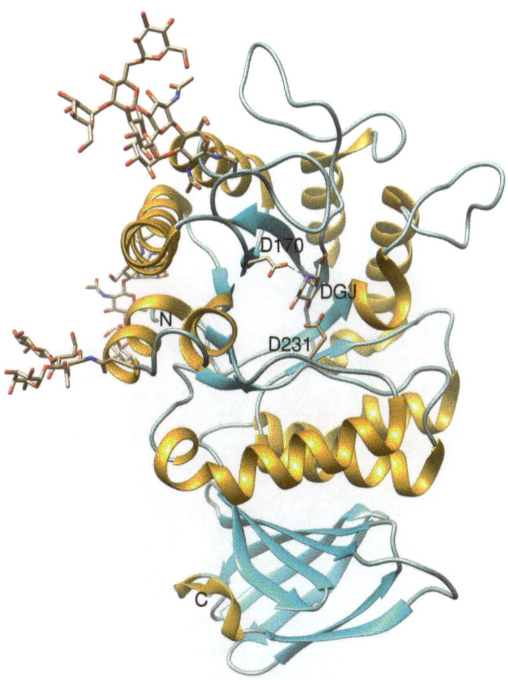

Fig. 4.7 Crystal structure of α-Gal A in complex with DGJ (PDB id 3s5y)

enzyme with ~400,000 better binding affinity for α-Gal A than galactose and stabilizes the enzyme according to circular dichroism, proteolysis and fluorescence experiments (Guce et al. 2011). The higher binding potency of DGJ has been attributed to a single ionic interaction of the heterocyclic nitrogen with the catalytic nucleophile D170. Indeed, a D170A mutation resulted in weaker binding of DGJ, thus confirming the key role of D170 in the binding. The overall structure of α-Gal A consists of an N-terminal (α/β)8 barrel that encompasses the active site and a C-terminal antiparallel domain (Fig. 4.7). A second sugar-binding site has been identified at the interface of the two domains far from the active site. It has been suggested that this site may be suited for targeting by a new class of pharmacological chaperones (Guce et al. 2010).

4.2.8 Krabbe Disease

Krabbe disease is an inherited neurodegenerative disorder known also as globoid cell leucodystrophy. It is caused by defects in the lysosomal enzyme β-galactocerebrosidase (GALC), which catalyzes the cleavage of the terminal β-D-galactose from the glycosphingolipids galactosylceramide and psychosine. The disease leads in rapid loss of the myelin sheath that surrounds nerve cells (demyelination). A number of missense mutations have been identified that result in misfolding of the protein and impair in its trafficking to the ER. Crystal structures of murine GALC in complex with substrate, product and covalent intermediate have revealed details of the catalytic mechanism and confirmed the role of two glutamate residues (E258, E182) as the nucleophile and the acid/base, respectively (Hill et al. 2013). The structures also showed significant conformational changes at the active site, including the non-catalytic R380 residue, which takes two alternative conformations during the catalytic cycle. The majority (~70%) of the disease-causing mutations involves modification of residues that are buried within the core of the structure. Treatment includes hematopoietic stem cell transplantation for the early infantile form of the disease. The use of ERT is unsuitable, as the administered enzyme will be unable to cross the blood-brain barrier to reach the central nervous system. PCT is currently being explored as an alternative strategy to overcome defects in GALC folding (Lee et al. 2010; Spratley and Deane 2016). In recent studies, several small-molecule inhibitors have been identified as promising chaperone candidates for GALC (Viuff and Jensen 2016). Amongst them, azasugars that bind at the active site of the enzyme have shown potency as stabilizing agents of GALC structure (Hill et al. 2015). Four compounds were tested and their crystal structures in complex with murine GALC showed that they all bind with the same orientation as β-D-galactose. No secondary binding sites were identified that could confer some allosteric effect. Structural analysis of the complexes combined with stability measurements and inhibition assays suggested that the position and the positive charge of the ring nitrogen could contribute to increased stabilization of GALC.

4.2.9 Hunter Syndrome

Known also as MPS type II (OMIM 309900), Hunter syndrome is a rare X-linked disease caused by a deficiency in iduronate-2-sulfatase (IDS; EC 3.1.6.13), a crucial enzyme involved in the lysosomal degradation of dermatan sulfate and heparan sulfate. Enzyme replacement therapy with idursulfase (brand name Elaprase, a recently approved orphan drug) is the first treatment for Hunter syndrome. It is a purified form of IDS and is produced by recombinant DNA technology in a human cell line. Elaprase has been found effective in the 'mild' form of the disease but not in the 'severe' form which is characterized by cardiac and neurological deterioration.

The crystal structure of human IDS with a bound sulphate ion has been reported at 2.3 Å resolution (Demydchuk et al. 2017). It shows similarities with other human sulfatases despite low sequence identity. The single polypeptide chain comprises two subdomains (SDs): an N-terminal SD1 (residues 34–443) known also as the 'heavy' chain and a C-terminal SD2 (residues 455–550) which corresponds to the 'light' chain (Figs. 4.8 and 4.9). SD1 and SD2 form a large hydrophobic interface. The catalytic site is found in SD1. There are six Cys residues, four of which are involved in disulphide bond formation. One free cysteine is found at the tip of a loop whilst the sixth Cys (Cys84) is post-translationally modified as a key residue in the active site. The thiol group of Cys84 is oxidized to formyl-

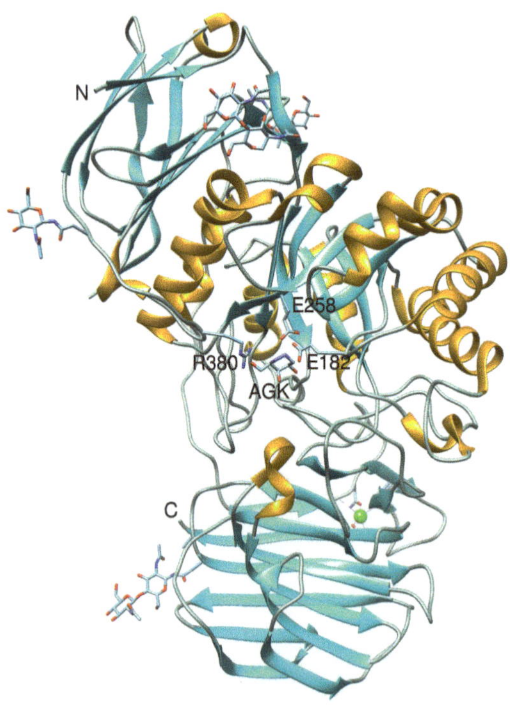

Fig. 4.8 Crystal structure of GALC in complex with azagalacto-fagomine (AGK; PDB id 4ufi)

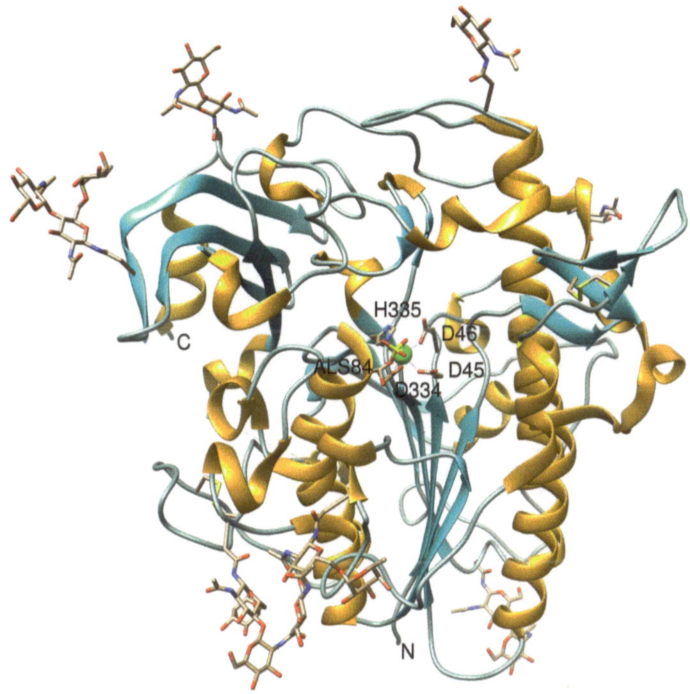

Fig. 4.9 Crystal structure of human IDS (PDB id 5fql). The modified Cys84 is shown (ALS84). Bound Ca^{2+} is shown as a green sphere

glycine (FGly) by a FGly-generating enzyme which recognizes the CxPSR motif. Of the 542 mutations known today, the majority (90%) involves residues at the core of the protein structure.

4.2.10 Hurler Syndrome

Hurler syndrome is the most severe form of MPS I, a rare lysosomal autosomal recessive storage disease caused by deficient activity of α-L-iduronidase (IDUA; EC3.2.1.76), an enzyme which removes α-L-iduronic acids from two GAGs: heparan sulfate and dermatan sulfate. The disease is characterized by skeletal abnormalities, cognitive impairment, heart disease, respiratory problems, enlarged liver and spleen, characteristic facies, and reduced life expectancy.

Human IDUA produced in CHO cells has been crystallized and the structure determined (PDB ID codes: 3w81 and 3w82). The overall architecture of human IDUA consists of three domains: a (β/α)8 triose phosphate isomerase (TIM) barrel domain, a β-sandwich domain with a short helix-loop-helix, and an immunoglobulin (Ig)-like domain (Maita et al. 2013). Interestingly, it was found that the mannose residue (Man7) of the N-glycan attached to N372 constitutes a part of the substrate-binding pocket and interacts directly with the substrate (Fig. 4.10). A deglycosylation study showed that enzyme activity was highly correlated with the N-glycan attached to N372. Similar results have been obtained with the structure of human IDUA although the enzyme lacked Man7 possibly owing to the *Arabidopsis thaliana* expression system used for its production (Bie et al. 2013). Structural analysis suggested that W306L, one of the more than 55 missense mutations associated with human IDUA may cause MPS I through a change in the conformation of the N372 glycan.

Computational modelling and molecular dynamics calculations based on the crystal structure of human IDUA showed a correlation between the structural changes and clinical phenotypes. Mutations responsible for severe and

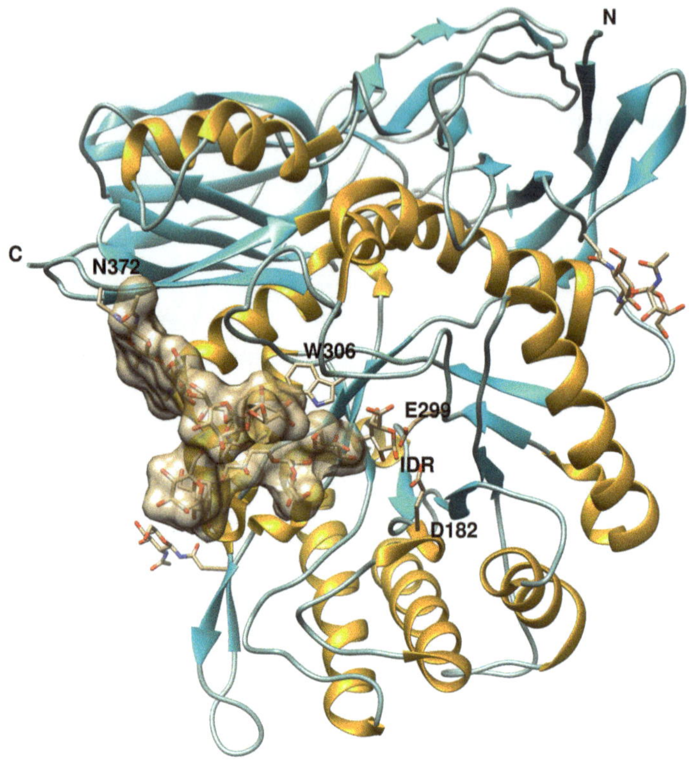

Fig. 4.10 Crystal structure of L-iduronidase in complex with iduronic acid (IDR). The molecular surface associated with the glycan at N372 is shown. IDR and the two catalytic glutamate residues are depicted (PDB id 3w82)

severe–intermediate groups (R533L/R, V620F, and R628P) appeared to induce larger conformational changes compared with those responsible for intermediate and attenuated ones (L535F, L578Q, F602I, and R619G) (Saito et al. 2014). Moreover, these structural changes were found in a region far from the active site of IDUA, suggesting that they may affect protein folding.

4.2.11 Sly Syndrome

Sly syndrome is an ultra-rare inherited metabolic condition known also as mucopolysaccharidosis type VII (MPS VII). It is caused by deficiency in β-glucuronidase (GUSB), an important lysosomal enzyme involved in the degradation of glucuronate-containing glycosaminoglycan (Naz et al. 2013). Phenotypes vary from mild skeletal abnormalities to more severe forms with hydrops fetalis, skeletal dysplasia and mental retardation (Montaño et al. 2016). In November 2017, Vestronidase alfa-vjbk (trade name Mepsevii), a recombinant form of the human β-glucuronidase was approved in the United States as the first drug for enzyme-replacement therapy of the disease. The structure of human β-glucuronidase comprises three domains with topologies similar to a jelly roll barrel, an immunoglobulin constant domain and a TIM barrel, respectively (Jain et al. 1996; Hassan et al. 2013). Residues 179–204 form a β-hairpin motif similar to the putative lysosomal targeting motif of cathepsin D in support of the view that lysosomal targeting has a structural basis. A large cleft at the interface of two monomers forms the active site with residues Glu451 and Glu540 proposed to be important for catalysis. Ninety mutations out of a total of 211 characterized deleterious mutations were predicted to be disease-causing mutations (Khan et al. 2016). Following molecular dynamics simulations, a number of mutations were predicted to destabilize the structure. However, other mutations were predicted to increase the stability of the protein. As these mutations were also associated with the disease, their role to the dysfunction of the enzyme was attributed to possible reduction of the substrate accessibility to the active site.

4.3 Cancer Treatment

The use of therapeutic enzymes in cancer involves two main strategies: amino acid deprivation of the tumor or enzyme-prodrug therapy. In the former strategy, enzymes are used to deplete amino acids in tumor cells which are auxotrophic to particular amino acids. In enzyme-prodrug therapy, antibody-conjugated enzymes are used to convert prodrug into cytotoxic drug at tumour cells and thereby killing tumor cells.

4.3.1 L-Asparaginase

L-asparagine is a nonessential amino acid in normal cells but in certain malignant cells it must be acquired through the blood. L-asparaginase is an anticancer agent that exploits this specific requirement of the malignant cells. As L-asparaginase catalyzes the conversion of L-asparagine to aspartic acid and ammonia, the use of L-asparaginase could deprive leukemic cells of their source of L-asparagine. Asparaginases (ASNases) have been used as anti-leukemic drugs in combination with vincristine and corticosteroids among others, to induce patients of acute lymphoblastic leukemia (ALL) into complete remission (Labrou et al. 2010).

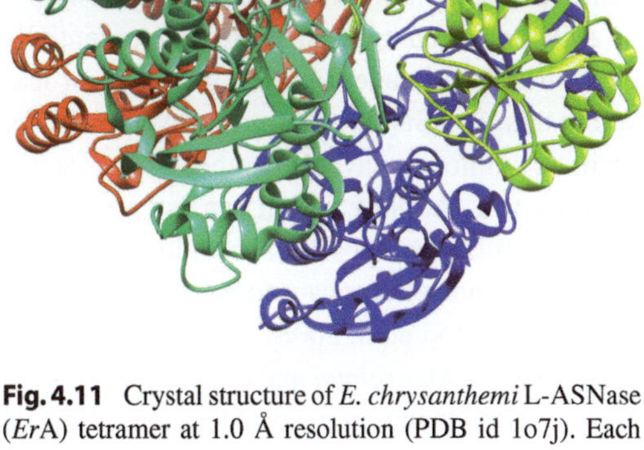

Fig. 4.11 Crystal structure of *E. chrysanthemi* L-ASNase (*Er*A) tetramer at 1.0 Å resolution (PDB id 1o7j). Each subunit of the tetramer is shown in different color

All bacterial L-ASNases are active as homo-tetramers composed of four subunits (A, B, C, and D) arranged in 222 symmetry, according to the nomenclature first appeared in the *E. coli* ASNase (EcAII) crystal structure (Fig. 4.11). Four identical non-cooperative active sites have been identified that are formed at the subunit interfaces: two between subunits A and C and two between B and D. The N- and C-terminal domains of the subunits are involved in the formation of each active site, thus the L-ASNase homotetramers are best described as dimers of intimate dimers. Residues responsible for ligand binding form the rigid part of the active site. The flexible part of the active site (residues 14–33 in *Ec*AII) controls the access to the binding pocket and carries the highly conserved for all L-ASNases catalytic nucleophile Thr12 (numbering according to *Ec*AII crystal structure (PDB code 4eca)). This flexible loop exists in different conformations depending on the presence or absence of a bound ligand in the active site. The active site is characterized by the presence of two catalytically indispensable Thr residues (Thr89 and Thr12 in *Ec*AII) in diametrically opposite locations above and below the CG atom of the substrate (Harms et al. 1991; Palm et al. 1996; Ortlund et al. 2000). The question, however, which of these two Thr residues acts as the primary nucleophile has been under continuous discussion. Based on the available data, a two-step ping-pong mechanism has been finally put forward involving both Thr residues, each acting at a different stage of the catalytic reaction. In this case, L-ASNases appear to differ from other enzymes by using two catalytic triads instead of one (Dodson and Wlodawer 1998).

The potential use of ASNase in cancer treatment goes back to the1950s when it was found that lymphomas in mice and rats were quickly regressed after the animals were injected with guinea pig serum. Later, it was concluded that L-asparaginase was the anticancer factor in guinea pig serum responsible for the observed effect (Avramis 2011). Therapeutic ASNases are fermentation native proteins, which originate from either *E. coli* or *Erwinia chrysanthemi*. To increase their efficacy and decrease the allergic reactions to the bacterial enzymes, pegylated products of both ASNase formulations have been made. Elspar® and Oncaspar® (Pegasparagase) are two commercially available forms of *E. coli* and *E. chrystanthemi* L-ASNases, respectively. Oncaspar® received USFDA approval in 2006 for treatment of ALL and is currently the most effective antileukemic biology oncology product available in the United States. Algal, plant and fungal L-asparaginases from different sources have been proposed as alternatives to currently used enzymes.

During L-ASNase therapy, repeated use of the drug is commonly needed owing to the enzyme's relatively short half-life due to low structural stability (Benjwal et al. 2006; Papageorgiou et al. 2008) and premature inactivation by serum proteases. Protein-engineering efforts towards new forms of L-ASNases have been undertaken to improve L-ASNase stability. *Ec*AII and ErA that are currently used in therapeutical settings have essentially different half-times in blood (van den Berg 2011). The recent identification of new L-ASNases has led to a closer look of their thermodynamic stabilities compared to the L-ASNases currently in the market. *Helicobacter pylori* L-ASNase shows a thermal stability between that of *Ec*AII and ErA (Dhavala and Papageorgiou 2009). The sensitivity of *Erwinia carotovora* ASNase (*Ew*A) to protease (e.g. trypsin) inactivation and its low half-life has hindered its development as a clinical therapeutic. An improved form of *Ew*A has been described by combining a structure-based molecular engineering approach with PEGylation (Kotzia et al. 2007). A R206His mutant was generated and treated with mPEG-SNHS [methoxypoly(ethylene glycol) succinate N-hydroxysuccinimide ester]. The resultant variant had seven PEGylated Lys residues, retained 82% of its original catalytic activity, exhibited enhanced resistance to trypsin degradation, and showed higher thermal stability compared to the wild-type enzyme (Kotzia et al. 2007).

In another work, the thermostability of *Ec*AII was enhanced by replacing Asp178 with Pro in a hydrogen-bonded turn ($D^{178}GR^{180}$) that contributes to the thermostability of the enzyme (Li et al. 2007). Notably, the kinetic constants K_m and k_{cat} for the mutant enzyme were not significantly affected by the mutation.

A thermostable variant of *Er*A that contains a single point mutation (D133V) has been produced through an in vitro directed evolution approach (Kotzia and Labrou 2009). The T_m of this variant was 55.8 °C whereas the wild-type enzyme has a T_m of 46.4 °C. Analysis of electrostatic potential of the wild-type enzyme showed that Asp133 is located at a neutral region on the enzyme surface and makes a significant and unfavorable electrostatic contribution to the overall stability.

Additional strategies to increase stability and reduce the immunogenicity of L-ASNase while still preserving its biological activity include entrapment of the enzyme into liposomes and microcapsules as previously described (Labrou et al. 2010).

4.3.2 Arginine Deiminase

Arginine deprivation of cancer cells has been also used as therapeutic strategy. Arginine is a non-essential amino acid synthesized by argininosuccinate synthetase (ASS) and argininosuccinate lyase from citrulline. Normal cells do not require arginine supply. However, certain cancer cells, such as melanoma, renal cell carcinomas and hepatocellular carcinomas lack ASS and they require circulating arginine. Arginine deiminase (ADI) is an enzyme from mycoplasma that converts arginine to citrulline, thereby reducing circulating arginine. This causes arginine deprivation in tumour cells that lack argininosuccinate synthetase, thereby inhibiting tumour progression.

ADI has been suggested to be a potentially better therapeutic agent for the treatment of leukemia than L-ASNase. ADI is highly specific for arginine without converting other amino acids, contrary to L-ASNase which can use asparagine and glutamine as substrates, resulting in some deleterious side effects caused by the degradation of glutamine (Ni et al. 2008).

ADIs from *M. arginini* and related species have been reported to inhibit the proliferation of several types of human cancer cells (Gong et al. 2000).

The structure of *M. arginini* ADI was determined in 2004 (Das et al. 2004) and resembles a 'clip-on fan' architecture with a 'clip' consisting of a five-helix bundle and a 'fan' of five blades. Each blade consists of a ββαβ module. The clip is inserted between the first and the second blade and is characterized by the presence of various hydrophobic and aromatic residues that form an extended hydrophobic core. Owing to large sequence variation amongst various ADIs, the role of the 'clip' region is unclear. It has been suggested that it may be responsible for initiating some cellular processes in the cell (Das et al. 2004).

ADIs from different organisms form either homodimers or homotetramers. X-ray crystal structures of ADI from *P. aeruginosa* (PaADI) (Galkin et al. 2005) (PDB id: 1rxx, 2abr, 2aci, 2a9g, 2aaf), *M. arginini* (MaADI) (Das et al. 2004) (PDB id: 1s9r, 1lxy) and *Streptococcus pyogenes* (PDB id 4bof) (Henningham et al. 2013) have been determined. An induced-fit mechanism has been proposed based on crystal structures of the apo-form of *Ma*ADI after comparison with previous 'closed' conformations in a covalent complex with two L- arginine substrate intermediates. The catalytic mechanism has been elucidated by a number of mutants (Galkin et al. 2005). A catalytic trial consisting of Cys-His-Glu/Asp has been identified (Cys406-His278-Glu224 in PaADI; Cys398- His269-Glu213 in MaADI).

The in vivo application of ADI as an anticancer agent faces two major challenges: strong antigenicity and short circulating half-life (around 4 h) in plasma. Modification of ADI with polyethylene glycols (PEGs) has shown promising results with an increase to the circulating half-life and a decrease in the immunogenicity. The best results have been obtained after PEGylation using PEG 20,000 (Han et al. 2016). Nevertheless, further modifications of the enzyme are required to improve its properties. Several regions and individual residues have been identified that could improve pH optimum, K_m, protein stability and solubility. Despite the efforts, however, there is still not a single mutant that could exhibit all the multiple property improvements required (Han et al. 2016).

ADI could also been used in other diseases, e.g. as vaccine against *Streptococcus pyogenes* (Henningham et al. 2013). Various catalytically inactive mutant forms of ADI showed a retained structure, recognition by antisera, and immunogenic epitopes, suggesting that they are ideal for inclusion in Group A Streptococcus (GAS) vaccine preparations.

4.3.3 Antibody-Directed Enzyme Prodrug Therapy (ADEPT)

This is another strategy of using enzymes as therapeutic agents in cancer. In this approach, an enzyme is conjugated to a monoclonal antibody that carries the enzyme to the site of the tumor (Sharma and Bagshawe 2017). A non-cytotoxic prodrug upon reaching the site of the tumor is converted to its active form by the enzyme, thus destroying the cancer cells but not normal cells. ADEPT with a modified carboxypeptidase A (CPA) specific to activate a methotrexate-phenylalanine (MTX-Phe) prodrug has been described (Deckert et al. 2004). Human β-glucuronidase is also an attractive enzyme in the ADEPT context for activation of glucuronide prodrugs (Chen et al. 2008).

4.4 Cystic Fibrosis (CF)

Dornase alfa is a recombinant human deoxyribonuclease (rhDNase I) that reduces the viscosity of sputum in the airways of CF patients by breaking down double-stranded DNA into lower-molecular mass fragments. Dornase alfa is commercially available as Pulmozyme®. The structure of rhDNase I produced in Chinese hamster ovary has been determined in the presence of phosphate and Mg^{2+} in the active site (Parsiegla et al. 2012). Combined structural and mutagenesis data suggested that Asn7 plays a key role in the catalysis by stabilizing the Mg^{2+} ion, thus providing details about the catalytic mechanism.

4.5 Cardiovascular Diseases

Fibrin is the main major protein formed during wound healing and the end product of the blood clotting cascade. Improper accumulation of fibrin in blood vessels leads to thrombolytic diseases, such as acute myocardial infarction (AMI; heart attacks) and stroke. Elimination and removal of the blood clot (thrombus) either with anti-coagulants (warfarin and heparin), antiplatelets (dipyridamole and aspirin), surgical treatment of the clot or fibrinolytic enzyme therapy are the main therapeutic approaches. Thrombolytic therapy using fibrinolytic enzymes has an advantage over anti-coagulants and antiplatelets, as the enzymes could act directly upon the existing clot. Two types of thrombolytic enzymes are currently employed: plasminogen activators and plasmin-like enzymes.

4.5.1 Plasminogen Activators

Tissue plasminogen activator (tPA) is a serine protease that consists of 527 amino acids with a molecular weight of 70 KDa. It binds to fibrin-rich clots via the fibronectin finger-like domain and the Kringle 2 domain. It catalyzes the conversion of plasminogen to plasmin by hydrolyzing a single Arg-Val in plasminogen. Alteplase (known commercially as Activase) is a recombinant tissue-plasminogen activator used as thrombolytic drug to break up blood clots in AMI and other severe conditions. tPA is rapidly inactivated by endogenous plasminogen activator inhibitor-1 (PAI-1). Because of that, high doses of tPA are required, leading to potential dangerous effects, such as intracranial hemorrhage and neurotoxicity. Current PAs differ in their source, antigenicity, half-life and haemorrhagic potential. The crystal structure of the tPA·PAI-1 Michaelis complex determined in 2015 (Gong et al. 2015) (Fig. 4.12) provided information that could lead to new thrombolytics agents with reduced PAI-1 inactivation.

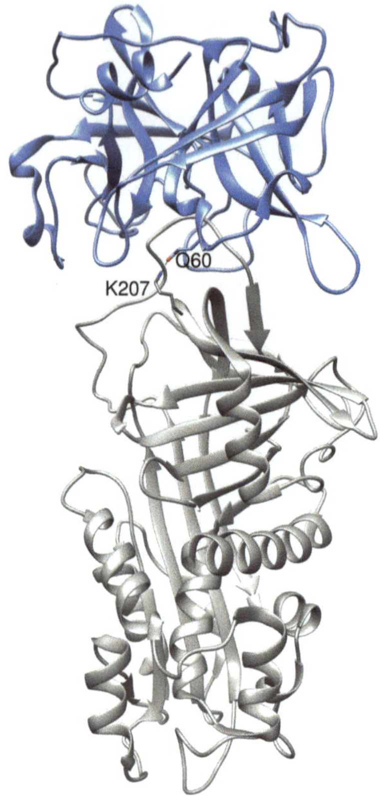

Fig. 4.12 Crystal structure of tPA-PAI-1 complex (PDB id 5brr). tPA is shown in cornflower blue and PAI-1 in gray. The Q60-K207 interaction is depicted

Reteplase is produced by recombinant techniques as a truncated non-glycosylated form of t-PA for use in thrombolytic treatment in AMI. It contains 355 of 527 amino acid residues of tPA. It lacks the finger-like domain and, as expected, it exhibits lower affinity for fibrin. However, clinical trials have shown faster reperfusion than t-PA without any increase in the rate of bleeding side-effects. Tenecteplase (TNK-tPA) is a variant of tPA showing slower inhibition by PAI-1 (Keyt et al. 1994). Compared to wild-type tPA, TNK-tPA has two mutations (T103N and N117Q), both within the kringle domain, and a tetra-Ala substitution at amino acids 296–299 in the protease domain. TNK possesses a longer half-life, increased resistance to plasminogen activator inhibitor, and improved fibrin specificity compared with alteplase (Semba et al. 2001).

4.5.1.1 Urokinase

Known also as urokinase-type plasminogen activator (uPA), urokinase is a 411-residue enzyme. The structure of uPA-PAI-1 Michaelis complex was determined in 2011 (Lin et al. 2011). Comparison with the structure of tPA·PAI-1 Michaelis complex (Gong et al. 2015) revealed significant differences between the two complexes that could explain the higher susceptibility of tPA to PAI-1 inhibition compared to that of uPA. The so called 37-loop (residues ^{36}KHRRSPGER^{39}) of tPA was found to have a contact area twice as large with PAI-1 (247 Å^2) than uPA (124 Å^2), suggesting the importance of this loop in its inhibition by PAI-1. Other structural differences include larger S1 and S2 β pockets in tPA for PAI-1 accommodation and stronger interactions of the tPA 60-loop with PAI-1 residues. Q60 in the tPA 60-loop is replaced by an isoleucine residue in uPA which is unable to make a strong interaction with K207 of PAI-1. In addition, residue 60a in uPA is an aspartate instead of glutamate in tPA and therefore no interaction with K107 of the inhibitor can occur in the uPA-PAI-1 complex. The structural information from these two complexes could lead to new tPA variants as thrombolytics with higher resistant to PAI-1. For example, tPA residues Gln-60 and Glu-60a of the 60-loop, Tyr- 151 of the 147-loop, and Arg-174 of the 169-loop have been suggested for mutagenesis studies.

4.5.2 Plasmin-Like Enzymes

Although tPA and urokinase are widely-used as thrombolytics, their high cost and side-effects including hemorrhage, have led to efforts for the development of newer and safer alternatives. Amongst them are plasmin-like enzymes that require no activation and are mainly produced by bacteria but also by earthworms (lumbrokinase) or snakes as venom components (fibrolase) (Pretzer et al. 1992).

4.5.2.1 Nattokinase

Nattokinase is a serine protease of the subtilisin family (Yanagisawa et al. 2010), with ~99.5% sequence similarity to subtilisins E and subtilisins J. It is used as a supplement because of its strong fibrinolytic activity. It is produced during fermentation of soybeans by *Bacillus subtilis natto* in the preparation of the Japanese food natto. The enzyme is now also produced by recombinant techniques (Cai et al. 2017). Although a single dose of nattokinase could help in reducing the risk of thrombosis in humans, further studies are required to establish potential clinical benefits of this enzyme (Kurosawa et al. 2015). The nattokinase structure (PDB id 4dww) is almost identical to that of subtilisin E.

4.5.2.2 Lumbrokinase

Earthworms have been used in traditional medicine in China, Japan and other Far East countries. Lumbrokinases (LKs) comprise a group of proteases isolated from different earthworm species, such as *Lumbricus rubellus* and *Eisenia fetida* (Wang et al. 2013). These proteases have the abilities not only to directly degrade fibrin and other proteins, but also to activate proenzymes such as plasminogen and prothrombin (Pan et al. 2010). LK is commercially available as Boluoke® and has been approved by the China Food and Drug Administration. It has also been used as oral supplement in many countries, such as USA, Canada and Japan. However, no approval has been received by FDA as a thrombolytic drug to date.

The earthworm fibrinolytic enzyme (EFE) component b from *Eisenia fetida* has the highest fibrinolytic activity in a mixture of seven EFE components in *E. fetida*. The structure of EFE-b (PDB id 1ym0) has revealed a trypsin-like protease with two chains: an N- terminal, pyroglutamated light chain and an N-glycosylated heavy chain (Wang et al. 2005a). The two chains are connected via a disulphide bridge. Detailed structural analysis has suggested that EFE-b may represent an intermediate stage in the evolution of vertebrate chymotrypsin and most non-digestive trypsin-like proteases (which both consist of two chains) from their ancestor (which consists of one chain). The increased stability of EFE-b compared to bovine trypsin or chymotrypsin has been attributed to several factors such as the presence of more charged amino acids and a higher ratio of Arg/Lys than in bovine trypsin or chymotrypsin, the pyroglutamate at the N-terminal of the light chain for protection against aminopeptidase, and increased rigidity at glycosylated residues. The crystal structure of EFE-a (PDB id 1m9u) revealed a chymotrypsin-like fold with specificity determinants of elastases at S1 subsite (Tang et al. 2002). EFE-a and EFE-b share a low sequence identity (~35%) and therefore have distinct structural features, especially in loop regions.

4.6 Aβ-Degrading Enzymes in Alzheimer Disease

Accumulation of the amyloid beta (Aβ) peptide in the brain is a central factor in the pathogenesis of Alzheimer disease (Miners et al. 2011). A number of strategies, including monoclonal antibodies and γ-secretase inhibitors to remove the Aβ peptide have attracted attention in recent years, without however too much success (Marr and Hafez 2014). Proteolytic cleavage of the peptide is another approach that has recently been proposed (Nalivaeva et al. 2012).

Neprilysin (NEP) is a membrane zinc endopeptidase that has received attention as a potential therapeutic enzyme owing to its ability to degrade the Aβ peptide. However, NEP has a wide range of peptide substrates, thus its therapeutic potential is currently limited. Efforts to create modified version of NEP with improved substrate specificity have been described (Webster et al. 2014). A G399V/G714K variant was found to have an approximately 20-fold improved activity on $A\beta_{1-40}$ and up to a 3200-fold reduction in activity on other peptides. The crystal structure (PDB id 4cth) revealed changes in the shape and size of the active site pocket that could explain the improved performance of the variant. In addition to NEP, NEP2, a NEP-like endopeptidase which cooperates with NEP to control Aβ levels in the brain has also been considered for Aβ clearance (Marr and Hafez 2014).

4.7 Nicotine Cessation

Tobacco smoking is the leading cause of many human diseases, such as cancer and cardiovascular disease. One of the major constituents of tobacco is nicotine, an addictive substance attributed to habitual smoking. Following its absorbance into the bloodstream through the lining of the mouth and the lungs, nicotine travels to the brain in a matter of seconds. Although vaccination antibodies to block the entry of nicotine to the brain have been used in immunopharmacotherapy, no satisfactory results have been produced so far and further pharmacokinetic data are required (Hoogsteder et al. 2014). As an alternative strategy, nicotine removal by enzymatic degradation has been suggested (Xue et al. 2018).

Pseudomonas putida NicA2, a flavin-containing enzyme, catalyzes the oxidation of nicotine to N-methyl-myosmine, which is subsequently hydrolyzed into a pseudooxynicotine product. NicA2 is a member of the flavoprotein amine oxidase family. Compared to other members of the amine oxidase family, NicA2 has an extension of 50 residues at the N-terminal. Fusion of an albumin domain to the N-terminal of a 50 a.a. truncated form of the enzyme has increased significantly the half-time of the enzyme without affecting its catalytic properties. The structure of full-length NicA2 has been determined at 2.2 Å resolution (Tararina et al. 2016). The 50 a.a. extension was found disordered in the NicA2 structure and it was suggested that it may play a role in localization. The NicA2 structure consists of two domains: a substrate-binding domain and a FAD-binding domain (Fig. 4.13). Comparison of the FAD-binding domain of NicA2 with that of other members of the flavin amine oxidase family showed high conservation. The substrate-binding domain, which consists of a large cavity and a long channel extending from the substrate-binding domain to the surface of the protein, was

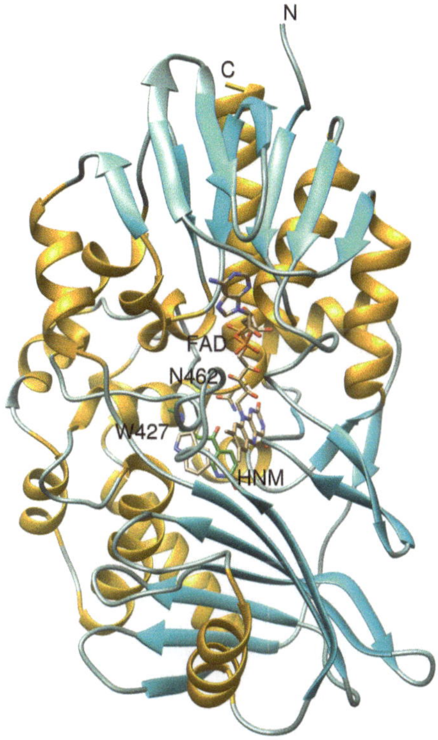

Fig. 4.13 Ribbon diagram of *Pseudomonas putida* NicA2 crystal structure (PDB id 3nho). Bound FAD, 'aromatic cage' residues W427 and N462, and bound 6-hydroxy-L-nicotine (HNM) are shown in stick representation

found to be more divergent. Interestingly, a docking exercise revealed no residues in hydrogen-bonding distance to the substrate's pyridine moiety, indicating possible conformational changes upon substrate binding. Two residues, Trp427 and Asn462, were found to form the so-called 'aromatic cage', a highly conserved region amongst flavin-dependent amine oxidases. In human monoamine oxidase B, the aromatic cage includes two Tyr residues and was proposed to assist substrate binding and to increase the nucleophilicity of the substrate amine. In NicA2, Tyr435 is replaced by Asn462 and therefore the presence of a polar residue may affect the role of the 'aromatic cage' in NicA2 regarding substrate specificity and catalytic mechanism. The detailed mapping of the active site could guide efforts for site-directed mutagenesis studies that may improve the catalytic efficiency of the enzyme.

4.8 Conclusions and Future Perspectives

In recent years, therapeutic enzymes have become major players in the pharmaceutical industry. Their use is expected to grow in the future as new targets have emerged for enzyme-based treatments. The use of enzymes in therapeutic settings is not, however, without limitations. The large size of the enzymes, their immunogenicity, their relatively short half-life and the high purity required are key limiting factors in their use. Although advancements in drug development and delivery over the past few decades have revolutionized enzyme therapy and newer drugs with improved stability and less antigenicity have been developed many challenges still remain. Progress in protein engineering methods offer new opportunities for tailoring the biochemical and biophysical properties of therapeutic enzymes to meet specific needs of clinical applications. Structural information has played a critical role in several cases to improve enzyme-based therapies and to develop small molecules as molecular chaperones and structure stabilizers. With the need of future optimization, structural information will continue to play a significant role in finding new strategies against diseases.

References

Aviezer D, Brill-Almon E, Shaaltiel Y, Hashmueli S, Bartfeld D, Mizrachi S, Liberman Y, Freeman A, Zimran A, Galun E (2009) A plant-derived recombinant human glucocerebrosidase enzyme – a preclinical and phase I investigation. PLoS One 4:e4792. https://doi.org/10.1371/journal.pone.0004792

Avramis VI (2011) Asparaginases: a successful class of drugs against leukemias and lymphomas. J Pediatr Hematol Oncol 33:573–579. https://doi.org/10.1097/MPH.0b013e31823313be

Bell SM, Wendt DJ, Zhang Y, Taylor TW, Long S, Tsuruda L, Zhao B, Laipis P, Fitzpatrick PA (2017) Formulation and PEGylation optimization of the therapeutic PEGylated phenylalanine ammonia lyase for the treatment of phenylketonuria. PLoS One 12:e0173269. https://doi.org/10.1371/journal.pone.0173269

Benjwal S, Verma S, Röhm K-H, Gursky O (2006) Monitoring protein aggregation during thermal unfold-

ing in circular dichroism experiments. Protein Sci 15:635–639. https://doi.org/10.1110/ps.051917406
Bennett LL, Mohan D (2013) Gaucher disease and its treatment options. Ann Pharmacother 47:1182–1193. https://doi.org/10.1177/1060028013500469
Bie H, Yin J, He X, Kermode AR, Goddard-Borger ED, Withers SG, James MNG (2013) Insights into mucopolysaccharidosis I from the structure and action of α-L-iduronidase. Nat Chem Biol 9:739–745. https://doi.org/10.1038/nchembio.1357
Bijvoet AG, Van Hirtum H, Kroos MA, Van de Kamp EH, Schoneveld O, Visser P, Brakenhoff JP, Weggeman M, van Corven EJ, Van der Ploeg AT, Reuser AJ (1999) Human acid alpha-glucosidase from rabbit milk has therapeutic effect in mice with glycogen storage disease type II. Hum Mol Genet 8:2145–2153
Bond CS, Clements PR, Ashby SJ, Collyer CA, Harrop SJ, Hopwood JJ, Guss JM (1997) Structure of a human lysosomal sulfatase. Structure 5:277–289
Brumshtein B, Wormald MR, Silman I, Futerman AH, Sussman JL (2006) Structural comparison of differently glycosylated forms of acid-beta-glucosidase, the defective enzyme in Gaucher disease. Acta Crystallogr D Biol Crystallogr 62:1458–1465. https://doi.org/10.1107/S0907444906038303
Cai D, Zhu C, Chen S (2017) Microbial production of nattokinase: current progress, challenge and prospect. World J Microbiol Biotechnol 33:84. https://doi.org/10.1007/s11274-017-2253-2
Calabrese JC, Jordan DB, Boodhoo A, Sariaslani S, Vannelli T (2004) Crystal structure of phenylalanine ammonia lyase: multiple helix dipoles implicated in catalysis. Biochemistry 43:11403–11416. https://doi.org/10.1021/bi049053
Chen K-C, Wu C-H, Chang C-Y, Lu W-C, Tseng Q, Prijovich ZM, Schechinger W, Liaw Y-C, Leu Y-L, Roffler SR (2008) Directed evolution of a lysosomal enzyme with enhanced activity at neutral pH by mammalian cell-surface display. Chem Biol 15:1277–1286. https://doi.org/10.1016/j.chembiol.2008.10.008
Das K, Butler GH, Kwiatkowski V, Clark AD, Yadav P, Arnold E (2004) Crystal structures of arginine deiminase with covalent reaction intermediates; implications for catalytic mechanism. Structure 12:657–667. https://doi.org/10.1016/j.str.2004.02.017
Deckert PM, Bornmann WG, Ritter G, Williams C, Franke J, Keilholz U, Thiel E, Old LJ, Bertino JR, Welt S (2004) Specific tumour localisation of a huA33 antibody – carboxypeptidase A conjugate and activation of methotrexate-phenylalanine. Int J Oncol 24:1289–1295
Demydchuk M, Hill CH, Zhou A, Bunkoczi G, Stein PE, Marchesan D, Deane JE, Read RJ (2017) Insights into Hunter syndrome from the structure of iduronate-2-sulfatase. Nat Commun 8:15786. https://doi.org/10.1038/ncomms15786
Desnick RJ, Schuchman EH (2012) Enzyme replacement therapy for lysosomal diseases: lessons from 20 years of experience and remaining challenges. Annu Rev Genomics Hum Genet 13:307–335. https://doi.org/10.1146/annurev-genom-090711-163739
Dhavala P, Papageorgiou AC (2009) Structure of helicobacter pylori L-asparaginase at 1.4 A resolution. Acta Crystallogr D Biol Crystallogr 65:1253–1261. https://doi.org/10.1107/S0907444909038244
Dodson G, Wlodawer A (1998) Catalytic triads and their relatives. Trends Biochem Sci 23:347–352
Dvir H, Harel M, McCarthy AA, Toker L, Silman I, Futerman AH, Sussman JL (2003) X-ray structure of human acid-beta-glucosidase, the defective enzyme in Gaucher disease. EMBO Rep 4:704–709. https://doi.org/10.1038/sj.embor.embor873
Galkin A, Lu X, Dunaway-Mariano D, Herzberg O (2005) Crystal structures representing the Michaelis complex and the thiouronium reaction intermediate of *Pseudomonas aeruginosa* arginine deiminase. J Biol Chem 280:34080–34087. https://doi.org/10.1074/jbc.M505471200
Gamez A, Sarkissian CN, Wang L, Kim W, Straub M, Patch MG, Chen L, Striepeke S, Fitzpatrick P, Lemontt JF, O'Neill C, Scriver CR, Stevens RC (2005) Development of pegylated forms of recombinant *Rhodosporidium toruloides* phenylalanine ammonia-lyase for the treatment of classical phenylketonuria. Mol Ther 11:986–989. https://doi.org/10.1016/j.ymthe.2005.02.013
Gong H, Zolzer F, von Recklinghausen G, Havers W, Schweigerer L (2000) Arginine deiminase inhibits proliferation of human leukemia cells more potently than asparaginase by inducing cell cycle arrest and apoptosis. Leukemia 14:826–829
Gong L, Liu M, Zeng T, Shi X, Yuan C, Andreasen PA, Huang M (2015) Crystal structure of the Michaelis complex between tissue-type plasminogen activator and plasminogen activators inhibitor-1. J Biol Chem 290:25795–25804. https://doi.org/10.1074/jbc.M115.677567
Guce AI, Clark NE, Salgado EN, Ivanen DR, Kulminskaya AA, Brumer H, Garman SC (2010) Catalytic mechanism of human alpha-galactosidase. J Biol Chem 285:3625–3632. https://doi.org/10.1074/jbc.M109.060145
Guce AI, Clark NE, Rogich JJ, Garman SC (2011) The molecular basis of pharmacological chaperoning in human α-galactosidase. Chem Biol 18:1521–1526. https://doi.org/10.1016/j.chembiol.2011.10.012
Han R-Z, Xu G-C, Dong J-J, Ni Y (2016) Arginine deiminase: recent advances in discovery, crystal structure, and protein engineering for improved properties as an anti-tumor drug. Appl Microbiol Biotechnol 100:4747–4760. https://doi.org/10.1007/s00253-016-7490-z
Harms E, Wehner A, Aung HP, Röhm KH (1991) A catalytic role for threonine-12 of *E. coli* asparaginase II as established by site-directed mutagenesis. FEBS Lett 285:55–58
Hassan MI, Waheed A, Grubb JH, Klei HE, Korolev S, Sly WS (2013) High resolution crystal structure of human β-glucuronidase reveals structural basis of

lysosome targeting. PLoS One 8:e79687. https://doi.org/10.1371/journal.pone.0079687
Henningham A, Ericsson DJ, Langer K, Casey LW, Jovcevski B, Chhatwal GS, Aquilina JA, Batzloff MR, Kobe B, Walker MJ (2013) Structure-informed design of an enzymatically inactive vaccine component for group A Streptococcus. MBio 4:e00509-13–e00509-13. https://doi.org/10.1128/mBio.00509-13
Hill CH, Graham SC, Read RJ, Deane JE (2013) Structural snapshots illustrate the catalytic cycle of β-galactocerebrosidase, the defective enzyme in Krabbe disease. Proc Natl Acad Sci USA 110:20479–20484. https://doi.org/10.1073/pnas.1311990110
Hill CH, Viuff AH, Spratley SJ, Salamone S, Christensen SH, Read RJ, Moriarty NW, Jensen HH, Deane JE (2015) Azasugar inhibitors as pharmacological chaperones for Krabbe disease. Chem Sci 6:3075–3086. https://doi.org/10.1039/c5sc00754b
Hoogsteder PHJ, Kotz D, van Spiegel PI, Viechtbauer W, van Schayck OCP (2014) Efficacy of the nicotine vaccine 3′-AmNic-rEPA (NicVAX) co-administered with varenicline and counselling for smoking cessation: a randomized placebo-controlled trial. Addiction 109:1252–1259. https://doi.org/10.1111/add.12573
Jain S, Drendel WB, Chen Z-W, Mathews FS, Sly WS, Grubb JH (1996) Structure of human β-glucuronidase reveals candidate lysosomal targeting and active-site motifs. Nat Struct Mol Biol 3:375–381. https://doi.org/10.1038/nsb0496-375
Keyt BA, Paoni NF, Refino CJ, Berleau L, Nguyen H, Chow A, Lai J, Peña L, Pater C, Ogez J (1994) A faster-acting and more potent form of tissue plasminogen activator. Proc Natl Acad Sci USA 91:3670–3674
Khan FI, Shahbaaz M, Bisetty K, Waheed A, Sly WS, Ahmad F, Hassan MI (2016) Large scale analysis of the mutational landscape in β-glucuronidase: a major player of mucopolysaccharidosis type VII. Gene 576:36–44. https://doi.org/10.1016/j.gene.2015.09.062
Kishnani PS, Beckemeyer AA (2014) New therapeutic approaches for Pompe disease: enzyme replacement therapy and beyond. Pediatr Endocrinol Rev 12(Suppl 1):114–124
Kotzia GA, Labrou NE (2009) Engineering thermal stability of L-asparaginase by in vitro directed evolution. FEBS J 276:1750–1761. https://doi.org/10.1111/j.1742-4658.2009.06910.x
Kotzia GA, Lappa K, Labrou NE (2007) Tailoring structure-function properties of L-asparaginase: engineering resistance to trypsin cleavage. Biochem J 404:337–343. https://doi.org/10.1042/BJ20061708
Kurosawa Y, Nirengi S, Homma T, Esaki K, Ohta M, Clark JF, Hamaoka T (2015) A single-dose of oral nattokinase potentiates thrombolysis and anti-coagulation profiles. Sci Rep 5:11601. https://doi.org/10.1038/srep11601
Labrou NE, Papageorgiou AC, Avramis VI (2010) Structure-function relationships and clinical applications of L-asparaginases. Curr Med Chem 17:2183–2195
Lee WC, Kang D, Causevic E, Herdt AR, Eckman EA, Eckman CB (2010) Molecular characterization of mutations that cause globoid cell leukodystrophy and pharmacological rescue using small molecule chemical chaperones. J Neurosci 30:5489–5497. https://doi.org/10.1523/JNEUROSCI.6383-09.2010
Li L-Z, Xie T-H, Li H-J, Qing C, Zhang G-M, Sun M-S (2007) Enhancing the thermostability of *Escherichia colil*-asparaginase II by substitution with pro in predicted hydrogen-bonded turn structures. Enzym Microb Technol 41:523–527. https://doi.org/10.1016/j.enzmictec.2007.04.004
Lin Z, Jiang L, Yuan C, Jensen JK, Zhang X, Luo Z, Furie BC, Furie B, Andreasen PA, Huang M (2011) Structural basis for recognition of urokinase-type plasminogen activator by plasminogen activator inhibitor-1. J Biol Chem 286:7027–7032. https://doi.org/10.1074/jbc.M110.204537
Lourenço CM, Giugliani R (2014) Evaluation of galsulfase for the treatment of mucopolysaccharidosis VI (Maroteaux-Lamy syndrome). Expert Opin Orphan Drugs 2:407–417. https://doi.org/10.1517/21678707.2014.885833
Maita N, Tsukimura T, Taniguchi T, Saito S, Ohno K, Taniguchi H, Sakuraba H (2013) Human α-L-iduronidase uses its own N-glycan as a substrate-binding and catalytic module. Proc Natl Acad Sci USA 110:14628–14633. https://doi.org/10.1073/pnas.1306939110
Marr RA, Hafez DM (2014) Amyloid-beta and Alzheimer's disease: the role of neprilysin-2 in amyloid-beta clearance. Front Aging Neurosci 6:187. https://doi.org/10.3389/fnagi.2014.00187
Miners JS, Barua N, Kehoe PG, Gill S, Love S (2011) Aβ-degrading enzymes: potential for treatment of Alzheimer disease. J Neuropathol Exp Neurol 70:944–959. https://doi.org/10.1097/NEN.0b013e3182345e46
Moffitt MC, Louie GV, Bowman ME, Pence J, Noel JP, Moore BS (2007) Discovery of two cyanobacterial phenylalanine ammonia lyases: kinetic and structural characterization. Biochemistry 46:1004–1012. https://doi.org/10.1021/bi061774g
Mohamed FE, Al-Gazali L, Al-Jasmi F, Ali BR (2017) Pharmaceutical chaperones and proteostasis regulators in the therapy of lysosomal storage disorders: current perspective and future promises. Front Pharmacol 8:448. https://doi.org/10.3389/fphar.2017.00448
Montaño AM, Lock-Hock N, Steiner RD, Graham BH, Szlago M, Greenstein R, Pineda M, Gonzalez-Meneses A, Çoker M, Bartholomew D, Sands MS, Wang R, Giugliani R, Macaya A, Pastores G, Ketko AK, Ezgü F, Tanaka A, Arash L, Beck M, Falk RE, Bhattacharya K, Franco J, White KK, Mitchell GA, Cimbalistiene L, Holtz M, Sly WS (2016) Clinical course of sly syndrome (mucopolysaccharidosis type VII). J Med Genet 53:403–418. https://doi.org/10.1136/jmedgenet-2015-103322
Nalivaeva NN, Beckett C, Belyaev ND, Turner AJ (2012) Are amyloid-degrading enzymes viable therapeutic targets in Alzheimer's dis-

ease? J Neurochem 120:167–185. https://doi.org/10.1111/j.1471-4159.2011.07510.x

Naz H, Islam A, Waheed A, Sly WS, Ahmad F, Hassan I (2013) Human β-glucuronidase: structure, function, and application in enzyme replacement therapy. Rejuvenation Res 16:352–363. https://doi.org/10.1089/rej.2013.1407

Ni Y, Schwaneberg U, Sun Z-H (2008) Arginine deiminase, a potential anti-tumor drug. Cancer Lett 261:1–11. https://doi.org/10.1016/j.canlet.2007.11.038

Ortlund E, Lacount MW, Lewinski K, Lebioda L (2000) Reactions of pseudomonas 7A glutaminase-asparaginase with diazo analogues of glutamine and asparagine result in unexpected covalent inhibitions and suggests an unusual catalytic triad Thr-Tyr-Glu. Biochemistry 39:1199–1204

Osellame LD, Duchen MR (2013) Defective quality control mechanisms and accumulation of damaged mitochondria link Gaucher and Parkinson diseases. Autophagy 9:1633–1635. https://doi.org/10.4161/auto.25878

Palm G, Lubkowski J, Derst C, Schleper S (1996) A covalently bound catalytic intermediate in *Escherichia coli* asparaginase: crystal structure of a Thr-89-Val mutant. FEBS Lett 390:211–216

Pan R, Zhang Z-J, He R-Q (2010) Earthworm protease. Appl Environ Soil 2010:13. https://doi.org/10.1155/2010/294258

Papageorgiou AC, Posypanova GA, Andersson CS, Sokolov NN, Krasotkina J (2008) Structural and functional insights into Erwinia carotovora L-asparaginase. FEBS J 275:4306–4316. https://doi.org/10.1111/j.1742-4658.2008.06574.x

Parenti G, Fecarotta S, la Marca G, Rossi B, Ascione S, Donati MA, Morandi LO, Ravaglia S, Pichiecchio A, Ombrone D, Sacchini M, Pasanisi MB, De Filippi P, Danesino C, Casa Della R, Romano A, Mollica C, Rosa M, Agovino T, Nusco E, Porto C, Andria G (2014) A chaperone enhances blood α-glucosidase activity in Pompe disease patients treated with enzyme replacement therapy. Mol Ther 22:2004–2012. https://doi.org/10.1038/mt.2014.138

Parsiegla G, Noguere C, Santell L, Lazarus RA, Bourne Y (2012) The structure of human DNase I bound to magnesium and phosphate ions points to a catalytic mechanism common to members of the DNase I-like superfamily. Biochemistry 51:10250–10258. https://doi.org/10.1021/bi300873f

Pettersen EF, Goddard TD, Huang CC, Couch GS, Greenblatt DM, Meng EC, Ferrin TE (2004) UCSF Chimera – a visualization system for exploratory research and analysis. J Comput Chem 25:1605–1612. https://doi.org/10.1002/jcc.20084

Pol-Fachin L, Siebert M, Verli H, Saraiva-Pereira ML (2016) Glycosylation is crucial for a proper catalytic site organization in human glucocerebrosidase. Glycoconj J 33:237–244. https://doi.org/10.1007/s10719-016-9661-7

Pretzer D, Schulteis B, Vander Velde DG, Smith CD, Mitchell JW, Manning MC (1992) Effect of zinc binding on the structure and stability of fibrolase, a fibrinolytic protein from snake venom. Pharm Res 9:870–877. https://doi.org/10.1023/A:1015840613799

Ran C, Brodin L, Forsgren L, Westerlund M, Ramezani M, Gellhaar S, Xiang F, Fardell C, Nissbrandt H, Söderkvist P, Puschmann A, Ygland E, Olson L, Willows T, Johansson A, Sydow O, Wirdefeldt K, Galter D, Svenningsson P, Belin AC (2016) Strong association between glucocerebrosidase mutations and Parkinson's disease in Sweden. Neurobiol Aging 45:212.e5–212.e11. https://doi.org/10.1016/j.neurobiolaging.2016.04.022

Rivera-Colón Y, Schutsky EK, Kita AZ, Garman SC (2012) The structure of human GALNS reveals the molecular basis for mucopolysaccharidosis IV A. J Mol Biol 423:736–751. https://doi.org/10.1016/j.jmb.2012.08.020

Roig-Zamboni V, Cobucci-Ponzano B, Iacono R, Ferrara MC, Germany S, Bourne Y, Parenti G, Moracci M, Sulzenbacher G (2017) Structure of human lysosomal acid α-glucosidase-a guide for the treatment of Pompe disease. Nat Commun 8:1111. https://doi.org/10.1038/s41467-017-01263-3

Saito S, Ohno K, Maita N, Sakuraba H (2014) Structural and clinical implications of amino acid substitutions in α-L-iduronidase: insight into the basis of mucopolysaccharidosis type I. Mol Genet Metab 111:107–112. https://doi.org/10.1016/j.ymgme.2013.10.005

Semba CP, Sugimoto K, Razavi MK, Society of Cardiovascular and Interventional Radiology (SCVIR) (2001) Alteplase and tenecteplase: applications in the peripheral circulation. Tech Vasc Interv Radiol 4:99–106

Sharma SK, Bagshawe KD (2017) Antibody Directed Enzyme Prodrug Therapy (ADEPT): trials and tribulations. Adv Drug Deliv Rev 118:2–7. https://doi.org/10.1016/j.addr.2017.09.009

Sidhu NS, Schreiber K, Propper K, Becker S, Uson I, Sheldrick GM, Gartner J, Kratzner R, Steinfeld R (2014) Structure of sulfamidase provides insight into the molecular pathology of mucopolysaccharidosis IIIA. Acta Cryst D70:1321–1335. 1–15. https://doi.org/10.1107/S1399004714002739

Spratley SJ, Deane JE (2016) New therapeutic approaches for Krabbe disease: the potential of pharmacological chaperones. J Neurosci Res 94:1203–1219. https://doi.org/10.1002/jnr.23762

Tang Y, Liang D, Jiang T, Zhang J, Gui L, Chang W (2002) Crystal structure of earthworm fibrinolytic enzyme component a: revealing the structural determinants of its dual fibrinolytic activity. J Mol Biol 321:57–68

Tararina MA, Janda KD, Allen KN (2016) Structural analysis provides mechanistic insight into nicotine oxidoreductase from Pseudomonas putida. Biochemistry 55:6595–6598. https://doi.org/10.1021/acs.biochem.6b00963

Thomas J, Levy H, Amato S, Vockley J, Zori R, Dimmock D, Harding CO, Bilder DA, Weng HH, Olbertz J, Merilainen M, Jiang J, Larimore K, Gupta S, Gu Z,

Northrup H (2018) Pegvaliase for the treatment of phenylketonuria: results of a long-term phase 3 clinical trial program (PRISM). Mol Genet Metab 124:27–38. https://doi.org/10.1016/j.ymgme.2018.03.006

Tomatsu S, Montaño AM, Dung VC, Ohashi A, Oikawa H, Oguma T, Orii T, Barrera L, Sly WS (2010) Enhancement of drug delivery: enzyme-replacement therapy for murine Morquio A syndrome. Mol Ther 18:1094–1102. https://doi.org/10.1038/mt.2010.32

van den Berg H (2011) Asparaginase revisited. Leuk Lymphoma 52:168–178. https://doi.org/10.3109/10428194.2010.537796

Van Hove JL, Yang HW, Wu JY, Brady RO, Chen YT (1996) High-level production of recombinant human lysosomal acid alpha-glucosidase in Chinese hamster ovary cells which targets to heart muscle and corrects glycogen accumulation in fibroblasts from patients with Pompe disease. Proc Natl Acad Sci USA 93:65–70

Viuff AH, Jensen HH (2016) Synthesis and evaluation of N-alkylated analogues of aza-galacto-fagomine – potential pharmacological chaperones for Krabbe disease. Org Biomol Chem 14:8545–8556. https://doi.org/10.1039/c6ob01309k

Wang F, Wang C, Li M, Zhang JP, Gui LL, An X-M, Chang WR (2005a) Crystal structure of earthworm fibrinolytic enzyme component B: a novel, glycosylated two-chained trypsin. J Mol Biol 348:671–685. https://doi.org/10.1016/j.jmb.2005.02.055

Wang L, Gamez A, Sarkissian CN, Straub M, Patch MG, Han GW, Striepeke S, Fitzpatrick P, Scriver CR, Stevens RC (2005b) Structure-based chemical modification strategy for enzyme replacement treatment of phenylketonuria. Mol Genet Metab 86:134–140. https://doi.org/10.1016/j.ymgme.2005.05.012

Wang L, Gamez A, Archer H, Abola EE, Sarkissian CN, Fitzpatrick P, Wendt D, Zhang Y, Vellard M, Bliesath J, Bell SM, Lemontt JF, Scriver CR, Stevens RC (2008) Structural and biochemical characterization of the therapeutic *Anabaena variabilis* phenylalanine ammonia lyase. J Mol Biol 380:623–635. https://doi.org/10.1016/j.jmb.2008.05.025

Wang KY, Tull L, Cooper E, Wang N, Liu D (2013) Recombinant protein production of earthworm lumbrokinase for potential antithrombotic application. Evid Based Complement Alternat Med 2013:783971–783978. https://doi.org/10.1155/2013/783971

Webster CI, Burrell M, Olsson L-L, Fowler SB, Digby S, Sandercock A, Snijder A, Tebbe J, Haupts U, Grudzinska J, Jermutus L, Andersson C (2014) Engineering neprilysin activity and specificity to create a novel therapeutic for Alzheimer's disease. PLoS One 9:e104001. https://doi.org/10.1371/journal.pone.0104001

Xue S, Kallupi M, Zhou B, Smith LC, Miranda PO, George O, Janda KD (2018) An enzymatic advance in nicotine cessation therapy. Chem Commun (Camb) 30:217. https://doi.org/10.1039/c7cc09134f

Yanagisawa Y, Chatake T, Chiba-Kamoshida K, Naito S, Ohsugi T, Sumi H, Yasuda I, Morimoto Y (2010) Purification, crystallization and preliminary X-ray diffraction experiment of nattokinase from Bacillus subtilis natto. Acta Crystallogr Sect F Struct Biol Cryst Commun 66:1670–1673. https://doi.org/10.1107/S1744309110043137

Application of Fluorescence in Studying Therapeutic Enzymes

5

Zhaoshuai Wang, Caihong Li, and Yinan Wei

Abstract

Fluorescence spectroscopy is one of the most important techniques in the study of therapeutic enzymes. The fluorescence phenomenon has been discovered and exploited for centuries, while therapeutic enzymes have been used in treatment of disease for only decades. This chapter provides a brief summary of the current applications of fluorescence methods in studying therapeutic enzymes to provide some insights on the selection of proper method tailored to the goal. First a brief introduction about therapeutic enzymes and history of fluorescence were provided, followed by discussions on how fluorescence was applied in the studies. Four popular fluorescence methods are discussed: fluorescence tracing, fluorescence resonance energy transfer (FRET), fluorescence quenching and fluorescence polarization. Selected application of the fluorescence methods in studying therapeutic enzymes are listed, and discussed in details in the following paragraphs.

Keywords

Therapeutic enzyme · Fluorescence tracing · FRET · Fluorescence quenching · Fluorescence polarization

Abbreviations

FRET	Fluorescence resonance energy transfer
MS	Mass spectroscopy
CD	Circular dichroism
FTIR	Fourier-transform infrared spectroscopy
US-FDA	Food and Drug Administration in USA
SCID	Severe combined immunodeficiency disease
mAbs	Monoclonal antibodies
IgG1	Immunoglobulin G1
GFP	Green Fluorescent Protein
GBA	Glucocerebrosidase
ABP	Activity-based probes
MCL	Mantle cell lymphoma
MESK	Microencapsulated streptokinase
FREE SK	Unencapsulated streptokinase

Z. Wang (✉)
Department of Pharmaceutical Science, University of Kentucky, Lexington, KY, USA
e-mail: zhaoshuai.wang@uky.edu

C. Li
Department of Educational, School, and Counseling Psychology, University of Kentucky, Lexington, KY, USA

Y. Wei
Department of Chemistry, University of Kentucky, Lexington, KY, USA

N. Labrou (ed.), *Therapeutic Enzymes: Function and Clinical Implications*, Advances in Experimental Medicine and Biology 1148, https://doi.org/10.1007/978-981-13-7709-9_5

CFP	Cyan fluorescent protein
YFP	Yellow fluorescent protein
Abl	Ableson
Bcr	Break point cluster
ECFP	Enhanced cyan fluorescent protein
HPMA	N-(2-hydroxypropyl) methacrylamide
ADA	Adenosine deaminase
FAM	6-carboxyfluo rescein
MWCNTs	Multi-walled carbon nanotubes
IgE	Immunoglobulin E
FcεRI	The IgE Fc receptor
IFN-γ	Interferon gamma

Therapeutic enzymes have been used as diagnostic probes and agents in the treatment of diseases for several decades. Comparing to small-molecule medicine, therapeutic enzymes have much higher molecular weight and more complicated structure which lead to their highly specific function. Therapeutic enzymes have several advantages such as superb specificity, well tolerated by the body, effective treatment, and versatility to modification and development (Leader et al. 2008). Many techniques help speed up the development of therapeutic enzymes, including mass spectroscopy (MS), circular dichroism (CD) spectroscopy, fluorescence spectroscopy, Fourier-transform infrared spectroscopy (FTIR), differential scanning calorimetry, etc. (Tatford et al. 2004; Wei et al. 2014). Herein, we'll discuss how fluorescence spectroscopy is applied in studying therapeutic enzymes, and mainly focus on four common fluorescence methods including fluorescence tracing, FRET, fluorescence quenching and fluorescence polarization.

5.1 A Brief Introduction of Therapeutic Enzymes

Therapeutic enzymes have been widely used in medicine as diagnostic probes and therapeutic agents for decades (Manchester 1995; Li et al. 2012; Goldberg 1992). The first therapeutic enzyme approved by Food and Drug Administration in USA (US-FDA) is human insulin, which is used to treat diabetes mellitus, diabetic ketoacidosis, and hyperkalaemia (Leader et al. 2008). A new era of enzymes begins following the approval of Adagen® (also known as pegademase bovine, product of ENZON Pharmaceuticals, Inc.) to treat severe combined immunodeficiency disease (SCID) (Vellard 2003). Until 2015, US-FDA has approved over 200 protein and peptide therapeutics to treat genetic-related (e.g., Fabry disease, Gaucher disease, Glycogen storage disease Type II) and non-genetic related disease (e.g., cancer, chronic inframammary diseases) (Carter and Lazar 2018; Goldberg 1992; Kinch 2015; Usmani et al. 2017).

Therapeutic enzymes can be classified into enzymes with regulatory activity and those with special targeting activities (Leader et al. 2008). With regulatory activity, a therapeutic enzyme replaces a protein that is deficient or abnormal, augments an existing pathway, or provides a novel function or activity (Leader et al. 2008). For example, by augmenting an existing pathway, insulin, the first protein medicine, could treat patients with Diabetes Mellitus (American Diabetes 2010). In addition to regulatory enzymes, enzymes with special targeting activities compose an important type of therapeutic enzymes. The most representative group of therapeutic enzymes with special targeting activities is monoclonal antibodies (mAbs), which work through ligand blockade, receptor blockade, receptor downregulation, depletion, and signaling induction (Chan and Carter 2010). The discovery of mAbs expanded the application of protein medicines to treatment of cancers or autoimmune diseases. For example, rituximab, formatted as chimeric Immunoglobulin G1 (IgG1), has been approved for treatment of non-hodgkin lymphoma, rheumatoid arthritis, etc. (McLaughlin et al. 1998). And ipilimumab, another drug formatted as human transgenic mouse-derived IgG1, has been used to treat melanoma (Hodi et al. 2010).

5.2 A Brief History About Fluorescence Spectroscopy

Fluorescence describes the phenomenon that a substance absorbs light with a certain wavelength and then emits photons at a different wavelength. The history of fluorescence starts in 1565, when the first observation of fluorescence emission was reported (Valeur and Berberan-Santos 2011). It was discovered later that fluorescence could also be observed in crystals of fluorite in 1819. Measurement of the fluorescence was not possible until 1940s when an instrument called the "Coleman filter fluorescence" was invented to determine the level of drugs in the plasma using fluorescence. In 1956, Dr. Robert Bowman invented the first fluorescence spectrophotometer, the tool that can vary the wavelength of exciting light and measure the intensity and wavelength of the emitted light. Nineteen sixty-two is a big milestone year for the development of fluorescence technology due to the discovery of the now widely used Green Fluorescent Protein (GFP), which has stimulated continuous efforts of engineering fluorescent proteins with different excitation and emission wavelengths. Nowadays fluorescence spectroscopy as a research technique has been extensively applied in many research fields including medical microbiology, genomics, proteomics, bioengineering, etc. (Shahzad et al. 2009).

5.3 Application of Fluorescence Spectroscopy in Studying Therapeutic Enzymes

As aforementioned, the main principle of fluorescence is that a substance absorbs light with a certain wavelength and then emits at a different wavelength. To explain the phenomena from a molecule level, the electrons of a fluorophore absorb energy from a radiation source, jump to a higher and unstable singlet state, then emit photons, and go back to their ground state. Based on the main principle, several fluorescence methods such as fluorescence tracing, FRET, fluorescence quenching, and fluorescence polarization have been developed.

Depending on the experimental objective, proper fluorescence method can be chosen. To study kinetic properties of a protein or a dynamic process in vivo, FRET is a good choice. To study protein-ligand binding, fluorescence polarization could be a good choice. In this review, the application of fluorescence in therapeutic enzymes will be discussed according to the fluorescence methods used. Selected examples are listed in the Table 5.1, and they will be discussed in details as follows.

5.3.1 Fluorescence Tracing

Fluorescence tracing is the labeling and monitoring of a target protein using a fluorescent probe, which is one of the most widely used fluorescence method in research. Fluorescent probe can be classified into two groups according to their size: fluorescent protein tag and fluorescence dye. First, fluorescent protein tag, discovered in 1960s, is extensively employed in biological chemistry research. Some examples of fluorescent proteins, mainly GFP and its variants, are shown in the Fig. 5.1. Second, as shown in Fig. 5.2, fluorescence dyes are smaller chemical compounds. Beside difference in size, the attachment methods to introduce them to the target proteins are different. Fluorescent proteins are attached to target proteins mainly through cloning, while the dyes are attached to target proteins by either covalent bonding or molecular interactions (Piston and Kremers 2007; Tyagarajan et al. 2003). Through monitoring the tracing fluorescent probe, researchers could track their target protein, even in live cells using a fluorescence microscope. Dr. Zhuang's group (Jones et al. 2011) reported 2D and 3D super-resolution imaging of lives cells using photo-switchable dyes such as cyanine dye, Alexa Fluor 647 and Atto655, and fluorescent proteins EOS and its derivatives mEOS2 and tdEOS.

Table 5.1 Summary for application of fluorescence techniques in studies of therapeutic enzymes

Fluorescence method	Fluorescence molecules used	Therapeutic enzyme studied	Medical use	References
Fluorescence tracing	Green ABP 5 and red ABP6	Glucocerebrosidase	Gaucher disease (GD)	Kallemeijn et al. (2017)
Fluorescence tracing	7-amino-4-methylcoumarin	L-asparaginase	Childhood acute lymphoblastic leukaemia (ALL)	Nath Christa et al. (2008)
Fluorescence tracing	N-hydroxysuccinimide ester rhodamine (piece) and Alexa Fluor 488 5-sulfodichlorophenol ester	Milatuzumab and rituximab	Mantle cell lymphoma	Alinari et al. (2011)
Fluorescence tracing	Alexa Fluor 594 and Alexa Fluor 488	Plasminogen activator	Acute myocardial infarction	Leach et al. (2004)
Fluorescence tracing	5-(acetamido)-fluorescein	Streptokinase	Myocardial infarction	Verhamme and Bock (2014)
Fluorescence tracing	Acridine orange/ethidium bromide	Rituximab	Refractory follicular CD20 + non-Hodgkin's lymphoma (NHL)	Li et al. (2011)
FRET	CFP and YFP	Imatinib	Chronic myelogenous leukemia (CML)	Tunceroglu et al. (2010)
FRET	Cy5 and Cy7	Doxorubicin-polymer conjugate	Cancer	Zhang et al. (2017)
Fluorescence quenching	(HiLyte Fluor647)–LPYPQPK(QXL670)	Prolyl endopeptidase	Gastrointestinal tract diseases	Schulz et al. (2016)
Fluorescence polarization	mCherry	Insulin	Diabetes	Yi et al. (2015)
FRET	PyMPO maleimide and fluorescein-5-maleimide	Factor VII	Haemophilia A	Wakabayashi and Fay (2013)
Fluorescence quenching	Synthesized substrates	β-Glucocerebrosidase	Gaucher's disease	Yadav et al. (2015)
Fluorescence polarization	Pyrenesulfonyl chloride	Interferon gamma	Chronic granulomatous disease and severe osteopetrosis	Falach et al. (1997)
Fluorescence quenching	A carbon nanotube-based fluorescent aptasensor	Adenosine deaminase	Severe immunodeficiency disease	Hu et al. (2014)
Fluorescence quenching	Alexa Fluor 488	Omalizumab	Severe allergic asthma	Kim et al. (2012)
Fluorescence polarization	Alexa Fluor 488 hydrazide	Hyaluronidase	Increase the absorption of injected drugs	Murai and Kawashima (2008)
Fluorescence tracing	Oregon Green 488	Imiglucerase	Gaucher disease	Piepenhagen et al. (2010)
Fluorescence tracing	Alexa Fluor 546	Alglucerase	Gaucher disease	Piepenhagen et al. (2010)

Applying fluorescence tracing, researchers studied therapeutic enzymes in different aspects. The first example is about the "Gaucher's Disease", in which glucocerebroside accumulates in cells due to the gene deficiency of Glucocerebrosidase (GBA). To date, there are

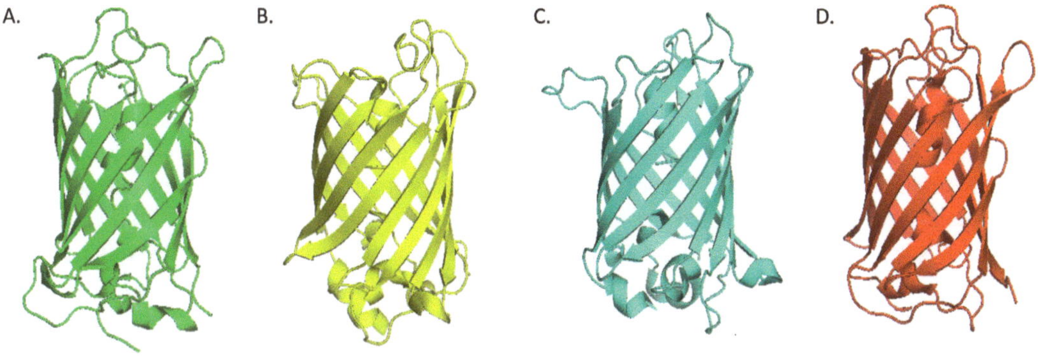

Fig. 5.1 GFP and its variants. (**a**) GFP. Excitation 488 nm, emission 507 nm. Protein Data Bank (PDB) ID 1GFL. (**b**) YFP. Excitation 514 nm, Emission 527 nm. The mutations relative to GFP: S65G, V68L, Q69K, S72A, T203Y, A206K. PDB ID: 1MYW. (**c**) CFP. Excitation 433 nm Emission 475 nm. The mutations relative to GFP: F64L, S65T, Y66W, N149I, M153T, V163A, A206K. PDB ID: 2WSN. (**d**) mCherry. Excitation 587 nm. Emission 610 nm. The mutations relative to GFP (~41 mutations): N6D, R17H, K194N, T195V, D196N, etc. PDB ID: 2H5Q (Shaner et al. 2004, 2005)

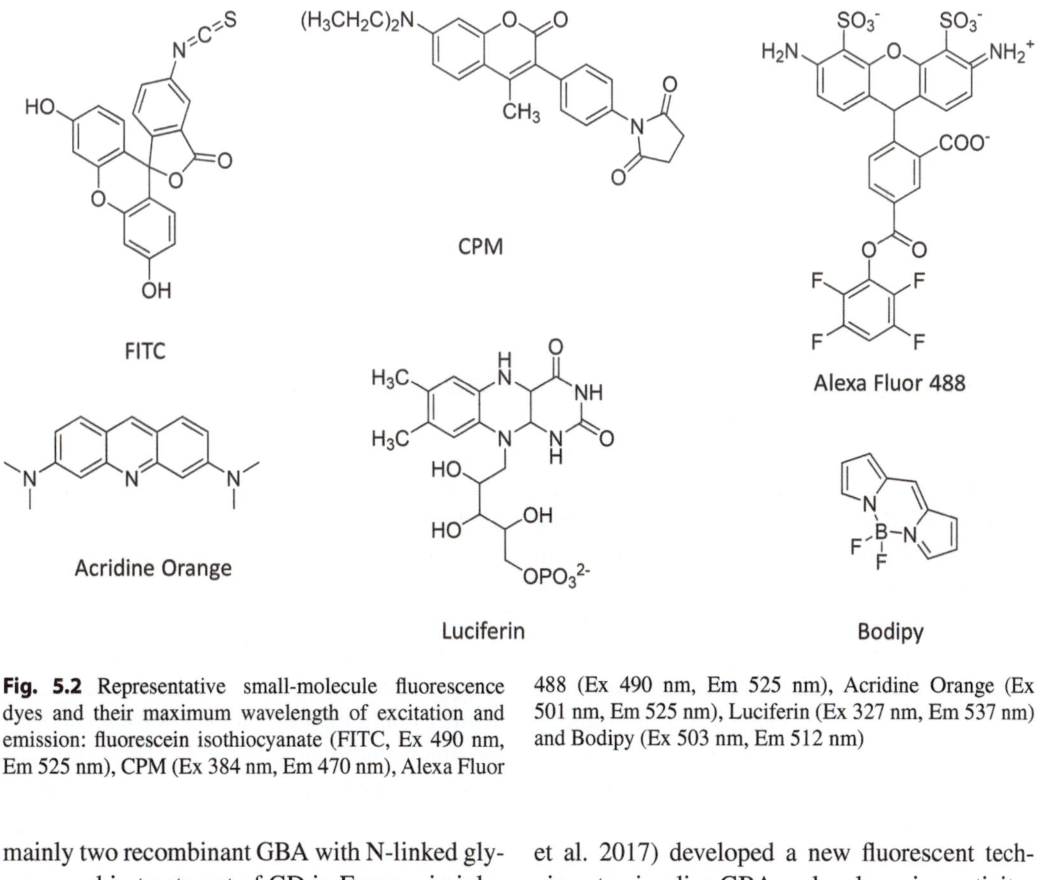

Fig. 5.2 Representative small-molecule fluorescence dyes and their maximum wavelength of excitation and emission: fluorescein isothiocyanate (FITC, Ex 490 nm, Em 525 nm), CPM (Ex 384 nm, Em 470 nm), Alexa Fluor 488 (Ex 490 nm, Em 525 nm), Acridine Orange (Ex 501 nm, Em 525 nm), Luciferin (Ex 327 nm, Em 537 nm) and Bodipy (Ex 503 nm, Em 512 nm)

mainly two recombinant GBA with N-linked glycans used in treatment of GD in Europe: imiglucerase (Cerezyme®, Genzyme Corporation) and velaglucerase alpha (VPRIV®, Shire PLC). Wouter W. Kallemeijn and colleagues (Kallemeijn et al. 2017) developed a new fluorescent technique to visualize GBA molecules using activity-based probes (ABP) green ABP5 and red ABP6. They comparatively studied the binding and uptake of ABP-labelled imiglucerase and velaglucerase in different cells including dendritic cells, cultured human macrophages, and live mice.(Kallemeijn et al. 2017) Besides imiglucerase and velaglucerase alpha, alglucerase (Ceredase®, Genzyme Corporation) can also be used in treatment of Gaucher's Disease. Peter A. Piepenhagen et al. investigated the biodistribution of two therapeutic enzymes Cerezyme® and Ceredase® using direct fluorescence labeling and confocal microscopy. In their study, the fluorescence dyes Oregon Green 488 and Alexa Fluor 546 were used to label the two therapeutic enzymes, which were administrated into mice through tail veins for animal study. Both therapeutic enzymes were detected in intact cells and organs, and the relative amounts of the enzymes were quantified.

The discovery and clinically use of antibody drugs is a big step forward in modern medicine (Carter and Lazar 2018). Using two different fluorescence dye N-hydroxysuccinimide ester rhodamine (Piece) and Alexa Fluor 488 5-sulfodichlorophenol ester (Invitrogen), the combination of two antibody drugs anti-CD74 (milatuzumab, Immunomedics) and anti-CD20 (rituximab, Genentech) in the treatment mantle cell lymphoma (MCL) were investigated. With the help of fluorescent tracing, the death of all MCL cells treated by the two antibody drugs was easily observed using a confocal microscope (Alinari et al. 2011). Rituximab was also investigated by Li et al. (2011), who studied the mechanism of apoptosis of B-lymphoma cells induced by rituximab using acridine orange/ethidium bromide double fluorescent staining. In their study, apoptosis of B-lymphoma cells could be clearly identified using dual fluorescence merge images obtained on a fluorescence microscope. In addition to studying the effect or mechanism of antibody drugs, some fluorescence-based assays useful in disease diagnosis have also been developed. Szittner and colleagues (2013) developed a fluorescence-based method in which antigen microarray was applied to identify serum antibodies. Then the antigen-antibody complexes were identified by the secondary antibody IgG labeled by a fluorescence dye FITC. This well-developed system could be applied in autoimmune disease diagnostics.

Streptokinase, which converts plasminogen to plasmin, is used in the treatment of several diseases such as acute evolving transmural myocardial infarction, pulmonary embolism, and deep vein thrombosis (Leader et al. 2008). Leach et al. (2004) explored the mechanism by which microencapsulated streptokinase (MESK) accelerates clot digestion compared to unencapsulated streptokinase (FREE SK) and reduces reperfusion times by as much as an order of magnitude in vivo. Alexa Fluor 594 and Alexa Fluor 488 were used to label fibrinogen and streptokinase, respectively. Images taken using Olympus Fluoview FV-500 Confocal Laser Scanning Biological Microscope showed that MESK had faster and higher penetration rate than FREE SK. Verhamme et al. (Verhamme and Bock 2014) studied the interaction between streptokinase and zymogen plasminogen or plasmin. They labelled the active sites of plasminogen or plasmin with 5-(iodoacetamido) fluorescein. Through stop-flow fluorescence spectrometry studies, they proposed a mechanism in which the streptokinase-plasminogen complex formed in a three-step process.

5.3.2 FRET

FRET is a powerful tool to study kinetic properties of proteins. The main principle is that when a "donor" fluorophore is close to a "acceptor" fluorophore, energy transfer can occur through non-radiative dipole-dipole coupling (Pietraszewska-Bogiel and Gadella 2011). In this process, the acceptor which absorbs the energy from the donor is excited and subsequently emits light at a longer wavelength than the donor fluorophore when it relaxes from the excited state to the ground state (Helms 2008). According to this principle, the intensity of FRET signal depends on the distance between the donor and accepter fluorophores, so that FRET is widely applied in studies involving conformational change of the protein or protein-protein interactions. For instance, using FRET between fluorescent protein pairs cyan fluorescent protein (CFP) and yellow fluorescent protein (YFP), Miyawaki et al (1997) investigated the interaction between two proteins (calmodulin and M13) in response to an increasing concentration of Ca^{2+}. Kurokawa et al. (2001) used this method to investigate the conformational change of phosphorylated adaptor protein, CrkII, in vivo and found epidermal growth factor-induced phosphorylation of CrkII was initiated at the peripheral plasma membrane.

Chronic myelogenous leukemia, a disease occurred in Hematopoietic stem cells, is mainly caused by the fusion between the Ableson (Abl) tyrosine kinase gene and the break point cluster (Bcr) gene. An active protein named Bcr-Abl, which is expressed from the fused gene, is the main target for medication such as imatinib (Ciftci et al. 2018; Nowell 2007). Tunceroglu et al. (2010) developed one rapid and sensitive FRET-based method using CFP and YFP as a FRET pair to determine the activity of Bcr-Abl and its inhibition by drug candidates. They fused YFP and CFP before and after the SH2-SH3 domain, respectively. FRET occurred if Bcr-Abl is active, while there is no FRET if imatinib or other effective drug candidate is present to inhibit the activity of Bcr-Abl. Therefore, the sensitivity to conformational change enabled the application of the FRET method in the screening of therapeutic enzymes. A sandwich construct in which CrkL, a major substrate of Bcr-Abl kinase, placed between Venus (a variant of YFP) and enhanced cyan fluorescent protein (ECFP) has been created as a FRET bio-sensor. The Venus-CrkL-ECFP construct displayed a high FRET signal while it is active.

Enzyme-responsive backbone degradable carriers, which can release drugs in tissue or cells, is widely used in the treatment of cancer (Yang et al. 2011). Zhang et al. (2017) investigated the enzyme responsive performance of N-(2-hydroxypropyl) methacrylamide (HPMA) copolymer-based drug delivery system in vivo and in vitro using FRET in which Cy5 and Cy7 were applied as donor and acceptor, respectively. Cy5 was attached to the HPMA backbone while Cy7 was attached to the termini of a peptide side chains. When the enzyme or other drug candidate cleaves the peptide, FRET signal is switched off. A notable FRET change was observed in an in vitro model in which Human ovarian cancer cells (A2780 and OVCAR-3) and mouse embryo fibroblast cells NIH/3T3 were used, and in in vivo model in which mice bearing the A2780 human ovarian tumor were used. This system is good for evaluating drug candidate including protein drug candidate at the levels of cells, tissue, and even in animal models.

5.3.3 Fluorescence Quenching

The third popular fluorescence method is fluorescence quenching, which is widely used to evaluate protein binding or folding. Fluorescence quenching could be observed either by taking advantage of the intrinsic fluorescent residues such as tryptophan, or introduced fluorescent probes. The intrinsic fluorescence quenching has been used to study the properties of proteins, such as structure or stability (Ye et al. 2014). An example is a study involving variants of calcium-regulatory protein calmodulin. Yuan et al. (1998) determined the orientation of the myosin light chain kinase and calmodulin kinase I bound to calmodulin. The following examples involves the quenching of the fluorescence dye, which is used more frequently in studies on therapeutic enzymes.

MX, the prolyl endopeptidase variant, is a potential therapy for celiac diseases and other gluten related disorders. Schulz and colleagues (2016) explored method to stabilize MX without comprising their activity by conjugating polymers at different locations. Then they measured the activity of conjugated polymer-MX complex in vitro and in vivo. To determine the activity of MX-polymer in rats, a fluorophore/quencher pair peptide (HiLyte Fluor647)–LPYPQPK(QXL670) was administered in to the gastrointestinal tract, followed by injection of the MX-polymer conjugates. Fluorescence signal cannot be observed until the peptide was cleaved by the conjugates which indicated the MX-polymer was active. They found that several PEG polymers performed well in stabilizing MX while maintaining its activity, which is a promising modification that might be useful in clinical applications.

Adenosine deaminase (ADA), which metabolizes adenosine to prevent the accumulation of adenosine due to the deficiency of the intrinsic ADA gene in patients, is used in treatment of severe combined immunodeficiency disease. Hu et al. (2014) developed a fluorescence-quenching based method to detect the activity of ADA using a fluorescence dye 6-carboxyfluo rescein (FAM). Multi-walled carbon nanotubes (MWCNTs) and two FAM-labeled anti-adenosine oligonucleotides were also involved in the method. Adenosine binds with FAM labeled oligonucleotides in the absence of ADA, forming a complex that cannot bind to MWCNTs, which lead to a high background fluorescence. Adenosine was converted into inosine when ADA was introduced into the system so that the oligonucleotides can be adsorbed onto MWCNTs, resulting in fluorescence quenching of FAM dye.

The binding of immunoglobulin E (IgE) with the IgE Fc receptor (FcεRI) is found on mast cells and basophils, triggering allergic response. Some medication like omalizumab has been developed to inhibit the binding to treat severe allergic asthma. However, the dissociation of the binding of IgE- FcεRI is extremely slow, even though the inhibitors are administrated. Herein, Kim et al. (2012) developed a protein inhibitor of IgE- FcεRI, FcDARPin E2_79, which acts through a non-classical inhibition mechanism to actively stimulating the dissociation of the ligand-receptor complex. In their study, a double mutant of IgE-Fc_{3-4} was labeled with Alexa Fluor 488 at residue 367, which is close to the binding site. Fluorescence quenching of Alexa Fluor488 labeled protein AF488-Fc was observed when inhibitors such as FcεRI or omalizumab (anti-IgE therapeutic enzyme) was present. The engineered protein DARPin E2_79 did not lead to quenching, which suggested a different inhibition mechanism by the protein. The new finding indicated that the complex IgE-FcεRI can be dissociated with a different mechanism from omalizumab, which could be used to block the allergic response and provide new solutions for treating allergy and asthma.

5.3.4 Fluorescence Polarization

Fluorescence polarization is a convenient and powerful technique to study protein-protein and protein-ligand interactions. The main principle is that when a fluorescent molecule is excited with plane-polarized light, the emitted light is depolarized due to the rapid movement of the fluorophores. When the fluorophore binds with a large molecule, its movement is slowed so that the emitted light is less depolarized. Fluorescence polarization has been applied to study molecular interactions such as DNA-protein interactions, protein-protein interactions and protein-ligand interactions. This method has been applied to study the dynamic properties of proteins and how the enzymes catalyze reactions. In addition, it has also been used to characterize how proteins interact with each other (Tompa et al. 1987).

Human insulin, the first recombinant therapeutic enzyme, has been used in the treatment of diabetes mellitus for more than 30 years (Leader et al. 2008). Some biosensor based on fluorescence polarization has been developed to track insulin dynamics in live cells and secretion in pancreatic β-cells. The fluorescent protein mCherry was attached to insulin, whose fluorescence polarization can reflect whether insulin was packaged into the granules (Yi et al. 2015). If insulin-Cherry was free and out of the granules, it showed high fluorescence polarization; while the polarization signal was decreased due to homo-FRET between molecules if many insulin-Cherry complexes were accumulated in the granules. The fluorescent biosensor provided a new way to assess protein-protein interaction or protein aggregation in live cells.

Interferon gamma (IFN-γ), which increases inflammatory and antimicrobial response, has been used to treat chronic granulomatous disease and severe osteopetrosis (Leader et al. 2008). Falach et al. (1997) developed a fluorescence polarization based method to study the initial interaction between IFN- γ and human amnion WISH cells. The enzyme was labeled using the fluorescence dye pyrenesulfonyl chloride. Fluorescence polarization result showed that the signal increased fourfold when IFN- γ bound to the human amnion WISH cells compared to the free IFN- γ. The fluorescence labeled enzyme provided a novel tool for studying the mechanism of interaction between the antiproliferative lymphokine and cancer cell receptors.

Hyaluronidase is routinely used as an adjuvant to increase the absorption of injected drugs by catalyzing the hydrolysis of hyaluronic acid to increase the tissue permeability (Leader et al. 2008). Murai et al. (Murai and Kawashima 2008) established a simple assay to measure degradation of hyaluronan by hyaluronidase based on fluorescence polarization. In the assay, hyaluronan was fluorinated with Alexa Fluor 488 hydrazide which can be detected by a fluorescence spectrometer. Fluorescence polarization measured at 519 nm emission decreased if hydaluronan was degraded by haluronidase, and the result was also confirmed by agarose gel electrophoresis. The developed method could provide a convenient way to detect the activity of cellular hyaluronidase and the binding with its substrates.

5.4 Conclusion

In this chapter, the history of therapeutic enzymes, the history of fluorescence, and the principles of the four fluorescence methods were reviewed. In addition, selected examples about how these fluorescence methods were applied in the studies provide some insight on choosing proper fluorescence method depending on the research target. However, there are not many studies of therapeutic enzymes in which fluorescence methods are employed, when we do the literature search. Since the development of therapeutic enzymes is faster than before, more efforts may be necessary for exploring how to apply fluorescence to facilitate those studies.

References

Alinari L, Yu B, Christian BA et al (2011) Combination anti-CD74 (milatuzumab) and anti-CD20 (rituximab) monoclonal antibody therapy has in vitro and in vivo activity in mantle cell lymphoma. Blood 117(17):4530–4541

American Diabetes A (2010) Diagnosis and classification of diabetes mellitus. Diabetes Care 33(Suppl 1):S62–S69

Carter PJ, Lazar GA (2018) Next generation antibody drugs: pursuit of the 'high-hanging fruit'. Nat Rev Drug Discov 17(3):197–223

Chan AC, Carter PJ (2010) Therapeutic antibodies for autoimmunity and inflammation. Nat Rev Immunol 10:301

Ciftci HI, Ozturk SE, Ali TFS et al (2018) The first pentacyclic triterpenoid gypsogenin derivative exhibiting anti-ABL1 kinase and anti-chronic myelogenous leukemia activities. Biol Pharm Bull 41(4):570–574

Falach A, Nathan I, Baram S et al (1997) Interaction of a novel fluorescent analog of interferon-γ with transformed cells. Bioconjug Chem 8(4):459–465

Goldberg DM (1992) Enzymes as agents for the treatment of disease. Clin Chim Acta 206(1):45–76

Helms V (2008) Principles of computational cell biology: from protein complexes to cellular networks. Wiley, Weinheim, pp 39–59

Hodi FS, O'Day SJ, McDermott DF et al (2010) Improved survival with ipilimumab in patients with metastatic melanoma. N Engl J Med 363:711–723

Hu K, Huang Y, Wang S et al (2014) A carbon nanotubes based fluorescent aptasensor for highly sensitive detection of adenosine deaminase activity and inhibitor screening in natural extracts. J Pharm Biomed Anal 95:164–168

Jones SA, Shim SH, He J et al (2011) Fast, three-dimensional super-resolution imaging of live cells. Nat Methods 8(6):499–508

Kallemeijn WW, Scheij S, Hoogendoorn S et al (2017) Investigations on therapeutic glucocerebrosidases through paired detection with fluorescent activity-based probes. PLoS One 12(2):e0170268

Kim B, Eggel A, Tarchevskaya SS et al (2012) Accelerated disassembly of IgE-receptor complexes by a disruptive macromolecular inhibitor. Nature 491:613

Kinch MS (2015) An overview of FDA-approved biologics medicines. Drug Discov Today 20(4):393–398

Kurokawa K, Mochizuki N, Ohba Y et al (2001) A pair of fluorescent resonance energy transfer-based probes for tyrosine phosphorylation of the CrkII adaptor protein in vivo. J Biol Chem 276(33):31305–31310

Leach JK, Patterson E, O'Rear EA (2004) Distributed intraclot thrombolysis: mechanism of accelerated thrombolysis with encapsulated plasminogen activators. J Thromb Haemost 2(9):1548–1555

Leader B, Baca QJ, Golan DE (2008) Protein therapeutics: a summary and pharmacological classification. Nat Rev Drug Discov 7:21

Li M, Liu L, Xi N et al (2011) Imaging and measuring the rituximab-induced changes of mechanical properties in B-lymphoma cells using atomic force microscopy. Biochem Biophys Res Commun 404(2):689–694

Li S, Yang X, Yang S et al (2012) Technology prospecting on enzymes: application, marketing and engineering. Comput Struct Biotechnol J 2:e201209017

Manchester KL (1995) Louis Pasteur (1822–1895) – chance and the prepared mind. Trends Biotechnol 13(12):511–515

McLaughlin P, Grillo-López AJ, Link BK et al (1998) Rituximab chimeric anti-CD20 monoclonal antibody therapy for relapsed indolent lymphoma: half of patients respond to a four-dose treatment program. J Clin Oncol 16(8):2825–2833

Miyawaki A, Llopis J, Heim R et al (1997) Fluorescent indicators for Ca^{2+}based on green fluorescent proteins and calmodulin. Nature 388:882–887

Murai T, Kawashima H (2008) A simple assay for hyaluronidase activity using fluorescence polarization. Biochem Biophys Res Commun 376(3):620–624

Nath CE, Dallapozza L, Eslick AE (2008) An isocratic fluorescence HPLC assay for the monitoring of l-asparaginase activity and l-asparagine depletion in children receiving *E. coli* l-asparaginase for the treatment of acute lymphoblastic leukaemia. Biomed Chromatogr 23(2):152–159

Nowell PC (2007) Discovery of the Philadelphia chromosome: a personal perspective. J Clin Invest 117(8):2033–2035

Piepenhagen PA, Vanpatten S, Hughes H (2010) Use of direct fluorescence labeling and confocal microscopy to determine the biodistribution of two protein therapeutics, Cerezyme® and Ceredase®. Microsc Res Tech 73(7):694–703

Pietraszewska-Bogiel A, Gadella TW (2011) FRET microscopy: from principle to routine technology in cell biology. J Microsc 241(2):111–118

Piston DW, Kremers GJ (2007) Fluorescent protein FRET: the good, the bad and the ugly. Trends Biochem Sci 32(9):407–414

Schulz JD, Patt M, Basler S et al (2016) Site-specific polymer conjugation stabilizes therapeutic enzymes in the gastrointestinal tract. Adv Mater 28(7):1455–1460

Shahzad A, Köhler G, Knapp M et al (2009) Emerging applications of fluorescence spectroscopy in medical microbiology field. J Transl Med 7:99

Shaner NC, Campbell RE, Steinbach PA et al (2004) Improved monomeric red, orange and yellow fluorescent proteins derived from Discosoma sp. red fluorescent protein. Nat Biotechnol 22:1567–1572

Shaner NC, Steinbach PA, Tsien RY (2005) A guide to choosing fluorescent proteins. Nat Methods 2:905–909

Szittner Z, Papp K, Sándor N et al (2013) Application of fluorescent monocytes for probing immune complexes on antigen microarrays. PLoS One 8(9):e72401

Tatford OC, Gomme PT, Bertolini J (2004) Analytical techniques for the evaluation of liquid protein therapeutics. Biotechnol Appl Biochem 40(1):67–81

Tompa P, Batke J, Ovadi J et al (1987) Quantitation of the interaction between citrate synthase and malate dehydrogenase. J Biol Chem 262(13):6089–6092

Tunceroglu A, Matsuda M, Birge RB (2010) Real-time fluorescent resonance energy transfer analysis to monitor drug resistance in chronic myelogenous leukemia. Mol Cancer Ther 9(11):3065–3073

Tyagarajan K, Pretzer E, Wiktorowicz JE (2003) Thiol-reactive dyes for fluorescence labeling of proteomic samples. Electrophoresis 24(14):2348–2358

Usmani SS, Bedi G, Samuel JS et al (2017) THPdb: database of FDA-approved peptide and protein therapeutics. PLoS One 12(7):e0181748

Valeur B, Berberan-Santos MN (2011) A brief history of fluorescence and phosphorescence before the emergence of quantum theory. J Chem Educ 88(6):731–738

Vellard M (2003) The enzyme as drug: application of enzymes as pharmaceuticals. Curr Opin Biotechnol 14(4):444–450

Verhamme IM, Bock PE (2014) Rapid binding of plasminogen to streptokinase in a catalytic complex reveals a three-step mechanism. J Biol Chem 289(40):28006–28018

Wakabayashi H, Fay PJ (2013) Molecular orientation of factor VIIIa on the phospholipid membrane surface determined by fluorescence resonance energy transfer. Biochem J 452(2):293–301

Wei H, Mo J, Tao L et al (2014) Hydrogen/deuterium exchange mass spectrometry for probing higher order structure of protein therapeutics: methodology and applications. Drug Discov Today 19(1):95–102

Yadav AK, Shen DL, Shan X et al (2015) Fluorescence-quenched substrates for live cell imaging of human glucocerebrosidase activity. J Am Chem Soc 137(3):1181–1189

Yang J, Luo K, Pan H et al (2011) Synthesis of biodegradable multiblock copolymers by click coupling of RAFT-generated HeterotelechelicPolyHPMA conjugates. React Funct Polym 71(3):294–302

Ye C, Wang Z, Lu W et al (2014) Unfolding study of a trimeric membrane protein AcrB. Protein Sci 23(7):897–905

Yi NY, He Q, Caligan TB et al (2015) Development of a cell-based fluorescence polarization biosensor using preproinsulin to identify compounds that alter insulin granule dynamics. Assay Drug Dev Technol 13(9):558–569

Yuan T, Weljie AM, Vogel HJ (1998) Tryptophan fluorescence quenching by methionine and selenomethionine residues of calmodulin: orientation of peptide and protein binding. Biochemistry 37(9):3187–3195

Zhang R, Yang J, Radford DC et al (2017) FRET imaging of enzyme-responsive HPMA copolymer conjugate. Macromol Biosci 17(1):1600125

Key Physicochemical Characteristics Influencing ADME Properties of Therapeutic Proteins

6

Xing Jing, Yan Hou, William Hallett, Chandrahas G. Sahajwalla, and Ping Ji

Abstract

Therapeutic proteins are a rapidly growing class of drugs in clinical settings. The pharmacokinetics (PK) of therapeutic proteins relies on their absorption, distribution, metabolism, and excretion (ADME) properties. Moreover, the ADME properties of therapeutic proteins are impacted by their physicochemical characteristics. Comprehensive evaluation of these characteristics and their impact on ADME properties are critical to successful drug development. This chapter summarizes all relevant physicochemical characteristics and their effect on ADME properties of therapeutic proteins.

Keywords

Protein therapeutics · Physicochemical characteristics · Pharmacokinetics (PK) · ADME

X. Jing (✉) · Y. Hou · W. Hallett
C. G. Sahajwalla · P. Ji
U.S. Food and Drug Administration,
Office of Clinical Pharmacology, DV II,
Silver Spring, MD, USA
e-mail: xing.jing@fda.hhs.gov

Abbreviations

ADA	Anti-drug antibody
ADC	Antibody drug conjugate
ADME	Absorption, distribution, metabolism, excretion
ASGPR	Asialoglycoprotein receptor
CD	Circular dichroism
cIEF	Capillary isoelectric focusing
DLS	Dynamic light scattering
DSC	Differential scanning calorimetry
DSF	Differential scanning fluorimetry
ECM	Extracellular matrix
Fab	Fragment antigen-binding domain
Fc	Fragment crystallizable domain
FcRn	Neonatal Fc receptor
FcγR	Fc gamma receptors
FDA	U.S. Food and Drug Administration
FRET	Förster resonance energy transfer
GlcNAc	N-acetylglucosamine
IgG	Immunoglobulin G
ITC	Isothermal titration calorimetry
mAb	Monoclonal antibody
ManR	Mannose receptor
MST	Microscale thermophoresis
MW	Molecular weight
PEG	Polyethylene glycol
pI	Isoelectric point
PK	Pharmacokinetics
SC	Subcutaneous
SPR	Surface plasmon resonance
$T_{1/2}$	Half-life

N. Labrou (ed.), *Therapeutic Enzymes: Function and Clinical Implications*, Advances in Experimental Medicine and Biology 1148, https://doi.org/10.1007/978-981-13-7709-9_6

T_{max}	Time to peak concentration
TMDD	Target mediated drug disposition
TNF	Tumor necrosis factor
V_d	Volume of distribution

6.1 Introduction

In recent decades, the clinical application of therapeutic proteins has revolutionized the treatment of many diseases. Since the approval of the recombinant human insulin in early 1980s, protein therapeutics have rapidly gained popularity in clinical use. To date, more than 140 therapeutic proteins have been granted approval from the U.S. Food and Drug Administration (FDA), for a wide range of indications, from alleviation of neuropathic pain to rheumatoid arthritis and lysosomal storage diseases.

Therapeutic proteins are biological products (biologics) produced from living organisms or contain component of living organisms (Dorai and Ganguly 2014). Over the past decades, the fast-evolving biotechnologies have also facilitated the development of numerous protein therapeutics such as cytokines, growth factors and replacement enzymes. The most remarkable milestone is the emergence of antibody therapeutics (Macielag 2012; Cavagna and Taylor 2014; Vidarsson et al. 2014; Wang et al. 2015).

Compared to small molecule drugs, therapeutic proteins have unique characteristics that underlie their pharmacokinetics and pharmacodynamics (PK/PD). Understanding protein characteristics that impact clinical performance of therapeutic proteins are essential for drug development of different phases. For example, product PK-influencing attributes could inform drug design and evaluation and dosing regimen selection in preclinical and clinical studies. This chapter describes all relevant physicochemical characteristics that impact ADME processes of therapeutic proteins.

Each section in this chapter briefly introduces the mechanisms governing the involvement of protein therapeutics in each ADME process. The subsequent subsections describe relevant protein physiochemical characteristics and their roles in the corresponding mechanisms. Furthermore, with data analytics for FDA-approved protein therapeutics as real examples, this chapter focuses on the established relationship between protein characteristics and PK performance, as well as the successful strategies for PK improvement. Finally, this chapter builds a comprehensive roadmap to summarize all established correlations between physiochemical characteristics of FDA approved therapeutic proteins and their ADME properties.

6.2 Impact of Physicochemical Characteristics on Absorption

The approved therapeutic proteins are mostly delivered through the subcutaneous (SC) or intravenous (IV) route of administration. As the intravenous administration route bypasses the absorption phase, this section primarily focuses on therapeutic proteins administered via the subcutaneous (SC) route. The rate and extent of absorption after subcutaneous dosing is dependent on factors such as the molecular structure, weight, size, and charge.

6.2.1 Molecular Weight and Size

The molecular weight of a therapeutic protein is its total molecular mass, which is composed of the mass of its amino acid sequence and the mass of modifications. The molecular size of a therapeutic protein, directly correlated with its molecular weight, represents its geometric dimensions. The primary structure of a therapeutic protein represents its amino acid sequence.

After SC administration, therapeutic proteins are delivered to the hypodermis that is primarily composed of negatively charged extracellular matrix (ECM) (Kinnunen and Mrsny 2014; Richter et al. 2012). From the ECM, therapeutic proteins enter systemic circulation through two

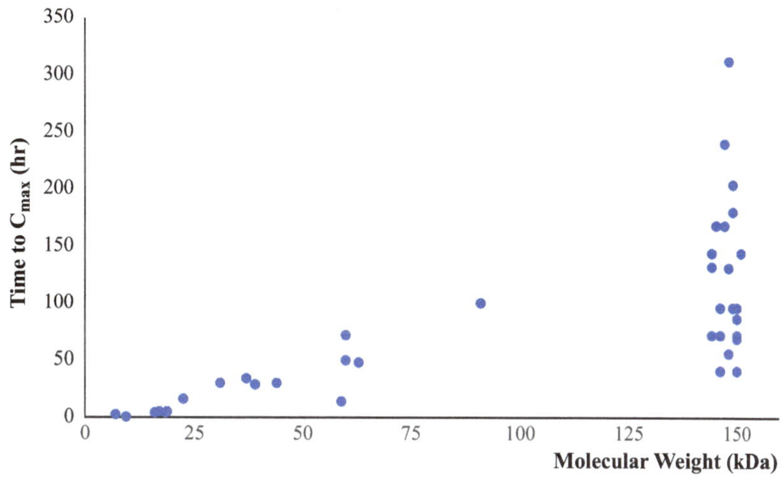

Fig. 6.1 Statistical analysis of the impact of molecular weight on T_{max} of FDA-approved therapeutic proteins

routes: blood capillaries via diffusion and lymphatic vessels via convection. The probability of absorption via convection increases with the increase in their molecular weight (MW) (Porter and Charman 2000; McLennan et al. 2005), with proteins larger than 15–20 kDa entering the circulation system primarily through the lymphatic system (Supersaxo et al. 1990). Studies have identified a correlation between T_{max} and MW (Porter and Charman 2000; Kagan 2014). Therapeutic proteins with MW from 9.4 to 18.8 kDa and 31–63 kDa generally reach T_{max} at 0.5–5 h and 14–72 h, respectively, after administration. Antibodies or derivatives with MW of 150 kDa have T_{max} values in the range of 40 h to 13 days (Fig. 6.1).

Molecular size also affects absorption (Tibbitts et al. 2016). Products with similar molecular weights but different molecular size may exhibit different T_{max} values (Fig. 6.1). Although the molecular weight of a protein correlates with its physical size, other physicochemical structural factors, such as modification and folding, could also impact its molecular size and thus the absorption profile.

6.2.1.1 Higher-Order Structure

The higher-order structure of a therapeutic protein means its three-dimension tertiary structure, which is affected by its primary structure. Moreover, higher-order structure also covers protein quaternary structure such as dimerization. Self-association of therapeutic proteins may affect absorption through its impact on MW and size. Having numerous analogs, insulin products are excellent examples illustrating the impact of dimerization on absorption (Fig. 6.2). Regular human insulin (i.e., Humulin R) is absorbed as hexamers (Palmieri et al. 2013) with a T_{max} of 4–8 h. By contrast, aspart insulin analogs (such as Novolog and Fiasp), bearing a mutation of proline28 to Aspartate in the B chain, have a T_{max} of 40–60 min. The difference in T_{max} between human insulin and aspart insulin analog is due to this single mutation that results in a change in oligomeric state (Brange and Volund 1999; Brange et al. 1988, 1990; Volund et al. 1991). Likewise, a series of other insulin mutants, identified by mutational analysis targeting the dimer interface, tend to form weaker oligomers and perform as fast-acting insulin drugs (Brange et al. 1988). For example, insulin (Humalog) with flipped lysine28/proline29 mutation and insulin glulisine (Apidra) with double mutations in B chain has a T_{max} of 30–90 min and 55 min, respectively. As another example, glargine insulin (Basaglar and Lantus) possesses two additional arginine residues and thus enhanced oligomerization, resulting in prolonged absorption with T_{max} about 12 h (Owens 2012; Hilgenfeld et al. 2014).

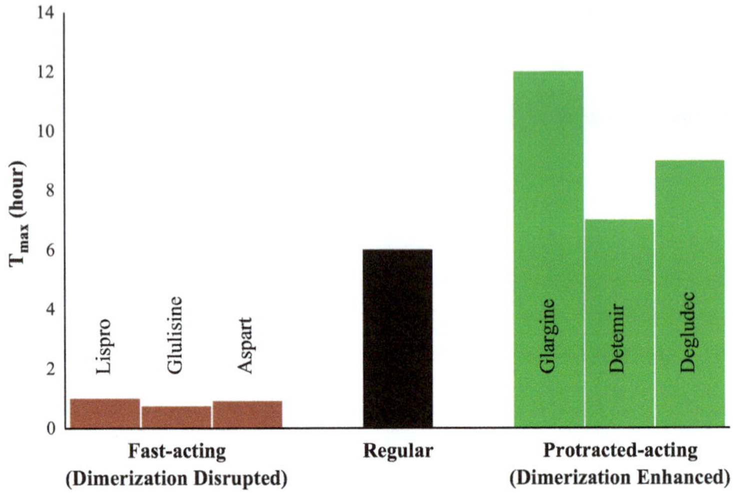

Fig. 6.2 Absorption rates of different types of approved insulin drugs with different dimerization affinity

6.2.2 Molecular Charge

The molecular charge of a therapeutic protein represents its net surface charge, commonly indicated by its isoelectric point (pI).

The negatively charged ECM may capture the positively charged portions of a therapeutic protein. Since most monoclonal antibodies (mAbs) or fusion proteins have a pI between seven and nine, the electrostatic interaction with ECM might delay their release to blood (Khawli et al. 2010; Bumbaca et al. 2012). For example, while Trulicity (dulaglutide) and Mircera (methoxy polyethylene glycol-epoetin beta) have similar MWs of ~60 kDa, they have different absorption rates (T_{max} of 48 and 72 h, respectively) which may be attributed to their difference in pIs (pI of 5.5 and 8.8, respectively).

6.2.3 Bioavailability

The extent of absorption after subcutaneous administration is dependent on pre-systemic clearance, which is dependent on multiple factors including MW, solubility and interaction with receptors, etc. (Datta-Mannan et al. 2012; Deng et al. 2012). Thus, no obvious trend could be observed between bioavailability and MW after SC administration (Fig. 6.3).

6.3 The Impact of Physicochemical Characteristics on Distribution

The volume of distribution of therapeutic proteins is affected by factors such as molecular weight, size, charge, and structure (Tabrizi et al. 2010; Tibbitts et al. 2016).

6.3.1 Molecular Weight and Size

Monoclonal antibodies (mAbs) exhibit limited distribution from the blood to the peripheral tissue primarily due to their large molecular size (Lobo et al. 2004). In general, the distribution of mAbs is restrained primarily in blood plasma and limited interstitial spaces (Boswell et al. 2010; Dostalek et al. 2013). The Fab fragments have larger volume of distribution because of smaller size (Thurber et al. 2008; Tabrizi et al. 2010). Full-length IgG therapeutics targeting TNF, on the other hand, are primarily distributed within the blood stream (Tabrizi et al. 2010).

6.3.2 Molecular Charge

Molecular charge of therapeutic proteins underlies their interactions with biological components that

Fig. 6.3 Evaluation of the impact of molecular weight on bioavailability of FDA-approved therapeutic proteins

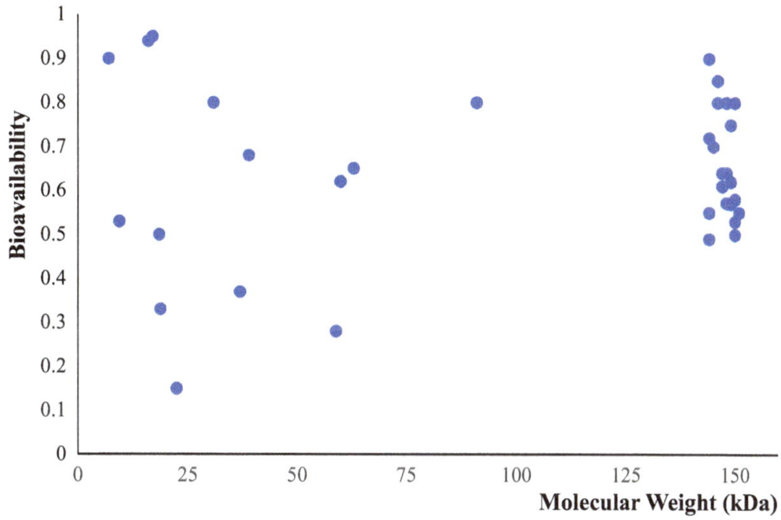

influence their distribution. For instance, one of the major mechanisms that govern distribution of antibody therapeutics to tumor tissues relies on receptor-mediated endocytosis and binding to the cellular antigens (Weinstein and van Osdol 1992; Tabrizi et al. 2010). Tighter binding between a mAb and tumor cells results in increased retention of an antibody at the periphery of tumor nodules, which form a binding-site barrier for antibody drugs (Graff and Wittrup 2003; Tabrizi et al. 2010). Playing a vital role in antibody elimination, the interaction between mAbs and neonatal Fc receptor (FcRn), which is essential to the FcRn-mediated endocytosis, also promotes tissue distribution of antibody therapeutics (Martins et al. 2016).

6.3.3 Primary and Higher-Order Structure

The primary structure of therapeutic proteins could also influence their distribution. For instance, compared to full-length mAbs, the Fab fragment alone may have larger volume of distribution because it has less potential for interaction with FcRn and thus can better overcome the binding-site barrier and penetrate deeper into tissues (Thurber et al. 2008; Tabrizi et al. 2010).

Additionally, the change in higher-order structure of a therapeutic protein can alter its distribution profile via change in apparent molecular size. For instance, the mean V_d of Humalog, an insulin lispro with disrupted dimerization as mentioned above, appeared to decrease with increase in dose (1.55 and 0.72 L/kg, respectively) in contrast to that of regular human insulin for which, V_d was comparable across the two dose groups (1.37 and 1.12 L/kg for 0.1 and 0.2 U/kg dose, respectively) (FDA database for labeling 2018).

6.3.4 Other Factors

Interactions with pharmacological targets on tissues or in plasma influence distribution of therapeutic proteins as well (Tabrizi et al. 2010). In addition to the abundance of the targets expressed, its interaction affinity with the drug can also have an impact on distribution. Structural factors, such as primary structure design and molecular charge, determine the interaction pattern between the drug and the drug target. More interestingly, for bispecific antibody-derived therapeutics, the relationship between binding affinity and distribution is not monotonic (List and Neri 2012; Kanodia et al. 2016). For instance, for the T-cell engagers, high affinity with T-cells will inhibit penetration to tumor tissues (List and Neri 2012). In another

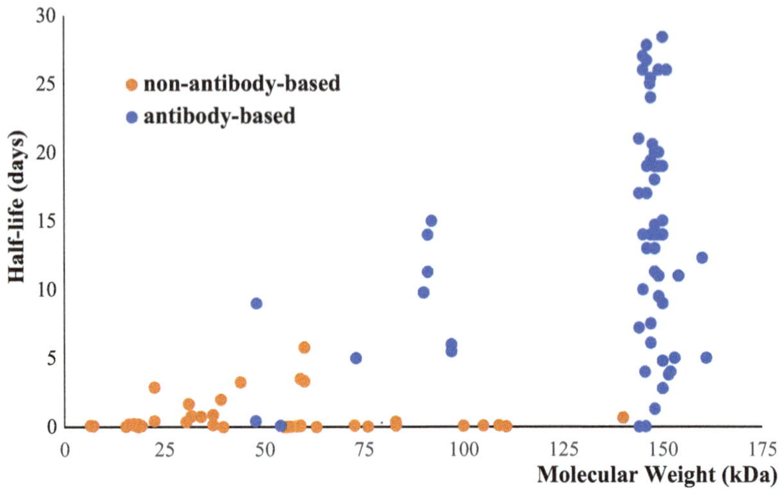

Fig. 6.4 The half-life of non-antibody (orange) vs full-length or mAb portion(s) (blue) therapeutic proteins

case, although anti-transferrin receptor (TfR)-based bispecific antibodies bind to TfR in order to pass the blood-brain-barrier, an intermediate affinity is required (Kanodia et al. 2016).

6.4 The Impact of Physicochemical Factors on Elimination

Therapeutic proteins are generally eliminated by two pathways: proteolytic catabolism and renal elimination (Shi 2014; Zhao et al. 2012), depending on its molecular weight. Kidneys may play a relevant role in the catabolism and elimination of only those biologics that have a size below the cut-off for glomerular filtration of approximately 60–70 kDa (Shi 2014; Vugmeyster et al. 2012; Tibbitts et al. 2016). For antibody-based therapeutics including monoclonal antibodies (mAbs), chimeric antibodies, antibody-drug conjugate (ADC), Fab and fusion proteins, the MW is generally much bigger than 60 kDa (Fig. 6.4), and proteolytic catabolism is the main elimination pathway.

6.4.1 Molecular Weight and Size

Clearance of therapeutic proteins can be modified through the change in weight and size such as dimerization and PEGylation. As an example, the therapeutic enzyme Fabrazyme (agalsidase beta) is a homodimer and has an apparent MW of 100 kDa. It is in equilibrium with its monomer of a MW of 50 kDa. The monomer is able to eliminate renally, promoting dissociation of the dimer. This process may help explain the short half-life of Fabrazyme (45–102 min).

PEGylation is a widely used to modify the physical size of small therapeutic proteins. It modifies biologics by covalent conjugation with polyethylene glycols (PEGs). In general, PEGylation can improve drug solubility, decrease immunogenicity, prolong residence time in body, and decrease degradation by metabolic enzymes, resulting in the improved PK and PD properties. A variety of biologics have been PEGylated with different PEGs to improve PK properties (Veronese and Pasut 2005; Hamidi et al. 2006; Jevsevar et al. 2010). First, attachment of PEG moieties greatly increases MW of therapeutic proteins. For instance, the PEGylaiton of asparaginase increases its MW from 34.5 to 380–450 kDa by attaching 69–82 molecules of mono-PEG (5 kDa each), resulting in a prolonged half-life of 5.8 days of Oncaspar (pegaspargase). The PEGylation of uricase (34 kDa) with approximately 40 PEG moieties increases the MW to 540 kDa and significantly increases its half-life (Richette et al. 2014). Cimzia (Certolizumab Pegol Injection), with a half-life of 14 days, is composed of an antibody

Fab fragment (50 kDa) conjugated to a 40 kDa PEG moiety. Second, PEGylation can increase the hydrodynamic radius. PEG exhibits a much greater molecular volume due to the extended conformation of the PEG polymer per unit of mass (Caliceti and Veronese 2003). Pegfilgrastim (39 kDa) consists of a single 20 kDa linear PEG molecule attached to filgrastim (18.8 kDa). The half-life of pegfilgrastim is much longer than filgrastim (15–80 h compared to 3.5–9 h, respectively) because the attached PEG moiety significantly increases the hydrodynamic radius (Yang and Kido 2011). Finally, PEGylation can increase the stability and reduce the catabolic elimination. The highly hydrated polyether backbone of the PEG moiety helps Pegasys (Peginterferon alfa-2a) forms a water shield, preventing the degradation by proteolytic enzymes and thus increasing the half-life of the parent protein (Jevsevar et al. 2010).

6.4.2 Charge

More negatively charged small proteins are less likely cleared by renal filtration because of the negatively charged framework of the kidney (Porter and Charman 2000). For instance, both Proleukin (Aldesleukin) (15.3 kDa) and Kalbitor (Ecallantide) (7.1 kDa) are cleared renally, but Aldesleukin has a shorter half-life compared to Ecallantide (13–85 min vs. 2 h, respectively). The small difference in half-life might be partly attributed to the difference in pI between Aldesleukin (pI: 6.83) and Ecallantide (pI: 5.58).

Total or local charge of therapeutic proteins can be modified by glycosylation. Aranesp (Darbepoetin alpha), for instance, is a 165-amino acid protein (37 kDa) that differs from Epogen/Procrit (Epoetin alfa) (30 kDa) by containing five N-linked oligosaccharide chains instead of three. While the two additional carbohydrate chains increase the MW of the glycoprotein by only 7 kDa, darbepoetin alpha has a threefold longer terminal half-life than epoetin alfa. The two additional carbohydrate molecules provide darbepoetin alpha significantly more negative charges as compared to epoetin alfa, thus elongating the half-life (Egrie and Browne 2002). Another case example of where glycosylation impacts clearance of a product is for Extavia (Interferon beta-1b) (18.5 kDa) and Rebif (interferon beta-1a) (22.5 kDa). The glycosylation of interferon beta-1a increases its solubility and stability in contrast to unmodified interferon beta-1b, resulting in a longer half-life for interferon beta-1a (69 h vs. 8 min to 4.3 h, respectively) (Song et al. 2014).

6.4.3 Elimination of Antibody-Based Therapeutics

Antibody-based therapeutics include mAbs, chimeric antibodies, antibody-drug conjugate (ADC), Fab and fusion proteins. As shown by Fig. 6.4, in general, therapeutic antibodies exhibit significantly longer half-lives than smaller non-antibody therapeutic proteins. The elimination of these therapeutics mainly occurs via intracellular catabolism (proteolysis) through two major catabolic pathways: Fc receptor mediated clearance and target mediated clearance (Lobo et al. 2004; Zhao et al. 2012).

The Fc receptor family is composed of cell surface receptors. Two major types of human Fc receptors that bind Fc domain of IgG are Fc gamma receptors (FcγR) and neonatal Fc receptor (FcRn) (Pechtner et al. 2017). The function of FcRn and the mechanism governing its protective role in antibody drug elimination have been well established. Briefly, FcRn protects IgG from lysosomal degradation and recycling IgG back into the circulation via specific binding to Fc domain of IgG. This recycling pathway preserves serum antibody level of IgG (Liu 2018; Lencer and Blumberg 2005; Sockolosky and Szoka 2015; Wang et al. 2008; Zhao et al. 2012; Pechtner et al. 2017). Changes in IgG-FcRn binding affinity resulted in altered clearance rate and half-lives of antibody drugs (Ghetie et al. 1997; Dall'Acqua et al. 2002, 2006; Vaccaro et al. 2005; Zalevsky et al. 2010). The structural properties of antibodies play an essential role in their interaction with FcRn and therefore FcRn-mediated recycling. Underlying the IgG-FcRn binding are two major structural factors of antibody therapeutics: surface charge and primary structure.

6.4.4 Primary Structure

Given that all FDA-approved antibody drugs are developed based on IgG1, IgG2 and IgG4 subtypes, these antibodies should potentially have an elimination half-life of approximately 21 days (Zhao et al. 2012; Dall'Acqua et al. 2002; Ghetie et al. 1997; Levêque et al. 2005). However, elimination half-lives of all marketed IgG1-based antibodies differ widely (Fig. 6.4). One of the major physiochemical features that underlie this diversity is the degree of humanization of antibody therapeutics, which include human IgG, human-murine chimeric and humanized IgG. Human and rat FcRn receptors share only 65% amino acid sequence homology, causing the differences in IgG-FcRn affinity and, in turn, FcRn-mediated elimination of mAbs between the two species (Kuo et al. 2010; Dostalek et al. 2013). Indeed, among all antibodies approved by FDA, there exists a correlation between half-life and the degree of antibody humanization. Elimination half-life increases with the increase in humanization of the antibody therapeutics (Fc absence < murine < chimeric < humanized < human), a trend consistent with the previously suggested order (Zhao et al. 2012; Dostalek et al. 2013). Abciximab and idarucizumab, for instance, have a short half-life of 0.5 and 10.3 h, respectively, because they contain only the Fab fragments of IgG thus lack of Fc-FcRn interaction. This shortened half-life is consistent with an animal study showing that IgG in mice without FcRn is catabolized significantly faster than wild-type mice (Lobo et al. 2004; Zhao et al. 2012). One FDA-approved bispecific antibody derivative, Blincyto (Blinatumomab), represents another striking case for the impact of Fc absence on elimination. Blinatumomab exhibits a half-life of only 2.1 h because it is comprised of only Fab domains.

Although possessing the human Fc domain could theoretically result in chimeric antibodies with similar elimination rates as humanized or human IgG, the observed difference in half-life indicates that the intact human IgG structure may be required to fully restore the native IgG-FcRn interaction. The antigen binding (Fab) portion of native IgG may either contribute to FcRn binding or reduce the elimination of IgG through other mechanisms. In addition to direct therapeutic functions, full-length IgG as well as IgG fragments are effective carriers of therapeutic agents such as small molecules or proteins. Small molecule therapeutics such as tumor toxins are covalently attached to mAbs to form antibody-drug conjugates (ADC) (Lambert 2005; Senter 2009; Wu and Senter 2005). Similarly, some protein therapeutics can be also linked to mAb platform by DNA recombinant technology to form fusion proteins (Strohl 2015; Pechtner et al. 2017). Even though ADCs and fusion proteins possess the Fc domain and thus capable to utilize FcRn-mediated recycling, some of these therapeutics exhibit much shorter half-life than expected. For example, Fc fusion drugs, abatacept, aflibercept, and etanercept, have elimination half-lives of 5–6 days, even though they are not cleared via renal filtration due to their molecular weights being greater than 70 kDa. Similarly, the currently marketed ADC drugs exhibit elimination half-lives of 1.3–6 days. These short half-lives indicate that the conjugated molecules or fused proteins may interfere with the Fc-FcRn interaction, therefore inhibiting FcRn mediated recycling pathway.

6.4.5 Impact of pI on FcRn-Dependent and FcRn-Independent Elimination

The molecular charge or pI of therapeutic IgGs also plays a significant role in their elimination. Firstly, the molecular charge is essential to the electrostatic interaction between IgG and FcRn and hence the FcRn-mediated elimination. FcRn-IgG interaction is strictly pH-dependent. Possessing a pI of 7–9, IgG binds FcRn in slightly acidic pH, but not in neutral pH, via electrostatic interaction between titratable histidine residues in

CH2-CH3 domains with acidic residues on the α2-domain of FcRn (Lencer and Blumberg 2005; Dall'Acqua et al. 2006; Qiu et al. 2016). This dependence upon physiological conditions underlies the IgG preservation mechanism that governs the release of IgG from FcRn-bound form within acidic lysosomes back to the systemic circulation (Ghetie and Ward 2000). Mutational analyses have mapped the IgG-FcRn binding sites. The pH-dependent electrostatic interaction of IgG is mainly contributed by histidine310 and histidine435, while isoleucine253 is required for the hydrophobic interaction with FcRn (Sockolosky and Szoka 2015; Martin et al. 2001).

Secondly, multiple studies have demonstrated that the pI of IgG therapeutics impacts their FcRn-independent elimination. For instance, within an animal study, lowering the pI of variable domain of an antibody significantly reduced the its clearance without affecting its binding affinity with FcRn (Igawa et al. 2010). Another mutational analysis targeting anti-hepatitis C antibody suggested that adding negative charges to the variable domains of antibodies with high pI increases their clearance, independent of FcRn binding (Li et al. 2014). These results suggest that the elimination reduction, by lowering pI, is due to the decrease in fluid phase pinocytosis, because the negative charge on the antibody inhibits its binding to cells. Alternatively, adding positive charges on the high pI antibody at the lower pH of lysosomes could cause a faster rate of degradation of the antibody when internalized via pinocytosis (Igawa et al. 2010; Li et al. 2014).

6.4.6 Target Mediated Drug Disposition (TMDD)

TMDD causes the non-linear clearance (Mager 2006; Keizer et al. 2010; Dostalek et al. 2013). If subject to TMDD upon binding to targets on cell surface, the therapeutic proteins are internalized into the cells and subsequently degraded in lysosomes (Mellman and Plutner 1984; Press et al. 1988; Coffey et al. 2004; Lammerts et al. 2006; Keizer et al. 2010). Therefore, the structural characteristics that impact the interaction between therapeutic proteins and their pharmacological targets will inevitably influence their TMDD mediated elimination. For antibodies, the relevant structural characteristics include the molecular chargewhich involves the electrostatic interaction, and proper glycosylation required for antigen or target recognition. During drug development for a product with a given mechanism of action, the principle of primary structure design is to provide proper higher-order structural properties such as hydrophobicity, molecular charge and correct glycosylation, in order to facilitate the interaction between the drug product and its target.

6.4.7 Glycosylation

Glycosylation on the Fc domain of antibodies may impact their clearance. Although human-like native glycosylation may not account for the long half-life of IgG, the absence of influence is only relevant when the glycosylation is buried (Liu 2015; Bumbaca et al. 2012; Higel et al. 2016). In certain situations, attached terminal carbohydrate moieties are exposed and available to bind glycan receptors, causing faster clearance through the glycan receptor mediated elimination pathway (Winkelhake and Nicolson 1976; Wright et al. 2000). Glycan receptors that are involved in the elimination of glycoproteins include mannose receptor (ManR) and asialoglycoprotein receptor (ASGPR), both of which are specific to certain glycan types. It has been demonstrated that high-mannose containing IgG or Fc-fusion proteins are cleared faster than those with other glycosylation patterns (Liu 2015; Wright and Morrison 1994; Kanda et al. 2007; Liu et al. 2011; Yu et al. 2012). Similarly, antibodies carrying the terminal N-acetylglucosamine (GlcNAc) or galactose also exhibit fast clearance facilitated by ASGPR that recognizes GlcNAc (Winkelhake and Nicolson 1976; Beck and Reichert 2011; Stefanich et al. 2008). The other type of glycan, sialic acid (NANA), on the other hand, is critical to reduce the clearance of antibodies or Fc-fusion proteins, because NANA is able to cap the galactose and to block the recognition by ASGPR (Schwartz 1991; Liu 2015).

6.5 Immunogenicity and Anti-drug Antibodies (ADA)

Immunogenicity, an unwanted immune response to therapeutic proteins, involves the generation of anti-drug antibodies (ADA) that compromise drug efficacy and raise safety concerns. Factors influencing the immunogenicity of biologics can be classified into disease-, patient-, or product-related. For example, the dose, route, frequency and duration of administration are important for the immunogenicity response and ADAs. ADA induction can affect PK profiles of therapeutic proteins via influencing their elimination (Lobo et al. 2004). Besides of the inherent characteristics (e.g., pI and glycosylation) of therapeutic proteins, ADA-binding represents a factor that influences the product interactions with biological components such as Fc receptors (Davies et al. 1993; Strohmeier et al. 1995). The effect of ADA-binding on elimination may depend on the number of antigenic sites available on a given therapeutic protein. If there are one or two binding sites found on a therapeutic protein, ADA can contribute to the increase in its half-life. If there are more than two binding sites, on the other hand, the ADA-binding may result in faster clearance of the product (Rehlaender and Cho 1998; Lobo et al. 2004). In addition, the degree of aggregation of manufactured therapeutic proteins also influences the induction of ADA (Ratanji et al. 2014).

6.6 Characterizing Physicochemical Properties Affecting ADME of Therapeutic Proteins

6.6.1 Structure

The primary structure of a therapeutic protein represents its amino acid sequence. The higher-order structure of therapeutic proteins means its three-dimension tertiary structure, which is affected by its primary structure. Moreover, higher-order structure also covers protein quaternary structure such as dimerization. Protein folding is the process in which a protein forms three-dimension structure from the primary structure. Protein stability means the ability of a protein to maintain its three-dimension structure.

Circular Dichroism (CD) is used extensively to evaluate protein structure. Even though CD does not provide 3-D structure information, it can monitor the extent and rate of structural variation and ligand binding. In pharmaceutical industry, CD is also used to assess the stability of the designed proteins (Kelly and Price 2000). Differential Scanning Calorimetry (DSC) and Differential Scanning Fluorimetry (DSF) are two fundamentals tools for assessing thermal stability of proteins that have been widely applied by the pharmaceutical industry. These two similar methods semi-quantitatively measure stability via determination of melting temperatures of proteins (Johnson 2013; Bernhards et al. 2009).

6.6.2 Size and Self-Association

The molecular size of a therapeutic protein, directly correlated with its molecular weight, represents its geometric dimension. Dynamic light scattering (DLS) analyses are routinely used in detection of protein aggregation, the size of proteins and complexes or to monitor the binding of ligands. Due to the limitation of currently protein manufacturing technologies, DLS has been used quite often to assess the quality of manufactured therapeutic proteins (Lorber et al. 2012). Analytical gel-filtration is a chromatography-based method to assess the hydrodynamic size of therapeutic proteins.

The dimerization of therapeutic proteins can be quantitatively measured with Förster resonance energy transfer (FRET) (Jing et al. 2010). This method can be extensively used for insulin drugs which have multiple analogs with different dimerization states.

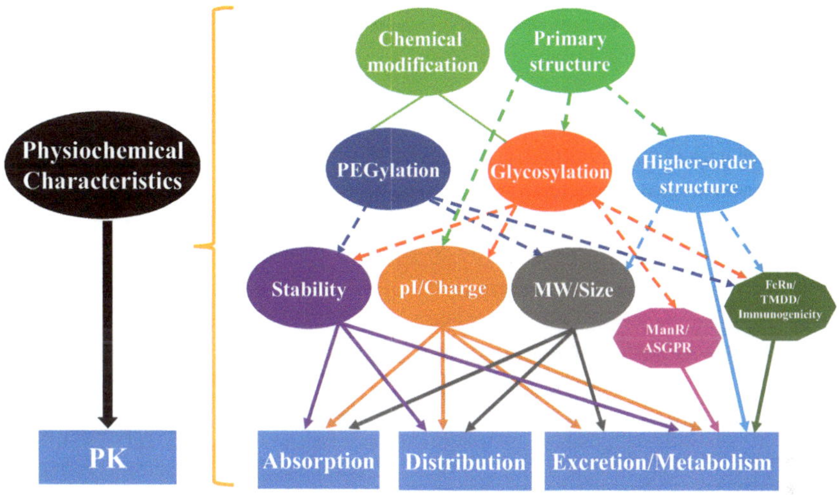

Fig. 6.5 Relationship between protein characteristics and their impact on ADME. A solid arrow line represents a direct influence of a protein characteristic or a mechanism on ADME. A dashed arrow line represents the influence of one protein characteristic to another or to a mechanism. The oval circles represent physiochemical features of therapeutic proteins. The two polygons represent mechanisms that correlate with physiochemical features and ADME properties of therapeutic proteins

6.6.3 Molecular Charge

The molecular charge of a therapeutic protein represents its net surface charge, commonly indicated by its isoelectric point. The simplest way to estimate the charge of a protein is to calculate pI from its amino acid sequence. However, this calculation may be inaccurate since it does not consider the folding of the target protein. Capillary isoelectric focusing (cIEF) is a high-resolution method that is widely applied to experimentally assess protein charge in pharmaceutical industry, especially for characterization of mAb drugs (Righetti 2004; Pergande and Cologna 2017).

6.6.4 Protein-Protein Interaction

Protein interactions are crucial to both therapeutic capabilities and PK of therapeutic proteins. Commonly used biophysical technique to characterize protein-protein interactions include Surface Plasmon Resonance (SPR) (Fabini and Danielson 2017), Isothermal Titration Calorimetry (ITC) (Pierce et al. 1999), Förster resonance energy transfer (FRET), and DLS (Hanlon et al. 2010).

6.7 Conclusions

In biological system, proteins are the major molecules responsible of executing biological functions because they possess advanced structures and biophysics. This concept underlies not only the therapeutic power of proteins in clinical therapy, but also more sophisticated correlation between their molecular characteristics and pharmacokinetic behaviors and clinical success.

This chapter describes structural characteristics and factors that impact either directly or indirectly the ADME properties of therapeutic proteins (Fig. 6.5). The molecular size of proteins, reflecting the combination of molecular weight and shape, determines not only their absorption route but also the elimination pathways. The molecular charge of therapeutic proteins, affecting their interactions with a wide range of biological components such as ECM framework, kidney, Fc receptors and pharmacological targets, has a significant impact on every step of ADME. The post-translational modification of therapeutic proteins is critical for their biological activities and, more importantly, affect a multitude of their structural characteristics such as molecular size, molecular charge, stability and protein-receptor interactions. Further advances in protein engineering technologies and development of more sophisticated bioanalytical tools will facilitate the understanding of the relationship between structure and ADME and promote successful development of novel therapeutic proteins.

Acknowledgments The authors sincerely thank the critical review of the chapter from Dr. Sarah Schrieber and Dr. Vitaliy Klimov from Office of Clinical Pharmacology, FDA.

References

Beck A, Reichert JM (2011) Therapeutic Fc-fusion proteins and peptides as successful alternatives to antibodies. MAbs 3:415–416

Bernhards RC, Jing X, Vogelaar NJ, Robinson H, Schubot FD (2009) Structural evidence suggests that antiactivator ExsD from *Pseudomonas aeruginosa* is a DNA binding protein. Protein Sci 18(3):503–513

Boswell CA, Tesar DB, Mukhyala K, Theil FP, Fielder PJ, Khawli LA (2010) Effects of charge on antibody tissue distribution and pharmacokinetics. Bioconjug Chem 21(12):2153–2163

Brange J, Volund A (1999) Insulin analogs with improved pharmacokinetic profiles. Adv Drug Deliv Rev 35:307–335

Brange J, Ribel U, Hansen JF, Dodson G, Hansen MT, Havelund S, Melberg SG, Norris F, Norris K, Snel L, Sorensen AR, Voigt HO (1988) Monomeric insulins obtained by protein engineering and their medical implications. Nature 333:679–682

Brange J, Owens DR, Kang S, Volund A (1990) Monomeric insulins and their experimental and clinical implications. Diabetes Care 13:923–954

Bumbaca D, Boswell CA, Fielder PJ, Khawli LA (2012) Physiochemical and biochemical factors influencing the pharmacokinetics of antibody therapeutics. AAPS J 14:554–558

Caliceti P, Veronese FM (2003) Pharmacokinetic and biodistribution properties of poly (ethylene glycol)-protein conjugates. Adv Drug Deliv Rev 55:1261–1277

Cavagna L, Taylor WJ (2014) The emerging role of biotechnological drugs in the treatment of gout. Biomed Res Int 2014:264859. https://doi.org/10.1155/2014/264859

Coffey GP, Stefanich E, Palmieri S, Eckert R, Padilla-Eagar J, Fielder PJ, Pippig S (2004) In vitro internalization, intracellular transport, and clearance of an anti-CD11a antibody (Raptiva) by human T-cells. J Pharmacol Exp Ther 310(3):896–904

Dall'Acqua WF, Woods RM, Ward ES, Palaszynski SR, Patel NK, Brewah YA, Wu H, Kiener PA, Langermann S (2002) Increasing the affinity of a human IgG1 for the neonatal Fc receptor: biological consequences. J Immunol 169(9):5171–5180

Dall'Acqua WF, Kiener PA, Wu H (2006) Properties of human IgG1s engineered for enhanced binding to the neonatal fc receptor (FcRn). J Biol Chem 281(33):23514–23524

Datta-Mannan A, Witcher DR, Lu J, Wroblewski VJ (2012) Influence of improved FcRn binding on the subcutaneous bioavailability of monoclonal antibodies in cynomolgus monkeys. MAbs 4(2):267–273

Davies KA, Erlendsson K, Beynon HL, Peters AM, Steinsson K, Valdimarsson H, Walport MJ (1993) Splenic uptake of immune complexes in man is complement-dependent. J Immunol 151(7):3866–3873

Deng R, Meng YG, Hoyte K, Lutman J, Lu Y, Iyer S, DeForge LE, Theil FP, Fielder PJ, Prabhu S (2012) Subcutaneous bioavailability of therapeutic antibodies as a function of FcRn binding affinity in mice. MAbs 4:101–109

Dorai H, Ganguly S (2014) Mammalian cell-produced therapeutic proteins: heterogeneity derived from protein degradation. Curr Opin Biotechnol 30:198–204

Dostalek M, Gardner I, Gurbaxani BM, Rose RH, Chetty M (2013) Pharmacokinetics, pharmacodynamics and physiologically-based pharmacokinetic modelling of monoclonal antibodies. Clin Pharmacokinet 52(2):83–124

Egrie JC, Browne JK (2002) Development and characterization of darbepoetin alfa. Oncology (Williston Park) 16:13–22

Fabini E, Danielson UH (2017) Monitoring drug-serum protein interactions for early ADME prediction through Surface Plasmon Resonance technology. J Pharm Biomed Anal 144:188–194

FDA Approved Drug Products Database of Labeling (2018) U.S. Food and Drug Administration. https://www.accessdata.fda.gov/drugsatfda_docs/label/2017/020563s176lbl.pdf. Accessed 01 Mar 2018

Ghetie V, Ward ES (2000) Multiple roles for the major histocompatibility complex class I-related receptor FcRn. Annu Rev Immunol 18:739–766

Ghetie V, Popov S, Borvak J, Radu C, Matesoi D, Medesan C, Ober RJ, Ward ES (1997) Increasing the serum persistence of an IgG fragment by random mutagenesis. Nat Biotechnol 15(7):637–640

Graff CP, Wittrup KD (2003) Theoretical analysis of antibody targeting of tumor spheroids: importance of dosage for penetration, and affinity for retention. Cancer Res 63(6):1288–1296

Hamidi M, Azadi A, Rafiei P (2006) Pharmacokinetic consequences of pegylation. Drug Deliv 13(6):399–409
Hanlon AD, Larkin MI, Reddick RM (2010) Free-solution, label-free protein-protein interactions characterized by dynamic light scattering. Biophys J 98(2):297–304
Higel F, Seidl A, Sörgel F, Friess W (2016) N-glycosylation heterogeneity and the influence on structure, function and pharmacokinetics of monoclonal antibodies and Fc fusion proteins. Eur J Pharm Biopharm 100:94–100
Hilgenfeld R, Seipke G, Berchtold H, Owens DR (2014) The evolution of insulin glargine and its continuing contribution to diabetes care. Drugs 74(8):911–927
Igawa T, Tsunoda H, Tachibana T, Maeda A, Mimoto F, Moriyama C, Nanami M, Sekimori Y, Nabuchi Y, Aso Y, Hattori K (2010) Reduced elimination of IgG antibodies by engineering the variable region. Protein Eng Des Sel 23(5):385–392
Jevsevar S, Kunstelj M, Porekar VG (2010) PEGylation of therapeutic proteins. Biotechnol J 5(1):113–128
Jing X, Jaw J, Robinson HH, Schubot FD (2010) Crystal structure and oligomeric state of the RetS signaling kinase sensory domain. Proteins 78(7):1631–1640
Johnson CM (2013) Differential scanning calorimetry as a tool for protein folding and stability. Arch Biochem Biophys 531(1–2):100–109
Kagan L (2014) Pharmacokinetic modeling of the subcutaneous absorption of therapeutic proteins. Drug Metab Dispos 42(11):1890–1905
Kanda Y, Yamada T, Mori K, Okazaki A, Inoue M, Kitajima-Miyama K, Kuni-Kamochi R, Nakano R, Yano K, Kakita S, Shitara K, Satoh M (2007) Comparison of biological activity among nonfucosylated therapeutic IgG1 antibodies with three different N-linked Fc oligosaccharides: the high-mannose, hybrid, and complex types. Glycobiology 17(1):104–118
Kanodia JS, Gadkar K, Bumbaca D, Zhang Y, Tong RK, Luk W, Hoyte K, Lu Y, Wildsmith KR, Couch JA, Watts RJ, Dennis MS, Ernst JA, Scearce-Levie K, Atwal JK, Ramanujan S, Joseph S (2016) Prospective design of anti-transferrin receptor bispecific antibodies for optimal delivery into the human brain. CPT Pharmacometrics Syst Pharmacol 5(5):283–291
Keizer RJ, Huitema AD, Schellens JH, Beijnen JH (2010) Clinical pharmacokinetics of therapeutic monoclonal antibodies. Clin Pharmacokinet 49(8):493–507
Kelly SM, Price NC (2000) The use of circular dichroism in the investigation of protein structure and function. Curr Protein Pept Sci 1(4):349–384
Khawli LA, Goswami S, Hutchinson R, Kwong ZW, Yang J, Wang X, Yao Z, Sreedhara A, Cano T, Tesar D, Nijem I, Allison DE, Wong PY, Kao YH, Quan C, Joshi A, Harris RJ, Motchnik P (2010) Charge variants in IgG1: isolation, characterization, in vitro binding properties and pharmacokinetics in rats. MAbs 2:613–624
Kinnunen HM, Mrsny RJ (2014) Improving the outcomes of biopharmaceutical delivery via the SC route by understanding the chemical, physical and physiological properties of the SC injection site. J Control Release 182:22–32
Kuo TT, Baker K, Yoshida M, Qiao SW, Aveson VG, Lencer WI (2010) Neonatal Fc receptor: from immunity to therapeutics. J Clin Immunol 30(6):777–789
Lambert JM (2005) Drug-conjugated monoclonal antibodies for the treatment of cancer. Curr Opin Pharmacol 5:543–549
Lammerts van Bueren JJ, Bleeker WK, Bøgh HO, Houtkamp M, Schuurman J, van de Winkel JG, Parren PW (2006) Effect of target dynamics on pharmacokinetics of a novel therapeutic antibody against the epidermal growth factor receptor: implications for the mechanisms of action. Cancer Res 66(15):7630–7638
Lencer WI, Blumberg RS (2005) A passionate kiss, then run-exocytosis and recycling of IgG by FcRn. Trends Cell Biol 15(1):5–9
Levêque D, Wisniewski S, Jehl F (2005) Pharmacokinetics of therapeutic monoclonal antibodies used in oncology. Anticancer Res 25(3c):2327–2344
Li B, Tesar D, Boswell CA, Cahaya HS, Wong A, Zhang J, Meng YG, Eigenbrot C, Pantua H, Diao J, Kapadia SB, Deng R, Kelley RF (2014) Framework selection can influence pharmacokinetics of a humanized therapeutic antibody through differences in molecule charge. MAbs 6(5):1255–1264
List T, Neri D (2012) Biodistribution studies with tumor-targeting bispecific antibodies reveal selective accumulation at the tumor site. MAbs 4(6):775–783
Liu L (2015) Antibody glycosylation and its impact on the pharmacokinetics and pharmacodynamics of monoclonal antibodies and Fc-fusion proteins. J Pharm Sci 104(6):1866–1884
Liu L (2018) Pharmacokinetics of monoclonal antibodies and Fc-fusion proteins. Protein Cell 9(1):15–32
Liu L, Stadheim A, Hamuro L, Pittman T, Wang W, Zha D, Hochman J, Prueksaritanont T (2011) Pharmacokinetics of IgG1 monoclonal antibodies produced in humanized Pichia pastoris with specific glycoforms: a comparative study with CHO produced materials. Biologicals 39:205–210
Lobo ED, Hansen RJ, Balthasar JP (2004) Antibody pharmacokinetics and pharmacodynamics. J Pharm Sci 93(11):2645–2668
Lorber B, Fischer F, Bailly M, Roy H, Kern D (2012) Protein analysis by dynamic light scattering: methods and techniques for students. Biochem Mol Biol Educ 40(6):372–382
Macielag MJ (2012) Chemical properties of antimicrobials and their uniqueness. In: Dougherty T, Pucci M (eds) Antibiotic discovery and development. Springer, Boston, pp 793–820
Mager DE (2006) Target-mediated drug disposition and dynamics. Biochem Pharmacol 72(1):1–10

Martin WL, West AP Jr, Gan L, Bjorkman PJ (2001) Crystal structure at 2.8Å of an FcRn/heterodimeric Fc complex: mechanism of pH-dependent binding. Mol Cell 7(4):867–877

Martins JP, Kennedy PJ, Santos HA, Barrias C, Sarmento B (2016) A comprehensive review of the neonatal Fc receptor and its application in drug delivery. Pharmacol Ther 161:22–39

McLennan DN, Porter CJ, Charman SA (2005) Subcutaneous drug delivery and the role of the lymphatics. Drug Discov Today Technol 2:89–96

Mellman I, Plutner H (1984) Internalization and degradation of macrophage Fc receptors bound to polyvalent immune complexes. J Cell Biol 98:1170–1177

Owens DR (2012) Optimizing treatment strategies with insulin glargine in type 2 diabetes. Expert Rev Endocrinol Metab 7(4):377–393

Palmieri LC, Favero-Retto MP, Lourenco D, Lima LM (2013) A T_3R_3 hexamer of the human insulin variant B28Asp. Physicochem Chem 173–174:1–7

Pechtner V, Karanikas CA, García-Pérez LE, Glaesner W (2017) A new approach to drug therapy: Fc-fusion technology. Prim Health Care 7:1

Pergande MR, Cologna SM (2017) Isoelectric point separations of peptides and proteins. Proteomes 5(1)

Pierce MM, Raman CS, Nall BT (1999) Isothermal titration calorimetry of protein-protein interactions. Methods 19(2):213–221

Porter CJ, Charman SA (2000) Lymphatic transport of proteins after subcutaneous administration. J Pharm Sci 89:297–310

Press OW, Hansen JA, Farr A (1988) Endocytosis and degradation of murine anti-human CD3 monoclonal antibodies by normal and malignant T-lymphocytes. Cancer Res 48(8):2249–2257

Qiu Y, Lv W, Xu M, Xu Y (2016) Single chain antibody fragments with pH dependent binding to FcRn enabled prolonged circulation of therapeutic peptide in vivo. J Control Release 229:37–47

Ratanji KD, Derrick JP, Dearman RJ, Kimber I (2014) Immunogenicity of therapeutic proteins: influence of aggregation. J Immunotoxicol 11(2):99–109

Rehlaender BN, Cho MJ (1998) Antibodies as carrier proteins. Pharm Res 15(11):1652–1656

Richette P, Frazier A, Bardin T (2014) Pharmacokinetics considerations for gout treatments. Expert Opin Drug Metab Toxicol 10(7):949–957

Richter WF, Bhansali SG, Morris ME (2012) Mechanistic determinants of biotherapeutics absorption following SC administration. AAPSJ 14(3):559–570

Righetti PG (2004) Determination of the isoelectric point of proteins by capillary isoelectric focusing. J Chromatogr A 1037(1–2):491–499

Schwartz AL (1991) Trafficking of asialoglycoproteins and the asialoglycoprotein receptor. Target Diagn Ther 4:3–39

Senter PD (2009) Potent antibody drug conjugates for cancer therapy. Curr Opin Chem Biol 13:235–244

Shi S (2014) Biologics: an update and challenge of their pharmacokinetics. Curr Drug Metab 15(3):271–290

Sockolosky JT, Szoka FC (2015) The neonatal Fc receptor, FcRn, as a target for drug delivery and therapy. Adv Drug Deliv Rev 91:109–124

Song K, Yoon IS, Kim NA, Kim DH, Lee J, Lee HJ, Lee S, Choi S, Choi MK, Kim HH, Jeong SH, Son WS, Kim DD, Shin YK (2014) Glycoengineering of interferon-β 1a improves its physicochemical and pharmacokinetic properties. PLoS One 9(5):e96967

Stefanich EG, Ren S, Danilenko DM, Lim A, Song A, Iyer S, Fielder PJ (2008) Evidence for an asialoglycoprotein receptor on nonparenchymal cells for O-linked glycoproteins. J Pharmacol Exp Ther 327:308–315

Strohl WR (2015) Fusion proteins for half-life extension of biologics as a strategy to make biobetters. BioDrugs 29(4):215–239

Strohmeier GR, Brunkhorst BA, Seetoo KF, Meshulam T, Bernardo J, Simons ER (1995) Role of the FCγR subclasses FcγRII and FcγRIII in the activation of human neutrophils by low and high valency immune complexes. J Leukoc Biol 58(4):415–422

Supersaxo A, Hein WR, Steffen H (1990) Effect of molecular weight on the lymphatic absorption of water-soluble compounds following subcutaneous administration. Pharm Res 7(2):167–169

Tabrizi M, Bornstein GG, Suria H (2010) Biodistribution mechanisms of therapeutic monoclonal antibodies in health and disease. AAPS J 12(1):33–43

Thurber GM, Schmidt MM, Wittrup KD (2008) Antibody tumor penetration: transport opposed by systemic and antigen-mediated clearance. Adv Drug Deliv Rev 60(12):1421–1434

Tibbitts J, Canter D, Graff R, Smith A, Khawli LA (2016) Key factors influencing ADME properties of therapeutic proteins: a need for ADME characterization in drug discovery and development. MAbs 8(2):229–245

Vaccaro C, Zhou J, Ober RJ, Ward ES (2005) Engineering the Fc region of immunoglobulin G to modulate in vivo antibody levels. Nat Biotechnol 23(10):1283–1288

Veronese FM, Pasut G (2005) PEGylation, successful approach to drug delivery. Drug Discov Today 10(21):1451–1458

Vidarsson G, Dekkers G, Rispens T (2014) IgG subclasses and allotypes: from structure to effector functions. Front Immunol 5:520

Volund A, Brange J, Drejer K, Jensen I, Markussen J, Ribel U, Sorensen AR, Schlichtkrull J (1991) In vitro and in vivo potency of insulin analogues designed for clinical use. Diabet Med 8(9):839–847

Vugmeyster Y, Xu X, Theil FP, Khawli LA, Leach MW (2012) Pharmacokinetics and toxicology of therapeutic proteins: advances and challenges. World J Biol Chem 3(4):73–92

Wang W, Wang EQ, Balthasar J (2008) Monoclonal antibody pharmacokinetics and pharmacodynamics. Clin Pharmacol Ther 84(5):548–558

Wang W, Erbe AK, Hank JA, Morris ZS, Sondel PM (2015) NK cell-mediated antibody-dependent cellular cytotoxicity in cancer immunotherapy. Front Immunol 6:368

Weinstein JN, van Osdol W (1992) The macroscopic and microscopic pharmacology of monoclonal antibodies. Int J Immunopharmacol 14(3):457–463

Winkelhake JL, Nicolson GL (1976) Aglycosylantibody. Effects of exoglycosidase treatments on autochthonous antibody survival time in the circulation. J Biol Chem 251:1074–1080

Wright A, Morrison SL (1994) Effect of altered CH2-associated carbohydrate structure on the functional properties and in vivo fate of chimeric mouse-human immunoglobulin G1. J Exp Med 180:1087–1096

Wright A, Sato Y, Okada T, Chang K, Endo T, Morrison S (2000) In vivo trafficking and catabolism of IgG1 antibodies with Fc associated carbohydrates of differing structure. Glycobiology 10:1347–1355

Wu AM, Senter PD (2005) Arming antibodies: prospects and challenges for immunoconjugates. Nat Biotechnol 23:1137–1146

Yang BB, Kido A (2011) Pharmacokinetics and pharmacodynamics of pegfilgrastim. Clin Pharmacokinet 50(5):295–306

Yu M, Brown D, Reed C, Chung S, Lutman J, Stefanich E, Wong A, Stephan JP, Bayer R (2012) Production, characterization, and pharmacokinetic properties of antibodies with N-linked mannose-5 glycans. MAbs 4:475–487

Zalevsky J, Chamberlain AK, Horton HM, Karki S, Leung IW, Sproule TJ, Lazar GA, Roopenian DC, Desjarlais JR (2010) Enhanced antibody half-life improves in vivo activity. Nat Biotechnol 28(2):157–159

Zhao L, Shang EY, Sahajwalla CG (2012) Application of pharmacokinetics-pharmacodynamics/clinical response modeling and simulation for biologics drug development. J Pharm Sci 101(12):4367–4382

Stability of Therapeutic Enzymes: Challenges and Recent Advances

7

Shubhrima Ghosh, Shahenvaz Alam, Anurag S. Rathore, and S. K. Khare

Abstract

Enzymes are biocatalysts that have found profound applications in the current biotherapeutic industry and play a crucial role in diagnosis, prevention, and biochemical analysis of major diseases. However, stability, protein degradation and immunogenicity in the body present unique challenges that are faced upon sustained use of such enzymes. The present chapter is an attempt to dissect the state-of-the-art in relation to the challenges of development of therapeutic enzymes and the recent advances to address them. At the very outset, diseases where enzymes have found effective applications and the various causes of enzyme instability have been discussed. In recent times, polymer or nano- conjugated resistant delivery methods, as well as mutagenesis have led to manifold increase in enzyme stability against thermal denaturation, acidic gut environment, proteolysis and immunogenicity. Further, methods of analytical characterization of proteins have been highlighted and explored to shape future research directions.

Keywords

Enzyme · Metabolic disorders · Immunogenicity · Protein denaturation · Mutagenesis

Abbreviations

US	United States
FDA	Food and drug administration
AML	Acute Myeloid Leukemia
ALL	Acute Lymphoblastic Leukemia
PEG	Polyethylene glycol
ADP	Adenosine Diphosphate
NAD	Nicotinamide Adenine Dinucleotide
ERT	Enzyme Replacement Therapy
GAG	Glucosaminoglycans
DNA	Deoxyribonucleic Acid
LAL	Lysosomal Acid Lipase
MPS	Mucopolysaccharidosis
LIPA	Lysosomal Acid Lipase
SCID	Severe Combined Immune Deficiency
ADA	Adenosine Deaminase
PPFE	Pleuroparenchymal Fibroelastosis
RNA	Ribonucleic Acid
DUB	Deubiquitinating Enzyme
GI	Gastrointestinal
BAb	Binding Antibodies
GD	Gaucher's Disease

S. Ghosh · S. Alam · S. K. Khare (✉)
Enzyme and Microbial Biochemistry Laboratory, Department of Chemistry, Indian Institute of Technology Delhi, New Delhi, India
e-mail: skkhare@chemistry.iitd.ac.in

A. S. Rathore
Department of Chemical Engineering, Indian Institute of Technology Delhi, New Delhi, India

N. Labrou (ed.), *Therapeutic Enzymes: Function and Clinical Implications*, Advances in Experimental Medicine and Biology 1148, https://doi.org/10.1007/978-981-13-7709-9_7

GALK1	Galactokinase 1
EC	Enzyme Commission
ATP	Adenosine Triphosphate
ABD	Albumin-Binding Domain
PVP	Poly(N-vinylpyrrolidone)
LbL	Layer-by-layer
MNP	Magnetic Nanoparticles
PAGE	Poly-Acrylamide Gel Electrophoresis
SDS	Sodium Dodecyl Sulfate
IEF	Isoelectric Focusing
2D	2-dimension
DIGE	Difference Gel Electrophoresis
HPLC	High-performance liquid chromatography
CD	Circular Dichroism
NMR	Nuclear Magnetic Resonance
XRD	X-ray Diffraction
FTIR	Fourier-Transform Infrared Spectroscopy
IR	Infrared
DLS	Dynamic Light Scattering
HT-DLS	High- Throughput Dynamic Light Scattering
UVRR	Ultraviolet Resonance Raman
ROA	Raman Optical Activity
ANS	8-Anilinonaphthalene-1-sulfonic acid
ThT	Thioflavin-T
TEM	Transmission Electron Microscopy
ELISA	Enzyme-Linked Immunosorbent Assay
SPR	Surface Plasmon Resonance
BLI	Bio-Layer Interferometry
ITC	Isothermal Titration Calorimetry
tPa	Tissue Plasminogen Activator
SKA	Streptokinase

7.1 Introduction

Proteins and peptides currently constitute a major share of therapeutics against several diseases. More than 239 proteins have been approved by US-FDA to be used as therapeutics while numerous others are undergoing clinical trials (Usmani et al. 2017). Recent developments in pharmaceutical biotechnology have increased the diversity of protein based therapeutics to include antibody-drug conjugates, vaccines, enzymes, cytokines, interferons and other recombinant protein medicines. Enzymes are a class of proteins which possess great potential as effective therapeutic drugs because of their specificity, high activity, and selectivity. However, due to their complex structures, they are prone to inactivation and unfolding. Therapeutic enzymes have found remarkable applications in rare disorders such as lysosomal storage diseases and severe combined immunodeficiency. Genetically engineered enzymes have been developed as successful drugs for chronic ailments such as cardiovascular diseases. Some crucial enzymes also find applications as diagnostics and markers for chronic disorders. The major differentiation of therapeutic enzymes based on their areas of application are discussed here and in Fig. 7.1.

7.1.1 Cancer

Treatment for cancer forms the most important of applications for therapeutic enzymes. L-Asparaginases, sourced from *E.coli* or *Erwinia chrysanthemi*, find wide applications in the therapy of acute myeloid leukemia (AML), acute lymphoblastic leukemia (ALL), and non-Hodgkin's lymphoma (Duval et al. 2002; Egler et al. 2016). It is also used in food manufacturing to decrease acrylamide content (Gökmen 2015). As a chemotherapeutic agent, asparaginase can be administered as an intramuscular, subcutaneous, or intravenous injection. Treatment by L-asparaginase leads to rapid depletion of asparagine in the cellular environment. Normal cells are able to synthesize asparagine while metabolic defect in asparagine synthesis inhibits malignant cells to do so. ALL patients may be hypersensitive to the native, unmodified L-asparaginase. Hence, PEGylated form of L-asparaginase has been recently approved for frontline use by FDA. Diptheria toxin (produced by *Corynebacterium diphtheria*) is another oncolytic enzyme still in experimental stages. This enzyme catalyzes transfer of the adenosine diphosphate ribose (ADP-ribose) moiety of nicotinamide adenine dinucleotide (NAD) to elongation factor 2,

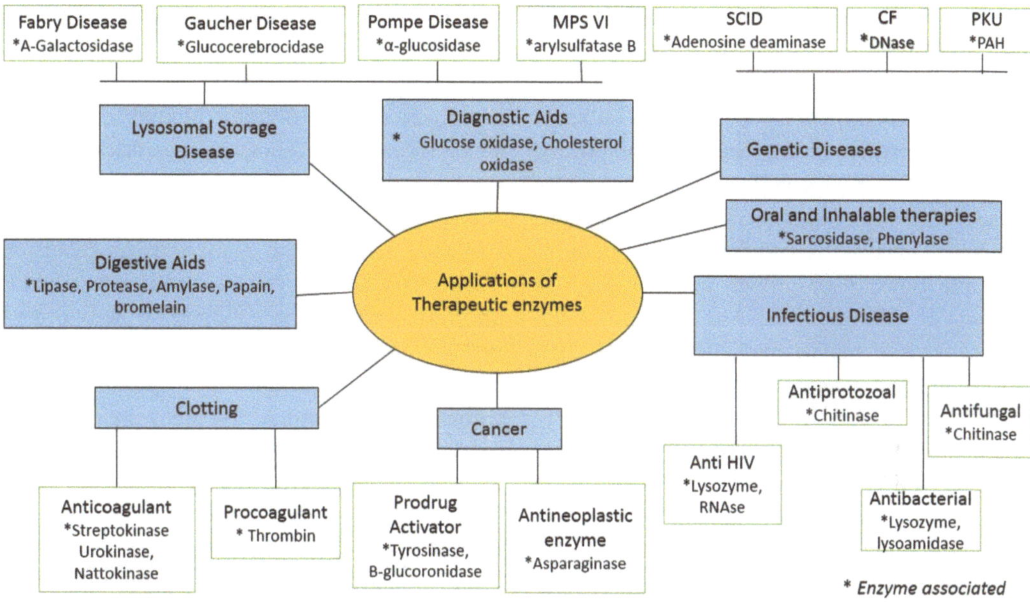

Fig. 7.1 Applications of therapeutic enzymes

thus halting protein synthesis. The protein synthesis machinery in malignant cells is more sensitive to this toxin as compared to normal cells, thus making it an effective anti-cancer drug. Denileukin diftitox (Ontak) is an antineoplatic agent comprising of engineered protein containing Diphtheria toxin and Interleukin-2 (Lansigan et al. 2010; Kaminetzky and Hymes 2008). Several other targeted toxins have undergone clinical trials in brain tumour models (Li et al. 2013).

Neuraminidase is an emerging target in sialidase mediated tumorigenesis regulation. This enzyme removes sialic acid residues from the surface of cancerous cells which alters their immunogenicity, and makes them sensitive to immune response (Haxho et al. 2016). Ribonucleases (Leland and Raines 2001; Shlyakhovenko 2016) and diverse group of proteolytic enzymes (Chabot et al. 2010) have also been explored as anti-neoplastic agents.

7.1.2 Inherited Metabolic Disorders

These diseases are recognized by an inherited metabolic deficiency to produce enzymes. Enzyme Replacement Therapy (ERT) is among the most effective and widely used treatments for such disorders. The most common of these deficiencies is that of the enzyme glucose-6-phosphate dehydrogenase which makes red blood cells more vulnerable to breaking down, and hemolytic anemia can occur. This is most frequently seen in African, Asian, Mediterranean or Middle-Eastern populations. The treatment is mostly symptomtic. Gaucher's disease is another autosomal recessive inherited deficiency of the enzyme glucocerebrosidase, resulting in the accumulation of a type of lipid called glucocerebrosides in the internal organs such as liver, spleen, bone marrow and the brain. Enzyme replacement therapy, with a modified form of the enzyme (intravenous infusion every 2 weeks) may help to reduce the skeletal anomalies and abnormal blood counts associated with certain types of this disorder (Kasturi and Amin Alpa 2001). The Hunter syndrome, characterized by stunted growth, coarse facial features, impaired intelligence, hearing loss, joint stiffness and thickening of heart valves, occurs in young males due to the deficiency of the enzyme iduronate-2-sulfatase that helps breakdown glycosaminoglycans (GAG). Idursulfase (Elaprase) (Muenzer et al. 2006; Da Silva et al. 2011) and idursulfase-beta (Hunterase) (Ngu et al. 2017) are the currently approved enzymes for ERT. However, the current preparations cannot cross the blood-brain barrier, thus making it unable to address intellectual impairment associated with Hunter's. Cystic fibrosis is another inherited autosomal recessive disorder which affects mostly the lungs, but also other organs such as liver, pancreas, intestine and kidneys. Symptoms include frequent lung infections, poor growth and infertility in most males. ERT using DNAse and lipase (pancreatic enzyme replacement therapy) has been successful in managing cystic fibrosis (Calvo-Lerma et al. 2017; Shak et al. 1990).

A major class of enzyme deficiency disorders is the Lysosomal Storage Diseases, including Fabry disease, Pompe disease, Maroteaux-Lamy Syndrome, Morquio A syndrome, and LAL Deficiency. Fabry disease is an X-linked lysosomal storage disorder caused by deficiency of the enzyme alpha-galactosidase A, which leads to the accumulation of globotriaosylceramide in skin, heart, kidney, eye, brain, and other organs. Patients are at an increased risk of kidney failure, heart attack, and stroke. Currently, the only treatment available is agalsidase alfa (marketed as Fabrazyme) which needs to be infused every 2 weeks (Alegra et al. 2012). Pompe disease occurs due to the genetic deficiency of the lysosomal acid α-glucosidase, leading to the accumulation of glycogen in cardiac, skeletal and smooth muscle, which results in debilitating muscle weakness. Recombinant human acid α-glucosidase has been used in infantile Pompe disease patients with good glycogen clearance in post treatment period (Thurberg et al. 2006; Angelini and Semplicini 2012). Maroteaux–Lamy syndrome, also known as mucopolysaccharidosis type VI, is caused by a deficiency in arylsulfatase B, causing neurological complications and dwarfism, while skeletal changes limit movement. A study of ERT with recombinant human arylsulfatase B, in two patients, improved their walking capability (But et al. 2011). A similar autosomal recessive disorder is Morquio syndrome (referred to as mucopolysaccharidosis IV, MPS IV). The two forms of the disease are distinguished by the deficiency of different gene products. A involves malfunction in galactosamine-6 sulfatase, while B involves a malfunction of the beta-galactosidase. ERT of MPS IV has been achieved by elosulfase alfa (Vimizim) (Lyseng-Williamson 2014). Lysosomal acid lipase (LAL) deficiency is another rare autosomal recessive lysosomal storage disease where the body does not produce enough active lysosomal acid lipase (LAL or LIPA) enzyme due to malfunctioning of the cholesterol metabolism. The lack of the LAL enzyme can lead to a build-up of fatty material in a number of body organs including the spleen, liver, and in the wall of blood vessels. Sebelipase alfa (Kanuma) is a recombinant human LAL protein which has been successfully used in long-term therapy in LAL affected individuals (Su et al. 2016).

ERT has been used successfully in the treatment of life-threating diseases such as severe combined immunodeficiency (SCID) for more than 20 years. Patients suffering from SCID have an impaired immune system, making them extremely vulnerable to infectious diseases. The second most common form of SCID arises from the deficiency of the enzyme Adenosine deaminase, which results in inhibition of lymphocyte proliferation. ERT with polyethyleneglycol-coupled adenosine deaminase metabolizes the toxic substrates of the ADA enzyme and prevents their accumulation (Chan et al. 2005; Hershfield et al. 1993). This restore T-cell functions for short term before the patient can undergo bone-marrow transplant.

7.1.3 Fibrinolytic and Anticoagulant Enzymes

Blood clots in arteries are hallmark of heart attacks and demand immediate removal of the clots to revive the patient. Clot formation is mediated by aggregation of platelets and formation of fibrin. In recent times, the thrombolytic enzymes such as Streptokinase (a protease derived from *Streptococcus*) along with Tissue Plasminogen Activator, Urokinase, Retavase, Tenectaplase have been used as "clot busters" to dissolve clots in the walls of the arteries. They involve the conversion of plasminogen into plasmin which finally degrades the fibrin and hence dissolving the clot (Sikri and Bardia 2007).

Anticoagulant enzymes are commonly used as blood thinning agents, prolonging the clotting time. Antiplatelets are closely related drugs which inhibit the aggregation of platelets, hence preventing the formation of clots. These enzymes are useful in the therapy of thrombotic disorders. PPFE-1 is an anticoagulant enzyme isolated from the bacterium *Paenibacillus polymyxa* (Lu et al. 2010). Similar enzymes such as phospholipase A2 have been screened from snake venoms (Kini 2005). Antithrombic peptides such as antithrombin, heparin and heparin mimics are considered as new classes of anticoagulants (Desai 2004).

7.1.4 Infectious Diseases

With increasing challenges of antibiotic resistance in microbes, the search is on for new drug targets and antibacterial/antiviral agents. Lysozyme (Griswold et al. 2014; Vereshchagin et al. 1985) and lysoamidase (Tishchenko et al. 2016) cause bacterial cell hydrolysis and thus are used in antibiotic therapy. Ribonuclease is used as an antiviral agent due to the fact that it causes hydrolysis of RNA (Ilinskaya and Mahmud 2014; Shah Mahmud et al. 2017). Enzymes such as trypsin, chymotrypsin and bromelain have been found effective against inflammatory responses, probably due to their ability to degrade protein-based inflammatory mediators (Jayachandran and Khobre 2017; Ghaffarinia et al. 2014). As a preventive measure, subtilisin has been found to degrade prion proteins in soil, thus reducing their transmission via soil or other environmental surfaces (Saunders et al. 2011). Fungal cell-wall lytic enzymes such as chitinases, glucanases, and proteases produced by actinomycetes and bacteria are found to be effective antifungal drugs (Choudhary et al. 2014)

Enzymes can also be potential drug targets. Deubiquitinating enzymes (DUBs) remove ubiquitin and ubiquitin-like modifications from proteins and they have been known to contribute to immune response pathways in microbial infections. Therapeutic inhibitors of DUBs have been designed, that can be used to limit the spread of infection (Nanduri et al. 2013). Various other enzymes such as DNA topoisomerases (Bakshi and Shapiro 2003), cysteine proteases, disulfide oxidoreductases, glutathione reductases and thioredoxin reductases have also been targeted against parasitic infections (Chibale 2002).

7.1.5 Burn Debridement

The removal of scar tissue in burn patients is of utmost importance for speedy recovery. Enzymatic debridement involves the controlled digestion and removal of necrotic tissues from a wound or a burn using exogenous proteolytic enzymes such as papain along with denaturing agents such as urea. This technique is considered advantageous than surgical debridement owing to non-selectivity and removal of surrounding healthy tissues (Langer et al. 2013). Extensive studies have proven the high efficacy of Debridase (obtained from crude Bromelain) in scar debridement in burn injuries (Rosenberg et al. 2004). However, side effects such as fever, pain and itching should also be considered for such applications. Additionally, trypsin from mammalian pancreas (which hydrolyses peptide bonds involving arg and lys) and collagenase sourced from *Clostridium* and culture extracts from animal cells are also used for debridement of necrotic tissues in wounds.

7.1.6 Digestive Aids

Most digestive enzymes are depolymerases which aid in the breakdown of complex proteins, polysaccharides and lipids. Such enzymatic preparations may consist of a single or multiple enzymes. Microbial α-amylases which hydrolyse α 1-4 glycosidic bonds are orally administered to aid in digestion and better absorption by the body. Lactase is another such enzyme which is widely used in patients having reduced inherent lactase activity and consequent lactose intolerance. Cellulase and xylanase are effective in digesting fruit and plant fibers. Other proteolytic enzymes such as papain or pepsin are also used to enable efficient protein digestion. They can be used as a supplement to alleviate autoimmune and inflammatory conditions and pain. Patients suffering from chronic pancreatitis, pancreatic carcinomas and cystic fibrosis are found to be deficient in pancreatic enzymes and can be successfully treated with pancreatin, a preparation extracted from pancreas containing various enzymes. Most of these enzymes are orally administered and are prone to gastric inactivation. Co-administration of inhibitors of gastric acid secretion, and enteric coated tablet or capsules help to overcome such inactivation problems.

Functions and applications of some key therapeutic enzymes are discussed in Table 7.1.

7.2 Why Are Administered Enzymes Unstable in the Body?

A major problem encountered with therapeutic enzymes is their instability and low circulation life in the human body. The therapeutic enzymes are generally administered to the patients via oral, subcutaneous, intramuscular or intravenous routes. Sourced mainly from bacterial sources, the temperature, pH and salinity of the administration route and destination site are often not optimal for the enzyme activity, while harsh environments can lead to protein denaturation. Presence of proteolytic enzymes and rejection by the immune system are among important factors affecting the stability of therapeutic enzymes,

Table 7.1 Some important therapeutic enzymes, their functions and applications

Enzyme	Disease	Function	References
L-glutaminase	Acute lymphoblastic leukemia	Blocking glutamine uptake, cell death due to starvation	Sarada (2013)
L-asparaginase	Acute lymphoblastic leukemia	Depletion of nutrient asparagine leads to tumor apoptosis	Goodsell (2005)
Serratiopeptidase	Inflammation and inflammatory disorder	Regulates recruitment of lymphocytes at the site of inflammation	Mouneshkumar Chappi et al. (2015)
Caspase	Cancer	Regulates death of cancerous cell via apoptosis	Perchellet et al. (2004)
Lipase	Cardiovascular diseases	Lowers the level of triglycerides	Alpers (2008)
Alginate lyase	Cystic fibrosis	Prevent mucoid formation and enhances antibiotic efficacy	Lamppa and Griswold (2013)
Chondroitinase	Acute spinal cord injury	Digests extracellular chondroitin sulphate proteoglycans, enhance axonal regeneration	Wang et al. (2011)
Superoxide dismutase	Oxidative stress and inflammation	Diminishes oxidative stress and prevent inflammation	Ishihara et al. (2009)
Penicillin acylase	β-lactam resistance	Production of 6-aminopenicillanic acid as β-lactam antibiotics	Illanes and Valencia (2017)
Laccase	Chronic lymphocytic leukemia	Possess cytotoxic activity against leukemia	Matuszewska et al. (2016)
Tissue plasminogen Activator (tPa)	Hepatic metastasis	Inhibits the growth of cancerous cells	Hayashi et al. (1999)
L-arabinofuranosidase	Adenocarcinoma	Inhibits DNA synthesis, cytotoxic to tumor cells	Sabu (2003)
Laronidase	Mucopolysaccharidosis I (genetic disorder)	Provide alternative function for IDUA enzyme	Liao et al. (2018)
Collagenase	Pyronie's disease	Treats skin ulcer and connective tissue disorders	Cocci et al. (2018)
Streptokinase	Coagulation and thrombosis	Activates plasminogen	Reed et al. (1998)

some of which are discussed below. These instability may lead to degradation via various processes like aggregation, denaturation, hydrolysis and oxidation of the enzymes.

7.2.1 Oral Administration

Oral administration of exogenous enzymes are currently undertaken for diseases such as exocrine pancreatic insufficiency, phenylketonuria and celiac disease. The aim of oral enzyme therapy is either to breakdown harmful substances or to supplement inherent digestive enzyme deficiencies (Fuhrmann and Leroux 2014). Most of the enzymes administered through the oral route face the potential possibility of degradation and inactivation owing to the harsh conditions in the gastrointestinal tract (GI). The difficulty arises due to the fact that the primary function of the GI tract is to enable protein digestion. The digestive system in mammals is divided into the stomach, small intestine (duodenum, jejunum, and ileum), and large intestine (cecum, colon, rectum, and anal canal). The primary obstacle is faced by oral enzymes in the stomach. Protein digestion begins in the stomach with the action of peptidases at a low acidic pH (1–2.5) (Cook et al. 2012; Fuhrmann and Leroux 2011). In this step, aided by protein unfolding, hydrolysis of peptide bonds at hydrophobic residues such as phenylalanine and leucine takes place (Hamuro et al. 2008). After the stomach, the ingested proteins enter the intestine, where they encounter neutral pH (Cook et al. 2012). However, they are still prone to the action of endopeptidases such as trypsin and chymotrypsin that preferentially cleave peptide bonds involving arginine, lysine and aromatic amino acids, respectively (Fuhrmann and Leroux 2011). Additional denaturation may occur due to the action of intestinal bile salts (Cook et al. 2012). The major strategies to improve stability of oral enzymes thus involves shielding the protein from the acidic environment of the GI tract.

7.2.2 Immunogenicity of Enzyme Drugs

In the context of therapeutic enzymes, immunogenicity refers to the immune response of the host against the enzyme which presents itself as an antigen. This leads to the activation of both cellular (T cell) and humoral (antibody) arms of the immune response. Antibodies directed against therapeutic enzymes may consist of IgM, IgG, IgE, and/or IgA isotypes. These antidrug antibodies (ADA) are of two types- (a) Binding Antibodies (BAbs), all isotypes of which are capable of binding to the therapeutic product, and (b) Neutralizing antibodies, which are a subpopulation of the total BAbs (Scagnolari et al. 2002). They inhibit functional activity of the therapeutic protein and are generally directed against the biologically active site of the drug. These antibodies are detected through immunoassays, cell-based bioassays or complementary ligand binding assays. In the treatment of many lysosomal storage diseases such as MPS and Pompe disease, ERT often results in the development of antibodies against the enzyme drug leading to a reduction in its efficacy (Harmatz 2015). In Gaucher's disease, it was found that 12.8% of patients develop antibody to glucocerebrosidase, while 90% of patients tolerize over time (Rosenberg et al. 1999). Many a time, cheaper, impure enzyme preparations with lacking relevant clinical studies, can lead to high immunogenicity. Higher anti-asparaginase antibodies titres was obtained for the biosimilar in comparison to the well-known Medac l-asparaginase in one study (Zenatti et al. 2018). Glycosylated proteins are common candidates for immunogenicity. Therapeutics containing non-human sugars produce specific antibodies against them. For example, taliglucerase alfa used in Gaucher's disease, produced *in planta*, elicitated a 13.5% response in healthy humans and 5% in glucocerebrosidase-deficient Gaucher disease (GD) patients (Zenatti et al. 2018). It is possible to predict the immunogenicity based on the following factors-

- Product derivation- Foreign, Self or Fusion
- Product specific attributes- Molecular structure, Purity, Formulation, Mechanism of action
- Patient specific factors- Population, Genetics, Age, Gender, Pre-existing antibodies
- Trial design attributes- Route of delivery, Dosage and frequency of administration, Immunomodulatory properties of the drug, Stage of product development

7.2.3 Temperature and Salt Concentration

Another factor that can influence the stability of therapeutic proteins is temperature. Mostly, extracted enzymes and other therapeutic proteins are stored at 4 °C for extended periods of time to retain their activity and structural integrity (Akash et al. 2015). Storage at room temperature degrades most therapeutic enzymes, which can be a problem for transport and storage in places where cooling devices are unavailable. During delivery, the human body temperature of 37 °C might not be optimal for the enzymes, and result in lower activity. Particular temperature sensitive enzymes may even undergo denaturation at this temperature. Additionally, diagnostic enzymes such as glucose oxidase, cholesterol oxidase need to work at room temperature (Yoo and Lee 2010; Pundir et al. 2012). Thus, current research on therapeutic enzymes has been increasingly focusing on temperature stable preparations. Protein aggregation is also influenced by salt concentrations of the plasma and extracellular fluids.

7.2.4 Protein Degradation

Therapeutic enzymes can lose their activity as a result of proteolysis, aggregation, and suboptimal conditions in cell compartments and extracellular fluids. Other routes of chemical degradation include oxidation, isomerization, deamidation, disulfide modifications, chemical cross-linking and fragmentation. Oxidation accounts for one of the major mechanisms of degradation under stress conditions as the amino acids Cys, Met, Trp, and His are all predisposed to oxidation (Schöneich 2000; Luo et al. 2011). These reactions can be triggered by light, presence of Reactive Oxygen Species, peroxide, or by the presence of oxygen during the manufacturing and storing process. The amino acid Asparagine is particularly susceptible to deamidation, and this can result in structural and functional modifications of proteins, owing to the introduction of negative charge (Kosky et al. 1999). Aggregates are mostly a result of chemical modifications inside the protein (Luo et al. 2011).

Proteolysis of therapeutic enzymes is a major drawback limiting their applications. Proteases are ubiquitous enzymes, present in almost all cells and tissues participating in digestion, post-translational processing, subcellular localization, fibrinolysis and hemostasis (Neurath 1984). The specific amino acid motifs in the peptide backbone render the therapeutic enzymes prone to degradation by proteases present in serum or plasma when delivered through injection routes. In blood coagulation, mostly serine proteases such as thrombin cleave sites C-terminal to lysine or arginine residues in P1 position of certain sequence motifs (Karstad et al. 2012). These proteases are also capable of degrading therapeutic proteins that contain residues important for cell entry or binding to the target via ionic interactions (Krizsan et al. 2014; Knappe et al. 2016). Through oral routes, digestive enzymes such as trypsin and chymotrypsin, and other non-specific proteases degrade proteins into peptides. Current efforts target removal or modification of amino acid motifs which can be cleaved by the proteases at the destination site or route of delivery of the enzyme.

7.2.5 Metabolism and Elimination

Hepatic first-pass metabolism and rapid elimination are a major problem in case of oral administration of therapeutic enzymes. This leads to reduced bioavailability and short biological half-life in circulation. Invasive delivery or use of encapsulation/polymer conjugation can help overcome this problem (Akash et al. 2015).

7.3 Chemical Basis of Stability

The molecular basis of denaturation of proteins is the disruption of secondary and tertiary structure. The calculation of the free energies of various determinants gives an idea of their contribution, and would be helpful in developing approaches for stable therapeutic preparations (Wong and Wong 1992). Of the various factors that contribute to a stable protein structure, hydrophobicity of amino acids is the most prominent. Hydrophobic amino acids are generally buried inside a protein, away from the surface, resulting in a tight packing. Solitary hydrophobic clusters may occur on the surface to facilitate interactions with lipids or other hydrophobic entities. Protecting the nonpolar groups from watery environment reduces the unfavorable entropy of the system, increasing protein stability (Tanford 1980). Other forces such as hydrogen bonding, van der Waals forces, salt bridges, and other electrostatic interactions also contribute to protein stability. Hydrogen bonding plays a major role in the secondary structure of proteins, giving rise to α-helices, β-sheets, random coils and turns (McDonald and Thornton 1994; Baker and Hubbard 1984). The energy contribution of salt bridges towards protein stability is also substantial (Hendsch and Tidor 1994). Significant contributions are also made by dipole-dipole interactions, electrostatic forces, along with binding with co-factors, ligands and metal ions which together maintain the integrity of the proteins (Burley and Petsko 1988).

7.4 Increasing Enzyme Stability

As is evident from the above factors, it is imperative to improve the stability of the enzymes during delivery and at the destination tissue. For attaining this target, it is essential to have a complete understanding of the type of protein, its structure-function relationships as well as requirement of formulation and delivery for their proper therapeutic effect. The strategies employed for stabilization of therapeutic enzymes can be broadly divided into (1) functional modification and (2) formulation.

7.4.1 Functional Modification

In this approach, tools are employed to alter the functionality of the enzymes by introducing structural modifications at the molecular level.

7.4.1.1 Mutagenesis and Protein Engineering

A commonly used approach that is applied to increase the contribution of various forces to the free energy of stabilization is protein engineering. It involves site-specific mutagenesis and cloning. In case of digestion of proteins in the GI tract, the replacement of leucine and phenylalanine residues by analogues which are not recognized by enzymes such as pepsin have been shown to increase their stability (Ehren et al. 2008). In another instance, *Thermomyces lanuginosus* lipase has been suggested as potential enzyme replacement candidate for pancreatic insufficiency, wherein the mutant was reported to be more stable at acidic pH and showed increased refolding capacity compared to the native enzyme (Wang et al. 2013). Galactokinase (GALK1; EC 2.7.1.6) is an enzyme which catalyses the ATP-dependent phosphorylation of galactose to form galactose-1-phosphate, as the first committed step of the Leloir pathway of galactose metabolism. Mutations in the human galactokinase gene results in the inherited recessive metabolic disorder, Type-II galactosaemia, characterized by decreasing ability to metabolize galactose, resulting in the early onset of cataracts. In addition to removal of galactokinase from diet, emerging treatment modalities resort to Enzyme Replacement Therapy for some of the inherited disorders. However a frequent problem witnessed is the short half-lives of proteins in cells or plasma. Researchers have identified six sites in galactokinase which deviate from the consensus sequences and restoring the consensus by combining all the six mutations increased the thermal stability of the enzyme without compromising the activity (McAuley et al. 2018).

As discussed above, Streptokinase is widely used to remove clots as it converts plasminogen to plasmin. However, plasmin cleaves peptide bonds between Lys and Arg residues, thus degrading Streptokinase, and requiring continuous infusions every 30–90 min. To make Streptokinase less susceptible, Lys at 59 and 386 were changed to Glu by site directed mutagenesis. The half-life of the mutants increased and the double mutant was 21-fold more resistant to proteolysis (Wu et al. 1998).

In another case, human variants of β-glucuronidase were identified, as a replacement for α-iduronidase, which is defective in MPS type- I patients. This would help overcome immune response which develops while using exogenous enzymes. Six residues were identified as hot spots in the catalytic domain of β-glucuronidase and were mutated to variable amino acids to enrich the library diversity. The screening of human β-glucuronidase library isolated variants which exhibited 100 to 290-fold greater activity against the α-iduronidase substrate 4-methylumbelliferyl α-L-iduronide and 7900 to 24,500-fold shift in enzymatic specificity as compared with the wild-type β-glucuronidase (Chuang et al. 2015).

7.4.1.2 Chemical Cross-Linking

Intra- and inter-molecular cross-linking is an important technique that is used to increase the proteolytic stability of enzymes and proteins, as they reinforce the active structure. Disulfide bonds are found to destabilize folded structures entropically, but stabilize them enthalpically to a greater extent, resulting in an overall stabilization of the protein molecule (Doig and Williams 1991). Various bifunctional reagents such as carbodiimides, chloroformates, disuccinimidyl suberate, dimethyl malonimidate, N-succinimidyl 3-maleimidopropionate and ethyl iodoacetimidate have been used to enhance enzymatic stability. These chemicals react with nucleophilic side chains of amino acids, such as the sulfhydryl group of cysteine, the amino groups of lysine, the carboxyl groups of aspartic and glutamic acids, the imidazolyl group of histidine, and the thioether group of methionine (Wong and Wong 1992). Enzymes such as catalase (Shaked and Wolfe 1988), amyloglucosidase (Tatsumoto et al. 1989) and α-chymotrypsin (Torchilin et al. 1978) have been stabilized using such agents. However, disulfide bridges are at risk of thiol-disulfide exchange or reduction during production, purification, or therapeutic use. As an alternative, thioether bridges are being explored for intermolecular cross-linking. Variants of the 46 aa albumin-binding domain (ABD) (which acts as a fusion partner for the human serum albumin), with single or double intramolecular thioether bridges were designed and synthesized. One promising variant, ABD_CL1, showed high thermal stability and dramatically increased stability in the presence of pepsin and trypsin/chymotrypsin, compared to the control protein (Lindgren and Eriksson Karlström 2014). The structural basis of cross-linking is derived from the fact that cross-linking reinforces active conformation. Intra- or intermolecular crosslinks reticulate the protein molecule to diminish the polypeptide entropy, which decreases the rate of denaturation. With the help of multiple cross-linked attachments, the molecule is held in its original active structure (Wong and Wong 1992).

7.4.2 Formulation

In this approach, instead of making structural modifications, the formulation of delivery of native therapeutic enzymes is altered.

7.4.2.1 Polymer Conjugation

Since the 1970s, the research in protein-polymer conjugation has been dominated by the versatile Poly(ethylene glycol). Owing to its biocompatibility and non-toxicity, PEGylation has been found to be one of the most successful techniques of improving the pharmacodynamics of injected proteins by increasing their circulating half-life and masking their immunogenicity (Harris and Chess 2003; Qi and Chilkoti 2015). Site-specific PEGylation in streptokinase was achieved by incorporation of cysteine at 'optimal' surface exposed sites within all the three domains in the enzyme, followed by PEGylation with 20 KDa PEG groups. The screened biologically active variants exhibited subdued immunoreactivity, proteolytic stability and 2 to 20-fold increase in half-life (Sawhney et al. 2016). PEGylation has also been successfully used in case of the enzyme phenylalanine hydroxylase, used as an ERT for phenylketonuria, providing more activity and stability to the enzyme (Gámez et al. 2004).

Some limitations of PEG such as non-degradability has led to the search for newer conjugation materials. Non-PEG alternatives such as Poly(N-vinylpyrrolidone) (PVP), polyglycerol, Poly(amino acid)-Based Hybrid Materials and recombinant polypeptides are finding increasing applications in protein conjugation (Qi and Chilkoti 2015). Many a time, gastro-resistant, cellulose acetate phthalate or methacrylate copolymers-coated tablets or microparticles are used to avoid digestion in the GI tract (Imrie et al. 2010). Microencapsulation is another technique which employs polymers for conjugation with enzymes. Catalytically active preparations of L-asparaginase, were made by microencapsulation. At the outset, a template of $CaCO_3$ template was generated by co precipitation with L-asparaginase. The polyelectrolyte microcapsules were then made by layer-by-layer (LbL) coating of two or three poly dextran/poly-L-arginine-based bilayers. Finally, the template material is dissolved, which ultimately gives rise to hollow capsules that contain the initially adsorbed enzyme. The resulting microcapsules showed increased stability in the presence of proteases such as trypsin and thrombin and resistance to thermal inactivation at 37 °C (Karamitros et al. 2013).

7.4.2.2 Nanoparticle Immobilization

Nanoparticles present the newest and one of the most versatile systems for conjugation and delivery of therapeutic enzymes, while ensuring stability. There are in fact, a multitude of nanocarriers and molecules available for selective tumour targeting in cancer. Gold nanoparticles have proven to be an inert, non-toxic carrier for many protein drugs. The interaction of gold nanoparticles with thiols, provides an effective and selective means of controlled intracellular release (Ghosh et al. 2008). Functionalized gold nanoparticles with long hydrocarbon ligand chains have been investigated to stabilize the delivery of DNA, RNA, peptides and proteins (DeLong et al. 2010). In another instance, Magnetic iron (II, III) oxide nanoparticles (MNP) have been employed to immobilize bacterial cholesterol oxidase with improved thermal and pH stability as well as reusability (Ghosh et al. 2018). Researchers have immobilized antioxidant enzymes such as, catalase and superoxide dismutase onto non-polymeric MNPs as a treatment for diseases involving increased production of reactive oxygen species and these nanoconjugates were found to exhibite resistance to proteolysis (Chorny et al. 2010). Other polymeric nanoparticle-protein conjugates have also been developed. In a recent study, biodegradable poly(lactic-co-glycolic acid) nanoparticles carrying rapamycin as an immunomodulatory agent, has shown its capability to induce immune tolerance to co-administered proteins. This is characterized by inhibition of antigen-specific hypersensitivity reactions, an increase in regulatory T cells, a reduction in B cell activation, and the induction of tolerogenic dendritic cells (Kishimoto et al. 2016). Polymeric nanoparticles (200–300 nm) composed of biodegradable block copolymers poly(ethylene glycol)-b-poly(lactic-glycolic acid), has been used to encapsulate catalase (Dziubla et al. 2005).

Some of the commonly used tools for improving the stability of enzymes used in various diseases are listed in Table 7.2.

7.5 Analytical Characterization of Therapeutic Enzymes

With the increase in the demand for biotherapeutics, quantity of production, quality control, safety and efficacy are important aspects of the manufacturing process. Several high-resolution analytical tools for the characterization of immunological, physicochemical and structural attributes of the product have been employed (Bui et al. 2015; Rathore 2009). The techniques employed can be classified into two broad categories-.

Table 7.2 Approaches for improving stability of therapeutic enzymes

Enzyme	Tools	Changes	References
Class A β-lactamases	Hybrid enzymes formed with small peptides attached at permissive position	Bifunctional proteins formed, determination of specific antibodies enhanced	Huynen et al. (2013)
L-asparaginase	Chemical modification with bovine serum albumin, ovalbumin by crosslinking using glutaraldehyde, N-bromosuccinimide, and mono-methoxy polyethylene glycol	High stability of L-asparaginase that was 8.5- and 7.62-fold more compared to native enzyme at 28 °C and 37 °C by the end of 24 h	(Mohan Kumar et al. (2014)
Serratiopeptidase	Immobilization with chitosan nanoparticles	Anti-inflammatory activity increased upto 32 h compared with plain enzyme	Mali et al. (2015)
Streptokinase (SKA)	Use of sRNA	FasX (sRNA) enhances ska transcript stability and tenfold increase in SKA activity	Ramirez Peña et al. (2010)
Alginate lyase	Site directed mutagenesis	Ca^{2+} binding activates enzyme, Ca^{2+} binding was enhanced by substitution of aspartate for glutamate in calcium binding site	Wang et al. (2018)
Superoxide dismutase	Chemically cross-linked multilayer polyion complex	Double coat nanozyme, half life in blood increased up to 60 min and enhanced enzymatic activity in vivo	Nukolova et al. (2018)
L-asparaginase	Site directed mutagenesis	Variants are designed by replacement of two positively charged residues (K139 and K207) on the surface loops with neutral and reverse charges; thermally stable	Vidya et al. (2014)
Lipase	Immobilization on silica and magnetite coated silica nanoparticles	Stability enhanced, lipase retained over 73% of the initial activity after 5 times reuse	Lee et al. (2016)
Chondroitinase	Site directed mutagenesis	Aromatic amino acid substitution at positions Asn 806 and Gln 810 with tyrosine improves activity and thermal stability of enzyme	Shahaboddin et al. (2017)
Penicillin acylase	Error prone PCR	Error prone PCR reactions applied to enzyme gene, activity (fourfold higher) and pH stability at pH 10 were improved	Balci et al. (2014)

7.5.1 Tools for Analytical Characterization

Poly-Acrylamide Gel Electrophoresis (PAGE) is perhaps one of the oldest analytical tool for investigating proteins that is commonly used for early characterization. It is used to estimate size and isoelectric point of the protein, based on the property of charged molecules to migrate in the presence of an electric field. Both non-denaturing (native) and reducing (SDS) PAGE are used to resolve proteins. Isoelectric focusing (IEF) is combined with PAGE to resolve proteins based on their isoelectric points in addition to size. Thus, 2D-PAGE involving IEF in the first dimension and PAGE on the second is considered to be a convenient method for qualitative evaluation of charge heterogeneity, stability and post-translational modifications (Nebija et al. 2014). More recently, a more advanced form of PAGE, 2D-Difference Gel Electrophoresis (DIGE), that involves direct labelling of proteins by fluorescent dyes and separation on 2D-PAGE, has been used for providing accurate identification of spots. Both PAGE and 2D-DIGE can be coupled with Mass Spectrometry for identification of specific gel bands or spots (Reichel and Thevis 2013; Fekete et al. 2015)

High Performace Liquid Chromatography (HPLC) involves separation using a liquid mobile phase and a solid stationary phase. It is widely considered and used as a fundamental tool for analysis of biotherapeutics owing to its high resolution, selectivity, robustness and diverse choice of phases. It is typically used for analysis of proteins, nucleic acids, lipids, and other small molecules in complex mixtures. Based on the type of intended interactions between the analyte and the stationary phase, there are a variety of modalities that are used for protein analysis including Ion- exchange, Size-exclusion, Hydrophic Interaction and Reversed-Phase Chromatography. HPLC can also be coupled with Mass Spectrometry for simultaneous identification of the separated proteins (Mitulović and Mechtler 2006).

Circular Dichronism (CD) Spectroscopy is based on the differential absorption of left and right-handed circularly polarized light in the presence of light absorbing chiral groups. In proteins, different structural elements such as α-helices and β-pleated sheets have a characteristic spectra. During denaturation in the presence of denaturants or high temperature, CD can be used to monitor the folding and unfolding behavior of proteins. This technique provides the advantage of rapid measurements of multiple samples containing 20 μg or less of proteins in physiological buffers in a few hours. However, residue-specific information can only be obtained through analytical tools such as NMR or XRD (Greenfield 2006).

Fourier Transform Infrared Spectroscopy (FTIR) is another widely used technique to unravel information about the secondary structure of proteins, mainly α-helices and β-sheets using infra-red (IR) incident beams. FTIR works on the principle of interferometry to record high-spectral-resolution information about molecules placed in the IR beam. A monochromatic beam is projected, the amount of light absorbed is measured and repeated for each different wavelength, thus giving rise to a spectra. FTIR offers an easy, non-destructive way to study the secondary structure of proteins with simple sample preparation methods and application over a wide variety of conditions. The polypeptide chains give rise to nine characteristic IR absorption bands, namely, amide A, B, and I–VII. Out of these, amide I spectroscopy has been used to study folding/unfolding and structural stability of protein molecules (Kong and Yu 2007).

Dynamic Light Scattering (DLS) is a technique for determining the average particle size distribution (mean effective diameter) of molecules in a suspension or biotherapeutic formulation. It involves a monochromatic light source, usually a laser, which is shot through a polarizer into the sample. The scattered light passes through a second polarizer, collected by a photoamplifier and projected onto a screen. The data generated depends on dispersion, particle concentration, and presence of ions in solution. Recently, high-through put platform in DLS (HT-DLS) has been used to quantify viscosity of mAb formulations (Some and Burge 2015).

Raman Spectroscopy records vibrational, rotational, and other low-frequency modes of a particular system and is generally employed to study the secondary structure of proteins by studying the Amide I bond between 1600 and 1700 cm^{-1} (Wen 2007). This technique is popular as it offers reduced susceptibility to interference by water on the scattered light as water is ineffective in scattering in this part of the spectrum. Other Raman techniques such as UVRR, ROA have been used for studying biotherapeutics including protein–protein interaction, protein aggregation and conformation (Wen 2007).

Fluorescence Spectroscopy is an important tool for studying changes in the tertiary structure of a protein, based on the intrinsic fluorescence of amino acid residues such as tryptophan and tyrosine (Vivian and Callis 2001). Upon excitation by an external source of light, these flurophores produce a characteristic emission spectrum. Changes in the tertiary structure due to protein denaturation can lead to alterations in either the intensity or the wavelength maxima (red or blue shifts) of the spectra, and are reflective of the local environment of the protein. In absence of an intrinsic fluorophore, extrinsic fluorophore dyes such as 1-anilino-8-naphthalenesulfonate (ANS) and Thioflavin-T (ThT) can be used to induce fluorescence. However, this technique is not efficient to predict positional changes and nature of the modifications.

X-Ray crystallography is a pioneering technique for studying the structure of proteins. Based on the diffraction pattern of the incident X-ray beam on the crystal, the three-dimensional structure of the protein is constructed. However, prior purification and crystallization of biotherapeutic proteins can prove challenging owing to the presence of inherent heterogeneity in the form of numerous post-translational modifications, which can be cumbersome for routine analysis, comparability and characterization (Berkowitz 2017).

Nuclear Magnetic Resonance (NMR) is based on the quantum mechanical properties of certain atomic nuclei. These properties depend on the local environment of the molecule, and their measurement provide information about chemical bonding and spatial arrangements. NMR parameters such as chemical shifts are most readily and accurately measurable and they reflect the conformations of native and nonnative states of proteins with great specificity (Cavalli et al. 2007). However, exploitation of this tool for routine characterization of therapeutic proteins has been limited due to factors including large size of protein, the low natural abundance of active nuclei, and the relatively low sensitivity of the NMR signals. The problems resolving overlapping peaks in larger proteins have been somewhat rectified by the introduction of isotope labelling and multidimensional experiments.

Transmission Electron Microscopy (TEM) is an analytical tool that has been lately being used in the field of biotherapeutics. It employs a beam of electrons which passes through a protein sample to form an image. Owing to the smaller wavelength of electrons as compared to photons, the resolution of images obtained is greater by many orders of magnitude, providing detailed internal structures of protein aggregates (Anderson and Webb 2011). Recent advances in TEM technology, especially with the development of Cryo-TEM has enabled in-depth studies of proteins (Tsuruta et al. 2015).

7.5.2 Tools for Functional Characterization

As discussed in the previous sections, immunogenicity is a major problem for a therapeutic product. **Enzyme-Linked-Immunosorbent-Assay (ELISA)** is a precursor, rapid technique to test the immunogenicity of a therapeutic protein, and has several applications in biotherapeutic characterization. In this assay, antigens (therapeutic protein) from a sample are attached to a surface. Enzyme-linked-antibodies specific to the antigen are then added. The signal can be detected by addition of a colorimetric substrate to the enzyme.

Surface Plasmon Resonance is a widely used label-free technique used for detection of biomolecular interactions, especially ligand-receptor kinetics (Nupur et al. 2018). It is an optical method which measures the refractive index of thin layers of materials deposited on metal surfaces. During SPR, polarised light strikes an electrically conducting surface, leading to the production of plasmons. These electron charged density waves reduce the intensity of reflected light at a specific angle known as the resonance angle, in proportion to the mass on a sensor surface. SPR has high sensitivity and is easy to use for real-time monitoring of association and dissociation of an interaction. The design data analysis methods of SPR have found wide applications in microbiology, virology, epitope mapping and protein, nucleic acid, cell and membrane based interactions (Pattnaik 2005).

Bio-Layer Interferometry (BLI) is another optical label-free technique based on the interference pattern of white light reflected from the surface of the immobilised enzyme on the biosensor tip and an internal reference layer. Changes in the interference pattern are observed based on the number of molecules on the tip of the biosensor and can be used in studying binding kinetics.

Isothermal Titration Calorimetry (ITC) is a quantitative technique for determining the thermodynamic parameters of biomolecular interactions, mostly in case of small molecules binding to larger macromolecules (such as protein or DNA) (Pierce et al. 1999). In this technique, the two interactants are mixed together and the heat released/absorbed during the process is measured via titration. ITC is typically used as a secondary screening technique for characterizing binding affinity of ligands for proteins.

7.6 Future Perspectives

With increasing demand of medical technologies, it is imperative to look for therapeutics with better stability, higher specificity, and greater circulation time with lesser immunogenicity and toxicity. Microbes are still the largest source of therapeutic proteins. However, due to physiochemical and functional divergence, the traditional method of culturing microorganisms has limited the exploration of around 90–99% of unculturable microbes. With integrated 'omics' approaches of Metagenomics, Transcriptomics, Metaproteomics, and Metabolomics, it has become easier to study unculturable microbes and mining for stable therapeutic enzymes from vast data sets, at the molecular level (Ghosh et al. 2017; Alain and Querellou 2009; Tiwari et al. 2018).

7.7 Conclusions

The market for enzyme based therapeutics is enormous in the current times and is projected to increase in the near future. Advancement in biotechnology have facilitated unprecedented increase in the production of significant enzymes. Factors such as proteolysis, chemical modifications, acidic environment, hepatic first pass, immunogenicity, temperature, and the route of administration as well as the target site, affect the stability of enzymes. However, these challenges can be overcome by functional and formulation based approaches through the techniques of site-specific mutagenesis and cloning/expression and conjugation with polymers or nanoparticles. An enhanced understanding into the routes of administration and the drug absorption mechanisms is necessary to streamline the development process. Further exploration of microbial sources for more stable enzyme forms and protein mining through 'omics' approaches could help in the development of therapeutics against myriad fatal disorders and syndromes.

References

Akash MSH, Rehman K, Tariq M, Chen S (2015) Development of therapeutic proteins: advances and challenges. Turk J Biol 39(3):343–358

Alain K, Querellou J (2009) Cultivating the uncultured: limits, advances and future challenges. Extremophiles 13(4):583–594

Alegra T, Vairo F, de Souza MV, Krug BRC, Schwartz IVD (2012) Enzyme replacement therapy for Fabry disease: a systematic review and meta-analysis. Genet Mol Biol 35(4):947–954

Alpers D (2008) Role of lipoprotein lipase in triglyceride metabolism: potential therapeutic target. Futur Lipidol 3(4):385–397

Anderson VL, Webb WW (2011) Transmission electron microscopy characterization of fluorescently labelled amyloid β 1-40 and α-synuclein aggregates. BMC Biotechnol 11(1):125

Angelini C, Semplicini C (2012) Enzyme replacement therapy for Pompe disease. Curr Neurol Neurosci Rep 12(1):70–75

Baker EN, Hubbard RE (1984) Hydrogen bonding in globular proteins. Prog Biophys Mol Biol 44(2):97–179

Bakshi RP, Shapiro TA (2003) DNA topoisomerases as targets for antiprotozoal therapy. Mini-Rev Med Chem 3(6):597–608

Balci H, Ozturk MT, Pijning T, Ozturk SI, Gumusel F (2014) Improved activity and pH stability of *E. coli* ATCC 11105 penicillin acylase by error-prone PCR. Appl Microbiol Biotechnol 98(10):4467–4477

Berkowitz SA (2017) Analytical characterization: structural assessment of biosimilarity. In: Biosimilar drug product development. CRC Press, Boca Raton, pp 15–82

Bui LA, Hurst S, Finch GL, Ingram B, Jacobs IA, Kirchhoff CF, Ng CK, Ryan AM (2015) Key considerations in the preclinical development of biosimilars. Drug Discov Today 20:3–15

Burley SK, Petsko GA (1988) Weakly polar interactions in proteins. In: Advances in protein chemistry. Elsevier, New York, pp 125–189

But WM, Wong MY, Chow JC, Chan WKY, Ko WT, Wu SP, Wong ML, Miu TY, Tse WY, Hung WWY (2011) Enzyme replacement therapy for mucopolysaccharidosis VI (Maroteaux-Lamy syndrome): experience in Hong Kong. Hong Kong Med J 17(4):317–324

Calvo-Lerma J, Martínez-Barona S, Masip E, Fornés V, Ribes-Koninckx C (2017) Pancreatic enzyme replacement therapy in cystic fibrosis: dose, variability and coefficient of fat absorption. Rev Esp Enferm Dig Organo Oficial Soc Esp Patol Dig 109(10):684–689

Cavalli A, Salvatella X, Dobson CM, Vendruscolo M (2007) Protein structure determination from NMR chemical shifts. Proc Natl Acad Sci 104(23):9615–9620

Chabot JA, Tsai WY, Fine RL, Chen C, Kumah CK, Antman KA, Grann VR (2010) Pancreatic proteolytic enzyme therapy compared with gemcitabine-based chemotherapy for the treatment of pancreatic cancer. J Clin Oncol 28(12):2058

Chan B, Wara D, Bastian J, Hershfield MS, Bohnsack J, Azen CG, Parkman R, Weinberg K, Kohn DB (2005) Long-term efficacy of enzyme replacement therapy for adenosine deaminase (ADA)-deficient severe combined immunodeficiency (SCID). Clin Immunol 117(2):133–143

Chibale K (2002) Towards broadspectrum antiprotozoal agents. Arkivoc 9:93–98

Chorny M, Hood E, Levy RJ, Muzykantov VR (2010) Endothelial delivery of antioxidant enzymes loaded into non-polymeric magnetic nanoparticles. J Control Release 146(1):144–151

Choudhary B, Nagpure A, Gupta RK (2014) Fungal cell-wall lytic enzymes, antifungal metabolite (s) production, and characterization from *Streptomyces exfoliatus* MT9 for controlling fruit-rotting fungi. J Basic Microbiol 54(12):1295–1309

Chuang HY, Suen CS, Hwang MJ, Roffler SR (2015) Toward reducing immunogenicity of enzyme replacement therapy: altering the specificity of human β-glucuronidase to compensate for α-iduronidase deficiency. Protein Eng Des Sel 28(11):519–530

Cocci A, Russo GI, Briganti A, Salonia A, Cacciamani G, Capece M, Falcone M, Timpano M, Cito G, Verze P (2018) Predictors of treatment success after collagenase clostridium histolyticum injection for Peyronie's disease. Development of a nomogram from a multicentre single-arm, non-placebo controlled clinical study. BJU Int 122:680–687

Cook MT, Tzortzis G, Charalampopoulos D, Khutoryanskiy VV (2012) Microencapsulation of probiotics for gastrointestinal delivery. J Control Release 162(1):56–67

Da Silva EM, Strufaldi MW, Andriolo RB, Silva LA (2011) Enzyme replacement therapy with idursulfase for mucopolysaccharidosis type II (Hunter syndrome). Cochrane Database Syst Rev 11(11)

DeLong RK, Reynolds CM, Malcolm Y, Schaeffer A, Severs T, Wanekaya A (2010) Functionalized gold nanoparticles for the binding, stabilization, and delivery of therapeutic DNA, RNA, and other biological macromolecules. Nanotechnol Sci Appl 3:53

Desai UR (2004) New antithrombin-based anticoagulants. Med Res Rev 24(2):151–181

Doig AJ, Williams DH (1991) Is the hydrophobic effect stabilizing or destabilizing in proteins?: The contribution of disulphide bonds to protein stability. J Mol Biol 217(2):389–398

Duval M, Suciu S, Ferster A, Rialland X, Nelken B, Lutz P, Benoit Y, Robert A, Manel A-M, Vilmer E (2002) Comparison of *Escherichia coli*-asparaginase with *Erwinia*-asparaginase in the treatment of childhood lymphoid malignancies: results of a randomized European Organisation for Research and Treatment of Cancer-Children's Leukemia Group phase 3 trial. Blood 99(8):2734–2739

Dziubla TD, Karim A, Muzykantov VR (2005) Polymer nanocarriers protecting active enzyme cargo against proteolysis. J Control Release 102(2):427–439

Egler RA, Ahuja SP, Matloub Y (2016) L-asparaginase in the treatment of patients with acute lymphoblastic leukemia. J Pharmacol Pharmacother 7(2):62

Ehren J, Govindarajan S, Morón B, Minshull J, Khosla C (2008) Protein engineering of improved prolyl endopeptidases for celiac sprue therapy. Protein Eng Des Sel 21(12):699–707

Fekete S, Guillarme D, Sandra P, Sandra K (2015) Chromatographic, electrophoretic, and mass spectrometric methods for the analytical characterization of protein biopharmaceuticals. Anal Chem 88(1):480–507

Fuhrmann G, Leroux JC (2011) In vivo fluorescence imaging of exogenous enzyme activity in the gastrointestinal tract. Proc Natl Acad Sci 108(22): 9032–9037

Fuhrmann G, Leroux JC (2014) Improving the stability and activity of oral therapeutic enzymes-recent advances and perspectives. Pharm Res 31(5):1099–1105

Gámez A, Wang L, Straub M, Patch MG, Stevens RC (2004) Toward PKU enzyme replacement therapy: PEGylation with activity retention for three forms of recombinant phenylalanine hydroxylase. Mol Ther 9(1):124–129

Ghaffarinia A, Jalili C, Riazi-Rad F, Mostafaie A, Parvaneh S, Pakravan N (2014) Anti-inflammatory effect of chymotrypsin to autoimmune response against CNS is dose-dependent. Cell Immunol 292(1–2):102–108

Ghosh P, Han G, De M, Kim CK, Rotello VM (2008) Gold nanoparticles in delivery applications. Adv Drug Deliv Rev 60(11):1307–1315

Ghosh S, Tiwari R, Hemamalini R, Khare SK (2017) Multi-omic approaches for mapping interactions among marine microbiomes. In: Understanding host-microbiome interactions-an omics approach. Springer, pp 353–368

Ghosh S, Ahmad R, Gautam VK, Khare SK (2018) Cholesterol-oxidase-magnetic nanobioconjugates for the production of 4-cholesten-3-one and 4-cholesten-3, 7-dione. Bioresour Technol 254:91–96

Gökmen V (2015) Acrylamide in food: analysis, content and potential health effects. Academic, Amsterdam

Goodsell DS (2005) The molecular perspective: L-asparaginase. Oncologist 10(3):238–239
Greenfield NJ (2006) Using circular dichroism spectra to estimate protein secondary structure. Nat Protoc 1(6):2876
Griswold KE, Bement JL, Teneback CC, Scanlon TC, Wargo MJ, Leclair LW (2014) Bioengineered lysozyme in combination therapies for *Pseudomonas aeruginosa* lung infections. Bioengineered 5(2):143–147
Hamuro Y, Coales SJ, Molnar KS, Tuske SJ, Morrow JA (2008) Specificity of immobilized porcine pepsin in H/D exchange compatible conditions. Rapid Commun Mass Spectrom 22(7):1041–1046
Harmatz P (2015) Enzyme replacement therapies and immunogenicity in lysosomal storage diseases: is there a pattern? Clin Ther 37(9):2130–2134
Harris JM, Chess RB (2003) Effect of pegylation on pharmaceuticals. Nat Rev Drug Discov 2(3):214
Haxho F, Neufeld RJ, Szewczuk MR (2016) Neuraminidase-1: a novel therapeutic target in multistage tumorigenesis. Oncotarget 7(26):40860
Hayashi S, Yokoyama I, Namii Y, Emi N, Uchida K, Takagi H (1999) Inhibitory effect on the establishment of hepatic metastasis by transduction of the tissue plasminogen activator gene to murine colon cancer. Cancer Gene Ther 6(4):380
Hendsch ZS, Tidor B (1994) Do salt bridges stabilize proteins? A continuum electrostatic analysis. Protein Sci 3(2):211–226
Hershfield MS, Chaffee S, Sorensen RU (1993) Enzyme replacement therapy with polyethylene glycol-adenosine deaminase in adenosine deaminase deficiency: overview and case reports of three patients, including two now receiving gene therapy. Pediatr Res 33(S4):S42
Huynen C, Filée P, Matagne A, Galleni M, Dumoulin M (2013) Class a-lactamases as versatile scaffolds to create hybrid enzymes: applications from basic research to medicine. Biomed Res Int 2013:1–6
Ilinskaya ON, Mahmud RS (2014) Ribonucleases as antiviral agents. Mol Biol 48(5):615–623
Illanes A, Valencia P (2017) Industrial and therapeutic enzymes: penicillin acylase. In: Current developments in biotechnology and bioengineering. Elsevier, Amsterdam, pp 267–305
Imrie CW, Connett G, Hall RI, Charnley RM (2010) Enzyme supplementation in cystic fibrosis, chronic pancreatitis, pancreatic and periampullary cancer. Aliment Pharmacol Ther 32:1–25
Ishihara T, Tanaka KI, Tasaka Y, Namba T, Suzuki J, Ishihara T, Okamoto S, Hibi T, Takenaga M, Igarashi R (2009) Therapeutic effect of lecithinized superoxide dismutase against colitis. J Pharmacol Exp Ther 328(1):152–164
Jayachandran S, Khobre P (2017) Efficacy of bromelain along with trypsin, rutoside trihydrate enzymes and diclofenac sodium combination therapy for the treatment of TMJ osteoarthritis-a randomised clinical trial. J Clin Diagn Res 11(6):ZC09
Kaminetzky D, Hymes KB (2008) Denileukin diftitox for the treatment of cutaneous T-cell lymphoma. Biol Targets Ther 2(4):717
Karamitros CS, Yashchenok AM, Möhwald H, Skirtach AG, Konrad M (2013) Preserving catalytic activity and enhancing biochemical stability of the therapeutic enzyme asparaginase by biocompatible multilayered polyelectrolyte microcapsules. Biomacromolecules 14(12):4398–4406
Karstad R, Isaksen G, Wynendaele E, Guttormsen Y, De Spiegeleer B, Brandsdal BO, Svendsen JS, Svenson J (2012) Targeting the S1 and S3 subsite of trypsin with unnatural cationic amino acids generates antimicrobial peptides with potential for oral administration. J Med Chem 55(14):6294–6305
Kasturi L, Amin Alpa S (2001) Enzyme replacement therapy in Gaucher's disease. Indian Pediatr 38(6):686–687
Kini RM (2005) Structure-function relationships and mechanism of anticoagulant phospholipase A2 enzymes from snake venoms. Toxicon 45(8):1147–1161
Kishimoto TK, Ferrari JD, LaMothe RA, Kolte PN, Griset AP, O'Neil C, Chan V, Browning E, Chalishazar A, Kuhlman W (2016) Improving the efficacy and safety of biologic drugs with tolerogenic nanoparticles. Nat Nanotechnol 11(10):890
Knappe D, Goldbach T, Hatfield MPD, Palermo NY, Weinert S, Sträter N, Hoffmann R, Lovas SN (2016) Proline-rich antimicrobial peptides optimized for binding to *Escherichia coli* chaperone DnaK. Protein Pept Lett 23(12):1061–1071
Kong J, Yu S (2007) Fourier transform infrared spectroscopic analysis of protein secondary structures. Acta Biochim Biophys Sin 39(8):549–559
Kosky AA, Razzaq UO, Treuheit MJ, Brems DN (1999) The effects of alpha-helix on the stability of Asn residues: deamidation rates in peptides of varying helicity. Protein Sci 8(11):2519–2523
Krizsan A, Volke D, Weinert S, Sträter N, Knappe D, Hoffmann R (2014) Insect-derived proline-rich antimicrobial peptides kill bacteria by inhibiting bacterial protein translation at the 70 S ribosome. Angew Chem Int Ed 53(45):12236–12239
Lamppa JW, Griswold KE (2013) Alginate lyase exhibits catalysis-independent biofilm dispersion and antibiotic synergy. Antimicrob Agents Chemother 57(1):137–145
Langer V, Bhandari PS, Rajagopalan S, Mukherjee MK (2013) Enzymatic debridement of large burn wounds with papain-urea: is it safe? Med J Armed Forces India 69(2):144–150
Lansigan F, Stearns DM, Foss F (2010) Role of denileukin diftitox in the treatment of persistent or recurrent cutaneous T-cell lymphoma. Cancer Manag Res 2:53
Lee HR, Kim MI, Hong SE, Choi J, Kim YM, Yoon KR, Lee S, Ha SH (2016) Effect of functional group on activity and stability of lipase immobilized on silica-coated magnetite nanoparticles with different functional group. Anal Sci Technol 29(3):105–113

Leland PA, Raines RT (2001) Cancer chemotherapy-ribonucleases to the rescue. Chem Biol 8(5):405–413

Li YM, Vallera DA, Hall WA (2013) Diphtheria toxin-based targeted toxin therapy for brain tumors. J Neuro-Oncol 114(2):155–164

Liao AY, Ghosh A, O'Leary C, Mercer J, Church HJ, Tylee KL, Goenka A, Holley R, Wynn RF, Jones SA (2018) Induction of immune tolerance to enzyme replacement therapy in mucopolysaccharidosis type I. Mol Genet Metab 123(2):S86

Lindgren J, Eriksson Karlström A (2014) Intramolecular thioether crosslinking of therapeutic proteins to increase proteolytic stability. ChemBioChem 15(14):2132–2138

Lu F, Lu Z, Bie X, Yao Z, Wang Y, Lu Y, Guo Y (2010) Purification and characterization of a novel anticoagulant and fibrinolytic enzyme produced by endophytic bacterium *Paenibacillus polymyxa* EJS-3. Thromb Res 126(5):e349–e355

Luo Q, Joubert MK, Stevenson R, Ketchem RR, Narhi LO, Wypych J (2011) Chemical modifications in therapeutic protein aggregates generated under different stress conditions. J Biol Chem 286(28): 25134–25144

Lyseng-Williamson KA (2014) Elosulfase Alfa: a review of its use in patients with mucopolysaccharidosis type IVA (Morquio A syndrome). BioDrugs 28(5):465–475

Mali N, Wavikar P, Vavia P (2015) Serratiopeptidase loaded chitosan nanoparticles by polyelectrolyte complexation: in vitro and in vivo evaluation. AAPS PharmSciTech 16(1):59–66

Matuszewska A, Karp M, Jaszek M, Janusz G, Osińska-Jaroszuk M, Sulej J, Stefaniuk D, Tomczak W, Giannopoulos K (2016) Laccase purified from Cerrena unicolor exerts antitumor activity against leukemic cells. Oncol Lett 11(3):2009–2018

McAuley M, Mesa-Torres N, McFall A, Morris S, Huang M, Pey A, Timson DJ (2018) Improving the activity and stability of human galactokinase for therapeutic and biotechnological applications. ChemBioChem 19:1088–1095

McDonald IK, Thornton JM (1994) Satisfying hydrogen bonding potential in proteins. J Mol Biol 238(5):777–793

Mitulović G, Mechtler K (2006) HPLC techniques for proteomics analysis-a short overview of latest developments. Brief Funct Genomics 5(4):249–260

Mohan Kumar NS, Kishore V, Manonmani HK (2014) Chemical modification of L-asparaginase from *Cladosporium* sp. for improved activity and thermal stability. Prep Biochem Biotechnol 44(5):433–450

Mouneshkumar Chappi D, Suresh KV, Patil MR, Desai R, Tauro DP, Shiva Bharani KNS, Parkar MI, Babaji HV (2015) Comparison of clinical efficacy of methylprednisolone and serratiopeptidase for reduction of postoperative sequelae after lower third molar surgery. J Clin Exp Dent 7(2):e197

Muenzer J, Wraith JE, Beck M, Giugliani R, Harmatz P, Eng CM, Vellodi A, Martin R, Ramaswami U, Gucsavas-Calikoglu M (2006) A phase II/III clinical study of enzyme replacement therapy with idursulfase in mucopolysaccharidosis II (Hunter syndrome). Genet Med 8(8):465

Nanduri B, Suvarnapunya AE, Venkatesan M, Edelmann MJ (2013) Deubiquitinating enzymes as promising drug targets for infectious diseases. Curr Pharm Des 19(18):3234–3247

Nebija D, Noe CR, Urban E, Lachmann B (2014) Quality control and stability studies with the monoclonal antibody, trastuzumab: application of 1D-vs. 2D-gel electrophoresis. Int J Mol Sci 15(4):6399–6411

Neurath H (1984) Evolution of proteolytic enzymes. Science 224(4647):350–357

Ngu LH, Peitee WO, Leong HY, Chew HB (2017) Case report of treatment experience with idursulfase beta (Hunterase) in an adolescent patient with MPS II. Mol Genet Metab Rep 12:28–32

Nukolova NV, Aleksashkin AD, Abakumova TO, Morozova AY, Gubskiy IL, Kirzhanova EA, Abakumov MA, Chekhonin VP, Klyachko NL, Kabanov AV (2018) Multilayer polyion complex nanoformulations of superoxide dismutase 1 for acute spinal cord injury. J Control Release 270:226–236

Nupur N, Chhabra N, Dash R, Rathore AS (2018) Assessment of structural and functional similarity of biosimilar products: rituximab as a case study. Translated by Taylor & Francis, pp 143–158

Pattnaik P (2005) Surface plasmon resonance. Appl Biochem Biotechnol 126(2):79–92

Perchellet EM, Wang Y, Weber RL, Lou K, Hua DH, Perchellet JPH (2004) Antitumor triptycene bisquinones induce a caspase-independent release of mitochondrial cytochrome c and a caspase-2-mediated activation of initiator caspase-8 and-9 in HL-60 cells by a mechanism which does not involve Fas signaling. Anti-Cancer Drugs 15(10):929–946

Pierce MM, Raman CS, Nall BT (1999) Isothermal titration calorimetry of protein-protein interactions. Methods 19(2):213–221

Pundir CS, Narang J, Chauhan N, Sharma P, Sharma R (2012) An amperometric cholesterol biosensor based on epoxy resin membrane bound cholesterol oxidase. Indian J Med Res 136(4):633

Qi Y, Chilkoti A (2015) Protein-polymer conjugation-moving beyond PEGylation. Curr Opin Chem Biol 28:181–193

Ramirez Peña E, Treviño J, Liu Z, Perez N, Sumby P (2010) The group A Streptococcus small regulatory RNA FasX enhances streptokinase activity by increasing the stability of the ska mRNA transcript. Mol Microbiol 78(6):1332–1347

Rathore AS (2009) Follow-on protein products: scientific issues, developments and challenges. Trends Biotechnol 27(12):698–705

Reed GL, Liu L, Houng AK, Matsueda LH, Lizbeth H (1998) Mechanisms of fibrin-independent and fibrin-dependent plasminogen activation by streptokinase. Circulation 98(17):41–41

Reichel C, Thevis M (2013) Gel electrophoretic methods for the analysis of biosimilar pharmaceuticals using the example of recombinant erythropoietin. Bioanalysis 5(5):587–602

Rosenberg M, Kingma W, Fitzpatrick MA, Richards SM (1999) Immunosurveillance of alglucerase enzyme therapy for Gaucher patients: induction of humoral tolerance in seroconverted patients after repeat administration. Blood 93(6):2081–2088

Rosenberg L, Lapid O, Bogdanov-Berezovsky A, Glesinger R, Krieger Y, Silberstein E, Sagi A, Judkins K, Singer AJ (2004) Safety and efficacy of a proteolytic enzyme for enzymatic burn debridement: a preliminary report. Burns 30(8):843–850

Sabu A (2003) Sources, properties and applications of microbial therapeutic enzymes. Indian J Biotechnol 2(2003):334–341

Sarada KV (2013) Production and applications of L-glutaminase using fermentation technology. Asia Pac J Res 1:1

Saunders SE, Bartz JC, Vercauteren KC, Bartelt-Hunt SL (2011) An enzymatic treatment of soil-bound prions effectively inhibits replication. Appl Environ Microbiol 77(13):4313–4317

Sawhney P, Kumar S, Maheshwari N, Singh Guleria S, Dhar N, Kashyap R, Sahni G (2016) Site-specific thiol-mediated PEGylation of streptokinase leads to improved properties with clinical potential. Curr Pharm Des 22(38):5868–5878

Scagnolari C, Bellomi F, Turriziani O, Bagnato F, Tomassini V, Lavolpe V, Ruggieri M, Bruschi F, Meucci G, Dicuonzo G (2002) Neutralizing and binding antibodies to IFN-β: relative frequency in relapsing-remitting multiple sclerosis patients treated with different IFN-β preparations. J Interf Cytokine Res 22(2):207–213

Schöneich C (2000) Mechanisms of metal-catalyzed oxidation of histidine to 2-oxo-histidine in peptides and proteins. J Pharm Biomed Anal 21(6):1093–1097

Shah Mahmud R, Müller C, Romanova Y, Mostafa A, Ulyanova V, Pleschka S, Ilinskaya O (2017) Ribonuclease from *Bacillus* acts as an antiviral agent against negative-and positive-sense single stranded human respiratory RNA viruses. BioMed Res Int 2017

Shahaboddin ME, Khajeh K, Maleki M, Golestani A (2017) Improvement of activity and stability of chondroitinase ABC I by introducing an aromatic cluster at the surface of protein. Enzym Microb Technol 105:38–44

Shak S, Capon DJ, Hellmiss R, Marsters SA, Baker CL (1990) Recombinant human DNase I reduces the viscosity of cystic fibrosis sputum. Proc Natl Acad Sci 87(23):9188–9192

Shaked ZE, Wolfe S (1988) Stabilization of pyranose 2-oxidase and catalase by chemical modification. In: Methods in enzymology, pp 599–615

Shlyakhovenko VO (2016) Ribonucleases. Possible new approach in cancer therapy. Exp Oncol 38(1):2–8

Sikri N, Bardia A (2007) A history of streptokinase use in acute myocardial infarction. Tex Heart Inst J 34(3):318

Some D, Burge R (2015) Innovations in automation-high-throughput dynamic light scattering for screening biotherapeutic formulations. Biopharm Int 28(11):23

Su K, Donaldson E, Sharma R (2016) Novel treatment options for lysosomal acid lipase deficiency: critical appraisal of sebelipase alfa. Appl Clin Genet 9:157

Tanford C (1980) The hydrophobic effect: formation of micelles and biological membranes, 2nd edn. Wiley, Brisbane

Tatsumoto KM, Oh KK, Baker JO, Himmel ME (1989) Enhanced stability of glucoamylase through chemical crosslinking. Appl Biochem Biotechnol 20(1):293–308

Thurberg BL, Maloney CL, Vaccaro C, Afonso K, Tsai AC-H, Bossen E, Kishnani PS, O'Callaghan M (2006) Characterization of pre-and post-treatment pathology after enzyme replacement therapy for Pompe disease. Lab Investig 86(12):1208

Tishchenko S, Gabdulkhakov A, Melnik B, Kudryakova I, Latypov O, Vasilyeva N, Leontievsky A (2016) Structural studies of component of lysoamidase bacteriolytic complex from *Lysobacter* sp. XL1. Protein J 35(1):44–50

Tiwari R, Nain L, Labrou NE, Shukla P (2018) Bioprospecting of functional cellulases from metagenome for second generation biofuel production: a review. Crit Rev Microbiol 44(2):244–257

Torchilin VP, Maksimenko AV, Smirknov VN, Berezin IV, Klibanov AM, Martinek K (1978) The principles of enzyme stabilization. III. The effect of the length of intra-molecular cross-linkages on thermostability of enzymes. Biochim Biophys Acta (BBA)-Enzymol 522(2):277–283

Tsuruta LR, Lopes dos Santos M, Moro AM (2015) Biosimilars advancements: moving on to the future. Biotechnol Prog 31(5):1139–1149

Usmani SS, Bedi G, Samuel JS, Singh S, Kalra S, Kumar P, Ahuja AA, Sharma M, Gautam A, Raghava GPS (2017) THPdb: database of FDA-approved peptide and protein therapeutics. PLoS One 12(7):e0181748

Vereshchagin IA, Vereshchagin SI, Zhuravleva OD (1985) Enhanced effectiveness of antibiotic therapy with an exogenous lysozyme in dysentery in children. Antibiot Med Biotekhnologiia Antibiot Med Biotechnol 30(12):927–931

Vidya J, Ushasree MV, Pandey A (2014) Effect of surface charge alteration on stability of l-asparaginase II from *Escherichia* sp. Enzym Microb Technol 56:15–19

Vivian JT, Callis PR (2001) Mechanisms of tryptophan fluorescence shifts in proteins. Biophys J 80(5):2093–2109

Wang D, Ichiyama RM, Zhao R, Andrews MR, Fawcett JW (2011) Chondroitinase combined with rehabilitation promotes recovery of forelimb function in rats with chronic spinal cord injury. J Neurosci 31(25):9332–9344

Wang H, Hagedorn J, Svendsen A, Borch K, Otzen DE (2013) Variant of the *Thermomyces lanuginosus* lipase with improved kinetic stability: a candidate for enzyme replacement therapy. Biophys Chem 172:43–52

Wang B, Ji SQ, Ma XQ, Lu M, Wang LS, Li FL (2018) Substitution of one calcium-binding amino acid strengthens substrate binding in a thermophilic alginate lyase. FEBS Lett 592(3):369–379

Wen Z-Q (2007) Raman spectroscopy of protein pharmaceuticals. J Pharm Sci 96(11):2861–2878

Wong SS, Wong L-JC (1992) Chemical crosslinking and the stabilization of proteins and enzymes. Enzym Microb Technol 14(11):866–874

Wu XC, Ye R, Duan Y, Wong S-L (1998) Engineering of plasmin-resistant forms of streptokinase and their production in *Bacillus subtilis*: streptokinase with longer functional half-life. Appl Environ Microbiol 64(3):824–829

Yoo EH, Lee SY (2010) Glucose biosensors: an overview of use in clinical practice. Sensors 10(5):4558–4576

Zenatti PP, Migita NA, Cury NM, Mendes-Silva RA, Gozzo FC, de Campos-Lima PO, Yunes JA, Brandalise SR (2018) Low bioavailability and high immunogenicity of a new brand of *E. coli*-asparaginase with active host contaminating proteins. EBioMedicine 30:158–166

8 Engineering Strategies for Oral Therapeutic Enzymes to Enhance Their Stability and Activity

Philipp Lapuhs and Gregor Fuhrmann

Abstract

Oral application of therapeutic enzymes is a promising and non-invasive administration that improves patient compliance. However, the gastrointestinal tract poses several challenges to the oral delivery of proteins, including harsh pH conditions and digestive proteases. A promising way to stabilise enzymes during their gastrointestinal route is by modification with polymers that can provide both steric shielding and selective interaction in different digestive compartments. We give an overview of modification technologies for oral enzymes ranging from functionalisation of native proteins, to site-specific mutation and protein-polymer engineering. We specifically focus on enzymes that are active directly in the gastrointestinal lumen and not systemically absorbed. In addition, we discuss examples of microparticle and nanoparticle encapsulated enzymes for improved oral delivery. The modification of orally administered enzymes offers a broad chemical variability and may be a promising tool for enhancing their gastrointestinal stability.

Keywords

Exogenous enzymes · Gastrointestinal tract · Oral delivery · Enzyme therapy · Protein-polymer conjugates · Non-invasive imaging · Gastro-resistant coating · Pharmaceutical formulation · Stomach-resistant coatings

Abbreviations

2-BIBB	2-bromoisobutyryl bromide
AP	Alkaline phosphatase
ATRP	Atom transfer radical polymerization
BCA	Bicinchoninic acid
BT*p*NA	Benzoyl-L-tyrosine *p*-nitroanilide
CAP	Cellulose acetate phthalate
CD	Circular dichroism
CLSM	Confocal laser scanning microscopy
CM	Carboxymethyl
CT	α_1-antichymotrypsin
DLS	Dynamic light scattering
DSC	Differential scanning calorimetry

P. Lapuhs
Helmholtz Institute for Pharmaceutical Research Saarland (HIPS), Helmholtz Centre for Infection Research (HZI), Biogenic Nanotherapeutics Group (BION), Saarbrücken, Germany

G. Fuhrmann (✉)
Helmholtz Institute for Pharmaceutical Research Saarland (HIPS), Helmholtz Centre for Infection Research (HZI), Biogenic Nanotherapeutics Group (BION), Saarbrücken, Germany

Department of Pharmacy, Saarland University, Saarbrücken, Germany
e-mail: gregor.fuhrmann@helmholtz-hips.de

N. Labrou (ed.), *Therapeutic Enzymes: Function and Clinical Implications*, Advances in Experimental Medicine and Biology 1148, https://doi.org/10.1007/978-981-13-7709-9_8

EDC	1-ethyl-3-(3-dimethyl-aminopropyl) carbodiimide
FDA	Food and drug administration USA
FITC-BSA	Fluorescein isothiocyanate conjugate - bovine serum albumin
FTIR	Fourier-transform infrared spectroscopy
GI	Gastrointestinal
GPC	Gel permeation chromatography
H/D	Hydrogen/deuterium
HA	Hyaluronic acid
HAP	Hydroxyapatite
HPLC	High performance liquid chromatography
HPMCP	Hydroxyl propyl methyl cellulose phthalate
HRP	Horseradish peroxidase
LC-MS	Liquid chromatography–mass spectrometry
LCST	Lower critical solution temperature
MALDI-TOF-MS	Matrix assisted laser desorption ionization time-of-flight mass spectrometry
MP	Microparticle
mPEG2-NHS	Branched PEG N-hydroxysuccinimide
MW	Molecular weight
NCC	Nanoceramic cores
NHS	N-hydroxysuccinimide
NHS-Br	*N*-Hydroxysuccinimide-bromide
NMR	Nuclear magnetic resonance
o-NP	Ortho-nitrophenol
o-NPG	Ortho-nitrophenyl-β-galactoside
PAMAM	Poly(amidoamine)
PBPE	Polymer-based protein engineering
pCBAm	Poly (carboxybetaine acrylamide)
PDMAEMA	Poly(2-(dimethylamino) ethyl methacrylate)
pDMAPS	Poly[N,N′-dimethyl (methacryloylethyl) ammonium propane sulfonate]
PEG	Polyethylene glycol
PEP	Proline-specific endopeptidase
pNIPAm	Poly (N-isopropylacrylamide)
pOEGMA	Poly(oligoethylene glycol monomethylethermethacrylate)
pQA	Poly-(quarternary ammonium methacrylate
pSMA	Poly-(sulfonate methacrylate)
RT	Room temperature
SDS-PAGE	Sodium dodecyl sulfate–polyacrylamide gel electrophoresis
SEC	Size exclusion chromatography
SEM	Scanning electron microscopy
SGC	Simulated gastric conditions
SGF	Simulated gastric fluid
SIC	Simulated intestinal tract conditions
SIF	Simulated intestinal fluid
Suc-Ala-Ala-Pro-Phe-pNA	N-succinyl-L-alanyl-L-alanyl-L-prolyl-L-phenyla-lanine4-nitroanilide
TEM	Transmission electron microscopy
T_m	Denaturation midpoint
TM-AvPAL	Triple mutant-Anabaena variabilis phenylalanine ammonia lyase
TNBSA	2,4,6-trinitrobenzene sulfonic acid
UCST	Upper critical solution temperature
UV-vis	Ultraviolet-visible α_1-anti
α-CT	α-chymotrypsin

8.1 Introduction

The oral administration of drugs is a non-invasive, patient-friendly and highly desirable application route (Vorselen et al. 2018). It is often associated with better therapy compliance because patients can administer the required drug dose themselves and they adhere to their therapy regimen more strictly which is overall clinically beneficial. Indeed, patients prefer an oral administration of their medicine over an intravenous application, not only for cancer therapy (Eek et al. 2016). A prominent example for a macromolecular compound for which oral administration would be an important breakthrough, is insulin. A formulation of insulin that could be simply ingested and would still reduce blood glucose concentrations would relieve millions of type 1 diabetes patients from their daily and repeatedly injections. Several academic approaches for enhancing the oral stability and bioavailability of insulin were developed, for example conjugation to a vitamin B12 receptor ligand for improved translocation (Clardy-James et al. 2012), conjugation of insulin to various natural polymers such as trehalose glycopolymer (Liu et al. 2017) and insulin encapsulated into nanocarriers (Chopra et al. 2017). The patients' preferences and advantages of oral (insulin) administration have been known already for several decades but clinically-translatable results and products are still limited (Vllasaliu et al. 2018). For a long time, chemical absorption enhancers have been considered the "holy grail" for oral delivery of macromolecules and proteins (Gupta et al. 2013), but clinical outcomes from these approaches are modest. A simple way to modify orally administered macromolecules is by conjugation to polymers of different size, architecture and morphology (Moroz et al. 2016). These polymers may transfer some of their physical-chemical properties onto the drug and could induce enhanced stability, activity or improved systemic absorption of the active principle (Fuhrmann and Fuhrmann 2017). In this chapter we are discussing recent approaches for polymer modification of orally administered enzymes, an important subclass of macromolecular drugs that are clinically relevant in gastrointestinal (GI) diseases such as lactose intolerance (Lomer et al. 2008), coeliac disease (Pinier et al. 2010), phenylketonuria (Blau et al. 2010), pancreatic insufficiency (Leeds et al. 2011) and many others. We focus on modification of enzymes that are active directly in the GI tract but these chemical and engineering principles are also relevant for systemically absorbed enzymes as these encounter comparable challenges upon oral administration.

8.2 Orally Administered Enzymes Are a Particular Class of Therapeutic Proteins

8.2.1 Oral Administration and the Physiology of the Gastrointestinal Tract

The GI tract is a difficult environment. It can be broadly compartmentalised into the oral cavity, oesophagus, stomach, duodenum, small intestine, large intestine, rectum, and anus (Fig. 8.1). It is a very complex combination of different organs, each holding a specific environment (e.g., pH, enzymes, digestive fluids) and various functions. Its physiological role is to disintegrate food components, isolate nutrients and make their absorption possible. This natural principle of digestion is the major challenge when developing macromolecules and especially enzymes as an orally administered formulation. In addition, orally given drugs may be influenced by food components, residual water content in each GI organ and the mean transit time which overall will enhance or reduce the absorption probability (Fuhrmann and Fuhrmann 2017). A detailed summary of these parameters, including GI fluid composition and retention times in the different organs are given in Table 8.1. For local oral application of enzymes, the most relevant GI compartments are stomach,

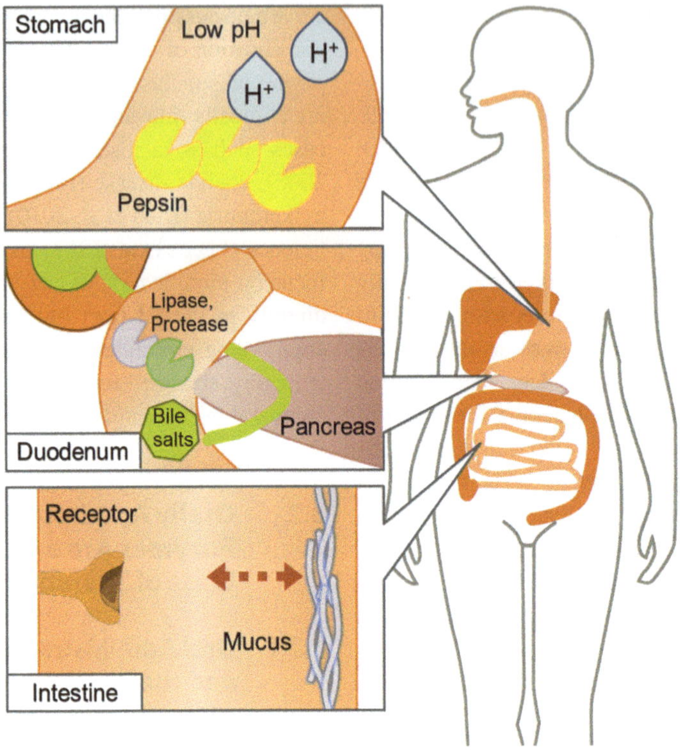

Fig. 8.1 Schematic overview of the human gastrointestinal tract and its most relevant organs for food digestion, i.e., stomach, duodenum and intestine. Each of these organs has a digestive fluid composed of specific enzymes, pH and other substances. These digestive fluids are the major challenge for oral administration of protein drugs, in particular enzymes. (Reproduced from Fuhrmann and Fuhrmann 2017, Copyright 2017 Elsevier)

Table 8.1 Composition of digestive fluids and retention time in the different GI compartments

Gastrointestinal compartment	pH range of digestive fluid	Important digestive enzymes	Average fluid volumes	Food residence time
Oral cavity	5–6	Amylase	0,6–1,7 L/day	<1 min
Stomach	1–3 (fasted)	Pepsin, amylase, lipase	Total volume: 20–50 mL (fasted)	1–8 h
	4–6 (fed)		2 L/day	
Duodenum	4–4,5	Trypsin, chymotrypsin, amylase, lipase		2–5 h
Jejunum	6,5		1 L/day	
Ileum	7,5		0,6 L/day	
Colon	6–8	Bacterial enzymes	–	4–70 h
Rectum	–	–	–	<1 min

small intestine and large intestine and these will be discussed in more detail in the next section. In the stomach, food digestion is initiated at low pH and presence of proteases, mostly pepsin. For exogenous enzymes, these conditions lead to unfolding of proteins and loss of enzymatic activity. Such gastric digestion is possibly the major challenge for oral administration of enzymes. When leaving the stomach, the digested food reaches the small intestine with a neutral pH and presence of pancreatic enzymes, mainly trypsin and chymotrypsin. Enzymes are often not stable against these proteases and in addition, bile salts can further inactivate the biocatalysts. The mucus lining in the GI tract also plays an important role when approaching the oral route. Mucus is the protective hydrogel matrix covering the GI walls (Bansil and Turner 2018). The mucus has a variable thickness and it is composed of a firmly adhered layer and a loosely adhered layer (Dean et al. 2017). While for example in the stomach the firmly and loosely adhered layer have comparable thicknesses with 150 and 120 μm, the firmly adherent layer in the duodenum is only about 15 μm thick and it is covered by 150 μm of loosely adherent layer (Dean et al. 2017). The loosely adhered layer is renewed every 4–6 h (Fuhrmann et al. 2013). Interestingly, the loosely mucus layer is the habitat for microbiota living in the GI tract and other mucin-degrading bacteria while the firmly adherent layer has a more restricted bacterial community including *Bacteroides fragilis* and *Acinetobacter* spp. (Donaldson et al. 2015). Mucus can trap pathogens, particles and other potentially harmful food components (Murgia et al. 2018). Ingested drugs may also interfere with mucus which may on one hand reduce their oral bioavailability but on the other hand could also be taken advantage of for enhanced retention of drugs in selected areas of the GI tract (Fuhrmann et al. 2013). Mucus penetration for orally administered nanoparticles is difficult to achieve because most carriers remain on the mucus surface and are readily cleaved by the GI motility. One of the few examples for mucus-penetrating nanocarriers presented was using biodegradable polymers poly(sebacic acid) and poly(ethylene glycol) (PSA-PEG) (Tang et al. 2009). Interestingly, PSA-PEG nanoparticles only showed a little reduction in diffusion speed between water and different types of mucus pointing to a promising system for in vivo evaluation.

Overall, the GI tract poses several barriers and challenges to the oral delivery of compounds, ranging from small molecules to large proteins (Fuhrmann and Fuhrmann 2017). On the other hand, some GI properties may also be taken advantage of, including mucus adhesion for sustained GI residence time (Fuhrmann et al. 2013), or stimuli-responsive release of molecules upon cleavage by native GI enzymes (Klinger and Landfester 2011; Aguado et al. 2018). Recent advances in nanomedicine for oral biologics application may be promising and indeed some potential has been demonstrated (Vllasaliu et al. 2018). In the following section, we will outline relevant challenges for the delivery of orally administered enzymes and aim at discussing any concurrently chances arising from the function and physiology of the GI system.

8.2.2 Challenges of Oral Administration of Therapeutic Enzymes

Among the various groups of orally administered drugs, macromolecules and proteins need to be examined differently compared to small molecule compounds (Mitragotri et al. 2014). For small molecules, biopharmaceutical properties including solubility in GI fluids and permeability across epithelial barriers are the key parameters when assessing their systemic bioavailability. For highly soluble small drugs their permeation through the GI membrane is the rate-limiting step for their absorption, while for well-permeable compounds their solubility will impact on the velocity of absorption. For these small molecule drugs, the biopharmaceutical classification system is creating a profound basis when estimating their ability to be orally absorbed and thus their therapeutic efficiency (Donaldson et al. 2015; Couvreur and Puisieux 1993). In contrast, proteins and peptides often need to be folded in their physiological state to be pharmacologically active. This important difference between macromolecular and small molecule drugs is extremely relevant when developing these compounds for an orally administered formulation (Fuhrmann and Fuhrmann 2017). Within the group of protein-based compounds, orally-administered enzymes are a special type of proteins as they require a unique three-dimensional folding of both their protein backbone and their active site. The active site is composed of a binding pocket to bind and orient the substrate and a catalytic domain which allows the enzymatic reaction to take place (Blocher et al. 1999). Small structural alterations in the binding and the catalytic site may result in substantial loss in overall enzymatic activity. This loss will inevitably influence the enzyme's therapeutic efficiency and will need to be addressed during pharmaceutical formulation.

The key players in this loss of enzymatic activity in the GI tract are: (i) pepsin and low pH in the stomach, (ii) pancreatic proteases, such as chymotrypsin and trypsin, and bile salts in the small intestine, and (iii) slightly basic pH and the microbial community in the large intestine. Low and high pH may lead to enzyme unfolding (Fuhrmann and Leroux 2011), while GI enzymes will digest enzymes and induce a loss in activity by degradation (Schulz et al. 2015a). Pepsin preferentially cleaves hydrophobic residues such as leucine and phenylalanine and often leaves amino acids histidine, lysine arginine and proline uncleaved (Hamuro et al. 2008). Trypsin and chymotrypsin prefer bonds between amino acids lysine, arginine and aromatic amino acids (Lundh 1957). Intestinal bile salts may additionally inactivate orally administered enzymes (Cook et al. 2012).

In this chapter, we will focus mostly on orally administered enzymes which are active directly in the GI tract, i.e., those that are not systemically absorbed, in contrast to enzymes that are supposed to function systemically after absorption. We are discussing engineering strategies for these enzymes using diseases such as coeliac disease (Tack et al. 2010), phenylketonuria (Kim et al. 2004), pancreatic insufficiency (Regan et al. 1977), including malabsorption in immunodeficient patients, sucrose and lactose intolerance and others as model disposition. In some cases, model enzymes are simply normal GI enzymes such as α-CT or pepsin. The presented examples are discussed in more technical and methodological detail to provide an in-depths understanding of modification strategies for oral enzymes. In all cases, the major focus is on assessing enzyme modifications to understand the different modification strategies, which include:

(i) Direct polymer modification of the enzyme backbone and
(ii) Encapsulation of orally administered enzymes into gastro-resistant carriers and formulation strategies.

We are specifically interested in the chemistry behind enzyme polymer modifications and are discussing biocompatibility aspects and challenges associated with large scale production of these enzyme conjugates (Gauthier and Klok 2010). Ultimately, an outlook on the pharmaceutical and clinical relevance of modification strategies is given.

8.3 Engineering Strategies for Oral Enzymes Which Are Effective Directly in the Gastrointestinal Tract

A potential avenue to enhance an enzyme's activity in the GI tract is by manipulation of its primary amino acid sequence using recombinant technologies. In doing so, it may be possible to reduce tendencies for enzyme unfolding under acidic conditions, such as in the stomach. It may also increase the enzyme stability towards pepsin or intestinal digestive enzymes to overall improve oral enzyme activity. It was shown that replacing hydrophobic residues such as phenylalanine and leucine by analogues that are less recognised by pepsin can enhance stability of oral enzymes under mild acidic conditions (Ehren et al. 2008). In that work, a combination of structure-based with sequence-based approaches and machine learning algorithms has proven useful to identify improved variants of an enzyme explored for oral coeliac disease therapy. Two rounds of iterative mutagenesis and analysis were applied leading to variants with 20% increased specific activity at pH 4.5 and a 200-fold greater pepsin resistance. A comparable example is the recombinant engineering of a ciliate-derived lipase that showed a >200-fold activity in comparison to the native lipase enzyme (Brock et al. 2016). Ciliates are a type of protozoans possessing hair-like organelles called cilia which are responsible for their nomenclature. This protozoan lipase was highly active compared to a pancreatin standard and also

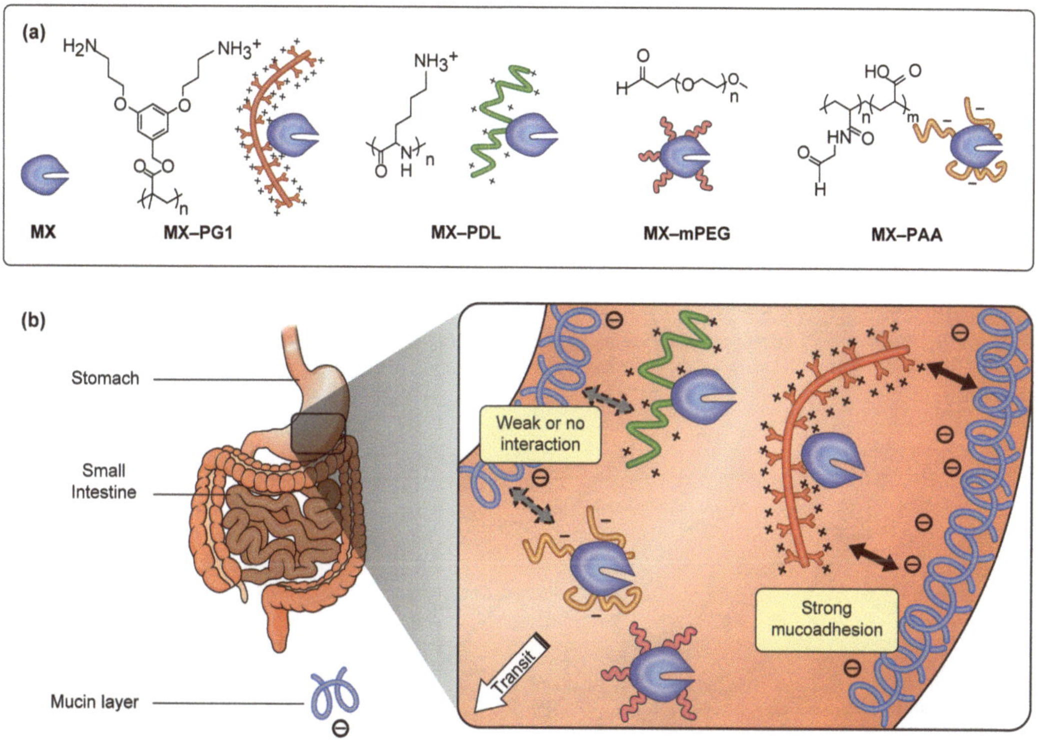

Fig. 8.2 Schematic overview of the general principle of polymer conjugation onto the surface of orally administered enzymes. The figure shows MX – an endopeptidase from *Myxococcus xanthus* (Shan et al. 2004; Fuhrmann and Leroux 2014) – as a model enzyme and (**a**) illustrates conjugation of various charged or neutral polymers onto its surface. (**b**) Upon oral administration the surface exposed polymers may convey specific interaction in the GI tract, including mucoadhesion or receptor binding. Depending on the chemical space the properties of enzyme conjugates may be tuned in different directions. (Reproduced from Fuhrmann et al. 2013, Copyright (2013) Macmillan Publishers)

when bile salts were present. Genetic engineering may also be a suitable avenue as shown by a modified yeast lipase that showed inherent stability under stomach in vivo conditions and also using a test meal evaluation in a mini pig model of pancreatic insufficiency (Aloulou et al. 2015). Conservative changes in protein backbone structure, and recombinant genetic engineering may indeed influence enzyme stability and functionality but the influence of such unnatural modifications on enzyme immunogenicity needs careful evaluation. Nevertheless, these examples nicely showed that engineering approaches can be useful for improving enzyme functionality.

The primary focus of this chapter is on direct polymer modification of orally administered enzymes. In doing so, chemical groups exposed on the surface of the enzymes are coupled via chemical conjugation to the polymer of choice. The schematic concept of this conjugation is outlined in Fig. 8.2. The available groups on the enzymes' surface may include amine groups (–NH_2) from lysine residues, sulphide bridges and free sulphide groups (-S-S and –SH) from cysteine residues, carboxyl groups (-COOH) from glutamic or aspartic acid and amide groups (-$CONH_2$) from glutamate residues. This list indicates the broad chemical diversity which is supplemented by the various types of polymers and polymer architectures that are conjugated onto these residues. Structures ranging from linear to comb-shaped architecture, from neutral to positively or negatively charged, and zwitterionic polymers to those that may selectively interact with GI mucus, endothelial surface proteins or even receptors (Fig. 8.3). The method of polymer stabilisation of enzymes is mostly based on steric hindrance and shielding of enzymes from the

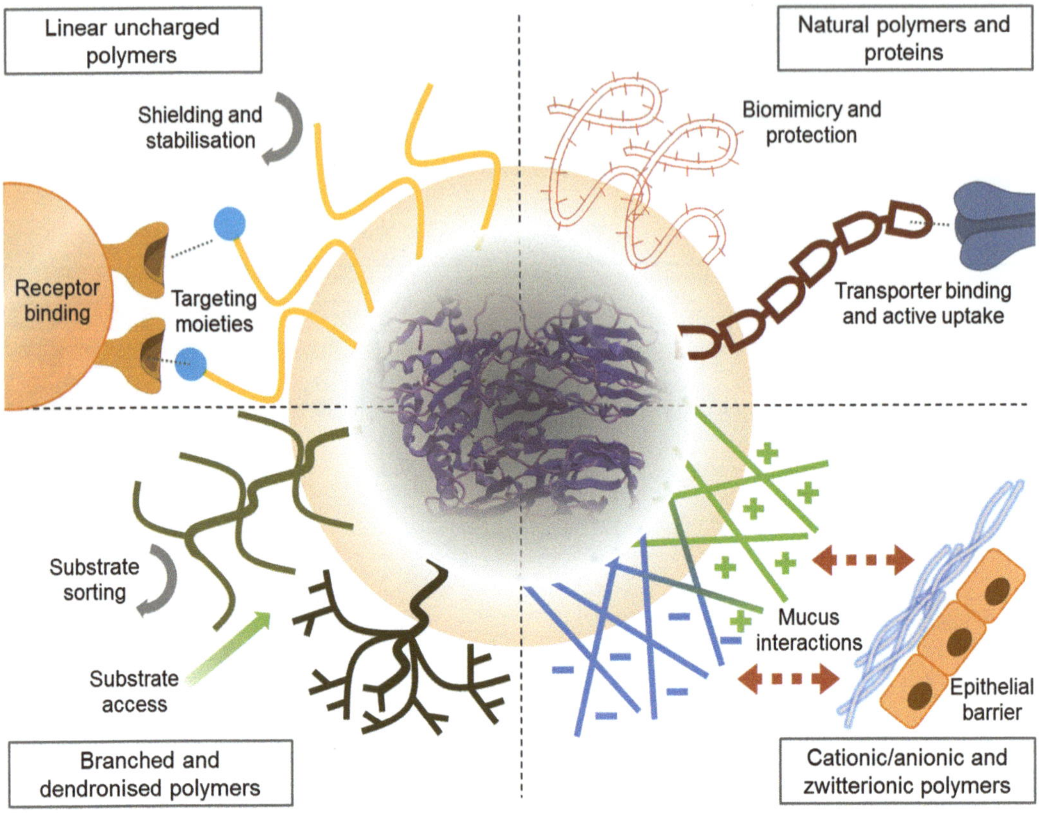

Fig. 8.3 Schematic overview of various polymer types and architecture for surface functionalisation of enzymes. A broad range of polymers is available for surface modification of enzymes. These are charged and neutral polymers, linear, branched or dendronised polymers, natural and biopolymers. This broad chemical variation allows tuning of specific enzymes to enhance their therapeutic selectivity and efficiency. (Reproduced from Fuhrmann and Fuhrmann 2017, Copyright (2017) Elsevier)

external GI environment. By protecting the enzyme from the harsh GI conditions, particularly in the stomach, it may be possible to retain a sufficient activity in the small intestine to cleave any potentially toxic substrate. Similar approaches have been successfully conducted to stabilise enzymes for systemic injection, for example polyethylene glycol (PEG)-Asparaginase (Graham 2003). Asparaginase is an enzyme that cleaves asparagine into aspartic acid and ammonia and it is used in cancer therapy. Upon injection, it degrades free asparagine making it inaccessible for cancer cells for which it is essential. The list of polymers that have been successfully coupled onto oral enzyme surfaces is not limited to PEG and oligo ethylene glycol derivatives but also includes poly lysine, chitosan, polyacrylic acid, various methacrylate polymers such as temperature-responsive poly(N-isopropylacrylamide) and quartenary ammonium methacrylate as we will show in the following. Overall, the chemical space when modifying oral enzymes is large which is a major advantage of this approach. It creates a multitude of modification options and increases the chances of finding clinically successful candidates. In the next examples, we will discuss types of polymer modification, types of enzymes, types of chemical linkage, and their overall impact on enzyme activity and stability in the GI tract.

8.3.1 PEGylation of Enzymes

PEGylation is the covalent attachment of poly (ethylene glycol) to any drugs, including small molecules, proteins and enzymes. PEG is a linear polymer approved by the FDA and used for several years for the development of clinically relevant conjugates (Rodríguez-Martínez et al. 2009). The first conjugates were developed in the 1990s, but they are mostly for systemic application (Hafner et al. 2014). Until now, at least nine PEG conjugates have reached the market, more are currently designed (Zelikin et al. 2016) which overall indicates the importance and timeliness of the PEGylation approach. For oral administration, PEG seems to be the most straightforward polymer as it is biocompatible and offers a broad chemical variance (Veronese and Pasut 2005). In the following, we are first discussing approaches with PEG alone, and then in comparison to other polymers.

To achieve PEGylation of a commercially available lactase, which is derived from Aspergillus oryzae, the enzyme was co-incubated with mPEG2-NHS in borate buffered medium (Turner et al. 2011). The NHS ester targeted 1 of the 16 lysine side chains found on the enzyme's surface and the TNBSA assay combined with SDS-PAGE indicated monoPEGylation. By combining SEC and HPLC, the separation of the enzyme-PEG conjugate from its educts was accomplished. The SEC elution profile showed the successful formation of the PEGylated enzyme. Further analyses of the modification indicated that PEGylation had no effect upon the enzyme's stability and activity at varying temperatures, whereas the activity and stability were increased at pH values between 2.5 and 4.5. By exposing the PEGylated enzyme to SGC and SIC, it could be shown that PEGylation protected the enzyme from proteolytic degradation by pepsin or pancreatin as well as from inactivation at lower pH values. Exposing both, PEGylated and unPEGylated enzyme to SGC followed by SIC resulted in retention of 25% and 12% of its activity for the PEGylated and unPEGylated enzyme, respectively. Surprisingly, an enhancement of the kinetic properties was found, indicated by slightly decreased k_m value and an increased V_{max} and k_{cat}.

Another model enzyme that is suited for assessing the impact of polymer modification on oral stability and activity is α-CT. Rodríguez-Martínez et al. conjugated α-CT with PEG of various MW, i.e., 700, 2000 and 5000 Da and at different grafting densities (Rodríguez-Martínez et al. 2009). The rate of digestion of the substrate Suc-Ala-Ala-Pro-Phe-pNA for both, α-CT and PEG-α-CT was determined by measuring the absorbance of p-nitroaniline. Compared to the native enzyme, PEGylation resulted in a decrease of activity of around 40–50% and the k_{cat} value decreased as the amount of PEG was increased when six or more molecules of PEG were attached. The main reason for the loss of activity might be the resulting increase of rigidity of the enzyme's core. The bioactivity did not depend on the MW of PEG, at least for α-CT. The k_m value increased depending on the amount of PEG attached: it doubled after the first PEG was attached, remained constant for up to six molecules and increased again after more than six PEGs were bound. The decrease in substrate affinity is a result of steric hindrance. To observe the effect of thermal stress on enzyme activity, α-CT and the conjugates were incubated at 45 °C prior to the activity measurement. The residual activity after incubating the enzyme at 45 °C increased with higher MW of PEG. Five thousand dalton PEG-conjugates showed 60% activity after 2.5 h of incubation, while the 700 Da conjugate retained about 30% and the native enzyme was inactivated. Overall, a higher MW PEG seemed to be beneficial for the enzyme, as there is no additional loss of activity while the stability is further increased.

In a continuation of this approach, it was studied in more detail which impact polymer modification may have on the structural protein dynamics. In doing so, similar PEG-α-CT conjugates with PEG of a various MW, i.e., 700, 2000 and 5000 Da and the degree of PEGylation was tuned between one and nine PEG molecules per α-CT (Rodríguez-Martínez et al. 2008). Different degrees of PEGylation were obtained by using molar ratios of activated mPEG to α-CT in the range of 1–13. To determine the degree of conjugation they titrated the free amino groups with TNBSA. By characterizing the PEG-α-CT conjugates via near-UV CD spectroscopy it could be shown that both, degree of PEGylation and molecular weight of PEG affects the structural integrity of the enzyme conjugate. Another aim was to study the influence of PEGylation on protein stability and structural dynamics. CD spectroscopy was used to estimate the effect on tertiary structure and DSC for thermal denaturation experiments. PEGylation resulted in increased T_m values by 6 °C after four PEG molecules were bound. Also, the free energy of protein unfolding and consequently the thermodynamic stability increased. Information on the structural dynamics of the conjugate could be received by FTIR H/D exchange experiments which indicated that PEGylation resulted in a more rigid protein core, as the rate of exchanging amide hydrogens is reduced. Again, the reduction of protein dynamics saturated after four PEG molecules were attached. As for the thermodynamic stability it was reasoned that the amount of PEG, not the MW of the attached PEG is crucial.

One of the clinically most advanced oral enzyme approaches is the development of phenylalanine cleaving enzymes for the treatment of phenylketonuria. In phenylketonuria a genetic defect in the phenylalanine cleaving enzyme phenylalanine hydroxylase – which catalyses the conversion of phenylalanine to tyrosine – leading to typical reduced cognitive development (Bélanger-Quintana et al. 2011). At the moment, the treatment comprises of reduced phenylalanine intake via the nutrition which is challenging to maintain. Phenylalanine hydroxylase used for replacement treatment might be promising but the enzyme requires certain co-factors which make a straightforward use difficult. A recombinant phenylalanine ammonia lyase (PAL) was developed as a robust alternative enzyme that can degrade phenylalanine without the need for cofactors (Kim et al. 2004). This PAL was produced in yeast and was successfully studied in an in vivo mouse model of phenylketonuria (Sarkissian et al. 1999) and has shown promising activity. Nevertheless, this PAL was applied parentally and it would be desirable to administer the enzyme by oral administration. When further developing this approach of orally administered PAL, a new variant, namely TM-AvPAL was investigated (Kang et al. 2010). This variant was modified using a silica sol-gel matrix and PEGylation of surface lysines. Silica particles were produced by adding the enzyme to a silica sol of pH > 4 and transferring this solution into liquid paraffin with 1% w/w Span 80, in which the gelation occurred. To achieve enzyme PEGylation with 5 kDa linear PEG via NHS-ester strategy, the enzyme was dropped into a stirring solution of PEG in potassium phosphate buffer of pH 8 and the reaction kinetics were assessed by SDS-PAGE. They found that a maximum protein load of around 13.8 mg/mL could be achieved without causing a negative effect on particle morphology or protease protection. The silica gel protected the enzyme from inactivation in acidic medium of pH 3.5 for 1 h. When incubating the enzyme containing silica particles in the presence of chymotrypsin or trypsin, the enzyme retained 50% of its activity after 2 h, while the free TM-AvPAL was inactivated after 15 min (Fig. 8.4). The effect of PEGylation on the enzyme stability was also evaluated and it was found that there was no protection from acidic conditions, as the enzyme also lost its activity after 15 min. PEG was able to protect the enzyme from digestion by trypsin and chymotrypsin at pH 6.8 for 4 h, however the protecting effect was more pronounced for trypsin, which suggests that PEG effectively blocked its potential cleaving sites. To measure the enzyme activity of PAL, it was added to a phenylalanine solution at pH 8.5, causing the formation of trans-cinnamic acid as digestion product. The substrate affinity was not affected by PEGylation. However, the enzyme encapsulated in silica particles had a lower specific substrate affinity. The oral formulation of this PEG-AvPAL needs further evaluation while an s.c. application has already reached clinical phase 2 trials (Pascucci et al. 2018).

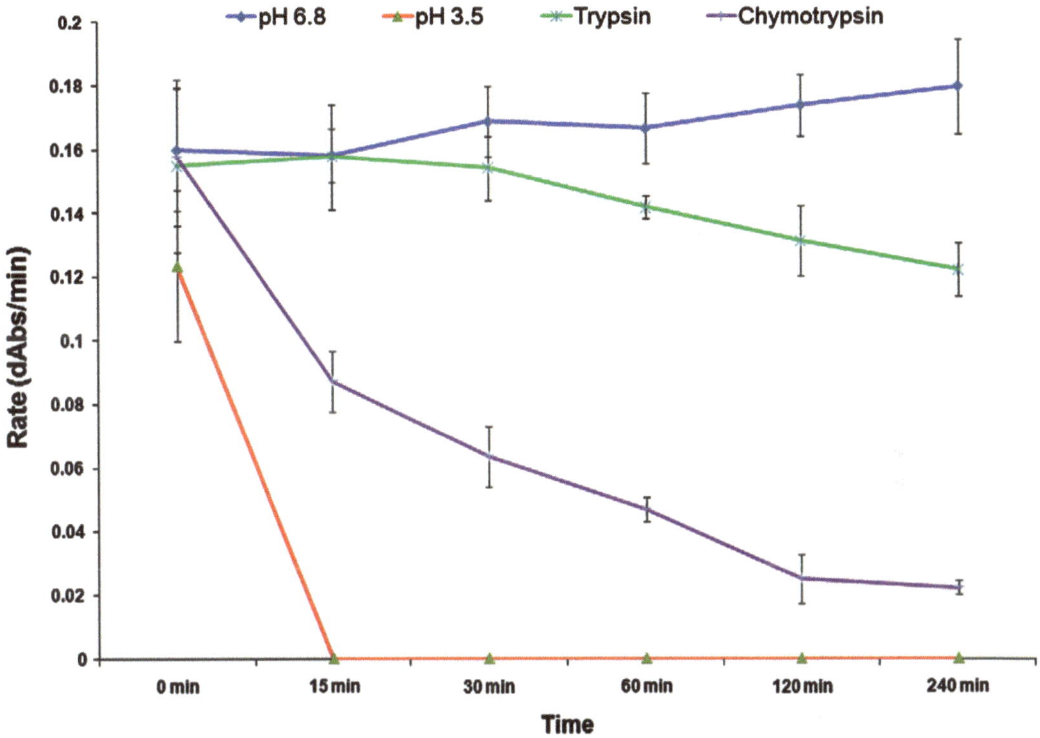

Fig. 8.4 The stability of PEGylated TM-AVPAL under different simulated GI conditions. Assay of 5 kDa PEG-TM-AvPAL by incubation with USP pH 3.5 buffer for 1 h, or USP pH 6.8 buffer, 5 mg/mL Chymotrypsin solution, or 5 mg/mL Trypsin solution for 2 h before assaying for catalytic activity. (Reproduced from Kang et al. 2010, Copyright (2010) Elsevier)

8.3.2 PEG in Comparison to Other Polymers

Using a non-invasive imaging method (Fuhrmann and Leroux 2011), the activity of an orally administered enzyme – PEP used in the treatment of coeliac disease – was approached in a systematic manner. PEP enzymes may potentially cleave deleterious food gluten, the major cause of coeliac disease, and could ameliorate clinical outcome. PEGylated PEPs showed improved cleavage of gluten model peptides and were more stable under simulated GI conditions (Robic 2007). In addition to PEG, PEP enzymes were modified with structurally and architecturally divers polymers, including poly lysine, poly acrylic acid and a dendronised polymer PG1 (Grotzky et al. 2012). It was shown that all polymers can enhance the stability of the enzymes in vivo. Most interestingly, the PG1 polymer induced an unprecedented enhancement of the PEP activity in the stomach which lasted about 4 h, i.e. the time by which the loosely adherent mucosal layer is shed off. PG1 was indeed mucoadhesive as shown both in vitro and in vivo and so was the PEP conjugate with this polymer.

All previously described approaches have used the enzyme in its natural conformation, which means they rely on chemically accessible functional groups on the enzyme surface, such as lysine-residues. The advantage of doing so is that the enzyme is used in its native state. A potential downside being accessible groups need to be present in significant abundance to allow coupling with the polymer of choice in the required density. In site-directed mutagenesis, the protein backbone of the enzyme is selectively engineered by introducing other residues and groups that allow a chemically selective coupling of the polymer of choice. One way of such modification of the PEP enzymes was followed by installing three cysteine residues onto the surface of the enzyme via site-directed mutagenesis of the producer bacterium, *Myxococcus xanthus* (Schulz et al. 2015b). To investigate the effect of polymer conjugation on the enzyme's stability and activity PEG was coupled via thiol-maleimide linkage by incubating the enzyme overnight with 100 molar equivalents of 5 kDa or 40 kDa mPEG-maleimide in sodium phosphate buffer and purification by repetitive ultrafiltration. By attaching one to three PEGs of different molar masses (5 kDa or 40 kDa) to the enzyme it was found that both position and total molar mass of polymer affect the stability and activity. In comparison, a generation three PAMAM dendrimer of 6.5 kDa was attached to the enzyme by triple-conjugation. Both, PEG and PAMAM conjugates were studied non-invasively in vivo (Fuhrmann and Leroux 2011) using a gluten-like substrate, carrying a fluorophore/quencher, enabling the detection of fluorescence at λex/em of 745/800 nm after its successful digestion. The fluorophore/quencher probe was given orally 5 min and 1–2 h after oral administration of the enzyme, to prove that the functionalised enzyme is stable even after longer incubation times in the GI tract. In vivo activity could be shown by the measurable fluorescence after release of the fluorophore due to digestion. While the 5 kDa PEG conjugate showed no statistical difference in activity compared to the negative control, 40 kDa PEG- and PAMAM-functionalisation resulted in a significant increase of enzyme activity. It is suggested that the cationic charge of PAMAM contributes to the protection, for example by complexing negatively charged mucin, making its smaller hydrodynamic diameter compared to the 5 kDa PEG negligible.

8.3.3 Methacrylate Polymers

As discussed in the previous section, albeit being more biocompatible and clinically approved PEG is not always superior to other polymers, such as charged or branched. In recent years, antibody formation against orally administered PEG is becoming clinically more relevant (Ishida et al. 2003; Zhao et al. 2012). Modification of oral enzymes without impairing their GI stability is tricky and may require the design of conceptually new polymer conjugates that are not based on simple PEG chains. A few of such new examples are discussed below where polymer chains are grown directly from the surface of the enzymes using a so-called polymer-based protein engineering (PBPE) avenue (Fig. 8.5).

In PBPE approaches, polymer chains are selectively grown from the surface of the enzyme, using a so-called "grafting from" approach. As example, an α-CT model enzyme was modified with the temperature responsive polymers pNIPAm and pDMAPS by ATRP (Cummings et al. 2013). MALDI-TOF analysis revealed that there were 12 ATRP initiators attached to the enzyme, from which the polymer chains could be grown. For polymer analysis, they were cleaved by hydrolysis and their molecular weight was determined to be 10 kDa for pDMAPS and 9 kDa for pNIPAm per chain. For further analyses, the critical solubility temperatures, i.e., UCST and LCST of free polymers and α-CT conjugates, k_{cat} and k_m were determined at 5 °C, 25 °C and 40 °C, and the effect of conjugation on enzyme stability were studied. The UCST of α-CT-pDMAPS was 13 °C and the LCST of α-CT-pNIPAm was 29.5 °C, both comparable to the isolated polymer with 12 °C respectively 30 °C. pNIPAm changes its conformation above the LCST, it collapses and becomes highly hydrophobic, while pDMAPS is hydrophobic below its UCST. Comparing the k_{cat} and k_m values at all temperatures, α-CT-pDMAPS showed lower k_m than the native α-CT, meaning it has a higher substrate affinity, possibly due to the interactions between the model substrate and pDMAPS. Even below the UCST its activity was higher. α-CT-pNIPAm collapsed at 40 °C followed by a decrease in activity, probably caused by the increased steric hindrance. Relating to the enzyme stability it could be shown that conjugation stabilises the enzyme from both, autolysis (main reason for inactivation at 25 °C) and protein denaturation (main reason for inactivation at 40 °C). The stabilizing effect of pNIPAm was higher than that of pDMAPS, because pNIPAm,

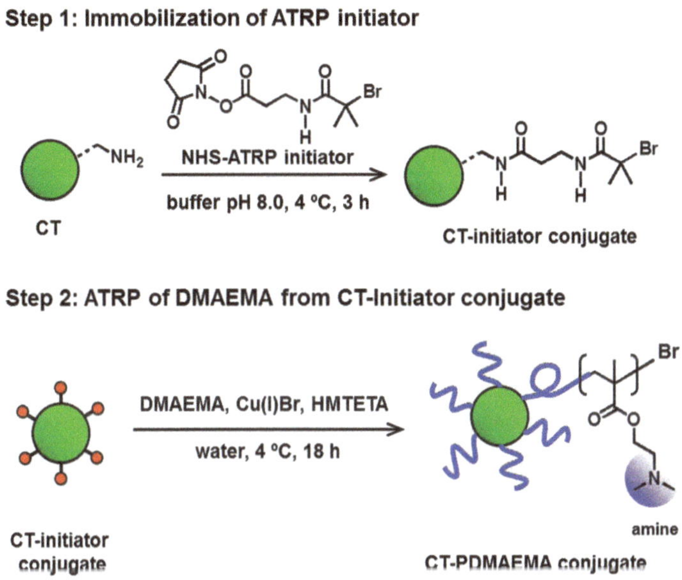

Fig. 8.5 Schematic overview of the principle of polymer-based protein engineering (PBPE). Overview of PBPE using ATRP with DMAEMA and Chymotrypsin. (Reproduced from Murata et al. 2013, Copyright (2013) American Chemical Society)

in its collapsed state at 40 °C, shows a higher degree of steric hindrance. By using PBPE the activity and stability of enzymes could be predictably modified by using temperature responsive polymers.

When extending the arsenal of polymers for α-CT modification, ATRP was applied to couple zwitterionic pCBAm, neutral pOEGMA, positive pQA, and negative pSMA on the enzyme's surface (Cummings et al. 2017). In doing so, NHS-Br initiator was immobilised on the surface and quantified via MALDI-TOF-MS. Then the reagent monomers CBAm, OEGMA475, QA and SMA were added in a suitable medium. The conjugates were purified by dialysis and the weight percent of chymotrypsin in the different conjugates defined by BCA assay. As discussed above, GI mucus plays a pivotal role in the interaction of orally administered enzymes and it is interesting to better understand the role of polymer type, charge and architecture in this complex system. The various α-CT-conjugates were tested for in vitro polymer mucoadhesion by adding them to mucin in different buffer systems and determining the biocatalytic activity of the conjugate using a cleavable substrate. Also, the gastric acid stability, the surface charge of the α-CT-initiator complex and the intrinsic tryptophan fluorescence of the α-CT conjugates was determined. It was found that each α-CT conjugate, depending on the polymer, had different characteristics regarding mucoadhesion, activity and stability at different pH values. When studying the free polymers for mucoadhesion, pQA showed significant mucin binding at all tested pH values, but pSMA and pOEGMA led to no adhesion to the negatively charged mucin at all. Electrostatic interaction was postulated to be the main driving force for mucin binding. Depending on the pH, pCBAm showed different degrees of mucin binding. α-CT-pQA and α-CT-pCBAm retained 70% activity, while α-CT-pSMA and α-CT-pOEGMA had less than 50% activity, compared to the native chymotrypsin. There was also a large increase in substrate affinity for the α-CT-pQA conjugate, while the affinity of the α-CT-pSMA conjugate decreased at pH 6, which can be explained by electrostatic attraction or repulsion of the negatively charged substrate. The stability at gastric conditions was dependent on the attached polymer. The positively charged α-CT-pCBAm (at lower pH values) and α-CT-pQA showed higher stability compared to the native chymotrypsin, while α-CT-pSMA and α-CT-pOEGMA were destabilised through functionalisation.

PDBE is seen as a potent alternative to site-directed mutagenesis because it may allow to predictably influence the functionality of catalytically active enzymes by polymer-engineering. For ATRP initiators, a common problem is that conventional initiators often lack sufficient water solubility to conduct modification in aqueous solvents, while most therapeutic enzymes are easily denatured in organic solvents. To overcome this issue, a water-soluble NHS-functionalised, amide containing ATRP initiator was synthesised for conjugation of PDMAEMA (Murata et al. 2013) (Fig. 8.5). This polymer was used as a model because of its strong hydrophilic or hydrophobic properties, depending on the tested pH values between pH 5 and 9. Four different chain lengths of PDMAEMA were evaluated and depending on the chain length of PDMEAMA a different structure was achieved. Shorter polymers resulted in a mushroom-like structure, while longer chains, appeared brush-like. A pH- and temperature-dependency on enzyme activity and stability were observed. Due to the protonated state of PDMEAMA at pH values below the pka, it was hypothesised that the positive charge of the polymer stabilises the negative charge of the aspartate in the catalytic triad of α-CT. As shown by the lower k_m, the substrate binding also improved below pH 8, while the k_m increased at higher pH values, probably due to hydrophobic interactions between the substrate and uncharged PDMEAMA. At pH 5, the catalytic efficiency increased tenfold. Regarding the temperature-dependency, the LCST was important. Below the LCST, which depends on the chain length, the chains were hydrated and expanded. Above the LCST the catalytic efficiency decreased significantly. Exposing the α-CT conjugate to pH 7 and 40 °C resulted in 80–90% retention of the initial activity, while the native enzyme retained 50%. Also, the half-life was increased by almost 30-fold, independent of the polymer length. It could be demonstrated that the activity and stability of the α-CT conjugate depended on charge and conformational state of the polymer. Thus, they could be influenced by pH value and temperature in the GI environment.

From an analytical point of view, it remains difficult to study the conformational changes that enzyme-polymer conjugates undergo in different environments. A NMR spectroscopy method was developed to further evaluate the conformational and permeability properties of protein-bound monolayers of pOEGMA as a biocompatible comb-polymer (Liu et al. 2012). The aim was to synthesise a variety of α-CT-pOEGMA conjugates and to analyse the behaviour of the polymer monolayer regarding the conformation and permeability by probing the polymer backbone flexibility via ^{1}H NMR spectroscopy and observation of the catalytic activity of the enzyme conjugate (Fig. 8.6). ATRP was again used to grow chains of the polymer from previously prepared α-CT macro-initiators, which were obtained by reaction of 2-BIBB with lysine residues. Between 2 and 12 polymer chains from α-CT, with a varying backbone length between 10 and 100 monomers and side-chain lengths between 2 and 21–23 oxyethylene units. Tryptic digestion LC-MS revealed a random distribution of initiator groups on α-CT and they found that the polymer chains of a given conjugate were comparable regarding their lengths. ^{1}H NMR spectroscopy suggested that the flexibility of pOEGMA decreased with an increasing number of chains per enzyme by transitioning from free over loose chains to a brush-like layer. An increasing backbone length caused the polymer to transition from a brush-ellipsoid to a brush-cylindrical state, which has an impact on the permeability of the polymer monolayer. There was no preference towards smaller or larger substrates in the state of loose chains, but they found a preferential activity towards the smaller substrates, both BT*p*NA and Suc-Ala-Ala-Pro-Phe-pNA, in a brush-ellipsoid conformation of pOEGMA. Comparing these two substrates, the smaller BT*p*NA had an even higher preferential activity. The authors indicate a pronounced selective permeability, so called "molecular sieving" caused by gaps between the polymers chains of the monolayer, which get smaller, the closer the polymer gets to the transition from ellipsoid to cylindrical state. When determining the enzyme

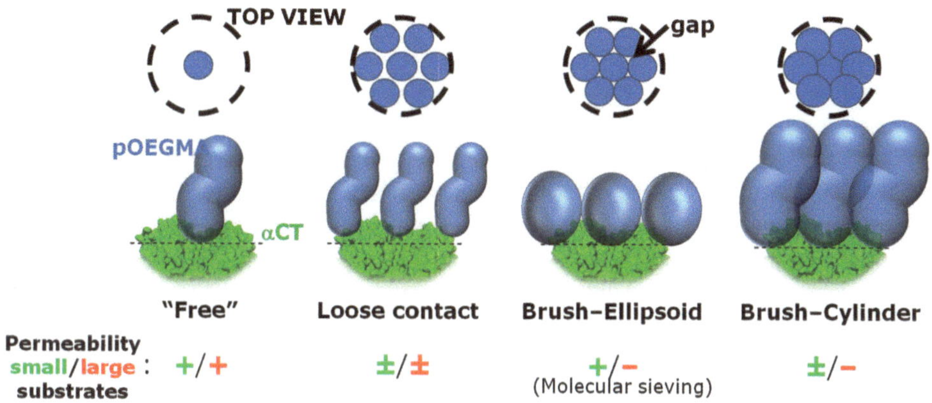

Fig. 8.6 Schematic illustration of the behaviour of architecturally different polymers conjugated onto the surface of a model enzyme. The idealised cartoon shows the organization and selective permeability characteristics of the four different regimes observed within the 3D parameter space of protein-bound pOEGMA examined. (Reproduced from Liu et al. 2012, Copyright (2012) Wiley)

activity in the presence of α1-antiCT depending on the state of the polymer, they found that the conjugates in free or loose contact state were inhibited by α1-antiCT, while the two brush-states were not affected. Due to the molecular sieving effect the brush-ellipsoid state was superior to the brush-cylindrical state. As a control, they tested for the activity of an α-CT-PEG conjugate with 8–9 linear chains and found that α1-antiCT caused a decrease in activity.

Summarising the different approaches for engineering orally administered enzymes to render optimal stability and activity, it may be useful to create a generalised rule to achieve successful clinical candidates. While PEGylation still remains the academic "gold-standard", it appears that other polymers than PEG may provide additional features to further enhance the stability of enzymes and enabling their selective interaction with GI components. In terms of modification strategies, there are three main avenues of intensive research activity, including (i) native enzyme modification taking advantage of the natural enzyme structure and conformation, (ii) site-specific mutation to introduce new chemically accessible residues and (iii) protein polymer engineering where enzymes are converted into a polymerisation initiator. All of these paths have their advantages and disadvantages and the decision what functionalisation is to follow will mostly depend on the natural enzyme conformation and stability and the availability of the appropriate polymer in the required molecular weight and architecture. Should new polymers be developed, their biocompatibility needs careful assessment, for example charged and dendronised polymers may interact with the GI cell barrier and induce instability which could allow foreign material to be translocated (Turner 2009). To bring these approaches forward, detail evaluation both in complex in vitro (Leonard et al. 2012) and in vivo models are required.

8.4 Formulation Strategies for Oral Enzymes Which Are Effective Directly in the Gastrointestinal Tract

GI enzymes and low pH in the stomach are one of the major issues for oral delivery of enzymes as discussed in previous sections. One of the most studied avenues to circumvent these drawbacks is the application of stomach-resistant and enteric coatings. These coatings are successfully applied for monolithic oral dosage forms tablets and capsules and they protect their cargo from the harsh GI environment and release (macromolecular) drugs under the milder conditions in the upper small intestine (Regan et al. 1977). Compared to single dosage forms (i.e., tablets and capsules), multiple dosage forms – due to their small size – provide a few advantages, such as reduced stomach residence and increased transition into the small intestine, specific interaction in different GI compartments and the option of combining them with selective release mechanisms (Zhang et al. 2015).

Here, we focus on newer systems such as microparticles and nanoparticles and we discuss strategies combining enzymes with biomaterials including hydrogels. Such combination with new materials may offer different ways of formulating enzymes taking advantage of the properties of the biomaterial itself (Fuhrmann 2018; Lee et al. 2018).

8.4.1 Microparticles, Macropores and Nanocores

Microparticle formulations for oral application have been widely assessed and studied for a long time, not only for protein-based drugs (Couvreur and Puisieux 1993). These systems are easily prepared with a good control over their size and morphology. An interesting example is a multi-particulate delivery system of pancreatin enzymes created by encapsulation with the acid-stable polymers CAP, Eudragit® L100 or HPMCP (Naikwade et al. 2009). During formulation development, solid dispersions with CAP in ratios of 1:1 and 1:2 of pancreatin to CAP were prepared by physical mixing, kneading and coevaporation methods. In case of Eudragit® L100 and HPMCP the dispersions were prepared in ratios of 1:2. Additionally, microspheres were prepared by solvent evaporation technique in a core-to-coat ratio of 1:0.25 and 1:0.5 and by nonsolvent addition technique in a ratio of 1:0.5, with a ratio of dispersion to vehicle of 1:5. They also prepared solid dispersions by spraying a solution of enzyme and polymer on the surface of sugar spheres. Protease, amylase and lipase content was determined spectrophotometrically or by titration. While some of the microspheres prepared by solvent evaporation or nonsolvent addition technique showed promising acid protection, enzyme activity and particle size, the solid dispersions failed due to insufficient protection against acid and the spraying on sugar spheres caused the enzymes to lose their activity. Pancreatin microspheres with Eudragit® L100 as coating material, prepared by solvent evaporation technique with a core to coat ratio of 1:0.5, were chosen for the comparison with marketed tablets. In vivo experiments with pancreatic enzyme deficient rats, showed the highest retained enzyme activity after 3 and 6 months storage. In vitro tests showed that the microspheres reduced fecal fat by 38%, which is 2.5 times more than by treatment with plain pancreatin. Safety of the formulation was confirmed by dose toxicity studies.

In a comparable approach, MPs were fabricated from Eudragit S100 via microemulsion technique using sonication which contained pH-responsive macropores (Kumar et al. 2017). Different sonication methods were tested to study the effect on pore formation. Lactase was encapsulated by suspending the MPs in a trehalose-containing solution of the enzyme. The MPs were freeze-dried to ensure sealing of the pores. As shown by SEM and TEM imaging, freeze-drying resulted in closure of the pores. When incubating the MPs in SGF and SIF to examine the pH responsiveness of the pores it was found that they remain closed in SGF but opened when the MPs were exposed to SIF. Regarding the release behaviour under different pH values, they found that fluorescent model drugs were not released at pH 2.0, while the measured fluorescence intensity increased rapidly when the MPs were exposed to pH 7.1, due to the opening of the pores under those conditions. Their investigation regarding the protective effect of the MPs towards lactase showed that the enzyme retained 50% activity, while the unprotected enzyme was completely inactive after 2 h of incubation in SGF. The MPs are a suitable carrier for the targeted release in the small intestine, which also protects the enzyme from acidic gastric environment, shown by the more than 15 times higher activity as commercially available lactase formulations.

Further reduction in size of enzyme loaded carriers, i.e. moving from microparticle to nanoparticle, may modulate the enzyme stability and interaction in the GI system. Nanoceramic cores (NCCs) are a glassy matrix of sugar oligomers providing an aqueous environment, which could improve the stability and activity of loaded enzymes (Parihar et al. 2017). HAP NCCs were prepared by self-precipitation method and then coated with cellobiose or trehalose, to increase the surface area as well as to ensure a protective environment for the adsorption of catalase. Encapsulation of the enzyme loaded NCs within alginate by adding $CaCl_2$ was to protect catalase from acidic conditions after oral application. A standard H_2O_2 degradation assay was used to determine the enzyme activity of encapsulated catalase. To determine the in vitro release, the loaded nanocores were incubated for 2 h in HCl buffer at pH 1.2, followed by 6 h incubation in alkaline phosphate buffer at pH 7.4, both at 37 °C and stirring at 100 rpm to mimic GIT passage. When comparing the adsorbed sugars, TEM pictures showed that cellobiose gave a well-packed structure, while trehalose aggregated to some extent. The results from adsorption isotherm suggested a higher total adsorption of catalase for the cellobiose system, which also had a better loading efficiency. Determination of enzyme activity showed, that the free catalase was completely inactive after treatment with acidic buffer, while the alginate encapsulated enzyme lost only 1.3–1.5% of its activity, using cellobiose respectively trehalose. Treatment with basic buffer resulted in activity retention of 81.0% for free catalase, 99.0% and 99.8% for cellobiose or trehalose as adsorbent. Regarding the in vitro release of the enzyme, they found that the alginate system kept its integrity during the incubation in acidic medium, due to its conversion to unionised alginic acid. The cumulative drug release of about 20% after 2 h is due to diffusive release. When entering the alkaline buffer, the alginate dissolved quickly, resulting in a rapid release of catalase, followed by a sustained release due to desorption from the core structure. The release in alkaline buffer followed first order kinetic. Storage stability studies revealed a loss of 1.5–1.9% of enzyme activity after 6 months.

8.4.2 Hydrogel Approaches

Hydrogels are a promising group of biomaterials with a high degree of functionality, ease of preparation and various preclinical applications. Hydrogels offer a broad chemical variance for modification, they are biocompatible and have been successfully applied for several applications, including tissue engineering (Parmar et al. 2015), regenerative medicine (Pashuck and Stevens 2012) and other therapeutic applications (Fuhrmann et al. 2018). For oral delivery, hydrogels can be obtained from synthetic and natural polymers which in addition to protection of cargoes can specifically interact in different regions of the GI tract, for example mucus adhesion of chitosan hydrogels. Polysaccharide hydrogels are well-studied and clinically evaluated which would allow a quick translation of results to preclinical assessment. Examples of these avenues are outlined below.

A recent approach was presented by preparing catalase-loaded HA-CM-chitosan hydrogel microspheres via emulsion cross-linking technique, using EDC/NHS as co-crosslinkers (Tang et al. 2014). The rate of enzymatic degradation was measured by determination of the mass loss after incubation in hyaluronidase-containing medium. Using the Bradford method, the encapsulation efficiency as well as the release behaviour was assessed. The activity of the entrapped catalase was compared to the activity of the native enzyme by determining the decay rate of hydrogen peroxide, after incubating both, loaded microspheres and native enzyme in trypsin containing buffer respectively in buffer solutions between 2.0 and 8.0 at 37 °C. Cross-linking of HA and CM-chitosan resulted in a higher stability against degrading enzymes such as hyaluronidases. When evaluating the activity of entrapped catalase as a function of encapsulation efficiency, the activity was lower than expected, probably due to the carboxylate anions of HA, which competitively block the binding site of catalase. An increase of CM-chitosan in the microspheres resulted in an increase of catalase activity. The encapsulated catalase was protected from acidic gastric conditions as well as from digestive enzymes as trypsin, as shown by the remaining activity of encapsulated enzyme compared to the native catalase. A HT-29 cell model was used to investigate the microsphere's capability of scavenging hydrogen peroxide and the resulting cell viability in the presence of catalase. The cell model showed that the HA-CM-chitosan microspheres with a ratio of HA to CM-chitosan of 1:3, successfully delivered catalase to the HT-29 cells, as demonstrated by the intracellular hydrogen peroxide scavenging ratio of 37.8%. They were able to demonstrate that encapsulated catalase had a significant protective effect on hydrogen peroxide induced stress, while native catalase failed to protect the HT-29 cells.

Another polysaccharide-based hydrogel was recently presented by preparing differently sized carrageenan-based hydrogel beads loaded with β-galactosidase and with or without $Mg(OH)_2$ as buffer (Zhang et al. 2017). Hydrogel beads of 255 μm and 2610 μm diameter were prepared and the effect of acidic conditions on the activity of lactase was studied by incubation in preheated SGF for 2 h. The pH inside the beads was determined by CLSM by using a fluorescence probe which bears a pH-dependent and a pH-independent fluorescence group. By calculating the ratio of intensities of both fluorescence groups, the pH inside the beads could be estimated. The pH inside the unbuffered beads dropped from 6.9 to below 4 after incubation, while it decreased slightly from 7.2 to 6.6 inside the buffer-containing beads, indicating the protecting effect of buffer-loaded beads against acidic conditions. Lactase in unbuffered beads completely lost its activity after 5 min incubation, independent of the size of the beads. As opposed to this, the enzyme activity depended on both, size of the beads and amount of buffer loaded. For smaller beads a higher amount of buffer was needed to retain a reasonable level of activity due to shorter diffusion distance for H^+ ions inside and higher amount of enzyme on the surface of the bead.

8.4.3 Other Avenues

Using a technique called emulsion electrospinning an interesting combination of enzymes with biomaterials was presented by loading the model enzymes HRP and AP into electrospun nanofibers made of Eudragit® L100 (Frizzell et al. 2017). Protein activity and loading-efficiency were optimised by varying electrospun voltage steady, the flow rate and the volume of the aqueous phase. By TEM imaging and by using FITC-BSA as loading material they showed that the electrospun nanofibers have a core-shell structure. Mimicking physiological pH conditions in the GI tract they demonstrated that less than 5% of the protein is released over 4 h in acidic conditions by the Eudragit® L100 coat. After 1 h at pH 6 nearly 100% of the cargo was released. They tested for the effect of flow rate and vol% of the aqueous phase on retention of enzyme stability after electrospinning. While HRP activity depended neither on flow rate nor volume of the aqueous phase, the activity of AP decreased with higher amounts of aqueous phase. By using PVA as stabiliser, activity for AP in the 20 vol% formulation could be increased from 60% to 75%. To increase the shelf life of this formulation, the nanofibers were lyophilised, which resulted in a 1.5 to 3.2-fold increase in recovery of initial bioactivity after 7 days of storage.

8.5 Clinical and Pharmaceutical Relevance of Enzyme Modifications

The oral administration of enzymes is currently studied for several GI dispositions, including coeliac disease, phenylketonuria and others. The administration via the oral route has many advantages, including non-invasive application, no need for trained personnel and increased patient compliance. The clinical use of enzymes is already indicated for pancreatic insufficiency, lactose and sucrose intolerance. In addition, oral administration of PEPs for coeliac disease is under investigation (Tack et al. 2013) with a dietary supplement on the market (Tolerase G).

Table 8.2 Successful engineering strategies for oral enzymes under preclinical evaluation

Oral enzyme	Delivery approach	Effect	References
Lactase	Polymer conjugation (PEG)	Improved in vitro stability in simulated gastric fluids	Turner et al. (2011)
	pH-responsive microparticles (Eudragit S100)	Improved in vitro stability and activity	Kumar et al. (2017)
	Buffer-containing hydrogel beads (carrageenan)	Protection from acidic conditions	Zhang et al. (2017)
Proline-specific endopeptidase	Polymer conjugation (PEG, PAMAM)	Improved in vivo stability and activity	Schulz et al. (2015b)
	Polymer conjugation (PEG)	Improved stability under GI conditions, improved substrate cleavage	Robic (2007)
α-Chymotrypsin	Polymer conjugation (PEG)	Decrease of activity, improved stability	Rodríguez-Martinez et al. (2009)
	Polymer conjugation (pDMAPS)	Improved substrate affinity and stability	Cummings et al. (2013)
	Polymer conjugation (pCBAm, pQA)	Improved in vitro stability under gastric conditions	Cummings et al. (2017)
	Polymer conjugation (PDMAEMA)	Improved substrate binding and catalytic efficiency	Murata et al. (2013)
	Polymer conjugation	NMR based tool to study the impact of surface engineering	Liu et al. (2012)
Phenylalanine ammonia lyase	Polymer conjugation (PEG)	Protection from digestive enzymes	Kang et al. (2010)
	Enteric coated particles	In vitro protection from acidic environment	
Pancreatic enzymes (Pancreatin)	Polymer encapsulation (Eudragit L100)	Improved in vivo efficacy	Naikwade et al. (2009)
Catalase	Enteric-coated nanocores (HAP)	Improved in vitro stability and activity	Parihar et al. (2017)
	Hydrogel microspheres (HA, CM-chitosan)	Protection from acidic conditions and degrading enzymes	Tang et al. (2014)

Nevertheless, all of these enzymes are unmodified and studied in their native state. Native, unmodified enzymes may easily be inactivated in the GI tract, mainly in the stomach and additional functionalisation strategies should be taken advantage of. Table 8.2 summarised the current status of engineered oral enzymes that are discussed in detail in this chapter.

Current approaches to stabilise enzymes in the GI tract and circumvent their inactivation under the harsh stomach conditions is by their incorporation into enteric coated capsules or tablets. Their clinical outcome is debatable given that for some diseases potentially deleterious compounds have to be inactivated in different compartments of the GI tract. In coeliac disease, toxic gluten needs to be cleaved in the stomach and before reaching the upper small intestine. Such enzyme activity would not be possible using enteric coated formulations. For these reasons, polymer conjugation of enzymes offers the possibility to engineer and tune enzymatic activity to the desired GI section (Fuhrmann et al. 2013). Polymer conjugation offers a broad toolset of methodologies to modify enzymes, including direct modification on the surface, site-specific mutation to introduce reactive groups, protein polymer engineering and an extensive chemical variability for polymers. Both, clinically accepted polymers and new polymers may be applied for enzyme modification. When developing these enzyme-conjugates there are two major issues to be addressed: (i) biocompatibility and toxicity aspects of enzyme modifications and polymers, and (ii) technological challenges when producing modified enzymes at larger scale. Biocompatibility evaluations need to be executed first using simple and complex cell models, followed by detailed in vivo evaluations. During these tests, pharmacokinetics of enzymes, stability and activity and potential immunogenicity reactions need to be studied. A scalable production process is required for advanced development of the enzyme-conjugates for clinical evaluations. Recombinant biotechnological production is possibly the way forward, as shown from PAL enzymes that have already reached clinical testing (Sarkissian et al. 2008). Further developments of new natural and bio-inspired polymers (Liu et al. 2017) will also help in developing well-tolerated and efficient conjugates with low toxicity. The combination of such systems with a triggered responsive release may further reduce any potential side-effects and could make the enzyme-polymer conjugates more selective and specific. Overall, polymer engineering of oral enzymes is a powerful tool to enhance their stability and activity in the GI tract. When bringing forward these systems, close cooperation between academic inventive talent and industrial expertise in the technological implementation will be inevitable.

References

Aguado BA, Grim JC, Rosales AM, Watson-Capps JJ, Anseth KS (2018) Engineering precision biomaterials for personalized medicine. Sci Transl Med 10(424)

Aloulou A, Schué M, Puccinelli D, Milano S, Delchambre C, Leblond Y, Laugier R, Carriere F (2015) *Yarrowia Lipolytica* lipase 2 is stable and highly active in test meals and increases fat absorption in an animal model of pancreatic exocrine insufficiency. Gastroenterology 149(7):1910–1919.e5

Bansil R, Turner BS (2018) The biology of mucus: composition, synthesis and organization. Adv Drug Deliv Rev 12:43–15

Bélanger-Quintana A, Burlina A, Harding CO, Muntau AC (2011) Up to date knowledge on different treatment strategies for phenylketonuria. Mol Genet Metab 104(0):S19–S25

Blau N, Van Spronsen FJ, Levy HL (2010) Phenylketonuria. Lancet 376(9750):1417–1427

Blocher M, Walde P, Dunn IJ (1999) Modeling of enzymatic reactions in vesicles: the case of alpha-chymotrypsin. Biotechnol Bioeng 62(1):36–43

Brock A, Aldag I, Edskes S, Hartmann M, Herzog T, Uhl W, Schnekenburger J (2016) Novel ciliate lipases for enzyme replacement during exocrine pancreatic insufficiency. Eur J Gastroenterol Hepatol 28(11):1305–1312

Chopra S, Bertrand N, Lim J-M, Wang A, Farokhzad OC, Karnik R (2017) Design of insulin-loaded nanoparticles enabled by multistep control of nanoprecipitation and zinc chelation. ACS Appl Mater Interfaces 9(13):11440–11450

Clardy-James S, Allis DG, Fairchild TJ, Doyle RP (2012) Examining the effects of vitamin B12 conjugation on the biological activity of insulin: a molecular dynamic and in vivo oral uptake investigation. MedChemComm 3(9):1054–1058

Cook MT, Tzortzis G, Charalampopoulos D, Khutoryanskiy VV (2012) Microencapsulation of probiotics for gastrointestinal delivery. J Control Release 162(1):56–67

Couvreur P, Puisieux F (1993) Nano- and microparticles for the delivery of polypeptides and proteins. Adv Drug Deliv Rev 10(2):141–162

Cummings C, Murata H, Koepsel R, Russell AJ (2013) Tailoring enzyme activity and stability using polymer-based protein engineering. Biomaterials 34(30):7437–7443

Cummings CS, Campbell AS, Baker SL, Carmali S, Murata H, Russell AJ (2017) Design of stomach acid-stable and mucin-binding enzyme polymer conjugates. Biomacromolecules 18(2):576–586

Dean SN, Turner KB, Medintz IL, Walper SA (2017) Targeting and delivery of therapeutic enzymes. Ther Deliv 8(7):577–595

Donaldson GP, Lee SM, Mazmanian SK (2015) Gut biogeography of the bacterial microbiota. Nat Rev Microbiol 14:20

Eek D, Krohe M, Mazar I, Horsfield A, Pompilus F, Friebe R, Shields AL (2016) Patient-reported preferences for oral versus intravenous administration for the treatment of cancer: a review of the literature. Patient Prefer Adherence 10:1609–1621

Ehren J, Govindarajan S, Moron B, Minshull J, Khosla C (2008) Protein engineering of improved prolyl endopeptidases for celiac sprue therapy. Protein Eng Des Sel 21(12):699–707

Frizzell H, Ohlsen TJ, Woodrow KA (2017) Protein-loaded emulsion electrospun fibers optimized for bioactivity retention and Ph-controlled release for peroral delivery of biologic therapeutics. Int J Pharm 533(1):99–110

Fuhrmann G (2018) Luminal coating of the intestine. Nat Mater 17:754–755

Fuhrmann K, Fuhrmann G (2017) Recent advances in oral delivery of macromolecular drugs and benefits of polymer conjugation. Curr Opin Colloid Interface Sci 31:67–74

Fuhrmann G, Leroux J-C (2011) In vivo fluorescence imaging of exogenous enzyme activity in the gastrointestinal tract. Proc Natl Acad Sci U S A 108(22):9032–9037

Fuhrmann G, Leroux J-C (2014) Improving the stability and activity of oral therapeutic enzymes –

recent advances and perspectives. Pharm Res 31(5):1099–1105

Fuhrmann G, Grotzky A, Lukic R, Matoori S, Luciani P, Yu H, Walde P, Schlüter AD, Gauthier MA, Leroux J-C (2013) Sustained gastrointestinal activity of dendronized polymer-enzyme conjugates. Nat Chem 5:582–589

Fuhrmann G, Chandrawati R, Pamar PA, Keane TJ, Maynard SA, Bertazzo S, Stevens MM (2018) Engineering extracellular vesicles with the tools of enzyme prodrug therapy. Adv Mater 30:1706616

Gauthier MA, Klok H-A (2010) Polymer-protein conjugates: an enzymatic activity perspective. Polym Chem 1(9):1352–1373

Graham ML (2003) Pegaspargase: a review of clinical studies. Adv Drug Deliv Rev 55(10):1293–1302

Grotzky A, Nauser T, Erdogan H, Schlüter AD, Walde P (2012) A fluorescently-labeled dendronized polymer-enzyme conjugate carrying multiple copies of two different types of active enzymes. J Am Chem Soc 134:11392–11395

Gupta V, Hwang BH, Doshi N, Mitragotri S (2013) A permeation enhancer for increasing transport of therapeutic macromolecules across the intestine. J Control Release 172(2):541–549

Hafner A, Lovrić J, Lakoš GP, Pepić I (2014) Nanotherapeutics in the Eu: an overview on current state and future directions. Int J Nanomed 9:1005–1023

Hamuro Y, Coales SJ, Molnar KS, Tuske SJ, Morrow JA (2008) Specificity of immobilized porcine pepsin in H/D exchange compatible conditions. Rapid Commun Mass Spectrom 22(7):1041–1046

Ishida T, Maeda R, Ichihara M, Irimura K, Kiwada H (2003) Accelerated clearance of pegylated liposomes in rats after repeated injections. J Control Release 88(1):35–42

Kang TS, Wang L, Sarkissian CN, Gámez A, Scriver CR, Stevens RC (2010) Converting an injectable protein therapeutic into an oral form: phenylalanine ammonia lyase for phenylketonuria. Mol Genet Metab 99(1):4–9

Kim W, Erlandsen H, Surendran S, Stevens RC, Gamez A, Michols-Matalon K, Tyring SK, Matalon R (2004) Trends in enzyme therapy for phenylketonuria. Mol Ther 10(2):220–224

Klinger D, Landfester K (2011) Dual stimuli-responsive poly(2-hydroxyethyl methacrylate-co-methacrylic acid) microgels based on photo-cleavable cross-linkers: Ph-dependent swelling and light-induced degradation. Macromolecules 44(24):9758–9772

Kumar A, Montemagno C, Choi H-J (2017) Smart microparticles with a Ph-responsive macropore for targeted oral drug delivery. Sci Rep 7(1):3059

Lee Y, Deelman TE, Chen K, Lin DSY, Tavakkoli A, Karp JM (2018) Therapeutic luminal coating of the intestine. Nat Mater 17:834–842

Leeds JS, Oppong K, Sanders DS (2011) The role of fecal elastase-1 in detecting exocrine pancreatic disease. Nat Rev Gastroenterol Hepatol 8(7):405–415

Leonard F, Ali H, Collnot EM, Crielaard BJ, Lammers T, Storm G, Lehr CM (2012) Screening of budesonide nanoformulations for treatment of inflammatory bowel disease in an inflamed 3d cell-culture model. Altex 29(3):275–285

Liu M, Tirino P, Radivojevic M, Phillips D, Gibson M, Leroux J-C, Gauthier MA (2012) Molecular sieving on the surface of a protein provides protection without loss of activity. Adv Funct Mater 23(16):2007–2015

Liu Y, Lee J, Mansfield KM, Ko JH, Sallam S, Wesdemiotis C, Maynard HD (2017) Trehalose glycopolymer enhances both solution stability and pharmacokinetics of a therapeutic protein. Bioconjug Chem 28(3):836–845

Lomer MCE, Parkes GC, Sanderson JD (2008) Review article: lactose intolerance in clinical practice – myths and realities. Aliment Pharmacol Ther 27(2):93–103

Lundh G (1957) Determination of trypsin and chymotrypsin in human intestinal content. Scand J Clin Lab Invest 9(3):229–232

Mitragotri S, Burke PA, Langer R (2014) Overcoming the challenges in administering biopharmaceuticals: formulation and delivery strategies. Nat Rev Drug Discov 13(9):655–672

Moroz E, Matoori S, Leroux J-C (2016) Oral delivery of macromolecular drugs: where we are after almost 100 years of attempts. Adv Drug Deliv Rev 101:108–121

Murata H, Cummings CS, Koepsel RR, Russell AJ (2013) Polymer-based protein engineering can rationally tune enzyme activity, Ph-dependence, and stability. Biomacromolecules 14(6):1919–1926

Murgia X, Loretz B, Hartwig O, Hittinger M, Lehr C-M (2018) The role of mucus on drug transport and its potential to affect therapeutic outcomes. Adv Drug Deliv Rev 124:82–97

Naikwade SR, Meshram RN, Bajaj AN (2009) Preparation and in vivo efficacy study of pancreatin microparticles as an enzyme replacement therapy for pancreatitis. Drug Dev Ind Pharm 35(4):417–432

Parihar AKS, Srivastava S, Patel S, Singh MR, Singh D (2017) Novel catalase loaded nanocores for the treatment of inflammatory bowel diseases. Artif Cells Nanomed Biotechnol 45(5):981–989

Parmar PA, Chow LW, St-Pierre J-P, Horejs C-M, Peng YY, Werkmeister JA, Ramshaw JAM, Stevens MM (2015) Collagen-mimetic peptide-modifiable hydrogels for articular cartilage regeneration. Biomaterials 54:213–225

Pascucci T, Rossi L, Colamartino M, Gabucci C, Carducci C, Valzania A, Sasso V, Bigini N, Pierigè F, Viscomi MT, Ventura R, Cabib S, Magnani M, Puglisi-Allegra S, Leuzzi V (2018) A new therapy prevents intellectual disability in mouse with phenylketonuria. Mol Genet Metab 124(1):39–49

Pashuck ET, Stevens MM (2012) Designing regenerative biomaterial therapies for the clinic. Sci Transl Med 4(160):160sr4

Pinier M, Fuhrmann G, Verdu E, Leroux J-C (2010) Prevention measures and exploratory pharmacologi-

cal treatments of celiac disease. Am J Gastroenterol 105:2551–2561
Regan PT, Malagelada J-R, Dimagno EP, Glanzman SL, Go VLW (1977) Comparative effects of antacids, cimetidine and enteric coating on the therapeutic response to oral enzymes in severe pancreatic insufficiency. N Engl J Med 297(16):854–858
Robic S (2007) Pegylated glutenase polypeptides. USA patent application WO/2007/047303. 26.04.2007
Rodríguez-Martínez JA, Solá RJ, Castillo B, Cintrón-Colón HR, Rivera-Rivera I, Barletta G, Griebenow K (2008) Stabilization of A-chymotrypsin upon pegylation correlates with reduced structural dynamics. Biotechnol Bioeng 101(6):1142–1149
Rodríguez-Martínez JA, Rivera-Rivera I, Solá RJ, Griebenow K (2009) Enzymatic activity and thermal stability of Peg-A-chymotrypsin conjugates. Biotechnol Lett 31(6):883–887
Sarkissian CN, Shao Z, Blain F, Peevers R, Su H, Heft R, Chang TMS, Scriver CR (1999) A different approach to treatment of phenylketonuria: phenylalanine degradation with recombinant phenylalanine ammonia lyase. Proc Natl Acad Sci U S A 96(5):2339–2344
Sarkissian CN, Gámez A, Wang L, Charbonneau M, Fitzpatrick P, Lemontt JF, Zhao B, Vellard M, Bell SM, Henschell C, Lambert A, Tsuruda L, Stevens RC, Scriver CR (2008) Preclinical evaluation of multiple species of pegylated recombinant phenylalanine ammonia lyase for the treatment of phenylketonuria. Proc Natl Acad Sci U S A 105(52):20894–20899
Schulz JD, Gauthier MA, Leroux J-C (2015a) Improving oral drug bioavailability with polycations? Eur J Pharm Biopharm 97(Part B):427–437
Schulz JD, Patt M, Basler S, Kries H, Hilvert D, Gauthier MA, Leroux J-C (2015b) Site-specific polymer conjugation stabilizes therapeutic enzymes in the gastrointestinal tract. Adv Mater 28(7):1455–1460
Shan L, Marti T, Sollid LM, Gray GM, Khosla C (2004) Comparative biochemical analysis of three bacterial prolyl endopeptidases: implications for coeliac sprue. Biochem J 383(2):311–318
Tack GJ, Verbeek WHM, Schreurs MWJ, Mulder CJJ (2010) The spectrum of celiac disease: epidemiology, clinical aspects and treatment. Nat Rev Gastroenterol Hepatol 7:204–213
Tack GJ, Van De Water JMW, Bruins MJ, Kooy-Winkelaar EMC, Van Bergen J, Bonnet P, Vreugdenhil ACE, Korponay-Szabo I, Edens L, Von Blomberg BME, Schreurs MWJ, Mulder CJ, Koning F (2013) Consumption of gluten with gluten-degrading enzyme by celiac patients: a pilot-study. World J Gastroenterol 19(35):5837–5847
Tang BC, Dawson M, Lai SK, Wang Y-Y, Suk JS, Yang M, Zeitlin P, Boyle MP, Fu J, Hanes J (2009) Biodegradable polymer nanoparticles that rapidly penetrate the human mucus barrier. Proc Natl Acad Sci 106(46):19268–19273
Tang D-W, Yu S-H, Wu W-S, Hsieh H-Y, Tsai Y-C, Mi F-L (2014) Hydrogel microspheres for stabilization of an antioxidant enzyme: effect of emulsion cross-linking of a dual polysaccharide system on the protection of enzyme activity. Colloids Surf B: Biointerfaces 113:59–68
Turner JR (2009) Intestinal mucosal barrier function in health and disease. Nat Rev Immunol 9(11):799–809
Turner KM, Pasut G, Veronese FM, Boyce A, Walsh G (2011) Stabilization of a supplemental digestive enzyme by post-translational engineering using chemically-activated polyethylene glycol. Biotechnol Lett 33(3):617–621
Veronese FM, Pasut G (2005) Pegylation, successful approach to drug delivery. Drug Discov Today 10(21):1451–1458
Vllasaliu D, Thanou M, Stolnik S, Fowler R (2018) Recent advances in oral delivery of biologics: nanomedicine and physical modes of delivery. Expert Opin Drug Deliv 15(8):759–770
Vorselen D, Van Dommelen SM, Sorkin R, Piontek MC, Schiller J, Döpp ST, Kooijmans SAA, Van Oirschot BA, Versluijs BA, Bierings MB, Van Wijk R, Schiffelers RM, Wuite GJL, Roos WH (2018) The fluid membrane determines mechanics of erythrocyte extracellular vesicles and is softened in hereditary spherocytosis. Nat Commun 9(1):4960
Zelikin AN, Ehrhardt C, Healy AM (2016) Materials and methods for delivery of biological drugs. Nat Chem 8(11):997–1007
Zhang S, Bellinger AM, Glettig DL, Barman R, Lee Y-AL, Zhu J, Cleveland C, Montgomery VA, Gu L, Nash LD, Maitland DJ, Langer R, Traverso G (2015) A Ph-responsive supramolecular polymer gel as an enteric elastomer for use in gastric devices. Nat Mater 14:1065–1071. advance online publication
Zhang Z, Zhang R, Mcclements DJ (2017) Lactase (B-galactosidase) encapsulation in hydrogel beads with controlled internal Ph microenvironments: impact of bead characteristics on enzyme activity. Food Hydrocoll 67:85–93
Zhao Y, Wang C, Wang L, Yang Q, Tang W, She Z, Deng Y (2012) A frustrating problem: accelerated blood clearance of pegylated solid lipid nanoparticles following subcutaneous injection in rats. Eur J Pharm Biopharm 81(3):506–513

9 Enzymes in Metabolic Anticancer Therapy

Maristella Maggi and Claudia Scotti

Abstract

Cancer treatment has greatly improved over the last 50 years, but it remains challenging in several cases. Useful therapeutic targets are normally unique peculiarities of cancer cells that distinguish them from normal cells and that can be tackled with appropriate drugs. It is now known that cell metabolism is rewired during tumorigenesis and metastasis as a consequence of oncogene activation and oncosuppressors inactivation, leading to a new cellular homeostasis typically directed towards anabolism. Because of these modifications, cells can become strongly or absolutely dependent on specific substrates, like sugars, lipids or amino acids. Cancer addictions are a relevant target for therapy, as removal of an essential substrate can lead to their selective cell-cycle arrest or even to cell death, leaving normal cells untouched. Enzymes can act as powerful agents in this respect, as demonstrated by asparaginase, which has been included in the treatment of Acute Lymphoblastic Leukemia for half a century. In this review, a short outline of cancer addictions will be provided, focusing on the main cancer amino acid dependencies described so far. Therapeutic enzymes which have been already experimented at the clinical level will be discussed, along with novel potential candidates that we propose as new promising molecules. The intrinsic limitations of their present molecular forms, along with molecular engineering solutions to explore, will also be presented.

Keywords

Cancer addictions · Asparaginase · Glutaminase · Arginine deaminase · Arginase

Abbreviations

ADC	L-arginine decarboxylase
ADI	Arginine deiminase
AGA	Aspartylglucosaminidase
ALL	Acute lymphoblastic leukaemia
ARGase	L-Arginase
ASL	Argininosuccinate lyase
ASNase	Asparaginase
ASNS	Asparagine synthetase
AspATs	Aspartate aminotransferases
ASRLG1	β-aspartyl peptidase/L-asparaginase

M. Maggi (✉) · C. Scotti
Department of Molecular Medicine, Unit of Immunology and General Pathology, University of Pavia, Pavia, Italy
e-mail: maristella.maggi01@universitadipavia.it; claudia.scotti@unipv.it

N. Labrou (ed.), *Therapeutic Enzymes: Function and Clinical Implications*, Advances in Experimental Medicine and Biology 1148, https://doi.org/10.1007/978-981-13-7709-9_9

ASS	Argininosuccinate synthase
ASS	Argininosuccinate synthetase
ATF4	Activating transcription factor 4
ATP	Adenosine triphosphate
BBB	Blood brain barrier
Bcl-2	B-cell lymphoma 2
BCT-100	Cysteines-mutated and PEGylated form of ARGase I
BFM	Berlin, Frankfurt, Münster
CGL	Cystathionine-γ-lyase
FAD	Flavin adenine dinucleotide
FAK	Focal adhesion kinase
GA	Glutaminase-asparaginase
GAC	Glutaminase C
GCS	Glycine cleavage system
GL	Bacterial glutaminase
GLS	Mammalian glutaminase
GO	Glycine oxidase
GRASPA	Erythrocytes encapsulating L-asparaginase
KGA	Kidney glutaminase
KRAS	Kirsten Rat sarcoma virus
Ma-ADI	*Mycoplasma arginini* derived ADI
MGL	L-methioninase
MTAP	Methylthioadenosine phosphorylase
mTORC1	Mammalian target of rapamycin complex 1
MTR	Methionine synthase
MYC and c-MYC	Myelocytomatosis gene
NADPH	Dihydronicotinamide-adenine dinucleotide Phosphate
NCI-60	National cancer institute-60 collection of cancer cell lines
Ntn	N-terminal nucleophile
OAT	Ornithine aminotransferase
OCT	Ornithine transcarbamyl transferase
P5C	Pyrroline-5-carboxylate
P5CS	Pyrroline-5-carboxylate synthase
PA	Plasminogen activator
PEG	Polyethylene glycole
PHGDH	Phosphoglycerate dehydrogenase
PIG-6	p53-induced gene 6
PLP	Pyrodoxal phosphate
PN	Pyridoxin
PpMGL	*Pseudomonas putida* L-methioninase
PRODH/POX	Proline oxidase
PSAT-1	Phosphoserine aminotransferase 1
PSPH	Phosphoserine phosphatase
PutA	Proline utilization A
PYCR1/2/L	Pyrroline-5-carboxylate reductase
Raf	Rapidly accelerated fibrosarcoma
Ras	Rat sarcoma
RBC	Red blood cell
RNAi	RNA interference
ROS	Radical oxygen species
RSV-BRL	Rous sarcoma virus-transformed buffalo-rat liver
SAM	S-adenosylmethionine
SC	Succinimidyl carbamate
SC-PEG	Calaspargase pegol
SDH	Serine dehydratase
SHMT	Serine hydroxymethyltransferase
TAT	L-tyrosine-2-oxoglutarate aminotransferase
TCA	Tricarboxylic acid cycle

9.1 Introduction

9.1.1 Cancer and Cancer Therapy

Cancer is the main cause of mortality worldwide, with 14.1 million new cancer cases, 8.2 million cancer deaths and 32.6 million people living with cancer within 5 years of diagnosis (Ferlay et al. 2015). The number of new cases is expected to rise by about 70% to about 22 million over the next two decades. Solid tumours deriving from epithelia are the most frequently diagnosed and

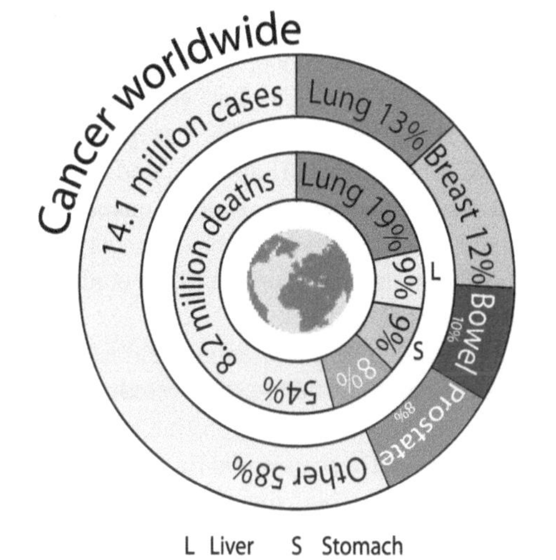

Fig. 9.1 Cancer incidence and mortality in 2012 (CRUK)

the most difficult to treat: in men, lung, prostate, colorectum, stomach, and liver cancers and, in women, breast, colorectum, lung, cervix, and stomach cancers are the top causes of death (Fig. 9.1). In fact, about 23% (59,364) of Phase I-III studies listed by ClinicalTrials.gov up to the 8th January 2017 are related to cancer, demonstrating a high unmet need for improved treatments (Aggarwal 2010).

Many advances in therapy in the last decades have led to a significant improvement in survival of patients affected by haematological tumours, but the therapy of solid tumours (more than 80% of the total) is still challenging (Ferlay et al. 2010). Good specificity of therapeutic drugs is still to be achieved, so that unacceptable toxicity on normal cells is a major drawback, with life-threatening side effects hampering treatment. Drug resistance selects for clones which survive and reiterate the disease either locally or at a distance from the original tumour, making treatment even more challenging and leading patients to death.

This review discusses cancer addictions and the consolidated or potential role of therapeutic enzymes in the removal of cellular metabolic substrates. New translational opportunities will be presented along with possible solutions to the intrinsic limitations of available molecules.

9.1.2 Cancer Metabolism and Addictions

9.1.2.1 Cancer Cell Metabolism

After the original observation by Otto Warburg in 1924 about the enhanced glucose consumption of cancer cells by glycolysis even in the presence of sufficient oxygen to support oxidative phosphorylation (Warburg effect) (Warburg et al. 1924), cancer metabolism has definitely emerged in the last 10 years as a potential target for therapy (Galluzzi et al. 2013; Vander Heiden 2011). The number of publications releated to this field has recently increased to more than 20,000 per year (source: PubMed, keywords: "cancer" and "metabolism"). It is now known that hyperactivation of oncogenes or suppression of oncosuppressor genes induce, both directly and indirectly, specific metabolic features in cancer cells ("metabolic rewiring") and even production of specific metabolites ("oncometabolites"), which are related to their need to move from a catabolic to an anabolic metabolism in order to produce new biomass (Galluzzi et al. 2013). In many cases, this generates a modification in cellular activity, providing a molecular signature and a dependency of the neoplasm on the supply of specific substrates.

Alterations of several branches of metabolism can be associated with malignant transformation, generating potential targets for therapy (Galluzzi et al. 2013). Examples of promising metabolic targets for therapy include enzymes of Krebs cycle, glycolysis, beta-oxidation and mitochondrial respiration, but also of lipid biosynthesis and of the pentose phosphate pathway (Galluzzi et al. 2013). Agents and approaches include mainly small molecules and RNAis aimed at blocking the activity or production of these enzymes. Their developmental stage is in some cases at the preclinical or even clinical level, but, in general, the toxicity of small molecule inhibitors and glycolitic pathways redundancy have thus far hampered their success (Vander Heiden 2011; Galluzzi et al. 2013). Definitely more promising seems to be the group of enzyme drugs targeting amino acid addictions, whose efficacy has already been shown to be a powerful tool in several cases.

9.1.2.2 Amino Acid Addictions

There is a massive amount of both clinical and experimental data regarding the usage of bacterial Asparaginases (ASNases, *Escherichia coli* and *Erwinia carotovora*) in the treatment of Acute Lymphoblastic Leukaemia (ALL), starting from the classical BFM (Berlin, Frankfurt, Münster) protocol, introduced by Hansjörg Riehm in the 1970s (Covini et al. 2014). ALL is indeed dependent on asparagine for survival and more recent evidence suggests also on glutamine (Chan et al. 2014). The recent case of glutamine is particularly relevant, as the vast majority of cancer cell lines shows a profound addiction towards this amino acid (DeBerardinis et al. 2007; Wise and Thompson 2010). These dependences rely on the metabolic modifications derived from oncogenic transformation and tumour progression. For example, among oncogenes, MYC activation induces glutamine dependence (Wise et al. 2008; Gao et al. 2009) and KRAS stimulates glutamine usage through mTORC1 activation (Laplante and Sabatini 2012; Csibi et al. 2013). Among oncosuppressors, p53 helps in the adaptation of cancer cells to serine and glutamine shortage (Reid et al. 2013; Maddocks et al. 2013).

Along these lines, new amino acid addictions have been, and are being, discovered: arginine, serine, glycine, methionine, proline and tyrosine have been found to be essential for several tumour types, especially solid tumours with very poor outcome (Table 9.1).

Specific amino acid dependences can grant therapeutic selectivity and provide a good therapeutic window. Indeed, in the field of amino acid addictions, pre-clinical and clinical work shows that enzymes can be used in order to successfully remove amino acids and lead cancer cells to

Table 9.1 Tumour amino acid addictions

Tumour type	Addiction substrate	Role	Enzymatic activity for therapy	References
CNS tumours (neuroblastoma, glioma)	Methionine	Biosynthesis of GSH and polyamines DNA methylation	L-methioninase	Cappelletti et al. (2008), Carreau et al. (2011), and Cavuoto and Fenech (2012)
ALL Ovarian carcinomas Hon-hodgkin lymphoma	Asparagine	Protein synthesis Cell cycle progression	Asparaginase	Bansal et al. (2012), Cellarier et al. (2003), Chan et al. (2014), and GRASPA (n.d.)
Ovarian cancer ALL Hepatoma	Glutamine	Protein synthesis Amino nitrogen transporter	Glutaminase	Chen and Cui (2015), Cluntun et al. (2017), and Covini et al. (2014)
Melanoma Hepatocellular carcinoma Small cell lung carcinoma Pancreatic carcinomas	Arginine	NO synthesis	L-arginine deiminase	CRUK (n.d.), Csibi et al. (2013), Das et al. (2004), Deberardinis et al. (2007), and Ding et al. (2017)
Breast cancer	Glycine	Purine synthesis	Glycine deoxidase	Douce et al. (2001)
Breast cancer	Serine	Glutamine synthesis	Serine deaminase	Dufour et al. (2012)
Prostate cancer Burkitt lymphoma	Proline	Glutamine synthesis	Proline dehydrogenase	El-Sayed (2010) and Elia et al. (2017)
Melanoma Prostate cancer	Tyrosine Phenylalanine	Melanin and hormones synthesis	Tyrosine transaminase	Equar et al. (2015) and Ferlay et al. (2010)

Table 9.2 Number of ongoing clinical trials testing anticancer enzymes. (Source: Clinicaltrials.gov)

Enzyme drug	Clinical trials (total number)
Asparaginase	243
Arginine deiminase	18
Arginase	9

The table reports only cancer-related trials where enzymes are used

death (Cantor et al. 2012). Most of the times, however, like in the case of ASNase, non-human enzymes, typically of bacterial origin, need to be adopted because of their superior efficiency.

9.1.2.3 Clinical Trials

The number of clinical trials (source: Clinicaltrials.gov) based on enzymes able to deplete the amino acids listed in Table 9.1 shows the situation reported in Table 9.2. It is clear the predominance of asparaginase-based studies (243) and the relevant number of those involving arginine deiminase and arginase (27). In the case of asparaginase, most investigations (87.8% of total trials) involve administration of the *E. coli* native (44.4%) or PEGylated (43.4%) form, or encapsulated in red blood cells (GRASPA, n.d., 3.7%). The *E. chrysanthemi* enzymes in their native (8.8%) or PEGylated (1.4%) form are the second main group of drugs used. Of interest the fact that a clinical trial (NCT01574274) regarding the comparison of Oncaspar with a new modified form of asparaginase (Calaspargase Pegol, SC-PEG) is under way. Calaspargase pegol (SC-PEG) replaces the SS linker in SS-PEG with a succinimidyl carbamate (SC) linker, creating a more stable PEGylated asparaginase molecule (Angiolillo et al. 2014). For arginine depletion, the two enzymes used (arginase and arginine deiminase) are being tested either in their native or PEGylated form.

The fact that other enzymes do not meet clinical interest at the moment likely reflects either previous unsuccessful attempts (see methioninase and glutaminase below) or the need to readdress the molecular and structural biology of the existing ones.

In this review, enzymes able to tackle amino acid addictions will be described. Along with molecules which have already been substantially experimented, new potential candidates will be proposed because of their specific features and targets.

9.2 Enzymatic activities

9.2.1 L-Asparaginase

L-asparagine is a non-essential amino acid with a relevant role in glycoproteins synthesis. It is also a secondary non-toxic ammonia carrier. In mammals, de novo synthesis of asparagine is catalyzed by asparagine synthetase (EC 6.3.5.4, ASNS), an ATP dependent amido-ligase that transfers the ammonia group from the glutamine donor to the aspartate acceptor to form asparagine and glutamate. In humans, ASNS expression is regulated by the amino acid and unfolded protein response mechanisms through increased levels of ATF4 (Sukhani and Umerah 1978). Given the "stress-related" expression of ASNS, its expression levels in different tissues vary. The relevance of exogenous asparagine supply for the proliferation and survival of lymphoma has been highlighted in the early 1950s of the last century when the use of guinea pig serum containing asparagine degrading enzyme had cytotoxic effect on lymphoma transplanted mice (Kidd 1953; Broome 1968). Guinea pig L-asparaginase (EC 3.5.1.1, ASNase) was identified as the asparagine degrading enzyme responsible for the observed cytotoxic effect and, since then, several studies have been performed on ASNases of different origins, with a particular interest in bacteria-derived ones (Zuo et al. 2015; Batool et al. 2016; Kotzia and Labrou 2007; Warangkar and Khobragade 2010; Bansal et al. 2012; Law and Wriston 1971; Pokrovskaya et al. 2012; Han et al. 2014; Maggi et al. 2015; Cappelletti et al. 2008; Moreno-Enriquez et al. 2012; Lubkowski et al. 1996).

ASNases (Fig. 9.2) are amidohydrolases which primarily catalyze the deamination of L-asparagine to L-aspartate releasing ammonia. With lower efficiency, ASNases are also capable of hydrolyzing L-glutamine to L-glutamate and ammonia. Bacteria can express two types of

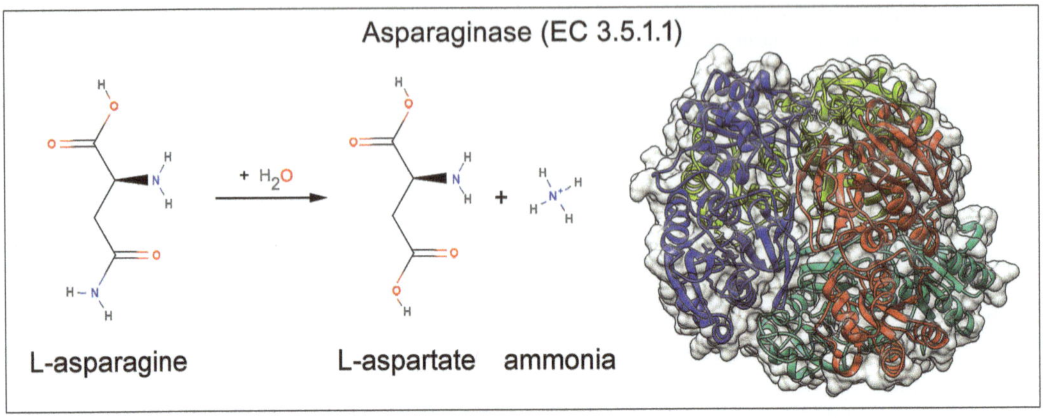

Fig. 9.2 L-asparaginase (ASNase): enzymatic reaction scheme (left) and molecular structure (right). The biological unit (tetrameric) structure (PDB ID: 3ECA) is represented as ribbons, colored by chain (red, blue, yellow and light blue), and surface (light grey)

L-Asparaginase: type I and type II. Type I L-ASNases are cytoplasmatic, constitutively expressed proteins. Kinetically, type I L-ASNases can carry out both L-asparagine and L-glutamine hydrolysis with higher efficiency for the former substrate. They show a cooperative behavior *vs.* asparagine with a low affinity (Km in the order of mM) and a hyperbolic kinetics *vs.* glutamine. Type II L-ASNases are periplasmic proteins expressed only in nutrient and oxygen stress conditions. Enzymes belonging to this class have a hyperbolic kinetics towards both substrates and they have a high affinity *vs.* asparagine (Km in the order of μM) and low to negligible activity towards glutamine (Michalska and Jaskolski 2006). Only type II L-ASNases have anti-cancer activity (Michalska and Jaskolski 2006). The enzyme reaction consists of two nucleophilic attacks involving two highly conserved catalytic triads (typically Thr-Lys-Asp and Thr-Tyr-Glu) in which two threonines play the main role. The enzyme reaction starts with a nucleophilic attack on the substrate amide C-atom operated by the first nucleophilic threonine, forming an acyl-intermediate and releasing a molecule of NH_3. The second step consists in a nucleophilic attack on the ester C atom of the product amide operated by a water molecule activated by the second triad, and results in the solving of the enzyme acyl-intermediate and in the release of the acid counterpart of the substrate (Lubkowski et al. 1994; Michalska and Jaskolski 2006).

Two bacterial ASNases are presently used in the clinics as drug for the treatment of pediatric Acute Lymphoblastic Leukemia (ALL). Both are type II asparaginases derived from *E. coli* and *Erwinia chrysanthemi*. The former, in its native (namely, Kidrolase and Spectrila) or PEGylated form (Oncaspar), has been used for more than five decades as first line drug for the treatment of ALL; *E. chrysanthemi* ASNase (Erwinase) is used as a second line drug in case of allergic reaction to Kidrolase and/or Oncaspar. The enzyme derived from *E. coli* has pronounced asparaginase activity and low glutaminase activity; Erwinase, at comparable asparaginase activity, has higher glutaminolytic activity if compared to Kidrolase and Oncaspar. On the other hand, the enzyme derived from *E. chrystanthemi* shows very poor stability in the blood stream if compared mainly to Oncaspar (6–16 h *vs.* 6 days, respectively and according to manufacturers data) and for this reason the former drug is preferred as a first therapeutic approach.

Treatment of ALL using ASNase in combination with vincristine and steroids, as established by the BFM protocol, is routinely used in the clinics worldwide and has an impressive success rate of roughly 90% 5-years survival in pediatric patients (Avramis and Panosyan 2005; Bhojwani et al. 2009). The drug efficacy is strictly linked to the incapability of leukemia cells to produce ASNS, the only enzyme capable of de novo asparagine synthesis. Interestingly, Su et al.

(2008) proved that asparaginase resistance mediated by ASNS expression correlates directly with ASNS protein content and not with coding mRNAs, suggesting that ASNS mRNA instability can be a mechanism of poor synthetase levels in cells sensitive to ASNase. In such class of cancer cells, asparaginase-mediated removal of the exogenous supply of the non-essential amino acid cannot be counteracted, causing protein synthesis inhibition through the mTOR pathway, cell cycle arrest and eventually cell death. Moreover, the secondary glutaminase activity of the enzyme is essential in favoring the drug effect in leukemia cells that express low levels of ASNS by removing the synthetase amino group donor glutamine. Glutaminase activity of asparaginases is a matter of discussion in the scientific community. Historically, it has been considered the main cause of the drug metabolic side effects, but recently its role in the drug efficacy has been re-evaluated and great importance has been given to concomitant asparagine and glutamine removal to counteract cancer cells resistance to treatment (Reinert et al. 2006; Chan et al. 2014; Maggi et al. 2015).

Given ASNases success in the treatment of pediatric ALL, the drug efficacy has been tested on several other tumours. Haematological malignancies that have shown sensitivity, though inconsistent, to L-ASNase-based treatment comprise acute myeloblastic leukemia, Hodgkin and non-Hodgkin lymphomas, myelosarcoma, myeloma, and the highly aggressive extranodal NK/T cell nasal type lymphoma (Willems et al. 2013; Perel et al. 2002; Yamaguchi et al. 2011). The most promising results on solid tumours have been obtained in in vitro and in vivo experimental models of pancreatic carcinoma, a cancer that is characterized by low to negligible expression of ASNS and, which, therefore, is sensitive to the asparaginase anti-proliferative mechanism (Dufour et al. 2012). Another solid tumour that shows in vitro sensitivity to ASNase is ovarian cancer, very likely because of the low expression levels of ASNS (Lorenzi et al. 2006, 2008).

Beside ASNS expression levels, also ATF4 is a good indicator of cancer cells sensitivity to ASNases: ATF4-deficient cells are strongly dependent on external asparagine supply as confirmed by in culture starvation experiments (Ye et al. 2010). ATF-4 is downstream in the amino acid response pathway and directly controls the expression of ASNS.

One of the main problems for the use of ASNase for the treatment of cancer in adult patients is its high immunogenicity that is linked to its bacterial origin and is more marked in the adult with respect to children. Recently, a less immunogenic and more stable formulation of ASNase has been tested with a positive outcome in adult pancreatic cancer patients, providing promising grounds for the use of ASNases as a new therapeutic strategy for pancreatic carcinoma (Dufour et al. 2012). This new formulation, namely GRASPA (n.d.), consists in the use of patient-derived red blood cells (RBCs) as a carrier for the drug, a strategy that is successful in reducing the host immune system exposure and, in turn, the reaction to the exogenous protein. An alternative strategy is chemical PEGylation of *E. coli* and *E. chrysanthemi* ASNases; the former is routinely used in the clinics and the latter has been tested in phase II clinical trials.

Finding a suitable asparagine-degrading enzyme of human origin would be the best solution to overcome immunogenicity of the drug. In humans, three asparagine-lytic enzymes have been identified: the 60 kDa lysophospholipase (EC 3.1.1.5), that can carry out both asparagine and lysophospholipids hydrolysis (Karamitros and Konrad 2014); the aspartylglucosaminidase (EC 3.5.1.26, AGA), a lysosomal enzyme that cleaves the amide bond between asparagine and N-acetylglucosamine in glycoproteins (Heiskanen et al. 1994); and the human β-aspartyl peptidase/L-asparaginase (EC 3.4.19.5, ASRLG1 (Nomme et al. 2014)).

The 60 kDa lysophospholipase has homology with bacterial type I and type II L-asparaginases in its N-domain; AGA and ASRLG1 are instead similar to type III L-asparaginases, a class of ASNases that is typically found in plants and is expressed as pro-enzymes belonging to the family of N-terminal nucleophile (Ntn) hydrolases. Type III asparaginases active form consists of two subunits (*e.g.*, α and β) generated by autoproteolytic cleavage that exposes catalytic threonines (Michalska and Jaskolski 2006).

Three are the main problems to be faced in order to consider clinical application of all three human enzymes: (i) secondary activities, (ii) need of activation for AGA and ASRGL1, and (iii) low affinity towards asparagine. Another issue to be considered is the lack of glutaminase activity that is, instead, becoming more and more relevant for the full therapeutic effect of asparaginases in some classes of leukemia. To our knowledge, several engineering studies are being undertaken in order to make human asparagine-degrading enzymes suitable for clinical application (Park et al. 1996; Tikkanen et al. 1996; Saarela et al. 1998, 2004a, b; Cantor et al. 2009; Hirano et al. 2009).

9.2.2 Arginine-Degrading Enzymes

Arginine is a conditionally essential amino acid, meaning that its endogenous production in normal conditions is sufficient to fulfill the organism metabolic demand, but in catabolic stress conditions the demand increases to such a level that endogenous production is not sufficient anymore (Fernandes et al. 2017; Morris et al. 2017). Cancer cells can become highly dependent on the extracellular supply of arginine, which is a key molecule in several pathways involved in metabolism, immunological response, neuro-signaling and stress-response (Wu and Morris 1998; Morris 2007). De novo synthesis of arginine is catalyzed in a two-step reaction by argininosuccinate synthase (EC 6.3.4.5, ASS) and argininosuccinate lyase (EC EC 4.3.2.1, ASL). ASS and/or ALS deficiency has been identified in several cancer cell lines dependent on arginine; among them carcinomas are the most represented tumours together with metastatic melanoma, Hodgkin's and non-Hodgkin's lymphoma, osteosarcoma and premyelocytic leukemia (Fernandes et al. 2017).

Arginine biosynthesis can also be obtained by the activation of a *salvage* pathway in which arginine is synthesized by ornithine transcarbamyl transferase (EC 2.1.3.6, OCT). OCT synthesis is typically downregolated in epithelial cells of different origin, therefore several tumours derived from epithelial tissues lack the arginine *salvage* pathway (Bobak et al. 2010).

Tumour cells deficient in ASS or OCT are auxotrophic for arginine, thus, targeting of its exogenous supply can be a suitable strategy for selective tumour cells killing. Out of the five enzymes capable of degrading arginine, three have been extensively investigated for their cytotoxic activity: L-Arginine decarboxylase (EC 4.1.1.19, ADC, Fig. 9.3), L-Arginase (EC 3.5.3.1, ARGase, Fig. 9.4), and arginine deiminase (EC 3.5.3.6, ADI, Fig. 9.5).

Beside plants and bacteria, ADC is also found in Mammals, where its expression is tissue specific (*i.e.*, liver and brain membranes). Human recombinant ADC has very high catalytic efficiency and affinity towards arginine, characteristics that make this enzyme an ideal candidate for enzymatic arginine deprivation. Unfortunately, the main product of ADC catalysis, agmantine, has strong cytotoxicity on healthy tissues if present in high doses in the extracellular space. Agmantine main physiological role is to regulate polyamines metabolism by increasing their catabolism, affecting in turn cell cycle progression. Actually, in vitro experiments using human recombinant ADC show that it has cytotoxic activity on several cancer cell lines, but its activity impairs also normal cells proliferation causing cell cycle arrest in G_1/S and eventually cell death (Wheatley and Campbell 2002; Philip et al. 2003).

ARGase is a binuclear Mn^{2+} amidinohydrolase that catalyzes the conversion of L-Arg into L-ornithine and urea. In humans, two isoforms have been identified: hARGase I, a cytosolic enzyme present only in liver and involved in urea cycle; hARGase II, a mitochondrial enzyme found in all tissues (Morris et al. 1997) and involved in mithocondrial synthesis of proline and glutamine (Patil et al. 2016). Isoform I main interest is arginase-starving agent for cancer therapy. The enzyme has mM affinity for arginine and a pH optimum of 9.3 (Glazer et al. 2011). The functional protein is an oligomer of 107 kDa characterized by the presence of a manganese binding cluster and a C-terminal S-shaped tail that is involved in inter-monomers interactions.

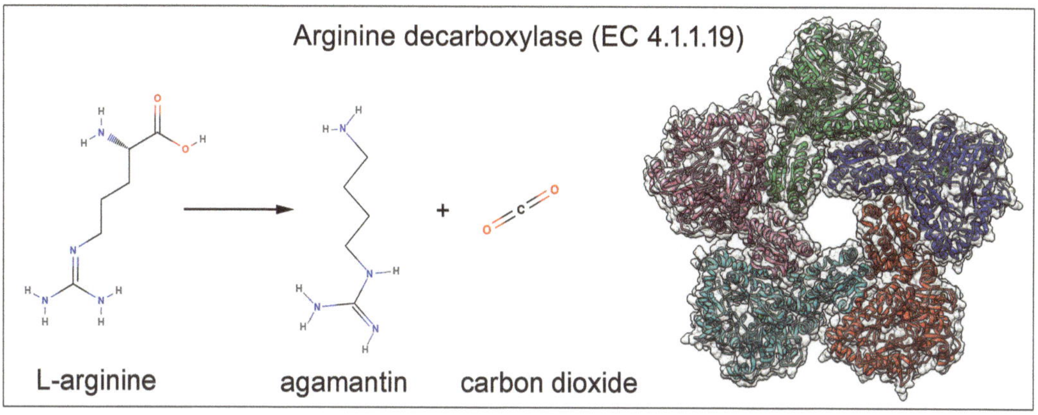

Fig. 9.3 Arginine decarboxylase (ADC): enzymatic reaction scheme (left) and molecular structure (right). The biological unit structure (PDB ID: 2VYC) is represented as ribbons, colored by chain (red, blue, green, pink and cyan), and surface (light grey)

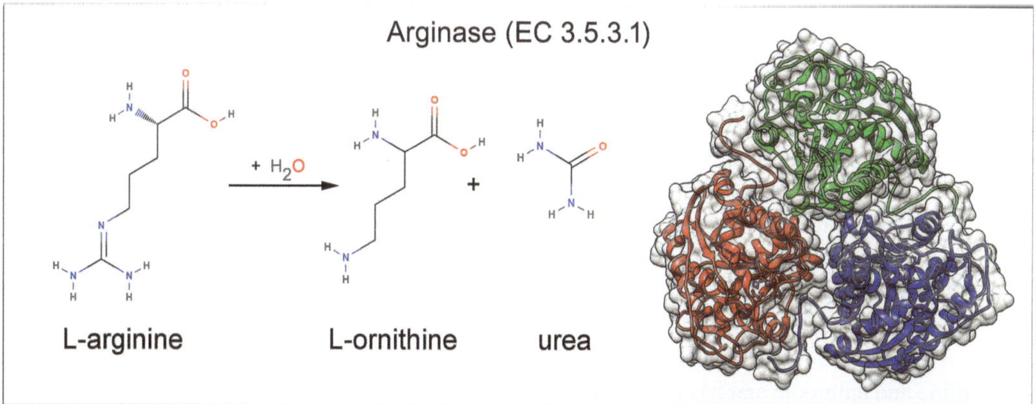

Fig. 9.4 Arginase I (ARGase I): enzymatic reaction scheme (left) and molecular structure (right). The biological unit structure (PDB ID: 4FCI) is represented as ribbons, colored by chain (red, blue and green), and surface (light grey)

Upon substrate binding to the catalytic pocket through interactions with Mn^{2+}, Arg undergoes a nucleophilic attack on the Cζ (step 1); subsequently, Asp128 operates an acidic protonation on the Nε nitrogen of the substrate (step 2) that causes the bond break with the Cζ resulting in the formation of ornithine and urea (step 3); the urea product is then released by the formation of a bridge between a water molecule and the Mn^{2+} (step 4); ornithine release from the active site is mediated by His141 resulting in the regeneration of the nucleophilic metal bringing a hydroxide anion (step 5) (Fernandes et al. 2017).

Human ARGase I has been tested on a variety of cancer cell lines and in animal models since 1950, but has showed some limitations for clinical application due to its catalytic properties (high Km at physiological pH), the requirement of Mn^{2+} for activity and its poor stability in vivo (Fung and Chan 2017). Engineered forms of the enzyme have allowed to improve on the drug limitations. Particularly, development of a cysteines-mutated and PEGylated form of ARGase I (namely, BCT-100) was successful in increasing the enzyme in vivo stability and, at the same time, in improving its catalytic properties (Fultang et al. 2016). Another mutated form of PEGylated ARGase I has been described and shows higher affinity for Co^{2+} than Mn^{2+} as a divalent ion co-factor and improved catalytic

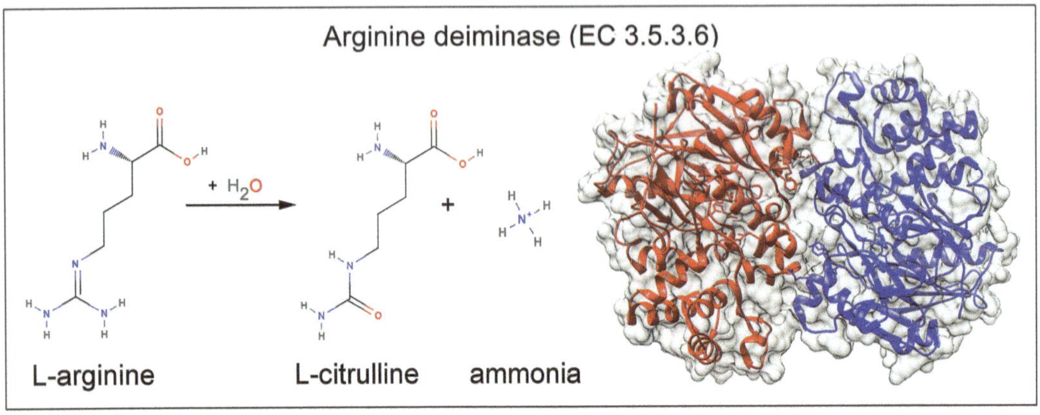

Fig. 9.5 Arginine deiminase (ADI): reaction scheme (left) and molecular structure (right). The biological unit structure (PDB ID: 1LXY) is represented as ribbons, colored by chain (red and blue), and surface (light grey)

properties at physiological pH (Tanios et al. 2013). The former modified ARGase I showed high toxicity on non-cancerous cells, a characteristic that has discouraged further studies for its clinical application (Agrawal et al. 2014; Khoury et al. 2015). BCT-100, instead, has been tested in several in vitro and in vivo models showing efficacy in reducing cancer cells growth mainly on leukemia and HCC models.

Another enzyme capable of catabolizing arginine is the procaryote derived arginine deiminase (ADI). The enzyme breaks down L-Arg into L-citrulline and ammonia and has μM affinity for Arg. The most extensively studied form of ADI is the one derived from *Mycoplasma arginini* (Ma-ADI), but several microbial sources of ADI have been described in the literature (Han et al. 2016). Ma-ADI is a homodimer of 90 kDa, and the enzyme can structurally acquire two conformations: an open one (apo-enzyme) and a closed one (enzyme-substrate complex). Upon substrate binding, in the closed conformation, a solvent channel is formed on the binding pocket to allow water access. The enzyme catalytic mechanism involves a conserved Cys-His-Glu triad and consists of two nucleophilic attacks: the first one is operated on the substrate Cζ by the activated sulfur of the thiol group of the catalytic Cys and results in the release of ammonia and formation of a reaction intermediate with L-citrulline covalently bound to the enzyme Cys. The intermediate is resolved by the second nucleophilic attack, again on the substrate Cζ, operated by a water molecule activated by the catalytic His and Glu. Following the second nucleophilic attack, the C-S covalent bond is cleaved and the L-citrulline product is released (Das et al. 2004).

Ma-ADI anti-cancer activity was first observed on Rous sarcoma virus-transformed buffalo-rat liver (RSV-BRL) (Miyazaki et al. 1990), and then confirmed on several other cancer cell lines deficient on argininosuccinate synthetase (ASS) (Patil et al. 2016; Fung and Chan 2017). In vivo studies highlighted the poor stability of the native enzyme, therefore further pre-clinical and clinical studies have been performed using a PEGylated form of the enzyme that showed higher stability in vivo and reduced immunogenicity with preserved catalytic and anti-cancer activities, and promising results for the treatment of HCC and melanomas (Miyazaki et al. 1990).

ARGase and ADI anti-cancer activity mechanism has not been fully revealed yet. In general, deprivation of amino acids leads to the activation of stress responses enhancing pro-survival authophagy and reducing protein synthesis. Cancer cells unable to produce endogenous Arg by the de novo (*e.g.*, cancer cells deficient for ASS and/or ALS) or *salvage* pathways (*e.g.*, cancer cells deficient for OCT) stay in a prolonged stress condition that eventually results in growth arrest and activation of pro-death pathways (Fultang et al. 2016). Arginine is also the only source of nitric oxide (NO) that, beside pro-inflammatory activity, is involved in tumour growth, mainly favoring adhesion and angiogenesis (Fukumura et al. 2006; Carreau et al. 2011). Solid tumours are well known to stimulate angiogenesis in order to sustain mass growth. Absence of arginine is related to poor angiogenesis that in turn causes tumour mass growth inhibition and, eventually, reduction due to hypoxia and poor metabolites supply (Patil et al. 2016).

9.2.3 L-Methioninase

In mammalian cells, the essential amino acid methionine plays several roles in cell metabolism and proliferation, of which three are the main ones: protein synthesis, biosynthesis of polyamines, and DNA methylation (Cavuoto and Fenech 2012). Indeed, S-adenosylmethionine (SAM), a derivative of methionine, is the primary methyl donor for DNA epigenetic modifications involved in gene expression regulation (Bleich et al. 2014). Upon methylation, SAM is converted into S-adenosylhomocystein (Sahu et al. 2016) that is promptly hydrolyzed to homocystein (Hcy) that, in turn, can be re-converted into methionine *via* the methionine de novo pathway in which methionine synthase (EC 2.1.1.13, MTR) plays the main role (Sharma et al. 2014). Beside the de novo pathway, methionine can be recovered by a *salvage* pathway in which methionine is generated by methylthioadenosine phosphorylase (EC 24.2.28, MTAP) (Sato and Nozaki 2009).

The first evidences of cancer cells dependency on methionine derive from starvation experiments in rodents transplated with Walker-256 carcinosarcoma cells (Sugimura et al. 1959). Afterwards, several tumoural cell lines have been treated with microbial L-methioninases and described as strongly dependent on the extracellular supply of methionine for survival and proliferation. Among them, central nervous system (CNS) tumours are the most represented together with other solid tumours (Fernandes et al. 2017).

L-methioninase (EC 4.4.1.11, MGL, Fig. 9.6) is a pyrodoxal phosphate (PLP) dependent γ-lyase found in all living organisms except mammals (Sharma et al. 2014). The enzyme catalyses the cleavage of methionine into methanethiol, α-ketobutyrate and ammonia. Bacterial MGLs, the most characterized and studied form of MGLs, are generally cytosolic homotetramers of molecular weight ranging from 149 to 173 kDa (Nakayama et al. 1984). Most of microbial MGLs show Km towards methionine in the mM range (El-Sayed 2010), with the exception of *Aspergillus* Rs-1a enzyme that shows a μM affinity for the substrate (Ruiz-Herrera and Starkey 1969). Microbial MGLs can catabolize other amino acids or amino acids-derived molecules like L-homocysteine, L-ethionine, L-selenomethionine, L-cystein,

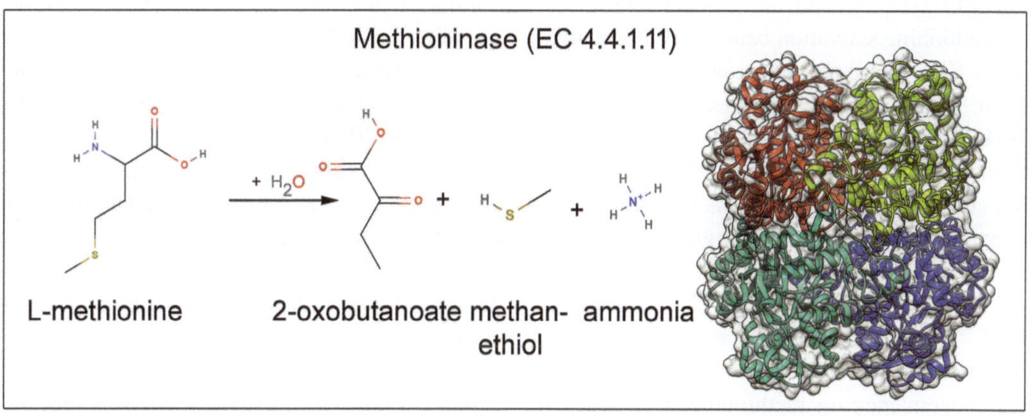

Fig. 9.6 L-Methioninase (MGL): enzymatic reaction scheme (left) and molecular structure (right). The biological unit structure (PDB ID: 5X2V) is represented as ribbons, colored by chain (red, yellow, blue and cyan), and surface (light grey)

S-methyl-L-cystein, homoserine, and L-vinylglycine (Sato and Nozaki 2009).

Pseudomonas putida L-methioninase (PpMGL) is the most extensively characterized microbial MGL and several papers have been published on its in vitro and in vivo anti-cancer activity (Tan et al. 1996; Yoshioka et al. 1998; Motoshima et al. 2000; Kudou et al. 2007, 2008; Fukumoto et al. 2012; Hoffman 2015). Structurally, PpMGL is a homotetramer, each monomer consisting of one N-terminal domain, two PLP binding domains and one C-domain. The enzyme functional unit consists of a dimer-dimer association and has four active sites. Structural and mutational analysis of PpMGL revealed the essential role of the Cys116 in substrate binding, a characteristic that distinguishes MGLs from other PLP-γ-family enzymes.

The general MGL catalytic mechanism is consistent with the one described for the PLP-γ-family of enzymes and consists of five steps: in the first two steps, L-Met is covalently associated to PLP through its amino acid primary amine, the association of methionine to PLP causes the cleavage of covalent bonds between the cofactor and the enzyme Lys residue (step 1) and the substrate α- and β-hydrogens are then transferred to PLP (step 2) forming an external aldimine; next, the methanethiol group is eliminated by the acidic attack of a tyrosine residue (steps 3 and 4); finally, the enzyme-product complex is resolved by the release of α-ketobutyrate and ammonia from PLP (step 5) (Sato and Nozaki 2009).

Methionine starvation causes cell cycle arrest in the S/G_2 phase that eventually results in cell death. Moreover, in vivo experiments using PpMGL alone or in combination with other chemotherapeutics like vincristine, 5-fluorouracil, and cisplatin, proved the efficacy of the drug as an antiproliferative agent with great potential for cancer therapy (Cellarier et al. 2003; Cavuoto and Fenech 2012; Sharma et al. 2014; Fernandes et al. 2017). Experimental evidences did not point to a specific mechanism underneath tumour cells dependency on methionine. Deficiency of proteins involved in the methionine de novo and/or *salvage* pathways is one of the possible mechanism involved in cancer cell inability to produce endogenous methionine, but it cannot be the only explanatory mechanism as some cancer cells dependent on methionine can produce normal amounts of it if cultured in the presence of homocystein. Therefore, it is hypothesized that cancer cells addiction to methionine depends on the huge demand of the essential amino acid more than on their incapability to produce it endogenously. Moreover, it has been reported that endogenous methionine is preferentially used for protein synthesis rather than for DNA methylation or polyamines synthesis (Martínez et al. 2017). Because of this, it can be stated that tumour cell lines dependent on methionine display a high demand of the amino acid more than an incapability to produce it and, in turn, can be defined as methionine-auxotrophic.

Besides mediating methionine-dependent tumour cells death, MGL can be useful also for the treatment of other solid tumours as a specific activator of the pro-drug selenomethionine. Indeed, MGL can use selenomethionine to produce methylselenol that catalyzes oxidation of thiols generating pro-apoptotic superoxides (Wang et al. 2002; Zhao et al. 2006; Kim et al. 2007; Park et al. 2007; Van Rite et al. 2013; Guillen et al. 2014). For such an application, targeting of MGL on to cancer is essential to reduce systemic toxicity. Two approaches in such a direction have been tried so far: targeting of MGL on tumour vasculature by fusion with annexin V, in order to localize the pro-drug activator only on the tumour mass (Van Rite et al. 2013); and tumour cells transfection with a viral construct encoding for MGL in order to obtain expression of the enzyme only in the cytosol of malignant cells (Miki et al. 2001; Yamamoto et al. 2003; Gupta et al. 2003).

Use of recombinant PpMGL in the clinics has been attempted (Hoffman 2015), but the drug presents some limitations mainly linked to its instability in the blood stream and to its immunogenicity (Stone et al. 2012). An attempt to overcome such limitations has been done by covalently binding the protein to polyethylene glycol (PEG-MGL), a strategy that was successful in increasing the drug half life (Yang et al. 2004a, b; Hoffman 2003). As an alternative strategy, MGL has been loaded into red blood cells (Gay et al. 2017). The system has been tested in in vivo models with promising results for increasing the drug half life. Microbial MGLs are strongly dependent on PLP for activity, but the availability of the cofactor in the blood stream is limited because of its prompt assimilation and degradation. Simultaneous loading of MGL and pyridoxin (PN) on RBCs can also overcome this limitation as erythrocytes are known as a PLP source and they can process its PN precursor efficiently, thus providing MGL with its cofactor.

Another route has been taken to improve MGL efficacy: engineering of primate and human MGL-structural homologous cystathionine-γ-lyase (CGL) in order to endow the mammalian enzyme with methioninase activity. The engineering of both human and primate CGL has been successful and the new molecules cytotoxic activity has been proved in both in vitro and in vivo models (Stone et al.; Tan et al. 1996).

9.2.4 L-Glutaminase

L-glutamine is a key amino acid involved in several metabolic and energetic pathways, moreover it is the main nitrogen transporter and the more abundant amino acid in mammals (Covini et al. 2014). Glutamine is a conditionally essential amino acid, in fact, even though cells can produce it, the demand of the amino acid during rapid proliferation increases to such a level that exogenous uptake is necessary to provide sufficient supply of the precursor for many energetic and metabolic reactions. Studies on cancer cell metabolism have lead to discover that many mutations on oncogenes or tumour suppressor genes increase the expression of proteins involved in glutamine metabolism and uptake, suggesting a strong correlation between glutamine supply and tumour proliferation (Cluntun et al. 2017). Further, it is well known that cancer cells in culture need addition of glutamine to survive and proliferate (Chen and Cui 2015). All these evidences point to a basal auxotrophy of cancer cells for glutamine, a characteristic independent from mutations involving expression of proteins related to the amino acid endogenous production. The essential role of glutamine for cell proliferation and metabolism can be linked to its utilization as a precursor for amino acids, nucleotides, the NADPH cofactor, and glutathione. Moreover, during cell proliferation, glutamine serves as an anaplerotic substrate to allow replenishment of the TCA cycle. Indeed, during proliferation, a large part of glutamine is uptaken in the cell mitochondria where it is converted into glutamate and ammonia by glutaminase (GLS). The obtained glutamate is then converted through a second deamination into α-ketoglutarate, a TCA cycle intermediate that becomes essential for the cell cycle progression during proliferation as citrate is shifted outside the mitochondrion to produce acetyl-CoA for lipid biosynthesis.

Given glutamine key energetic and biosynthetic role for cancer development, glutamine deprivation has become a hot topic for metabolic cancer therapy.

Mammals express two isoforms of glutaminase (EC 3.5.1.2), GLS1 and GLS2. GLS1 is found in two splicing isoforms, KGA and GAC, that are expressed in different tissues, being KGA predominant in the kidney. GLS2 is also known as liver glutaminase for its main organ of expression. Both isoforms are phosphate-dependent mitochondrial enzymes involved in glutamine catabolism and several evidences have shown that GLS1 is mainly over-expressed in cancers, favoring tumour growth (Chen and Cui 2015). Human GLSs have millimolar affinity for Gln and low catalytic efficiency if compared to microbial glutaminases. A mitochondria-isolated KGA has been reported to be effective for tumour burden reduction in mice bringing ascitic and solid tumours (Maity et al. 1999).

More extensive studies on glutaminase anticancer activity have been conducted on microbial-derived enzymes (Nandakumar et al. 2003). Two types of glutaminases are found in microorganisms: true glutaminase (EC 3.5.1.2, GL, Fig. 9.7) and glutaminase-asparaginase (EC 3.5.1.38, GA, Fig. 9.8). The two classes of enzymes differ for substrate specificity and affinity: GLs are highly specific for glutamine as a substrate and catalyze its conversion to glutamate and ammonia with a millimolar affinity for the substrate; GAs, beside converting glutamine into glutamate and ammonia, are also capable of converting asparagine into aspartate and ammonia with similar

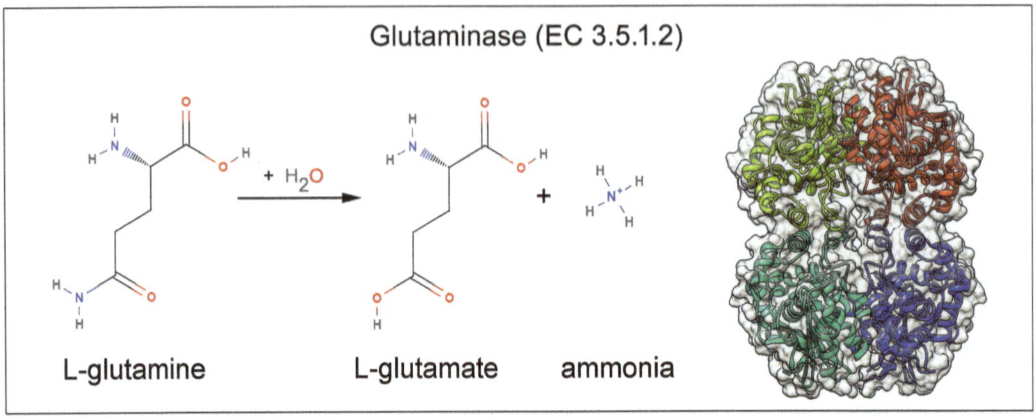

Fig. 9.7 Glutaminase (GL): enzymatic reaction scheme (left) and molecular structure (right). The biological unit structure (PDB ID: 3AGF) is represented as ribbons, colored by chain (red, yellow, cyan and blue), and surface (light grey)

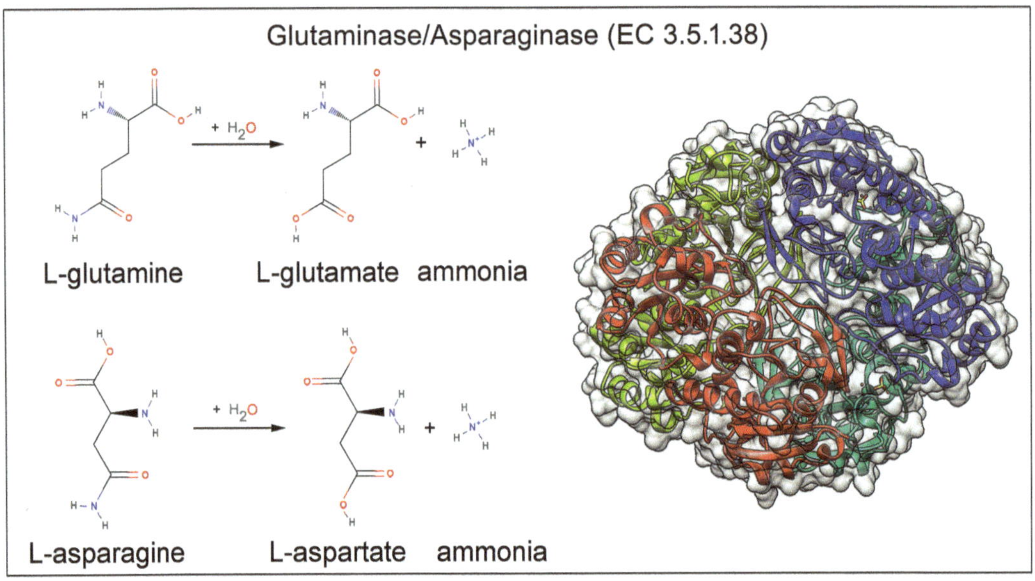

Fig. 9.8 Glutaminase/Asparaginase (GA): enzymatic reaction scheme (left) and molecular structure (right). The biological unit structure (PDB ID: 4PGA) is represented as ribbons, colored by chain (red, blue, cyan and yellow), and surface (light grey)

efficiency and μM affinity for both substrates (Nandakumar et al. 2003).

GLs catalytic mechanism is similar to the one of chymotrypsin and consists of a two-step acylation-deacylation: in the first step, a γ-glutamyl intermediate is formed between the substrate and the Oγ of the primary catalytic serine in the active site, and, upon intermediate formation, the product (ammonium) is released; the glutamyl-intermediate is then resolved by a second nucleophilic attack operated by a water molecule activated by a Glu or Tyr residue in the catalytic pocket and the glutamate product is released (Brown et al. 2008). *Pseudomonas* 7A glutaminase-asparaginase crystal structure revealed that for this class of enzyme the first nucleophilic attack on the amide carbon of glutamine is operated by the nucleophilic Thr34 (instead of a Ser) activated by Tyr34 and Glu294. The resulting acyl-intermediated is resolved by a water molecule activated by Thr100, Asp101 and Lys173 (Lubkowski et al. 1994). The residues involved in the catalytic mechanism are divided into two catalytic triads, corresponding to the ones described for true asparaginases (EC 3.5.1.1) as described in the previous section.

GLs and GAs from different microbial sources have been tested for their cytotoxic activity on ascitis and solid tumours both in in vitro and in vivo experimental models (Riley et al. 1974; Roberts et al. 1970). Pure glutaminases did not result effective in inhibiting tumour growth and proliferation both in vitro and in vivo, very likely because of the enzyme poor stability and high Km (Riley et al. 1974; Roberts et al. 1970). Quite promising results have been obtained, instead, using GA from *Acinobacter glutaminasificans*, *Achromobacter* sp. and *Pseudomonas* 7A; all three enzymes have been tested in patients in Phase I clinical trials, but none of them was capable of inducing full or partial remission both in adult and pediatric patients and, more relevant, therapy was accompanied by several side effects mainly affecting the central nervous system (Warrell et al. 1980, 1982; Riley et al. 1974; Roberts et al. 1970).

High toxicity associated with glutaminases-based therapy represents a huge drawback in the development of enzyme mediated glutamine-starving therapy using microbial enzymes with high affinity for the substrate. Other strategies have been employed for targeting of glutamine metabolism in cancer cells: specific GLS inhibitors, inhibition of glutamine uptake by targeting membrane amino acids transporters, and suicidal glutamine mimetics (Higashitani et al. 1990). Moreover, the clinical success of the use of asparaginase for the treatment of Acute Lymphoblastic Leukemia (ALL) shows that minimal enzymatic removal of glutamine is beneficial in the treatment of some tumours (for details, see the Asparaginase section).

9.2.5 Glycine Oxidase

Glycine is the smallest amino acid and represents a structural precursor to the synthesis of glutathion, proteins, lipids, nucleic acids and heme groups, either as is or as a source of one-carbon units (Amelio et al. 2014). Its metabolism is very complex. Sources for the cell are represented both by extracellular glycine and by intracellular biosynthesis, mainly, but not only, from serine through the action of serine hydroxymethyltransferase (SHMT). The reaction represents a major source of methyl groups for de novo synthesis of nucleotides and DNA methylation (Amelio et al. 2014).

Glycine metabolism is associated with tumourigensis and cancer cell proliferation. Particularly, Jain and collaborators (Jain et al. 2012) have shown that both glycine consumption and expression of enzymes in the mitochondrial glycine biosynthetic pathway correlate with the rate of proliferation of cancer cells in the NCI-60 panel of cell lines (Jain et al. 2012). Antagonising glycine uptake and its mitochondrial biosynthesis preferentially impairs rapidly proliferating cells, and upregulation of glycine metabolism correlates with cell proliferation and poor prognosis in several tumours. Glycine deprivation may therefore be a new route for human cancer chemotherapy. Removal of glycine from the diet is a possible approach, along with serine removal, because of the close interconnection between the metabolism of the two amino acids (see below and (Amelio et al. 2017; Maddocks et al. 2017)).

An alternative approach is removal of glycine by exploiting enzymes able to cleave the molecule. The glycine cleavage system (GCS) is a multienzyme complex comprising four different components: P, H, T and L proteins. In humans, a mutation in the GCS encoding genes can lead to nonketotic hyperglycinemia (Tada and Kure 1993). In vivo, the stoichiometry of the complex is impressive: 2P:27H:9T:1L. This system catalyses glycine cleavage in almost all organisms through a multistep reaction, where the four proteins act in a highly coordinated way (Douce et al. 2001). Initially, the P-protein catalyses the decarboxylation of the glycine molecule, while the residual aminomethyl groups is transferred to a sulphur atom on the lipoyl group of the oxidized H-protein, generating the aminomethylated H protein. The methylene group is then transferred from the H protein to a tetrahydrofolate molecule, resulting in the release of ammonia and the generation of a reduced H protein. Finally, the

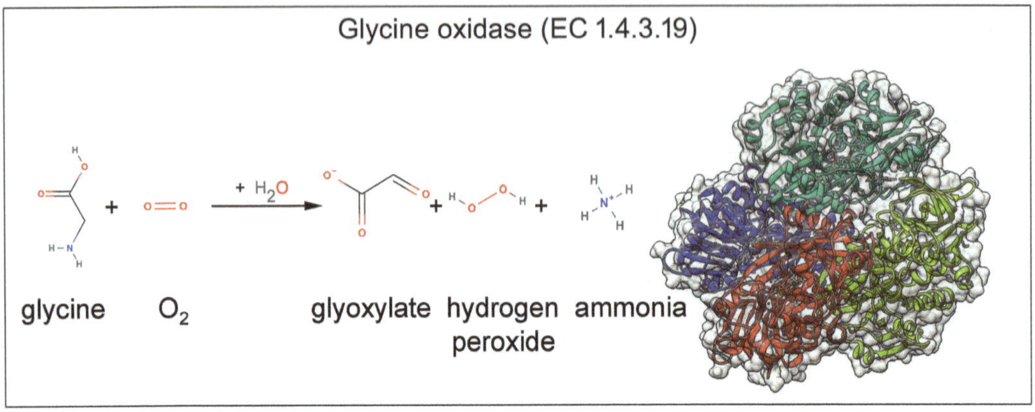

Fig. 9.9 Glycine oxidase (GO): enzymatic reaction scheme (left) and molecular structure (right). The biological unit structure (PDB ID: 1NG4) is represented as ribbons, colored by chain (red, blue, cyan and yellow), and surface (light grey)

dihydrolipoyl group of the reduced H protein is oxidised by the L protein and the lipoyl group of the H protein is regenerated, completing the catalytic cycle. Only recently the structure of the P protein from *Thermus thermophylus* has been determined (Nakai et al. 2005). Because of the high complexity of this reaction, of the need of the whole GCS system to be present and correctly assembled to be functional, we consider it unlikely useful, at least in its native form, as a drug for cancer therapy. We are not aware at the moment of attempts to reduce the size of the complex and symplify its structure in order to build a molecule more apt to become a drug for glycine removal in vivo.

In contrast, a better candidate could be represented in our opinion by a glycine oxidase activity. Glycine oxidase (EC 1.4.3.19, GO, Fig. 9.9) catalyses the oxidative deamination of glycine, generating glyoxylate, ammonium and hydrogen peroxide. One such activity was isolated in 1998 (Nishiya and Imanaka 1998), by studying the *yjbR* open reading frame of *Bacillus subtilis*. Initially characterized for its weak similarity to bacterial sarcosine oxidases, it turned out to be a deaminating oxidase, specifically a glycine oxidase. The protein is a homotetrameric flavoenzyme, with each subunit theoretically weighing 40.9 kDa, and an activity of 0.46 U/mg towards glycine. The N-terminus includes the flavin adenine binding site and the C-terminus the substrate binding site. Further characterization was later performed (Job et al. 2002), showing how the FAD flavin cofactor is present in a 1:1 stoichiometric ratio per monomer and not covalently bound to the enzyme. Compared to similar amino acid flavoprotein oxidases, the activity and kinetic efficiency are low (Pollegioni et al. 2008, 2013), but GO variants with improved kinetic properties have already been generated both for *Bacillus subtilis* (Rosini et al. 2014) and *Bacillus licheniformis* (Zhang et al. 2016). Other bacterial sources of glycine oxidases are also under investigation. For the *Bacillus cereus* variant, it has been demonstrated a role for Asn336 in determining substrate affinity (Wu et al. 2016). The crystal structure of the enzyme from *Geobacillus kaustophilus* has been determined in 2015 (Shiono et al. 2015), when the one from *Pseudomonas putida* KT2440 was also purified and characterised (Equar et al. 2015).

The enzyme from the Gram-negative marine bacterium *Marinomonas mediterranea* is very different from the ones described so far. It is more specific for glycine than the other enzymes, which can also catalyse reactions on sarcosine and N-ethylgycine (Campillo-Brocal et al. 2013). Moreover, it does not use FAD as a cofactor, but there is evidence supporting the presence of a quinone cofactor, generated by a post-translational modification of amino acid residues (Campillo-Brocal et al. 2013). The discovery of these enzymes paves the way to new opportunities to treat cancer glycine addiction.

9.2.6 Serine Deaminase

Serine is part of an essential pathway in cancer (Kalhan and Hanson 2012; DeBerardinis et al. 2007), which is very closely interconnected to glycine metabolism (Amelio et al. 2014; Locasale 2013). Serine can be synthetised starting from 3-phosphoglycerate (produced in the glycolytic cycle) which is converted by phosphoglycerate dehydrogenase (PHGDH), PSAT-1 and PSPH into serine. Serine can also be imported from the extracellular environment and it donates the carbon atom from its side chain to folate, converting it into glycine. Because of its key role, serine has been defined as the "central hub of cancer metabolism" (Amelio et al. 2014). Several studies demonstrate that de novo serine synthesis through PHGDH is dramatically enhanced in a subset of breast cancers and melanomas (Possemato et al. 2011; Locasale et al. 2011). An increase in PHGDH copy number correlated with a dependence on serine production. Enhanced serine production by bone metastatic breast cancer cells stimulates osteoclastogenesis (Pollari et al. 2011). A dietary requirement for serine and glycine was demonstrated for in vivo tumour growth (Maddocks et al. 2013) and tumour adaptation to the deprivation of these two amino acids requires p53 functionality. Modulation of serine metabolism has been studied in rat hepatoma 3924A cells (Snell et al. 1987).

At the moment, there has not been a systematic attempt to tackle this addiction from the pharmacologic point of view. We propose that serine deaminase (serine dehydratase or serine ammonia lyase, EC 4.3.1.17, SDH, Fig. 9.10), which catalyses the transformation of serine into pyruvate and ammonium, could be a good enzyme for removal of serine in serine-addicted cancers. SDH is widely distributed in nature, with physico-chemical features variable from species to species. For example, the rat enzyme is a dimer (Xue et al. 1999), while the yeast and *Escherichia coli* ones (able to catalyse deamination of threonine as well) are tetramers. A significant advantage is that a human serine dehydratase from liver has been isolated and characterised (Sun et al. 2005), which would mean the advantage of avoiding the immune response in the host upon administration. The human enzyme is a pyridoxal-phosphate dependent enzyme, with a homodimeric structure. Each subunit includes 322 residues and consists of two domains, a small one and a large one, the latter being the catalytic domain and containing the cofactor.

9.2.7 Proline Deaminase

Proline is a unique amino acid, as its alpha-amino group is blocked into a pyrrolidine ring. It is therefore the sole secondary (imino) amino acid and has its own metabolic pathways (Phang et al.

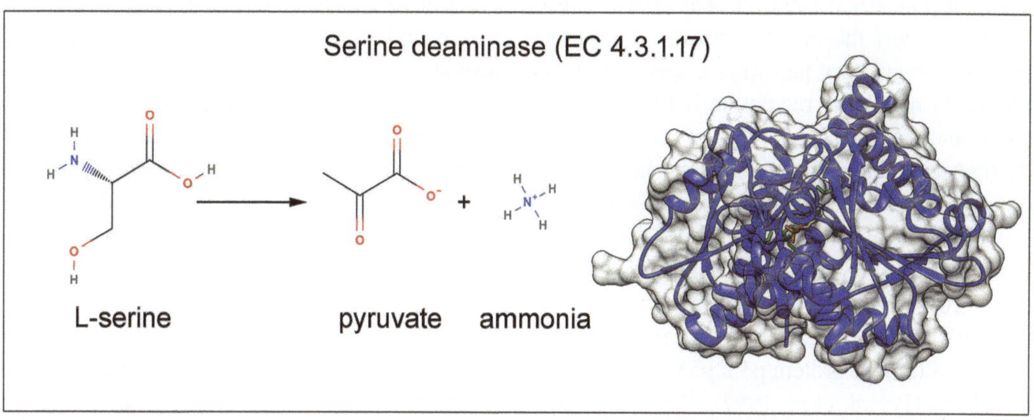

Fig. 9.10 Serine deaminase (SDH): enzymatic reaction scheme (left) and molecular structure (right). The biological unit structure (PDB ID: 5C3U) is represented as ribbons (blue) and surface (light grey)

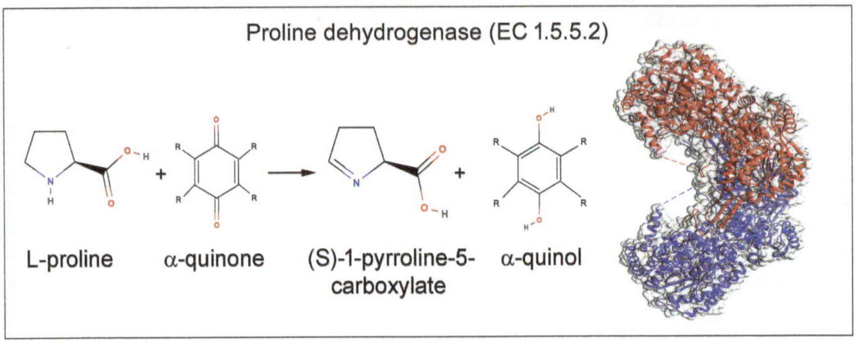

Fig. 9.11 Proline dehydrogenase (PRODH/POX): enzymatic reaction scheme (left) and molecular structure (right). The biological unit structure (PDB ID: 4NMA) is represented as ribbons, coloured by chain (blue and red), and surface (light grey)

2015). Proline can be synthethised starting from glutamate by pyrroline-5-carboxylate (P5C) synthase (P5CS) and ornithine by ornithine aminotransferase (OAT). P5C is also the intermediate of the proline degradation pathway, mediated only by proline dehydrogenase, also known as proline oxidase (EC 1.5.5.2, PRODH/POX, Fig. 9.11). Interestingly, this enzyme, which can open the proline ring, is strategically associated to the inner mitochondrial membrane, close to the electron transport chain proteins (Phang et al. 2008). The reaction leads to ATP production and proline can indeed be used as the single source of energy in several organisms (Nagata et al. 2003). Radical oxygen species (ROS) are the main by-product of the reaction and they seem to work as secondary messengers through complex III especially in apoptosis (Liu et al. 2006) and in autophagy (Liu et al. 2012; Zabirnyk et al. 2010). This is the reason why the proline pathway is regarded as a main regulatory pathway: a special role for proline in metabolic regulation is now accepted (Ahn and Metallo 2015; Phang et al. 2008). Proline metabolism has also been shown to be critical in cancer reprogramming (Sahu et al. 2016; Elia et al. 2017; Liu et al. 2015) and its clinical relevance has been established (Ding et al. 2017). An important discovery was that PRODH/POX was encoded by a gene under the control of tumour protein p53: p53-induced gene 6 (PIG-6) (Polyak et al. 1997). The enzymes of the proline biosynthetic pathway (P5CS, PYCR1/2/L) are markedly upregulated by c-MYC (Liu et al. 2012). Tumours sensitive to proline metabolism are melanoma (Kardos et al. 2015), lung cancer (Liu et al. 2012) and breast cancer (Elia et al. 2017).

As yet, proline removal has not been attempted as a therapeutic strategy, but there are good reasons to follow this route, especially according to the recent developments described above. First, proline addiction can be tackled in a very specific way, thanks to the uniqueness of proline molecular structure; second, appropriate enzymes exist, able to degrade it; finally, enzyme mediated proline degradation would occur in the extracellular space, where ROS could be easily quenched by plasma redox systems, therefore not interfering with intracellular signaling of healthy cells.

The human PRODH-POX enzyme (EC 1.5.99.B2) can be a good candidate for the purpose, as this would minimize the risk of the immune response by the host. The structure of the enzyme, however, is not yet available, which could represent a limitation in its usage. Only the structure of the PutA homologue enzyme from *Brayrhizobium japonicum* was described in 2010 (Srivastava et al. 2010). PutA (proline utilization A), however, is bifunctional and catalyses the oxidation of proline to glutamate via the sequential activities of FAD-dependent PRODH and NAD-dependent P5C dehydrogenase domains. Even this enzyme could be an option for human treatment, though the immune reponse would for sure represent an issue.

A potential hurdle in fighting proline cancer addiction by enzymatic methods is that collagen, forming 25% of total body proteins, has been described as a reservoir of proline, wherefrom the amino acid can be reobtained by cancer cells. This issue would need careful consideration in the complex dynamics of each cancer type.

9.2.8 Tyrosine Aminotransferase

Melanomas and prostate cancer have been described as being dependent on tyrosine and phenylalanine and, in some cases, methionine. These studies are still in their infancy, but it has been shown that removal of tyrosine and phenylalanine can lead human prostate cancer DU145 cells to apoptosis without affecting normal cells. This effect is associated with inhibition of expression and phosphorylation of focal adhesion kinase (FAK) and of extracellular-regulated kinase (Maddocks et al.; Fu et al. 2003). This effect is however not visible in prostate cancer PC3 cells, which are instead sensitive to methionine deprivation (Fu et al. 2003).

Restriction of tyrosine and phenylalanine from human (A375) and murine (B16) cell lines generates alterations in metabolic and signaling pathways in a concentration dependent and time-ordered fashion (Fu and Meadows 2007). Early events include modulation of FAK/G protein pathways and of the plasminogen activator (PA)/PA inhibitor pathway. Late events lead to apoptosis by affecting the FAK/Ras/Raf and Bcl-2 pathways. This threshold effect has been described both in melanoma and in other sold tumours (Fu et al. 2003; Pelayo et al. 2001) and does not seem to be present in normal cells (Fu et al. 1999).

Enzymatic removal of amino acids is an as yet unattempted potential approach. Aminotransferases (EC 2.6.1.x) are part of the α-superfamily of vitamin B6-dependent enzymes (Alexander et al. 1994) and they are classifed into four families, with family I including seven subfamilies (Jensen and Gu 1996). The most thoroughly studied enzyme of the aminotransferase superfamily are the aspartate aminotransferases (AspATs), most of which belong to subfamily Iα. They include both highly specific enzymes, like cytosolic enzymes from higher vertebrates, and broad substrate specificity aminotransferases, which can utilise different substrates with different affinities. Crystallographic studies on the first group have disclosed their catalytic mechanism (Kirsch et al. 1984). The Iγ subfamily enzymes are less well characterised and include tyrosine aminotransferase (L-tyrosine-2-oxoglutarate aminotransferase, EC 2.6.1.5, TAT, Fig. 9.12), a pyridoxal phosphate dependent enzyme which catalyses the transamination reaction included in the catabolic pathways of aromatic amino acids. In this reaction, the α-amino group of L-tyrosine, or L-phenylalanine, is transferred to α-ketoglutarate. While mammalian TAT, however, favours these substrates, others have a broader specifity, like the one from *Trypanosoma cruzi* (Blankenfeldt et al. 1999), which uses all three aromatic amino acids and alanine as substrates, along with three α-ketoacids as acceptors (pyruvate, α-ketoglutarate, oxalacetate).

Not many TAT structures have been described: the first was the one from *T. cruzi* (Blankenfeldt et al. 1999). The structure of the human enzyme (hTAT) is also available (PDB ID: 3DYD, released in 2008), along with those of the mouse (Mehere et al. 2010), *E. coli* (Ko et al. 1999) and *Leishmania infantum* (Moreno et al. 2014).

Typically, the enzymes share a similar architecture and include two domains: a major one and a minor one. The major domain binds the PLP group (in mTAT and hTAT to Lys280) and includes the active site, while the minor domain binds the substrate (Blankenfeldt et al. 1999). In hTAT, for example, the residues critical for tyrosine binding are Phe169, Phe410 and Arg417. The catalytic mechanism seems to involve a ping-pong mechanism, similar to the one of AspAT (Kirsch et al. 1984), where the pyridoxal group forms an intermediate with the aromatic amino acid, then resolved with the release of the corresponding α-ketoacid.

Ideally, the human enzyme, which is mainly expressed in the liver, would be the best candidate

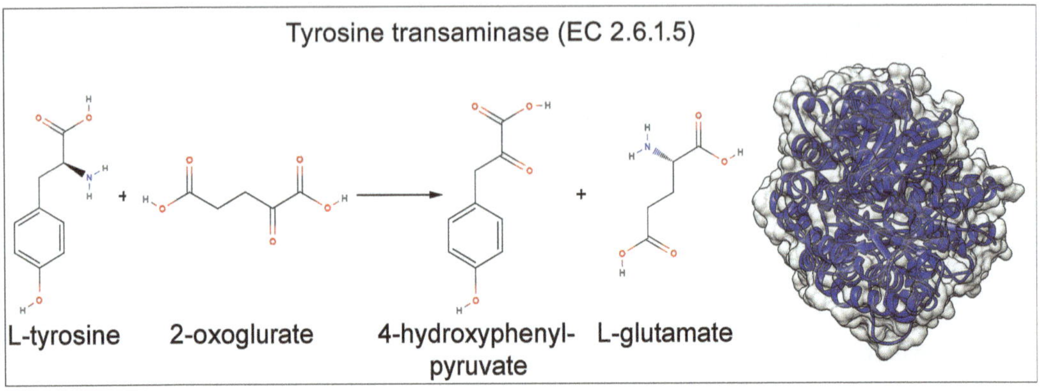

Fig. 9.12 Tyrosine transaminase (TAT): enzymatic reaction scheme (left) and molecular structure (right). The biological unit structure (PDB ID: 1BW0) is represented as ribbons (blue) and surface (light grey)

for therapy. However, it shows the lowest Km for tyrosine among those described (mM range, (Sivaraman and Kirsch 2006)). It is therefore likely that engineering of activity and specificity of the existing enzymes, or evaluation of alternative sources, must be considered in the near future for therapeutic applications.

9.3 Open Issues

Several limitations are intrinsic to the non-self nature of most of the enzymes described above. The problems shared by almost all of them are the immune response induced in the human host, the side effects due to systemic depletion of amino acids, and the inability to overcome the blood-brain barrier (BBB), behind which either minimal residual disease can cause relapse or the primary tumour is located.

In the case of ASNase, several attempts have been undertaken to reduce the immune response of the host. For example, shielding the *E. coli* protein with polyethylene glycol (PEGaspargase, Oncaspar) has lead to a reduction, but not a total disappearance, of the cases where a second line therapy replacement with *E. chrysanthemi* had to be activated (Cantor et al. 2012). Good results are also being obtained by loading red blood cells of the host with the enzyme before injection (GRASPA (n.d.)) and one observation has been performed on the possibility to make ASNase reach the surface of lung cells in the form of a nanobiocomposite by targeting the folate receptor (Muthukumar et al. 2014).

Arginase is the only enzyme of human origin under clinical evaluation. A review about deimmunization strategies for therapeutic enzymes by Cantor et al. (2012) indicates essentially two approaches. On one side, in silico analysis allows the prediction of the main epitopes, which can be followed by site-directed mutagenesis. On the other hand, engineering of substrate specificities into a human enzyme is a possible alternative to limit immunogenicity only to the catalytic site.

9.4 Perspectives

Systemic depletion of a specific amino acid necessarily has systemic consequences and, hence, side effects. A thus far unexplored approach is based on targeting of enzymes onto tumoural cells: as mentioned above, L-asparaginase, for example, has been targeted to lung tumour cells (Muthukumar et al. 2014), but much more can be done into this direction. Depletion of amino acids in close proximity of cancer cells could prove as successful as, or more, than systemic depletion. It is also expected that intracellular control of amino acid addiction could be even more effective on tumour control. Finding the way to deliver enzymes into the cell and into the right cell compartment to lead to cell death is another challenge that could be put on the agenda of researchers interested in tackling cancer metabolism.

Considering the success of combination therapy in the case of asparaginase, this strategy for sure needs to be explored also for other enzymes, but even interconnections between amino acid pathways need to be explored more deeply in relation to cancer and to possible multiple addictions.

9.5 Conclusions

In this review we have presented an outline of the enzymes potentially useful to tackle cancer metabolic addictions, presenting their general and specific limitations and proposing solutions to overcome them. We also suggest a targeted, in place of a systemic, approach in order to reduce side effects and optimise specific cell destruction.

In our view, the future of enzymes for cancer therapy will entail significant molecular engineering work, especially for novel candidates, and it is likely to reserve relevant progress in the treatment of tumours.

References

Aggarwal S (2010) Targeted cancer therapies. Nat Rev Drug Discov 9:427–428

Agrawal V, Woo JH, Mauldin JP, Stone EM, Meininger CJ, Jo C, Kleypas K, Frenkel EP, Frankel AE (2014) In-vivo evaluation of human recombinant co-arginase against A375 melanoma xenografts. Melanoma Res 24:556–567

Ahn CS, Metallo CM (2015) Mitochondria as biosynthetic factories for cancer proliferation. Cancer Metab 3:1

Alexander FW, Sandmeier E, Mehta PK, Christen P (1994) Evolutionary relationships among pyridoxal-5′-phosphate-dependent enzymes. Regio-specific alpha, beta and gamma families. Eur J Biochem 219:953–960

Amelio I, Cutruzzola F, Antonov A, Agostini M, Melino G (2014) Serine and glycine metabolism in cancer. Trends Biochem Sci 39:191–198

Amelio I, Melino G, Frezza C (2017) Exploiting tumour addiction with a serine and glycine-free diet. Cell Death Differ 24:1311–1313

Angiolillo AL, Schore RJ, Devidas M, Borowitz MJ, Carroll AJ, Gastier-Foster JM, Heerema NA, Keilani T, Lane AR, Loh ML, Reaman GH, Adamson PC, Wood B, Wood C, Zheng HW, Raetz EA, Winick NJ, Carroll WL, Hunger SP (2014) Pharmacokinetic and pharmacodynamic properties of calaspargase pegol *Escherichia coli* L-asparaginase in the treatment of patients with acute lymphoblastic leukemia: results from Children's Oncology Group Study AALL07P4. J Clin Oncol 32:3874–3882

Avramis VI, Panosyan EH (2005) Pharmacokinetic/pharmacodynamic relationships of asparaginase formulations: the past, the present and recommendations for the future. Clin Pharmacokinet 44:367–393

Bansal S, Srivastava A, Mukherjee G, Pandey R, Verma AK, Mishra P, Kundu B (2012) Hyperthermophilic asparaginase mutants with enhanced substrate affinity and antineoplastic activity: structural insights on their mechanism of action. FASEB J 26:1161–1171

Batool T, Makky EA, Jalal M, Yusoff MM (2016) A comprehensive review on L-asparaginase and its applications. Appl Biochem Biotechnol 178:900–923

Bhojwani D, Howard SC, Pui CH (2009) High-risk childhood acute lymphoblastic leukemia. Clin Lymphoma Myeloma 9(Suppl 3):S222–S230

Blankenfeldt W, Nowicki C, Montemartini-Kalisz M, Kalisz HM, Hecht HJ (1999) Crystal structure of Trypanosoma cruzi tyrosine aminotransferase: substrate specificity is influenced by cofactor binding mode. Protein Sci 8:2406–2417

Bleich S, Semmler A, Frieling H, Thumfart L, Muschler M, Hillemacher T, Kornhuber J, Kallweit U, Simon M, Linnebank M (2014) Genetic variants of methionine metabolism and DNA methylation. Epigenomics 6:585–591

Bobak YP, Vynnytska BO, Kurlishchuk YV, Sibirny AA, Stasyk OV (2010) Cancer cell sensitivity to arginine deprivation in vitro is not determined by endogenous levels of arginine metabolic enzymes. Cell Biol Int 34:1085–1089

Broome JD (1968) Studies on the mechanism of tumor inhibition by L-asparaginase. Effects of the enzyme on asparagine levels in the blood, normal tissues, and 6C3HED lymphomas of mice: differences in asparagine formation and utilization in asparaginase-sensitive and -resis. J Exp Med 127:1055–1072

Brown G, Singer A, Proudfoot M, Skarina T, Kim Y, Chang C, Dementieva I, Kuznetsova E, Gonzalez CF, Joachimiak A, Savchenko A, Yakunin AF (2008) Functional and structural characterization of four glutaminases from Escherichia coli and Bacillus subtilis. Biochemistry 47:5724–5735

Campillo-Brocal JC, Lucas-Elio P, Sanchez-Amat A (2013) Identification in Marinomonas mediterranea of a novel quinoprotein with glycine oxidase activity. Microbiology 2:684–694

Cantor JR, Stone EM, Chantranupong L, Georgiou G (2009) The human asparaginase-like protein 1 hASRGL1 is an Ntn hydrolase with beta-aspartyl peptidase activity. Biochemistry 48:11026–11031

Cantor JR, Panayiotou V, Agnello G, Georgiou G, Stone EM (2012) Engineering reduced-immunogenicity enzymes for amino acid depletion therapy in cancer. Methods Enzymol 502:291–319

Cappelletti D, Chiarelli LR, Pasquetto MV, Stivala S, Valentini G, Scotti C (2008) Helicobacter pyloril-asparaginase: a promising chemotherapeutic agent. Biochem Biophys Res Commun 377:1222–1226
Carreau A, Kieda C, Grillon C (2011) Nitric oxide modulates the expression of endothelial cell adhesion molecules involved in angiogenesis and leukocyte recruitment. Exp Cell Res 317:29–41
Cavuoto P, Fenech MF (2012) A review of methionine dependency and the role of methionine restriction in cancer growth control and life-span extension. Cancer Treat Rev 38:726–736
Cellarier E, Durando X, Vasson MP, Farges MC, Demiden A, Maurizis JC, Madelmont JC, Chollet P (2003) Methionine dependency and cancer treatment. Cancer Treat Rev 29:489–499
Chan WK, Lorenzi PL, Anishkin A, Purwaha P, Rogers DM, Sukharev S, Rempe SB, Weinstein JN (2014) The glutaminase activity of L-asparaginase is not required for anticancer activity against ASNS-negative cells. Blood 123:3596–3606
Chen L, Cui H (2015) Targeting glutamine induces apoptosis: a cancer therapy approach. Int J Mol Sci 16:22830–22855
Cluntun AA, Lukey MJ, Cerione RA, Locasale JW (2017) Glutamine metabolism in cancer: understanding the heterogeneity. Trends Cancer 3:169–180
Covini D, Maggi M, Tardito S, Bussolati O, Chiarelli LR, Pasquetto MV, Vecchia L, Valentini G, Scotti C (2014) Expanding targets for a metabolic therapy of cancer: L-asparaginase. Top Anti-Cancer Res 3:418–445
CRUK (n.d.). Available: http://www.cancerresearchuk.org/cancer-info/cancerstats/world/. Accessed
Csibi A, Fendt SM, Li C, Poulogiannis G, Choo AY, Chapski DJ, Jeong SM, Dempsey JM, Parkhitko A, Morrison T, Henske EP, Haigis MC, Cantley LC, Stephanopoulos G, Yu J, Blenis J (2013) The mTORC1 pathway stimulates glutamine metabolism and cell proliferation by repressing SIRT4. Cell 153:840–854
Das K, Butler GH, Kwiatkowski V, Clark AD, Yadav P, Arnold E (2004) Crystal structures of arginine deiminase with covalent reaction intermediates; implications for catalytic mechanism. Structure 12:657–667
Deberardinis RJ, Mancuso A, Daikhin E, Nissim I, Yudkoff M, Wehrli S, Thompson CB (2007) Beyond aerobic glycolysis: transformed cells can engage in glutamine metabolism that exceeds the requirement for protein and nucleotide synthesis. Proc Natl Acad Sci U S A 104:19345–19350
Ding J, Kuo ML, Su L, Xue L, Luh F, Zhang H, Wang J, Lin TG, Zhang K, Chu P, Zheng S, Liu X, Yen Y (2017) Human mitochondrial pyrroline-5-carboxylate reductase 1 promotes invasiveness and impacts survival in breast cancers. Carcinogenesis 38:519–531
Douce R, Bourguignon J, Neuburger M, Rebeille F (2001) The glycine decarboxylase system: a fascinating complex. Trends Plant Sci 6:167–176
Dufour E, Gay F, Aguera K, Scoazec JY, Horand F, Lorenzi PL, Godfrin Y (2012) Pancreatic tumor sensitivity to plasma L-asparagine starvation. Pancreas 41:940–948
Elia I, Broekaert D, Christen S, Boon R, Radaelli E, Orth MF, Verfaillie C, Grunewald TGP, Fendt SM (2017) Proline metabolism supports metastasis formation and could be inhibited to selectively target metastasizing cancer cells. Nat Commun 8:15267
El-Sayed AS (2010) Microbial L-methioninase: production, molecular characterization, and therapeutic applications. Appl Microbiol Biotechnol 86:445–467
Equar MY, Tani Y, Mihara H (2015) Purification and properties of glycine oxidase from pseudomonas putida KT2440. J Nutr Sci Vitaminol (Tokyo) 61:506–510
Ferlay J, Shin HR, Bray F, Forman D, Mathers C, Parkin DM (2010) Estimates of worldwide burden of cancer in 2008: GLOBOCAN 2008. Int J Cancer 127:2893–2917
Ferlay J, Soerjomataram I, Dikshit R, Eser S, Mathers C, Rebelo M, Parkin DM, Forman D, Bray F (2015) Cancer incidence and mortality worldwide: sources, methods and major patterns in GLOBOCAN 2012. Int J Cancer 136:E359–E386
Fernandes HS, Silva Teixeira CS, Fernandes PA, Ramos MJ, Cerqueira NM (2017) Amino acid deprivation using enzymes as a targeted therapy for cancer and viral infections. Expert Opin Ther Pat 27:283–297
Fu YM, Meadows GG (2007) Specific amino acid dependency regulates the cellular behavior of melanoma. J Nutr 137:1591S–1596S. discussion 1597S–1598S
Fu YM, Yu ZX, Pelayo BA, Ferrans VJ, Meadows GG (1999) Focal adhesion kinase-dependent apoptosis of melanoma induced by tyrosine and phenylalanine deficiency. Cancer Res 59:758–765
Fu YM, Yu ZX, Li YQ, Ge X, Sanchez PJ, Fu X, Meadows GG (2003) Specific amino acid dependency regulates invasiveness and viability of androgen-independent prostate cancer cells. Nutr Cancer 45:60–73
Fukumoto M, Kudou D, Murano S, Shiba T, Sato D, Tamura T, Harada S, Inagaki K (2012) The role of amino acid residues in the active site of L-methionine γ-lyase from Pseudomonas putida. Biosci Biotechnol Biochem 76:1275–1284
Fukumura D, Kashiwagi S, Jain RK (2006) The role of nitric oxide in tumour progression. Nat Rev Cancer 6:521–534
Fultang L, Vardon A, De Santo C, Mussai F (2016) Molecular basis and current strategies of therapeutic arginine depletion for cancer. Int J Cancer 139:501–509
Fung MKL, Chan GC (2017) Drug-induced amino acid deprivation as strategy for cancer therapy. J Hematol Oncol 10:144
Galluzzi L, Kepp O, Vander Heiden MG, Kroemer G (2013) Metabolic targets for cancer therapy. Nat Rev Drug Discov 12:829–846
Gao P, Tchernyshyov I, Chang TC, Lee YS, Kita K, Ochi T, Zeller KI, De Marzo AM, Van Eyk JE, Mendell JT, Dang CV (2009) c-Myc suppression of miR-23a/b enhances mitochondrial glutaminase expression and glutamine metabolism. Nature 458:762–765

Gay F, Aguera K, Senechal K, Tainturier A, Berlier W, Maucort-Boulch D, Honnorat J, Horand F, Godfrin Y, Bourgeaux V (2017) Methionine tumor starvation by erythrocyte-encapsulated methionine gamma-lyase activity controlled with per os vitamin B6. Cancer Med 6:1437–1452

Glazer ES, Stone EM, Zhu C, Massey KL, Hamir AN, Curley SA (2011) Bioengineered human arginase I with enhanced activity and stability controls hepatocellular and pancreatic carcinoma xenografts. Transl Oncol 4:138–146

GRASPA (n.d.). Available: http://erytech.com/erytech-reports-positive-top-line-phase-iii-results-from-clinical-study-with-graspa-in-acute-lymphoblastic-leukemia/. Accessed

Guillen KP, Kurkjian C, Harrison RG (2014) Targeted enzyme prodrug therapy for metastatic prostate cancer – a comparative study of L-methioninase, purine nucleoside phosphorylase, and cytosine deaminase. J Biomed Sci 21:65

Gupta A, Miki K, Xu M, Yamamoto N, Moossa AR, Hoffman RM (2003) Combination efficacy of doxorubicin and adenoviral methioninase gene therapy with prodrug selenomethionine. Anticancer Res 23:1181–1188

Han S, Jung J, Park W (2014) Biochemical characterization of L-asparaginase in NaCl-tolerant Staphylococcus sp. OJ82 isolated from fermented seafood. J Microbiol Biotechnol 24:1096–1104

Han RZ, Xu GC, Dong JJ, Ni Y (2016) Arginine deiminase: recent advances in discovery, crystal structure, and protein engineering for improved properties as an anti-tumor drug. Appl Microbiol Biotechnol 100:4747–4760

Heiskanen T, Tollersrud OK, Zhao M, Peltonen L (1994) Large-scale purification of human aspartylglucosaminidase: utilization of exceptional sodium dodecyl sulfate resistance. Protein Expr Purif 5:205–210

Higashitani A, Tabata S, Ogawa T, Ogawa H, Shibata M, Hotta Y (1990) ATP-independent strand transfer protein from murine spermatocytes, spermatids, and spermatozoa. Exp Cell Res 186:317–323

Hirano T, Kishi M, Sugimoto H, Taguchi R, Obinata H, Ohshima N, Tatei K, Izumi T (2009) Thioesterase activity and subcellular localization of acylprotein thioesterase 1/lysophospholipase 1. Biochim Biophys Acta 1791:797–805

Hoffman RM (2003) PEG-methioninase. Adv Exp Med Biol 519:69–79

Hoffman RM (2015) Development of recombinant methioninase to target the general cancer-specific metabolic defect of methionine dependence: a 40-year odyssey. Expert Opin Biol Ther 15:21–31

Jain M, Nilsson R, Sharma S, Madhusudhan N, Kitami T, Souza AL, Kafri R, Kirschner MW, Clish CB, Mootha VK (2012) Metabolite profiling identifies a key role for glycine in rapid cancer cell proliferation. Science 336:1040–1044

Jensen RA, Gu W (1996) Evolutionary recruitment of biochemically specialized subdivisions of family I within the protein superfamily of aminotransferases. J Bacteriol 178:2161–2171

Job V, Marcone GL, Pilone MS, Pollegioni L (2002) Glycine oxidase from Bacillus subtilis. Characterization of a new flavoprotein. J Biol Chem 277:6985–6993

Kalhan SC, Hanson RW (2012) Resurgence of serine: an often neglected but indispensable amino acid. J Biol Chem 287:19786–19791

Karamitros CS, Konrad M (2014) Human 60-kDa lysophospholipase contains an N-terminal L-asparaginase domain that is allosterically regulated by L-asparagine. J Biol Chem 289:12962–12975

Kardos GR, Wastyk HC, Robertson GP (2015) Disruption of proline synthesis in melanoma inhibits protein production mediated by the GCN2 pathway. Mol Cancer Res 13:1408–1420

Khoury O, Ghazale N, Stone E, El-Sibai M, Frankel AE, Abi-Habib RJ (2015) Human recombinant arginase I (Co)-PEG5000 [HuArgI (Co)-PEG5000]-induced arginine depletion is selectively cytotoxic to human glioblastoma cells. J Neuro-Oncol 122:75–85

Kidd JG (1953) Regression of transplanted lymphomas induced in vivo by means of normal guinea pig serum. II. Studies on the nature of the active serum constituent: histological mechanism of the regression: tests for effects of guinea pig serum on lymphoma cells in vitro. J Exp Med 98:583–606

Kim A, Oh JH, Park JM, Chung AS (2007) Methylselenol generated from selenomethionine by methioninase downregulates integrin expression and induces caspase-mediated apoptosis of B16F10 melanoma cells. J Cell Physiol 212:386–400

Kirsch JF, Eichele G, Ford GC, Vincent MG, Jansonius JN, Gehring H, Christen P (1984) Mechanism of action of aspartate aminotransferase proposed on the basis of its spatial structure. J Mol Biol 174:497–525

Ko TP, Wu SP, Yang WZ, Tsai H, Yuan HS (1999) Crystallization and preliminary crystallographic analysis of the *Escherichia coli* tyrosine aminotransferase. Acta Crystallogr D Biol Crystallogr 55:1474–1477

Kotzia GA, Labrou NE (2007) L-Asparaginase from Erwinia Chrysanthemi 3937: cloning, expression and characterization. J Biotechnol 127:657–669

Kudou D, Misaki S, Yamashita M, Tamura T, Takakura T, Yoshioka T, Yagi S, Hoffman RM, Takimoto A, Esaki N, Inagaki K (2007) Structure of the antitumour enzyme L-methionine gamma-lyase from Pseudomonas putida at 1.8 A resolution. J Biochem 141:535–544

Kudou D, Misaki S, Yamashita M, Tamura T, Esaki N, Inagaki K (2008) The role of cysteine 116 in the active site of the antitumor enzyme L-methionine gamma-lyase from Pseudomonas putida. Biosci Biotechnol Biochem 72:1722–1730

Laplante M, Sabatini DM (2012) mTOR signaling in growth control and disease. Cell 149:274–293

Law AS, Wriston JC (1971) Purification and properties of *Bacillus coagulans* L-asparaginase. Arch Biochem Biophys 147:744–752

Liu Y, Borchert GL, Surazynski A, Hu CA, Phang JM (2006) Proline oxidase activates both intrinsic and extrinsic pathways for apoptosis: the role of ROS/superoxides, NFAT and MEK/ERK signaling. Oncogene 25:5640–5647

Liu W, Le A, Hancock C, Lane AN, Dang CV, Fan TW, Phang JM (2012) Reprogramming of proline and glutamine metabolism contributes to the proliferative and metabolic responses regulated by oncogenic transcription factor c-MYC. Proc Natl Acad Sci U S A 109:8983–8988

Liu W, Hancock CN, Fischer JW, Harman M, Phang JM (2015) Proline biosynthesis augments tumor cell growth and aerobic glycolysis: involvement of pyridine nucleotides. Sci Rep 5:17206

Locasale JW (2013) Serine, glycine and one-carbon units: cancer metabolism in full circle. Nat Rev Cancer 13:572–583

Locasale JW, Grassian AR, Melman T, Lyssiotis CA, Mattaini KR, Bass AJ, Heffron G, Metallo CM, Muranen T, Sharfi H, Sasaki AT, Anastasiou D, Mullarky E, VOKES NI, Sasaki M, Beroukhim R, Stephanopoulos G, Ligon AH, Meyerson M, Richardson AL, Chin L, Wagner G, Asara JM, Brugge JS, Cantley LC, Vander Heiden MG (2011) Phosphoglycerate dehydrogenase diverts glycolytic flux and contributes to oncogenesis. Nat Genet 43:869–874

Lorenzi PL, Reinhold WC, Rudelius M, Gunsior M, Shankavaram U, Bussey KJ, Scherf U, Eichler GS, Martin SE, Chin K, GRAY JW, Kohn EC, Horak ID, Von Hoff DD, Raffeld M, Goldsmith PK, Caplen NJ, Weinstein JN (2006) Asparagine synthetase as a causal, predictive biomarker for L-asparaginase activity in ovarian cancer cells. Mol Cancer Ther 5:2613–2623

Lorenzi PL, Llamas J, Gunsior M, Ozbun L, Reinhold WC, Varma S, Ji H, Kim H, Hutchinson AA, Kohn EC, Goldsmith PK, Birrer MJ, Weinstein JN (2008) Asparagine synthetase is a predictive biomarker of L-asparaginase activity in ovarian cancer cell lines. Mol Cancer Ther 7:3123–3128

Lubkowski J, Wlodawer A, Ammon HL, Copeland TD, Swain AL (1994) Structural characterization of pseudomonas 7A glutaminase-asparaginase. Biochemistry 33:10257–10265

Lubkowski J, Palm GJ, Gilliland GL, Derst C, Röhm KH, Wlodawer A (1996) Crystal structure and amino acid sequence of Wolinella succinogenes L-asparaginase. Eur J Biochem 241:201–207

Maddocks OD, Berkers CR, Mason SM, Zheng L, Blyth K, Gottlieb E, Vousden KH (2013) Serine starvation induces stress and p53-dependent metabolic remodelling in cancer cells. Nature 493:542–546

Maddocks ODK, Athineos D, Cheung EC, Lee P, Zhang T, Van den Broek NJF, Mackay GM, Labuschagne CF, Gay D, Kruiswijk F, Blagih J, Vincent DF, Campbell KJ, Ceteci F, Sansom OJ, Blyth K, Vousden KH (2017) Modulating the therapeutic response of tumours to dietary serine and glycine starvation. Nature 544:372–376

Maggi M, Chiarelli LR, Valentini G, Scotti C (2015) Engineering of Helicobacter pylori L-asparaginase: characterization of two functionally distinct groups of mutants. PLoS One 10:e0117025

Maity P, Chakraborty S, Bhattacharya P, Sarkar R (1999) Isolation and purification of phosphate dependent glutaminase from sarcoma-180 tumor and its antineoplastic effects on murine model system. J Exp Clin Cancer Res 18:475–480

Martínez Y, Li X, Liu G, Bin P, Yan W, Más D, Valdivié M, Hu CA, Ren W, Yin Y (2017) The role of methionine on metabolism, oxidative stress, and diseases. Amino Acids 49:2091–2098

Mehere P, Han Q, Lemkul JA, Vavricka CJ, Robinson H, Bevan DR, Li J (2010) Tyrosine aminotransferase: biochemical and structural properties and molecular dynamics simulations. Protein Cell 1:1023–1032

Michalska K, Jaskolski M (2006) Structural aspects of L-asparaginases, their friends and relations. Acta Biochim Pol 53:627–640

Miki K, Xu M, Gupta A, BA Y, Tan Y, Al-Refaie W, Bouvet M, Makuuchi M, Moossa AR, Hoffman RM (2001) Methioninase cancer gene therapy with selenomethionine as suicide prodrug substrate. Cancer Res 61:6805–6810

Miyazaki K, Takaku H, Umeda M, Fujita T, Huang WD, Kimura T, Yamashita J, Horio T (1990) Potent growth inhibition of human tumor cells in culture by arginine deiminase purified from a culture medium of a mycoplasma-infected cell line. Cancer Res 50:4522–4527

Moreno MA, Alonso A, Alcolea PJ, Abramov A, De Lacoba MG, Abendroth J, Zhang S, Edwards T, Lorimer D, Myler PJ, Larraga V (2014) Tyrosine aminotransferase from Leishmania infantum: a new drug target candidate. Int J Parasitol Drugs Drug Resist 4:347–354

Moreno-Enriquez A, Evangelista-Martinez Z, Gonzalez-Mondragon EG, Calderon-Flores A, Arreguin R, Perez-Rueda E, Huerta-Saquero A (2012) Biochemical characterization of recombinant L-asparaginase (AnsA) from Rhizobium etli, a member of an increasing rhizobial-type family of L-asparaginases. J Microbiol Biotechnol 22:292–300

Morris SM (2007) Arginine metabolism: boundaries of our knowledge. J Nutr 137:1602S–1609S

Morris SM, Bhamidipati D, Kepka-Lenhart D (1997) Human type II arginase: sequence analysis and tissue-specific expression. Gene 193:157–161

Morris CR, Hamilton-Reeves J, Martindale RG, Sarav M, Ochoa Gautier JB (2017) Acquired amino acid deficiencies: a focus on arginine and glutamine. Nutr Clin Pract 32:30S–47S

Motoshima H, Inagaki K, Kumasaka T, Furuichi M, Inoue H, Tamura T, Esaki N, Soda K, Tanaka N, Yamamoto M, Tanaka H (2000) Crystal structure of the pyridoxal 5′-phosphate dependent L-methionine gamma-lyase from Pseudomonas putida. J Biochem 128:349–354

Muthukumar T, Chamundeeswari M, Prabhavathi S, Gurunathan B, Chandhuru J, Sastry TP (2014) Carbon nanoparticle from a natural source fabricated for folate receptor targeting, imaging and drug delivery application in A549 lung cancer cells. Eur J Pharm Biopharm 88:730–736

Nagata K, Nagata Y, Sato T, Fujino MA, Nakajima K, Tamura T (2003) L-serine, D- and L-proline and alanine as respiratory substrates of Helicobacter pylori: correlation between in vitro and in vivo amino acid levels. Microbiology 149:2023–2030

Nakai T, Nakagawa N, Maoka N, Masui R, Kuramitsu S, Kamiya N (2005) Structure of P-protein of the glycine cleavage system: implications for nonketotic hyperglycinemia. EMBO J 24:1523–1536

Nakayama T, Esaki N, Sugie K, Beresov TT, Tanaka H, Soda K (1984) Purification of bacterial L-methionine gamma-lyase. Anal Biochem 138:421–424

Nandakumar R, Yoshimune K, Wakayama M, Moriguchi M (2003) Microbial glutaminase: biochemistry, molecular approaches and applications in the food industry. J Mol Catal B Enzym 23:87–100

Nishiya Y, Imanaka T (1998) Purification and characterization of a novel glycine oxidase from Bacillus subtilis. FEBS Lett 438:263–266

Nomme J, Su Y, Lavie A (2014) Elucidation of the specific function of the conserved threonine triad responsible for human L-asparaginase autocleavage and substrate hydrolysis. J Mol Biol 426:2471–2485

Park H, Vettese-Dadey M, Aronson NN (1996) Glycosylation and phosphorylation of lysosomal glycosylasparaginase. Arch Biochem Biophys 328:73–77

Park JM, Kim A, Oh JH, Chung AS (2007) Methylseleninic acid inhibits PMA-stimulated pro-MMP-2 activation mediated by MT1-MMP expression and further tumor invasion through suppression of NF-kappaB activation. Carcinogenesis 28:837–847

Patil MD, Bhaumik J, Babykutty S, Banerjee UC, Fukumura D (2016) Arginine dependence of tumor cells: targeting a chink in cancer's armor. Oncogene 35:4957–4972

Pelayo BA, Fu YM, Meadows GG (2001) Decreased tissue plasminogen activator and increased plasminogen activator inhibitors and increased activator protein-1 and specific promoter 1 are associated with inhibition of invasion in human A375 melanoma deprived of tyrosine and phenylalanine. Int J Oncol 18:877–883

Perel Y, Auvrignon A, Leblanc T, Vannier JP, Michel G, Nelken B, Gandemer V, Schmitt C, Lamagnere JP, De Lumley L, Bader-Meunier B, Couillaud G, Schaison G, Landman-Parker J, Thuret I, Dalle JH, Baruchel A, Leverger G, Immunology, G. L. O. T. F. S. O. P. H. A (2002) Impact of addition of maintenance therapy to intensive induction and consolidation chemotherapy for childhood acute myeloblastic leukemia: results of a prospective randomized trial, LAME 89/91. Leucámie Aiqüe Myéloïde Enfant. J Clin Oncol 20:2774–2782

Phang JM, Donald SP, Pandhare J, Liu Y (2008) The metabolism of proline, a stress substrate, modulates carcinogenic pathways. Amino Acids 35:681–690

Phang JM, Liu W, Hancock CN, Fischer JW (2015) Proline metabolism and cancer: emerging links to glutamine and collagen. Curr Opin Clin Nutr Metab Care 18:71–77

Philip R, Campbell E, Wheatley DN (2003) Arginine deprivation, growth inhibition and tumour cell death: 2. Enzymatic degradation of arginine in normal and malignant cell cultures. Br J Cancer 88:613–623

Pokrovskaya MV, Aleksandrova SS, Pokrovsky VS, Omeljanjuk NM, Borisova AA, Anisimova NY, Sokolov NN (2012) Cloning, expression and characterization of the recombinant Yersinia pseudotuberculosis L-asparaginase. Protein Expr Purif 82:150–154

Pollari S, Kakonen SM, Edgren H, Wolf M, Kohonen P, Sara H, Guise T, Nees M, Kallioniemi O (2011) Enhanced serine production by bone metastatic breast cancer cells stimulates osteoclastogenesis. Breast Cancer Res Treat 125:421–430

Pollegioni L, Molla G, Sacchi S, Rosini E, Verga R, Pilone MS (2008) Properties and applications of microbial D-amino acid oxidases: current state and perspectives. Appl Microbiol Biotechnol 78:1–16

Pollegioni L, Motta P, Molla G (2013) L-amino acid oxidase as biocatalyst: a dream too far? Appl Microbiol Biotechnol 97:9323–9341

Polyak K, Xia Y, Zweier JL, Kinzler KW, Vogelstein B (1997) A model for p53-induced apoptosis. Nature 389:300–305

Possemato R, Marks KM, Shaul YD, Pacold ME, Kim D, Birsoy K, Sethumadhavan S, Woo HK, Jang HG, Jha AK, Chen WW, Barrett FG, Stransky N, Tsun ZY, Cowley GS, Barretina J, Kalaany NY, Hsu PP, Ottina K, Chan AM, Yuan B, Garraway LA, Root DE, Mino-Kenudson M, Brachtel EF, Driggers EM, Sabatini DM (2011) Functional genomics reveal that the serine synthesis pathway is essential in breast cancer. Nature 476:346–350

Reid MA, Wang WI, Rosales KR, Welliver MX, Pan M, Kong M (2013) The B55alpha subunit of PP2A drives a p53-dependent metabolic adaptation to glutamine deprivation. Mol Cell 50:200–211

Reinert RB, Oberle LM, Wek SA, Bunpo P, Wang XP, Mileva I, Goodwin LO, Aldrich CJ, Durden DL, Mcnurlan MA, Wek RC, Anthony TG (2006) Role of glutamine depletion in directing tissue-specific nutrient stress responses to L-asparaginase. J Biol Chem 281:31222–31233

Riley V, Spackman D, Fitzmaurice MA, Roberts J, Holcenberg JS, Dolowy WC (1974) Therapeutic properties of a new glutaminase-asparaginase preparation and the influence of the lactate dehydrogenase-elevating virus. Cancer Res 34:429–438

Roberts J, Holcenberg JS, Dolowy WC (1970) Antineoplastic activity of highly purified bacterial glutaminases. Nature 227:1136–1137

Rosini E, Piubelli L, Molla G, Frattini L, Valentino M, Varriale A, D'auria S, Pollegioni L (2014) Novel biosensors based on optimized glycine oxidase. FEBS J 281:3460–3472

Ruiz-Herrera J, Starkey RL (1969) Dissimilation of methionine by a demethiolase of Aspergillus species. J Bacteriol 99:764–770

Saarela J, Laine M, Tikkanen R, Oinonen C, Jalanko A, Rouvinen J, Peltonen L (1998) Activation and oligomerization of aspartylglucosaminidase. J Biol Chem 273:25320–25328

Saarela J, Oinonen C, Jalanko A, Rouvinen J, Peltonen L (2004a) Autoproteolytic activation of human aspartylglucosaminidase. Biochem J 378:363–371

Saarela J, Von schantz C, Peltonen L, Jalanko A (2004b) A novel aspartylglucosaminuria mutation affects translocation of aspartylglucosaminidase. Hum Mutat 24:350–351

Sahu N, Dela Cruz D, Gao M, Sandoval W, Haverty PM, Liu J, Stephan JP, Haley B, Classon M, Hatzivassiliou G, Settleman J (2016) Proline starvation induces unresolved ER stress and hinders mTORC1-dependent tumorigenesis. Cell Metab 24:753–761

Sato D, Nozaki T (2009) Methionine gamma-lyase: the unique reaction mechanism, physiological roles, and therapeutic applications against infectious diseases and cancers. IUBMB Life 61:1019–1028

Sharma B, Singh S, Kanwar SS (2014) L-methionase: a therapeutic enzyme to treat malignancies. Biomed Res Int 2014:506287

Shiono T, Nomura T, Nishiya Y, Arai R (2015) Crustal structure of glycine oxidase from *Geobacillus kaustophilus*. Photon Fact Act Rep 2014:32

Sivaraman S, Kirsch JF (2006) The narrow substrate specificity of human tyrosine aminotransferase – the enzyme deficient in tyrosinemia type II. FEBS J 273:1920–1929

Snell K, Natsumeda Y, Weber G (1987) The modulation of serine metabolism in hepatoma 3924A during different phases of cellular proliferation in culture. Biochem J 245:609–612

Srivastava D, Schuermann JP, White TA, Krishnan N, Sanyal N, Hura GL, Tan A, Henzl MT, Becker DF, Tanner JJ (2010) Crystal structure of the bifunctional proline utilization A flavoenzyme from Bradyrhizobium japonicum. Proc Natl Acad Sci U S A 107:2878–2883

Stone E, Chantranupong L, Gonzalez C, O'neal J, Rani M, Vandenberg C, Georgiou G (2012) Strategies for optimizing the serum persistence of engineered human arginase I for cancer therapy. J Control Release 158:171–179

Su N, Pan YX, Zhou M, Harvey RC, Hunger SP, Kilberg MS (2008) Correlation between asparaginase sensitivity and asparagine synthetase protein content, but not mRNA, in acute lymphoblastic leukemia cell lines. Pediatr Blood Cancer 50:274–279

Sugimura T, Birnbaum SM, Winitz M, Greenstein JP (1959) Quantitative nutritional studies with water-soluble, chemically defined diets. VIII. The forced feeding of diets each lacking in one essential amino acid. Arch Biochem Biophys 81:448–455

Sukhani S, Umerah BC (1978) Wilson-Mikity syndrome – a case report. Med J Zambia 12:58–60

Sun L, Bartlam M, Liu Y, Pang H, Rao Z (2005) Crystal structure of the pyridoxal-5′-phosphate-dependent serine dehydratase from human liver. Protein Sci 14:791–798

Tada K, Kure S (1993) Non-ketotic hyperglycinaemia: molecular lesion, diagnosis and pathophysiology. J Inherit Metab Dis 16:691–703

Tan Y, Xu M, Guo H, Sun X, Kubota T, Hoffman RM (1996) Anticancer efficacy of methioninase in vivo. Anticancer Res 16:3931–3936

Tanios R, Bekdash A, Kassab E, Stone E, Georgiou G, Frankel AE, Abi-Habib RJ (2013) Human recombinant arginase I(Co)-PEG5000 [HuArgI(Co)-PEG5000]-induced arginine depletion is selectively cytotoxic to human acute myeloid leukemia cells. Leuk Res 37:1565–1571

Tikkanen R, Riikonen A, Oinonen C, Rouvinen R, Peltonen L (1996) Functional analyses of active site residues of human lysosomal aspartylglucosaminidase: implications for catalytic mechanism and autocatalytic activation. EMBO J 15:2954–2960

Van Rite BD, Krais JJ, Cherry M, Sikavitsas VI, Kurkjian C, Harrison RG (2013) Antitumor activity of an enzyme prodrug therapy targeted to the breast tumor vasculature. Cancer Investig 31:505–510

Vander Heiden MG (2011) Targeting cancer metabolism: a therapeutic window opens. Nat Rev Drug Discov 10:671–684

Wang Z, Jiang C, Lu J (2002) Induction of caspase-mediated apoptosis and cell-cycle G1 arrest by selenium metabolite methylselenol. Mol Carcinog 34:113–120

Warangkar SC, Khobragade CN (2010) Purification, characterization, and effect of thiol compounds on activity of the Erwinia carotovora L-asparaginase. Enzym Res 2010:165878

Warburg O, Posener K, Negelein E (1924) Ueber den Stoffwechsel der Tumoren. Biochem Z 152:319–324

Warrell RP Jr, Arlin ZA, Gee TS, Chou TC, Roberts J, Young CW (1982) Clinical evaluation of succinylated Acinetobacter glutaminase-asparaginase in adult leukemia. Cancer Treat Rep 66:1479–1485

Warrell RP, Chou TC, Gordon C, Tan C, Roberts J, Sternberg SS, Philips FS, Young CW (1980) Phase I evaluation of succinylated Acinetobacter glutaminase-asparaginase in adults. Cancer Res 40:4546–4551

Wheatley DN, Campbell E (2002) Arginine catabolism, liver extracts and cancer. Pathol Oncol Res 8:18–25

Willems L, Jacque N, Jacquel A, Neveux N, Maciel TT, Lambert M, Schmitt A, Poulain L, Green AS, Uzunov M, Kosmider O, Radford-Weiss I, Moura IC, Auberger P, Ifrah N, Bardet V, Chapuis N, Lacombe C, Mayeux P, Tamburini J, Bouscary D (2013) Inhibiting glutamine uptake represents an attractive new strategy for treating acute myeloid leukemia. Blood 122:3521–3532

Wise DR, Thompson CB (2010) Glutamine addiction: a new therapeutic target in cancer. Trends Biochem Sci 35:427–433

Wise DR, Deberardinis RJ, Mancuso A, Sayed N, Zhang XY, Pfeiffer HK, Nissim I, Daikhin E, Yudkoff M,

Mcmahon SB, Thompson CB (2008) Myc regulates a transcriptional program that stimulates mitochondrial glutaminolysis and leads to glutamine addiction. Proc Natl Acad Sci U S A 105:18782–18787

Wu G, Morris SM (1998) Arginine metabolism: nitric oxide and beyond. Biochem J 336(Pt 1):1–17

Wu G, Zhan T, Guo Y, Kumar A, Liu Z (2016) Asn336 is involved in the substrate affinity of glycine oxidase from *Bacillus cereus*. Electron J Biotechnol 22:26–30

Xue HH, Sakaguchi T, Fujie M, Ogawa H, Ichiyama A (1999) Flux of the L-serine metabolism in rabbit, human, and dog livers. Substantial contributions of both mitochondrial and peroxisomal serine:pyruvate/alanine:glyoxylate aminotransferase. J Biol Chem 274:16028–16033

Yamaguchi M, Kwong YL, Kim WS, Maeda Y, Hashimoto C, Suh C, Izutsu K, Ishida F, Isobe Y, Sueoka E, Suzumiya J, Kodama T, Kimura H, Hyo R, Nakamura S, Oshimi K, Suzuki R (2011) Phase II study of SMILE chemotherapy for newly diagnosed stage IV, relapsed, or refractory extranodal natural killer (NK)/T-cell lymphoma, nasal type: the NK-Cell Tumor Study Group study. J Clin Oncol 29:4410–4416

Yamamoto N, Gupta A, Xu M, Miki K, Tsujimoto Y, Tsuchiya H, Tomita K, Moossa AR, Hoffman RM (2003) Methioninase gene therapy with selenomethionine induces apoptosis in bcl-2-overproducing lung cancer cells. Cancer Gene Ther 10:445–450

Yang Z, Sun X, Li S, Tan Y, Wang X, Zhang N, Yagi S, Takakura T, Kobayashi Y, Takimoto A, Yoshioka T, Suginaka A, Frenkel EP, Hoffman RM (2004a) Circulating half-life of PEGylated recombinant methioninase holoenzyme is highly dose dependent on cofactor pyridoxal-5′-phosphate. Cancer Res 64:5775–5778

Yang Z, Wang J, Lu Q, Xu J, Kobayashi Y, Takakura T, Takimoto A, Yoshioka T, Lian C, Chen C, Zhang D, Zhang Y, Li S, Sun X, Tan Y, Yagi S, Frenkel EP, Hoffman RM (2004b) PEGylation confers greatly extended half-life and attenuated immunogenicity to recombinant methioninase in primates. Cancer Res 64:6673–6678

Ye J, Kumanova M, Hart LS, Sloane K, Zhang H, De Panis DN, Bobrovnikova-Marjon E, Diehl JA, Ron D, Koumenis C (2010) The GCN2-ATF4 pathway is critical for tumour cell survival and proliferation in response to nutrient deprivation. EMBO J 29:2082–2096

Yoshioka T, Wada T, Uchida N, Maki H, Yoshida H, Ide N, Kasai H, Hojo K, Shono K, Maekawa R, Yagi S, Hoffman RM, Sugita K (1998) Anticancer efficacy in vivo and in vitro, synergy with 5-fluorouracil, and safety of recombinant methioninase. Cancer Res 58:2583–2587

Zabirnyk O, Liu W, Khalil S, Sharma A, Phang JM (2010) Oxidized low-density lipoproteins upregulate proline oxidase to initiate ROS-dependent autophagy. Carcinogenesis 31:446–454

Zhang K, Guo Y, Yao P, Lin Y, Kumar A, Liu Z, Wu G, Zhang L (2016) Characterization and directed evolution of BliGO, a novel glycine oxidase from Bacillus licheniformis. Enzym Microb Technol 85:12–18

Zhao R, Domann FE, Zhong W (2006) Apoptosis induced by selenomethionine and methioninase is superoxide mediated and p53 dependent in human prostate cancer cells. Mol Cancer Ther 5:3275–3284

Zuo S, Zhang T, Jiang B, Mu W (2015) Recent research progress on microbial L-asparaginases. Appl Microbiol Biotechnol 99:1069–1079

Alkaline Phosphatase Replacement Therapy

10

Maria Luisa Bianchi and Silvia Vai

Abstract

Hypophosphatasia (HPP) is a rare genetic disease, characterized by the defective production of tissue-non-specific alkaline phosphatase (TNSALP). Six subtypes of the disease – affecting neonates (beginning in utero), infants, children, or adults – are recognized: perinatal lethal, prenatal benign, infantile, childhood, adult, and odontohypophosphatasia. The clinical presentation of these subtypes is very different and the severity ranges from mild to lethal. This chapter, after an overview of the genetics, epidemiology, classification, and clinical presentation of the different forms of HPP, will review the current experience with enzyme replacement therapy (ERT).

Keywords

Asfotase alfa · Alkaline phosphatase · Bone · Enzyme replacement therapy · Fractures · Hypophosphatasia · Hypomineralization · Teeth

Abbreviations

ALP	Alkaline phosphatase
BMD	Bone mineral density
BSALP	Bone-specific alkaline phos phatase
CPAP	Continuous positive airway pressure
DXA	Dual-energy X-ray absorptiometry
ERT	Enzyme replacement therapy
GABA	Gamma-aminobutyric acid
HPP	Hypophosphatasia
NPP1	Nucleoside pyrophos phohydrolase-1
NSAIDs	Non steroidal anti-inflammatory drugs
PEA	Phosphoethanolamine
PHOSPHO1	Phosphatase orphan 1
Pi	Inorganic phosphate
PL	Pyridoxal
PLP	Pyridoxal-5′-phosphate
PPi	Inorganic pyrophosphate
PTH	Parathyroid hormone
RGI-C	Radiographic global impression of change
TNSALP	Tissue-non-specific alkaline phosphatase

M. L. Bianchi (✉) · S. Vai
Experimental Laboratory for Children's Bone Metabolism Research, Bone Metabolism Unit, Istituto Auxologico Italiano IRCCS, Milan, Italy
e-mail: ml.bianchi@auxologico.it

N. Labrou (ed.), *Therapeutic Enzymes: Function and Clinical Implications*, Advances in Experimental Medicine and Biology 1148, https://doi.org/10.1007/978-981-13-7709-9_10

10.1 Hypophosphatasia

Hypophosphatasia (HPP), first described in 1948 (Rathbun 1948), is a rare inherited disorder of bone and mineral metabolism, caused by loss-of-function mutations in the *ALPL* gene, leading to reduced activity of the tissue-non-specific isoenzyme of alkaline phosphatase (TNSALP). Over 360 *ALPL* mutations are known.

The clinical presentation of HPP is very variable, depending on age of appearance and, possibly, on type of mutation and mechanism of inheritance. The most severe symptoms and signs are observed in neonates, infants and children, and a worse prognosis is usually associated with earlier manifestations. Defective bone and tooth mineralization, respiratory insufficiency, seizures, chronic pain, altered calcium and phosphate metabolism, renal problems, impaired growth and mobility, developmental delay, recurrent fractures, premature tooth loss are typical clinical findings. Biochemical tests show persistent hypophosphatasemia (low levels of serum alkaline phosphatase (ALP) and TNSALP, adjusted for age and gender) and increased levels of ALP substrates (inorganic pyrophosphate [PPi]; pyridoxal-5′-phosphate [PLP], the active metabolite of vitamin B6; and phosphoethanolamine [PEA]). Among them, PPi is a potent inhibitor of mineralization, and its accumulation explains the bone and tooth complications that are a hallmark of HPP.

The clinical classification is essentially dependent on age at onset of first signs and symptoms, and six subtypes of the disease are currently recognized, with severity ranging from mild to lethal (see Sect. 10.1.4) (Hofmann et al. 2013; Linglart and Biosse-Duplan 2016; Mornet 2018; Whyte 2016).

In the last decade, the efficacy and safety of enzyme replacement therapy with asfotase alfa, a recombinant TNSALP, has been demonstrated in newborns, infants, and children with severe HPP. In 2015, asfotase alfa has been approved by regulatory agencies in several countries, including the European Union (Hofmann et al. 2016; Mornet 2018; Whyte 2017a).

10.1.1 Alkaline Phosphatase

Alkaline phosphatase (ALP) is a membrane-bound phosphomonoesterase that catalyzes dephosphorylation reactions (cleavage of phosphoester bonds with release of a hydroxyl group and phosphate) and is essential in the catabolism of PPi and other substances. It is present in all animal species. In humans, four ALP isoenzymes, encoded by different genes, are known: three are tissue-specific ALPs (intestinal ALP, placental ALP, placenta-like or germ cell ALP), and the fourth is the ubiquitous non-specific TNSALP, that accounts for about 95% of the total serum ALP activity and is most abundant in liver, bone, and kidney (Caswell et al. 1991; Cole 2008; Mornet 2008; Hofmann et al. 2013; Hoshi et al. 1997; Whyte 2010). In bone, TNSALP is found in pre-osteoblasts, osteoblasts, matrix vesicles, osteoid in areas of new bone formation, trabecular lining cells, newly embedded osteocytes, subperiosteal, endosteal, and bone marrow cells. TNSALP is also present in growth plates (chondrocytes and cartilage matrix), articular cartilages and teeth (ameloblasts and odontoblasts) (Hoshi et al. 1997; Miao and Scutt 2002). TNSALP activity is essential for bone and teeth mineralization, and its deficiency leads to skeletal hypomineralization, a clinical hallmark of HPP. The bone TNSALP isoform is often called bone-specific ALP (BSALP). TNSALP may also have a role in inflammatory processes, as suggested by the occurrence of muscle, bone, and cartilage damage, including osteomyelitis and arthritis, in HPP (Hofmann et al. 2013).

TNSALP function in tissues other than bone is still incompletely known. Reviews of current knowledge are provided by Buchet et al. (2013) and Mornet (2018). TNSALP is certainly involved in the metabolism of vitamin B6, whose active metabolite (PLP) has a key role in several physiological processes, including the synthesis of major neurotransmitters (serotonin, dopamine, epinephrine, norepinephrine, GABA). Studies on knockout mice have suggested that the defective dephosphorylation of PLP to pyridoxal (PL) could explain the seizures observed in some perinatal/infantile HPP forms. Such seizures are the only known clinical sign of altered vitamin B6 metabolism in HPP, although, for unclear reasons, a positive response to vitamin B6 treatment in these cases seems to be an ominous sign. It must be underlined that such seizures are only observed in very young subjects. Older children or adults have no signs of vitamin B6 deficiency or toxicity (Baumgartner-Sigl et al. 2007; Mornet 2008; Whyte 2010).

TNSALP seems also involved in the development of central nervous system and in some brain functions (Fonta et al. 2005; Hofmann et al. 2013; Kermer et al. 2010), and these actions might explain the neurological and psychological signs and symptoms (neonatal seizures, chronic pain, anxiety, restlessness, depression) that are often observed in HPP.

10.1.2 Genetics

The *ALPL* gene is located on the short (p) arm of chromosome 1, bands 1p36.1-p34 (Weiss et al. 1988; Smith et al. 1988). Human TNSALP contains 524 aminoacid residues (Silvent et al. 2014). HPP is caused by any mutation in the *ALPL* gene leading to decreased TNSALP activity and increased levels of its substrates. As of August 2018, the *Tissue Nonspecific Alkaline Phosphatase Gene Mutations Database* (http://www.sesep.uvsq.fr/03_hypo_mutations.php) includes 365 *ALPL* mutations. The inheritance mechanism can be autosomal recessive or dominant. The most severe forms of HPP are autosomal recessive and are characterized by nearly total suppression of TNSALP activity. The milder forms can be either autosomal recessive or autosomal dominant and are characterized by reduced, but not suppressed, TNSALP activity. Some severe mutations may cause mild forms of HPP in heterozygotes, due to a dominant negative effect (Fauvert et al. 2009; Mornet and Nunes 2011; Mornet 2017). The phenotypic presentation of the disease may be significantly different even within the same family (Hofmann et al. 2014).

10.1.3 Epidemiology

HPP is present worldwide, but its prevalence in different populations is highly variable. Canadians have a particularly high prevalence of the severe form, estimated at 1:100,000 (Fraser 1957). In particular, Canadian Mennonites show a very high prevalence (up to 1/2500), and 1:25 subjects may be a carrier (Orton et al. 2008). In African-Americans, HPP seems to be particularly rare (Whyte et al. 2006). In Europe, the prevalence of severe HPP has been estimated at 1:300,000 (Mornet et al. 2011). In Japan, the prevalence of the perinatal lethal form is about 1:900,000 and seems often due to homozygosis for *ALPL* mutation c.1559delT, that is found only among the Japanese, with an estimated carrier frequency of 1:480 (Watanabe et al. 2011).

The prevalence of the milder forms of HPP, including the adult forms, is extremely difficult to estimate, as many cases probably remain undiagnosed. According to a genetic model proposed by Mornet et al., the prevalence of dominant mild HPP in the European population might be about 1:6370 (Mornet et al. 2011).

10.1.4 Classification

The clinical presentation of HPP is very variable, ranging from lethal perinatal forms (death in utero or soon after birth, with almost total absence of bone mineralization) to mild adult forms (dental problems and osteopenia) (Bianchi 2015).

The original classification of HPP, essentially unchanged since 1957, described five HPP subtypes: "perinatal lethal HPP", "infantile HPP", "childhood HPP", "adult HPP", and "odonto-HPP". A sixth subtype "prenatal (or perinatal) benign HPP" was subsequently identified and added (Wenkert et al. 2011).

In 2015, reappraising their 25-year experience with 173 pediatric HPP patients, Whyte et al. proposed to distinguish two forms of "childhood HPP", mild and severe (Whyte et al. 2015). Accordingly, an updated classification would now recognize seven "major forms" of HPP, that is, in order of increasing severity: "odonto HPP", "adult HPP", "mild childhood HPP", "severe childhood HPP", "infantile HPP", "perinatal HPP", plus the time-limited form "prenatal benign HPP" (Whyte 2017a) (Table 10.1).

In practice, there can be considerable overlapping of clinical features, particularly in the infantile and childhood forms, and some cases can be very difficult to classify. Both children and adults can present with pain, stiffness, and weakness of the lower limbs (possibly due to bone microfractures). Also, anomalies of tooth color, eruption and exfoliation, shape and structure, enamel, dentin, cementum are observed in both childhood and adult forms. The clinical severity of HPP is essentially linked to its skeletal complications, and a worse prognosis is usually associated with earlier signs and symptoms (Cole 2008; Fallon et al. 1984; Hofmann et al. 2013; Whyte et al. 2015; Whyte 2017a).

Very recently, Whyte et al. (2018), re-evaluating a cohort of 165 preteenager HPP patients, highlighted the four biochemical hallmarks of HPP in children. All patients had subnormal serum total and bone-specific ALP, and elevated plasma PLP, and nearly all had high urinary PPi excretion. The mean levels of these four biomarkers correlated with HPP severity, ranked according to the proposed HPP nosology ("odonto HPP", "mild childhood HPP", "severe childhood HPP", "infantile HPP"), with data overlapping among the four patient groups. The authors conclude that "these four biochemical hallmarks represent both a sensitive and reliable tool for diagnosing children with HPP", and "validate our expanded clinical nosology for pediatric HPP" as an "improved framework for conceptualizing and working with this disorder's remarkably broad-ranging severity".

10.1.5 Bone and Dental Problems in HPP

Defective mineralization of bone and teeth is the most characteristic and severe consequence of insufficient TNSALP activity (Figs. 10.1, 10.2, and 10.3) (Hofmann et al. 2013; Linglart and Biosse-Duplan 2016; Whyte 2017a). The mineralization of cartilage, bone, and teeth is the result of physicochemical and biochemical processes regulated by several promoting and inhibiting factors. Among them, three phosphatases – TNSALP, phosphatase orphan 1 (PHOSPHO1), and nucleoside pyrophosphohydrolase-1 (NPP1) – play an essential role in maintaining an optimal Pi/PPi ratio. This ratio is crucial in the mineralization process, where Pi is needed for the formation of hydroxyapatite crystals, while PPi inhibits it (Millán 2013).

In an early phase of mineralization, Ca^{2+} and Pi accumulate within the bone matrix vesicles, forming hydroxyapatite crystals. In a second phase, the vesicles' membranes break down and the hydroxyapatite crystals are deposited in the collagenous matrix (Anderson 1995; Golub 2009; Hessle et al. 2002; Schinke et al. 1999).

In HPP, insufficient TNSALP activity leads to the accumulation of excess PPi, which inhibits the calcium/phosphate crystal nucleation and hampers the growth of hydroxyapatite crystals in the matrix vesicles (Hessle et al. 2002; Harmey et al. 2004; Anderson et al. 2004; Orimo 2010). Even if osteoblasts seem to be unaffected (Wennberg et al. 2000), the whole process of bone mineralization is seriously impaired, and hypomineralized bone, spontaneous fractures, tooth loss, as well as soft tissue calcifications (osteoarthritis, arterial calcification), are characteristic manifestations of the disease (Millán 2013).

Regarding dental problems, premature exfoliation and structural alterations of the primary (deciduous) teeth and an increased occurrence of severe dental caries is observed in all forms of HPP (Fig. 10.2). Alveolar bone and dental cementum are poorly mineralized, and premature tooth loss is common. The anterior teeth (incisors) are more frequently affected. Impaired dentinogenesis, revealed by large pulp chambers, and enamel hypoplasia, predisposing to dental caries, are also described. The primary teeth are most often affected, but permanent teeth are not spared (Atar and Körperich 2010; Beumer et al. 1973; Olsson et al. 1996; van den Bos et al. 2005).

Table 10.1 Different forms of hypophosphatasia

Form	Inheritance	Clinical course and features	Dental features
Perinatal lethal	AR	Most severe form	–
		Stillbirth or death within days/weeks after birth	
		Severe hypomineralization (deformed and short limbs)	
		Derangements in calcium/phosphate metabolism	
		Osteochondral spurs in forearms, legs	
		Severe lung hypoplasia (chest deformities, rib fractures)	
		Seizures, dependent by pyridoxine (B6)	
Prenatal (or perinatal) benign	AR/AD	Benign evolution (but long-term course unknown)	–
		Limb shortening and bowing of long bones observed in utero	
		Spontaneous improvement of skeletal defects after birth	
Infantile	AR	First symptoms during the first 6 months of life	Premature loss of deciduous teeth
		Poor prognosis during the 1st year of life (mortality has been estimated to be 50% during infancy)	
		Severe hypomineralization (rachitic ribs)	
		Premature craniosynostosis (Chiari I malformation, hydrostatic hydrocephalus, hydrosyringomyelia)	
		Rickets-like signs of hypomineralization	
		Swallowing disorders, irritability, seizures	
		Severe muscular hypotonia	
		Hypercalciuria, nephrocalcinosis	
Childhood ("mild" or "severe")	AR/AD	First symptoms after 6th month of life	Premature loss of deciduous teeth
		Rickets-like signs of hypomineralization	
		Short stature, failure to thrive	Caries
		Delayed walking	
		Repeated fractures	
		Waddling gait due to bone deformities	
		Chronic bone pain (lower extremities)	
		Muscular hypotonia	
		Lack of appetite, nausea, gastrointestinal problems	
Adult	AR/AD	Presents in middle age	Loss of permanent teeth at 40–60 years of age
		Stress fractures of metatarsals, tibia. Femur pseudofractures	
		Fragility fractures	
		Osteomalacia, osteoporosis	
		History of delayed fracture healing in childhood; often mild rickets	
		Chondrocalcinosis, osteoarthritis	
		Myopathy, weakness	
		Renal abnormalities, reduced GFR nephrocalcinosis and kidney stones	
		Psychiatric symptoms (insomnia, restlessness, anxiety, depression)	
Odonto HPP	AR/AD	Not associated with bone, articular, or muscular problems	Premature (i.e. age <5 years) exfoliation of deciduous and/or permanent teeth (mainly incisors)
			Severe dental caries
			Reduced thickness of dentin Enlarged pulp chambers
			Alveolar bone loss

Legend: *AD* autosomal dominant, *AR* autosomal recessive, *GFR* glomerular filtration rate

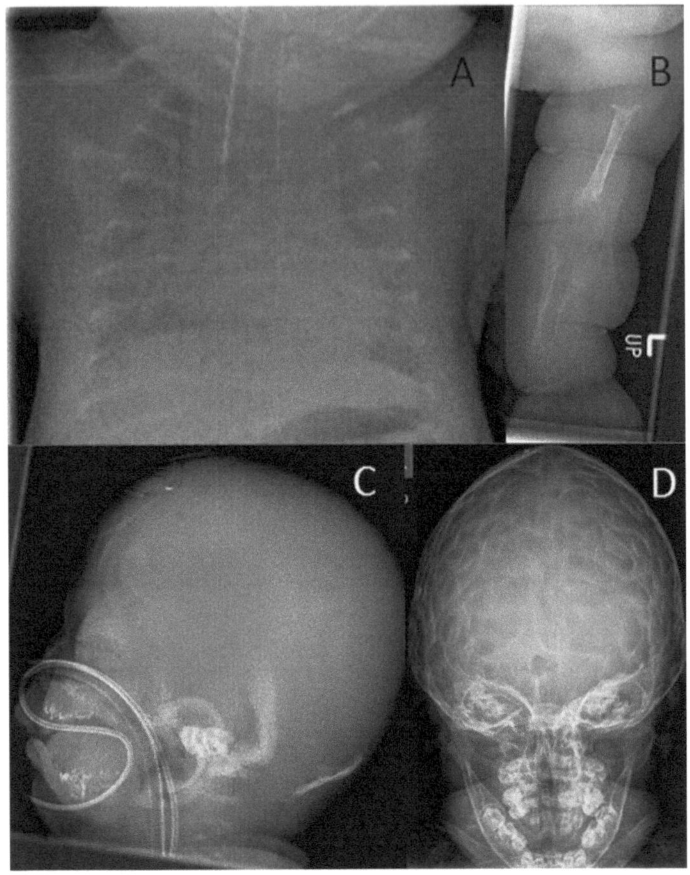

Fig. 10.1 Radiographies of a case of perinatal and infantile HPP. (**a**) Chest of an infant with thin ribs, lung hypoplasia, endotracheal intubation and ventilation. (**b**) Upper limb showing extreme hypomineralization. (**c**) Lateral skull showing severe hypomineralization, and ossification only at frontal bone, base and occiput. (**d**) Antero-posterior skull showing a copper beaten skull due to craniosynostosis with increased intracranial pressure. (From Hofmann et al. 2013, reproduced with permission)

10.1.6 Diagnosis

The diagnosis of HPP must be based on a consistent set of medical history and physical examination data, biochemical tests, and skeletal radiographic findings (that can be pathognomonic in perinatal, infantile, and severe childhood HPP) (Bianchi 2015; Hofmann et al. 2013; Kishnani et al. 2017; Linglart and Biosse-Duplan 2016; Whyte 2016, 2017a, b). As all rare diseases, HPP (the mild forms in particular) can be difficult to recognize and the diagnosis is often delayed. HPP is associated with a very wide range of presenting signs, symptoms, and complications (defective skeletal mineralization, respiratory insufficiency, seizures, altered calcium and phosphate metabolism, renal problems, pain, delay in growth and development, impaired mobility, recurrent fractures, tooth loss, etc.), that can be misinterpreted as indicators of other more common diseases. The clinical findings, and consequently the differential diagnosis, are very different depending on age at presentation. For example, the presence of short bowed legs revealed by ultrasonography in a fetus might suggest osteogenesis imperfecta or other skeletal dysplasias. In a neonate, HPP should be differentiated from congenital rickets, neonatal hyperparathyroidism, osteogenesis imperfecta, mucolipidosis II. In an infant, HPP might be misinterpreted as any cause of failure to thrive, nutritional rickets, or nephrocalcinosis due to idiopathic hypercalcemia. In a child, the presence of short stature, bone pain, fractures, motor delay might suggest mild osteogenesis imperfecta, myopathies, arthritis, chronic recurrent multifocal osteomyelitis, fibrous dysplasia. In an adult, especially in the presence of fractures, HPP can easily be confused with osteoporosis (Bishop 2015; Hofmann et al. 2013; Linglart and Biosse-Duplan 2016).

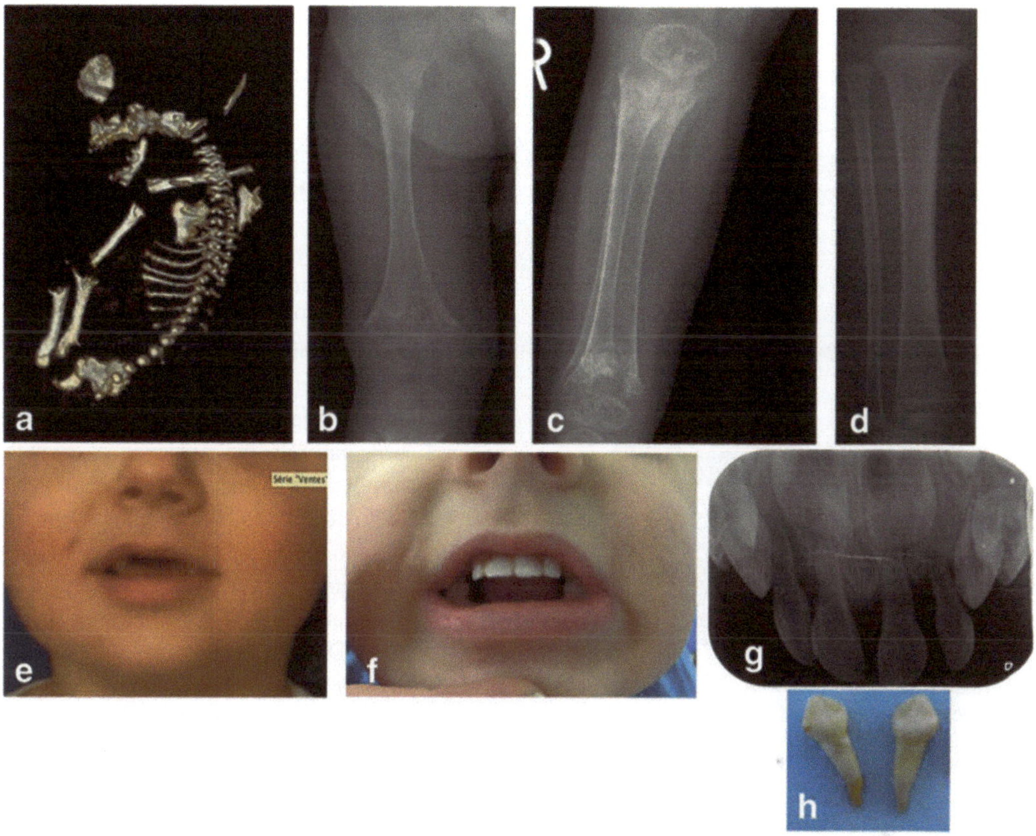

Fig. 10.2 Images showing clinical characteristics of HPP in children. (**a**) Scannographic image at 30 weeks of pregnancy in a fetus affected with hypophosphatasia and a homozygous mutation in the ALPL gene. Note the lack of mineralization of the skull, the thin ribs, but most of all, the abnormal distal femoral metaphyses with eperons and irregularities. (**b**) Femur and (**c**) tibia X-rays in an 18-month-young boy with HPP with a compound heterozygous mutation in the ALPL gene. (**d**) For comparison, an X-ray of an unaffected 20-month-old boy is shown. (**e**) Premature loss of the anterior upper and lower teeth before the age of 2 years. (**f**) Anicesmile with a fixed partial denture in the same child 1 year later. (**g**) Retroalveolar radiograph of the same patient before he lost upper incisors and canines during the following year. Note the reduced alveolar bone level around the incisors. (**h**) Pictures of exfoliated teeth (decidual canines) with their entire roots. (From Linglart and Biosse-Duplan 2016, reproduced with permission)

The typical laboratory findings of HPP are persistently low serum levels of ALP for age and gender, and increased levels of its substrates PPi, PLP, and PEA. However, low ALP activity alone, even if correctly based on sex- and age-specific reference ranges, cannot establish a diagnosis of HPP, as it may depend on many different causes (use of certain drugs, celiac disease, hypothyroidism, osteogenesis imperfecta type II, milk-alkali syndrome, vitamin D toxicity, etc.). High serum levels of PLP are a more specific marker of HPP, and are usually correlated with disease severity (Whyte 2016). It can also be noted at this point that a standardized assay for serum PPi, a potentially very useful laboratory test in the diagnosis and follow-up of HPP, is currently not available.

Genetic analysis of *TNSALP* mutations is often unnecessary for a diagnosis of HPP, but is obviously helpful to clarify inheritance patterns, evaluate the risk of HPP recurrence in a future pregnancy, and for prenatal assessment. However, even a positive mutation analysis is not sufficient for a diagnosis of HPP, if not accompanied by clinical, biochemical and radiologic evidence of disease (Whyte 2017a).

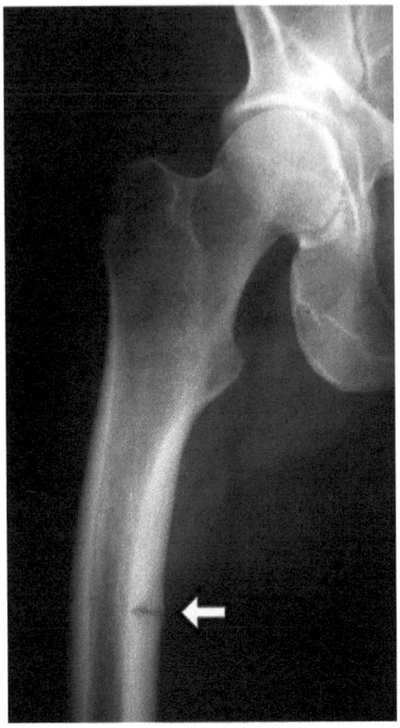

Fig. 10.3 HPP in an adult with severe childhood type. X-ray shows a pseudofrature (arrow) medially in the proximal femur who presented also bowing and cortical thickening. (From Whyte 2017a, reproduced with permission)

Whyte also underlines the importance of a correct diagnosis of HPP before making the decision to treat a patient with asfotase alfa, because mistakenly using the drug in conditions other than HPP could result in excessive mineralization with negative consequences (Whyte 2017a). In an adult, the misdiagnosis of HPP as osteoporosis may lead to the erroneous prescription of a bisphosphonate, which is contraindicated in HPP (Wüster and Ziegler 1992; Sutton et al. 2012; Whyte 2009).

10.2 General Management of HPP

Until the availability of enzyme replacement therapy (ERT, see below), therapeutic interventions in HPP addressed its different clinical problems (chronic pain, musculo-articular-skeletal alterations, fractures, psychological and neurological problems, renal problems, dental problems), aiming to alleviate symptoms and prevent complications (Rockman-Greenberg 2013). The wide spectrum of HPP manifestations ideally required the collaboration of an experienced multispecialty team.

10.2.1 Nutrition

In the presence of hyperphosphatemia, restriction of dietary phosphate (Pi) and/or pharmacologic binding of dietary Pi have been proposed, since Pi is known to inhibit TNSALP activity and gene expression (Wenkert et al. 2002).

In children, adequate nutrition should be ensured, especially in the presence of growth retardation. Vitamin D insufficiency should be corrected with standard supplementation, taking care to avoid too high doses that could aggravate the hypercalcemia/hypercalciuria in infantile HPP (Mornet and Nunes 2011). Active metabolites of vitamin D (such as 1α-hydroxy-vitamin D or 1,25-dihydroxyvitamin D) should not be used, to avoid increase of intestinal Pi absorption and serum Pi levels. Calcium supplementation is also to be avoided except in carefully selected cases, considering the risk of hypercalcemia and kidney stones. In adults, supplementation with calcium and vitamin D can be used to prevent secondary hyperparathyroidism (Mornet and Nunes 2011).

10.2.2 Physical Activity

Moderate but regular physical exercise is recommended at any age, first to build and then to maintain bone mineral mass and bone strength. Only low-impact physical activities are suitable, especially in the severe forms of HPP, to minimize the risk of fragility fractures.

10.2.3 Inflammatory Manifestations

Patients with bone and joint pain and inflammatory symptoms may benefit from the judicious intermittent use of non steroidal anti-inflammatory drugs (NSAIDs), with significant subjective and functional improvement, that may last for some weeks after withdrawal. NSAIDs have also been successfully used for bone and joint pain in childhood HPP (Girschick et al. 2006). In the presence of impaired renal function, NSAID use is risky and requires close monitoring (Hofmann et al. 2013).

10.2.4 Osteoporosis and Fractures

Osteoporosis and bone fragility are a major problem in the severe forms of HPP. The prevention of fractures through the avoidance of risky physical activities is very important, since fractures in HPP patients heal with great difficulty and in a long time, requiring prolonged casting or even orthopedic interventions.

Bisphosphonates, the classical anti-osteoporosis drugs, have not been seriously studied in HPP (Mornet and Nunes 2011). There are theoretical reasons against them, however, as they are structurally related to PPi and can further suppress TNSALP activity (Wüster and Ziegler 1992). Moreover, there is a major difference between the bone fragility of HPP and that of classical osteoporosis. In HPP, bone fragility is caused by deficient bone mineralization, while in osteoporosis it derives from excess bone resorption. For this reason, the anti-resorptive drugs like bisphosphonates and denosumab that are so effective in classical osteoporosis are possibly contraindicated in HPP and may have untoward side effects. It has been reported that adults with undiagnosed HPP treated with bisphosphonates may have an increase in and/or worsening of fractures, or may sustain atypical (lateral subtrochanteric) femur fractures (Fig. 10.4) (Cundy et al. 2015; Mornet and Nunes 2011; Sutton et al. 2012; Whyte 2009).

Since PTH stimulates TNSALP synthesis in osteoblasts, teriparatide (recombinant human parathyroid hormone 1–34) has been used in some adults with HPP. Whyte et al. (2007) reported the healing of metatarsal stress fractures and proximal femur pseudofractures with a 18-month teriparatide treatment. Camacho et al. (2008) reported the positive effects of a 24-month treatment in a 75-year-old woman with multiple low-trauma fractures: skeletal mineralization and bone turnover markers improved, and there were no more fractures. Doshi et al. (2009) described an interesting case of bisphosphonate-related atypical femur fractures in a 50-year-old woman, successfully treated with teriparatide for 16 months after the correct diagnosis of HPP was

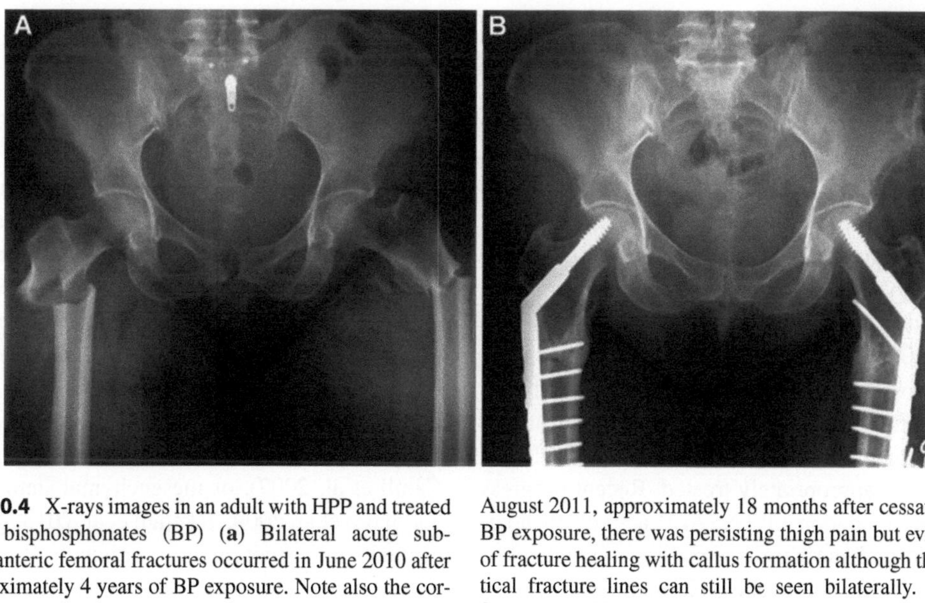

Fig. 10.4 X-rays images in an adult with HPP and treated with bisphosphonates (BP) (**a**) Bilateral acute subtrochanteric femoral fractures occurred in June 2010 after approximately 4 years of BP exposure. Note also the cortical thickening of both upper femoral shafts. (**b**) In August 2011, approximately 18 months after cessation of BP exposure, there was persisting thigh pain but evidence of fracture healing with callus formation although the cortical fracture lines can still be seen bilaterally. (From Sutton et al. 2012, reproduced with permission)

made. There are other case reports of successful treatment (Schalin-Jäntti et al. 2010; Camacho et al. 2016), but also cases in which the benefits did not last (Gagnon et al. 2010), or no significant effect was observed (Laroche 2012). Considering these inconsistent results, Gagnon et al. (2010) suggested that the response might depend on the specific gene mutations. Teriparatide should not be used for a total duration of more than 2 years in a single patient, and must not be used in young patients with open epiphyses, due to a potentially increased risk of osteosarcoma (observed in animal studies) (Mornet and Nunes 2011).

Calcitonin followed by a bisphosphonate (clodronate) was unsuccessfully tried in a 7-month-old girl with infantile HPP (Deeb et al. 2000).

Anti-sclerostin treatment might stimulate bone formation, and there is a report that eight adult patients with HPP showed increases in bone formation markers and bone mineral density (Seefried et al. 2017).

Bone marrow and stem cell transplantation has been tried in some infants and children with life-threatening HPP, with some improvement in skeletal mineralization and prolonged survival (until 3–7 years), but without consistent improvement in ALP activity (Cahill et al. 2007; Taketani et al. 2015; Whyte et al. 2003).

10.2.5 Neurological and Psychological Symptoms

Seizures, apparently due to defective dephosphorylation of PLP, are reported in severe perinatal and infantile HPP. For unclear reasons, their response to vitamin B6 or pyridoxine treatment seems not a positive sign and may on the contrary suggest a lethal prognosis (Baumgartner-Sigl et al. 2007; Mornet 2008; Whyte 2010).

Chronic pain, insomnia, anxiety, irritability, restlessness, depression can be very disturbing symptoms in the less severe forms of HPP and should be appropriately treated. Recent progress in the elucidation of TNSALP activity in nociception and purinergic signaling might lead to more specific approaches (Hofmann et al. 2013).

10.2.6 Neurosurgical Problems

In the infantile and childhood forms, the premature fusion of cranial sutures is common, and radiological, neurological, and ophtalmoscopic assessments should be regularly made until adolescence. Complications like secondary ectopia of cerebellar tonsils (Chiari I malformation) or hydrosyringomyelia may appear, and in these cases lifetime surveillance is needed (Collmann et al. 2009). Neurosurgical interventions may be required in the presence of neurologic signs and symptoms of increased intracranial pressure (headache, seizures, papilledema, numbness of extremities, paralysis).

10.2.7 Renal Problems

Kidneys can be involved in HPP in several ways. High arterial pressure is common and must be treated. Kidney stones can be a complication of hypercalcemia and hypercalciuria. Hyperphosphatemia may require treatment with phosphate binders. Nephrocalcinosis can also develop, with progressive worsening of renal function. The prudent, short-term use of NSAIDs might be helpful, possibly with the supervision of a nephrologist (Rockman-Greenberg 2013).

10.3 Enzyme Replacement Therapy with Asfotase Alfa

The first attempts at ERT were made with intravenous administration of ALP-enriched plasma from patients with Paget bone disease, purified human liver ALP, or purified human placental ALP in infants with severe HPP, but the results were disappointing (Weninger et al. 1989; Whyte et al. 1982, 1992). Some years later, transplantation of T-cell-depleted bone marrow (Whyte et al. 2003), bone fragments and cultured osteoblasts (Cahill et al. 2007), or mesenchymal stem-cells (Tadokoro et al. 2009; Taketani et al. 2013, 2015) resulted in encouraging, although limited, improvement. These first studies suggested that increasing enzyme activity in plasma was not sufficient, while TNSALP-expressing bone cells could help correct the genetic defect of HPP.

In 2004, Enobia Pharma Inc. (a biotech company based in Montreal, Canada) filed a patent application for a bioengineered substitute of TNSALP (original documentation available at https://patentimages.storage.googleapis.com/cf/e0/bd/7ecc5788ba1853/US7763712.pdf). This product (sALP-FcD10, first named ENB-0040; currently called asfotase alfa) is a recombinant fusion glycoprotein made of two identical polypeptide chains (connected by disulfide bonds), that contain the catalytic ectodomain of human TNSALP (i.e. the soluble part of TNSALP, cut at the C-terminal membrane-bound region), with the addition of the human IgG1 Fc region and a C-terminal deca-aspartate motif. The Fc region was added to facilitate cromatographic purification of the drug, and the deca-aspartate motif, which has high affinity for hydroxyapatite crystals, for specific bone targeting (Fig. 10.5) (Hofmann et al. 2016).

In 2011, Enobia Pharma was acquired by Alexion Pharmaceuticals Inc. (New Haven, CT, USA), and ENB-0040 was renamed asfotase alfa. Since 2008, ENB-0040/asfotase alfa has been investigated in many studies, first in animals and then in children with severe HPP. The positive results led to its approval by the regulatory agencies of Japan (PMDA), European Union (EMA), Canada (Health Canada), and USA (FDA) in 2015. Asfotase alfa, currently marketed as Strensiq® by Alexion Pharmaceuticals, is currently the only approved treatment for HPP, with the indication "*treatment of patients with perinatal-, infantile- and juvenile-onset hypophosphatasia*". The European Medicines Agency product information is available online (Strensiq® EMA product information 2018).

Strensiq® (asfotase alfa) is a clear, colorless, aqueous solution for subcutaneous injections, now available in single use vials at different doses and concentrations (18 mg/0.45 mL, 28 mg/0.7 mL, 40 mg/1.0 mL, 80 mg/0.8 mL). The recommended dosage is 2 mg/kg body weight three times a week or 1 mg/kg body weight six times a week. The maximum injection volume should not exceed 1 mL, and if necessary, multiple subcutaneous injections can be administered at the same time. The safety and efficacy of asfotase alfa in HPP patients with renal or hepatic problems have not been studied (Hofmann et al. 2016; Whyte 2017a; see also Strensiq® Alexion product information 2018; Strensiq® Alexion Prescription Information 2018). Table 10.2 resumes the clinical trials on HPP treatment with asfotase alfa, as listed at *https://clinicaltrials.gov*.

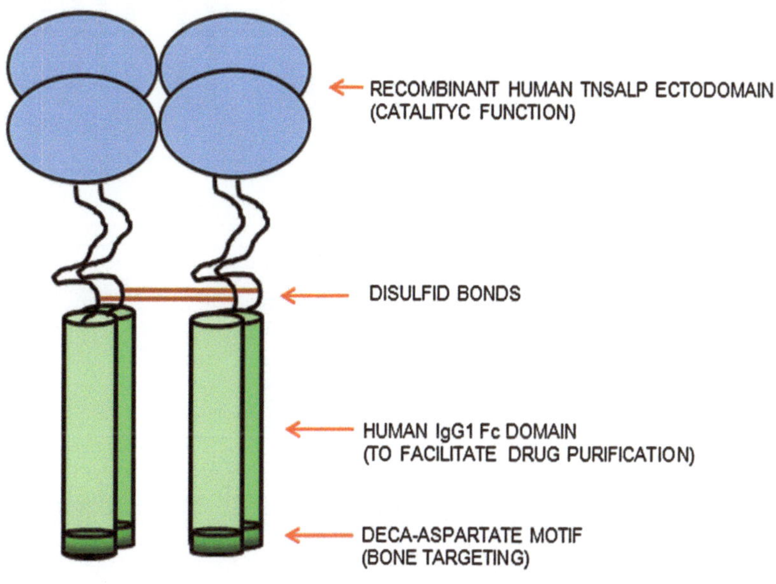

Fig. 10.5 The structure of asfotase alfa (drawn as a homotetramer): a recombinant fusion glycoprotein, containing the human TNSALP ectodomain with the addition of the human IgG1 Fc region and a C-terminal deca-aspartate motif. (Redrawn from http://www.ema.europa.eu/docs/en_GB/document_library/EPAR_-_Public_assessment_report/human/003794/WC500194340.pdf)

Table 10.2 Clinical trials on hypophosphatasia and asfotase alfa

NCT number	Titles	Recruitment	Characteristics	Population	Sponsor	Dates and Results	Locations
NCT03418389	Evaluate and monitor physical performance of adults treated with asfotase alfa for HPP	Not yet recruiting	Observational	Enrollment:	Alexion Pharm.	Start: July 2018	Würzburg University Hospital, Würzburg, Germany
			Cohort time				
			Perspective	Age: 18 years and older (adult, older adult)		First posted: Feb. 1 2018	
			Outcome measures:				
			Physical performance				
			Body composition	Sex: M and F		Last update posted: May 1, 2018	
			Health related quality of life				
			And 4 more outcomes				
NCT02797821	Pharmacokinetic, and dose response study of asfotase alfa in adult patients with pediatric-onset HPP	Completed	Interventional phase: 2	Enrollment: 27	Alexion Pharm.	Study start: Jun. 2016	Shriners Hospitals for Children, Saint Louis, MO, USA
			Randomized	Age: 18 years and older (adult, older adult)		First posted: Jun. 14, 2016	Duke University Medical Center, Durham, NC, USA
			Parallel assignment				
			Masking: none (open label)	Sex: M and F		Last update posted: Jul. 10, 2018	Vanderbilt Medical Center, Nashville, TN, USA
			Primary purpose: treatment				
			Outcome measures:			RESULTS not yet available	University of Würzburg, Würzburg, Germany
			Change in PPi from baseline to pre-3rd dose in week 9				
			Change in PLP from baseline to pre-3rd dose in week 9				
			And 11 more outcomes				

NCT02496689	Expanded access program for asfotase alfa treatment for patients with infantile- or juvenile-onset HPP	Approved for marketing	Expanded access	Enrollment:	Alexion Pharm.	Study start: NA	Colorado Center for Bone Research, Lakewood, CO, USA
				Age: child, adult, older adult		First posted: Jul. 14, 2015	
				Sex: M and F		Last update posted: Oct. 25, 2017	Children's Hospital of Pittsburg, Pittsburg, PA, USA
							Hôpital Bicêtre, Le Kremlin- Bicêtre, France
							And 2 more
NCT02456038	Safety and efficacy of asfotase alfa in patients with HPP	Completed	Interventional phase 2	Enrollment: 13	Translational Research Center for Medical Innovation, Kobe, Hyogo, Japan	Study start: August 2014	Kurume University Hospital, Kurume, Fukuoka, Japan
			Single group assignment	Age: child, Adult, older adult		First posted: May 28, 2015	
			Masking: none (open label)			Last update posted: March 31, 2016	National Hospital Organization, Nagara Medical Center, Gifu, Japan
			Primary purpose: treatment				
			Outcome measures:	Sex: M and F	Osaka University Graduate School of Medicine, Osaka, Japan	*RESULTS*	Hiroshima University Hospital, Hiroshima, Japan
			Number of subjects AEs as an assessment of safety of repeated SC injections of asfotase alfa			Improved skeletal, respiratory and clinical symptoms	And 9 more
			Overall survival			No serious AEs (Kitaoka et al. 2017)	
			Effect of asfotase alfa on skeletal manifestations of HPP as measured by X-rays RGI-C scale				
			And 4 more outcomes				

(continued)

Table 10.2 (continued)

NCT number	Titles	Recruitment	Characteristics	Population	Sponsor	Dates and Results	Locations
NCT01205152 (ENB-003-08)	Extension study of protocol ENB-002-08 (NCT00744042) – study of asfotase alfa in infants and young children with HPP	Completed	Interventional phase 2	Enrollment: 10	Alexion Pharm.	Study start: Apr. 2009	DuPont Hospital for Children, Wilmington, DE, USA
			Single group assignment	Age: 24 weeks to 42 months (child)		First posted: Sept. 20, 2010	St. John's Medical Research Institute, Springfield, MO, USA
			Masking: none (open label)	Sex: M and F		Last update posted: Nov.17, 2017	University of Nebraska Medical Center, Omaha, NE, USA
			Primary purpose: treatment			*RESULTS*	And 5 more
			Outcome measures:			Asfotase alfa mineralizes skeleton, including the ribs, improves respiratory function and survival in life-threatening perinatal and infantile HPP (Whyte et. al. 2016a, b; Whyte 2017a)	
			Long-term tolerability of SC asfotase alfa				
			Long-term efficacy of asfotase alfa in treating rickets in infants and young children				
			Long-term pharmacodynamics of asfotase alfa: plasma PPi and PLP levels				
			And 4 more outcomes				

NCT01203826	Extension study of protocol ENB-006-09 (NCT00952484) – Study of asfotase alfa in children with HPP	Completed	Interventional	Enrollment: 12	Alexion Pharm.	Study start: Apr. 2010	Shriners Hospital for Children, Saint Louis, MO, USA
			Phase: phase 2				
			Single group assignment	Age: 5–12 years (child)		First posted: Sept. 16, 2010	Children's Hospital Health Sciences Centre, Winnipeg, Manitoba, Canada
			Masking: none (open label)			Last update posted: Nov. 9, 2017	
			Primary purpose: treatment	Sex: M and F		*RESULTS*	
			Outcome measures:			With asfotase alfa significant X-rays and significant growth, strength, motor function, agility, and quality of life improvements, reduction in pain and disability (Whyte et al. 2016a)	
			Skeletal radiograph RGI-C scale compared to baseline (pre-treatment) in study				
			ENB-006-09				

(continued)

Table 10.2 (continued)

NCT number	Titles	Recruitment	Characteristics	Population	Sponsor	Dates and Results	Locations
NCT01176266 (ENB-010-10)	Open-label study of asfotase alfa in infants and children ≤5 years of age with HPP	Completed	Interventional	Enrollment: 69	Alexion Pharm.	Study start: Jul. 2010	Children's Hospital & Research Center Oakland, Oakland, CA, USA
			Phase 2 and 3	Age: up to 5 years (child)		First posted: Aug 5, 2010	
			Single group assignment	Sex: M and F		Last update posted: Feb 26, 2018	Cincinnati Children's Hospital Medical Center, Cincinnati, OH, USA
			Masking: none (open label)				
			Primary purpose: treatment				
			Outcome measures:			*RESULTS*	Children's Hospital of Pittsburg, Pittsburg, PE, USA
			Effect of asfotase alfa on skeletal manifestations of HPP			Improved survival rates (Whyte et al. 2016b)	
			Safety and tolerability of repeated SC injections of asfotase alfa				And 20 more
			Effect of asfotase alfa on ventilator-free survival (week 312)				
			And 11 more outcomes				
NCT01163149	Safety and efficacy study of asfotase alfa in adolescents and adults with HPP	Completed	Interventional phase 2	Enrollment: 19	Alexion Pharm.	Study start: Jun. 2010	Shriner's Hospital for Children, Saint Louis, MO, USA
			Randomized	Age: 13–65 years (child, adult, older adult)		First posted: Sept. 18, 2017	Duke University Medical Center, Durham, NC, USA
			Parallel assignment			Last update posted: Sept. 18, 2017	
			Masking: none (open label)				
			Primary purpose: treatment	Sex: M and F		RESULTS unpublished	Health Sciences Centre, Winnipeg University, Winnipeg, Manitoba, Canada
			Outcome measures:				
			Change from baseline to week 24 for plasma PLP and PPi				
			Safety and tolerability of asfotase alfa				
			And 6 more outcomes				

NCT00952484 (ENB 006-09)	Safety and efficacy of asfotase alfa in juvenile patients with HPP	Completed	Interventional phase 2	Enrollment: 13	Alexion Pharm.	Study start: Sept. 2009	Shriners Hospital for Children, St. Louis, MO, USA
			Randomized	Age: 5–12 years (child)		First posted: Aug. 6, 2009	The University of Manitoba Health Services Centre, Winnipeg, Manitoba, Canada
			Parallel assignment			Last update posted: May 2, 2016	
			Masking: none (open label)				
			Primary purpose: treatment	Sex: M and F			
			Outcome measures:			*RESULTS*	
			Change in rickets severity on skeletal radiographs from baseline to week 24 measured by RGI-C scale			With asfotase alfa significant X-rays and significant growth, strength, motor function, agility, and quality of life improvements, reduction in pain and disability (Whyte et al. 2016a)	
			Change in osteomalacia: osteoid thickness, osteoid volume/bone volume and mineralization lag time, measured by trans-iliac crest bone biopsy				
			And 8 more outcomes				

(continued)

Table 10.2 (continued)

NCT number	Titles	Recruitment	Characteristics	Population	Sponsor	Dates and Results	Locations
NCT00744042 (ENB002-08)	Safety and efficacy study of asfotase alfa in severely affected infants with HPP	Completed	Interventional phase 1 and 2	Enrollment: 11	Alexion Pharm.	Study start: Sept. 2008	Arkansas Children's Hospital, Little Rock, AR, USA
			Single group assignment	Age: up to 36 months (child)		First posted: Aug. 29, 2008	Du Pont Hospital for Children, Wilmington, DE, USA
			Masking: none (open label)	Sex: M and F		Last update posted: Feb. 22, 2016	St. John's Hospital, Springfield, MO, USA
			Primary purpose: treatment			*RESULTS*	And 7 more
			Outcome measures:			Improved skeletal mineralization, respiratory and physical function (Whyte et al. 2012), improved survival rates (Whyte et al. 2016a, b; White 2017a) (including extension NCT01205152)	
			Change in rickets severity from baseline to week 24 based on change in RGI-C				
			Maximum serum concentration of asfotase alfa (Cmax)				
			Time at maximum serum concentration of asfotase alfa (Tmax)				
			Area Under serum concentration-time curve to last measurable concentration of asfotase alfa (AUCt)				

NCT00739505	Safety study of human recombinant tissue non-specific alkaline phosphatase fusion protein asfotase alfa in adults with HPP adults with HPP	Completed	Interventional phase 1 Non-randomized Single group assignment Masking: none (open label) Primary purpose: treatment Outcome measures: Safety and tolerability of asfotase alfa given intravenously (IV) and given SC Pharmacokinetics (PK) of asfotase alfa given IV and SC to assess bioavailability of the SC asfotase alfa	Enrollment: 6 Age: 18–80 years (adult, older adult) Sex: M and F	Alexion Pharm.	Study start: Aug. 2008 First posted: Aug. 21, 2008 Last update posted: May 29, 2014 RESULTS unpublished	Barnes Jewish Hospital-Washington University School of Medicine, St. Louis, MO, USA Duke University Medical Center, Durham, NC, USA Department of Pediatrics & Child Health Sciences Centre Winnipeg University of Manitoba, Winnipeg, Manitoba, Canada
NTC02531867	Post-approval clinical study of asfotase alfa treatment for patients with HPP in Japan	Completed	Interventional Phase 4 Single group assignment Masking: none (open label) Primary purpose: treatment Outcome measures: Number of participants with AEs including injection site reactions (ISRs) and injection associated reactions (IARs)	Enrollment: 13 Age: child, adult, older adult Sex: M and F	Alexion Pharm.	Study start: June 2015 First posted: May 6, 2015 Last update posted: May 8, 2017 RESULTS unpublished	

(continued)

Table 10.2 (continued)

NCT number	Titles	Recruitment	Characteristics	Population	Sponsor	Dates and Results	Locations
NCT00894075	Safety and efficacy study of ENB-0040 in juvenile patients with HPP	Withdrawn	Interventional	Enrollment: 0	Alexion Pharm.	Study start: July 2019	
			Phase 2	Age: 5–12 years (child)		First posted: Aug. 25, 2015	
			Single group assignment			Last update posted: Jan. 25 2013	
			Masking: none (open label)	Sex: M and F			
			Primary purpose: treatment				
			Outcome measures:				
			Skeletal radiographs using CGI-C scoring system				
			PK using serum peak and trough levels and PD of PPi and PLP as biomarkers for HPP				
NCT02306720	An observational, longitudinal, prospective, long-term registry of patients with HPP	Enrolling by invitation	Observational	Enrollment: 500	Alexion Pharm.	Study start: Dec. 2014	Little Rock, AR, USA
			Case-control	Age: child, adult, older adult		First posted: Dec. 3 2014	Centennial, CO, USA
			Prospective			Last update posted: Aug. 17 2017	Lakewood, CO, USA
			Outcome measures:				
			Patients-reported outcomes	Sex: M and F			And 52 more
NCT02751801	Health Burden of HPP	Recruiting	Observational	Enrollment: 17	Sheffield Teaching Hospitals NHS Foundation Trust	Study start: Nov. 2016	Academic Unit of Bone Metabolism, Sheffield, South Yorks, UK
			Case-only	Age: up to 100 years, child, adult, older adult		first posted: Apr. 26 2016	
			Retrospective				
			Outcome measures:	Sex: M and F		Last update posted: Nov. 24 2017	
			Description of healthcare use				
			Healthcare use appointments				
			Healthcare cost				

From ClinicalTrials.gov, as of August 23, 2018

HPP hypophosphatasia, *Alexion Pharm* Alexion Pharmaceuticals, *AEs* adverse events, *NA* not available, *SC* subcutaneous, *RGI-C scale* qualitative radiographic global impression of change, *PPi* inorganic pyrophosphate, *PLP* pyridoxal-5-phosphate, *PD* pharmacodynamics, *PK* pharmacokinetics

10.3.1 Preclinical Studies

In the first published preclinical study, TNSALP-null *Akp2*$^{-/-}$ mice (i.e. mice deprived of the murine TNSALP gene), a good animal model for infantile HPP (Fedde et al. 1999) were treated with daily subcutaneous injections of ENB-0040 at different doses (1, 2, or 8.2 mg/kg/day) for up to 52 days after birth. Seventy-five percent of those receiving the highest doses survived and showed normal physical activity, normal growth, no skeletal or dental alterations, and no epilepsy, while the median survival of untreated mice was only 18.5 days. Plasma levels of calcium, PPi and pyridoxal remained within normal range (Millán et al. 2008).

Another study on the same animal model investigated the dose-response relationship with different doses (0.5, 2.0, or 8.2 mg/kg/day) of subcutaneous ENB-0040 for 43 days after birth. Bone defects were prevented in 80% of mice with doses of 2.8–3.2 mg/kg/day, depending on the skeletal site. Long bones apparently responded positively to lower doses. Overall, bone mineralization, bone length, body weight and median survival improved with increasing doses. Interestingly, the authors observed that urinary PPi (a good marker for the diagnosis of HPP) is probably not useful in the follow-up of ERT, since it remained elevated in all treated groups (Yadav et al. 2011). Further studies on *Akp2*$^{-/-}$ mice reported that ENB-0040 can prevent hypomineralization of alveolar bone, dentin, and cementum, as well as enamel defects (McKee et al. 2011; Foster et al. 2013).

10.3.2 Studies in Infants and Children

Following these preliminary positive results in animal models, ENB-0040 was investigated in infants and children with very severe forms of HPP.

In 2012, Rodriguez et al. used ENB-0040 in a preterm 3-week-old infant with HPP, with severe respiratory insufficiency requiring mechanical ventilation since birth. After 12 weeks of treatment, lung function and chest wall mechanics were improved with respect to a baseline evaluation at age 8 weeks (after 5 weeks of treatment), and oxygen flux could be reduced. At age 32 weeks, the infant could be discharged from the hospital, but unfortunately, only 2 weeks later, had to be rehospitalized because of fever, hypoxemia, pneumonia and shock, and in a few days he died from septic shock with multi-organ failure, disseminated intravascular coagulopathy, hypoxemic respiratory failure and acute respiratory distress syndrome (Rodriguez et al. 2012).

In the same year, the first results of a Phase 2 multinational open-label study of ERT treatment with ENB-0040 in infants and young children with life-threatening HPP (ENB-002-08; ClinicalTrials.gov number: NCT00744042) were published. The study enrolled 11 patients (7 girls and 4 boys, aged 2 weeks to 3 years; 5 with perinatal HPP and 6 with infantile HPP). All children presented typical HPP symptoms (failure to thrive, severe hypomineralization, fractures, substantial motor delay or regression) within the first 6 months of life. Those with erupted teeth had lost dentition, and all but two had nephrocalcinosis or renal stones. Ten of the 11 patients required some respiratory support (5 at baseline and 5 at a later time) during the study. They were treated with a single intravenous infusion (2 mg/kg) of ENB-0040 initially, followed by subcutaneous injections three times a week (1 mg/kg, to be increased up to 3 mg/kg in case of worsening pulmonary function, failure to thrive or absence of skeletal improvement). One child was withdrawn after the initial intravenous dose, because of disturbing symptoms. Ten children completed 6 months of treatment and nine completed 1 year. One of them died from sepsis (unrelated to treatment) during the 7th month of treatment. Skeletal healing became apparent within the first 24 weeks of treatment in all patients but one, with increased BMC (bone mineral content), improved endochondral and membranous bone formation, fracture healing, reduced deformity, resolution of radiolucencies and sclerosis, improved bone modeling and remodeling. One patient who had no visible bone at baseline did not have an early response to treatment, but showed some remineralization at 48 weeks. After 6 months of therapy, nine patients showed healing of rickets, as well as improvement in pulmonary function and developmental milestones. Serum PTH had an increasing trend in parallel with bone mineralization, allowing liberalization of dietary calcium intake. Serum calcium and phosphate levels showed only transient minor fluctuations. Plasma levels of TNSALP substrates (PPi and PLP) decreased; in particular, PPi (measured in five patients) significantly decreased over the 48 weeks of treatment. Apparently, ERT had no effect on craniosynostosis and two patients required cranial surgery. The most common ERT-related adverse event was mild, transient erythema localized at the injection site. Nephrocalcinosis did not progress after the first 6 months of treatment and even improved in some patients. No ectopic calcifications were observed. The increased bone mineralization process did not lead to hypocalcemia ("hungry bones"). There were no signs of drug hypersensitivity or tachyphylaxis, and only four patients developed low titers of anti-ENB-0040 antibodies, with no evident clinical, biochemical, or autoimmune abnormalities at 48 weeks of treatment. Three serious adverse events in three patients (one each of respiratory distress, craniosynostosis, conductive hearing loss), however, were considered as possibly related to the treatment (Whyte et al. 2012a).

In 2014, the results of an extension (NCT01205152) of the previous study were presented at a medical conference in Canada. The nine patients who had completed the 1st year of treatment, continued for two more years (until month 36). Asfotase alfa (the ENB-0040 was given the new name in the meantime) was well tolerated, and the improvement in skeletal mineralization continued. Only one patient continued to need supplemental oxygen at the last assessment after 3 years. Survival at 3 years was 90%, while in previous studies in comparable populations it was less than 50% (Whyte et al. 2014). Still later, Whyte communicated that after more than 5 years of treatment, these patients were showing further improvements and none was requiring respiratory support (Whyte 2017a), and that similarly positive results were being observed in an ongoing study on a larger number of comparable patients (Liese et al. 2016).

At another 2014 medical conference, Rockman-Greenberg et al. presented the results of an ongoing Phase II, open-label, global, multicenter trial (NCT01176266) on 15 children with HPP (age range 0.1–304 weeks; N = 8 age <22 weeks; N = 7 age 168–304 weeks). The patients received asfotase alfa as a subcutaneous injection (initial dose 2 mg/kg, three times per week; with adjustments permitted). The 13 patients with at least 6 months of follow-up data were evaluated. Significantly improved bone mineralization was observed at week 12 in 6/14 patients (42.9%), at week 24 in 8/12 patients (66.7%), at week 36 in 10/11 patients (90.9%) and at week 48 in 10/10 patients (100%). Eight of the 15 patients required some respiratory support during the study (5 at baseline and 3 at a later time); 4 of them still needed respiratory support at the last assessment. The overall survival was 93% (one patient withdrew on day 16 and died on day 22, from disease-related complications, not attributed to the drug). There were no serious drug-related adverse events, the most common events being mild to moderate reactions at injection site (Rockman-Greenberg et al. 2014).

After 2014, the results of further Phase 2 clinical trials of asfotase alfa in infants and adolescents with perinatal, infantile, or childhood HPP were presented at medical conferences and published in medical journals, confirming the positive effects of ERT on bone mineralization, growth, respiratory function, and mobility.

The first of these trials (NCT00952484, extended by NCT01203826) evaluated the efficacy and safety of asfotase alfa treatment in children aged 6–12 years, with rickets and impaired physical function due to HPP. Thirteen children were initially enrolled, one withdrew, 12 continued treatment for up to 5 years. Significant improvement in bone mineralization was observed after 6 months of treatment, then further improvements occurred and persisted after 5 years. In particular, 9/13 patients (69%) could be classified as "responders" (i.e. obtaining a

Radiographic Global Impression of Change (RGI-C) score ≥2) after 6 months of therapy, compared to 1/16 (6%) in the control group. The percentage of treated patients classified as "responders" increased to 75% after 2 years of treatment, 88% after 3 years, and 92% after 5 years. Overall, growth, strength, motor function, agility, and quality of life improved, in many cases achieving normal values for age- and sex-matched peers, and were sustained at 5 years. Pain and disability resolved in most cases. The most common adverse effects were mild to moderate reactions at injection site. No clinically important ectopic calcification or treatment resistance were observed. Low titers of anti–asfotase alfa antibody appeared in all patients. These results confirmed that ERT with asfotase alfa has good and sustained efficacy and a good safety profile for children suffering from HPP (Fig. 10.6) (Whyte et al. 2016a).

In another article (Whyte et al. 2016b), the survival rates observed in two ongoing, multicenter, open-label, phase 2 interventional studies of asfotase alfa treatment in neonates and infants with severe HPP (NCT00744042/NCT01205152 and NCT01176266) were compared with those observed in a retrospective, multinational, non-interventional, natural history study (NCT01419028) of patients with perinatal or infantile HPP with rachitic chest deformity, respiratory compromise, or vitamin B6–dependent seizures, used as controls. Thirty-seven patients were included in the two ERT studies (median treatment duration, 2.7 years) and 48 (with similar age and HPP severity) in the historical control study. The main outcome measures were survival, skeletal health (radiographically evaluated), and ventilatory status. Ninety-five percent (35/37) of treated patients were alive at 1 year of age versus 42% (20/48) of the historical controls, 84% (31/37) versus 27% (13/48) at 5 years of age. The median survival for the historical controls was 8.9 months, with respiratory complications or failure as the primary cause of death. The better respiratory outcome was accompanied by radiographically demonstrated improvements in skeletal mineralization. In particular, only 5% (1/20) of the historical controls who required respiratory assistance survived, while 76% (16/21) of the ventilated treated patients survived, and 75% of them (12/16) could be weaned from respiratory support. These results confirmed that asfotase alfa significantly improves skeletal mineralization, respiratory

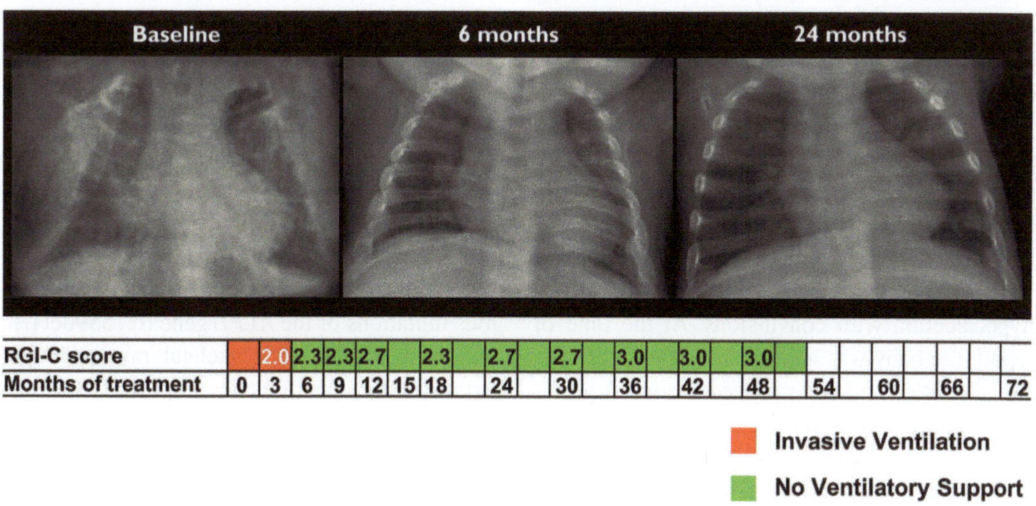

Fig. 10.6 Radiographic changes with asfotase alfa treatment in a patient with perinatal/infantile HPP. Better rib mineralization, chest structure, and thoracic volume with improved ventilatory status in an infant with hypophosphatasia (5.1 weeks of age at treatment baseline). RGI-C, Radiographic Global Impression of Change scale. (From Whyte et al. 2016b, reproduced with permission)

function and prolong survival in life-threatening perinatal and infantile HPP.

Another 1-year open-label, multicentre, prospective trial (NCT02456038) included 13 patients (9 females, 4 males; age 0 days-34 years) with *ALPL* gene mutations: 6 patients had perinatal HPP, 5 infantile HPP, 1 childhood HPP, 1 adult HPP. All but one were treated with asfotase alfa (2 mg/kg three times weekly). The primary outcome measure was safety, determined on the basis of observed adverse events (AEs), and the secondary outcome measure was efficacy, determined on the basis of overall survival, respiratory status, rickets severity and gross motor development. The more severe AEs (convulsion and hypocalcaemia, possibly related to treatment) were observed in one patient with perinatal HPP. Three patients with infantile HPP had hypercalcemia and/or hyperphosphatemia, requiring low-calcium and/or low-phosphate formula. The most frequent AEs were injection site reactions. Regarding efficacy, all patients survived, with improvements in radiographic findings, respiratory function, and development (Kitaoka et al. 2017).

To complete this survey, a few interesting case reports on ERT in perinatal or infantile HPP deserve mention.

Okazaki et al. (2016) report a very interesting case of a female infant with perinatal lethal HPP diagnosed in utero. At birth, she had severe skeletal hypomineralization and severe respiratory insufficiency requiring invasive ventilation and deep sedation. She was treated with asfotase alfa, starting the day after birth, and improvements in bone mineralization were visible at 3 weeks of age. She required calcium supplementation for the first 3 months of treatment, to correct serious hypocalcemia with convulsions. At the time of report, the baby was no more requiring mechanical ventilation and had survived for over 1 year.

In another case report, a dramatic clinical improvement was observed in a girl with infantile HPP, treated with asfotase alfa for more than 5 years (Fig. 10.7) (Simm and Savarirayan 2017).

A similar case was very recently reported by Oyachi et al. (2018). A newborn girl presented with respiratory insufficiency and seizures. Low ALP activity and high pyridoxal phosphate levels were measured in the umbilical cord blood, and severe rickets-like bone alterations were observed radiologically. She was treated with mechanical ventilation and pyridoxine hydrochloride. Perinatal severe HPP was diagnosed, and asfotase alfa treatment was started at 6 days of age. Genetic analysis revealed compound heterozygous mutations of the *ALPL* gene (c.1559delT/p. Ser188Pro). With ERT, skeletal mineralization and respiratory insufficiency improved. Seizures ceased and pyridoxine hydrochloride was tapered off at age 1 year. There were no remarkable side-effects.

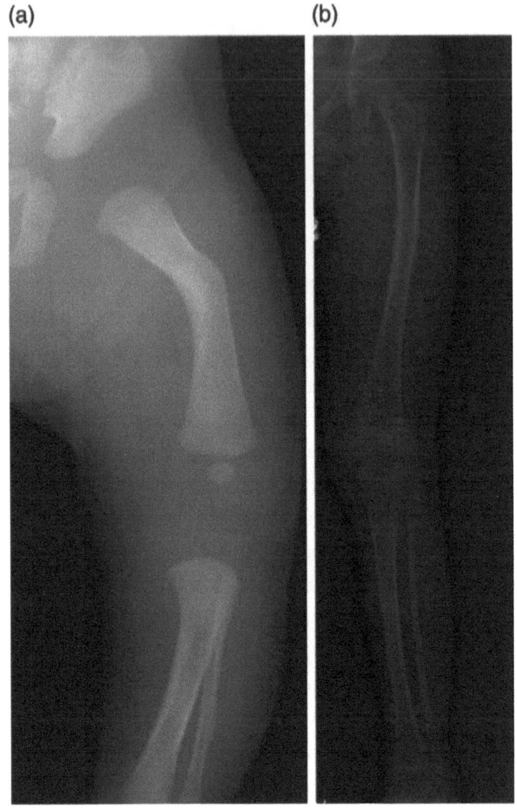

Fig. 10.7 Comparison of left femur, from (**a**) 2 months of age, before use of enzyme replacement therapy and (**b**) at age 5 years 10 months during asfotase alfa, demonstrating significant reduction in femoral bowing. (From Simm and Savarirayan 2017, reproduced with permission)

Finally, on the negative side, Costain et al. (2017) describe in great detail the case of a female baby with perinatal HPP for whom ERT with asfotase alfa was not successful. The baby was born with signs of poor bone mineralization and severe respiratory insufficiency, that required ventilation by continuous positive airway pressure (CPAP). On day 13, after a diagnosis of perinatal HPP, she was put on ERT with asfotase alfa. In the following months, notwithstanding some radiological improvement in bone mineralization, her chest wall and lungs failed to grow, she could never be weaned from mechanical ventilatory assistance and her general condition remained very poor. Eventually, the multidisciplinary healthcare team and the parents, after a difficult evaluation of the potential benefits versus the ongoing harms, made a decision to stop ERT, extubate, and provide only supportive care. On day 100, the baby died.

10.3.3 Studies in Adolescents and Adults

Regarding the less severe forms of HPP, there are only very few studies and the benefits of treatment with asfotase alfa have not yet been demonstrated in mild childhood HPP, adult HPP, and odontoHPP.

A poster (Kishnani et al. 2012) and a short communication (Whyte et al. 2012b) is all that we could find about an open-label, multicenter, randomized, controlled Phase II study of the safety and efficacy of asfotase alfa in 6 adolescents and 13 adults with HPP (mean age 42 years, range 14–68 years) (NCT01163149). Thirteen subjects received daily subcutaneous injection of asfotase alfa (dose 2.1 (N = 7) or 3.5 (N = 6) mg/kg/week) and six received no treatment (controls). After 24 weeks of treatment with asfotase alfa, TNSALP substrate levels (serum PPi and serum PLP) were significantly decreased, motor function (6-min walk test) was also improved, although by only 26 m on average, and treatment was well tolerated, with no serious adverse events related to treatment. Hofmann et al. (2016) cited an extension of this study in which all these subjects (including the non-treated controls) were treated with 3.5 mg/kg/week of asfotase alfa for another 24 weeks, then with 1 mg/kg/day 6 days/week for an additional 48 weeks (or until regulatory approval of the drug), but no data on this extension have apparently been published.

There are very few other data on adults. Remde et al. (2017) described a hemodialyzed 59-year old woman with childhood-onset HPP and a history of multiple fractures and orthopedic procedures that after 13 months of treatment with asfotase alfa had a dramatic increase in quality of life, increased mobility, reduction in pain drugs, and significant improvement in bone mineralization, with consolidation of existing fractures and no new fractures.

At the time of writing, asfotase alfa has not been approved for use in adult HPP, although some possible benefits could be expected, at least in selected cases, on the basis of the experience in children and the few data available for older patients.

Shapiro and Lewiecki (2017) discuss the problems and perspectives of ERT with asfotase alfa in adults, and present some suggestions about the possible criteria for treatment. The extent of HPP-related disability should always be a major criterion. Most important is a history of childhood involvement (before age 18 years), such as early loss of primary or secondary teeth, craniosynostosis, gait disturbance, developmental delays, skeletal deformity, bone pain or fractures. In the absence of childhood symptoms, these other criteria are suggested: musculoskeletal pain requiring prescription pain medications; disabling polyarthropathy and/or chondrocalcinosis; major low-trauma fractures (e.g., spine, hip, humerus, pelvis) attributable to HPP; delayed or incomplete fracture healing or fracture non-union, especially if requiring orthopedic surgery; disabling functional impairment (e.g., mobility, gait, activities of daily living) assessed by validated measures; low bone mineral density (BMD) (T-score ≤ -2.5 in postmenopausal women and men aged ≥ 50 years; or Z-score ≤ -2.0 in younger adults) by dual-energy X-ray absorptiometry (DXA); radiological evidence of nephrocalcinosis.

The benefits and risks of asfotase alfa ERT in adults are still undetermined, as well as the optimal dose, need for dose adjustments, and duration of therapy. Adverse effects of long-term ERT are still unknown. The currently high cost of therapy is obviously an important aspect and might justify the use of asfotase alfa only in severe cases of HPP.

10.3.4 Adverse Reactions to Asfotase Alfa

The adverse reactions to asfotase alfa (Strensiq®) reported in clinical trials (on a total of 79 patients with perinatal or infantile-onset HPP, and 20 patients with juvenile-onset HPP) are presented in detail in the Strensiq® prescribing information document (Strensiq® Alexion Prescription Information 2018) and are discussed in some review articles (Hofmann et al. 2016; Kishnani et al. 2017).

The most common reactions were injection site reactions (in 63% of patients). Other common adverse reactions included lipodystrophy (28%), ectopic calcifications (14%), and hypersensitivity reactions (12%). Injection site reactions, lipodystrophy and ectopic calcification were higher in patients with juvenile-onset HPP than in patients with perinatal/infantile-onset. Most injection-site reactions and injection-associated reactions were nonserious, mild to moderate, and resolved within a week. They were more common in patients receiving six injections per week compared with those receiving only three injections per week. Localized lipodystrophy (atrophy or hypertrophy) was observed after several months of injections. Rotation of injection site seems to reduce the risk of site reactions. Ectopic calcifications were observed in eyes (including cornea and conjunctiva) and kidneys (nephrocalcinosis), without evidence of visual disturbances or changes in renal function. There was insufficient information to determine whether these manifestations were disease- or drug-related.

Hypersensitivity reactions (12% of patients) included vomiting, erythema, fever, irritability, nausea, pain, rigor/chills, oral hypoesthesia, headache, flushing, and anaphylaxis (1%). Less common adverse reactions (reported in <1%) were hypocalcemia, renal stones, chronic hepatitis, decreased vitamin B6.

Since increased levels of parathyroid hormone (PTH) and decreased levels of calcium have been reported during treatment with asfotase alfa, particularly in the first 12 weeks, PTH and calcium levels should be monitored during treatment and supplementation with calcium and/or vitamin D may be indicated (Hofmann et al. 2016).

10.3.5 Immunogenicity

Being a protein, asfotase alfa has a potential for immunogenicity. In clinical trials, antibodies to asfotase alfa have been found in 76 (78%) out of 98 patients tested. Of these, 34 (45%) had neutralizing antibodies. However, the development of antibodies has not been associated with clinical resistance to the drug (Hofmann et al. 2016; Whyte 2017a).

10.3.6 Interactions

There are no data on the interactions of asfotase alfa with other therapeutic drugs (Hofmann et al. 2016).

10.4 The Goals of Therapy with Asfotase Alfa and Its Monitoring

The treatment of HPP patients with asfotase alfa poses different problems depending on the disease subtype and severity as well as on the patient's history and clinical manifestations. In younger patients, the main treatment goals are improved skeletal mineralization, prevention and/or control of complications (seizures, respiratory insufficiency, renal failure, neurosurgical problems), reduction of pain, improved growth and development, improved mobility. In particular, in perinatal or infantile HPP, where survival is the primary goal, respiratory function and seizure control are urgent problems, and ventilatory support is often required. In childhood HPP, depending on disease severity, longer-term goals, such as growth, mobility, and bone mineralization, can be pursued. In adults, HPP-related fractures are the most important problem, and treatment should aim at improving fracture healing and reducing the risk of new fractures. Other treatment goals are functional improvement (strength, endurance, fatigue, pain). Oral health is another important treatment goal for patients of all ages, and overall quality of life should of course be the guiding light of the caregivers.

In the absence of published guidelines, the monitoring of asfotase alfa therapy in the different forms of HPP and the different age groups was discussed by an international panel of experienced physicians, convened in 2016 by the producer of asfotase alfa, Alexion Pharmaceuticals Inc. The conclusions and recommendations of this important meeting (regarding the type and frequency of clinical and radiological evaluations, laboratory tests, and assessments of efficacy/safety to be performed during the course of treatment) are presented in a consensus report by Kishnani et al. (2017).

10.5 Challenges and Future Perspectives

Asfotase alfa treatment for HPP is now widely recognized, and hopefully, the accumulation of experience with it will not only increase the benefits for patients, but will also help us develop a better understanding of the pathophysiology of HPP and its extremely variable picture and severity. There are, however, many aspects that require further investigations, and as happens with all new treatments, problems and challenges have emerged.

10.5.1 Problems and Challenges

Treatment with asfotase alfa is a delicate matter, and before starting it, a correct diagnosis of HPP must be made (for diagnostic challenges, see above Sect. 10.1.6). The inappropriate use of this enzyme in conditions other than HPP could result in serious adverse consequences from excessive mineralization.

When asfotase alfa therapy is indicated, as in perinatal, infantile, and severe childhood HPP, a timely initiation is important to minimize the risks from respiratory and other complications and the need for intensive care. Perhaps in utero therapy of perinatal HPP will be possible in the future (Whyte 2017a).

Monitoring and follow-up are important. The dose of asfotase alfa, which is strictly based on the patient's weight, must be regularly adjusted because, should it become insufficient, clinical and radiological worsening may soon recur (Whyte 2017a).

Ectopic calcifications are a theoretically possible adverse effect of treatment with asfotase alfa. Should future clinical experience demonstrate the appropriateness of long-term or even life-long treatment regimens, the risk of vascular or cardiac valve calcifications will have to be considered as the patients' age increases, and deserves specific investigations (Hofmann et al. 2016; Whyte 2017a). There is also some evidence that high alkaline phosphatase activity could be associated with degenerative diseases of the central nervous system, and this risk must also be taken into account (Hofmann et al. 2016).

Throughout life, the physiological production of ALP varies according to skeletal growth and development, bone modeling and remodeling activity, and fracture repair: whether dose adaptation according to the physiological needs is appropriate has to be determined by further studies. Also, the possibility of intermittent treatment regimens in milder forms of HPP, which might require ERT only in challenge situations like fractures or implant surgery, should be investigated (Hofmann et al. 2016).

10.5.2 Future Perspectives

Currently, asfotase alfa treatment requires three or six administrations per week, with up to two subcutaneous injections per administration if high doses are needed. The half-life of asfotase alfa following subcutaneous administration is 5 days (Strensiq® Alexion Prescription Information 2018). Some future drug modification might allow to increase the interval between injections (Orimo 2016).

Bone marrow stem-cell transplantation has been attempted in the past, with little success (Whyte et al. 2003; Taketani et al. 2015), but the efficacy and safety of combination treatment of asfotase alfa and stem-cell transplantation may deserve evaluation (Orimo 2016).

Finally, ex-vivo gene therapy using bone marrow stem cells is also being evaluated, and has already been successfully attempted in TNSALP-null *Akp2*$^{-/-}$ mice (Yamamoto et al. 2011; Matsumoto et al. 2011). Use of a viral vector containing the *ALPL* gene is also being explored, and an experiment with TNSALP-null *Akp2*$^{-/-}$ mice has already been made with positive results (Iijima et al. 2015). Of course, this is only a very preliminary step, because the safety of vector administration must be demonstrated before its use in humans, and the vector with the highest transduction efficacy must be identified (Orimo 2016).

10.6 Conclusions

HPP is a very complex, multi-systemic disease, with a very variable, multi-faceted presentation and a wide range of severity, from very mild to lethal. In its different forms, HPP can reveal itself at different ages, and may involve different organs (bone, tooth, muscle, lung, kidney, gastrointestinal tract, peripheral and central nervous system). Its pathophysiology is still not completely understood in all its aspects and the diagnosis can be difficult and is often delayed.

Once a diagnosis of HPP is established, ERT with asfotase alfa is the only effective and promising treatment currently available, especially for the severe forms of infancy and childhood. The first clinical trials with asfotase alfa have demonstrated clinical, radiographic, and biochemical improvements in infants and children with perinatal, infantile or childhood HPP, persisting after over 5 years of treatment in many cases.

Notwithstanding the inconvenience of frequent injections (with the related adverse reactions), this therapy has demonstrated a good safety profile, but its long-term efficacy and safety are unknown. The treatment must be carefully tailored to individual requirements and its management and follow-up are highly demanding tasks, requiring the involvement of an experienced and dedicated multidisciplinary team. Finally, the high cost of therapy is a major issue, that cannot be addressed here.

Some positive results with asfotase alfa have also been reported in adolescents and adults with milder forms of HPP, but the benefits of asfotase alfa have not been definitely demonstrated in these cases and alternative treatments could eventually prove preferable.

As a last note, considering the rarity of the disease, the development of medical centers with the required multidisciplinary expertise is a major issue, and underlines the necessity for a regular, timely sharing of clinical observations and experiences via national and international "rare diseases" networks, in order to build a sound evidence base on the natural history of HPP and the long-term impact of treatment.

References

Anderson HC (1995) Molecular biology of matrix vesicles. Clin Orthop Relat Res 314:266–280

Anderson HC, Sipe JB, Hessle L, Dhanyamraju R, Atti E, Camacho NP, Millán JL (2004) Impaired calcification around matrix vesicles of growth plate and bone in alkaline phosphatase- deficient mice. Am J Pathol 164(3):841–847

Atar M, Körperich EJ (2010) Systemic disorders and their influence on the development of dental hard tissues: a literature review. J Dent 38(4):296–306

Baumgartner-Sigl S, Haberlandt E, Mumm S, Scholl-Bürgi S, Sergi C, Ryan L, Ericson KL, Whyte MP, Högler W (2007) Pyridoxine-responsive seizures as the first symptom of infantile hypophosphatasia caused by two novel missense mutations (c.677T.C, p.M226T; c.1112C.T, p.T371I) of the tissue-nonspecific alkaline phosphatase gene. Bone 40(6):1655–1661

Beumer J III, Trowbridge HO, Silverman S Jr, Eisenberg E (1973) Childhood hypophosphatasia and the premature loss of teeth. A clinical and laboratory study of seven cases. Oral Surg Oral Med Oral Pathol 35(5):631–640

Bianchi ML (2015) Hypophosphatasia: an overview of the disease and its treatment. Osteoporos Int 26(12):2743–2757

Bishop N (2015) Clinical management of hypophosphatasia. Clin Cases Miner Bone Metab 12(2):170–173
Buchet R, Millán JL, Magne D (2013) Multisystemic functions of alkaline phosphatase. In: Millán JL (ed) Phosphatase modulators. Methods in molecular biology. Springer Science Business Media, Humana Press, New York, pp 27–51
Cahill RA, Wenkert D, Perlman SA, Steele A, Coburn SP, McAlister WH, Mumm S, Whyte MP (2007) Infantile hypophosphatasia: transplantation therapy trial using bone fragments and cultured osteoblasts. J Clin Endocrinol Metab 92(8):2923–2930
Camacho PM, Painter S, Kadanoff R (2008) Treatment of adult hypophosphatasia with teriparatide. Endocr Pract 14(2):204–208
Camacho PM, Mazhari AM, Wilczynski C, Kadanoff R, Mumm S, Whyte MP (2016) Adult hypophosphatasia treated with teriparatide: report of two patients and review of the literature. Endocr Pract 22(8):941–950
Caswell AM, Whyte MP, Russell RG (1991) Hypophosphatasia and the extracellular metabolism of inorganic pyrophosphate: clinical and laboratory aspects. Crit Rev Clin Lab Sci 28(3):175–232
Cole DEC (2008) Hypophosphatasia update: recent advances in diagnosis and treatment. Clin Genet 73(3):232–235
Collmann H, Mornet E, Gattenlohner S, Beck C, Girschick H (2009) Neurosurgical aspects of childhood hypophosphatasia. Childs Nerv Syst 25:217–223
Costain G, Moore AM, Munroe L, Williams A, Zlotnik Shaul R, Rockman-Greenberg C, Offringa M, Kannu P (2017) Enzyme replacement therapy in perinatal hypophosphatasia: case report of a negative outcome and lessons for clinical practice. Mol Genet Metab Rep 14:22–26
Cundy T, Michigami T, Tachikawa K, Dray M, Collins JF, Paschalis EP, Gamsjaeger S, Roschger A, Fratzl-Zelman N, Roschger P, Klaushofer K (2015) Reversible deterioration in hypophosphatasia caused by renal failure with bisphosphonate treatment. J Bone Miner Res 30(9):1726–1737
Deeb AA, Bruce SN, Morris AAM, Cheetham TD (2000) Infantile hypophosphatasia: disappointing results of treatment. Acta Paedriatr 89(6):730–743
Doshi KB, Hamrahian AH, Licata AA (2009) Teriparatide treatment in adult hypophosphatasia in a patient exposed to bisphosphonate: a case report. Clin Cases Miner Bone Metab 6(3):266–269
Fallon MD, Teitelbaum SL, Weinstein RS, Goldfischer S, Brown DM, Whyte MP (1984) Hypophosphatasia: clinicopathologic comparison of the infantile, childhood, and adult forms. Medicine (Baltimore) 63(1):12–24
Fauvert D, Brun-Heath I, Lia-Baldini AS, Bellazi L, Taillandier A, Serre JL, de Mazancourt P, Mornet E (2009) Mild forms of hypophosphatasia mostly result from dominant negative effect of severe alleles or from compound heterozygosity for severe and moderate alleles. BMC Med Genet 10:51
Fedde KN, Blair L, Silverstein J, Coburn SP, Ryan LM, Weinstein RS, Waymire K, Narisawa S, Millán JL, MacGregor GR, Whyte MP (1999) Alkaline phosphatase knock-out mice recapitulate the metabolic and skeletal defects of infantile hypophosphatasia. J Bone Miner Res 14(12):2015–2026
Fonta C, Negyessy L, Renaud L, Barone P (2005) Postnatal development of alkaline phosphatase activity correlates with the maturation of neurotransmission in the cerebral cortex. J Comp Neurol 486(2):179–196
Foster BL, Nagatomo KJ, Tso HW, Tran AB, Nociti FH Jr, Narisawa S, Yadav MC, McKee MD, Millán JI, Somerman MJ (2013) Tooth root dentin mineralization defects in a mouse model of hypophosphatasia. J Bone Miner Res 28(2):271–282
Fraser D (1957) Hypophosphatasia. Am J Med 22(5):730–746
Gagnon C, Sims NA, Mumm S, McAuley SA, Jung C, Poulton IJ, Ng KW, Ebeling PR (2010) Lack of sustained response to teriparatide in a patient with adult hypophosphatasia. J Clin Endocrinol Metab 95(3):1007–1012
Girschick HJ, Schneider P, Haubitz I, Hiort O, Collmann H, Beer M, Shin YS, Seyberth HW (2006) Effective NSAID treatment indicates that hyperprostaglandinism is affecting the clinical severity of childhood hypophosphatasia. Orphanet J Rare Dis 1:24
Golub EE (2009) Role of matrix vesicles in biomineralization. Biochim Biophys Acta 1790(12):1592–1598
Harmey D, Hessle L, Narisawa S, Johnson KA, Terkeltaub R, Millan JL (2004) Concerted regulation of inorganic pyrophosphate and osteopontin by akp2, enpp1, and ank: an integrated model of the pathogenesis of mineralization disorders. Am J Pathol 164(4):1199–1209
Hessle L, Johnson KA, Anderson HC, Narisawa S, Sali A, Goding JW, Terkeltaub R, Millan JL (2002) Tissue-nonspecific alkaline phosphatase and plasma cell membrane glycoprotein-1 are central antagonistic regulators of bone mineralization. Proc Natl Acad Sci U S A 99(14):9445–9449
Hofmann C, Girschick HJ, Mentrup B, Graser S, Seefried L, Liese J, Jakob F (2013) Clinical aspects of hypophosphatasia: an update. Clin Rev Bone Miner Metab 11(2):60–70
Hofmann C, Girschick H, Mornet E, Schneider D, Jakob F, Mentrup B (2014) Unexpected high intrafamilial phenotypic variability observed in hypophosphatasia. Eur J Hum Genet 22(10):1160–1164
Hofmann C, Seefried L, Jakob F (2016) Asfotase alfa: enzyme replacement for the treatment of bone disease in hypophosphatasia. Drugs Today (Barc) 52(5):271–285
Hoshi K, Amizuka N, Oda K, Ikehara Y, Ozawa H (1997) Immunolocalization of tissuc non-specific alkaline phosphatase in mice. Histochem Cell Biol 107(3):183–191
Iijima O, Miyake K, Watanabe A, Miyake N, Igarashi T, Kanokoda C, Nakamura-Takahashi A, Kinoshita H, Noguchi T, Abe S, Narisawa S, Millán JL, Okada T,

Shimada T (2015) Prevention of lethal murine hypophosphatasia by neonatal ex vivo gene therapy using lentivirally transduced bone marrow cells. Hum Gene Ther 26(10):801–812

Kermer V, Ritter M, Albuquerque B, Leib C, Stanke M, Zimmermann H (2010) Knockdown of tissue nonspecific alkaline phosphatase impairs neural stem cell proliferation and differentiation. Neurosci Lett 485(3):208–211

Kishnani PS, Rockman-Greenberg C, Whyte MP, Weber T, Mhanni A, Madson K, Reeves A, Mack K, Plotkin H, Kreher N, Landy H (2012) Hypophosphatasia: enzyme replacement therapy (Asfotase alfa) decreases TNSALP substrate accumulation and improves functional outcome in affected adolescents and adults. Poster presented at the 2012 American College of Medical Genetics and Genomics Annual Meeting, Charlotte, NC, USA, March 27–31, 2012

Kishnani PS, Rush ET, Arundel P, Bishop N, Dahir K, Fraser W, Harmatz P, Linglart A, Munn CF, Nunes ME, Saal HM, Seefried L, Ozono K (2017) Monitoring guidance for patients with hypophosphatasia treated with asfotase alfa. Mol Genet Metab 122(1–2):4–17

Kitaoka T, Tajima T, Nagasaki K, Kikuchi T, Yamamoto K, Michigami T, Okada S, Fujiwara I, Kokaji M, Mochizuki H, Ogata T, Tatebayashi K, Watanabe A, Yatsuga S, Kubota T, Ozono K (2017) Safety and efficacy of treatment with asfotase alfa in patients with hypophosphatasia: results from a Japanese clinical trial. Clin Endocrinol 87(1):10–19

Laroche M (2012) Failure of teriparatide in treatment of bone complications of adult hypophosphatasia. Calcif Tissue Int 90(3):250

Liese J, Hofmann C, Harmatz P, et al (2016) Efficacy and safety of asfotase alfa in patients with infantile hypophosphatasia treated for up to 3.5 years: results from a phase II, open-label, uncontrolled study. Proceedings of the 98th Annual Endocrine Society Meeting; Boston, MA; April 1–4, 2016

Linglart A, Biosse-Duplan M (2016) Hypophosphatasia. Curr Osteoporos Rep 14(3):95–105

Matsumoto T, Miyake K, Yamamoto S, Orimo H, Miyake N, Odagaki Y, Adachi K, Iijima O, Narisawa S, Millán JL, Fukunaga Y, Shimada T (2011) Rescue of severe infantile hypophosphatasia mice by AAV-mediated sustained expression of soluble alkaline phosphatase. Hum Gene Ther 22(11):1355–1364

McKee MD, Nakano Y, Masica DL, Gray JJ, Lemire I, Heft R, Whyte MP, Crine P, Millán JL (2011) Enzyme replacement therapy prevents dental defects in a model of hypophosphatasia. J Dent Res 90(4):470–476

Miao D, Scutt A (2002) Histochemical localization of alkaline phosphatase activity in decalcified bone and cartilage. J Histochem Cytochem 50(3):333–340

Millán JL (2013) The role of phosphatases in the initiation of skeletal mineralization. Calcif Tissue Int 93(4):299–306

Millán JL, Narisawa S, Lemire I, Loisel TP, Boileau G, Leonard P, Gramatikova S, Terkeltaub R, Camacho NP, McKee MD, Crine P, Whyte MP (2008) Enzyme replacement therapy for murine hypophosphatasia. J Bone Miner Res 23(6):777–787

Mornet E (2008) Hypophosphatasia. Best Pract Res Clin Rheumatol 22(1):113–127

Mornet E (2017) Genetics of hypophosphatasia. Arch Pediatr 24(5S2):5S51–5S56

Mornet E (2018) Hypophosphatasia. Metabolism 82:142–155

Mornet E, Nunes ME (2011) Hypophosphatasia. In: Pagon RA, Adam MP, Ardinger HH, Bird TD, Dolan CR, Fong CT, Smith RJH, Stephens K (eds) GeneReviews® [Internet] 1993–2014. Seattle. http://www.ncbi.nlm.nih.gov/books/NBK1150/

Mornet E, Yvard A, Taillandier A, Fauvert D, Simon-Bouy B (2011) A molecular-based estimation of the prevalence of hypophosphatasia in the European population. Ann Hum Genet 75(3):439–445

Okazaki Y, Kitajima H, Mochizuki N, Kitaoka T, Michigami T, Ozono K (2016) Lethal hypophosphatasia successfully treated with enzyme replacement from day 1 after birth. Eur J Pediatr 175(3):433–437

Olsson A, Matsson L, Blomquist HK, Larsson A, Sjodin B (1996) Hypophosphatasia affecting the permanent dentition. J Oral Pathol Med 25(6):343–347

Orimo H (2010) The mechanism of mineralization and the role of alkaline phosphatase in health and disease. J Nippon Med Sch 77(1):4–12

Orimo H (2016) Pathophysiology of hypophosphatasia and the potential role of asfotase alfa. Ther Clin Risk Manag 12:777–786

Orton NC, Innes AM, Chudley AE, Bech-Hansen NT (2008) Unique disease heritage of the Dutch-German Mennonite population. Am J Med Genet A 146A(8):1072–1087

Oyachi M, Harada D, Sakamoto N, Ueyama K, Kondo K, Kishimoto K, Izui M, Nagamatsu Y, Kashiwagi H, Yamamuro M, Tamura M, Kikuchi S, Akiyama T, Michigami T, Seino Y, Namba N (2018) A case of perinatal hypophosphatasia with a novel mutation in the ALPL gene: clinical course and review of the literature. Clin Pediatr Endocrinol 27(3):179–186

Rathbun JC (1948) Hypophosphatasia: a new developmental anomaly. Am J Dis Child 75:822–831

Remde H, Cooper MS, Quinkler M (2017) Successful Asfotase alfa treatment in an adult dialysis patient with childhood-onset hypophosphatasia. J Endocr Soc 1(9):1188–1193

Rockman-Greenberg C (2013) Hypophosphatasia. Pediatr Endocrinol Rev 10(Suppl 2):380–388

Rockman-Greenberg C, Vockley J, Harmatz P, Vallée M, Bedrosian CL, Hofmann C, Liese J (2014) Asfotase alfa improves skeletal mineralization and respiratory function in infants and young children with hypophosphatasia: results from up to 12 months' treatment. Poster presented at the 2014 American College of Medical Genetics & Genomics Annual Meeting, Nashville, TN, USA, March 25–29, 2014

Rodriguez E, Bober MB, Davey L, Zamora A, Li Puma AB, Chidekel A, Shaffer TH (2012) Respiratory mechanics in an infant with perinatal lethal hypo-

phosphatasia treated with human recombinant enzyme replacement therapy. Pediatr Pulmonol 47(9):917–922
Schalin-Jäntti C, Mornet E, Lamminen A, Välimäki MJ (2010) Parathyroid hormone treatment improves pain and fracture healing in adult hypophosphatasia. J Clin Endocrinol Metab 95(12):5174–5149
Schinke T, McKee MD, Karsenty G (1999) Extracellular matrix calcification: where is the action? Nat Genet 21(12):150–151
Seefried L, Baumann J, Hemsley S, Hofmann C, Kunstmann E, Kiese B, Huang Y, Chivers S, Valentin MA, Borah B, Roubenoff R, Junker U, Jakob F (2017) Efficacy of anti-sclerostin monoclonal antibody BPS804 in adult patients with hypophosphatasia. J Clin Invest 127(6):2148–2158
Shapiro JR, Lewiecki EM (2017) Hypophosphatasia in adults: clinical assessment and treatment considerations. J Bone Miner Res 32(10):1977–1980
Silvent J, Gasse B, Mornet E, Sire JY (2014) Molecular evolution of the tissue-nonspecific alkaline phosphatase allows prediction and validation of missense mutations responsible for hypophosphatasia. J Biol Chem 289(35):24168–24179
Simm PJ, Savarirayan R (2017) Successful use of enzyme replacement therapy in infantile hypophosphatasia. J Paediatr Child Health 53:925–926. (Letter to the Editor)
Smith M, Weiss MJ, Griffin CA, Murray JC, Buetow KH, Emanuel BS, Henthorn PS, Harris H (1988) Regional assignment of the gene for human liver/bone/kidney alkaline phosphatase to chromosome 1p36.1–p34. Genomics 2(2):139–143
Strensiq® Alexion Prescription Information (2018). Available at: http://alexion.com/Documents/Strensiq_USPI.aspx. Accessed 27 Aug 2018
Strensiq® Alexion Product Information (2018). Available at: http://strensiq.com/hcp/. Accessed 27 Aug 2018
Strensiq® EMA Product Information (2018). Available at: http://www.ema.europa.eu/ema/index.jsp?curl=pages/medicines/human/medicines/003794/human_med_001901.jsp&mid=WC0b01ac058001d124. Accessed 27 Aug 2018
Sutton RA, Mumm S, Coburn SP, Ericson KL, Whyte MP (2012) "Atypical femoral fractures" during bisphosphonate exposure in adult hypophosphatasia. J Bone Miner Res 27(5):987–994
Tadokoro M, Kanai R, Taketani T, Uchino Y, Yamaguchi S, Ohgushi H (2009) New bone formation by allogenic mesenchymal stem cell transplantation in a patient with perinatal hypophosphatasia. J Pediatr 154(6):924–930
Taketani T, Kanai R, Abe M, Mishima S, Tadokoro M, Katsube Y, Yuba S, Ogushi H, Fukuda S, Yamaguchi S (2013) Therapy related Ph+ leukemia after both bone marrow and mesenchymal stem cell transplantation for hypophosphatasia. Pediatr Int 55(3):e52–e55
Taketani T, Oyama C, Mihara A, Tanabe Y, Abe M, Hirade T, Yamamoto S, Bo R, Kanai R, Tadenuma T, Michibata Y, Yamamoto S, Hattori M, Katsube Y, Ohnishi H, Sasao M, Oda Y, Hattori K, Yuba S, Ohgushi H, Yamaguchi S (2015) Ex vivo expanded allogeneic mesenchymal stem cells with bone marrow transplantation improved osteogenesis in infants with severe hypophosphatasia. Cell Transplant 24(10):1931–1943
van den Bos T, Handoko G, Niehof A, Ryan LM, Coburn SP, Whyte MP, Beertsen W (2005) Cementum and dentin in hypophosphatasia. J Dent Res 84(11):1021–1025
Watanabe A, Karasugi T, Sawai H, Naing BT, Ikegawa S, Orimo H, Shimada T (2011) Prevalence of c.1559delT in *ALPL*, a common mutation resulting in the perinatal (lethal) form of hypophosphatasia in Japanese and effects of the mutation on heterozygous carriers. J Hum Genet 56(2):166–168
Weiss MJ, Ray K, Henthorn PS, Lamb B, Kadesch T, Harris H (1988) Structure of the human liver/bone/kidney alkaline phosphatase gene. J Biol Chem 263(24):12002–12010
Weninger M, Stinson RA, Plenk H Jr, Böck P, Pollak A (1989) Biochemical and morphological effects of human hepatic alkaline hosphatase in a neonate with hypophosphatasia. Acta Paediatr Scand Suppl 360:154–160
Wenkert D, Podgornik MN, Coburn SP, Ryan LM, Mumm S, Whyte MP (2002) Dietary phosphate restriction therapy for hypophosphatasia: preliminary observations. J Bone Miner Res 17(suppl 1):S384
Wenkert D, McAlister WH, Coburn SP, Zerega JA, Ryan LM, Ericson KL, Hersh JH, Mumm S, Whyte MP (2011) Hypophosphatasia: nonlethal disease despite skeletal presentation in utero (17 new cases and literature review). J Bone Miner Res 26(10):2389–2398
Wennberg C, Hessle L, Lundberg P, Mauro S, Narisawa S, Lerner UH, Millán JL (2000) Functional characterization of osteoblasts and osteoclasts from alkaline phosphatase knockout mice. J Bone Miner Res 15(10):1879–1888
Whyte MP (2009) Atypical femoral fractures, bisphosphonates, and adult hypophosphatasia. J Bone Miner Res 24(6):1132–1134
Whyte MP (2010) Physiological role of alkaline phosphatase explored in hypophosphatasia. Ann N Y Acad Sci 1192:190–200
Whyte MP (2016) Hypophosphatasia – aetiology, nosology, pathogenesis, diagnosis and treatment. Nat Rev Endocrinol 12(4):233–246
Whyte MP (2017a) Hypophosphatasia: enzyme replacement therapy brings new opportunities and new challenges. J Bone Miner Res 32(4):667–675
Whyte MP (2017b) Hypophosphatasia: an overview for 2017. Bone 102(9):15–25
Whyte MP, Valdes R Jr, Ryan LM, McAlister WH (1982) Infantile hypophosphatasia: enzyme replacement therapy by intravenous infusion of alkaline phosphatase-rich plasma from patients with Paget bone disease. J Pediatr 101(3):379–386
Whyte MP, Habib D, Coburn SP, Tecklenburg F, Ryan L, Fedde KN, Stinson RA (1992) Failure of hyperphosphatasemia by intravenous infusion of purified placental alkaline phosphatase (ALP) to correct severe

hypophosphatasia: evidence against a role for circulating ALP in skeletal mineralization. J Bone Miner Res 7(Suppl1):S155
Whyte MP, Kurtzberg J, McAlister WH, Mumm S, Podgornik MN, Coburn SP, Ryan LM, Miller CR, Gottesman GS, Smith AK, Douville J, Waters-Pick B, Armstrong RD, Martin PL (2003) Marrow cell transplantation for infantile hypophosphatasia. J Bone Miner Res 18(4):624–636
Whyte MP, Essmyer K, Geimer M, Mumm S (2006) Homozygosity for TNSALP mutation 1348c>T (Arg433Cys) causes infantile hypophosphatasia manifesting transient disease correction and variably lethal outcome in a kindred of black ancestry. J Pediatr 148(6):753–758
Whyte MP, Mumm S, Deal C (2007) Adult hypophosphatasia treated with teriparatide. J Clin Endocrinol Metab 92(4):1203–1208
Whyte MP, Greenberg CR, Salman NJ, Bober MB, McAlister WH, Wenkert D, Van Sickle BJ, Simmons JH, Edgar TS, Bauer ML, Hamdan MA, Bishop N, Lutz RE, McGinn M, Craig S, Moore JN, Taylor JW, Cleveland RH, Cranley WR, Lim R, Thacher TD, Mayhew JE, Downs M, Millán JL, Skrinar AM, Crine P, Landy H (2012a) Enzyme-replacement therapy in life-threatening hypophosphatasia. N Engl J Med 366(10):904–913
Whyte MP, Kishnani PS, Greenberg CR, Madson K, Mack K, Weber T, Mhanni A, Plotkin H, Kreher N, Landy H, Kreher N (2012b) Asfotase alfa decreases TNSALP substrate accumulation and improves functional outcomes in affected adolescents and adults. Bull Group Int Rech Sci Stomatol Odontol 51:35
Whyte MP, Simmons JH, Bishop N, Lutz RE, Vallée M, Melian A, Odrljin T, for the Study 003-08 Investigators (2014) Asfotase alfa: sustained efficacy and tolerability in infants and young children with life-threatening hypophosphatasia. Poster presented at the 2014 Pediatric Academic Societies and Asian Society for Pediatric Research Joint Meeting, Vancouver, BC, Canada, May 3–6, 2014
Whyte MP, Zhang F, Wenkert D, McAlister WH, Mack KE, Benigno MC, Coburn SP, Wagy S, Griffin DM, Ericson KL, Mumm S (2015) Hypophosphatasia: validation and expansion of the clinical nosology for children from 25 years experience with 173 pediatric patients. Bone 75(6):229–239
Whyte MP, Madson KL, Phillips D, Reeves A, McAlister WH, Yakimoski A, Mack KE, Hamilton K, Kagan K, Fujita KP, Thompson DD, Moseley S, Odrljin T, Rockman-Greenberg C (2016a) Asfotase alfa therapy for children with hypophosphatasia. JCI Insight 1(9):e85971
Whyte MP, Rockman-Greenberg C, Ozono K, Riese R, Moseley S, Melian A, Thompson DD, Bishop N, Hofmann C (2016b) Asfotase alfa treatment improves survival for perinatal and infantile hypophosphatasia. J Clin Endocrinol Metab 101(1):334–342
Whyte MP, Coburn SP, Ryan LM, Ericson KL, Zhang F (2018) Hypophosphatasia: biochemical hallmarks validate the expanded pediatric clinical nosology. Bone 110(5):96–106
Wüster C, Ziegler R (1992) Reduced bone mineral density and low parathyroid hormone levels in patients with the adult form of hypophosphatasia. Clin Investig 70(7):560–565
Yadav MC, Lemire I, Leonard P, Boileau G, Blond L, Beliveau M, Cory E, Sah RL, Whyte MP, Crine P, Millán JL (2011) Dose response of bone-targeted enzyme replacement for murine hypophosphatasia. Bone 49(2):250–256
Yamamoto S, Orimo H, Matsumoto T, Iijima O, Narisawa S, Maeda T, Millán JL, Shimada T (2011) Prolonged survival and henotypic correction of Akp2(−/−) hypophosphatasia mice by lentiviral gene therapy. J Bone Miner Res 26(1):135–142

Enzybiotics: Enzyme-Based Antibacterials as Therapeutics

11

Dorien Dams and Yves Briers

Abstract

Antibiotics have saved millions of lives. However, the overuse and misuse of antibiotics have contributed to a rapid emergence of antibiotic resistance worldwide. In addition, there is an unprecedented void in the development of new antibiotic classes by the pharmaceutical industry since the first introduction of antibiotics. This antibiotic crisis underscores the urgent and increasing necessity of new, innovative antibiotics. Enzybiotics are such a promising class of antibiotics. They are derived from endolysins, bacteriophage-encoded enzymes that degrade the bacterial cell wall of the infected cell at the end of the lytic replication cycle. Enzybiotics are featured by a rapid and unique mode-of-action, a high specificity to kill pathogens, a low probability for bacterial resistance development and a proteinaceous nature. (Engineered) endolysins have been demonstrated to be effective in a variety of animal models to combat both Gram-positive and Gram-negative bacteria and have entered different phases of preclinical and clinical trials. In addition, mycobacteriophage-encoded endolysins have been successfully used to inhibit mycobacteria in vitro. In this chapter we focus on the (pre)clinical progress of enzybiotics as potent therapeutic agent against human pathogenic bacteria.

Keywords

Endolysin · Enzybiotics · Clinical trials · Animal models · Multidrug-resistance

Abbreviations

AG	Arabinogalactan
Art	Artilysin
AMP	Antimicrobial peptide
CBD	Cell wall binding domain
CM	Cytoplasmic membrane
CoNS	Coagulase-negative staphylococci
DAP	Daptomycin
EAD	Enzymatically active domain
EDTA	Ethylenediaminetetraacetic acid
GlcNAc	N-acetylglucosamine
GLP	Good laboratory practice
HHP	High hydrostatic pressure
LAL	*Limulus amoebocyte* lysate
LPS	Lipopolysaccharide
MA	Mycolic acids
MIC	Minimal inhibitory concentration
MRSA	Methicillin-resistant *Staphylococcus aureus*
MSSA	Methicillin-sensitive *Staphylococcus aureus*
MurNAc	N-acetylmuramic acid

D. Dams · Y. Briers (✉)
Laboratory of Applied Biotechnology, Department of Biotechnology, Ghent University, Ghent, Belgium
e-mail: Yves.Briers@UGent.be

N. Labrou (ed.), *Therapeutic Enzymes: Function and Clinical Implications*, Advances in Experimental Medicine and Biology 1148, https://doi.org/10.1007/978-981-13-7709-9_11

NA	Not applicable
ND	Not defined
OM	Outer membrane
OMP	Outer membrane permeabilizing
PAE	Post-antibiotic effect
PA-SME	Post-antibiotic sub-MIC effect
PG	Peptidoglycan
SME	Sub-MIC effect

11.1 Introduction

Multidrug-resistant pathogens represent a major cause of morbidity and mortality and pose a serious threat on healthcare, both for humans and animals. In the United States alone, more than two million people are infected every year with bacteria that are resistant to conventional antibiotics. A significant proportion of 23,000 people dies (reported by the Centres for Disease Control and Prevention (US) 2013). Besides resistance development limiting the therapeutic potential of currently used antibiotics, there is an increasing awareness that the collateral damage of broad-spectrum antibiotics, i.e. killing of the benign commensal microbiome, causes harmful long-term effects such as an increased risk of secondary infections, asthma, obesity, diabetes type I and II (Langdon et al. 2016). Therefore, the development of therapeutics to treat (polymicrobial) drug-resistant infections without affecting the commensal microbiota is essential. However, since the 1980s, no new classes of antibiotics have entered the market, which further underlines the global need for novel strategies to combat multidrug- and pandrug-resistant pathogens.

Enzybiotics are a promising new class of enzyme-based antibacterials that may offer a response to this global call. They are safe, effective, fast-acting and highly specific. In addition, they weaken biofilms and have a low probability to provoke resistance development. They can be used alone or in combination with traditional antibiotics. Enzybiotics are derived from endolysins, bacteriophage-encoded enzymes that lyse the infected bacterial cell at the end of the lytic replication cycle. They enzymatically degrade the peptidoglycan (PG) until the infected cell can no longer withstand the internal osmotic pressure and lyses, followed by dispersion of newly formed viral particles (i.e. "lysis form within"). Given the increasing incidence of antibiotic resistance and the lack of new alternative therapies, this unique and well-known feature of endolysins has spurred the idea to apply them exogenously to kill human pathogens (Nelson et al. 2001). Indeed, purified recombinant endolysins induce rapid osmotic lysis of Gram-positive bacteria through degradation of externally accessible PG with consequent cell death (i.e. "lysis from without"). The efficacy of endolysins has been demonstrated in numerous animal models of human disease (Nelson et al. 2001; Schmelcher et al. 2012a; Roach and Donovan 2015; Gerstmans et al. 2017). In addition, extensive engineering efforts have been done during the last two decades to enhance the potential of endolysins as therapeutic enzymes against Gram-positive bacteria and to extend their application to Gram-negative pathogens, which have a protective outer membrane. These endeavors have paved the path for recombinant endolysins to enter different phases in preclinical and clinical trials. Currently, different lead enzybiotics have entered the clinical phase with humans. The accelerating clinical advances and their high technical feasibility make phage endolysins as therapeutics the highest ranked alternatives to replace conventional antibiotics according to a recent pipeline investigation (Czaplewski et al. 2016).

11.2 Bacteriophages

11.2.1 From Phage Discovery Over Phage Therapy to Endolysin Therapy

In the early 20th century bacteriophages (or phages) were independently discovered by Twort (1915) and Felix d'Herelle (1917), but D'Herelle was the first to describe bacteriophages as bacterial viruses which have the ability to infect bacteria followed by bacterial lysis. Not long after their discovery, the therapeutic potential of phages to treat human and animal bacterial infections was recognized (Twort 1915; Sulakvelidze et al. 2001; Salmond and Fineran 2015). The first attempt to use bacteriophages therapeutically was the treatment of a 12-year-old boy suffering from severe dysentery. When the phage preparation was regarded as safe, a single dose was administered to the child. The boy, together with three additional patients fully recovered within a few days after a single administration (Sulakvelidze et al. 2001). Despite the successful outcome, the results of these studies were not published. In 1921, Richard Bruynoghe and Joseph Maisin have published the first study about the treatment of staphylococcal skin disease with phages. A regression of the infection was reported 24–48 h after treatment (Lavigne and Robben 2012). Several similar studies about phage therapy have followed and in early 1930 the commercialization of phages against an array of pathogens started. Despite these efforts, phage therapy for humans was abandoned in the Western medicine soon after the introduction of broad-spectrum antibiotics after World War II. Nowadays, the emergence of multidrug- and even pandrug-resistant pathogens has revitalized the interest in phages (Hanlon 2007). Nevertheless, phage therapy needs to overcome several technical hurdles such as regulation, narrow host range, bacterial resistance to phages, manufacturing, side effects of bacterial lysis, difficulties in the delivery of purified phage preparations, and the complicated pharmacokinetics of phages because of their self-replicating nature (Sandeep 2006; Hermoso et al. 2007). Using phage-encoded endolysins (or lysins) offer the potential to circumvent these obstacles typically related to intact phages (López et al. 2004; Hermoso et al. 2007).

11.2.2 Biological Function of Endolysin-Mediated Bacteriolysis

The biological role of endolysins is lysis of the infected bacterial cell at the end of the lytic cycle of bacteriophages in order to release progeny phage particles. The release is complicated by the presence of the bacterial cell wall, acting as a strong barrier. As phages co-evolved with their bacterial hosts, they have developed a variety of strategies to overcome this barrier (Young 2014). The most prevalent mechanism is the two component holin-endolysin system specific to the *Caudovirales* (tailed phages with dsDNA) and regulated by a timing mechanism to achieve lysis at an optimal time (Fig. 11.1). Endolysins accumulate in the cytoplasm as they cannot migrate through the cytoplasmic membrane (CM). The access to the PG in the periplasm is regulated by holins. Holins are small hydrophobic proteins that accumulate as homodimers in the CM in a uniformly distributed manner. At a genetically predetermined holin concentration, the holins aggregate into homo-oligomers and partially depolarize the CM by forming pores. The accumulated endolysins take advantage of these pores to gain access to the PG to degrade it. In addition, phage-encoded spanins weaken the outer membrane. When the impaired cell wall can no longer withstand the internal osmotic pressure, osmotic lysis and the release of the progeny ("lysis from within") occur (Wang et al. 2000; White et al. 2011; Young 2014).

11.3 The Structural and Biochemical Diversity of Endolysins

Although all endolysins have a conserved biological function, the constant evolutionary struggle between bacteriophage and bacterium has resulted in a huge biochemical and structural diversity among endolysins (Loessner 2005). Along with the high global abundance of phages, this diversity creates an enormous reservoir of specific endolysins that all have a medical potential.

11.3.1 Catalytic Specificity of Endolysins

The PG layer is a highly preserved layer that ensures the structural rigidity and integrity of the

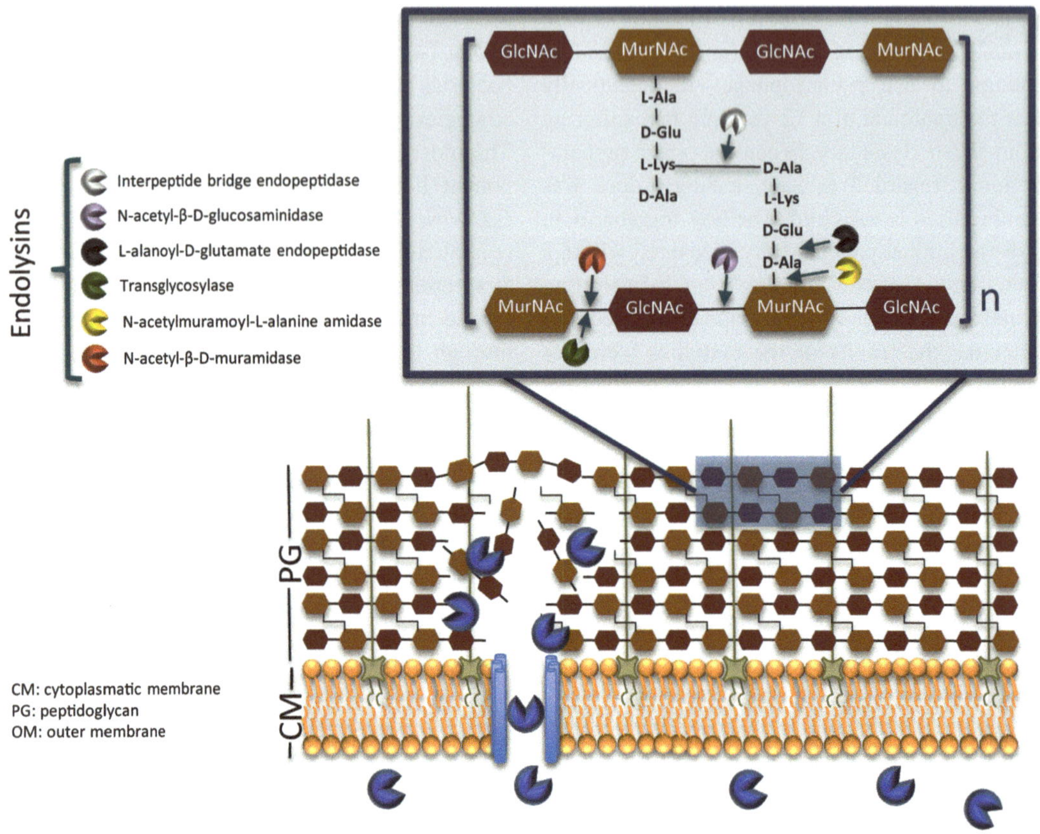

Fig. 11.1 Schematic representation of the PG structure of *S. aureus*, the different cleavage specificities of endolysins and the simplified holin/endolysin lysis mechanism. (Bottom) When the critical holin concentration is achieved, holins depolarize the CM by forming pore lesions through which the endolysins accumulated in the cytoplasm gain access to PG. As a consequence, the PG is degraded and the cell wall cannot longer withstand the internal osmotic pressure and finally undergoes lysis. (Top) The glycan chains consist of β-1,4-linked, alternating monomers of N-acetylglucosamine (GlcNAc) and N-acetylmuramic acid (MurNAc) residues and are further cross-linked via stem peptides that are attached to the MurNAc residues. The endolysins can be differentiated in three groups based on their cleavage specificity: (1) endopeptidases – targeting peptide bonds – including L-alanoyl-D-glutamate endopeptidase and interpeptide bridge endopeptidase, (2) amidases – targeting amide bonds – including the N-acetylmuramoyl-L-alanine amidases and (3) glycosidases – targeting glyosidic bonds – including N-acetyl-β-D-glucosaminidases, N-acetyl-β-D-muramidases and transglycosylases. (Hermoso et al. 2007)

bacterial cells. PG consists of a polysaccharide of alternating N-acetylmuramic acid (MurNAc) and N-acetylglucosamine (GlcNAc) residues, linked by a β1-4 glycosidic bond. The glycan polymer is further stabilized by cross-links of the stem peptides, which are attached to MurNAc residues. Endolysins can recognize and digest a specific chemical bond of PG. They are classified into three different groups, according to the targeted bond (i.e. the amide, peptide or glycosidic bond) (Fig. 11.1) (Madigan et al. 2008).

(I) The first group of endolysins (N-acetylmuramoyl-L-alanine amidases; E.C 3.5.1.28) hydrolyses the amide bond between N-acetylmuramoyl residues and L-alanine, the first amino acid of the stem peptide. (II) The second group (endopeptidases; E.C 3.4.X.X) cleaves the peptide bond between two amino acids. L-alanoyl-D-glutamate endopeptidase cleaves the bond between L-alanine and D-glutamate in contrast to an interpeptide bridge-specific endopeptidase such as D-alanyl-glycyl endopeptidase that targets the cross-link between adjacent stem peptides. (III) Glycosidases are the final group and hydrolyze glyosidic bonds. Three subgroups can be differentiated. The first subgroup (N-acetyl-β-D-glucosaminidases; E.C 3.2.1.52) cleaves the N-acetylglucosaminyl-β-1-4-N-acetylmuramine bond on the reducing side of GlcNac. Both other subgroups, N-acetyl-β-D-muramidases (E.C 3.2.1.17, generically also termed lysozymes) and lytic transglycosylases (E.C. 4.2.2.n1, cleave the N-acetylmuramoyl-β-1-4-N-acetylmuramine bond on the reducing side of MurNAc. In contrary to all other groups, lytic transglycosylases are no hydrolases and cleave the β-1,4-glycosidic bond by an intramolecular reaction, resulting in a N-acetyl-1,6-anhydro-muramyl moiety (Hermoso et al. 2007; Nelson et al. 2012; Schmelcher et al. 2012a).

11.3.2 The Structural Properties and Function of Endolysins

Next to catalytic diversity, endolysins have a remarkable structural diversity, which has a large impact on both enzyme kinetics and specificity (Oliveira et al. 2013). The structural variety of endolysins is strongly related with the cell wall structure and differs for endolysins from phages infecting Gram-positive and Gram-negative species (Fig. 11.2). The Gram-negative cell wall is composed of a CM covered by one to three layers of PG and an additional outer membrane (OM). The OM is a complex asymmetric membrane featured by lipopolysaccharide (LPS) molecules composed of lipid A, core polysaccharide and O-specific polysaccharide. Lipid A generally consists of phosphorylated glucosamine disaccharide with fatty acid chains attached, providing hydrophobic stabilization of the OM. The anionic phosphates of the lipid A and the core polysaccharide moieties account for the negative charge of the OM and are stabilized by ionic interactions with

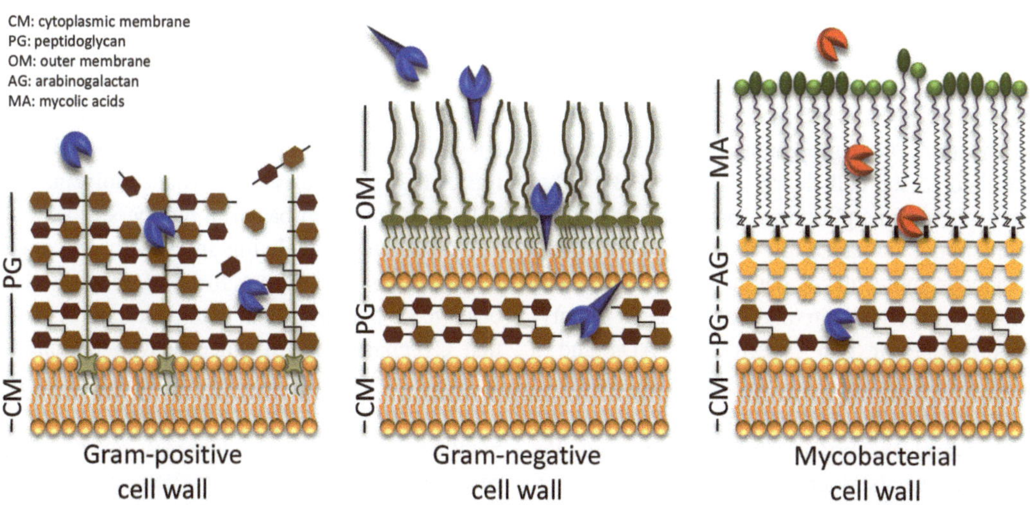

Fig. 11.2 Antimicrobial activity of endolysins and Artilysin®s on different cell wall types. Left: The Gram-positive bacterial cell wall contains a CM and a thick PG layer. As a result, PG of the Gram-positive cell wall is readily accessible to exogenously added endolysins (blue) making these phage enzymes suitable for the treatment of Gram-positive infections. Middle: The Gram-negative cell wall is composed of a CM covered by one to three layers of PG, and an additional protective outer membrane (OM) hindering the access of endolysins to the PG. Artilysin®s (blue) can reach the PG through the fusion of an endolysin with an OMP-peptide. First, the OMP-peptide acts as a wedge by local and transient membrane destabilization, whereafter the Artilysin® acts similar as a native endolysin. Right: Mycobacteria consists of a mycolyl-arabinogalactan-PG complex. Mycobacterium phages produces two lytic enzymes, a PG hydrolase LysA (blue) and a lipase LysB (red) targeting the mycolic acids. (This figure was originally published in Biochem Soc Trans. Gerstmans et al. (2016) © Portland Press Limited)

divalent cations (Mg^{2+} and Ca^{2+}). The LPS layer serves thus as a barrier for both hydrophobic and hydrophilic molecules larger than 600 Da. In contrary, Gram-positive bacteria only contain a thick PG layer decorated with (lipo)teichoic acids, which are linked with the PG and are responsible for the overall negative charge of the cell surface (reviewed by Nelson et al. 2012; Schmelcher et al. 2012a). The cell wall of mycobacteria is structurally different from Gram-positive and Gram-negative bacteria as it comprises an additional layer composed of mycolyl-arabinogalactan-PG complexes (Brennan 2003).

Endolysins can be divided in globular or modular enzymes. Modular endolysins are generally encoded by phages infecting Gram-positive bacteria and mycobacteria, but also by some phages infecting Gram-negative bacteria. They are composed of multiple functional domains with separate activities: (1) an enzymatically active domain (EAD) and (2) a cell wall binding domain (CBD). The EAD is responsible for the catalytic action of the enzyme, digesting a specific chemical bond of PG. A CBD directs an endolysin to its substrate, i.e. the PG layer or a specific cell wall ligand such as (lipo)teichoic acids, resulting in an increased affinity and substrate proximity (Schmelcher et al. 2012a; Payne et al. 2013). Even after PG degradation, endolysins remain attached to the cell debris via their CBD. This may prevent neighboring, potential host cells from being damaged by diffusion of released endolysin molecules (Loessner et al. 2002). The most common architecture of modular endolysins is one or two N-terminal EADs and a C-terminal CBD, but the specific composition or order of the domains is not universally conserved and several exceptions have been reported: the CBD can be also located at the N-terminus (e.g. most modular endolysins from phages infecting Gram-negative bacteria), the CBD can be squeezed between two EADs, or can be absent. The domains are usually connected by a flexible inter-domain linker sequence, which can vary in size and ensures an autonomous function of both domains (Schmelcher et al. 2012a; Roach and Donovan 2015). A number of 89 different architectural modular organizations have been described, illustrating the high evolutionary diversity among endolysins (Oliveira et al. 2013).

Globular endolysins originating from Gram-negative infecting phages do not have a CBD and consist of a single domain functioning as an EAD (Briers et al. 2007). The presence of an OM preventing exogenous cleavage of the PG layer by released endolysins and the thin PG layer may eliminate the need for a CBD that directs endolysins to the cell wall. Notwithstanding, several Gram-negative infecting phages producing modular enzymes have already been discovered (Walmagh et al. 2012), especially in so-called large jumbo phages. These modular endolysins consist of N-terminal CBD specific for the conserved A1γ PG (conserved domains PG_binding_1 and PG_binding_3) fused to a C-terminal EAD (Briers et al. 2009) or two C-terminal EADs (Oliveira et al. 2013).

11.4 Preclinical Analysis of Enzybiotics as Antibacterials

The highly lytic nature of endolysins spurred the idea to use them as enzyme-based antibiotics (or enzybiotics). Nelson et al. (2001) were the first to report that purified recombinant endolysins are able to reduce high bacterial numbers in an animal infection model. In this study, complete eradication of bacteria from the oral mucosa was observed 2 h after an oral administration of purified C1 phage lysin to mice heavily colonized with group A streptococci. Since then, numerous preclinical studies have demonstrated their potential as therapeutic, whereas several concerns related to their proteinaceous nature have been rebutted. Whereas initially only Gram-positive pathogens were targeted because their PG is easily accessible, Gram-negative pathogens can meanwhile also be killed by engineered endolysins or endolysins with intrinsic antibacterial activity (Table 11.1).

Enzybiotics have a **high specificity**, often at genus, species or even serovar level. Specificity

Table 11.1 Preclinical aspects and observations of enzybiotics related to their mode-of-action and proteinaceous nature

Aspect	Observation
Spectrum of activity	Enzybiotics have generally a high specificity to genus, species or even serovar level. The specificity can be adapted by protein engineering
Killing rate	Enzybiotics are featured by a rapid mode-of-action with generally killing upon contact depending on the dose. This high killing rate is explained by the enzymatic mode-of-action
Resistance development	No resistance development has been observed for endolysins targeting the conserved PG layer. The low resistance development profile of enzybiotics can also be explained by the high specificity and rapid action of endolysins. One exception are endopeptidases that act on the species-specific interpeptide cross-bridge
Antibacterial synergetic effect and resensitization of antibiotics	Synergistic effect between two endolysins with different catalytic specificities or between a endolysin and an antibiotic has been observed, resulting in a faster degradation of the substrate or a facilitated antibiotic uptake. In some combinations, enzybiotics resensitize bacteria to antibiotics against which they are resistant
Immunogenicity of proteins	Neutralizing antibodies can be raised against enzybiotics but do not completely neutralize in vivo or have no significant impact at all. This is currently explained by the high affinity of CBDs to its substrate and the fast kinetics of enzybiotics
Allergenicity	The proteinaceous nature of endolysins may potentially induce an allergic reaction, but no allergic reaction have been observed yet
Half-life	Enzybiotics have a short half-life, but are functional in in vivo models in this short frame due to the rapid mode-of-action
Inflammatory response	A pro-inflammatory response due to cellular debris was not observed after single and interval administration of endolysin. However, an optimal dosing regime is required since a continuous administration can lead to an increased pro-inflammatory response
Killing of intracellular bacteria	Some native endolysins and chimeric endolysins are able to kill intracellular pathogens. Alternatively, endolysins fused with protein transduction domains penetrate mammalian cells and kill intracellular bacteria

can be inferred from both domains. E.g. CBDs can bind specific ligands such as choline in the pneumococcal cell wall (Hermoso et al. 2003) or specific substituents of serotype-specific teichoic acids of *Listeria monocytogenes* (Eugster and Loessner 2012), whereas some EADs specifically cleave the pentaglycine crossbridge that only occurs in the PG of *Staphylococcus aureus* (Callewaert et al. 2011). Therefore, enzybiotics generally have a narrow spectrum of antibacterial activity, leaving commensal flora unaffected in contrast to the current, broad-spectrum antibiotics (e.g. penicillin and tetracycline) (Nelson et al. 2012). Moreover, endolysins have a unique and **rapid mode-of-action** holding several advantages compared to conventional antibiotics. Their ability to actively degrade the PG in seconds or minutes, depending on the concentration, makes them much faster than the existing antibiotics. The active degradation of the cell wall renders them also effective against persister cells (i.e. metabolically inactive cells) (Briers et al. 2014a; Gutierrez et al. 2014; Defraine et al. 2016). Enzybiotics do not require an active metabolism in contrast to antibiotics that target essential metabolic steps within actively growing bacteria. Compared to the rapid mode-of-action of enzybiotics, inhibition of such metabolic steps by small molecule antibiotics is usually a slow process, eventually leading to cell death after 24–48 h (Allison et al. 2011; Briers et al. 2014a).

Endolysin resistance has not been observed among strains from diverse sources and resistant strains can generally not be selected during in vitro experiments in which strains were repeatedly exposed to subinhibitory concentrations of (engineered) phage endolysins (Fischetti 2005; Pastagia et al. 2011; Briers et al. 2014a; Defraine et al. 2016). The high specificity and rapid action of endolysins, and the immutable nature of the PG layer may explain this **low probability of resistance development against phage endolysins**. This beneficial feature supports a prolonged use of enzybiotics, even under regimes of repeated use. The use of many broad-range antibiotics in contrary has led to the selection of resistant strains of both target pathogen and commensal bacteria, which is often accelerated by the distribution of resistant genes via horizontal gene transfer (Johnsborg and Håvarstein 2009). In spite of this overall low probability of resistance development, an exception of this rule has been reported for endolysins that target the pentaglycine crossbridge in *S. aureus*. The strain *S. aureus* Newman developed resistance against both endolysin LysK and bacteriocin lysostaphin. Serial exposure (i.e. ten rounds) *of S. aureus* Newman to subinhibitory doses of LysK and lysostaphin in liquid culture resulted in a 42-fold and 585-fold increases of the minimal inhibition concentration (MIC), respectively (Becker et al. 2016). Resistance is acquired by substitution of glycine residues of the pentaglycine bridge by serines (Climo et al. 1998). This indicates that endolysins targeting species-specific interpeptide crossbridges are susceptible to resistance development in contrast to endolysins targeting highly conserved bonds in the PG layer (i.e. the polysaccharide backbone and amide bond between the polysaccharide backbone and the stem peptide). Amidases and muramidases for example are the most prevalent endolysins targeting these highly conserved bonds.

Antibacterial synergistic effects have been demonstrated between two or more endolysins or between endolysins and other antibacterial agents like traditional antibiotics. This synergetic effect implies an enhanced and faster degradation of PG resulting in an improved antibacterial efficiency, a requirement of smaller doses and a potentially reduced risk of resistance development (Schmelcher et al. 2012b). Synergy between two enzymes with different catalytic specificities (e.g. pneumococcal phage endolysins CpI-1 and Pal, and staphylococcal phage endolysins LysK and lysostaphin) can be explained by two possible mechanisms: either two endolysins simultaneously digest a different bond of the PG network, resulting in a more extensive degradation or one endolysin cleaves the first bond, concurrently improving the accessibility for the other endolysin (Loeffler and Fischetti 2003; Becker et al. 2008; Schmelcher et al. 2012b). The synergy mechanism between endolysins and antibiotics (e.g. the pneumococcal phage lysin Cpl-1 and gentamicin) remains unclear but it is proposed that partial PG degradation by the endolysin facilitates antibiotic uptake. In addition, it has been reported that the pneumococcal phage endolysin Cpl-1 can **resensitize** *Streptococcus pneumoniae* to penicillin against which they are resistant. As such, endolysins can slow down the emergence of antibiotic-resistant strains (Djurkovic et al. 2005; Viertel et al. 2014).

Since PG is a exclusively present in bacteria and not in mammalian cells, the risk on cytotoxic effects for humans and animals is minimized. However, the proteinaceous character of enzybiotics may provoke an **immune response**. In vitro and in vivo studies have shown that neutralizing antibodies can indeed be raised against endolysins after repeated exposure. These antibodies reduce the antibacterial activity of endolysins but do not completely neutralize them, while in other studies they do not have a significant impact at all (Fischetti 2010). Thus, endolysins can be used repeatedly to treat the same bacterial infection (Jado et al. 2003; Loeffler and Fischetti 2003; Hermoso et al. 2007; Rashel et al. 2007; Zhang et al. 2016). These observations may be explained by the high affinity of the CBD for its substrate (nanomolar range) that exceeds the affinity of endolysin-specific antibodies and the fast kinetics of these enzymes that outperform the hosts immune response (Loessner et al. 2002; Jado et al. 2003; Schmelcher et al. 2010). The proteinaceous nature of endolysins may potentially also induce an allergic reaction, but so far **no allergic reactions** have been reported against endolysins in two clinical phase I studies with SAL200 (ClinicalTrials.gov NCT01855048; Jun et al. 2017) and CF-301 (ClinicalTrials.gov NCT02439359; Cassino 2016).

Proteins, including endolysins, generally have a **short half-life**, which is estimated to be approximately between 4 and 40 min for endolysins (Loeffler and Fischetti 2003; Jun et al. 2017). In spite of the short half-life, the efficacy of endolysins has been demonstrated in various animal models with intravenous administration. Again, the fast mode-of-action of enzybiotics appears to be of pivotal importance to compensate the short half-life. Loeffler et al. (2003) has reported that due to the narrow window of action a repeated administration was required to make a therapeutic treatment successful.

The release of cellular debris of lysed bacteria upon systemic administration of enzybiotics in humans or animals may induce a **pro-inflammatory response**. This bacterial cell debris includes lipopolysaccharides (LPS), (lipo) teichoic acids and PG through membrane fragmentation and may provoke serious complications such as a septic shock (Fischetti 2010). Entenza et al. (2005) found that rats treated with a continuous intravenous infusion of Cpl-1 lysin (originating of *Streptococcus pneumoniae* phage) show an increased pro-inflammatory cytokine concentration in comparison with untreated rats. Witzenrath et al. (2009) instead reported that administration in 12-h intervals of the same enzyme reduces cytokine concentrations compared with untreated animals. The latter observation indicates that there is an optimal dosing for endolysins. The optimal dose should suffice to digest the PG and to kill the bacterial pathogen without additional fragmentation of the PG layer. An optimization of the dosing regimen when using endolysins in therapeutic applications is therefore inevitable (Entenza et al. 2005; Witzenrath et al. 2009; Fischetti 2010).

Initially, it was thought that pathogens that propagate and survive intracellularly to evade the immune system would be inaccessible for enzybiotics. However, recently, both native endolysins (PlyC) and chimeric endolysins (K-L) have been shown to **eradicate intracellular pathogens** such as *Streptococcus pyogenes* and *S. aureus*. Alternatively, the fusion of endolysins with protein transduction domains led to functional endolysins that get into mammalian cells and kill intracellular bacteria (Becker et al. 2016; Shen et al. 2016).

11.5 Endolysins as Therapeutics

11.5.1 Endolysins as Therapeutic Agents Against Gram-Positive Bacterial Infections

In 2001 the group of Vincent Fischetti demonstrated that purified recombinant endolysins can be used as a preventive and curative agent of streptococcal infections in mice. Since then, extensive efforts have been done to expand their potential as therapeutics including the administration of enzybiotics in complex environments (e.g., blood stream, mucous membranes,...) (Schmelcher et al. 2012a). Various in vitro and in vivo animal infection models of human diseases demonstrated that the administration of purified endolysin to animals infected by Gram-positive pathogens (such as *Staphylococcus aureus*, *staphylococcus* sp., *streptococcus* sp., *Bacillus anthracis* and *Enterococcus* sp.) rescues them from an otherwise deadly infection (Nelson et al. 2001, 2012; Schuch et al. 2002; see Haddad et al. 2017 for an overview of all animal models). In addition, several protein engineering strategies such as mutagenesis, truncation or domain swapping (i.e. chimeric endolysins) are applied to improve the antibacterial activity and/or modify the specificity and other features (reviewed by Gerstmans et al. 2017). This collection of successful reports on the use of recombinant endolysins to combat pathogens inspired to the term 'enzybiotics' (Hermoso et al. 2007). Below we focus on the in vitro and in vivo models of all (engineered) endolysins that are currently in the clinical development stage (Table 11.2).

Table 11.2 In vitro and in vivo models of enzybiotics in clinical development, and completed clinical trials

Company	(Engineered) Endolysin	Target	Delivery method	Model type	Observations	Reference/ identifier
ContraFect	CF-301	*S. aureus* bacteremia	NA	In vitro	CF-301 showed a potent and rapid antibacterial effect, anti-biofilm activity and synergy with daptomycin	Schuch et al. (2014)
			ND	In vivo (mouse)	Reduction of bacterial counts is bloodstream of mice with bacteremia; the combination CF-301 and antibiotics increased survival in bacteremia	Schuch et al. (2014)
			NA	In vitro	A strong PAE, PA-SME and SME were observed against 14 staphylococcal strains	Schuch (2016)
			NA	In vitro	Exposure to subinhibitory levels increased daptomycin susceptibility, reduced growth rates, decreased biofilm formation and inhibited a virulence phenotype	Oh and Schuch (2017)
			ND	In vivo (mouse)	Exposure to subinhibitory levels reduced growth rates in a neutropenic mouse thigh infection model	Oh and Schuch (2017)
			Intravenous	In vivo (human)	Clinical phase I trial evaluating safety and tolerability of single dose injection, CF-301 has a good safety profile and is well tolerated	NCT02439359 Jandourek et al. (2017a, b)
Intron biotechnology	SAL200 (SAL-1)	*S. aureus*	Rat serum	In vitro	Rapid and effective antibacterial activity of formulated SAL200	Jun et al. (2013)
			Intravenous	In vivo (mouse)	Reduced the mortality of mice and reduction of bacterial counts in the bloodstream	Jun et al. (2013)
			Intravenous	In vivo (rat, dog)	No adverse clinical effects upon administration of endolysin	Jun et al. (2014)
			Intravenous	In vivo (monkey)	Single high dose injection or 5-day multiple-dose administration were well tolerated	Jun et al. (2016)
			Intravenous	In vivo (human)	Phase I clinical trial evaluating safety, pharmacokinetics and pharmacodynamics of SAL200 in healthy man, no severe adverse events were observed	NCT01855048 Jun et al. (2017)

Gangagen	P128	*S. aureus*	NA	In vitro	Effective against panel of clinical strains	Paul et al. (2011)
			NA	In vitro	Stable activity in body fluids and no cytotoxic effects on mammalian cells	George et al. (2012)
			Intranasal	In vivo (rat)	Complete nasal decolonization or reduced cell numbers	Paul et al. (2011)
			Topical	In vivo (dog)	Complete healing of dogs with canine pyoderma	Junjappa et al. (2013)
			Intravenous	In vivo (rat)	No anaphylaxis and type III hypersensitivity was observed after single dose injection, protection of rat from bacteremia and formation of renal abscesses	Channabasappa et al. (2017)
			Intravenous	In vivo (mouse)	Evaluation of pharmacokinetics and efficiency of P128 in mice with bacteremia, P128 showed rapid antibacterial effect	Sriram et al. (2017)
			Intranasal	In vivo (human)	Clinical phase I/II trial evaluating safety and efficacy in healthy human and chronic kidney disease patients, P128 was well tolerated and effective in reducing nasal carriage in humans	NCT01746654
Micreos	Staphefekt™ SA.100	Staphylococcal infections	Topical	In vivo (human)	Clinical relevant reduction of *S. aureus* on skin of three humans with dermatoses	Totté et al. (2017)

11.5.1.1 Enzybiotics in the Clinical Pipeline

Enzybiotics are increasingly developed for human and veterinary pathogens as pharmaceutical product (clinical phase II trials are going on by ContraFect, Intron Biotechnology, GangaGen, Micreos), cosmetic (GladSkin series with StaphEfekt™ as functional compound, commercialized by Micreos) and wound care spray (Medolysin® based on Artilysins announced by Lysando).

CF-301

Endolysin CF-301 from the New York-based company Contrafect has a potent anti-staphylococcal activity for the treatment of *S. aureus* bloodstream infections including endocarditis and bacteremia. This endolysin, also referred to as PlySs2, was identified from a *Streptococcus suis* prophage (Gilmer et al. 2013). Schuch et al. (2014) demonstrated that CF-301 has a rapid antibacterial activity against *S. aureus* strains, an anti-biofilm activity, and acts synergistically in combination with standard antibiotics. CF-301 shows bactericidal activity in vitro against 250 *S. aureus* strains, including 120 MRSA isolates and 27 multidrug-resistant strains. A more than 3-log reduction in cell number within 30 min was observed in vitro in contrary to antibiotics that required 6–12 h to reach similar reduction levels. In vivo, a CF-301 treatment of mice with MRSA bacteremia resulted in a 2-log CFU reduction in bloodstream within 1 h. In addition, CF-301 (1x MIC) was able to eradicate a *S. aureus* biofilm in 2 h, whereas high doses of antibiotics (1000x MIC) fail. CF-301 also exhibits a potent in vitro synergy with daptomycin (DAP) (64- to 256-fold increased daptomycin susceptibility) and increased survival in bacteremia when combined with vancomycin (from 7–31% to 82–90% survival) or DAP (from 3% to 67% survival) (Schuch et al. 2014). Further, a strong post-antibiotic effect (PAE, time period to resume normal growth after the antibiotic treatment has stopped and the serum concentration is below the MIC), post-antibiotic sub-MIC effect (PA-SME, effect of sub-MICs on bacteria during PAE phase) and sub-MIC effect (SME, effect of sub-MICs on bacteria without previous exposure to suprainhibitory concentrations) has been demonstrated for CF-301. An in vitro experiment against a panel of 14 staphylococcal strains in human serum indicated a PAE of 4.8 h, a PA-SME up to 7.5 h and a SME up to 7.8 h. The in vivo PAE tested in a neutropenic mouse thigh model significantly exceeds the in vitro PAE, reaching 23–26 h (Schuch 2016). A follow-up study of the CF-301 SME reported that the exposure to subinhibitory levels of CF-301 as low as 0.004x MIC increases antibiotic susceptibility (DAP), reduces growth rates (tested in vitro and in a neutropenic mouse thigh infection model model), decreases biofilm formation and inhibits a virulence phenotype (Oh and Schuch 2017).

CF-301 is formulated for intravenous injection. This endolysin has completed a Phase I trial in healthy human volunteers. In the phase I clinical trial (clinicaltrials.gov NCT02439359), a placebo-controlled, dose-escalating study (from 0.04 to 0.4 mg/kg/dose) has been conducted to examine the safety and tolerability of single intravenous dose of CF-301 in healthy human subjects. The clinical data have demonstrated that CF-301 is well tolerated and has a good safety profile. The latter implies that adverse clinical safety signals (hypersensitivity or serious adverse events), acute cardiovascular and inflammatory responses were absent (Cassino 2016; Jandourek et al. 2017a, b). To investigate the inflammatory response on CF-301, a range of inflammatory markers such as the high-sensitivity C-reactive protein, the erythrocyte sedimentation rate and complement factors Bb, C3a, C5a and CH50 were analyzed. No differences between placebo and CF-301 injected human subjects have been reported (Jandourek et al. 2017b). In addition, no clinically relevant changes in systolic and diastolic blood pressure, heart rate and QT were found in PK/PD models based on the clinical phase I data (Ghahramani et al. 2017).

N-Rephasin®SAL200

SAL200 with anti-staphylococcal activity (commercial name: N-Rephasin®) is developed by Intron Biotechnology. SAL200 is the recombinant variant of endolysin SAL-1 that is derived from the staphylococcus phage SAP-1. A stabilizing formulation containing 0.01 M L-histidine (pH 6.0), 5% (w/v) sorbitol, 10 mM $CaCl_2$ and 0.1% (w/v) Poloxamer 188 was developed for human application. Initial in vitro experiments with the formulated SAL200 demonstrated a rapid and effective antibacterial activity against a panel of clinical and biofilm-forming *S. aureus* isolates. A daily intravenous administration of formulated SAL200 during 3 days in mice infected with MRSA significantly increased their survival rate. No bacteria could be detected in blood and splenic tissue of mice treated with SAL200 in contrary to the control group (Jun et al. 2013). In addition, GLP-compliant safety evaluation studies of intravenously administered SAL200 have been executed on rats and dogs, including toxicity, central nervous system, respiratory, and cardiovascular function tests. Both, single-dose and repeated-dose (one dose per day for a period of 4 weeks) toxicity tests in rats does not result in adverse clinical effects related with the administration. Similar to rats, repeated dose toxicity tests were conducted in dogs. After a 2 week treatment, dogs showed no changes in body weight, food consumption, ophthalmology, electrocardiography, hematology, serum biochemistry, organ weight or urinalysis. However, from 10 days after the first treatment, clinical signs such as subdued behavior, prone position, irregular respiration, and vomiting have been reported in dogs but were transient and mild. These clinical signs were resolved 30 min to 1 h after each new injection. In the safety pharmacology studies, no adverse effects were observed in the central nervous, respiratory and function tests. Also in the cardiovascular function tests there were no adverse events or laboratory abnormalities observed by the first and second administration. However, mild and transient changes were observed upon the third and fourth injection but again, the signs were resolved 6 h after administration. Further investigation has demonstrated that a repeated administration of SAL200 elicit an immune response in both dogs (14 days) and rats (28 days), observed by the presence of anti-SAL-1 antibodies (Jun et al. 2014). In addition, a reduction in blood C3 complement was observed in the exposed dogs. C3 complement proteins support antibodies and phagocytic cells by killing foreign invaders and thus play an important role in the innate immune system. However, it is unclear whether this response was due to the residual lipopolysaccharide endotoxin in the recombinant protein preparation or to the enzyme itself. After these preclinical tests, the pharmacokinetics of SAL200 have been studied by an intravenous administration in monkeys. SAL200 was well tolerated and no adverse events or laboratory abnormalities were observed when injected as a single dose administration (up to 80 mg/kg body weight) or as a 5-day multiple-dose administration (up to 40 mg/kg per day) (Jun et al. 2016).

Based on these data, SAL200 entered phase 1 clinical trial (ClinicalTrials.gov NCT01855048, results published in Jun et al. 2017) to evaluate the safety, pharmacokinetics and pharmacodynamics of intravenous medication SAL200 in healthy men. After an intravenous infusion of single ascending doses, SAL 200 was well tolerated and no severe adverse events or clinically significant values were observed and all present clinical signs including fatigue, rigors, headache, and myalgia were transient, self-limiting, and mild. However, as expected the humoral immune response was induced and antibodies ranging from 2 to 12 μg/ml were formed against the recombinant endolysin SAL200. Also the pharmacokinetics and pharmacodynamics analysis support the potential use of SAL200 as new endolysin-based therapeutic drug. The antibacterial activity in blood has been assessed with an ex vivo blood assay using blood samples collected from all active pharmaceutical ingredient-treated participants 1 h after injection. These blood samples were spotted on a lawn of *S. aureus* bacteria and compared with a standard series of SAL200. These studies ensure that a dosing regimen of more than 1 mg/kg of SAL200 is a viable treatment because of the following observations: (I) The blood SAL200 concentration was greater than 0.078 μg/ml in all collected blood samples 1 h after injection, which is the minimum bactericidal concentration to kill a bacterial population of 1×10^6 CFU/ml in serum environment. (II) The time to reduce the optical density of the initial bacterial suspension with 50% (TOD_{50}, equivalent to one-half log drop in initial viable bacteria) was less than 10 min (Jun et al. 2017).

P128

The company Gangagen (India) created the engineered enzybiotic P128 for intranasal use against *S. aureus*. P128 is a chimeric protein that combines the phage tail-associated catalytic domain Lys16 of staphylococcus phage K with the well-known staphylococcal cell wall binding SH3b domain from lysostaphin. In an initial in vitro experiment, *S. aureus* was found to be effective (>99% reduction of cell numbers) against a panel of *S. aureus* clinical strains, including MRSA, methicillin-sensitive *S. aureus* (MSSA), and a mupirocin-resistant *S. aureus* (Paul et al. 2011; Vipra et al. 2012). George et al. (2012) confirmed the stable activity of P128 in body fluids such as blood, plasma and normal and hyperimmune sera and demonstrated that P128 has no cytotoxic effects on mammalian cells, indicating the potency of P128 as antibacterial agent. To evaluate the in vivo efficiency, P128 was formulated as a hydrogel and tested in a nasal rat colonization model using MRSA USA300. Rats treated with the P128 hydrogel were either completely decolonized (four out of the nine rats) or the bacterial cell numbers were significantly reduced (Paul et al. 2011). A study on dogs diagnosed with a canine staphylococcal pyoderma skin infection further demonstrated the clinical efficiency of P128. Nearly 5–6 log reduction was seen in vitro upon P128 treatment on canine pyoderma isolates. Further, a case study with 17 dogs suffering from canine pyoderma were treated twice daily for 8 days with hydrogel P128. All lesions of the dogs under treatment healed completely after treatment and no recurrence of the symptoms occurred within 2 months (Junjappa et al. 2013). In addition, two studies have reported P128 as effective antibiofilm agent against sinus-derived clinical *S. aureus* isolates (Drilling et al. 2016) and coagulase-negative staphylococci (CoNS), the major cause of catheter-related bloodstream infections (Poonacha et al. 2017). The latter study revealed that P128 has a potent efficiency against both planktonic cells, biofilms and persister cells of three CoNS species *S. epidermidis*, *S. haemolyticus*, and *S. lugdunensis*. Moreover, the combination of P128 and the selected antibiotics (vancomycin, daptomycin, linezolid) showed a high synergistic inhibition of these staphylococcal strains (Poonacha et al. 2017). A renal abscess rat model was used to evaluate the efficiency of P128 against *S. aureus* bacteremia and its potential hypersensitivity reactions. Rats injected with a single intravenous dose of P128 up to 12.0 mg/kg and re-injected after a 15 day resting period showed no abnormal clinical signs of Type I hypersensitivity (anaphylaxis) in contrary to the positive control group injected with ovalbumin, an agent that causes anaphylaxis. In addition, no P128-related tissue injury (vasculitis or glomerulonephritis) typical for type III hypersensitivity was observed. As expected, low titers of anti-P128 antibodies were raised in the P128-dosed animals. Notwithstanding this, the concentration of a single intravenous bolus dose of 2.5 mg/kg in rats remains above the minimal inhibitory concentration (4 μg/mL) for 15 min. In addition, a single dose of P128 (2.5 mg/kg) was efficacious in rescuing animals from fatal MRSA USA300 bacteremia and prevents formation of renal abscesses (Channabasappa et al. 2017). Based on these data, a second preclinical study was started to evaluate pharmacokinetics and efficacy of P128 in a neutropenic mouse model of bacteremia. A single bolus (10, 30 and 60 mg/kg) of P138 was intravenously administered and caused a rapid and dose-dependent antibacterial activity. A maximum bactericidal effect was detected for all test dose levels after 30 mins and the cell numbers remain low after 24 h. Finally, the half-life was determined between 5.2 h (30 mg/kg dose) – 5.6 h (60 mg/kg dose) (Sriram et al. 2017).

Staphefekt™

In 2013, Micreos Human Health BV (the Netherlands) launched Staphefekt™, a recombinant staphylococcal phage endolysin, which is the functional compound in the cetomacrogol-based cream and the gel-based Gladskin series. These products are promoted by Micreos for the treatment of various inflammatory skin infections provoked by *S. aureus* by humans such as eczema, rosacea, skin irritation and inflammatory acne. The Gladskin products are currently registered as a (class I) medical device and are prescription-free available on the market in Europe. In a case study, three human objects suffering from *S. aureus*-related dermatoses were treated with Staphefekt™ resulting in a clinically relevant reduction of *S. aureus* on the skin. However, the clinical symptoms quickly recurred after stopping the treatment with Staphefekt™ (Totté et al. 2017).

11.5.2 Enzybiotics Against Gram-Negative Bacteria

Gram-negative bacteria have a protective outer membrane (OM) that shields the access to the cell for both hydrophobic and hydrophilic molecules, including endolysins. The outer membrane comprises a lipopolysaccharide (LPS) layer which is typically stabilized by electrostatic interactions between divalent cations and anionic phosphates and hydrophobic stacking of the fatty acids of the lipid A moiety of the LPS molecules. The presence of this membrane generally excludes Gram-negative bacteria from being killed by exogenously added endolysins.

Notwithstanding this largely impermeable barrier, some endolysins can permeate to a certain extent the OM of Gram-negative bacteria. Especially *Acinetobacter baumannii* appears to be sensitive to a number of such endolysins including LysAB3 and LysAB4 (Lai et al. 2013), PlyAB1 (Huang et al. 2014), PlyF307 (Lood et al. 2015), LysABP-01 (Thummeepak et al. 2016), ABgp46 (Oliveira et al. 2016) and LysPA26 (Guo et al. 2017). However, higher concentrations of those endolysins appear to be needed to kill Gram-negative bacteria (100–500 μg/ml) compared to Gram-positive bacteria (<10 μg/ml), indicating that this OM still hinders the passage of endolysin (Lim et al. 2014; Thummeepak et al. 2016; Shavrina et al. 2016). PlyF307 has a high bactericidal activity (>5 log reduction in cell number) against all tested clinical *A. baumannii* strains. Further, a treatment with PlyF307 (100 μg/ml) resulted in a significant reduction of planktonic and biofilm *A. baumannii* (~2-log reduction in colonizing bacteria) both in vitro and in vivo. Finally, PlyF307 increased survival of mice with lethal *A. baumannii* bacteremia with 50% (Lood et al. 2015). Recent studies focusing on antimicrobial peptides (AMPs) reported that several endolysin-derived peptides have physicochemical properties to disrupt and penetrate the Gram-negative OM and function as potent antibacterials (Thandar et al. 2016). Further, different compounds facilitate the passage of endolysins through the OM. Aromatic essential oils (carvacrol) disintegrate the OM by inducing LPS release (Diez-Martinez et al. 2013), chelating agents (EDTA) capture the stabilizing divalent cations from their binding sites (Briers et al. 2007) and polycationic compounds (polymyxin E) competitively displace the divalent cations (Mg^{2+} and Ca^{2+}) (Vaara 1993; Thummeepak et al. 2016). Another approach is the use of high hydrostatic pressure (HHP) to permeabilize the OM for endolysins (Briers et al. 2008). However, this approach may only find application in the food sector for products that cannot be pasteurized (e.g. oysters, guacamole) and are mostly not suitable for therapeutic applications.

We reported the development of Artilysin®s (Briers et al. 2014a). Artilysins are protein engineered endolysins that kill Gram-negative bacteria. The Artilysin® structure is based on a fusion of a selected endolysin and a specific outer membrane permeabilizing (OMP-)peptide. This OMP-peptide has physicochemical properties that interferes with the stabilizing forces of the LPS layer in the OM. OMP-peptides may comprise cationic and hydrophobic amino acids, giving them a polycationic or amphipathic nature. The OMP-peptide locally and transiently destabilizes the outer membrane for endolysin passage to the periplasm. Once the Artilysin® reaches the periplasm, they act similarly to a native endolysin: the CBD binds its target (specifically, the PG_1 and PG_3 CBDs bind PG with chemotype A1γ) and the EAD cleaves specific bonds of the PG, eventually resulting in osmotic lysis of the bacterial cell.

OM permeabilizers, such as EDTA, can further enhance the antibacterial activity of Artilysin®s (Briers et al. 2014a, b).

The best described Artilysin® is Artilysin-175 (Art-175). Art-175 is a fusion of the broad-spectrum sheep myeloid-29 acid (SMAP-29) peptide to the Gram-negative specific endolysin KZ144. The endolysin moiety has been further modified by site-specific mutations of three cysteines to serines to create a more stable and active Artilysin®. Art-175 has a high bactericidal effect (>4 log reduction in cell number after 30 min) against all tested *P. aeruginosa* strains, including environmental, clinical and multidrug-resistant strains. Art-175 is also highly bactericidal against multidrug-resistant *Acinetobacter baumannii*, including their persisters, resulting in a complete eradication of bacterial cultures (up to 8 log reduction) (Defraine et al. 2016). To assess potential resistance development, three different *P. aeruginosa* strains were serially exposed to subinhibitory doses of Art-175. Similar to endolysins, the highly selective pressure of the subinhibitory doses did not lead to the recovery of resistant variants after 20 cycles in contrast to when control antibiotics are used. In addition, no cross-resistance against Art-175 could be observed with thirteen prevalent resistance mechanisms, including colistin (Schirmeier et al. 2018). Art-175 shows no cytotoxicity against mouse connective tissue fibroblasts. These data indicate that Art-175 is well suited for a broad range of applications in hygiene, veterinary and humane medicine, including persister-driven chronic infections. Persisters are insensitive to traditional antibiotics, because they have shut down essential metabolic processes targeted by those antibiotics. Metabolically inactive persisters cells likely remain susceptible to Art-175 because of its active mode of action (i.e. enzymatic PG degradation), while traditional antibiotics rely on a passive inhibition mechanism.

In vivo efficiency of Art-085 (a fusion of OMP-peptide SMAP-29 and endolysin KZ144, similar to Art-175 but without mutations) has been reported in two dogs with otitis that not could be healed with standard antibiotics. Three different *P. aeruginosa* strains and one β-hemolytic *Streptococcus* sp. were identified in both ears of the first dog diagnosed with otitis externa. The dog recovered completely after a systematic treatment of Art-085 (three doses of Art-085 the 1st day followed by one daily dose during 6 days) without relapse. Similar results have been observed with the second dog suffering from media purulenta. Two different *P. aeruginosa* strains and one *Proteus mirabilis* strain were identified in both ears and a 3-week treatment with marbofloxacin showed no observable effect. Again, systematic treatment of Art-085 (seven treatments with Art-085 within 1 day and administration of three additional doses at day 7 and 8) was effective. After 2 weeks no relapse was observed (Briers and Lavigne 2015).

LoGT-008, which comprises a polycationic nonapeptide fused to the endolysin PVP-SE1gp146, was evaluated in two infection models. Human keratinocytes were cultured in vitro and infected by *P. aeruginosa*, mimicking barrier-disrupted, infected skin wounds as in burns. LoGT-008 could fully protect the cultured human keratinocytes monolayer infected with the highly virulent *P. aeruginosa* strain PA14, which is otherwise lethal. No cytotoxic effects were observed. Antibacterial killing of *P. aeruginosa* strain PA14 was confirmed using a simple in vivo *Caenorhabditis elegans* infection model. Treatment with LoGT-008 led to an increased survival rate (63%) in comparison with the untreated worms (10%) and ciprofloxacin (45%) (Briers et al. 2014b).

Lysando AG commercializes Artilysin®s and reports about the development of a wound care spray. The spray supports the healing process by creating an optimum moisture and a protective film and has bactericidal activity against pathogens. In a study to determine the efficacy of the wound spray, a group of human subjects was daily treated for a long period (>30 days). For 90% of the patients, the wound healing process started immediately after application and the wound sizes were reduced. Within 30 days, complete healing of the wounds was detected for 40% of the human subjects. One of the patients that has been treated with the wound spray was a coma patient with severe decubitus that was chronically infected by MRSA. After a few days, signs of healing were described and complete wound healing appeared after 10 months (Lysando 2017).

Artilysation or modifying the properties of an endolysin by fusion with a specific peptide can also be used to improve the properties of Gram-positive-specific endolysins. Artilysin-240 (Art-240), a fusion of a polycationic nonapeptide fused to the C-terminus of endolysin λSa2lys has an increased in vitro antibacterial activity against stationary streptococci cells. A 0.5–1.5 higher log reduction in bacterial cell number was observed compared to λSa2lys. Moreover, Art-240 has an increased enzymatic activity over a broader range of pH values and NaCl conditions compared to endolysin λSa2lys. Further experiments demonstrate that Art-240 has a twofold higher killing rate causing a 1.7-log reduction in cell number after 5 min whereas λSa2lys needs 30 min to cause the same reduction in cell number. In addition, a 4 to 12-fold reduced dose of Art-240 is required to achieve the same bactericidal effect as the native endolysin (e.g. 12.5 nM of Art-240 and 150 nM of λSa2lys are needed for a 2-log reduction). The positive charges of the fused polycationic nonapeptide, acting as an additional and local positive anchor, likely strengthen the interactions of Art-240 with the anionic phosphate groups in teichoic acids present on the cell surface (Rodríguez-Rubio et al. 2016).

11.5.3 Endolysins Active Against Mycobacteria

Mycobacterium tuberculosis, the causative agent of tuberculosis, poses a major threat because of the emergence of drug-resistant mycobacteria and the lack of effective therapies. The mycobacterial outer membrane possesses a mycolyl-arabinogalactan-PG complex composed of an inner PG layer (first barrier) that is covalently attached to arabinogalactan (AG), which in turn is esterified with a mycolic acid rich layer. The latter layer provides mycobacteria of a second, lipid barrier (Brennan 2003). In nature, mycobacteriophages attack both barriers from within by producing two cell wall hydrolytic enzymes, LysA and LysB. LysA is a PG hydrolase that degrades the PG layer and LysB is an enzyme with lipolytic activity that completes lysis by cleaving the linkage between the mycolic acids and the arabinogalactan layer (Gil et al. 2008; Payne et al. 2009). The presence of this mycolyl-arabinogalactan layer partly restricts access of endolysins (LysA) to the PG layer when added exogenously (Payne et al. 2009; Grover et al. 2014).

The past few years, mycobacteriophage-derived endolysins were isolated and tested against cell wall components of different mycobacteria strains. Catalão et al. (2011) examined the endolysin LysA derived from mycobacteriophage Ms6. Close inspection of the *lysA* gene revealed a shorter open reading frame entirely embedded in the same reading frame, encoding a second functional PG hydrolase. *E. coli* crude extracts that contain one of both lysins were spotted onto a bacterial lawn of test strains and inhibited bacterial growth of Gram-positive bacteria and mycobacteria. Grover et al. (2014) have demonstrated the bacteriostatic activity of LysB (isolated from two different mycobacteriophages Bxz2 and Ms6) against *M. smegmatis* in combination with surfactants. The presence of surfactants eliminates artifacts due to cell aggregation and facilitates the activity of LysB. The bacteriostatic activity of LysB was highest in the presence of Tween 80. This additional anti-bacterial effect can be explained by the oleic acid release due to LysB-mediated hydrolysis of Tween 80 (Grover et al. 2014).

11.6 Conclusion

The rise of multidrug-resistant and pandrug-resistant pathogens and the lack of new antibiotics has accelerated the research focus towards enzybiotics as potent antibiotic alternatives. Their high antibacterial activity against multidrug-resistant strains and persisters, low resistance development profile, anti-biofilm activity and synergy with antibiotics makes enzybiotics a promising candidate to use as novel therapeutics for human and animal health. Numerous preclinical studies have already demonstrated the potential of (engineered) endolysins to combat Gram-positive and Gram-negative pathogens and initial research has been done to extend enzybiotics towards mycobacteria. Currently, different lead enzybiotics are in the clinical development stage. Considering the fact that different clinical trials are still in progress, we expect that new clinical developments will be revealed in the near future.

References

Allison KR, Brynildsen MP, Collins JJ (2011) Metabolite-enabled eradication of bacterial persisters by aminoglycosides. Nature 473:216–220. https://doi.org/10.1038/nature10069

Becker SC, Foster-Frey J, Donovan DM (2008) The phage K lytic enzyme LysK and lysostaphin act synergistically to kill MRSA. FEMS Microbiol Lett 287:185–191. https://doi.org/10.1111/j.1574-6968.2008.01308.x

Becker SC, Roach DR, Chauhan VS et al (2016) Triple-acting lytic enzyme treatment of drug-resistant and intracellular *Staphylococcus aureus*. Sci Rep 6:25063. https://doi.org/10.1038/srep25063

Brennan PJ (2003) Structure, function, and biogenesis of the cell wall of *Mycobacterium tuberculosis*. Tuberculosis 83:91–97. https://doi.org/10.1016/S1472-9792(02)00089-6

Briers Y, Lavigne R (2015) Breaking barriers: expansion of the use of endolysins as novel antibacterials against gram-negative bacteria. Future Microbiol 10:377–390. https://doi.org/10.2217/fmb.15.8

Briers Y, Volckaert G, Cornelissen A et al (2007) Muralytic activity and modular structure of the endolysins of *Pseudomonas aeruginosa* bacteriophages phiKZ and EL. Mol Microbiol 65:1334–1344. https://doi.org/10.1111/j.1365-2958.2007.05870.x

Briers Y, Cornelissen A, Aertsen A et al (2008) Analysis of outer membrane permeability of Pseudomonas aeruginosa and bactericidal activity of endolysins KZ144 and EL188 under high hydrostatic pressure. FEMS Microbiol Lett 280:113–119. https://doi.org/10.1111/j.1574-6968.2007.01051.x

Briers Y, Schmelcher M, Loessner MJ et al (2009) The high-affinity peptidoglycan binding domain of Pseudomonas phage endolysin KZ144. Biochem Biophys Res Commun 383:187–191. https://doi.org/10.1016/j.bbrc.2009.03.161

Briers Y, Walmagh M, Grymonprez B et al (2014a) Art-175 is a highly efficient antibacterial against multidrug-resistant strains and persisters of *Pseudomonas aeruginosa*. Antimicrob Agents Chemother 58:3774–3784. https://doi.org/10.1128/AAC.02668-14

Briers Y, Walmagh M, Van Puyenbroeck V et al (2014b) Engineered endolysin-based "Artilysins" to combat multidrug-resistant gram-negative pathogens. MBio 5:e01379–e01314. https://doi.org/10.1128/mBio.01379-14

Callewaert L, Walmagh M, Michiels CW, Lavigne R (2011) Food applications of bacterial cell wall hydrolases. Curr Opin Biotechnol 22:164–171. https://doi.org/10.1016/j.copbio.2010.10.012

Cassino C (2016) Results of the first in human study of lysin CF-301 evaluating the safety, tolerability and pharmacokinetic 1248 profile in healthy volunteers. In: 26th ECCMID, April 9. http://c.eqcdn.com/_aa148792491891c86783e37af306cef5/contrafect/db/257/1148/pdf/ContraFect+CF-301+ECCMID+2016+Poster.pdf. Accessed 20 Dec 2017

Catalão MJ, Milho C, Gil F, et al (2011) A second endolysin gene is fully embedded in-frame with the lysA gene of mycobacteriophage Ms6. PLoS One 6:e20515. doi: 10.1371/journal.pone.0020515

Centres for Disease Control and Prevention (US) (2013) Antibiotic resistance threats in the United States, 2013. Centres for Disease Control and Prevention, US Department of Health and Human Services

Channabasappa S, Chikkamadaiah R, Durgaiah M, et al (2017) Preclinical studies of anti-staphylococcal ectolysin P128 for potential systemic hypersensitivity and evaluation of efficacy in *Staphylococcus aureus* bacteremia with renal abscesses in rats. In: ASM Microbe, June. http://www.gangagen.com/GangaGen-News/presentation_poster/ASM 2017 Rat final.pdf. Accessed 28 Dec 2017

Climo MW, Patron RL, Goldstein BP, Archer GL (1998) Lysostaphin treatment of experimental methicillin-resistant *Staphylococcus aureus* aortic valve endocarditis. Antimicrob Agents Chemother 42:1355–1360

Czaplewski L, Bax R, Clokie M et al (2016) Alternatives to antibiotica – a pipeline portfolio review. Lancet Infect Dis 16:239–251. https://doi.org/10.1016/S1473-3099(15)00466-1

Defraine V, Schuermans J, Grymonprez B et al (2016) Efficacy of artilysin Art-175 against resistant and persistent Acinetobacter baumannii. Antimicrob Agents Chemother 60:3480–3488. https://doi.org/10.1128/AAC.00285-16

D'Herelle F. (1917) Sur un microbe invisible antagoniste des bacilles dysentériques. C. R. Acad. Sci. 165:373–375

Diez-Martinez R, de Paz HD, Bustamante N et al (2013) Improving the lethal effect of cpl-7, a pneumococcal phage lysozyme with broad bactericidal activity, by inverting the net charge of its cell wall-binding module. Antimicrob Agents Chemother 57:5355–5365. https://doi.org/10.1128/AAC.01372-13

Djurkovic S, Loeffler JM, Fischetti VA (2005) Synergistic killing of *Streptococcus pneumoniae* with the bacteriophage lytic enzyme Cpl-1 and penicillin or gentamicin depends on the level of penicillin resistance. Antimicrob Agents Chemother 49:1225–1228. https://doi.org/10.1128/AAC.49.3.1225-1228.2005

Drilling AJ, Cooksley C, Chan C et al (2016) Fighting sinus-derived *Staphylococcus aureus* biofilms in vitro with a bacteriophage-derived muralytic enzyme. Int Forum Allergy Rhinol 6:349–355. https://doi.org/10.1002/alr.21680

Entenza JM, Loeffler JM, Grandgirard D et al (2005) Therapeutic effects of bacteriophage Cpl-1 lysin against *Streptococcus pneumoniae* endocarditis in rats. Antimicrob Agents Chemother 49:4789–4792. https://doi.org/10.1128/AAC.49.11.4789-4792.2005

Eugster MR, Loessner MJ (2012) Wall teichoic acids restrict access of bacteriophage endolysin Ply118, Ply511, and PlyP40 cell wall binding domains to the *Listeria monocytogenes* peptidoglycan. J Bacteriol 194:6498–6506. https://doi.org/10.1128/JB.00808-12

Fischetti VA (2005) Bacteriophage lytic enzymes: novel anti-infectives. Trends Microbiol 13:491–496. https://doi.org/10.1016/j.tim.2005.08.007

Fischetti VA (2010) Bacteriophage endolysins: a novel anti-infective to control gram-positive pathogens. Antimicrob Agents Chemother 300:357–362. https://doi.org/10.1016/j.ijmm.2010.04.002

George SE, Chikkamadaiah R, Durgaiah M et al (2012) Biochemical characterization and evaluation of cytotoxicity of antistaphylococcal chimeric protein P128. BMC Res Notes 5:280. https://doi.org/10.1186/1756-0500-5-280

Gerstmans H, Rodríguez-Rubio L, Lavigne R, Briers Y (2016) From endolysins to Artilysin®s: novel enzyme-based approaches to kill drug-resistant bacteria. Biochem Soc Trans 44:123–128. https://doi.org/10.1042/BST20150192

Gerstmans H, Criel B, Briers Y (2017) Synthetic biology of modular endolysins. Biotechnol Adv. https://doi.org/10.1016/j.biotechadv.2017.12.009

Ghahramani P, Khariton T, Jones S, et al (2017) Population pharmacokinetic-pharmacodynamic assessment of cardiac safety endpoints for CF-301, a first-in-class antibacterial lysin. In: ASM Microbe, June 3. https://www.contrafect.com/news/posters-and-presentations

Gil F, Catalao MJ, Moniz-Pereira J et al (2008) The lytic cassette of mycobacteriophage Ms6 encodes an enzyme with lipolytic activity. Microbiology 154:1364–1371. https://doi.org/10.1099/mic.0.2007/014621-0

Gilmer DB, Schmitz JE, Euler CW, Fischetti VA (2013) Novel bacteriophage lysin with broad lytic activity protects against mixed infection by Streptococcus pyogenes and methicillin-resistant Staphylococcus aureus. Antimicrob Agents Chemother 57:2743–2750. doi: 10.1128/AAC.02526-12

Grover N, Paskaleva EE, Mehta KK et al (2014) Growth inhibition of Mycobacterium smegmatis by mycobacteriophage-derived enzymes. Enzym Microb Technol 63:1–6. https://doi.org/10.1016/j.enzmictec.2014.04.018

Guo M, Feng C, Ren J et al (2017) A novel antimicrobial endolysin, LysPA26, against *Pseudomonas aeruginosa*. Front Microbiol 8:293. https://doi.org/10.3389/fmicb.2017.00293

Gutierrez D, Ruas-Madiedo P, Martinez B et al (2014) Effective removal of staphylococcal biofilms by the endolysin LysH5. PLoS One 9:e107307. https://doi.org/10.1371/journal.pone.0107307

Haddad KH, Schmelcher M, Sabzalipoor H, et al (2017) Recombinant endolysins as potential therapeutics against antibiotic-resistant *Staphylococcus aureus*: current status of research and novel delivery strategies. Clin Microbiol Rev 31. doi: https://doi.org/10.1128/CMR.00071-17

Hanlon GW (2007) Bacteriophages: an appraisal of their role in the treatment of bacterial infections. Int J Antimicrob Agents 30:118–128. https://doi.org/10.1016/j.ijantimicag.2007.04.006

Hermoso JA, Monterroso B, Albert A et al (2003) Structural basis for selective recognition of pneumococcal cell wall by modular endolysin from phage Cp-1. Structure 11:1239–1249. https://doi.org/10.1016/j.str.2003.09.005. showArticle Info

Hermoso JA, García JL, García P (2007) Taking aim on bacterial pathogens: from phage therapy to enzybiotics. Curr Opin Microbiol 10:461–472

Huang G, Shen X, Gong Y et al (2014) Antibacterial properties of Acinetobacter baumannii phage Abp1 endolysin (PlyAB1). BMC Infect Dis 14:681. https://doi.org/10.1186/s12879-014-0681-2

Jado I, López R, García E et al (2003) Phage lytic enzymes as therapy for antibiotic-resistant *Streptococcus pneumoniae* infection in a murine sepsis model. J Antimicrob Chemother 52:967–973. https://doi.org/10.1093/jac/dkg485

Jandourek A, Boyle J, Cassino C, et al (2017a) Long term immunology results of a phase 1 placebo controlled dose escalating study to examine the safety of CF-301 in human volunteers. In: 27th ECCMID, April 22. http://c.eqcdn.com/_aa148792491891c86783e37af306cef5/contrafect/db/257/1164/pdf/ECCMID+2017+Immunogenicity.pdf. Accessed 16 Dec 2017

Jandourek A, Boyle J, Murphy G, Cassino C (2017b) Inflammatory markers in a phase 1 placebo controlled dose escalating study of intravenous doses of CF-301 in human subjects. In: ASM Microbe, June 2. http://c.eqcdn.com/_aa148792491891c86783e37af306cef5/contrafect/db/257/1167/pdf/ASM+2017+CF-301-102+Phase+1+Inflammatory+Markers+Poster+FINAL+POSTER.pdf. Accessed 16 Dec 2017

Johnsborg O, Håvarstein LS (2009) Regulation of natural genetic transformation and acquisition of transforming DNA in *Streptococcus pneumoniae*. FEMS Microbiol Rev 33:627–642

Jun SY, Jung GM, Yoon SJ et al (2013) Antibacterial properties of a pre-formulated recombinant phage endolysin, SAL-1. Int J Antimicrob Agents 41:156–161. https://doi.org/10.1016/j.ijantimicag.2012.10.011

Jun SY, Jung GM, Yoon SJ et al (2014) Preclinical safety evaluation of intravenously administered SAL200 containing the recombinant phage endolysin SAL-1 as a pharmaceutical ingredient. Antimicrob Agents Chemother 58:2084–2088. https://doi.org/10.1128/AAC.02232-13

Jun SY, Jung GM, Yoon SJ et al (2016) Pharmacokinetics of the phage endolysin-based candidate drug

SAL200 in monkeys and its appropriate intravenous dosing period. Clin Exp Pharmacol Physiol 43:1013–1016. https://doi.org/10.1111/1440-1681.12613

Jun SY, Jang IJ, Yoon S et al (2017) Pharmacokinetics and tolerance of the phage endolysin-based candidate drug SAL200 after a single intravenous administration among healthy volunteers. Antimicrob Agents Chemother 61:e02629–e02616. https://doi.org/10.1128/AAC.02629-16

Junjappa RP, Desai SN, Roy P et al (2013) Efficacy of anti-staphylococcal protein P128 for the treatment of canine pyoderma: potential applications. Vet Res Commun 37:217–228. https://doi.org/10.1007/s11259-013-9565-y

Lai M-J, Soo P-C, Lin N-T et al (2013) Identification and characterisation of the putative phage-related endolysins through full genome sequence analysis in Acinetobacter baumannii ATCC 17978. Int J Antimicrob Agents 42:141–148. https://doi.org/10.1016/j.ijantimicag.2013.04.022

Langdon A, Crook N, Dantas G (2016) The effects of antibiotics on the microbiome throughout development and alternative approaches for therapeutic modulation. Genome Med 8:39. https://doi.org/10.1186/s13073-016-0294-z

Lavigne R, Robben J (2012) Professor Dr. Richard Bruynoghe: a 1951 overview of his bacteriophage research spanning three decades. Bacteriophage 2:1–4. https://doi.org/10.4161/bact.20024

Lim J-A, Shin H, Heu S, Ryu S (2014) Exogenous lytic activity of SPN9CC endolysin against gram-negative bacteria. J Microbiol Biotechnol 24:803–811. https://doi.org/10.4014/jmb.1403.03035

Loeffler JM, Fischetti VA (2003) Synergistic lethal effect of a combination of phage lytic enzymes with different activities on penicillin-sensitive and-resistant *Streptococcus pneumoniae* strains. Antimicrob Agents Chemother 47:375–377

Loeffler JM, Djurkovic S, Fischetti VA (2003) Phage lytic enzyme Cpl-1 as a novel antimicrobial for pneumococcal bacteremia. Infect Immun 71:6199–6204. https://doi10.1128/iai.71.11.6199-6204.2003

Loessner MJ (2005) Bacteriophage endolysins – current state of research and applications. Curr Opin Microbiol 8:480–487

Loessner MJ, Kramer K, Ebel F, Scherer S (2002) C-terminal domains of *Listeria monocytogenes* bacteriophage murein hydrolases determine specific recognition and high-affinity binding to bacterial cell wall carbohydrates. Mol Microbiol 44:335–349

Lood R, Winer BY, Pelzek AJ et al (2015) Novel phage lysin capable of killing the multidrug-resistant gram-negative bacterium Acinetobacter baumannii in a mouse bacteremia model. Antimicrob Agents Chemother 59:1983–1991. https://doi.org/10.1128/AAC.04641-14

López R, Garcíia E, García P (2004) Enzymes for anti-infective therapy: phage lysins. Drug Discov Today Ther Strateg 1:469–474

Lysando (2017) Topical applications in humans. https://www.lysando.com/topical-applications.html. Accessed 14 Jan 2017

Madigan MT, Martinko JM, Dunlap PV, Clark DP (2008) Brock biology of microorganisms 12th edn. Int Microbiol 11:65–73

Nelson D, Loomis L, Fischetti VA (2001) Prevention and elimination of upper respiratory colonization of mice by group A streptococci by using a bacteriophage lytic enzyme. Proc Natl Acad Sci U S A 98:4107–4112. https://doi.org/10.1073/pnas.061038398

Nelson DC, Schmelcher M, Rodriguez-Rubio L et al (2012) Endolysins as antimicrobials. Adv Virus Res 83:299. https://doi.org/10.1016/B978-0-12-394438-2.00007-4

Oh J, Schuch R (2017) The sub-MIC effect of lysin CF-301 on *Staphylococcus aureus* (*S. aureus*). In: ASM Microbe, June 2. http://c.eqcdn.com/_aa148792491891c86783e37af306cef5/contrafect/db/257/1169/pdf/A%0A1548 SM+2017+PAE+Final+Version+Poster.pdf. Accessed 20 Dec 2017

Oliveira H, Melo LDR, Santos SB et al (2013) Molecular aspects and comparative genomics of bacteriophage endolysins. J Virol 87:4558–4570. https://doi.org/10.1128/JVI.03277-12

Oliveira H, Vilas Boas D, Mesnage S et al (2016) Structural and enzymatic characterization of ABgp46, a novel phage endolysin with broad anti-gram-negative bacterial activity. Front Microbiol 7:208. https://doi.org/10.3389/fmicb.2016.00208

Pastagia M, Euler C, Chahales P et al (2011) A novel chimeric lysin shows superiority to mupirocin for skin decolonization of methicillin-resistant and-sensitive *Staphylococcus aureus* strains. Antimicrob Agents Chemother 55:738–744. https://doi.org/10.1128/AAC.00890-10

Paul VD, Rajagopalan SS, Sundarrajan S et al (2011) A novel bacteriophage Tail-Associated Muralytic Enzyme (TAME) from phage K and its development into a potent antistaphylococcal protein. BMC Microbiol 11:226. https://doi.org/10.1186/1471-2180-11-226

Payne K, Sun Q, Sacchettini J, Hatfull GF (2009) Mycobacteriophage lysin B is a novel mycolylarabinogalactan esterase. Mol Microbiol 73:367–381. https://doi.org/10.1111/j.1365-2958.2009.06775.x

Payne CM, Resch MG, Chen L et al (2013) Glycosylated linkers in multimodular lignocellulose-degrading enzymes dynamically bind to cellulose. Proc Natl Acad Sci 110:14646–14651. https://doi.org/10.1073/pnas.1309106110

Poonacha N, Nair S, Desai S, et al (2017) Efficient killing of planktonic and biofilm-embedded coagulase-negative staphylococci by bactericidal protein P128. Antimicrob Agents Chemother 61. doi: https://doi.org/10.1128/AAC.00457-17

Rashel M, Uchiyama J, Ujihara T et al (2007) Efficient elimination of multidrug-resistant *Staphylococcus aureus* by cloned lysin derived from bacteriophage φMR11. J Infect Dis 196:1237–1247. https://doi.org/10.1086/521305

Roach DR, Donovan DM (2015) Antimicrobial bacteriophage-derived proteins and therapeutic applications. Bacteriophage 5:e1062590. https://doi.org/10.1080/21597081.2015.1062590

Rodríguez-Rubio L, Chang W-L, Gutiérrez D et al (2016) "Artilysation" of endolysin λSa2lys strongly improves its enzymatic and antibacterial activity against streptococci. Sci Rep 6:35382. https://doi.org/10.1038/srep35382

Salmond GPC, Fineran PC (2015) A century of the phage: past, present and future. Nat Rev Microbiol 13:777–786. https://doi.org/10.1038/nrmicro3564

Sandeep K (2006) Bacteriophage precision drug against bacterial infections. Curr Sci 90:631–633

Schirmeier E, Zimmermann P, Hofmann V et al (2018) Inhibitory and bactericidal effect of Artilysin®Art-175 against colistin-resistant mcr-1-positive *Escherichia coli* isolates. Int J Antimicrob Agents. https://doi.org/10.1016/j.ijantimicag.2017.08.027

Schmelcher M, Shabarova T, Eugster MR et al (2010) Rapid multiplex detection and differentiation of Listeria cells by use of fluorescent phage endolysin cell wall binding domains. Appl Environ Microbiol 76:5745–5756. https://doi.org/10.1128/AEM.00801-10

Schmelcher M, Donovan DM, Loessner MJ (2012a) Bacteriophage endolysins as novel antimicrobials. Future Microbiol 7:1147–1171. https://doi.org/10.2217/fmb.12.97

Schmelcher M, Powell AM, Becker SC et al (2012b) Chimeric phage lysins act synergistically with lysostaphin to kill mastitis-causing *Staphylococcus aureus* in murine mammary glands. Appl Environ Microbiol 78:2297–2305. https://doi.org/10.1128/AEM.07050-11

Schuch R (2016) Post-antibiotic effects of lysin CF-301 against *Staphylococcus aureus* in human serum. http://c.eqcdn.com/_aa148792491891c86783e37af306cef5/contrafect/db/257/1150/pdf/Post-Antibiotic+Effects+of+Lysin+CF-301+Against+Staphylococcus+aureus+in+Human+Serum.pdf. Accessed 15 Jan 2017

Schuch R, Nelson D, Fischetti VA (2002) A bacteriolytic agent that detects and kills *Bacillus anthracis*. Nature 418:884–889. https://doi.org/10.1038/nature01026

Schuch R, Lee HM, Schneider BC et al (2014) Combination therapy with lysin CF-301 and antibiotic is superior to antibiotic alone for treating methicillin-resistant *Staphylococcus aureus*-induced murine bacteremia. J Infect Dis 209:1469–1478. https://doi.org/10.1093/infdis/jit637

Shavrina MS, Zimin AA, Molochkov NV et al (2016) In vitro study of the antibacterial effect of the bacteriophage T5 thermostable endolysin on *Escherichia coli* cells. J Appl Microbiol 121:1282–1290. https://doi.org/10.1111/jam.13251

Shen Y, Barros M, Vennemann T et al (2016) A bacteriophage endolysin that eliminates intracellular streptococci. Elife 5:e13152

Sriram B, Channabasappa S, Chikkamadaiah R, et al (2017) Pharmacokinetics and efficacy of ectolysin P128 in a mouse model of systemic Methicillin Resistant *Staphylococcus aureus* (MRSA) infection. In: ASM Microbe, June. http://www.gangagen.com/GangaGen-News/presentation_poster/ASM2017-Mousefinal.pdf. Accessed 18 Dec 2017

Sulakvelidze A, Alavidze Z, Morris JG (2001) Bacteriophage therapy. Antimicrob Agents Chemother 45:649–659. https://doi.org/10.1128/AAC.45.3.649-659.2001

Thandar M, Lood R, Winer BY et al (2016) Novel engineered peptides of a phage lysin as effective antimicrobials against multidrug-resistant Acinetobacter baumannii. Antimicrob Agents Chemother 60:2671–2679. https://doi.org/10.1128/AAC.02972-15

Thummeepak R, Kitti T, Kunthalert D, Sitthisak S (2016) Enhanced antibacterial activity of Acinetobacter baumannii bacteriophage ØABP-01 endolysin (LysABP-01) in combination with colistin. Front Microbiol 7:1402. https://doi.org/10.3389/fmicb.2016.01402

Totté JEE, van Doorn MB, Pasmans SGMA (2017) Successful treatment of chronic *Staphylococcus aureus*-related dermatoses with the topical endolysin Staphefekt SA. 100: a report of 3 cases. Case Rep Dermatol 9:19–25. https://doi.org/10.1159/000473872

Twort FW (1915) An investigation on the nature of ultra-microscopic viruses. Lancet 186:1241–1243

Vaara M (1993) Outer membrane permeability barrier to azithromycin, clarithromycin, and roxithromycin in gram-negative enteric bacteria. Antimicrob Agents Chemother 37:354–356

Viertel TM, Ritter K, Horz H-P (2014) Viruses versus bacteria-novel approaches to phage therapy as a tool against multidrug-resistant pathogens. J Antimicrob Chemother 69:2326–2336. https://doi.org/10.1093/jac/dku173

Vipra AA, Desai SN, Roy P et al (2012) Antistaphylococcal activity of bacteriophage derived chimeric protein P128. BMC Microbiol 12:41. https://doi.org/10.1186/1471-2180-12-41

Walmagh M, Briers Y, Dos Santos SB et al (2012) Characterization of modular bacteriophage endolysins from Myoviridae phages OBP, 201φ2-1 and PVP-SE1. PLoS One 7:e36991. https://doi.org/10.1371/journal.pone.0036991

Wang IN, Smith DL, Young R (2000) Holins: the protein clocks of bacteriophage infections. Annu Rev Microbiol 54:799–825. https://doi.org/10.1146/annurev.micro.54.1.799

White R, Chiba S, Pang T et al (2011) Holin triggering in real time. Annu Rev Microbiol 108:798–803. https://doi.org/10.1073/pnas.1011921108

Witzenrath M, Schmeck B, Doehn JM et al (2009) Systemic use of the endolysin Cpl-1 rescues mice with fatal pneumococcal pneumonia. Crit Care Med 37:642–649. https://doi.org/10.1097/CCM.0b013e31819586a6

Young R (2014) Phage lysis: three steps, three choices, one outcome. J Microbiol 52:243–258. https://doi.org/10.1038/srep29344

Zhang L, Li D, Li X et al (2016) LysGH15 kills *Staphylococcus aureus* without being affected by the humoral immune response or inducing inflammation. Sci Rep 6:29344. https://doi.org/10.1038/srep29344

Clinical Applications of Hyaluronidase

12

Gregor Cornelius Weber, Bettina Alexandra Buhren, Holger Schrumpf, Johannes Wohlrab, and Peter Arne Gerber

Abstract

Hyaluronidases are enzymes that degrade hyaluronic acid, which constitutes an essential part of the extracellular matrix. Initially discovered in bacteria, hyaluronidases are known to be widely distributed in nature and have been found in many classes including insects, snakes, fish and mammals. In the human, six different hyaluronidases, HYAL1-4, HYAL-P1 and PH-20, have been identified. PH-20 exerts the strongest biologic activity, is found in high concentrations in the testicles and can be localized on the head and the acrosome of human spermatozoa. Today, animal-derived bovine or ovine testicular hyaluronidases as well as synthetic hyaluronidases are clinically applied as adjuncts to increase the bioavailability of drugs, for the therapy of extravasations, or for the management of complications associated with the aesthetic injection of hyaluronic acid-based fillers. Further applications in the fields of surgery, aesthetic medicine, immunology, oncology, and many others can be expected for years to come. Here, we give an overview over the molecular and cellular mode of action of hyaluronidase and the hyaluronic acid metabolism, as well as over current and potential future clinical applications of hyaluronidase.

Keywords

Hylase · Hyaluronic acid · Hyaluronan · Filler · Spreading factor · Extravasation · Bioavailability

Abbreviations

BEL	Belotero®
BTH	Bovine testicular hyaluronidase
CHO	Chinese hamster ovary
CPM	Cohesive polydensified matrix
DMSO	Dimethylsulfoxide
ECM	Extracellular matrix
EMV	Emervel®
FDA	Food and Drug Administration
GAG	Glycosaminoglycan
HA	Hyaluronan
HAS	Hyaluronan synthase
HYAL	Hyaluronidase

G. C. Weber · B. A. Buhren · H. Schrumpf
P. A. Gerber (✉)
Department of Dermatology, Medical Faculty, Heinrich Heine University Düsseldorf, Düsseldorf, Germany
e-mail: peterarne.gerber@med.uni-duesseldorf.de

J. Wohlrab
Department of Dermatology and Venereology, Martin Luther University Halle-Wittenberg, Halle (Saale), Germany

Institute of Applied Dermatopharmacy, Martin Luther University Halle-Wittenberg, Halle (Saale), Germany

N. Labrou (ed.), *Therapeutic Enzymes: Function and Clinical Implications*, Advances in Experimental Medicine and Biology 1148, https://doi.org/10.1007/978-981-13-7709-9_12

Ig	Immunoglobulin
IGSC	Subcutaneous immunoglobulins
IU	International units
JUV	Juvederm®
kDA	Kilodalton
l	Liter
LA	Local anesthetic
mAb	Monoclonal antibody
ml	Milliliter
µm	Micrometer
OTH	Ovine testicular hyaluronidase
PEGPH20	PEGylated recombinant human hyaluronidase
PG	Proteoglycan
RES	Restylane®
rHuPH20	Recombinant human hyaluronidase
S. aureus	*Staphylococcus aureus*

12.1 Introduction

In 1942 Duran-Reynals published the first comprehensive review on the interplay of hyaluronidase (hyase, HYAL) and hyaluronic acid (hyaluronan, HA) entitled "Tissue permeability and the spreading factors in infection" (Duran-Reynals 1942). Five years later Karl Meyer published a physiological review on "The biological significance of hyaluronic acid and hyaluronidase" (Meyer 1947). Thereafter more than 10,000 scientific publications can be found when the search term "hyaluronidase" is entered in the US-National Institute of Health's PubMed.gov. Duran-Reynals as well as Meyer precisely described the fundamental function and interactions of hyaluronan and hyaluronidases that have remained the basis for all findings published until today (McClean 1941; Meyer 1947; Duran-Reynals 1942; Kreil 1995): Hyaluronan is a proteoglycan of high molecular weight (102–104 kDa) with 2000–25,000 repeating disaccharide units and an extended length of 2–25 µm (Fraser et al. 1997) (Fig. 12.1). It is an essential part of the extracellular matrix (ECM), which is composed of a complex network of fibrous proteins such as collagen and elastin and proteoglycans (PGs) to which characteristic glycosaminoglycan (GAG) chains are attached. Together the structural framework and components of the extracellular matrix play an important role in the regulation of cellular proliferation, adhesion and migration (Bissell et al. 1982). Via the secretion of various chemokines, cytokines and growth factors there is a constant physicochemical interaction between the extracellular matrix and involved cells, constituting a dynamic micromilieu with various functional states (Wohlrab et al. 2014). The metabolism of hyaluronan is tightly regulated by the interplay of synthetizing enzymes, hyaluronan synthases (HAS), on the one hand and degrading enzymes, hyaluronidases (HYAL), on the other hand (Meyer 1947; Menzel and Farr 1998). The enzyme hyaluronidase was initially identified in autolysates of pneumococci. Moreover, hyaluronidases were found in *Staphylococcus aureus* (*S. aureus*) and other bacteria, in venoms of bees and snakes, including rattlesnake and copperhead, as well as in venom and sperm of various other species (Meyer 1947). The high concentration of hyaluronidase in the acrosome and membrane of spermatozoa correlates to the discovery of the activity of hyaluronidase as a so called "spreading factor" through analysis of testicular extracts by Duran-Reynals and others (Martin-Deleon 2011; Duran-Reynals 1942).

Today, animal-derived as well as synthetic hyaluronidases are clinically applied as adjuncts to increase the spreading or delivery of drugs, including local anesthetics (LA) or immunoglobulins (Igs). In addition, hyaluronidases are used for extravasation management. Finally, hyaluronidases are considered a standard of care for the management of complications associated with the aesthetic injection of hyaluronan-based fillers (Lv et al. 2015; Buhren et al. 2016). This review aims to give an overview of the molecular and cellular mode of action as well as the current and potential future clinical applications of hyaluronidase.

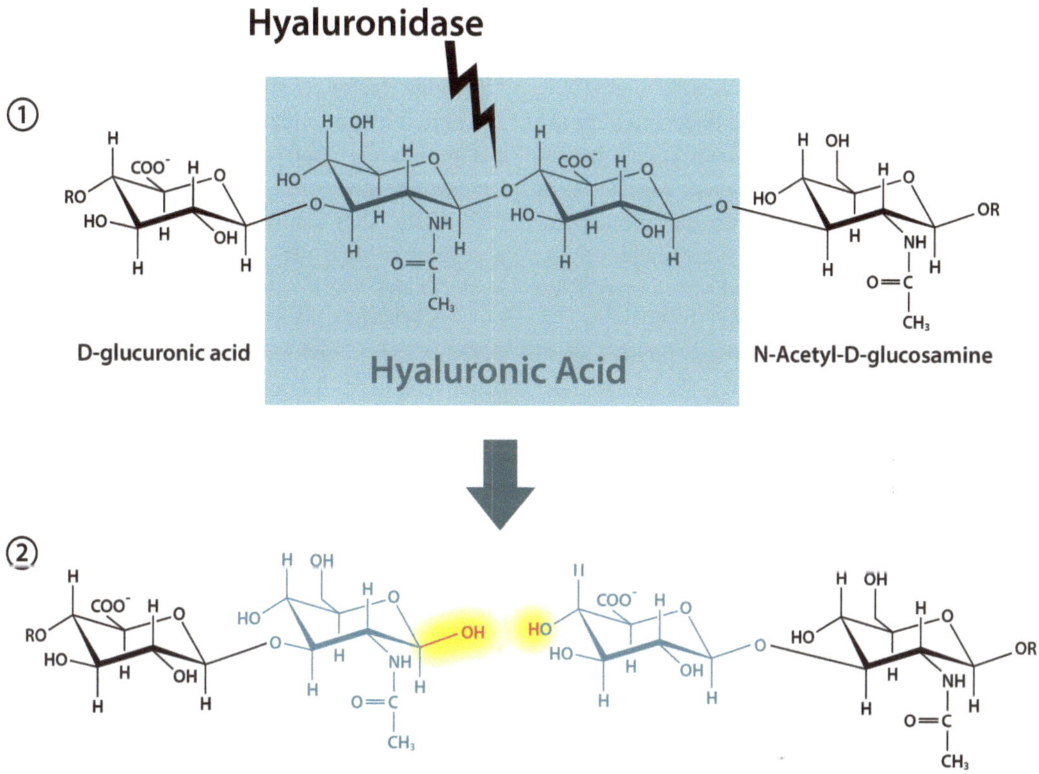

Fig. 12.1 Structure of hyaluronic acid and cleavage by hyaluronidase. Hyaluronic acid is a polysaccharide that consists of the repeating disaccharide D-glucuronic acid and N-Acetyl-D-glucosamine. Hyaluronidase cleaves its β1,4-glycosidic bonds

12.2 The Hyaluronan Metabolism

Hyaluronan is a polysaccharide that consists of the repeating disaccharide N-Acetyl-D-glucosamine and D-glucuronic acid: (-3)-*β*-D-*N*-acetylglucosamine-*β*(1,4)-D-glucuronic acid-*β*(1-) (Fig. 12.1). With a polymer length of up to 25 μm it is the largest polysaccharide found in vertebrates and is characterized by a high degree of polymerization (Fraser et al. 1997; Hascall and Esko 2015; Bohaumilitzky et al. 2017). This polymerization contributes to a high intrinsic viscosity of HA-solutions (Menzel and Farr 1998). It is estimated that the adult human body contains a total of 12–15 g of hyaluronan with more than 50% occurring in the skin (Bohaumilitzky et al. 2017). Moreover, hyaluronan is found in high concentrations in synovial fluids and joint tissues, as well as in the eye (Hascall and Esko 2015; Buhren et al. 2016; Meyer 1947). In the skin hyaluronan is primarily found in the dermis (Tammi et al. 1988; Juhlin 1997; Laurent and Fraser 1992; Buhren et al. 2016). It is extremely hygroscopic. Each disaccharide unit can bind 15 molecules of water, allowing 1 g of hyaluronan to bind an estimated 16 l of water (Bohaumilitzky et al. 2017). This high hydration capacity significantly contributes to the viscoelastic properties of the skin (Buhren et al. 2016; Juhlin 1997; Wohlrab et al. 2014). Hyaluronan is involved in a large number of physiological and pathophysiological processes such as mechanical tissue characteristics, the regulation of proliferation and differentiation of various cell types, as well as metabolism and cell transport. The cellular effects of hyaluronan are to a great extend mediated by the binding of exogenous hyaluronan or hyaluronan fragments to cell surface receptors, most notably CD44 or RHAMM (Csoka et al. 1997; Menzel and Farr 1998; Lokeshwar et al. 2001; Aruffo et al. 1990; Kreil 1995). Upon binding CD44 can mediate the endocytosis of hyaluronan, which is then hydrolysed by lysosomal hyaluronidase. Activation of CD44 can mediate cellular proliferation, cytokine production or cellular attachment on the one, but also metastatic spread of tumor cells on the other hand (Menzel and Farr 1998).

The half-life of non-stabilized, physiological hyaluronan in skin is about 24 h (Bohaumilitzky et al. 2017). Its degradation is mediated either by radical scission through reactive oxygen species (ROS) or by enzymatic cleavage. Regarding enzymatic synthetization in mammals three different hyaluronan synthases, HAS1, HAS2 and HAS3, are known (Weigel 2015). Concerning the hyaluronan-degrading hyaluronidases, six relevant gene loci on two chromosomes have been identified so far: the HYAL1, HYAL2 and HYAL3 genes on chromosome 3q21.3 and (Hyal-) PH20, HYAL4 and HYAL-P1 on chromosome 7q31.3 (Bertolami and Donoff 1982). For the human metabolome mainly HYAL1 (plasma hyaluronidase), HYAL2 (lysosomal hyaluronidase) and PH20 (sperm hyaluronidase) are of functional relevance (Kreil 1995).

12.3 Types and Functions of Hyaluronidases

Corresponding to their enzymatic properties and modus of splitting hyaluronan, as well as to their descent, hyaluronidases can be classified into three main groups: (i) hyaluronoglucosaminidase or "testicular-type hyaluronidase" (hyaluronate 4-glycanohydrolase, EC 3.2.1.35), (ii) hyaluronate glycanohydrolase or "leech-type hyaluronidase" (EC 3.2.1.36), and (iii) hyaluronate lyase or "bacterial-type hyaluronidase" (EC 4.2.99.1) (Kreil 1995; Menzel and Farr 1998). As noted before, mammalian, "testicular-type" hyaluronate 4-glycanohydrolase is present in mammalian spermatozoa but also in venoms of hymenoptera or snakes (Kreil 1995). The high enzyme activity in animal testicles has been described as early as 1928 by Duran-Reynals and has led to the definition of hyaluronidase as "spreading factor" (McClean 1941).

Human hyaluronidases are all β,1-4 endoglucosaminidases that cleave the β-N-acetylhexosamine-glycosidic bonds of hyaluronan (Fig. 12.1). They can furthermore be categorized according to their pH-dependent activity. Acidic hyaluronidases are active at a pH of 3–4 and include HYAL1, HYAL2, HYAL3 and HYAL4. PH20 is considered a neutral hyaluronidase and is active at pH 5–8 (Bohaumilitzky et al. 2017).

The most active human hyaluronidase is PH20 (sperm adhesion molecule 1, SPAM 1) which is involved in the process of fertilization and which is located on the head and the acrosome of human spermatozoa. PH20 can degrade the hyaluronan-rich matrix of the cumulus oophorus and enable sperm adhesion and penetration (Lv et al. 2015; Kreil 1995). Human plasma hyaluronidase (HYAL1) is ubiquitous in somatic tissues. It is the main hyaluronidase for the catabolism of hyaluronan in the extracellular matrix. Furthermore, it is associated with tumor biology, lysosomal storage disorders, and wound healing (Lv et al. 2015). Lysosomal hyaluronidase (HYAL2) is also widely expressed. It is localized in lysosomes or on the cell membrane. HYAL2 has been associated with tumor biology and angiogenesis (Bohaumilitzky et al. 2017; Lv et al. 2015). Hyaluronidases mediate a wide range of their biological effects indirectly via the generation of bioactive hyaluronan fragments of different sizes. In general, high molecular weight hyaluronan of >1000 kDa is proposed to be rather homeostatic. It has been shown to act rather anti-inflammatory, anti-proliferative and anti-angiogenic, and to be associated to wound healing processes. HA-breakdown products of medium (250–1000 kDa), low (10–250 kDa) or oligomeric size (<10 kDa) can be regarded as cellular alarm signals and have been associated with diverse effects including a pro- as well as anti-tumor activity (Bohaumilitzky et al. 2017; Stern et al. 2006). With regard to the central role of hyaluronidases for the generation of HA-breakdown fragments data on the controversial roles of HYAL1 and HYAL2 in tumorigenesis are complex and have extensively been reviewed by Lokeshwar and others (Lokeshwar and Selzer 2008). As compared to HYAL1, HYAL2 and PH20 the proposed biological activity and functional importance of HYAL3 and HYAL4 are low. HYAL3 is regarded as a non-enzymatic regulator of HYAL1 (Atmuri et al. 2008), whereas HYAL4 is proposed to be a chondroitinase without any activity against hyaluronate (Bohaumilitzky et al. 2017).

To conclude, hyaluronan, hyaluronan metabolizing enzymes and hyaluronan fragments have been associated with various physiological and pathological processes such as mechanical tissue characteristics, hydration and aging, fertilization and embryogenesis, proliferation, differentiation and wound healing as well as tumor biology (Hyde and Old 1999; Bohaumilitzky et al. 2017; Dai et al. 2007; Stern 2008; Fronza et al. 2014). Accordingly, different modes of action can be of interest for the therapeutic use of hyaluronidase (Herman and Castellot 1987). These could be categorized as follows: (1) targeted degradation of hyaluronan in order to manipulate structural tissue characteristics, (2) targeted degradation of hyaluronan in order to alter the diffusion properties of tissue, and (3) targeted reduction of the chain lengths of hyaluronan in order to exploit metabolic effects.

12.4 Therapeutic Use of Hyaluronidase

Succeeding the early discovery of the biological activity of hyaluronidase the enzymatic protein has been identified, characterized and developed for therapeutic use (Schoog 1951; Soldi 1951; Perrault and Housset 1952; Schonenberg 1952). To begin with, inspired by the diffusion enhancing effect of hyaluronidase in snake venoms, its degrading effects on hyaluronan in the extracellular matrix have been utilized to enhance the diffusion kinetics of intra- or subcutaneously applied drugs (Thorpe 1951; Lewis-Smith 1986). Here, hyaluronidase is simultaneously applied with local anesthetics and other injectable drugs (Landsman and Mandy 1991; Courtiss et al. 1995). Later, the diffusion enhancing effects have been objectified and evidenced in clinical studies (Wohlrab et al. 2012a, b). At the same time, applications in ophthalmology have been developed (Fanning 2001; Dieleman et al. 2012; Gul et al. 2015). Here, hyaluronidase is used in peribulbar anesthesia and vitreolysis (Schwartz et al. 1996; Narayanan and Kuppermann 2009; Zahl et al. 1991; Crawford and Kerr 1994; Dempsey et al. 1996, 1997). Further applications are the acceleration of the resorption of infiltrated toxic medication such as vinblastine and etoposide in case of extravasation during intravenous cytostatic therapy (Smith et al. 1997).

Commercial formulations of hyaluronidases include hyaluronidase from bovine testicles (BTH), ovine testicles (OTH) and also recombinant human hyaluronidase (rHuPH20) (Wohlrab et al. 2014; Meyer 1937; McClean 1941; Farr et al. 1997; Pirrello et al. 2007; Lee et al. 2010; Connolly et al. 2011). Purified hyaluronidases that have been obtained from other mammalian species show similar mechanisms of action as compared to human hyaluronidase (Bertolami and Donoff 1982; Garvin and Chipman 1974; Zaneveld et al. 1973). Bovine hyaluronidase (e.g. Hylase Dessau®, Riemser Pharma GmbH, Greifswald, Germany) has been used in several medical fields for many years including ophthalmo- and dermatosurgery as well as aesthetic medicine. It consists of a tetramer with four equal subunits with a molecular mass of 60 kDa (Khorlin et al. 1973; Barsukov et al. 2003). In comparison to bovine hyaluronidase, ovine hyaluronidase (e.g. Vitrase®, Bausch&Lomb, Rochester, NY, USA) is proposed to have a higher purity with a molecular mass of approximately 55 kDa and is used for comparable indications (Silverstein et al. 2012). Since animal-derived hyaluronidases can be contaminated with immunoglobulins and proteases they are proposed to potentially increase IgE-mediated immunogenic response like hypersensitivity reactions or anaphylaxis (Benditt et al. 1951; Kind and Roffler 1961; Schwartzman 1951). In 2005 a highly purified soluble form of the naturally occurring human hyaluronidase was approved by the US Food and Drug Administration (FDA). This enzyme, rHuPH20 (e.g. ENHANZE®, Halozyme Therapeutics, Inc., San Diego, CA, USA), exerts a higher specific activity as compared to commercially available animal-derived hyaluronidases. Enzyme activity is correlated to the concentration of active hyaluronidase protein (U) per total protein (mg), and is approximately 500–700 U/mg for unprocessed, animal-derived hyaluronidase, approximately 18,000 U/mg for pharmaceutically prepared animal-derived hyaluronidase, and approximately 120,000 U/mg for rHuPH20 (Silverstein et al. 2012). rHuPH20 has a molecular weight of 61 kDa and is produced by genetically engineered Chinese Hamster Ovary (CHO) cells which contain a DNA plasmin encoding PH20 (Yocum et al. 2007). Subsequently rHuPH20 is purified from non-hyaluronidase impurities and potential pathogens during processing steps. The resulting purity of almost 99% is associated with a significant lower risk of immunogenic reactions providing a superior safety profile (Silverstein et al. 2012; Spandorfer et al. 2012). With a half-life of shorter than 30 min rHuPH20 is short acting making it difficult to detect in plasma after subcutaneous administration (Frost 2007). rHuPH20 is used in combination with immunoglobulins to compensate deficiency syndromes or for subcutaneous rehydration therapy (Connolly et al. 2011; Wasserman 2017; Wasserman et al. 2016; Carne et al. 2016; Pirrello et al. 2007; Lee et al. 2010). Further applications in connection with the use of therapeutic antibodies and proteins in chronic inflammatory, infectious and oncological diseases can be expected for years to come (Stebliuk 1972; Zaoli 1958; Warnery et al. 1954; Martinez-Quintanilla et al. 2015; Guedan et al. 2010).

12.4.1 Hyaluronidase for Local Infiltration Anesthesia

The efficacy of intracutaneously applied drugs is limited due to specified bioavailability of the active substance within the respective target compartment. Hyaluronidases can catalyse the degradation of hyaluronan in the extracellular matrix, thereby enabling enhanced tissue permeability and increasing the bioavailability of injected drugs (Fig. 12.2) (Lokeshwar et al. 2001; Wohlrab et al. 2012a, b, 2014). This principle was first applied for dermatosurgery in 1951 when Thorpe showed that adjuvant hyaluronidase promotes the diffusion of an injected anesthetic (Thorpe 1951). The subcutaneous co-application of hyaluronidases as an adjuvant to local infiltration anesthetics has been shown to significantly increase the effectiveness of local anesthesia especially in the first minutes after injection and enhanced the safety of the surgical procedure (Remy et al. 2008; Wohlrab et al. 2012b). Of note, the rate of diffusion is proportional to the amount of hyaluronidase given (Lee et al. 2010).

The use of hyaluronidase as an adjunct to local anesthetics has been described for ophthalmology, nail surgery, traumatology, dentistry, and others. In 1986 Lewis-Smith demonstrated that a local anesthetic combined with hyaluronidase was associated with an immediate analgesia and a more than doubled anesthetized area as compared to the local anesthetic alone (Lewis-Smith 1986). Later Landsman and Mandy reported that the co-application of hyaluronidase with epinephrine subcutaneously for plastic surgery of the head improved the diffusion of infiltrated anesthetics, enhanced anesthesia, and improved the ease of dissection (Landsman and Mandy 1991). In 1994 Connolly et al. noticed while using hyaluronidase in combination with bupivacaine in surgery for ingrown toenail, that hyaluronidase leads to a more rapid dissolution of the local anesthetic and an earlier relief of pain as compared to bupivacaine alone (Connolly et al. 1994). In 2001 a prospective, randomized, double-blind trial examined the anesthetic effect of lidocaine compared to a combination of lidocaine plus hyaluronidase for inferior alveolar nerve blocks. Surprisingly, it was concluded that the addition of hyaluronidase only had a neglectable effect on the efficacy of the nerve block, whereas its property to loosen structures of the connective tissue was associated with a higher incidence of severe

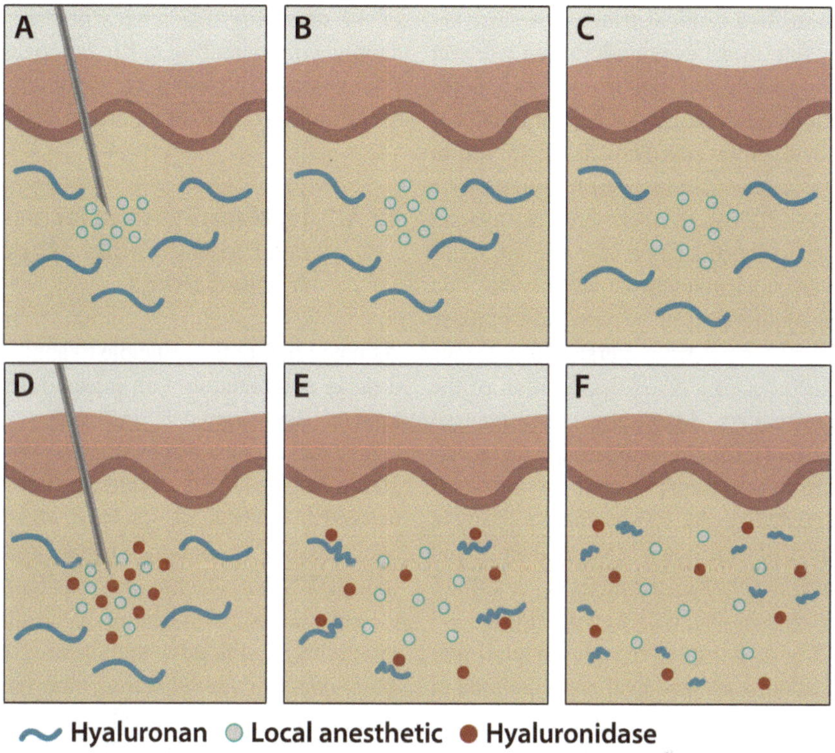

Fig. 12.2 Hyaluronidase increases the bioavailability of injected drugs. Hyaluronidase can catalyse the degradation of hyaluronan in the extracellular matrix, thereby enabling enhanced tissue permeability and increasing the bioavailability of injected drugs, e.g. local infiltration anesthetics. (**a**–**c**) The efficacy of conventional local infiltration anesthesia is limited by the fact that the dense extracellular matrix constrains a rapid dispersion of the drug within the tissue. (**d**–**f**) Hyaluronidase is co-injected with the local anesthetic to promote dispersion of the drug as a "spreading factor". In the case of local infiltration anesthetics this results in (**f**) a faster and more widespread anesthesia and hence a higher efficacy

postoperative pain (Ridenour et al. 2001). Ahmed et al. administered a local anesthetic in combination with hyaluronidase to harvest large split thickness skin grafts using only a single injection instead of multiple injections without hyaluronidase (Ahmed and Ahmed 2004). Furthermore, the co-application of hyaluronidase and local anesthetics was described for the treatment of chronic pain syndrome, resulting in a reduction of oedema and detachments of epidural adhesions (Dunn et al. 2010). Today, hyaluronidase in combination with various local anesthetics is most commonly used in ophthalmology. The co-administration of hyaluronidase improves the efficacy of retrobulbar anesthetic blocks and peribulbar infiltrations. Lindley-Jones used a mix of 0.5% bupivacaine, 4% lignocaine and hyaluronidase for peribulbar anesthesia during phacoemulsification surgery and represented a less painful procedure compared to topical anesthesia (Lindley-Jones 2000). Khandwala et al. showed in a prospective randomised double-blind study on 19 patients that the addition of hyaluronidase significantly augmented the dispersal of local anesthetics. Herein, 5 ml of 2% lidocaine alone, or with 15 IU/ml hyaluronidase were applied via infero-nasal sub-Tenon's injection. Subsequently at 3 and 5 min B-scan sonography was performed, recording significantly less depth of local anesthetic fluid in the group that had received hyaluronidase, which correlated to a significantly higher dispersal rate of the local anesthetic (Khandwala et al. 2008). Also Remy et al. evaluated the efficacy and safety of hyaluronidase in a prospective double-blind, placebo-controlled clinical trial as an adjuvant to mepivacaine in 40 eyes undergoing cataract surgery. The addition of hyaluronidase to mepivacaine enhanced the safety of the surgical procedure due to more complete akinesia within 5 min and a faster onset of complete anesthesia (Remy et al. 2008). Furthermore, Narayanan and Kuppermann described the use of hyaluronidase for pharmacologic vitreolysis. The intravitreal application of ovine hyaluronidase in patients suffering from persistent vitreous hemorrhage induced a rapid degradation of the extracellular matrix, leading to an increased diffusion rate of vitreous hemorrhage and a statistical significant improvement in best-corrected visual acuity (BCVA) as compared to the application of saline (Narayanan and Kuppermann 2009; Kuppermann et al. 2005). In 2012 Wohlrab and colleagues enrolled 44 participants in a prospective, randomized, placebo-controlled study and evaluated the area of anesthetized skin of local anesthetics with a solution of 75 IU of bovine testicular hyaluronidase (Hylase® Dessau, Riemser Pharma GmbH, Greifswald, Germany) and 0.5% lidocaine versus 0.5% lidocaine alone. The results show that the co-application of hyaluronidase with local anesthetics leads to an approximately 50% larger anesthetized area than local anesthetics alone and that this effect occurs fast within the first 5–15 min (Wohlrab et al. 2012b). Another analysis of the same group addressed the question whether hyaluronidase may impair cutaneous wound healing. This prospective, single-center, placebo-controlled, double-blind, intraindividual comparison included 20 participants. The authors created two superficial wounds in each participant by removing the epidermis via induced suction blisters. After excision of the blister tegmen a solution of 150 IU of bovine testicular hyaluronidase (Hylase® Dessau, Riemser Pharma GmbH, Greifswald, Germany) in combination with 0.5% lidocaine was injected sublesional to one wound and 0.5% lidocaine alone was injected sublesional to the other wound. Over a course of 14 days any differences neither in macroscopic wound healing nor in hemovascular perfusion measured via laser-Doppler flow were identified (Wohlrab et al. 2012a).

12.4.2 Hyaluronidase to Increase the Bioavailability of Locally Applied Drugs

As noted before the principle to apply hyaluronidase in combination with injected drugs is best known for surgical infiltration anesthesia (Fig. 12.2) (Lokeshwar et al. 2001; Wohlrab et al. 2012a, b, 2014). Yet, besides the use of animal-derived hyaluronidase for local anesthetics, the application of recombinant human hyaluronidase (rHuPH20) has recently gained importance for the subcutaneous application of insulin, volume, morphine, antibiotics, therapeutic monoclonal antibodies, immunoglobulins, and other indications (Bookbinder et al. 2006).

In 2009 Vaughn and colleagues reported for the first time that co-administration of rHuPH20 significantly accelerated pharmacokinetics of subcutaneously injected insulin by means of increased peak serum insulin concentrations, C(max), reduced time to C(max), and t(max). Clinically these effects resulted in significantly improved glucodynamic parameters (Vaughn et al. 2009). Considering the studies performed by Vaughn and others it can be stated that hyaluronidase can accelerate the absorption of subcutaneous applied human insulin allowing for a control of the glycodynamic effect similar to rapid-acting insulin analogues without an increased risk of hypoglycaemia (Hompesch et al. 2011; Garg et al. 2014; Vaughn et al. 2009; Muchmore and Vaughn 2010). Similar positive effects of rHuPH20 on C(max) and t(max) were observed for subcutaneous morphine as reported by Thomas et al. and for the antibiotic ceftriaxone as reported by Harb et al. (2010) and Thomas et al. (2009). Furthermore, Allen et al. demonstrated in a phase IV, multicenter, single-arm study that rHuPH20 facilitated the safe and effective rehydration in children of 2 months to 10 years of age. Herein, the subcutaneous injection of rHuPH20 was followed by subcutaneous infusion of 20 ml/kg isotonic fluid. A successful rehydration was achieved in more than 80% of the patients. In children the subcutaneous application of medications and most notably large volumes is of particular advantage as compared to the intravenous route, as the compliance with painful interventions is low and vessels are often small and more difficult to puncture (Allen et al. 2009). Recently, also the effective combination of rHuPH20 and a C1 inhibitor (C1 INH) for the treatment of patients with hereditary angioedema with C1 INH deficiency (HAE) was reported (Weller et al. 2017).

In 2012 Wassermann et al. reported for the first time the successful administration of rHuPH20-facilitated subcutaneous immunoglobulin (IGSC) in patients with primary immunodeficiency (PI). With regard to efficacy, safety and pharmacokinetics rHuPH20-facilitated IGSC was equivalent to the gold standard, intravenous immunoglobulin (IVIG). The tolerability of rHuPH20-facilitated IGSC was good at a lower frequency of systemic reactions (Wasserman et al. 2012). In line with these results, more recently Speth et al. used high dose hyaluronidase-facilitated IGSC in five patients with steroid-refractory juvenile dermatomyositis. Again, it was demonstrated that subcutaneous application of IgG in combination with hyaluronidase resulted in similar peak serum levels of IgG as compared to IVIG, resulting in mostly clinically inactive disease without any relevant local or systemic adverse effect (Speth et al. 2016).

Concerning the intention to combine hyaluronidase with therapeutic monoclonal antibodies (mAbs), it can be stated that monoclonal antibodies are typically given by slow intravenous infusion and that therapy is hence associated with long infusion times. Furthermore, intravenous applications are associated with the risk of extravasation and other infusion-related complications like infections or thromboembolic events. In 2013 Morcos and colleagues reported significantly improved pharmacokinetic and pharmacodynamic parameters for the subcutaneous co-administration of the monoclonal antibody trastuzumab (Herceptin®, Hoffmann-La Roche Ltd., Basel, Switzerland) and rHuPH20. Similar observations were reported for rituximab (MabThera®, Hoffmann-La Roche Ltd., Basel, Switzerland) (Morcos et al. 2013; Shpilberg and Jackisch 2013; Hamizi et al. 2013; Kagan and Mager 2013). Trastuzumab is a humanized monoclonal anti-HER2-antibody that is used for the therapy of HER2-positive breast cancer. Rituximab is a chimeric monoclonal anti-CD20-antibody used to treat autoimmune diseases (e.g. rheumatoid arthritis) or cancer (e.g. non-Hodgkin's lymphoma, chronic lymphocytic leukaemia) (Kagan and Mager 2013). It was concluded that subcutaneous administration of monoclonal antibodies co-formulated with rHuP20 may enable a novel effective, well-tolerated, cost-effective, and convenient pharmacologic therapy, thereby improving patient's satisfaction (Shpilberg and Jackisch 2013).

Finally, most recently PEGylated rHuPH20 (PEGPH20) in combination with gemcitabine and nab-paclitaxel is being assessed for the oncologic therapy of hyaluronic acid-high advanced pancreatic cancer. Hyaluronic acid-high pancreatic cancer is characterized by the marked production of hyaluronan by the tumor itself, which results in a modulation of the tumormicromilieu and an increase of interstitial tumor pressure, eventually limiting the perfusion of tumor tissue. Accordingly, systemically applied anticancer drugs may not sufficiently enough penetrate into the targeted tumor compartment to reach cytotoxic concentrations. PEGPH20 may break down this burden to re-establish tumor perfusion and enable the accumulation of chemotherapeutics in order to increase the antineoplastic effect (Doherty et al. 2018).

12.4.3 Hyaluronidase for Extravasation Management

Extravasation can be defined as the accidental leakage of intravenous fluids or medications into the perivascular or subcutaneous space, potentially resulting in significant tissue damage. The estimated incidence of extravasation may range from 0.001% to 7% (Boulanger et al. 2015). The extent of potential damage depends on the properties, concentration, and the amount of the infused substance. Severe extravasation injury is induced most frequently by certain chemotherapeutic agents, vasoactive substances that have the ability to induce ischaemia, concentrated electrolytes, iodinated contrast media, and other hyperosmolar solutions (Schummer et al. 2003). Since the extent of deep soft tissue damage is difficult to predict, the degree of severity of extravasation is often underestimated. Most extravasations have comparably minor, transient consequences, such as pain, swelling, localized erythema or hyperpigmentation. However, in severe cases extensive tissue damage, ulceration, necrosis and compartment syndrome may occur, potentially necessitating surgical debridement or even amputation (MacCara 1983; Rowlett 2012; Montgomery and Budreau 1996; Wiegand and Brown 2010; Dougherty 2008).

Because of the scarcity of evidence, available algorithms for the management of extravasation remain a matter of debate. Specific interventions depend on the type of infused drug as well as the anatomic location and the extent of extravasation (Boulanger et al. 2015). Regardless, the initial step for managing a suspected vesicant or irritating extravasation is to stop the infusion (Schrijvers 2003; Kassner 2000; Bertelli 1995; Albanell and Baselga 2000; Dunagin 1982; Goolsby and Lombardo 2006). Additional measures include aspiration of as much drug and blood as possible, marking of the affected area, the elevation of affected limbs, warm or cold compresses, as well as the application of systemic analgesics, if necessary (Boulanger et al. 2015). To date various medications and substances, including antidotes, depending on the agent that has caused extravasation have been tested and found to have beneficial effects to vesicant extravasation injuries. Sodium thiosulfate, for example, has been shown to be useful for the management of extravasations of nitrogen mustard or cisplatin (Dorr 1990; Boulanger et al. 2015). Topical dimethylsulfoxide (DMSO) is suggested for the management of extravasation of a wide range of chemotherapeutics including anthracyclines, cisplatin, carboplatin, mitomycin and others (Langer et al. 2000; El-Saghir et al. 2004; Alberts and Dorr 1991; Olver et al. 1988; Bertelli et al. 1995).

With regard to hyaluronidase, its application has been suggested in order to dilute local toxic concentrations in particular of cytostatic drugs like vinca alkaloids or paclitaxel (Boulanger et al. 2015). For vinorelbine, vinblastine or vincristine the local injection of 250 IU of hyaluronidase was reported to result in pain relief within several days after the extravasation event (Bertelli et al. 1994). In cases of paclitaxel extravasation the benefit of hyaluronidase is discussed controversially (Stanford and Hardwicke 2003). Whereas Bertelli and colleagues reported in a series of five patients with paclitaxel extravasation that the infiltration of hyaluronidase was effective and resulted in complete remission of local symptoms in all cases, Dubois et al. found in a series of four patients that hyaluronidase delayed healing in two patients as compared to the sole application of cold compresses in the other two patients (Bertelli et al. 1997; Dubois et al. 1996).

Apart from the extravasation of antineoplastic agents the extravasation of diagnostic substances is a relatively frequent occurrence during radiographic procedures. The ability of iodinated contrast media to produce severe extravasation injury largely depends on the volume and height of osmolality. Studies and clinical experience demonstrated that contrast media with osmolalities between 1025 and 1420 mOsm/kg facilitates a higher risk for severe extravasation injury (Wang et al. 2007; Bellin et al. 2002). Today, the use of contrast media agents with lower osmolality has largely reduced the risk of complications (Reynolds et al. 2014). Yet, in particular rapid bolus injections of iodinated contrast media are associated with more severe clinical manifestations, such as blistering and ulcerations (Federle et al. 1998). Small volumes of contrast media extravasation on the other hand may cause minor complications such as pain upon injection of the agent, swelling, oedema, and erythema (Belzunegui et al. 2011; Sbitany et al. 2010;

Selek et al. 2007). Extravasation of diagnostic substances is generally managed successfully conservative with warm or cold compresses and elevation alone (Doellman et al. 2009; Elam et al. 1991), in more severe cases, surgical consultation is recommended. With regard to hyaluronidase, the successful use has been reported in some cases (Bookbinder et al. 2006; Cochran et al. 2002). One exemplary case published by Rowlett reported complete clinical resolution after subcutaneous administration of recombinant human hyaluronidase following the extravasation of 100 ml of iodinated contrast media in the upper limb. Herein, 750 IU of hyaluronidase were injected in a circle around the extravasation site resulting to marked improvement after 4 h (Rowlett 2012). Finally, Fox et al. described in a case report a good clinical response after administration of intradermal hyaluronidase following the extravasation of amiodarone with decreased expansion of erythema and warmth as well as a decrease in pain (Fox et al. 2017).

12.4.4 Hyaluronidase for the Management of Hyaluronic Acid Filler Associated Complications

Skin changes are among the most visible signs of aging. One of the hallmarks of skin aging is a decrease in the skin's hyaluronan content (Papakonstantinou et al. 2012; Meyer and Stern 1994). This decrease of hyaluronan is mediated by both extrinsic and intrinsic factors. Extrinsic factors (extrinsic skin aging), such as the scission of hyaluronan by the build-up of reactive oxygen species (ROS) as a by-product of cellular metabolism or generated by UV-exposure (photoaging) or other pro-oxidative stressors, on the one hand, as well as intrinsic factors (intrinsic skin aging), such as an age-related decline in hyaluronan-neosynthesis by dermal fibroblasts or hormonal changes, on the other hand, contribute simultaneously to a progressive loss of skin integrity (Glogau 1997). These degenerative changes that affect hyaluronan and other components of the extracellular matrix, such as collagen or elastin, result in a loss of volume and hydration as well as a diminished elasticity, resulting in sagging skin, and fine and coarse wrinkles. Over the past decades various substances, including autologous fat, silicone, collagen, calcium hydroxylapatite, poly-L-lactic acid and most notably hyaluronic acid have been applied to augment and diminish the signs of skin and facial aging.

Today, the injection of gels containing hyaluronic acid, HA-based reversible dermal fillers (HA-fillers), is regarded as gold standard for non-surgical minimal invasive facial rejuvenation and is one of the most commonly performed procedures in cosmetic dermatology and plastic surgery. Briefly, HA-fillers are injected for facial recontouring, tissue augmentation and deep skin hydration (Narins et al. 2010, 2011). If performed properly, the injection of HA-fillers is safe and effective and can be synergistically combined with additional minimal invasive aesthetic procedures. Figure 12.3 shows the case of a patient that presented for the treatment of the signs of her facial aging at initial presentation and in the course after a multimodal therapy including laser skin resurfacing, as well as botulinum toxin A and HA-filler injections (Fig. 12.3). Figure 12.4 displays the successful augmentation of a patient's tear-trough with 0.2 ml of an HA-filler for each side (Fig. 12.4).

Potential complications of filler injections range from pain, edema and hematomas, over unaesthetic overcorrections, the Tyndall-effect – a bluish discoloration caused by too superficially injection of HA -, lower eyelid edema following tear-trough augmentation (Fig. 12.5), up to granulomas (Fig. 12.6), infections, and finally tissue necrosis with ulcerations and scarring or even blindness due to vascular occlusions (Figs. 12.7 and 12.8) (Buhren et al. 2016; DeLorenzi 2013; Myung et al. 2017). Vascular complications are most severe and will likely lead to necrosis, when the occluded arteries are responsible for the main blood supply to certain areas and collaterals are sparse or absent. Also more superficial facial arteries are at a higher risk of iatrogenic occlusion. It is proposed that it is not even necessary to inject the filler directly into the lumen of the vessel but that also too large volumes of filler placed

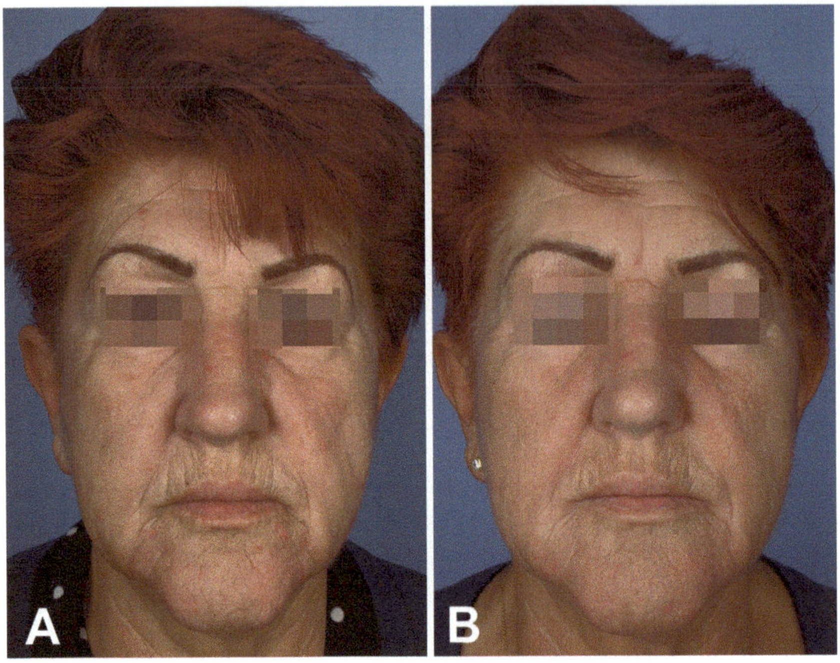

Fig. 12.3 Multimodal aesthetic treatment combining laser resurfacing, botulinum and HA-filler injections. (**a**) The 58-year old patient presented for a treatment of her marked signs of facial and skin aging including loss of volume, sagging skin, as well as fine and coarse wrinkles. Her personal anamnesis was positive for >40 years of intense UV-exposure and cigarette smoking. Panel (**b**) shows the patient after a multimodal facial aesthetic treatment combining fractional laser resurfacing (2.940 nm Er:YAG laser), botulinum toxin A and HA-filler injections

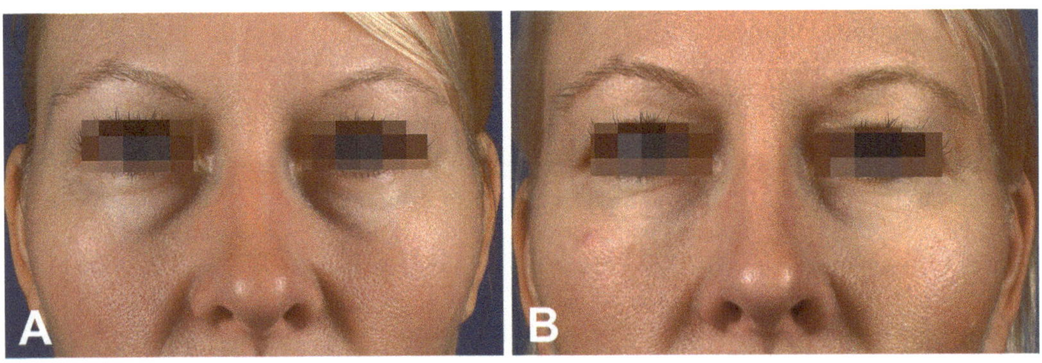

Fig. 12.4 Successful tear-trough augmentation using an HA-filler. Thirty five-year old patient presenting for the treatment of a moderate tear-trough deformity (**a**) prior to therapy and (**b**) 3 weeks after injection of 0.2 ml of an HA-filler to each side using a blunt cannula

in the perivascular space may compress the vessel from the outside eventually leading to a functional occlusion. "High risk" or "expert areas" with a higher probability of vascular complications following the injection of fillers include the glabellar region (supratrochlear artery) (Figs. 12.7 and 12.8), the nasociliar region (dorsal nasal artery), the temporal region (temporal arteries), the superior nasolabial fold and the tear-trough (angular artery), or the nasolabial sulcus (facial artery)

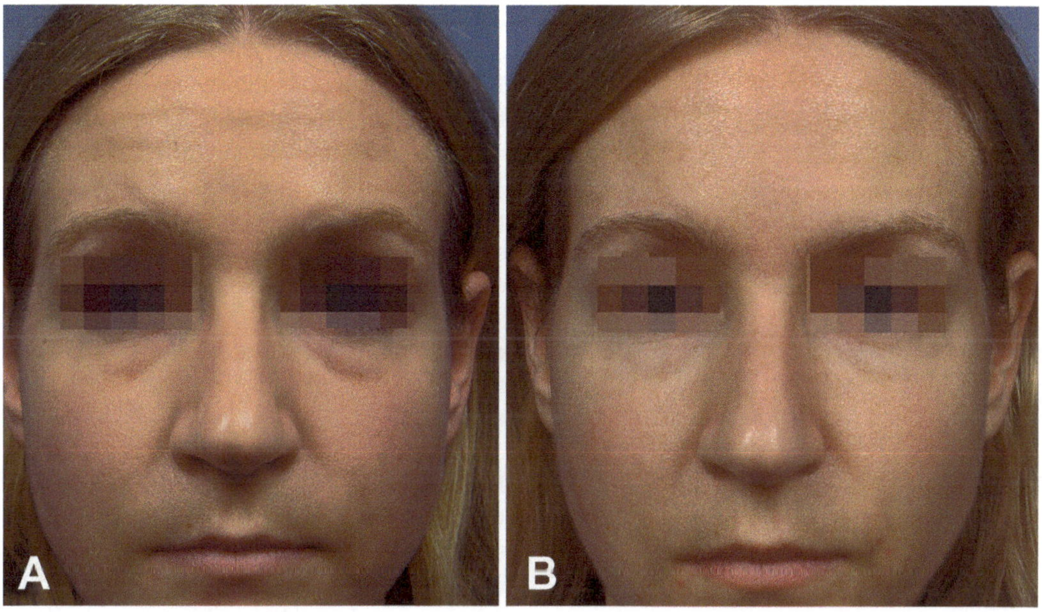

Fig. 12.5 Hyluronidase injection resolves eyelid edema associated with augmentation of the tear-trough with an HA-filler. Twenty eight-year old patient presenting with persistent lower eyelid edema 3 months after augmentation of the tear-trough with an HA-filler. The treatment of a moderate tear trough deformity (**a**) prior to therapy and (**b**) 2 weeks after injection of 50 IU of bovine hyaluronidase into each lower eyelid

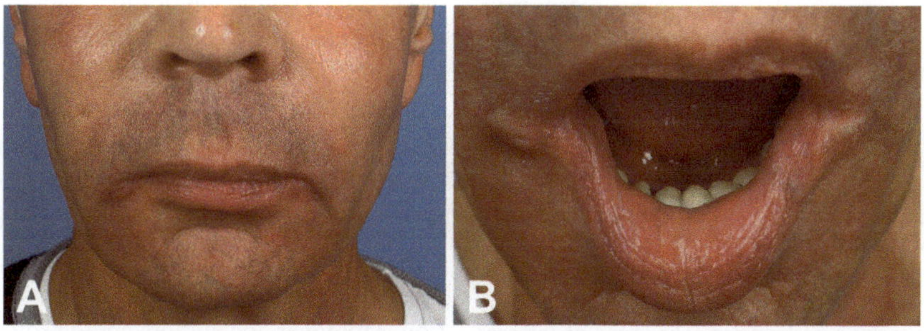

Fig. 12.6 Foreign body granulomas after augmentation of the corners of mouth and lips with an unknown filler substance. (**a**) Overview and (**b**) detail of a 54-year old patient presenting with multiple visible and palpable firm nodes of the lips and perioral region months after injection of an unknown filler substance for augmentation of the corners of mouth and lips

(Beleznay et al. 2015; Ozturk et al. 2013). Early clinical signs of vascular occlusion include immediate and severe pain as well as blanching of the supplied area. Later, pain may increase and reddish to bluish discolorations of the skin can occur (Fig. 12.7) which may then progress to tissue necrosis and atrophic scarring (Fig. 12.8). Preventive measures include aspiration, a slow injection, careful attention for patient's notices of immediate pain, and the avoidance of too large injection volumes particular in high risk areas.

Taking in account the potentially severe complications associated with aesthetic filler injections, the reasons why hyaluronan is preferred over other filler substances include not only degradability, biocompatibility, and hygroscopic properties but most notably also the availability of a specific antidote, hyaluronidase (Fig. 12.9)

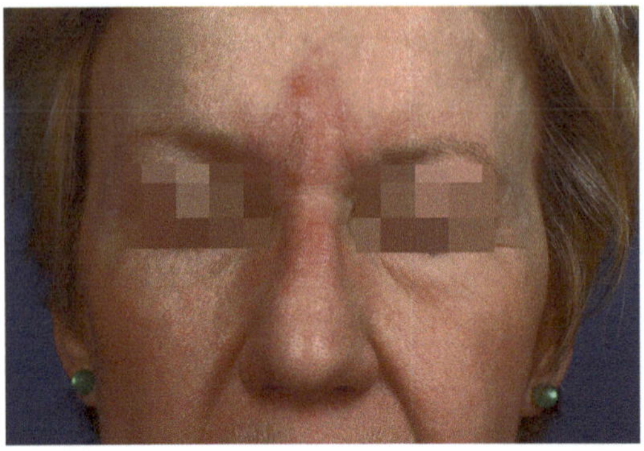

Fig. 12.7 Erythema and ulcerations 3 weeks after injection of an HA-filler into the glabella region. Fifty seven-year old patient presenting with erythema and ulcerations 3 weeks after injection of an HA-filler into the glabella region and probable occlusion of the supratrochlear and dorsal nasal artery. The ulcerations will likely progress to atrophic scars

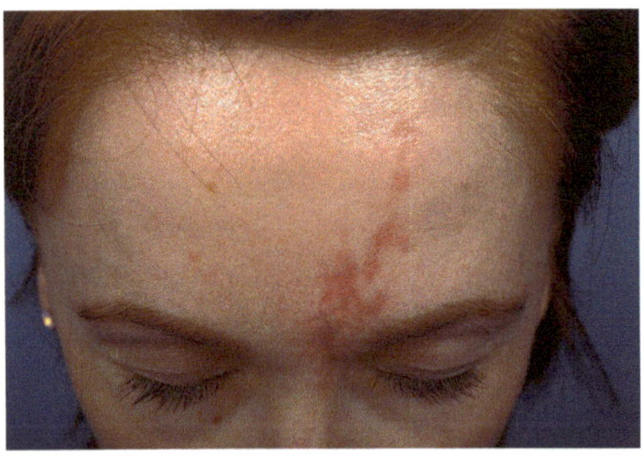

Fig. 12.8 Atrophic scar 2 months after injection of an HA-filler for augmentation of glabella folds by a cosmetician. Forty two-year old patient presenting with an atrophic scar 2 months after injection of an HA-filler into the glabella region by a cosmetician. The shape of the scar resembles the anatomical course of the supratrochlear artery

(Hilton et al. 2014; Buhren et al. 2016; Hirsch et al. 2007a, b). The injection of hyaluronidase may not only correct unaesthetic overcorrections but may also reverse chronic edema or the accidental vascular occlusion after HA-filler injections and may thereby prevent skin necrosis or blindness. In 2011 Kim et al. used an elegant in vivo rabbit ear model to assess whether early (<4 h after occlusion) or late subcutaneous injections of hyaluronidase (24 h after occlusion) could prevent skin necrosis following vascular occlusion of an artery with an HA-filler (Kim et al. 2011). Briefly, the authors injected an HA-filler directly into the auricular arteries of a rabbit. Next, hyaluronidase or a negative control of saline where injected into the ear either early, 4 h after injection of HA or late at 24 h after injection of HA. In the course the developing skin necrosis was analyzed. In both groups the ears infiltrated with saline solution developed widespread skin necrosis. Interestingly, for the ears infiltrated with hyaluronidase, those ears treated after 24 h also developed widespread necrosis, whereas ears infiltrated with hyaluronidase after 4 h did not develop any necrosis. Hence, it can be concluded that in cases of vascular complications following HA-filler injections the injection of hyaluronidase has a higher likeliness of success if performed early within the first 4 h after vascular occlusion. Most recently, Wang et al. used the same rabbit ear model to address the question, whether hyaluronidase can effectively penetrate into occluded vessels to degrade HA-fillers that may occlude the vascular lumen or whether hyaluronidase has to be injected directly into the lumen of the vessel (Wang et al. 2017). First, the HA-filler was injected into a main artery of the rabbit's ear. Immediately after embolism one group was injected with hyaluronidase subcutaneously, whereas another group received intravascular hyaluronidase. Two weeks later the authors found that the subcutaneous injection of hyaluronidase had resulted in less necrosis as compared to the intravascular injection. Moreover, the authors could prove that perivascular hyaluronidase could effectively degrade intravascular HA-fillers within 1 h. Hence, from the clinician's perspective in cases of definitive or even only suspected vascular occlusion the immediate infiltration of the entire area using large volumes of hyaluronidase is recommended. Besides the management of vascular occlusions and overcorrections, hyaluronidase has also been shown to be beneficial for the therapy of chronic edema following HA-filler injections. This complication correlated to the hygroscopic effect of hyaluronan and is most frequent for the augmentation of the so called tear-trough (Fig. 12.5). In a retrospective analysis, Hilton et al. demonstrated

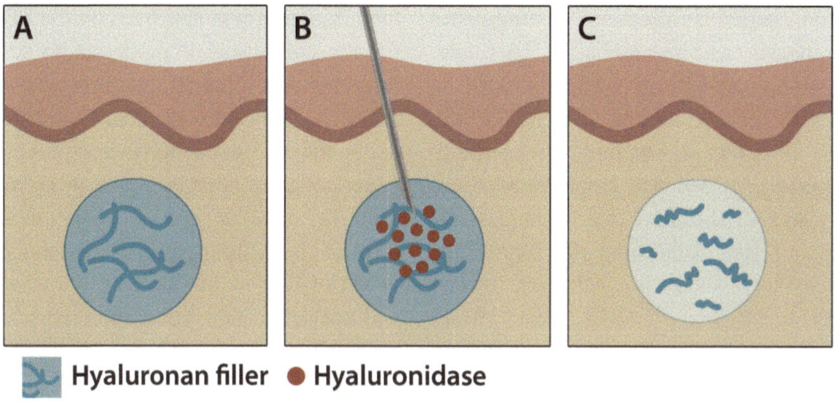

Fig. 12.9 Degradation of an HA-filler by hyaluronidase in vivo. (**a**) Injected dermal HA-fillers can last for several months up to years. (**b** and **c**) The infiltration of hyaluronidase can effectively and rapidly degrade these HA-fillers. The proven degradability of HA-fillers by hyaluronidase represents a safety-feature for respective products

that a single injection of bovine hyaluronidase effectively resolved eyelid edema of respective patients (Hilton et al. 2014). For the clinical practice the injection of smaller volumes of hyaluronidase was recommended, as larger volumes of the potent enzyme would not only resolve the edema but would completely degrade the injected HA-filler and would thereby abrogate the effect of augmentation. To conclude, it can be stated that the immediate availability of hyaluronidase is regarded a necessity for every physician who injects HA-fillers (Hirsch et al. 2007a; Kim et al. 2011; Jahn et al. 2014)

Yet, despite the availability of hyaluronidase it is controversially discussed whether all HA-fillers can be degraded by the enzyme assimilably effective. A difference in or even resistance to "degradability" may be attributed to the concentration of HA in the filler, the degree of cross-linking, gel viscosity, gel hardness, its cohesive properties, and finally the duration of the filler in the skin (Jones et al. 2010; Landau 2015; Rao et al. 2014). With regard to the chemical structure it can be stated that most modern HA-fillers are chemically modified using cross-linker molecules, such as 1,4-butanediol diglycidal ether (BDDE), to create three-dimensional network-structures with intermolecular cross-linked bonds and bridges to enhance the gel strength and longevity (Fig. 12.10). Through the cross-linking, the half-life of HA-fillers and clinical duration in the skin can reach up to 1 year and longer (De Boulle et al. 2013; Fraser et al. 1981).

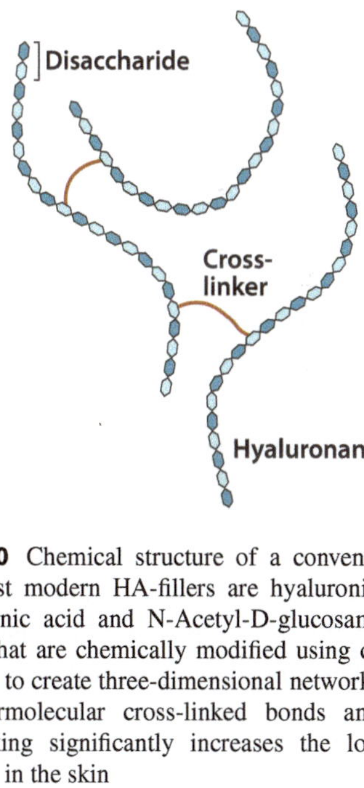

Fig. 12.10 Chemical structure of a conventional HA-filler. Most modern HA-fillers are hyaluronic acid-like D-glucuronic acid and N-Acetyl-D-glucosamine disaccharides that are chemically modified using cross-linker molecules to create three-dimensional network-structures with intermolecular cross-linked bonds and bridges. Cross-linking significantly increases the longevity of HA-fillers in the skin

Modern HA-fillers can be differentiated with regard to the following properties: presence and degree of cross-linkers between the hyaluronan polymer chains, polymer chain length, size of hyaluronan gel particles, hyaluronan concentration and purity of raw material (Stocks et al. 2011; Sundaram et al. 2010; Edsman et al. 2012).

A common superclassification distinguishes between biphasic and monophasic HA-fillers. Biphasic (non-cohesive) HA-fillers, like the Emervel® (EMV) or Restylane® (RES) line (Galderma Pharma, Lausanne, Switzerland; manufactured by Q-Med AB, Uppsala, Sweden) contain a range of different microsphere sizes, whereas monophasic (cohesive) HA-fillers, like the Juvederm® (JUV; Pharm-Allergan, Irvine, CA, USA) or Belotero® (BEL) line (Merz Pharmaceuticals GmbH, Frankfurt, Germany; manufactured by ANTEIS SA, Geneva, Switzerland) contain homogeneous microspheres (Attenello and Maas 2015; Buntrock et al. 2013; Flynn et al. 2011; Kontis 2013; Ballin et al. 2013; Tran et al. 2014). RES is a non-animal stabilized, partially cross-linked HA gel (NASHA). It is cross-linked with BDDE with a low level of 1%. RES is produced from the fermentation of equine streptococci, and was the first HA-filler to receive FDA approval in December 2003 for the correction of moderate to severe facial wrinkles and folds (Narins and Bowman 2005). Within the RES line of products there are various filler formulations which all have identically concentrations of hyaluronan (20 mg/ml) but differ in the size of the HA particles, expressed in number of gel particles per ml (Narins and Bowman 2005). BEL is a bacterially fermented, monophasic double cross-linked HA gel filler that is based on the so called Cohesive Polydensified Matrix (CPM) technology and that creates a low-viscosity. BEL was approved by the FDA in November 2011 (Tran et al. 2014). The BEL product line includes various filler formulations with total concentrations of hyaluronan ranging from 20.0 to 26.0 mg/ml that are indicated for the treatment of superficial lines up to volumizing of the face. It is produced by two steps of cross-linking. First the cross-linking of a determined amount of hyaluronan is performed with BDDE. In the second step, a new amount of hyaluronan is inserted and additionally cross-linked. The resulting variable densities of cross-linked HA zones is proposed to enable an optimal spreading of the gel into surrounding tissue (Lorenc et al. 2013). The JUV product line, finally, includes non–animal-based fillers with concentrations of 22.0–26.0 mg/ml of hyaluronan. It was approved by the FDA in 2006 for the correction of moderate to severe facial wrinkles and folds (Romagnoli and Belmontesi 2008). JUV is derived from *Streptococcus equi* using the so called "Hylacross technology" to generate a monophasic gel of highly BDDE-cross-linked (11%) cohesive molecules. The high level of cross-linking is proposed to increase longevity of JUV in the skin up to 12 months (Dayan and Bassichis 2008; Tezel and Fredrickson 2008). This is likely correlated to the fact that the strong cross-linking of monophasic JUV may limit the access of hyaluronidase to its HA substrate. On the other hand, the biphasic nature of EMV/RES and its distinct particles offer the enzyme a greater surface to attack (Sall and Ferard 2007).

With regard to longevity, like natural hyaluronan, also cross-linked hyaluronan hydrogels are degraded via radical scission through ROS or by enzymatic cleavage (Kenne et al. 2013; De Boulle et al. 2013). An increase in ROS, and hence oxidative stress, may occur as a consequence of foreign body reactions to biomaterial or through other external environmental stressors like mechanical tissue injury, which can cause a transient inflammatory reaction with an interaction of macrophages and foreign body giant cells (Harman 1992; Valko et al. 2007; Henson 1971a, b). Inflammation-induced ROS eventually degrade physiologic and synthetic, cross-linked hyaluronan-chains through cleavage of glycosidic bonds (Cowman 2017; Volpi et al. 2009; Rees et al. 2004; Soltes et al. 2006).

In line with physiologic hyaluronan, cross-linked HA-fillers are also cleaved by hyaluronidases. The increase in longevity of HA-fillers is, in part, related to the fact that cross-linking results in stronger resistance against enzymatic degradation via hyaluronidase (Yui et al. 1993). Yet, taking in account the potential risks associated with filler injections, a proven, effective degradability of an HA-filler by injected hyaluronidase can be regarded as "safety-feature" (Fig. 12.11). To this end, several studies have systematically assessed the degradability of HA-fillers by hyaluronidase (Juhasz et al. 2017; Jones et al. 2010; Rao et al. 2014; Hwang and Song 2017; Sall and Ferard 2007; Buhren

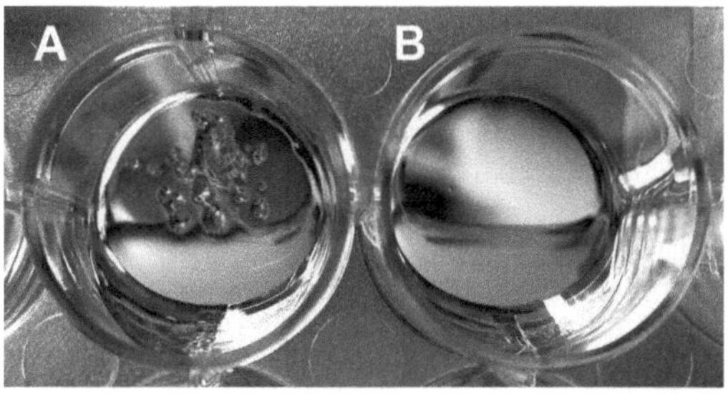

Fig. 12.11 Degradability of an HA-filler by hyaluronidase in vitro. Incubation of a conventional, cross-linked HA-filler with (**a**) medium control and (**b**) medium plus bovine hyaluronidase over 20 h at 37 °C and 5% CO_2 in vitro. Whereas the cohesive gel structure of the (**a**) HA-filler incubated with medium is still intact, the (**b**) HA-filler incubated with hyaluronidase is completely dissolved

et al. 2018). In 2007 Sall et al. used a test based on the colorimetric determination of N-acetyl-D-glucosamine released from 11 different HA-fillers to evaluate the sensitivities of different cross-linked HA-hydrogels to the action of bovine hyaluronidase (Sall and Ferard 2007). Three years later Jones et al. conducted an in vitro comparison of three HA-based fillers with different hyaluronan contents treated with ovine hyaluronidase via analysis of the degradation products using size-exclusion chromatography coupled with a multiangle lightscattering detector (Jones et al. 2010). Both studies reported that highly cross-linked HA-fillers exhibited the strongest resistance to enzymatic degradation (Jones et al. 2010; Sall and Ferard 2007), as compared to biphasic particular HA-fillers (Sall and Ferard 2007). In 2014 Rao et al. used an in vitro photographic approach to visualize the physical properties of four different HA-fillers incubated with rHuPH20 (Rao et al. 2014). Clinical results showed again that biphasic, low-level cross-linked EMV was degraded by hyaluronidase most effectively in a dose-dependent manner followed by JUV. Interestingly, this study reported that BEL retained its to the greatest extent. In 2016 Menzinger et al. reported that ovine testicular hyaluronidase effectively and dose-dependently degraded EMV in a murine model in vivo (Menzinger et al. 2016). In 2017, Juhasz et al. conducted an in vivo human blinded randomized study to assess the degradation of HA-fillers by palpation using a 5-point palpation scale during 14 day follow up period. In total 7 different HA-fillers, including RES/EMV, BEL and JUV, were injected into the back skin of 15 participants. Subsequently a second injection of ovine hyaluronidase (20 or 40 IU; Vitrase®, Valeant Pharmaceuticals, Laval, Canada) or saline control was performed. In conclusion, all fillers treated with ovine hyaluronidase showed a significant decrease in volume as compared to saline. BEL was found to be the fastest to dissolve (Juhasz et al. 2017). Most recently and finally, Buhren and Gerber et al. published their results of a highly standardized in vitro analysis of the degradability of RES/EMV, BEL and JUV by bovine hyaluronidase (Buhren et al. 2018).

Taken together, with regard to analyzed HA-fillers, applied hyaluronidases, treatment protocols, experimental setups and modes of analysis, these cited studies show a significant heterogeneity. This may explain also the heterogeneity with regard to the results and conclusions for the good or bad degradability of certain HA-fillers by different hyaluronidases. As different study outcomes result in uncertainty amongst physicians injecting HA-fillers, future studies should aim to assess the degradability of respective products at a high level of standardization.

12.5 Conclusion

Starting from the 1940s the enzyme hyaluronidase has evolved from a "spreading factor" for local infiltration anesthesia, over the gold standard for the management of HA-filler associated complications to a general adjunct that can significantly increase the bioavailability of a broad variety of different drugs. A glance at the number of yearly scientific publications on "hyaluronidase" shows an exponential rise in recent years pointing toward the rising clinical importance of hyaluronidase. Further indications in surgery, aesthetic medicine, immunology, oncology, and many others can be expected for years to come.

References

Ahmed S, Ahmed OA (2004) Hyaluronidase revisited-a single injection technique for harvesting split thickness skin grafts under local anaesthesia. Br J Plast Surg 57:589–591

Albanell J, Baselga J (2000) Systemic therapy emergencies. Semin Oncol 27:347–361

Alberts DS, Dorr RT (1991) Case report: topical DMSO for mitomycin-C-induced skin ulceration. Oncol Nurs Forum 18:693–695

Allen CH, Etzwiler LS, Miller MK et al (2009) Recombinant human hyaluronidase-enabled subcutaneous pediatric rehydration. Pediatrics 124:e858–e867

Aruffo A, Stamenkovic I, Melnick M et al (1990) CD44 is the principal cell surface receptor for hyaluronate. Cell 61:1303–1313

Atmuri V, Martin DC, Hemming R et al (2008) Hyaluronidase 3 (HYAL3) knockout mice do not display evidence of hyaluronan accumulation. Matrix Biol 27:653–660

Attenello NH, Maas CS (2015) Injectable fillers: review of material and properties. Facial Plast Surg 31:29–34

Ballin AC, Cazzaniga A, Brandt FS (2013) Long-term efficacy, safety and durability of Juvederm(R) XC. Clin Cosmet Investig Dermatol 6:183–189

Barsukov AK, Kozhevnikova OV, Khokhriakova AV (2003) Isolation and purification of bovine testicular hyaluronidase. Prikl Biokhim Mikrobiol 39:625–629

Beleznay K, Carruthers JD, Humphrey S et al (2015) Avoiding and treating blindness from fillers: a review of the world literature. Dermatol Surg 41:1097–1117

Bellin MF, Jakobsen JA, Tomassin I et al (2002) Contrast medium extravasation injury: guidelines for prevention and management. Eur Radiol 12:2807–2812

Belzunegui T, Louis CJ, Torrededia L et al (2011) Extravasation of radiographic contrast material and compartment syndrome in the hand: a case report. Scand J Trauma Resusc Emerg Med 19:9

Benditt EP, Schiller S, Matthews MB et al (1951) Evidence that hyaluronidase is not the factor in testicular extract causing increased vascular permeability. Proc Soc Exp Biol Med 77:643–646

Bertelli G (1995) Prevention and management of extravasation of cytotoxic drugs. Drug Saf 12:245–255

Bertelli G, Dini D, Forno GB et al (1994) Hyaluronidase as an antidote to extravasation of Vinca alkaloids: clinical results. J Cancer Res Clin Oncol 120:505–506

Bertelli G, Gozza A, Forno GB et al (1995) Topical dimethylsulfoxide for the prevention of soft tissue injury after extravasation of vesicant cytotoxic drugs: a prospective clinical study. J Clin Oncol 13:2851–2855

Bertelli G, Cafferata MA, Ardizzoni A et al (1997) Skin ulceration potential of paclitaxel in a mouse skin model in vivo. Cancer 79:2266–2269

Bertolami CN, Donoff RB (1982) Identification, characterization, and partial purification of mammalian skin wound hyaluronidase. J Invest Dermatol 79:417–421

Bissell MJ, Hall HG, Parry G (1982) How does the extracellular matrix direct gene expression? J Theor Biol 99:31–68

Bohaumilitzky L, Huber AK, Stork EM et al (2017) A trickster in disguise: Hyaluronan's ambivalent roles in the matrix. Front Oncol 7:242

Bookbinder LH, Hofer A, Haller MF et al (2006) A recombinant human enzyme for enhanced interstitial transport of therapeutics. J Control Release 114:230–241

Boulanger J, Ducharme A, Dufour A et al (2015) Management of the extravasation of anti-neoplastic agents. Support Care Cancer 23:1459–1471

Buhren BA, Schrumpf H, Hoff NP et al (2016) Hyaluronidase: from clinical applications to molecular and cellular mechanisms. Eur J Med Res 21:5

Buhren BA, Schrumpf H, Bolke E et al (2018) Standardized in vitro analysis of the degradability of hyaluronic acid fillers by hyaluronidase. Eur J Med Res 23:37

Buntrock H, Reuther T, Prager W et al (2013) Efficacy, safety, and patient satisfaction of a monophasic cohesive polydensified matrix versus a biphasic nonanimal stabilized hyaluronic acid filler after single injection in nasolabial folds. Dermatol Surg 39:1097–1105

Carne E, Ponsford M, El-Shanawany T et al (2016) Five years of self-administered hyaluronidase facilitated subcutaneous immunoglobulin (fSCIg) home therapy in a patient with primary immunodeficiency. J Clin Pathol 69:87–88

Cochran ST, Bomyea K, Kahn M (2002) Treatment of iodinated contrast material extravasation with hyaluronidase. Acad Radiol 9(Suppl 2):S544–S546

Connolly AA, Meyer LC, Tate JJ (1994) Local anaesthetic agents in surgery for ingrown toenail. Br J Surg 81:425–426

Connolly S, Korzemba H, Harb G et al (2011) Techniques for hyaluronidase-facilitated subcutaneous fluid administration with recombinant human hyaluronidase: the increased flow utilizing subcutaneously enabled administration technique (INFUSE AT) study. J Infus Nurs 34:300–307

Courtiss EH, Ransil BJ, Russo J (1995) The effects of hyaluronidase on local anesthesia: a prospective, randomized, controlled, double-blind study. Plast Reconstr Surg 95:876–883

Cowman MK (2017) Hyaluronan and hyaluronan fragments. Adv Carbohydr Chem Biochem 74:1–59

Crawford M, Kerr WJ (1994) The effect of hyaluronidase on peribulbar block. Anaesthesia 49:907–908

Csoka TB, Frost GI, Stern R (1997) Hyaluronidases in tissue invasion. Invasion Metastasis 17:297–311

Dai G, Freudenberger T, Zipper P et al (2007) Chronic ultraviolet B irradiation causes loss of hyaluronic acid from mouse dermis because of down-regulation of hyaluronic acid synthases. Am J Pathol 171:1451–1461

Dayan SH, Bassichis BA (2008) Facial dermal fillers: selection of appropriate products and techniques. Aesthet Surg J 28:335–347
De Boulle K, Glogau R, Kono T et al (2013) A review of the metabolism of 1,4-butanediol diglycidyl ether-crosslinked hyaluronic acid dermal fillers. Dermatol Surg 39:1758–1766
Delorenzi C (2013) Complications of injectable fillers, part I. Aesthet Surg J 33:561–575
Dempsey GA, Barrett PJ, Kirby J (1996) The effect of hyaluronidase on peribulbar block. Anaesthesia 51:515
Dempsey GA, Barrett PJ, Kirby IJ (1997) Hyaluronidase and peribulbar block. Br J Anaesth 78:671–674
Dieleman M, Bettink-Remeijer MW, Jansen J et al (2012) High incidence of adverse reactions to locoregional anaesthesia containing hyaluronidase after uneventful ophthalmic surgery. Acta Ophthalmol 90:e245–e246
Doellman D, Hadaway L, Bowe-Geddes LA et al (2009) Infiltration and extravasation: update on prevention and management. J Infus Nurs 32:203–211
Doherty GJ, Tempero M, Corrie PG (2018) HALO-109-301: a phase III trial of PEGPH20 (with gemcitabine and nab-paclitaxel) in hyaluronic acid-high stage IV pancreatic cancer. Future Oncol 14:13–22
Dorr RT (1990) Antidotes to vesicant chemotherapy extravasations. Blood Rev 4:41–60
Dougherty L (2008) IV therapy: recognizing the differences between infiltration and extravasation. Br J Nurs 17:896. 898–901
Dubois A, Fehr M, Bochtler H et al (1996) Clinical course and management of paclitaxel extravasation. Oncol Rep 3:973–974
Dunagin WG (1982) Clinical toxicity of chemotherapeutic agents: dermatologic toxicity. Semin Oncol 9:14–22
Dunn AL, Heavner JE, Racz G et al (2010) Hyaluronidase: a review of approved formulations, indications and off-label use in chronic pain management. Expert Opin Biol Ther 10:127–131
Duran-Reynals F (1942) TISSUE PERMEABILITY AND THE SPREADING FACTORS IN INFECTION: a contribution to the host: parasite problem. Bacteriol Rev 6:197–252
Edsman K, Nord LI, Ohrlund A et al (2012) Gel properties of hyaluronic acid dermal fillers. Dermatol Surg 38:1170–1179
Elam EA, Dorr RT, Lagel KE et al (1991) Cutaneous ulceration due to contrast extravasation. Experimental assessment of injury and potential antidotes. Investig Radiol 26:13–16
El-Saghir N, Otrock Z, Mufarrij A et al (2004) Dexrazoxane for anthracycline extravasation and GM-CSF for skin ulceration and wound healing. Lancet Oncol 5:320–321
Fanning GL (2001) Hyaluronidase in ophthalmic anesthesia. Anesth Analg 92:560
Farr C, Menzel J, Seeberger J et al (1997) Clinical pharmacology and possible applications of hyaluronidase with reference to Hylase "Dessau". Wien Med Wochenschr 147:347–355
Federle MP, Chang PJ, Confer S et al (1998) Frequency and effects of extravasation of ionic and nonionic CT contrast media during rapid bolus injection. Radiology 206:637–640
Flynn TC, Sarazin D, Bezzola A et al (2011) Comparative histology of intradermal implantation of mono and biphasic hyaluronic acid fillers. Dermatol Surg 37:637–643
Fox AN, Villanueva R, Miller JL (2017) Management of amiodarone extravasation with intradermal hyaluronidase. Am J Health Syst Pharm 74:1545–1548
Fraser JR, Laurent TC, Pertoft H et al (1981) Plasma clearance, tissue distribution and metabolism of hyaluronic acid injected intravenously in the rabbit. Biochem J 200:415–424
Fraser JR, Laurent TC, Laurent UB (1997) Hyaluronan: its nature, distribution, functions and turnover. J Intern Med 242:27–33
Fronza M, Caetano GF, Leite MN et al (2014) Hyaluronidase modulates inflammatory response and accelerates the cutaneous wound healing. PLoS One 9:e112297
Frost GI (2007) Recombinant human hyaluronidase (rHuPH20): an enabling platform for subcutaneous drug and fluid administration. Expert Opin Drug Deliv 4:427–440
Garg SK, Buse JB, Skyler JS et al (2014) Subcutaneous injection of hyaluronidase with recombinant human insulin compared with insulin lispro in type 1 diabetes. Diabetes Obes Metab 16:1065–1069
Garvin JH Jr, Chipman DM (1974) Subunit structure of testicular hyaluronidase. FEBS Lett 39:157–159
Glogau RG (1997) Physiologic and structural changes associated with aging skin. Dermatol Clin 15:555–559
Goolsby TV, Lombardo FA (2006) Extravasation of chemotherapeutic agents: prevention and treatment. Semin Oncol 33:139–143
Guedan S, Rojas JJ, Gros A et al (2010) Hyaluronidase expression by an oncolytic adenovirus enhances its intratumoral spread and suppresses tumor growth. Mol Ther 18:1275–1283
Gul A, Can E, Niyaz L et al (2015) Hyaluronidase in ophthalmic evisceration surgery. Trop Dr 45:100–104
Hamizi S, Freyer G, Bakrin N et al (2013) Subcutaneous trastuzumab: development of a new formulation for treatment of HER2-positive early breast cancer. Onco Targets Ther 6:89–94
Harb G, Lebel F, Battikha J et al (2010) Safety and pharmacokinetics of subcutaneous ceftriaxone administered with or without recombinant human hyaluronidase (rHuPH20) versus intravenous ceftriaxone administration in adult volunteers. Curr Med Res Opin 26:279–288
Harman D (1992) Role of free radicals in aging and disease. Ann N Y Acad Sci 673:126–141
Hascall V, Esko JD (2015) Hyaluronan. In: Varki A, Cummings RD, Esko JD, Stanley P, Hart GW, Aebi M, Darvill AG, Kinoshita T, Packer NH, Prestegard JH, Schnaar RL, Seeberger PH (eds) Essentials of glycobiology. Cold Spring Harbor Laboratory Press, Cold Spring Harbor

Henson PM (1971a) The immunologic release of constituents from neutrophil leukocytes. II. Mechanisms of release during phagocytosis, and adherence to nonphagocytosable surfaces. J Immunol 107:1547–1557
Henson PM (1971b) The immunologic release of constituents from neutrophil leukocytes. I. The role of antibody and complement on nonphagocytosable surfaces or phagocytosable particles. J Immunol 107:1535–1546
Herman IM, Castellot JJ Jr (1987) Regulation of vascular smooth muscle cell growth by endothelial-synthesized extracellular matrices. Arteriosclerosis 7:463–469
Hilton S, Schrumpf H, Buhren BA et al (2014) Hyaluronidase injection for the treatment of eyelid edema: a retrospective analysis of 20 patients. Eur J Med Res 19:30
Hirsch RJ, Brody HJ, Carruthers JD (2007a) Hyaluronidase in the office: a necessity for every dermasurgeon that injects hyaluronic acid. J Cosmet Laser Ther 9:182–185
Hirsch RJ, Cohen JL, Carruthers JD (2007b) Successful management of an unusual presentation of impending necrosis following a hyaluronic acid injection embolus and a proposed algorithm for management with hyaluronidase. Dermatol Surg 33:357–360
Hompesch M, Muchmore DB, Morrow L et al (2011) Accelerated insulin pharmacokinetics and improved postprandial glycemic control in patients with type 1 diabetes after coadministration of prandial insulins with hyaluronidase. Diabetes Care 34:666–668
Hwang E, Song YS (2017) Quantitative correlation between hyaluronic acid filler and hyaluronidase. J Craniofac Surg 28:838–841
Hyde CE, Old RW (1999) Expression pattern of a novel hyaluronidase during Xenopus embryogenesis. Mech Dev 82:213–217
Jahn K, Homey B, Gerber PA (2014) Management of complications after aesthetic hyaluronic acid injections. Hautarzt 65:851–853
Jones D, Tezel A, Borrell M (2010) In vitro resistance to degradation of hyaluronic acid dermal fillers by ovine testicular hyaluronidase. Dermatol Surg 36:804–809
Juhasz MLW, Levin MK, Marmur ES (2017) The kinetics of reversible hyaluronic acid filler injection treated with hyaluronidase. Dermatol Surg 43:841–847
Juhlin L (1997) Hyaluronan in skin. J Intern Med 242:61–66
Kagan L, Mager DE (2013) Mechanisms of subcutaneous absorption of rituximab in rats. Drug Metab Dispos 41:248–255
Kassner E (2000) Evaluation and treatment of chemotherapy extravasation injuries. J Pediatr Oncol Nurs 17:135–148
Kenne L, Gohil S, Nilsson EM et al (2013) Modification and cross-linking parameters in hyaluronic acid hydrogels – definitions and analytical methods. Carbohydr Polym 91:410–418
Khandwala M, Ahmed S, Goel S et al (2008) The effect of hyaluronidase on ultrasound-measured dispersal of local anaesthetic following sub-Tenon injection. Eye (Lond) 22:1065–1068
Khorlin AY, Vikha IV, Milishnikov AN (1973) Subunit structure of testicular hyaluronidase. FEBS Lett 31:107–110
Kim DW, Yoon ES, Ji YH et al (2011) Vascular complications of hyaluronic acid fillers and the role of hyaluronidase in management. J Plast Reconstr Aesthet Surg 64:1590–1595
Kind LS, Roffler S (1961) Allergic reactions to hyaluronidase. Proc Soc Exp Biol Med 106:734–735
Kontis TC (2013) Contemporary review of injectable facial fillers. JAMA Facial Plast Surg 15:58–64
Kreil G (1995) Hyaluronidases – a group of neglected enzymes. Protein Sci 4:1666–1669
Kuppermann BD, Thomas EL, De Smet MD et al (2005) Pooled efficacy results from two multinational randomized controlled clinical trials of a single intravitreous injection of highly purified ovine hyaluronidase (Vitrase) for the management of vitreous hemorrhage. Am J Ophthalmol 140:573–584
Landau M (2015) Hyaluronidase caveats in treating filler complications. Dermatol Surg 41(Suppl 1):S347–S353
Landsman L, Mandy SH (1991) Adjuncts to scalp reduction surgery. Intraoperative tissue expanders and hyaluronidase. J Dermatol Surg Oncol 17:670–672
Langer SW, Sehested M, Jensen PB (2000) Treatment of anthracycline extravasation with dexrazoxane. Clin Cancer Res 6:3680–3686
Laurent TC, Fraser JR (1992) Hyaluronan. FASEB J 6:2397–2404
Lee A, Grummer SE, Kriegel D et al (2010) Hyaluronidase. Dermatol Surg 36:1071–1077
Lewis-Smith PA (1986) Adjunctive use of hyaluronidase in local anaesthesia. Br J Plast Surg 39:554–558
Lindley-Jones MF (2000) Topical anaesthesia for phacoemulsification surgery. Clin Exp Ophthalmol 28:287–289
Lokeshwar VB, Selzer MG (2008) Hyalurondiase: both a tumor promoter and suppressor. Semin Cancer Biol 18:281–287
Lokeshwar VB, Rubinowicz D, Schroeder GL et al (2001) Stromal and epithelial expression of tumor markers hyaluronic acid and HYAL1 hyaluronidase in prostate cancer. J Biol Chem 276:11922–11932
Lorenc ZP, Fagien S, Flynn TC et al (2013) Review of key Belotero Balance safety and efficacy trials. Plast Reconstr Surg 132:33S–40S
Lv SH, Rong SF, Cai BG et al (2015) Property and current clinical applications of mammal hyaluronidase. Eur Rev Med Pharmacol Sci 19:3968–3976
Maccara ME (1983) Extravasation: a hazard of intravenous therapy. Drug Intell Clin Pharm 17:713–717
Martin-Deleon PA (2011) Germ-cell hyaluronidases: their roles in sperm function. Int J Androl 34:e306–e318
Martinez-Quintanilla J, He D, Wakimoto H et al (2015) Encapsulated stem cells loaded with hyaluronidase-expressing oncolytic virus for brain tumor therapy. Mol Ther 23:108–118
Mcclean D (1941) Studies on diffusing factors: the hyaluronidase activity of testicular extracts, bacterial culture filtrates and other agents that increase tissue permeability. Biochem J 35:159–183

Menzel EJ, Farr C (1998) Hyaluronidase and its substrate hyaluronan: biochemistry, biological activities and therapeutic uses. Cancer Lett 131:3–11

Menzinger S, Kaya A, Saurat JH et al (2016) Injected hyaluronidase reduces the volume of exogenous hyaluronate fillers in mice and results in clinical improvement in a patient with pretibial myxedema. Dermatopathology (Basel) 3:61–67

Meyer AW (1937) The hunters in embryology: part III. Cal West Med 46:38–40

Meyer K (1947) The biological significance of hyaluronic acid and hyaluronidase. Physiol Rev 27:335–359

Meyer LJ, Stern R (1994) Age-dependent changes of hyaluronan in human skin. J Invest Dermatol 102:385–389

Montgomery LA, Budreau GK (1996) Implementing a clinical practice guideline to improve pediatric intravenous infiltration outcomes. AACN Clin Issues 7:411–424

Morcos PN, Zhang X, Mcintyre C et al (2013) Pharmacokinetics and pharmacodynamics of single subcutaneous doses of tocilizumab administered with or without rHuPH20. Int J Clin Pharmacol Ther 51:537–548

Muchmore DB, Vaughn DE (2010) Review of the mechanism of action and clinical efficacy of recombinant human hyaluronidase coadministration with current prandial insulin formulations. J Diabetes Sci Technol 4:419–428

Myung Y, Yim S, Jeong JH et al (2017) The classification and prognosis of periocular complications related to blindness following cosmetic filler injection. Plast Reconstr Surg 140:61–64

Narayanan R, Kuppermann BD (2009) Hyaluronidase for pharmacologic vitreolysis. Dev Ophthalmol 44:20–25

Narins RS, Bowman PH (2005) Injectable skin fillers. Clin Plast Surg 32:151–162

Narins RS, Coleman WP 3rd, Donofrio LM et al (2010) Improvement in nasolabial folds with a hyaluronic acid filler using a cohesive polydensified matrix technology: results from an 18-month open-label extension trial. Dermatol Surg 36(Suppl 3):1800–1808

Narins RS, Brandt FS, Dayan SH et al (2011) Persistence of nasolabial fold correction with a hyaluronic acid dermal filler with retreatment: results of an 18-month extension study. Dermatol Surg 37:644–650

Olver IN, Aisner J, Hament A et al (1988) A prospective study of topical dimethyl sulfoxide for treating anthracycline extravasation. J Clin Oncol 6:1732–1735

Ozturk CN, Li Y, Tung R et al (2013) Complications following injection of soft-tissue fillers. Aesthet Surg J 33:862–877

Papakonstantinou E, Roth M, Karakiulakis G (2012) Hyaluronic acid: a key molecule in skin aging. Dermatoendocrinology 4:253–258

Perrault M, Housset E (1952) Hyaluronidase and its therapeutic applications. Therapie 7:196–206

Pirrello RD, Ting Chen C, Thomas SH (2007) Initial experiences with subcutaneous recombinant human hyaluronidase. J Palliat Med 10:861–864

Rao V, Chi S, Woodward J (2014) Reversing facial fillers: interactions between hyaluronidase and commercially available hyaluronic-acid based fillers. J Drugs Dermatol 13:1053–1056

Rees MD, Hawkins CL, Davies MJ (2004) Hypochlorite and superoxide radicals can act synergistically to induce fragmentation of hyaluronan and chondroitin sulphates. Biochem J 381:175–184

Remy M, Pinter F, Nentwich MM et al (2008) Efficacy and safety of hyaluronidase 75 IU as an adjuvant to mepivacaine for retrobulbar anesthesia in cataract surgery. J Cataract Refract Surg 34:1966–1969

Reynolds PM, Maclaren R, Mueller SW et al (2014) Management of extravasation injuries: a focused evaluation of noncytotoxic medications. Pharmacotherapy 34:617–632

Ridenour S, Reader A, Beck M et al (2001) Anesthetic efficacy of a combination of hyaluronidase and lidocaine with epinephrine in inferior alveolar nerve blocks. Anesth Prog 48:9–15

Romagnoli M, Belmontesi M (2008) Hyaluronic acid-based fillers: theory and practice. Clin Dermatol 26:123–159

Rowlett J (2012) Extravasation of contrast media managed with recombinant human hyaluronidase. Am J Emerg Med 30:2102 e1–2102 e3

Sall I, Ferard G (2007) Comparison of the sensitivity of 11 crosslinked hyaluronic acid gels to bovine testis hyaluronidase. Polym Degrad Stab 92:915–919

Sbitany H, Koltz PF, Mays C et al (2010) CT contrast extravasation in the upper extremity: strategies for management. Int J Surg 8:384–386

Schonenberg H (1952) Therapeutic use of hyaluronidase in pediatrics. Kinderarztl Prax 20:223–228

Schoog M (1951) Therapeutic use of hyaluronidase. Dermatol Wochenschr 124:1033–1037

Schrijvers DL (2003) Extravasation: a dreaded complication of chemotherapy. Ann Oncol 14(Suppl 3):iii26–iii30

Schummer W, Schummer C, Muller A et al (2003) Extravasation: a rare complication of central venous cannulation? Case report of an imminent erosion of the common carotid artery. Anaesthesist 52:711–717

Schwartz DM, Shuster S, Jumper MD et al (1996) Human vitreous hyaluronidase: isolation and characterization. Curr Eye Res 15:1156–1162

Schwartzman J (1951) Hyaluronidase in pediatrics. N Y State J Med 51:215–221

Selek H, Ozer H, Aygencel G et al (2007) Compartment syndrome in the hand due to extravasation of contrast material. Arch Orthop Trauma Surg 127:425–427

Shpilberg O, Jackisch C (2013) Subcutaneous administration of rituximab (MabThera) and trastuzumab (Herceptin) using hyaluronidase. Br J Cancer 109:1556–1561

Silverstein SM, Greenbaum S, Stern R (2012) Hyaluronidase in ophthalmology. J Appl Res 12(1):1–13

Smith KJ, Skelton HG, Turiansky G et al (1997) Hyaluronidase enhances the therapeutic effect of vin-

blastine in intralesional treatment of Kaposi's sarcoma. Military Medical Consortium for the Advancement of Retroviral Research (MMCARR). J Am Acad Dermatol 36:239–242
Soldi A (1951) Hyaluronidase and its therapeutic applications. Farmacol Sci Tec 6:765–791
Soltes L, Mendichi R, Kogan G et al (2006) Degradative action of reactive oxygen species on hyaluronan. Biomacromolecules 7:659–668
Spandorfer PR, Mace SE, Okada PJ et al (2012) A randomized clinical trial of recombinant human hyaluronidase-facilitated subcutaneous versus intravenous rehydration in mild to moderately dehydrated children in the emergency department. Clin Ther 34:2232–2245
Speth F, Haas JP, Hinze CH (2016) Treatment with high-dose recombinant human hyaluronidase-facilitated subcutaneous immune globulins in patients with juvenile dermatomyositis who are intolerant to intravenous immune globulins: a report of 5 cases. Pediatr Rheumatol Online J 14:52
Stanford BL, Hardwicke F (2003) A review of clinical experience with paclitaxel extravasations. Support Care Cancer 11:270–277
Stebliuk PN (1972) Combined effect of hyaluronidase, ribonuclease and antibiotics on pathogenous Staphylococci. Vrach Delo 4:151
Stern R (2008) Hyaluronidases in cancer biology. Semin Cancer Biol 18:275–280
Stern R, Asari AA, Sugahara KN (2006) Hyaluronan fragments: an information-rich system. Eur J Cell Biol 85:699–715
Stocks D, Sundaram H, Michaels J et al (2011) Rheological evaluation of the physical properties of hyaluronic acid dermal fillers. J Drugs Dermatol 10:974–980
Sundaram H, Voigts B, Beer K et al (2010) Comparison of the rheological properties of viscosity and elasticity in two categories of soft tissue fillers: calcium hydroxylapatite and hyaluronic acid. Dermatol Surg 36(Suppl 3):1859–1865
Tammi R, Ripellino JA, Margolis RU et al (1988) Localization of epidermal hyaluronic acid using the hyaluronate binding region of cartilage proteoglycan as a specific probe. J Invest Dermatol 90:412–414
Tezel A, Fredrickson GH (2008) The science of hyaluronic acid dermal fillers. J Cosmet Laser Ther 10:35–42
Thomas JR, Wallace MS, Yocum RC et al (2009) The INFUSE-Morphine study: use of recombinant human hyaluronidase (rHuPH20) to enhance the absorption of subcutaneously administered morphine in patients with advanced illness. J Pain Symptom Manag 38:663–672
Thorpe JN (1951) Procaine with hyaluronidase as local anesthetic. Lancet 1:210–211
Tran C, Carraux P, Micheels P et al (2014) In vivo bio-integration of three hyaluronic acid fillers in human skin: a histological study. Dermatology 228:47–54
Valko M, Leibfritz D, Moncol J et al (2007) Free radicals and antioxidants in normal physiological functions and human disease. Int J Biochem Cell Biol 39:44–84
Vaughn DE, Yocum RC, Muchmore DB et al (2009) Accelerated pharmacokinetics and glucodynamics of prandial insulins injected with recombinant human hyaluronidase. Diabetes Technol Ther 11:345–352
Volpi N, Schiller J, Stern R et al (2009) Role, metabolism, chemical modifications and applications of hyaluronan. Curr Med Chem 16:1718–1745
Wang CL, Cohan RH, Ellis JH et al (2007) Frequency, management, and outcome of extravasation of non-ionic iodinated contrast medium in 69,657 intravenous injections. Radiology 243:80–87
Wang M, Li W, Zhang Y et al (2017) Comparison of intra-arterial and subcutaneous testicular hyaluronidase injection treatments and the vascular complications of hyaluronic acid filler. Dermatol Surg 43:246–254
Warnery, Dumon G, Brin et al (1954) Antibiotics and hyaluronidase combined in aerosols in the treatment of lesions of pulmonary tuberculosis; 20 months' trial. Rev Tuberc 18:37–45
Wasserman RL (2017) Recombinant human hyaluronidase-facilitated subcutaneous immunoglobulin infusion in primary immunodeficiency diseases. Immunotherapy 9:1035–1050
Wasserman RL, Melamed I, Stein MR et al (2012) Recombinant human hyaluronidase-facilitated subcutaneous infusion of human immunoglobulins for primary immunodeficiency. J Allergy Clin Immunol 130:951–7 e11
Wasserman RL, Melamed I, Kobrynski L et al (2016) Recombinant human hyaluronidase facilitated subcutaneous immunoglobulin treatment in pediatric patients with primary immunodeficiencies: long-term efficacy, safety and tolerability. Immunotherapy 8:1175–1186
Weigel PH (2015) Hyaluronan synthase: the mechanism of initiation at the reducing end and a pendulum model for polysaccharide translocation to the cell exterior. Int J Cell Biol 2015:367579
Weller K, Maurer M, Fridman M et al (2017) Health-related quality of life with hereditary angioedema following prophylaxis with subcutaneous C1-inhibitor with recombinant hyaluronidase. Allergy Asthma Proc 38:143–151
Wiegand R, Brown J (2010) Hyaluronidase for the management of dextrose extravasation. Am J Emerg Med 28:257 e1–257 e2
Wohlrab J, Finke R, Franke WG et al (2012a) Clinical trial for safety evaluation of hyaluronidase as diffusion enhancing adjuvant for infiltration analgesia of skin with lidocaine. Dermatol Surg 38:91–96
Wohlrab J, Finke R, Franke WG et al (2012b) Efficacy study of hyaluronidase as a diffusion promoter for lidocaine in infiltration analgesia of skin. Plast Reconstr Surg 129:771e–772e

Wohlrab J, Wohlrab D, Wohlrab L et al (2014) Use of hyaluronidase for pharmacokinetic increase in bioavailability of intracutaneously applied substances. Skin Pharmacol Physiol 27:276–282

Yocum RC, Kennard D, Heiner LS (2007) Assessment and implication of the allergic sensitivity to a single dose of recombinant human hyaluronidase injection: a double-blind, placebo-controlled clinical trial. J Infus Nurs 30:293–299

Yui N, Nihira J, Okano T et al (1993) Regulated release of drug microspheres from inflammation responsive degradable matrices of crosslinked hyaluronic acid. J Control Release 25:133–143

Zahl K, Jordan A, Mcgroarty J et al (1991) Peribulbar anesthesia. Effect of bicarbonate on mixtures of lidocaine, bupivacaine, and hyaluronidase with or without epinephrine. Ophthalmology 98:239–242

Zaneveld LJ, Polakoski KL, Schumacher GF (1973) Properties of acrosomal hyaluronidase from bull spermatozoa. Evidence for its similarity to testicular hyaluronidase. J Biol Chem 248:564–570

Zaoli G (1958) Hyaluronidase-antibiotic association in the treatment of acute otitis media of the infant & child. Minerva Otorinolaringol 8:354–358

13 Alkaline Phosphatase Replacement Therapy for Hypophosphatasia in Development and Practice

S. A. Bowden and B. L. Foster

Abstract

Hypophosphatasia (HPP) is an inherited disorder that affects bone and tooth mineralization characterized by low serum alkaline phosphatase. HPP is caused by loss-of-function mutations in the *ALPL* gene encoding the protein, tissue-nonspecific alkaline phosphatase (TNSALP). TNSALP is expressed by mineralizing cells of the skeleton and dentition and is associated with the mineralization process. Generalized reduction of activity of the TNSALP leads to accumulation of its substrates, including inorganic pyrophosphate (PP_i) that inhibits physiological mineralization. This leads to defective skeletal mineralization, with manifestations including rickets, osteomalacia, fractures, and bone pain, all of which can result in multi-systemic complications with significant morbidity, as well as mortality in severe cases. Dental manifestations are nearly universal among affected individuals and feature most prominently premature loss of deciduous teeth. Management of HPP has been limited to supportive care until the introduction of a TNSALP enzyme replacement therapy (ERT), asfotase alfa (AA). AA ERT has proven to be transformative, improving survival in severely affected infants and increasing overall quality of life in children and adults with HPP. This chapter provides an overview of TNSALP expression and functions, summarizes HPP clinical types and pathologies, discusses early attempts at therapies for HPP, summarizes development of HPP mouse models, reviews design and validation of AA ERT, and provides up-to-date accounts of AA ERT efficacy in clinical trials and case reports, including therapeutic response, adverse effects, limitations, and potential future directions in therapy.

Keywords

Hypophosphatasia · Alkaline phosphatase · Asfotase alfa · Bone mineralization · Rickets · Osteomalacia · Teeth

S. A. Bowden (✉)
Division of Endocrinology, Department of Pediatrics, Nationwide Children's Hospital/The Ohio State University College of Medicine, Columbus, OH, USA
e-mail: Sasigarn.Bowden@nationwidechildrens.org

B. L. Foster
Division of Biosciences, College of Dentistry, The Ohio State University, Columbus, OH, USA

Abbreviations

HPP	Hypophosphatasia
ALPL	Alkaline phosphatase gene
TNSALP	Tissue-nonspecific isoenzyme of alkaline phosphatase

N. Labrou (ed.), *Therapeutic Enzymes: Function and Clinical Implications*, Advances in Experimental Medicine and Biology 1148, https://doi.org/10.1007/978-981-13-7709-9_13

ALP	Alkaline phosphatase enzyme activity, circulating
IAP	Intestinal alkaline phosphatase protein
ALPI	Intestinal alkaline phosphatase gene
PLAP	Placental alkaline phosphatase protein
ALPP	Placental alkaline phosphatase gene
GCAP	Germ cell alkaline phosphatase protein
ALPPL2	Germ cell alkaline phosphatase gene
AA	Asfotase alfa
ERT	Enzyme replacement therapy
P_i	Inorganic phosphate
PP_i	Inorganic pyrophosphate
PLP	Pyridoxal 5′-phosphate
PEA	Phosphoethanolamine
GPI	Glycosylphosphatidylinositol
HAP	Hydroxyapatite
CHO	Chinese hamster ovary cells
BSP	Bone sialoprotein
OPN	Osteopontin
DSP	Dentin sialoprotein
PHEX	Phosphate-regulating endopeptidase
XLH	X-linked hypophosphatemia
PDB	Paget's disease of the bone
PDL	Periodontal ligament
OMIM	Online Mendelian Inheritance in Man compendium
ANK/ANKH	Progressive ankylosis protein
ENPP1	Ectonucleotide pyrophosphatase phosphodiesterase 1 protein
PTH	Parathyroid hormone
TPTD	Teriparatide
BMT	Bone marrow transplantation
SCB	Stromal cell boost
IV	Intravenous
SC	Subcutaneous
IP	Intraperitoneal

13.1 Introduction

Hypophosphatasia (HPP) is an inherited disorder (OMIM #146300, #241500, #241510) characterized by low serum alkaline phosphatase. HPP is caused by loss-of-function mutations in the *ALPL* gene encoding the protein, tissue-nonspecific alkaline phosphatase (TNSALP). TNSALP is expressed by mineralizing cells of the skeleton and dentition and is associated with the mineralization process. Generalized reduction of activity of the TNSALP leads to accumulation of substrates of TNSALP including inorganic pyrophosphate (PP_i), a potent inhibitor of mineralization. This leads to defective skeletal and dental mineralization, resulting in multisystemic complications with significant morbidity and mortality in the most severe cases. Management of HPP has been limited to supportive care until the introduction of a TNSALP enzyme replacement therapy (ERT), asfotase alfa (AA). AA ERT has proven to be transformative, improving survival in severely affected infants and increasing overall quality of life in children and adults with HPP. This chapter provides an overview of TNSALP expression and functions, summarizes HPP clinical types and pathologies, discusses early attempts at therapies for HPP, summarizes development of HPP mouse models, reviews design and validation of AA ERT, and provides up-to-date accounts of AA ERT efficacy in clinical trials and case reports, including therapeutic recommendations and response, adverse effects, and limitations.

13.2 Tissue Non-specific Alkaline Phosphatase: An Enzyme Essential for Mineralization

The enzyme we now know as tissue-nonspecific alkaline phosphatase (TNSALP, TNALP, or TNAP) was first reported by Dr. Robert Robison in 1923 (Robison 1923 and reviewed in Siller and Whyte 2018). Robison hypothesized that the enzymatic activity he detected in extracts from rat and rabbit bones that liberated inorganic phosphate (P_i) from hexosemonophosphoric acid was involved in calcium phosphate deposition in the skeleton. Robison explored properties of the enzyme further in subsequent studies and terminology regarding the enzyme shifted from phosphoric esterase to bone enzyme, phosphatase, and bone phosphatase, finally being referred to as alkaline phosphatase based on non-physiologic alkaline conditions used to assay the enzyme activity in vitro, as well as to differentiate it from an acidic prostate phosphatase discovered shortly afterwards (Goodwin and Robison 1924; Gutman and Gutman 1938; Kay and Robison 1924; Martland and Robison 1924, 1926, 1927, 1929; Robison et al. 1930; Robison and Soames 1924, 1925; Siller and Whyte 2018).

TNSALP is encoded by the *ALPL* gene on chromosome 1 in humans (NM_000478.5). Three additional alkaline phosphatase isozymes exist in humans: intestinal (IAP encoded by *ALPI* on chromosome 2; NM_001631.4), placental (PLAP encoded by *ALPP* on chromosome 2; NM_001632.4), and germ cell (GCAP or ALPG encoded by *ALPPL2* on chromosome 2; NM_031313.2) (Millán 2006; Millan and Whyte 2016). In humans, the *ALPL* gene encodes 12 exons and 11 introns, adding up to a total of 69,034 base pairs, though exons Ia and Ib are noncoding and separated from the ATG initiation site in exon II. Human *ALPL* mRNA is translated into the TNSALP protein comprising 524 amino acids (Fig. 13.1a).

TNSALP includes a number of important conserved amino acid sites, motifs, and domains that can be delineated in the two-dimensional sequence (Fig. 13.1b), though these function in the context of the three-dimensional protein structure, that includes a dimeric structure composed of two TNSALP monomers (Fig. 13.1c, d). TNSALP is bound to cell plasma membrane surfaces by a glycosylphosphatidylinositol (GPI) anchor that can be cleaved to release the enzyme into circulation, where circulating alkaline phosphatase activity (ALP) can be detected in plasma. The enzyme active site is located in the extracellular domain making TNSALP an ectoenzyme. Additional important structural and functional motifs in the TNSALP amino acid sequence include a hydrophobic domain involved in monomer-monomer interactions to assemble the functional enzyme homodimer, three metal cation (Zn^{2+} and Mg^{2+}) binding sites critical for enzyme activity, a flexible crown domain involved in interactions with collagen matrix and inhibitors, an N-terminal α-helix that (along with the crown domain) contributes to stabilization of the dimeric structure, and N-linked glycosylation sites that affect catalytic activity and kinetic properties of the enzyme. Across nearly its entire amino acid sequence, TNSALP is extremely highly evolutionarily conserved (Fig. 13.1e), suggesting functional importance for the majority of the protein structure.

While a broad substrate specificity has been demonstrated in vitro, natural substrates indicated by TNSALP loss-of-function include inorganic pyrophosphate (PP_i), phosphoethanolamine (PEA), and pyridoxal 5′-phosphate (PLP), described in more detail in the next section. TNSALP is highly expressed in bones, teeth, liver, and kidney (and at lower levels in fibroblasts, endothelial cells, and nervous system), thus its nomenclature as a "non-specific" enzyme. In the skeleton, bone forming and mineralizing osteoblasts are the primary cells expressing TNSALP, while in teeth and their supporting tissues, ameloblasts, odontoblasts, cementoblasts, and other cells of the PDL all express TNSALP (Fig. 13.2a–l).

TNSALP is likely the most critical enzyme for mineralization of bones and teeth. Upon its discovery, Robison presciently hypothesized that TNSALP was associated with skeletal mineralization, possibly by locally increasing P_i through dephosphorylation of substrates (Robison 1923; Siller and Whyte 2018). The ability for TNSALP to hydrolyze and thus inactivate PP_i came to be understood as a possibly more important function of TNSALP in mineralization. PP_i is a potent inhibitor of calcium phosphate (hydroxyapatite;

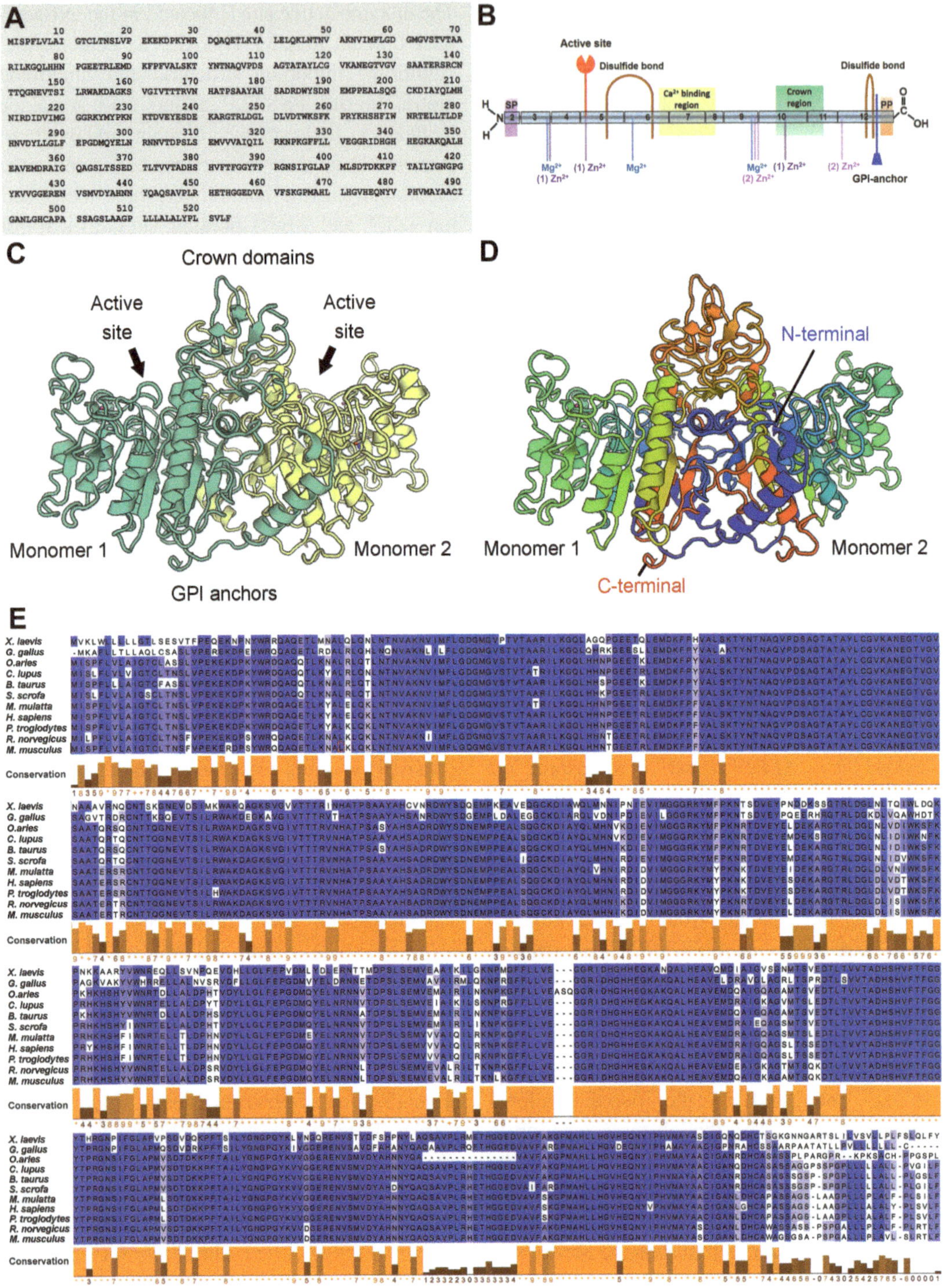

Fig. 13.1 Human TNSALP sequence and structure. (**a**) Human TNSALP is comprised of 524 amino acids, shown here by their one-letter codes. (**b**) Human TNSALP 2D protein structure superimposed over the 12 exons and demarcating conserved functional sites and domains. (**c**) Human TNSALP 3D structure showing the functional dimer with monomer 1 (green, on the left) and monomer 2 (yellow, on the right). Functional sites and domains are indicated for both monomers. 3D images are based on crystalized structure of PLAP and were imaged through Swiss-Model (www.swissmodel.expasy.org) and UniProt (www.uniprot.org). (**d**) Human TNSALP 3D structure

HAP) crystal growth that constitutes the inorganic mineralized component of bones and teeth (Bisaz et al. 1968; Fleisch and Bisaz 1962a, b; Fleisch et al. 1965, 1966; Meyer 1984; Meyer and Fleisch 1984). While TNSALP acts as a pro-mineralization enzyme decreasing local levels of PP_i, counter-regulatory proteins, including progressive ankylosis protein (ANK/ANKH) and ectonucleotide pyrophosphatase phosphodiesterase 1 (ENPP1), increase PP_i production, altogether comprising a complex feedback system of regulators that control the location and extent of mineralization in the body (Fig. 13.2m) (Foster et al. 2012; Gurley et al. 2006; Harmey et al. 2004; Ho et al. 2000; Johnson et al. 2000; Millan 2013; Murshed et al. 2005; Nociti et al. 2002; Rutsch et al. 2001, 2003; Terkeltaub 2001; Zweifler et al. 2015).

13.3 Hypophosphatasia

HPP is characterized by reduced ALP activity and extracellular accumulation of PP_i, resulting in defective mineralization of bones and teeth. HPP is caused by loss-of-function mutations in the *ALPL* gene that encodes TNSALP. At this writing, 377 such mutations have been reported (http://www.sesep.uvsq.fr/03_hypo_mutations.php), with the majority (~70%) being missense mutations, but also including nonsense mutations, deletions, and alterations in *ALPL* regulatory regions. The mode of inheritance can be either autosomal recessive or autosomal dominant (Mornet 2017; Thakker et al. 2017). The prevalence is highest among Mennonites in Manitoba, Canada, where approximately 1 in 2500 neonates manifests lethal form of HPP (Greenberg et al. 1993). The prevalence of severe and moderate HPP in Europe has been estimated to be 1 in 300,000 and 1 in 6370, respectively (Mornet et al. 2011). A critical function of TNSALP in skeletal tissues is to hydrolyze and thus reduce levels of PP_i, allowing physiological mineralization to proceed. Deficiency of TNSALP in HPP leads to increased PP_i levels that inhibit HAP crystal nucleation and growth in the extracellular matrix, thereby impairing skeletal and dental mineralization. This condition also secondarily leads to disturbances of calcium and phosphorus homeostasis. Elevation of serum calcium or phosphorus levels sometimes occurs and is thought to be the result of a combination of normal gut absorption of these ions and the inability to effectively incorporate them into bone HAP. In addition to increased PP_i, HPP also leads to increased extracellular accumulation of two other known physiological substrates of TNSALP, PLP and PEA.

The clinical spectrum of HPP is broad and highly variable, even within families. Clinical manifestations range from perinatal death to severe bone deformities in early childhood, to primary tooth loss with little or no other clinically detectable systemic or skeletal manifestations. The severity typically correlates with earlier disease onset. HPP is classified into six different clinical forms, based on the age onset of clinical symptoms: perinatal, benign prenatal, infantile, childhood, adult, and odontohypophosphatasia (odonto-HPP) (Table 13.1). This classification delineates disease severity and may correlate with prognosis and ALP levels, although there is considerably overlap in clinical phenotypes and biochemical hallmarks in HPP (Whyte et al. 2018).

Fig. 13.1 (continued) showing both monomers in rainbow colors to indicate N-terminal (blue) to C-terminal (red) sequence. N- and C-termini are indicated for monomer 1 (left side). (**e**) TNSALP multiple sequence comparisons between human (*Homo sapiens*; UniProt record P05186), chimpanzee (*Pan troglodytes*; K7B4Y6), Rhesus macaque monkey (*Macaca mulatta*; A0A1D5R5B1), rat (*Rattus norvegicus*; P08289) mouse (*Mus musculus*; P09242); dog (*Canis lupus familiaris*; F1PF95), cow (*Bos taurus*; P09487), sheep (*Ovis aries*; W5PFB8), pig (*Sus scrofa*; A0A287BSC3), chicken (*Gallus gallus*; Q92058), and African clawed frog (*Xenopus laevis*; Q7ZYJ4). Regions well conserved across species are indicated by darker blue coloration of the amino acids in the sequence rows. Relative conservation is also indicated by a histogram below the sequence comparisons, with numerical scores from 0 to 9, where higher scores indicate better conservation (also shown by lighter orange color) and * indicating perfect conservation across all species. Sequence comparison performed by Clustal Omega (www.clustal.org/omega) and viewed by Jalview (www.jalview.org)

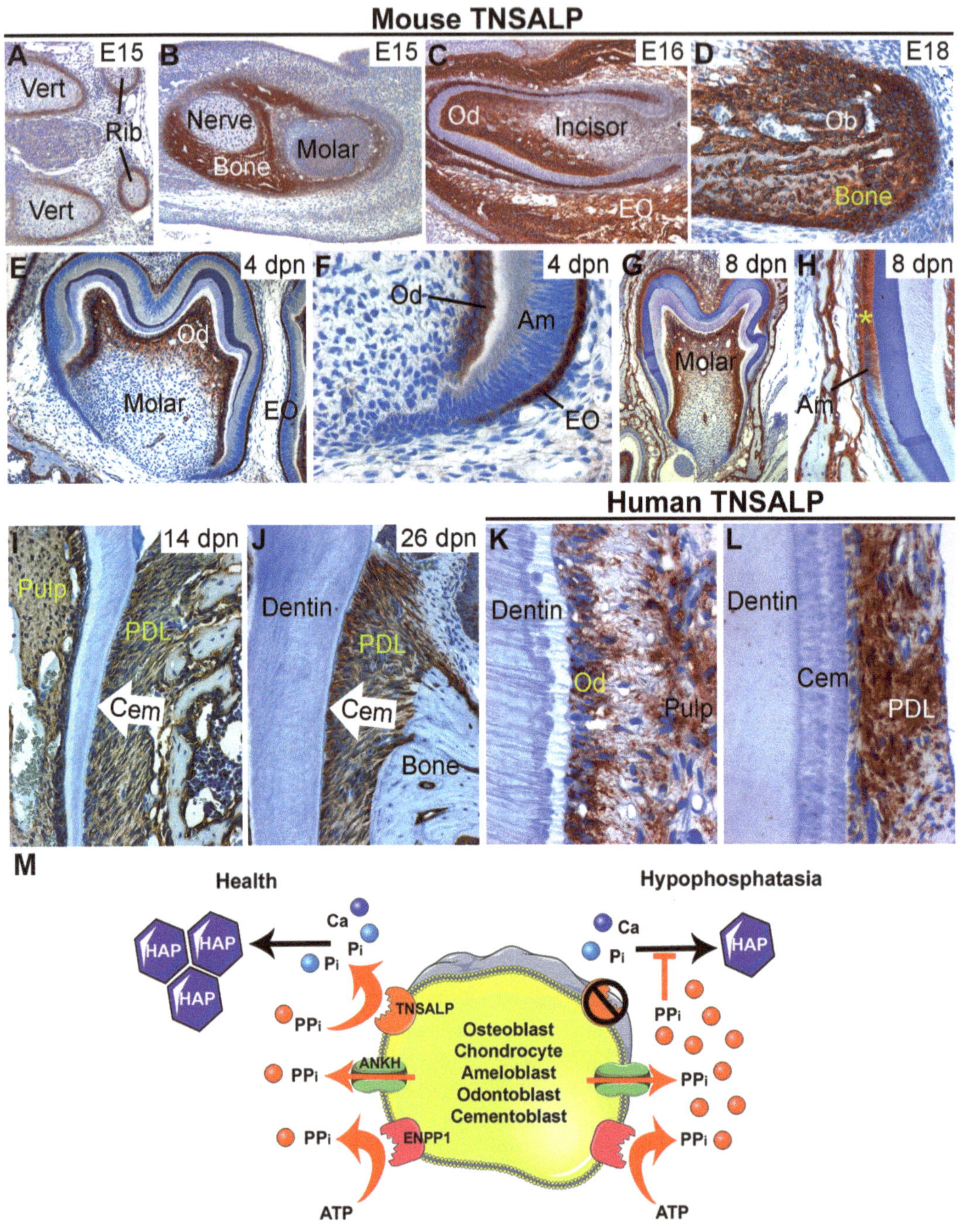

Fig. 13.2 Mineralizing cells express TNSALP. Immunohistochemistry in mouse (**a**–**j**) and human (**k** and **l**) tissues shows the expression of TNSALP (red color). TNSALP is expressed around (**a**) forming vertebrae (Vert) and ribs at mouse embryonic day 15 (E15), and is strongly expressed in the mineralizing bone of the mandible, (**b**) though is not yet expressed in the molar tooth that is undergoing morphogenesis but not yet mineralizing. (**c**) The incisor tooth is mineralizing by E16, and TNSALP is found in the enamel organ (EO) and in odontoblasts (Od) forming dentin. (**d**) Osteoblasts (Ob) of the mouse jaw bone at E18 strongly express TNSALP. (**e** and **f**) At 4 days postnatal (dpn), Od and EO of the molar strongly express TNSALP, though ameloblasts (Am) do not. (**g** and **h**) By 8 dpn, Am entering the maturation stage (*) of enamel mineralization begin expressing TNSALP. (**i** and **j**) At later ages of 14 and 26 dpn, after periodontal tissues have formed, the entire periodontal ligament (PDL) between

13.3.1 Clinical Classification of HPP and the Clinical/Radiologic Findings

13.3.1.1 Perinatal HPP

Perinatal HPP is the most severe form of HPP and is typically lethal as a result of an almost complete absence of skeletal mineralization. Severely affected infants often die shortly after birth due to respiratory complications arising from hypoplastic lungs and skeletal deformities of the thorax. The key radiographic and sonographic features that are characteristic of and unique to severe lethal perinatal HPP are absent ossification of whole bones at or after 11 weeks' gestation (Offiah et al. 2018). Other prenatal imaging findings, characteristic of perinatal HPP and not unique to HPP, include shortening, bowing and angulation of the long bones, mid-diaphyseal ("Bowdler") spurs, slender and poorly ossified thin ribs, metaphyseal lucencies, and deficient ossification in the skull observed as wide sutures and fontanelles (Fig. 13.3a–c) (Offiah et al. 2018; Zankl et al. 2008). These features also arise from other skeletal dysplasias (e.g. osteogenesis imperfecta, cleidocranial dysplasia, campomelic dysplasia, and achondrogenesis subtypes), which should be in the differential diagnosis. HPP can be distinguished and confirmed by prenatal ALP measurements.

13.3.1.2 Benign Prenatal HPP

Benign prenatal HPP manifests in utero, with abnormal imaging findings similar to the perinatal HPP. Affected fetuses exhibit skeletal deformities including poorly mineralized bone or short, severely bowed legs, which can sometimes be diagnosed as the perinatal lethal form of HPP (described above) (Fig. 13.3d, e). However, in cases of the benign prenatal form of HPP, the skeletal phenotype can be less severe and spontaneous improvement is observed beginning in the third trimester of pregnancy and continuing after birth (Offiah et al. 2018; Wenkert et al. 2011). Postnatal clinical outcomes range in severity from infantile to odonto-HPP phenotypes, therefore, abnormal prenatal ultrasound findings before the third trimester are not predictive of perinatal lethal HPP.

13.3.1.3 Infantile HPP

Infantile HPP presents before 6 months of age and is associated with approximately 50% mortality due to respiratory failure due to severe hypomineralization and mechanical weakness of the chest wall. Affected infants can appear normal after birth until emergence of poor feeding, failure to thrive, and hypotonia with delayed motor milestones develop. Radiologic findings include generalized hypomineralization with severe skeletal deformities, including rachitic defects of the long bones and chest (Fig. 13.3f). Infants may also have muscle pain and weakness from a static myopathy, possibly related to accumulation of TNSALP substrate, PP_i (Seshia et al. 1990). Craniosynostosis and other skull abnormalities occur in about 40% of infants with infantile HPP (Fig. 13.3g, h), and these complications may require neurosurgical intervention due to intracranial hypertension (Collmann et al. 2009). Proptosis, mild hypertelorism, and brachycephaly can develop. Unlike other forms of hereditary

Fig. 13.2 (continued) the root surface cementum (Cem) and bone becomes strongly positive for TNSALP. (**k** and **l**) Human TNSALP dental expression patterns parallel those in mice, with strong expression in Od and in the PDL. (**m**) Model of TNSALP function in mineralizing cells. Levels of the mineralization inhibitor, PP_i, are controlled by activities of ENPP1, ANKH/ANK, and TNSALP, all expressed by mineralizing cells. ENPP1 enzymatically cleaves nucleotide triphosphates (e.g. adenosine triphosphate, ATP) to generate PP_i, while ANKH/ANK directs PP_i transport to the extracellular space, both increasing pericellular PP_i levels. TNSALP hydrolyzes PP_i to allow inorganic phosphate (P_i) and calcium (Ca) to precipitate as hydroxyapatite (HAP), the inorganic component of bones and teeth. When TNSALP activity is lost in HPP, excess PP_i inhibits HAP crystal initiation and growth, causing mineralization defects in the skeleton and dentition. Figure designed with images from Servier Medical Art (https://smart.servier.com/) under a Creative Commons Attribution 3.0 Unported License. (Images in panels **e**, **f** reused with permission from McKee et al. J Dent Res 92(8): 721–727, 2013. Images in panels **g**, **h** reused with permission from Yadav et al. 2012. Images in panels **i**–**l** reused with permission from Zweifler et al. 2015. Image in panel **m** reused with permission from Bowden and Foster. Drug Des Devel Ther 12: 3147–3161, 2018)

Table 13.1 Clinical forms of HPP and their characteristics

Form	Onset	Clinical features	Imaging findings
Perinatal lethal	*In utero*/at birth	Profound hypomineralization Short and deformed extremities Polyhydraminos Respiratory insufficiency at birth to the first week of life Pyridoxine-responsive seizures Stillbirth or death within days/weeks after birth	Absence of mineralization of the roof of the skull and bones Thin ribs Fractures
Benign prenatal	*In utero*/at birth	Severe hypomineralization with skeletal deformities Spontaneous improvement in the skeletal disease after birth	Same as perinatal lethal form
Infantile	First 6 months of life	Severe hypomineralization with severe skeletal deformities (rachitic ribs, *genu varum*) Craniosynostosis Pyridoxine-responsive seizures Failure to thrive and delayed developmental milestones Hypercalcemia and hypercalciuria Premature loss of primary teeth	Absent bones or severe mineralization defects; Gracile bones; "Tongues" radiolucency in metaphyses; Patchy areas of osteosclerosis; Copper beaten appearance of skull radiograph; Wedging of the lower thoracic and upper lumbar vertebrae; Nephrocalcinosis on renal ultrasound.
Childhood	After 6 months of life	Rickets (bowed legs with bony enlargement near joints due to widened metaphyses) Chronic skeletal pain Recurrent fractures Short stature Abnormal ambulation or gait Premature loss of primary teeth	Same as infantile form
Adult	After 18 years	Fragility fractures Stress fractures of metatarsals, tibias Delayed fracture healing Osteomalacia Chondrocalcinosis Osteoarthropathy	Recurrent metatarsal fractures; Vertebral crush fractures; Femoral pseudofractures.
Odonto-HPP	Before 4–5 years of age	Premature loss of primary teeth prior to age 5 years Loss of permanent teeth Abnormal dentin Thin roots with wide pulp chambers Delayed eruption	Alveolar bone loss; Lack of other skeletal or radiologic manifestations.

and nutritional rickets, serum calcium levels are generally high at diagnosis. Hypercalciuria and nephrocalcinosis can also occur as consequences of hypercalcemia. Irritability and vomiting are also common, arising from hypercalcemia or increased intracranial pressure with papilledema secondary to craniosynostosis.

13.3.1.4 Childhood HPP

Childhood HPP presents after age 6 months, with wide ranging clinical manifestations that were suggested to be subdivided into "mild" and "severe" types in 2015 (Whyte et al. 2015a, 2018). Individuals with mild childhood HPP can maintain good physical function with minimal symptoms or skeletal changes. Severe childhood-onset HPP can feature cranial hypomineralization or craniosynostosis (Fig. 13.3i), however skeletal rickets is typically the main feature (Whyte et al. 2015a), manifesting as bowed legs and bony enlargement near joints due to widened metaphyses, and radiographic findings including characteristic 'tongues' of radiolucency projecting from growth plates into metaphyses (Fig. 13.3j, k). Chronic skeletal pain, recurrent fractures, short stature, muscle weakness, abnormal ambulation or gait, and premature loss of deciduous teeth are common signs or symptoms (Fig. 13.3l). Rarely, childhood HPP can present with chronic multifocal non-bacterial osteomyelitis mimicking malignancy thought to be due to marrow edema secondary to PP_i crystal deposition (Girschick et al. 2007; Whyte et al. 2009). Craniosynostosis can cause chronic increased intracranial pressure, papilledema and impaired visual acuity (Libby Kosnik-Infinger et al. 2015). Individuals exhibiting more severe manifestations of childhood HPP may present more obvious dental defects, including enamel hypoplasia and discoloration (Fig. 13.3m).

13.3.1.5 Adult HPP

Adult HPP is typically diagnosed in middle age based on presentation of chondrocalcinosis, osteoarthropathy, and/or recurrent stress fractures that are often poorly healing (Fig. 13.3n). Femoral pseudofractures and atypical subtrochanteric femoral fractures have been reported in adult HPP; the latter can occur without or following exposure to bisphosphonates (Berkseth et al. 2013; Genest and Seefried 2018; Lawrence et al. 2017; Sutton et al. 2012). Individuals with lower ALP levels and higher PLP and PEA levels tend to exhibit more fractures. There is a diverse spectrum of clinical manifestations within adult HPP, ranging from minimal symptoms such as dental abnormalities with no pain and normal bone mineralization on bone biopsy, to significant pain and fractures, with osteomalacia on bone biopsy (Fig. 13.3o, p). Asymptomatic or minimally symptomatic adults with HPP are identified based on low ALP levels or family history of HPP, and may report a history of 'childhood rickets' or fractures (Berkseth et al. 2013). Some affected adults recall a history of premature loss of their deciduous teeth, and may also have early loss of the adult dentition (Whyte 2016). Bone mineral density is low in adult HPP with severe skeletal manifestations, but is normal in asymptomatic adult HPP or those with mild manifestations. However, despite the absence of clinical skeletal abnormalities or decreased BMD, low bone turnover or low bone remodeling has been described in individuals with mild adult HPP, consistent with persistent hypophosphatasemia (Lopez-Delgado et al. 2018). Adult HPP can become debilitating, with severe disability secondary to recurrent fractures, muscle weakness, musculoskeletal pain and restricted range of motion from chondrocalcinosis (Berkseth et al. 2013; Lawrence et al. 2017; Weber et al. 2016).

13.3.1.6 Odonto-HPP

Odonto-HPP features dental defects as the primary manifestations, with biochemical characteristics of HPP and mild skeletal abnormalities or lack of clinically apparent skeletal changes. Dental defects are nearly universal among individuals affected by HPP regardless of the clinical form (Bloch-Zupan 2016; Feeney et al. 2018; Reibel et al. 2009). The most common dental sign is premature loss of primary/deciduous teeth prior to age 5 years, an important diagnostic criterion for HPP (Fig. 13.3q). In a large cohort of pediatric patients, about 98% exhibited premature loss of primary teeth (Whyte et al. 2015b). Anterior teeth (maxillary and mandibular incisors and canines) are the most frequently lost, typically require no or very mild trauma to extrude them from the sockets, and have an unusual and characteristic appearance of being fully rooted, rather than featuring partially resorbed roots typical of physiologically exfoliated primary teeth (Fig. 13.3r). In cases of odonto-HPP or mild childhood HPP, premature tooth loss is often the first recognized sign that something is amiss, and is thus a critical diagnostic criterion that places the general or pediatric dentist in a position to refer the patient to an endocrinologist. Tooth loss is the direct result of elevated PP_i inhibiting formation and mineraliza-

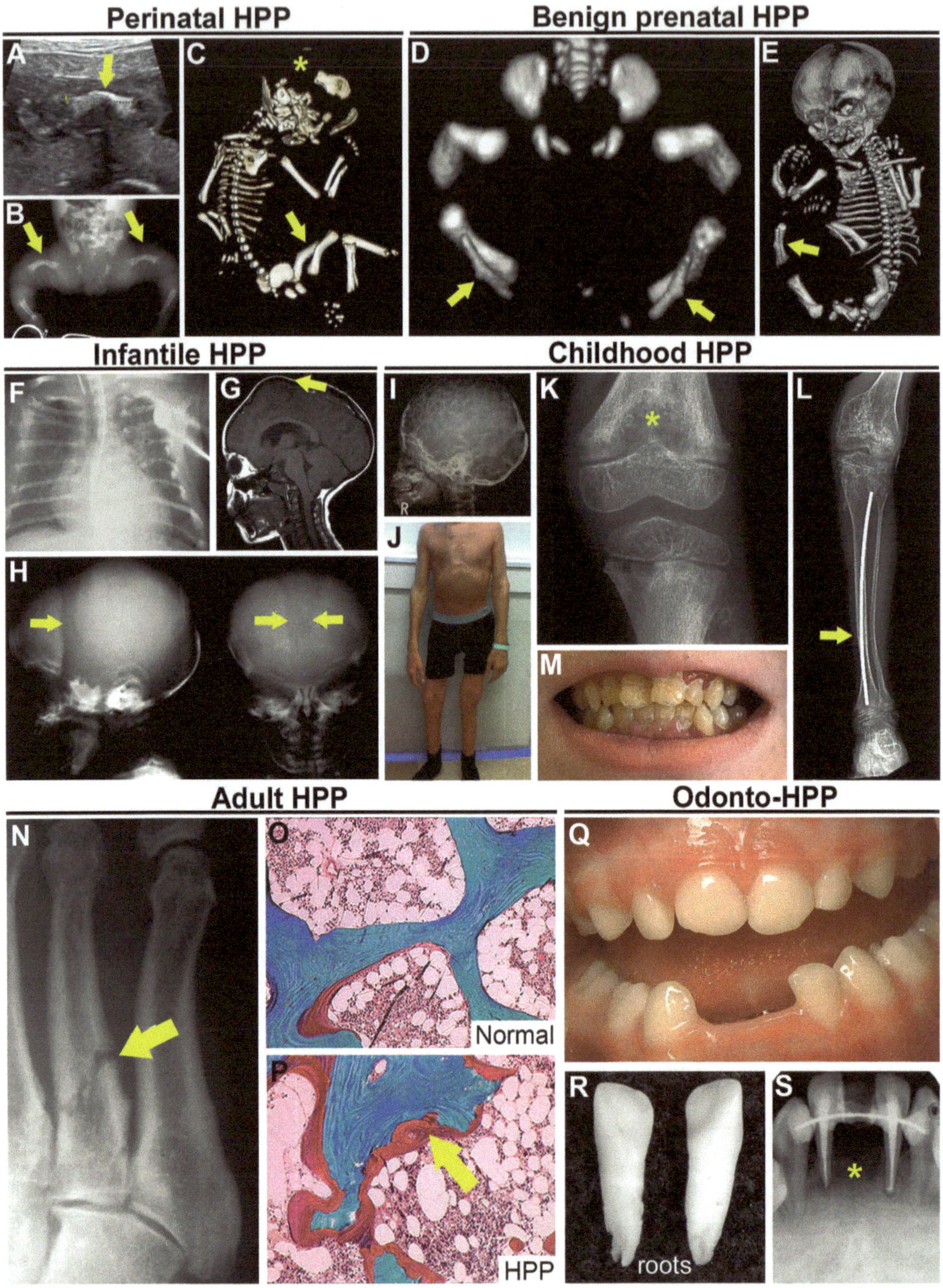

Fig. 13.3 Skeletal and dental defects associated with clinical forms of HPP. Overview of characteristic features of perinatal (**a**–**c**), benign prenatal (**d**, **e**), infantile (**f**–**h**), childhood (**i**–**m**), adult (**n**–**p**), and odonto-HPP (**q**–**s**) forms. (**a**) A fetus with perinatal HPP at 18 weeks-of-gestation displays a short and angulated femur (yellow arrow) by ultrasound. (**b**) Postmortem radiograph of a fetus with perinatal HPP at 38 weeks of gestation reveals bowed femora (yellow arrows) and metaphyseal tongues of radiolucency. (**c**) CT image of the same fetus

tion of tooth root cementum (Bruckner et al. 1962; van den Bos et al. 2005). Additional dental abnormalities reported in association with HPP include loss of secondary/permanent teeth, tooth mobility, abnormal or thin dentin, large pulp space, abnormal tooth root shapes, periodontal disease or alveolar bone loss, malocclusion, and enamel defects (Fig. 13.3s). However, there is presently not a consensus on how common these manifestations are, or how they relate to genotype, biochemical findings, or musculoskeletal effects of HPP, or why a small percentage of affected individuals do not lose their teeth prematurely. It is important to note that a subset of patients diagnosed with odonto-HPP have been observed to develop mild to moderate skeletal manifestations such as fractures and bone pain later in life, therefore, long-term follow-up is recommended (Mori et al. 2016).

13.3.2 Diagnosis of HPP

Diagnosis of HPP can be made with confidence when the clinical history, and physical and radiographic skeletal findings (as described in detail above) are consistent with this diagnosis, and when serum ALP is below the normal range for the patient's age. Patients with perinatal or infantile onset HPP may be misdiagnosed with severe form of osteogenesis imperfecta, however, HPP can be distinguished by low serum ALP. Circulating ALP is elevated in osteogenesis imperfecta and other forms of rickets. Other pediatric skeletal disorders with low serum ALP levels that can mimic HPP and should be considered as a differential diagnosis of HPP include a rare lethal form of osteogenesis imperfecta (Royce et al. 1988), neonatal lethal osteochondrodysplasia (Wyckoff et al. 2005), and severe cleidocranial dysplasia (El-Gharbawy et al. 2010; Unger et al. 2002). Evaluation for high substrates

Fig. 13.3 (continued) with perinatal HPP at 25 weeks-of-gestation shows bowed femurs (yellow arrow), deficient rib ossification, and widened cranial sutures (yellow ∗). (**d**) CT image of a fetus with benign prenatal HPP shows bowed long bones with normal metaphyses, in addition to mid-diaphyseal spurs on the fibulas (yellow arrows). (**e**) CT image of the same infant with benign prenatal HPP shows bowed leg bones and normal cranial ossification. (**f**) Radiograph of a 23-week-old child with infantile HPP shows gracile, deformed, and fractured ribs. (**g**) MRI of the skull of a 6-year-old child with infantile HPP exhibiting craniosynostosis and the resulting bregmatic bump (yellow arrow). (**h**) Lateral and anterior radiograph of a 14-week-old child with infantile HPP showing hypomineralization of the calvarium, giving the appearance of widened sutures (yellow arrows). (**i**) Radiograph of a 4-year-old individual with childhood HPP reveals hypomineralization of the cranial vault as seen by the "copper beaten" appearance of the skull. (**j**) Deformities of lower extremities with joint widening at knees and elbows in a 15-year-old boy with severe childhood hypophosphatasia. (**k**) Radiograph of the knee of a child with severe childhood HPP reveals tongues of radiolucency (yellow ∗) extending from the distal metaphysis of the femur. (**l**) Knee radiograph of the same individual from (panel **j**) reveals hypomineralized bone, coarsened trabeculae, and an intramedullary rod in the tibia. A fracture line is seen in the diaphysis of the tibia (yellow arrow). (**m**) Oral photograph of a boy with severe childhood HPP showing enamel discoloration and hypoplasia, manifested as horizontal bands and irregular crown appearance and texture. (**n**) Radiograph of an individual with adult HPP showing right fourth metatarsal fracture (yellow arrow). (**o**) Goldner trichrome stain of normal iliac crest biopsy compared to (**p**) the same from an individual with adult HPP, showing accumulation of excessive osteoid (red layer indicated by yellow arrow) on the surface of the mineralized bone (green). (**q**) Oral photograph of a 2.5-year-old child with HPP exhibiting premature loss of primary lower incisors. (**r**) Primary incisors that spontaneously exfoliated from a child with HPP. (**s**) Oral radiograph of a 20-year-old individual diagnosed with odonto-HPP showing loss of secondary incisor, endodontic treatment after fracture, splinting to try and stabilize remaining anterior teeth, and generalized alveolar bone loss (yellow ∗). (Images in panels **a–e** reproduced from Offiah et al. Pediatr Radiol 1–20, 2018, and used under the terms of the Creative Commons CC BY license. Images in panel **f**, **h** reproduced from Millán and Whyte 2016, and used under the terms of the Creative Commons CC BY license. Images in panels **g**, **i** reproduced with permission from Collmann et al. 2009. Images in panels **j**, **l** reproduced with permission from Bowden and Foster, Drug Des Devel Ther 12: 3147–3161, 2018. Image in panel **k** reproduced with permission from Whyte 2017. Image in panel **n** reproduced with permission from Whyte et al. 2007. Images in panels **o**, **p** reproduced with permission from Berkseth et al. 2013. Image in panel **q** reproduced from Reibel et al. 2009, in accordance with BMC's open access policy. Image in panel **r** reproduced with permission from Whyte 2017. Image in panel **s** reproduced with permission from Rodrigues et al. 2012)

(PP_i, PLP, and PEA) or molecular genetic testing can be important to make a correct diagnosis.

Serum ALP levels in parents, as a noninvasive diagnostic tool, may aid in prenatal differential diagnosis when bone dysplasia is detected by fetal imaging; when ALP is low, a high index of suspicion for perinatal HPP is raised (Castells et al. 2018).

For less severe forms of HPP, diagnosis can be challenging, and requires further laboratory evaluation to assess TNSALP substrate levels, as described below in next section. Low serum ALP can be found in several clinical situations (Table 13.2) where a distinct cause may be apparent. For example, a boy with Duchene muscular dystrophy on chronic corticosteroid treatment would feature low serum ALP due to the long-term suppression of bone turnover resulting from chronic corticosteroid treatment. In this clinical setting, further laboratory workup to rule out HPP is not required. In some clinical situations, where the reason for low ALP level cannot be ascertained, clinicians may need to obtain additional laboratory tests including serum zinc, magnesium, thyroid function test, complete blood count, parathyroid hormone (PTH), vitamin B_{12}, vitamin C, vitamin D, celiac antibodies, serum ceruloplasmin, and renal and liver function tests, in order to rule out conditions outlined in Table 13.2 (Saraff et al. 2016).

13.3.3 Laboratory Workup for HPP

The key to correct diagnosis of HPP is a low ALP level. A critical caveat is that ALP levels must be interpreted within the context of age- and gender-appropriate reference ranges. ALP levels are considerably much higher in healthy children compared to adults. Serum ALP is especially high during the growth spurt of adolescence, which occurs earlier in girls than in boys. Some clinical laboratories still report values only for adults. Therefore, a child's serum ALP level may be incorrectly interpreted to be normal if using an adult reference range, when it is actually remarkably low if using the correct and much higher pediatric reference range.

Table 13.2 Differential diagnosis of hypophosphatasia (HPP) from low serum alkaline phosphatase (ALP)

Conditions with low ALP
Hypophosphatasia (HPP)
Hypothyroidism
Cushing syndrome or chronic corticosteroid treatment
Profound anemia
Wilson's disease
Celiac disease
Starvation
Milk-alkali syndrome
Cardiac-bypass surgery
Zinc or magnesium deficiency
Vitamin C deficiency
Vitamin D intoxication
Radioactive heavy metals
Improper collection of blood specimen (e.g., EDTA, oxalate)
Inappropriate reference range
Conditions with low serum ALP that can mimic HPP
Osteogenesis imperfecta (OI) type II
Cleidocranial dysplasia

Patients with persistently low ALP levels require further diagnostic evaluation for HPP, even in the absence of other clinical symptoms (Saraff et al. 2016). To confirm diagnosis of HPP, clinicians should document elevation of TNSALP substrates resulting from markedly reduced ALP activity. Three substrate markers elevated in HPP are plasma/serum PLP and PP_i, and urine PEA. Serum PLP typically is ordered as "vitamin B_6". Ingestion of vitamin supplements containing vitamin B_6 can result in false positive values, therefore, vitamin supplements must be avoided for 1 week before laboratory testing. Serum PLP concentrations are associated with phenotype severity (Akiyama et al. 2018), and the occurrence of fractures and multiple symptoms in adult HPP (Schmidt et al. 2017). A medical laboratory test for serum PP_i is not commercially available and is performed only in research laboratory settings. Elevated urine PEA supports a diagnosis of HPP, but is not pathognomonic. When low serum ALP levels are associated with elevated PLP and/or PEA concentrations, radiographs of wrist and knee joints should be obtained. Antero-posterior and lateral view of skull may also need to be evaluated. Patients with suspected HPP should be referred to endocrinologists or bone specialists.

13.3.4 Genetic Testing

Mutational analysis of the *ALPL* gene can be performed to establish the diagnosis of HPP, although it may not be necessary in straightforward cases when clinical, radiographic and biochemical laboratory findings are consistent with HPP. *ALPL* gene analysis in affected individuals and family members can provide genetic information to help understand inheritance pattern and recurrence risks for genetic counselling to the family. In some cases, *ALPL* mutations are difficult to identify through the most common sequencing methods.

13.4 Attempted Treatments for Severe Hypophosphatasia Prior to Enzyme Replacement Therapy

Prior to the development and approval of AA ERT described below, management of HPP was limited to supportive care that included pain relief, orthopedic surgeries for fractures, treatment of hypercalcemia with hydration and low-calcium diet, or ventilator support for severely affected infants with respiratory insufficiency. While this range of support remains the standard of care for the majority of affected individuals, novel therapeutic interventions have been attempted in cases of severely affected infants and children. Despite the lack of long-term success of these attempted therapies, they provided invaluable insights into the etiopathology of HPP.

13.4.1 Blood Transfusion

Recognition that HPP caused low circulating ALP levels prompted interventions to attempt to ameliorate effects of severe HPP. Paget's disease of bone (PDB; OMIM# 167250, 239000, 602080, 616833), a metabolic disease of increased and disorganized bone remodeling that typically strikes in adulthood, causes dramatically increased ALP, likely due to increased numbers and activity of osteoblasts (Vallet and Ralston 2016). In a novel and clever approach to treat a metabolic disorder, Whyte and colleagues collected ALP-rich plasma from adult subjects with PDB and administered this by intravenous (IV) infusion to four subjects with severe infantile HPP (Whyte et al. 1982, 1984). Weekly treatment over the course of five or more weeks elevated ALP into the normal range for all subjects. Hypercalcemia was better controlled in subjects, one subject showed radiographic stabilization of rickets, and another exhibited histological improvement of skeletal mineralization. However, urinary PP_i and PEA remained elevated, and hypomineralization and skeletal deformities persisted and worsened over time. After failure of PDB plasma to improve the manifestations of HPP, one subject was additionally serially administered parathyroid hormone (PTH), prednisone, and then infused with normal plasma over several months [mimicking another report indicating improvement after infusing an HPP subject with normal plasma (Albeggiani and Cataldo 1982). These combined interventions failed to make a substantial improvement and all four patients died from pneumonia secondary to HPP developmental defects (Whyte et al. 1986). Based on these disappointing outcomes, the authors speculated that perhaps ALP in circulation is not physiologically active or is not accessing the tissues where its activity is required, an observation supported by previous in vivo studies (Clubb et al. 1965; Jung et al. 1970).

13.4.2 Bone Marrow Transplantation

In 2003, Whyte and colleagues took a different tack, administering bone marrow cell transplantation (BMT) to an 18-month-old female subject with severe infantile HPP (Whyte et al. 2003). With this intervention, T-cell depleted marrow from a healthy sister was successfully engrafted and provided substantial benefit to the subject, notably reversing rickets and improving skeletal mineralization at 3 months post-BMT. Improvements proved transient when host hematopoiesis returned, accompanied by worsening rickets, scoliosis, and fractures by 6 months post-BMT. This reversal prompted a follow-up "stromal cell boost" (SCB) treatment. This second round again appeared to promote skeletal improvements, though severe disabilities remained and biochemical features of HPP were not corrected. This therapy was also accompanied by use of cyclosporine, glucocorticoids, and calcitonin, all of which can affect skeletal remodeling, though engraftment of donor mesenchymal cells likely contributed the largest improvement. In a modified approach, Whyte and colleagues treated a 9.5-month-old female with severe infantile HPP with BMT using her father's marrow, followed by implantation of bone fragments (also from the father) subcutaneous (SC) and intraperitoneally (IP), and IV administration of primary osteoblasts harvested and expanded ex vivo from additional bone fragments (Cahill et al. 2007). Though minimal engraftment was detectable over time, the subject showed decreased severity of rickets and scoliosis and improvement in skeletal mineralization, described by the authors as a shift from infantile to a milder phenotype more consistent with childhood HPP. Intriguingly, the subject was reported to retain her deciduous teeth after BMT and bone transfer.

13.4.3 Other Interventions

Shortcomings of these attempted therapies have been hypothesized to arise from relatively short treatment times, insufficiently elevated steady state ALP in circulation, and/or inability for soluble TNSALP to access or be retained at sites of skeletal mineralization, i.e. the mineralization front. Other than attempts to directly or indirectly supply functional ALP, few treatments have been attempted and none have been very successful. Bisphosphonates, a group of drugs that prevent bone loss by inhibiting osteoclast function, have been given to some individuals with HPP, usually due to misunderstanding of pathological mechanisms of HPP or without proper diagnosis (Cundy et al. 2015; Deeb et al. 2000; Doshi et al. 2009; Righetti et al. 2018; Sutton et al. 2012). As bisphosphonates do not promote bone formation, and the first generation of the drugs functioned much like synthetic PP_i to disrupt mineralization, these drugs typically worsen the hypomineralization caused by HPP and are strongly contraindicated for use in HPP patients.

Teriparatide (TPTD), a recombinant peptide based on the N-terminal portion of human parathyroid hormone (PTH), is an anabolic agent used to treat osteoporosis. Administration of TPTD to an individual with adult HPP and several stress fractures was able to increase ALP levels to within normal range (albeit at the lower edge), reduce bone pain, and was associated with good fracture healing (Whyte et al. 2007). Additional case reports on TPTD use in subjects with HPP have indicated potentially positive effects on reducing osteomalacia or accelerating bone healing (Cundy et al. 2015; Doshi et al. 2009; Righetti et al. 2018). Relative success of TPTD treatment may depend on the phenotype and biochemistry of each individual HPP patient as the increased numbers and activity of osteoblasts will still produce defective TNSALP enzyme.

13.5 Animal Models of Hypophosphatasia

13.5.1 *Alpl* Knock-Out Mice

The experimental interventions described above were performed with consent on patients with severe HPP and a high likelihood of lethality. While these therapeutic attempts were mostly unsuccessful at promoting long-term improvement in individuals with severe HPP, they revealed important clues about the pathology of HPP and what a successful therapeutic intervention would require. However, an animal model of HPP was necessary to systematically test the safety and efficacy of new therapies prior to use in human patients. With the discovery that loss-of-function mutations in the gene *ALPL* caused

HPP (Weiss et al. 1988, 1989a, b), researchers now had a target for developing clinical interventions. Based on breakthroughs in mouse genetics in the 1980s (Doetschman et al. 1987; Evans and Kaufman 1981; Mak 2007; Skoultchi et al. 1987; Thomas and Capecchi 1987), the ability to genetically inactivate specifically targeted genes in mice emerged in the 1990s as a novel and powerful strategy for understanding the functional importance of the encoded proteins (Hall et al. 2009). As they would for many other genes and diseases, these advances paved the way for development of mouse models to study HPP.

In the first reported mouse model of HPP from the laboratory of Dr. Grant MacGregor (sometimes referred to as the EM strain, for their creation at Emory University), exons 2–6 of mouse *Alpl* were targeted by homologous recombination in embryonic stem (ES) cells, reducing serum ALP by more than 90% (Waymire et al. 1995). Interestingly, authors reported that homozygous *Alpl* knockout ($Alpl^{-/-}$) mice at early postnatal ages did not display the expected skeletal defects. However, by 2 weeks of age, $Alpl^{-/-}$ mice developed spontaneous seizures, leading to early lethality. $Alpl^{-/-}$ mice featured increases in proposed TNSALP substrates, PP_i, PEA, and PLP. Increased PLP associated with decreased γ-aminobutyric acid (GABA) in the brain was found to be responsible for seizures, and pyridoxal injections combined with soft diet (chewable even with hypomineralized teeth and jaws) rescued some mice from seizures and early lethality, establishing the importance of TNSALP in vitamin B6 metabolism. In knockout mice where lifespan was extended by rescue, authors identified apparent defects in incisors consistent with dental enamel phenotypes reported in some human subjects with HPP. The second mouse model of HPP was reported shortly thereafter from the laboratory group of Dr. Jose Luis Millán (the so-called LJ strain, for creation at the Burnham Institute in La Jolla), targeting *Alpl* exons 5–8 through homologous recombination in ES cells (Narisawa et al. 1997). Like the previous model, these $Alpl^{-/-}$ mice displayed severe seizures and died before weaning. While body sizes and skeletons of $Alpl^{-/-}$ mice appeared similar to controls at early ages, analysis of bones at 8 days postnatal (dpn) or later revealed numerous examples of hypomineralization and fractures, as well as long bone growth plate abnormalities and hypomineralized and abnormal vertebrae (Fig. 13.4a–c). A follow-up report directly comparing the EM and LJ mouse strains determined that by 10 dpn, both models featured substrate accumulation and prominent bone hypomineralization, followed by progressive rachitic changes in long bones, accumulation of osteoid, and occurrence of bone fractures (Fedde et al. 1999). While minor differences in HPP severity were documented (possibly due to genetic background), both models replicated clinical manifestations of severe infantile HPP. The Millán lab employed the LJ HPP mouse model extensively over the next two decades, exploring the role of TNSALP in bone mineralization, vitamin B metabolism, and other organ systems, as well as elucidating interactions of TNSALP with other mineralization regulators including ANK and ENPP1 (Anderson et al. 2004, 2005; Cruz et al. 2017; Harmey et al. 2004, 2006; Hessle et al. 2002; Johnson et al. 2000; Narisawa et al. 2001, 2003; Sebastian-Serrano et al. 2016; Shao et al. 2000; Street et al. 2013; Wennberg et al. 2000).

While an earlier publication confirmed that $Alpl^{-/-}$ mice featured cementum hypoplasia consistent with descriptions in the HPP case report literature (Beertsen et al. 1999), a series of additional reports focused in greater detail on other developmental dental and craniofacial defects in $Alpl^{-/-}$ mice, finding disturbed enamel mineralization, dentin hypomineralization, and defective cranial base mineralization, abnormal cranial shape, and craniosynostosis (Fig. 13.4d–g) (Durussel et al. 2016; Foster et al. 2013; Liu et al. 2014; Nam et al. 2017; Yadav et al. 2012). In particular, it became clear through studies in HPP (and other mouse) models and that the acellular cementum critical for tooth attachment was especially sensitive to disturbances in local PP_i metabolism (Fig. 13.4h) (Foster et al. 2012; Nociti et al. 2002; Rodrigues et al. 2011; Zweifler et al. 2015), an insight made possible through the initial clues provided by the natural experiment of *ALPL* loss-of-function mutations.

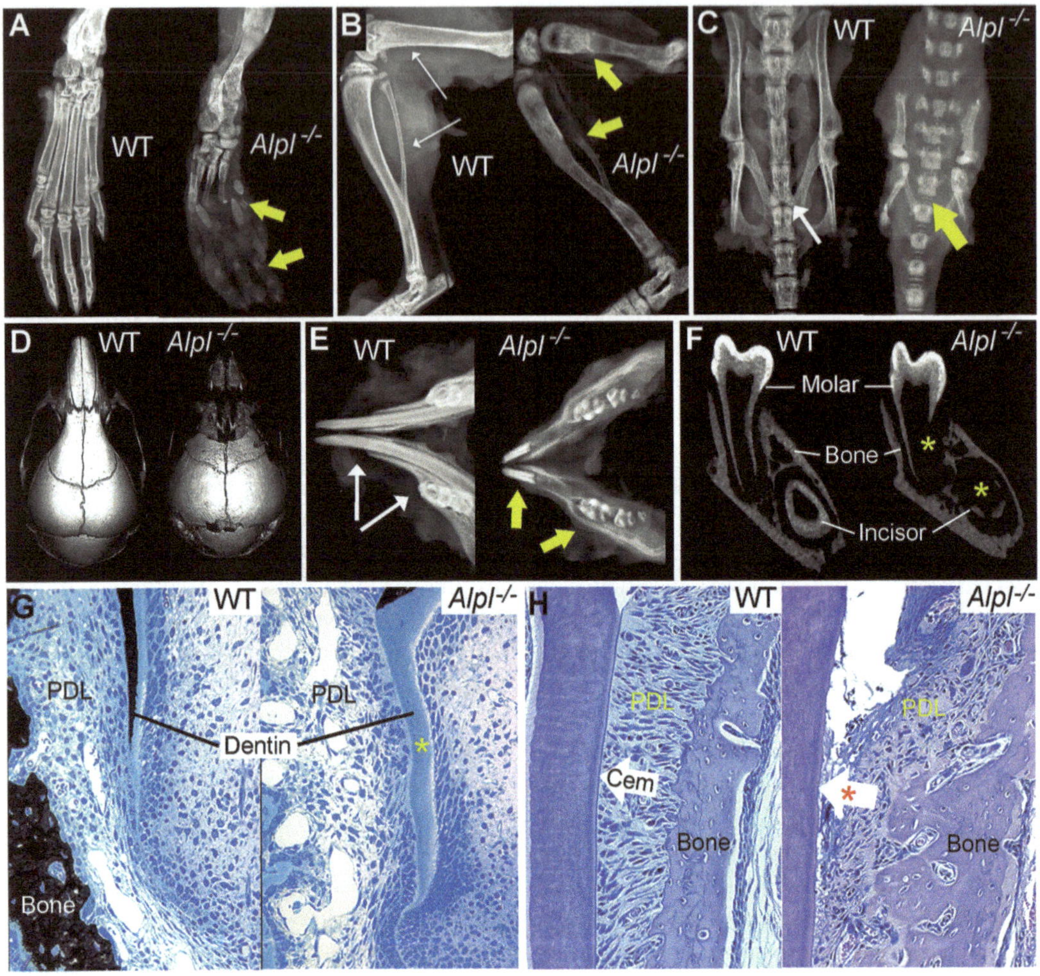

Fig. 13.4 The *Alpl* knockout mouse model of severe infantile HPP. (**a**) Radiographs of hind paws of wild type (WT) and $Alpl^{-/-}$ mice at 22 dpn. $Alpl^{-/-}$ mouse phalanges and metatarsals exhibit hypomineralization and deformities (yellow arrows). (**b**) Radiographs of hind limbs of WT and $Alpl^{-/-}$ mice at 22 dpn. $Alpl^{-/-}$ mouse femurs, tibias, and fibulas (white arrows) exhibit reduced mineralization, bowing, fracturing, and growth plate defects (yellow arrows). (**c**) Radiographs of caudal vertebrae (white arrow) of WT and $Alpl^{-/-}$ mice at 22 dpn. $Alpl^{-/-}$ mice show enlarged spaces between hypomineralized vertebrae (yellow arrow). (**d**) Radiographs of skulls of WT and $Alpl^{-/-}$ mice at 15 dpn. $Alpl^{-/-}$ mouse cranial bones feature severe hypomineralization and altered craniofacial shape. (**e**) Radiographs of mandibles with molars and incisors of WT (white arrows) and $Alpl^{-/-}$ mice at 22 dpn. $Alpl^{-/-}$ mouse mandibles show reduced radiopacity in molars and incisors (yellow arrows). (**f**) Micro-CT of WT and $Alpl^{-/-}$ mouse mandibles of at 14 dpn. $Alpl^{-/-}$ mouse molars, incisors, and alveolar bone show radiolucency (yellow ∗) indicative of severe hypomineralization. (**e**) Von Kossa stained undecalcified tissue sections of WT and $Alpl^{-/-}$ mouse mandibles at 12 dpn. Compared to well mineralized molar dentin in WT (indicated by black stain), $Alpl^{-/-}$ mouse molar roots featured hypomineralized dentin matrix (lack of black stain). (**f**) Hematoxylin and eosin (H&E) tissue sections of WT and $Alpl^{-/-}$ mouse mandibles at 22 dpn. Compared to the organized and functional periodontal complex in WT, $Alpl^{-/-}$ mouse molars lack acellular cementum (Cem) (red ∗), causing detachment of the periodontal ligament (PDL) and disorganized PDL and alveolar bone. (Images in panels **a–c**, and **e** reproduced with permission from Yadav et al. 2011. Images in panel **d** reproduced with permission from Liu et al. 2014. Images in panel **g** reproduced with permission from Foster et al. 2013. Images in **h** reproduced with permission from Bowden and Foster, Drug Des Devel Ther 12:3147–3161, 2018)

13.5.2 *Alpl* Knock-In Mice

Alpl$^{-/-}$ mice have been invaluable for providing insights into HPP pathology and treatment (as detailed in the following sections), however, the severity of the disease and resulting early lethality have been limitations for understanding the less severe end of the HPP spectrum experienced by many patients, and have additionally prevented long-term studies on therapeutic interventions at later ages. Therefore, attempts have been made to develop novel and less severe HPP mouse models. One approach was to genetically knock into mice an autosomal dominant mutation from a well-characterized kindred of HPP subjects reporting primarily dental defects and identified as falling within the odonto-HPP clinical type (Hu et al. 2000). Heterozygous *Alpl*$^{+/A116T}$ mice featured 50% decreased ALP and no apparent developmental, structural, or mechanical long bone phenotype by 120 dpn, considered to be adulthood in mice (Foster et al. 2015). Alterations were described in parietal bones of the skull, as well as alveolar bone of the jaw, where accumulation of osteoid and increased bone resorption were noted. Based on this phenotype, *Alpl*$^{+/A116T}$ mice were described as a mouse model of odonto-HPP, though prominent defects in cementum, dentin, and enamel were not found, making it difficult to use these mice in therapeutic rescue experiments.

13.5.3 *Alpl* Conditional Knock-Out Mice

A second approach at creating a less severe manifestation of HPP in mice employed conditional ablation of the *Alpl* gene through the Cre/lox genetic system. In this strategy, two short loxP nucleotide sequences were inserted into mouse *Alpl* allele introns surrounding exons 3 and 4. Conditional deletion of *Alpl* was achieved by crossing floxed *Alpl* (*Alpl*$^{fl/fl}$) mice with mouse lines carrying a *Cre* recombinase transgene under either the *Col1a1* promoter (to delete *Alpl* in osteoblasts and dental cells) or *Prx1* promoter (to delete *Alpl* in limb buds, chondrocytes, osteoblasts, and craniofacial mesenchyme) (Foster et al. 2017). Both conditional knockout lines lacked seizures and early lethality prominent in *Alpl*$^{-/-}$ mice, but displayed 75% reduced ALP and profound skeletal defects including rachitic changes, osteomalacia, deformations, and signs of multiple fractures at the advanced age of 180 dpn. Key aspects of HPP-associated dental defects were recapitulated, and these conditional *Alpl* knockouts were the first to demonstrate periodontal breakdown and alveolar bone loss, likely in part due to their longer lifespan allowing sufficient time for this manifestation. *Alpl*$^{fl/fl}$ mice may be crossed with any number of other Cre recombinase-carrying mouse lines, allowing targeted *Alpl* ablation in tissue- or time-specific manner that makes this a powerful approach for understanding pathological mechanisms and investigating potential therapies.

13.5.4 *Alpl* Knock-In Sheep

In 2018, the first large animal model of HPP was established. Gaddy, Suva, and colleagues knocked into sheep the same A116T *ALPL* exon 10 mutation used to engineer *Alpl*$^{+/A116T}$ mice, as described above. *Alpl*$^{+/A116T}$ heterozygous sheep exhibited approximately 30% decrease in ALP activity, decreased vertebrae size, metaphyseal flaring, altered gait, primary incisors with thin and short roots, reduced alveolar bone levels, and abnormal muscle histology (Williams et al. 2018). This model may provide novel insights into pathology and therapies because unlike in mice, sheep bone organization and remodeling is highly analogous to humans and sheep are diphiodont with primary and secondary dentitions.

13.6 Development of Recombinant Mineral-Targeting TNSALP for ERT

13.6.1 Origins of the Mineral-Targeted ERT Concept

Based on unsuccessful attempts to treat HPP with transfusions of high TNSALP blood and cell transplants, a new goal was set to develop a recombinant TNSALP enzyme that could be effectively targeted to where PP_iase activity was required, i.e. the mineralization front of developing bones and teeth. Dr. Philippe Crine had begun exploring this concept in the context of another enzyme, PHEX, and its associated hereditary metabolic disorder, X-linked hypophosphatemia (XLH; OMIM# 307800) (Boileau et al. 2001; Campos et al. 2003). Along with scientists at Enobia Pharma (Montreal, Canada) Dr. Crine turned his attention to HPP.

As a first step to engineer a soluble secreted TNSALP that could be expressed in vitro, the hydrophilic GPI-anchor was removed from the C-terminus, and the Fc region of human IgG antibody was added to allow column chromatography purification (Millan et al. 2008). In order to enhance delivery of recombinant TNSALP to bone, a highly negatively charged sequence of ten sequential aspartic acid residues, the so-called deca-aspartate or D_{10} extension, was added to the C-terminus of the protein (Fig. 13.5a). This type of acidic amino acid "tail" had been previously demonstrated to significantly improve in vivo delivery and retention of TNSALP to bones in mice (Nishioka et al. 2006). The high negative charge density of the D_{10} tail mimicks naturally occurring bone-associated proteins with acidic amino acid motifs, e.g. members of the Small Integrin-Binding Ligand N-Linked Glycoprotein (SIBLING) protein family including osteopontin (OPN), dentin phosphoprotein (DPP), and bone sialoprotein (BSP) (Fisher and Fedarko 2003; Staines et al. 2012). Recombinant TNSALP-D_{10} expressed in vitro in Chinese hamster ovary (CHO) cells was purified and molecular mass was confirmed by Western blotting to be consistent with homodimer formation (Millan et al. 2008). Importantly, TNSALP-D_{10} bound to HAP surfaces 32-fold more efficiently than the unmodified TNSALP control, and exhibited enzymatic catalytic activity in the bound fraction.

As a first in vivo test for TNSALP-D_{10}, pharmacokinetics and tissue distribution were investigated in adult and newborn mice. A single intravenous (IV) bolus of 5 mg/kg in adult mice provided proof-of-concept when prolonged retention of radiolabeled enzyme was detected in bone, but sustained accumulation was not found in other tissues (Millan et al. 2008). Repeated daily subcutaneous injection of 10 mg/kg enzyme in newborns reproducibly elevated circulating ALP levels to about 50-fold higher than normal levels, also increasing enzyme catalytic activity in bone.

13.6.2 Preclinical Studies of Recombinant TNSALP Enzyme Replacement Therapy

Efficacy of asfotase alfa was investigated using the LJ *Alpl*$^{-/-}$ mouse line, a model of severe infantile HPP developed in the lab of Dr. Jose Luis Millán (described in detail above). Based on daily subcutaneous injections to newborns over the course of 15 days, 2 mg/kg was determined to be the minimal efficacious dose, as indicated by normalized growth rate, increased vertebral bone mineral density (BMD), and positive changes in cortical and trabecular bone (Millan et al. 2008). The higher dose of 8.2 mg/kg injected subcutaneously to newborns increased lifespans and prevented skeletal defects and fractures in *Alpl*$^{-/-}$ mice over both short and long term experiments (Fig. 13.5b, c). Untreated knockout mice died by a median 18.5 days, whereas 75% of treated *Alpl*$^{-/-}$ mice lived to 52 dpn (the preset conclusion of the experiment), displaying normal physical activity and dramatic improvement in long bone length, appearance of secondary ossification centers, and lack of fractures. Further dose-response experiments established that the dose preventing 80% of bone defects in mice (ED_{80}) was 3 mg/kg for skeletal sites including feet, lower limbs, ribs, and jaws (Yadav et al. 2011). High dose AA ERT prevented craniosynostosis and largely normalized cranial shape and mineralization of craniofacial bones (Fig. 13.5d) (Durussel et al. 2016; Liu et al. 2015; Nam et al. 2017). AA treatment significantly improved dental mineralization and function in *Alpl*$^{-/-}$ mice, preventing enamel defects, significantly improving dentin mineralization, and allowing formation of functional acellular cementum to normalize tooth attachment (Fig. 13.5e) (Foster et al. 2013; McKee et al. 2011; Millan et al. 2008; Yadav et al. 2012).

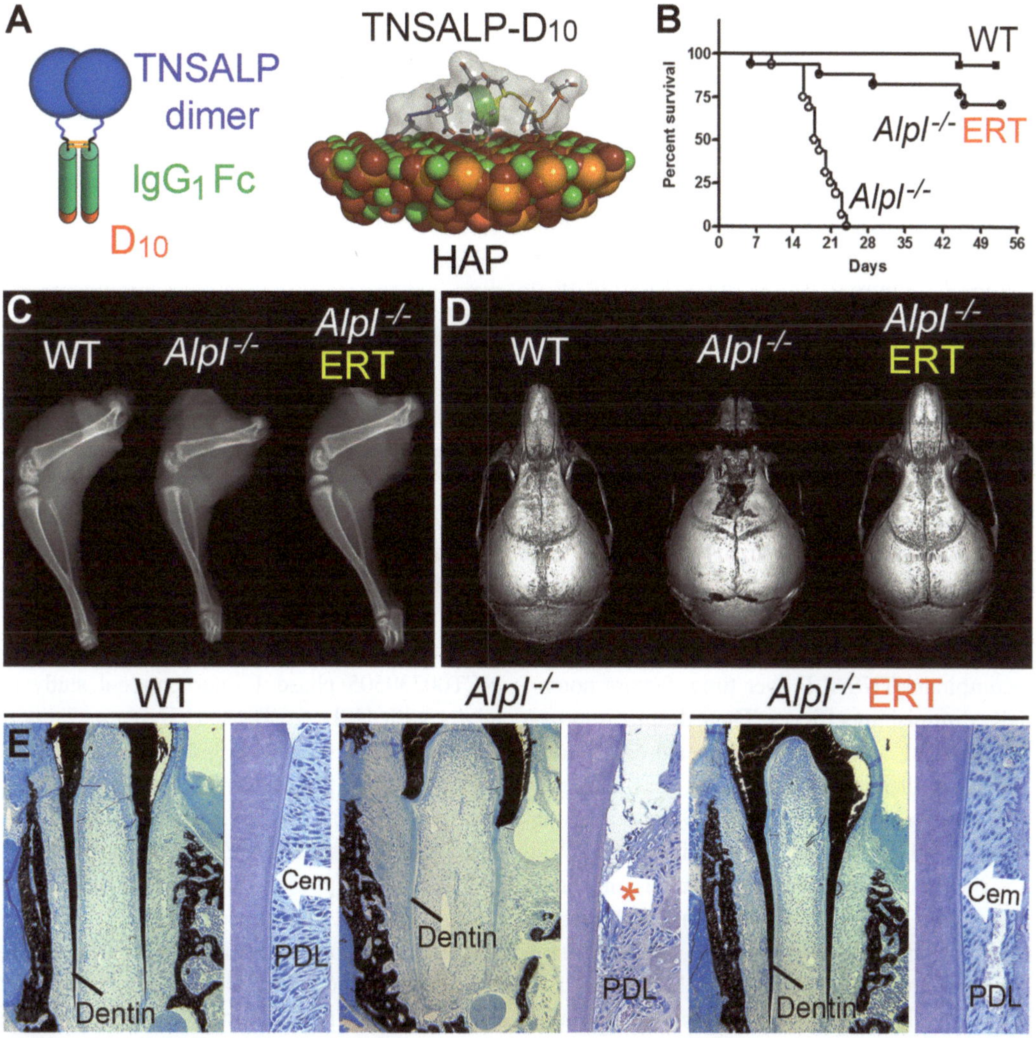

Fig. 13.5 TNSALP enzyme replacement therapy in a mouse model of HPP. (**a**) Model of recombinant asfotase alfa enzyme showing the TNSALP dimer (blue), human IgG_1 Fc domain (green), and D_{10} deca-aspartate tail (red). On the right, a simulated model shows predicted interaction of the highly negatively charged D_{10} tail with the positively charged calcium plane in the hydroxyapatite (HAP) crystal. (**b**) Percent survival of wild type (WT), untreated *Alpl*$^{-/-}$, and *Alpl*$^{-/-}$ mice receiving 8.2 mg/kg TNSALP ERT over the course of the study. (**c**) Radiographs of hind limbs at 16 dpn shows improvements in long bone lengths, shape, and appearance of secondary ossification centers in *Alpl*$^{-/-}$ mice receiving ERT. (**d**) Micro-CT of skulls at 15 dpn shows that ERT produces improvements in size, shape, and mineralization of craniofacial bones in *Alpl*$^{-/-}$ mice. (**e**) Von Kossa stained undecalcified tissue sections (left panels for each group) indicate that ERT rescues root dentin mineralization in *Alpl*$^{-/-}$ mice, as indicated by black stained appearance of mineralized tissues. Hematoxylin and eosin (H&E) stained sections (right panels for each group) reveal that ERT prevents acellular cementum (AC) hypoplasia (*) in *Alpl*$^{-/-}$ mouse molars, allowing periodontal ligament (PDL) attachment and normal periodontal architecture. (Images in panel **a** reproduced with permission (via republication of material within the agreed-upon thresholds between STM Permissions Guidelines signatories) from McKee et al. 2011. Graph in panel **b**, images in panel **c**, and von Kossa images in panel **e** reproduced with permission from Millán et al. 2008. Images in panel **d** reproduced with permission from Liu et al. 2015. Images in panel **e** reproduced with permission from Bowden and Foster, Drug Des Devel Ther 12:3147–3161, 2018)

The importance of the mineral-targeting aspect of TNSALP-D_{10} was tested by another group that attempted to use recombinant TNSALP lacking the C-terminal GPI anchor, effectively making this a soluble form of the enzyme (Oikawa et al. 2014). Purified enzyme produced by CHO cells was administered to 1 dpn *Alpl*$^{-/-}$ mice by IV infusion of 10 U/g TNSALP, followed by SC administration of a larger dose of 20 U/g from 3 to 28 dpn and finally with 10 U/g delivered IV every 3 days until mice were 6 month old. The lifespans of treated *Alpl*$^{-/-}$ mice were extended, however, decreased body weight compared to WT control mice persisted from about 30 dpn until the end of the study. Treated *Alpl*$^{-/-}$ mice exhibited shorter body length, reduced bone length, hypomineralization, and incisor malocclusion indicating cementum defects. Pharmacokinetic studies indicated rapid clearance and development of antibodies against recombinant TNSALP over time, factors potentially diminishing efficacy. Thus, despite improving survival and eliminating seizures, soluble TNSALP therapy proved much less effective than mineral-targeted TNSALP-D_{10} at correcting skeletal and dental disorders in this study.

These initial *Alpl*$^{-/-}$ mouse studies indicated tremendous potential for enzyme replacement therapy in treating HPP. However, one important limitation in the study designs and their interpretation was that enzyme replacement was initiated prior to onset of the majority of HPP skeletal and dental manifestations, as mouse pups were injected starting at early postnatal ages. Therefore, AA was demonstrated to prevent HPP-associated pathology in mice, but not necessarily reverse or ameliorate already existing pathology. Because the skeleton is in a constant state of remodeling through the actions of osteoblasts, osteocytes, and osteoclasts, introduction of functional TNSALP-D_{10} to mice or humans with HPP would be expected to promote mineralization of osteoid, resulting in replacement of poor quality bone with much improved bone tissue with superior mechanical properties. However, enamel, dentin, and cementum of teeth do not remodel, and dentin and cementum have limited ability for repair, therefore timing of therapeutic intervention is likely critical to correct formation and improve function of the teeth. This lesson on early intervention of treatment has been learned through other endocrine disorders and metabolic disorders affecting teeth, such as nutritional vitamin D deficiency and X-linked hypophosphatemia (XLH; OMIM 307800) (Biosse Duplan et al. 2017; Davit-Beal et al. 2014; Foster and Hujoel 2018).

13.7 Therapeutic Efficacy of TNSALP ERT in Clinical Trials and Case Reports

Following the successful preclinical studies in *Alpl*$^{-/-}$ mice described in the preceding section, in 2008, the first clinical trials for TNSALP-D_{10} in infants and young children with perinatal or infantile HPP began (NCT00744042 phases 1 and 2 interventional study in infants). Additional clinical trials that followed included: NCT00739505 phase 1 interventional study in adults with HPP, NCT01205152 phase 2 interventional study in infants and children in 2009, NCT00952484 phase 2 interventional study in juveniles in 2009, NCT01203826 phase 2 interventional extension study in children in 2010, NCT01163149 phase 2 interventional study in adolescents and adults in 2010, NCT01176266 phases 2 and 3 open label interventional study in infants and young children in 2010, and NCT02797821 phase 2 interventional study in adults with pediatric-onset HPP in 2016, all of which now completed in the US, Canada, Europe, and Australia. In Japan, clinical trials completed include NCT02456038 phase 2 interventional study in children and adults in 2014, and NCT02531867 phase 4 interventional study in children and adults in 2015. Ongoing and recruiting clinical trials as of this writing include: NCT02496689 expanded access trial for children and adults with HPP (in the U.S. and France); NCT03418389 observational trial in adults with pediatric-onset HPP treated with AA (Germany); NCT02306720 prospective long-term observational trial in children and adults who have received AA (multiple countries); and NCT02751801 observational retrospective trial to evaluate the personal and economic burden of HPP to determine whether a clinical trial for less severe clinical forms is warranted (United Kingdom). TNSALP-D_{10} was renamed as asfotase alfa (AA) (under Alexion Pharmaceuticals, Inc., New Haven, CT, USA; also known as ALXN1215 or previously ENB-0040 under Enobia Pharma, Inc.)

In 2015, AA (Strensiq) was approved by regulatory agencies in Japan, then Canada, the European Union, and the United States for pediatric-onset HPP. Treatment outcomes of AA ERT from clinical trials and case reports published to date are summarized in Table 13.3.

Table 13.3 Treatment outcomes of asfotase alfa in clinical trials and case reports

Reference and clinical trial number	Study type, patient number (N)	Age at treatment	Dosage and duration of therapy	Main clinical outcomes (skeletal, respiratory, and survival)	Comments
Perinatal and infantile HPP (treatment initiation before 3 years of age)					
Whyte et al. (2012); NCT00744042	Open-label, multinational clinical trial; N = 10 (4 perinatal, 6 infantile)	Mean age: 13.1 months (Range 0.6–36 months)	Single IV infusion at 2 mg/kg, followed by SC 1 mg/kg 3 times/ week; Dose could be increased up to 3 mg/kg if no improvement; Duration: 1 year	Healing of rickets and improved mineralization at 6 months. Most patients were off ventilator by 48 weeks of treatment. Improvement in motor function. Improved survival rate	Progressive craniosynostosis requiring neurosurgery in 2 patients. Nephrocalcinosis did not progress after 6 months of treatment and even improved in some patients. No hypocalcemia, no ectopic calcification detected. One patient died from sepsis at age 7.5 months.
Whyte et al. (2016b) NCT00744042 NCT01176266 NCT01205152 NCT01419028	Open-label, multinational, multicenter, phase 2 clinical trial; N = 37 vs. 48 historical controls	Mean age: 23 months (Range 0–71 months)	SC 1 mg/kg 6 times/week or 2 mg/kg 3 times/ week; Duration: 2.7 years	Improved skeletal mineralization. 75% weaned off ventilator support. Improved survival rate in treated patients vs. historical controls: 95% vs. 42% at age 1 year and 84% vs. 27% at age 5 years.	2 deaths: 1 from pneumonia and 1 from neurological complications of craniosynostosis
Kitaoka et al. (2017) NCT02456038	Open-label, multicenter clinical trial; N = 10	Mean age: 3 months (Range birth – 7.6 months)	SC 2 mg/kg 3 times/week; Duration: 1 year (Range 0.1–2.4 years)	Improved mineralization of bones including ribs. Two patients weaned off respiratory support during the first month of AA treatment. 100% survival rate	Hypocalcemia with seizure in 1 patient. No progression of nephrocalcinosis in 3 patients.

(continued)

Table 13.3 (continued)

Reference and clinical trial number	Study type, patient number (N)	Age at treatment	Dosage and duration of therapy	Main clinical outcomes (skeletal, respiratory, and survival)	Comments
Rodrigues et al. (2012)	Case report; Perinatal lethal HPP	3 weeks old	Single IV 2 mg/kg infusion, then SC 1 mg/kg 3 times/week for 8 weeks, then 2 mg/kg 3 times/week; Duration: 31 weeks	Improved bone mineralization. Decreased oxygen and pressure on mechanical ventilation at 10 weeks. Discharged home with a portable ventilator without oxygen at 32 weeks. The patient died at 34 weeks of sepsis and lung infections.	The patient was born at 33 weeks gestational age and developed bronchopulmonary dysplasia
Okazaki et al. (2016)	Case report; Perinatal lethal HPP	1 day old	SC 2 mg/kg 3 times per week; Duration: 1.5 years (at time of report)	Visible improvement of bone mineralization by 3 weeks of age. Weaned off ventilation at 7 months of age. Discharged from hospital at 10 months of age and still on tracheostomy and oxygen at 18 months of age.	No craniosynostosis. Hypocalcemia and convulsion after AA treatment, requiring calcium supplement for 3 months. Hearing loss improved.
Oyachi et al. (2018)	Case report; Perinatal lethal HPP	6 days old	SC 2 mg/kg 3 times/week; Duration: 0.5 year (at time of report)	Rickets disappeared by 2 months of age. Discharged home at 6 months of age.	Pyridoxine treatment for seizures was discontinued after AA treatment. Transient hypocalcemia within 1 week of AA treatment, not requiring calcium supplement.

(continued)

Table 13.3 (continued)

Reference and clinical trial number	Study type, patient number (N)	Age at treatment	Dosage and duration of therapy	Main clinical outcomes (skeletal, respiratory, and survival)	Comments
Costain et al. (2018)	Case report; Perinatal lethal HPP	13 days old	SC 2 mg/kg 3 times/week; Dose increased to a maximum of 9 mg/kg/dose on day 67 due to lack of clinical improvement	Improved bone mineralization but no improvement in respiratory function and persistently small chest size; suspected pulmonary hypoplasia. Patient died on day 100 after AA treatment and respiratory support were withdrawn.	Hypocalcemia requiring calcium supplementation
Hacihamdioglu et al. (2018)	Case report; Perinatal lethal HPP	21 days old	SC 2 mg/kg 3 times/week; Duration: 1 year (at time of report)	Discharged home at 7 months of age. Intubated at birth, though after 12 months of treatment, ventilation via tracheostomy was needed only during sleep.	No hypocalcemia during AA treatment. No sign of craniosynostosis at 12 months.
Rougier et al. (2018)	Case report; Perinatal lethal HPP	22 days old	SC 2 mg/kg 3 times/week for 1 month, then 3 mg/kg 3 times per week, then back to 2 mg/kg 3 times per week at 34 weeks of age; Duration: 3 years	Improvement in mineralization of bones and ribs. Intubated at birth, though taken off ventilation at 41 weeks of age. Improved muscle tone.	Neurosurgery for craniosynostosis at age 34 weeks. Severe nephrocalcinosis discovered at age 5 months.
Ucakturk et al. (2018)	Case report; Perinatal lethal HPP	40 days old	SC 2 mg/kg 3 times/week; Duration: 8 weeks	Minimal improvement in bone mineralization. Intubated at birth, and no significant improvement in respiratory function. The patient died at 97 days due to ventilator-associated pneumonia and sepsis.	AA treatment had not been long enough to detect improvement in bone mineralization and respiratory function.

(continued)

Table 13.3 (continued)

Reference and clinical trial number	Study type, patient number (N)	Age at treatment	Dosage and duration of therapy	Main clinical outcomes (skeletal, respiratory, and survival)	Comments
Children with HPP					
Whyte et al. (2016a); NCT00952484 NCT01203826	Open-label study in infantile or childhood HPP; N = 13 (5 infantile, 8 childhood)	Mean age 8.8 years (Range 6–12 years)	Initially randomized to SC 2 or 3 mg/kg 3 times/week for 6 months, then 6 mg/kg/week; Duration: 5 years	Significant healing of the skeletal manifestations. Significant improvements in growth, strength, motor function, agility, and quality of life. Pain and disability resolved.	Mild to moderate injection-site reactions. Localized lipohypertrophy persisted in 6 of 8 patients after 3 years of treatment.
Adolescents and adults with HPP					
Kishnani et al. (2017a, b)	Phase II, randomized, dose-ranging, open-label, multi-center study; N = 19	6 children aged 13–18 years, and 13 adults aged 18–65years.	0.3 (n = 7) or 0.5 mg/kg/day (n = 6) asfotase alfa for 6 months, then all received 0.5 mg/kg/day, increased 6–12 months later to 1 mg/kg 6 times/ week; Duration: 5 years	Improved physical function as assessed by 6 min walk test and Bruininks-Oseretsky Test of Motor Proficiency.	3 patients withdrew: 2 for injection site reactions at 2 years and 3.5 years, 1 for poor compliance at 2.5 years.
Bowden and Adler (2018a)	Case report; Severe childhood HPP	15 years old	SC 2 mg/kg 3 times/week; Duration: 1 year	Improved growth and physical function. Chronic pain resolved.	Worsening scoliosis from 60 to 110 degrees.
Remde et al. (2017)	Case report; Adult with childhood-onset HPP	57 years old	SC 80 mg 3 times/ week (given after hemodialysis for chronic renal failure); Duration: 13 months	Reduced bone pain; Able to walk independently for 3 m; Improved bone mineralization; No new fractures.	Prior to AA, patient received bisphosphonate therapy for a year and teriparatide for 4 months.
Klidaras et al. (2018)	Case report; Adult with infantile-onset HPP	41 years old	SC 1 mg/kg 6 times/week; Duration: 14 months at time of report	Healing of femoral pseudofracture of 17 years.	
	Case report; Adult with HPP	61 years old	AA treatment for 16 months at time of report	Complete healing of nonhealing femoral fracture of 8 years. Significantly less pain with ambulation. Improvement in gait.	Prior to AA, patient was treated with teriparatide for 6 months to assist with fracture healing; discontinued due to side effect and no response in fracture healing.

(continued)

Table 13.3 (continued)

Reference and clinical trial number	Study type, patient number (N)	Age at treatment	Dosage and duration of therapy	Main clinical outcomes (skeletal, respiratory, and survival)	Comments
Freitas et al. (2018)	Case report; Adult with childhood-onset HPP	36 years old	SC 2 mg/kg 3 times/week; Duration: 12 months at time of report	Able to walk without assistive devices. Improved bone mineral density. Improved bone microarchitecture detected by high-resolution peripheral quantitative computed tomography (HR-pQCT).	Also had stage III chronic kidney disease. Prior to AA, patient was treated with alendronate 70 mg weekly for 8 years until 1 year before AA treatment.

13.7.1 Treatment of Perinatal and Infantile HPP

AA ERT substantially improved bone mineralization, leading to improvement in respiratory function and survival in neonates and infants with HPP (Table 13.3) (Costain et al. 2018; Hacihamdioglu et al. 2018; Kitaoka et al. 2017; Oyachi et al. 2018; Rougier et al. 2018; Ucakturk et al. 2018; Whyte et al. 2012, 2016b). Improvements in skeletal radiographs were apparent as early as week 3 (Okazaki et al. 2016) and were dramatic by 24 weeks of treatment. Specifically, radiographs indicated improved mineralization, healing of rickets, resolution of radiolucencies and sclerosis, fracture healing, and reduced deformity (Fig. 13.6a–f). Patients with gross motor delay were bearing weight or walking by 48 weeks of treatment. The survival rate in 37 infants and young children (≤5 years) with perinatal or infantile HPP treated with AA had significantly improved compared to the historical control group (95% vs. 42% at age 1 year and 84% vs. 27% at age 5 years, respectively) (Whyte et al. 2016b). Another study in Japan reported 100% survival rate in perinatal and infantile HPP treated with AA (Kitaoka et al. 2017), which might have been associated with early treatment with AA (mean age at start of treatment 3 months vs. 13–23 months in cohorts described by Whyte et al). AA therapy started on day one of life resulted in survival in an infant with lethal perinatal HPP as well as dramatic improvement in skeletal mineralization (Okazaki et al. 2016). The window of survival with AA treatment maybe narrow in lethal perinatal HPP, as one neonate with this severe life-threatening form died on day 4 of life without AA treatment (Castells et al. 2018). Prompt diagnosis and expeditious initiation of AA therapy is crucial to survival as it will likely improve ventilation and minimize intensive care.

13.7.2 Treatment of Children (6–12 Years) with HPP

In 2016, Whyte and colleagues reported results of 5 years of AA ERT in a cohort of 11 children aged 6–12 years (Table 13.3) (Whyte et al. 2016a). Similar to the cohort with the more severe form of infantile HPP, substantial radiographic improvements were noted by 6 months, and persisted through 5 years of the study. Significant increase in weight Z scores was noted at 6 weeks of ERT, followed by significant increase in height Z scores (from mean −1.26 at

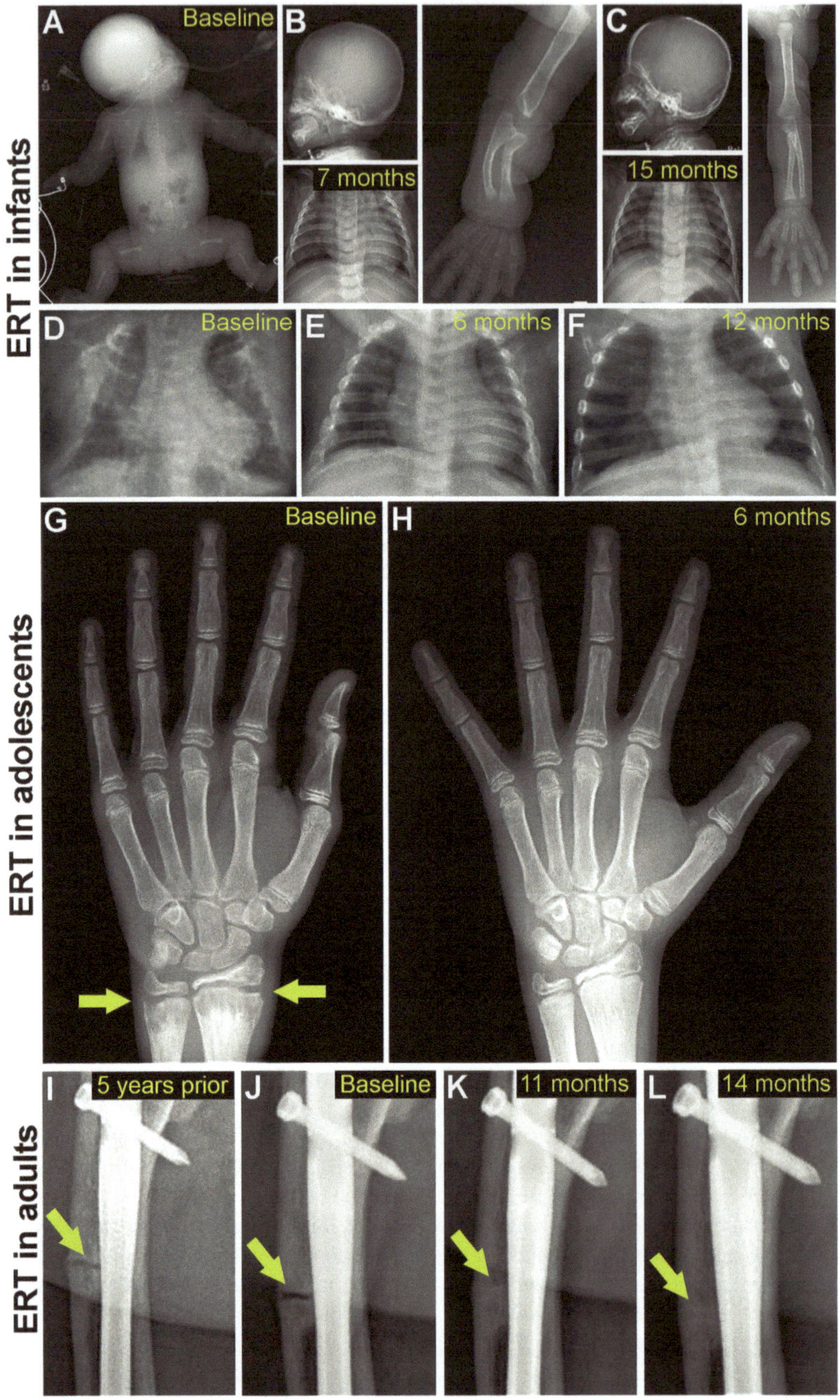

Fig. 13.6 TNSALP enzyme replacement therapy in human subjects. (**a–c**) Whole body, cranial, rib, and arm radiographs of an infant with perinatal HPP at baseline and 7 and 15 months after AA ERT. In addition to allowing survival, ERT was associated with substantial improvements in mineralization and reduction of rachitic

baseline to −0.8) after 1.5 years of AA treatment. Most children receiving ERT displayed increased muscle strength and motor skills on par with healthy children, had normal ambulation and reduced disability, and reported reduced pain. Case reports of children administered AA ERT reported similarly dramatic skeletal improvements (Kitaoka et al. 2017).

13.7.3 Treatment of Adolescents and Adults with HPP

Similar to the study in children, after 5 years of AA treatment, adolescents and adults with HPP had significant improvement in physical function with increases in 6-min walking test, and improvement in health-related quality of life (HRQOL) as measured by the Childhood Health Assessment Questionnaire Disability Index and the Pediatric Outcomes Data Collection Instrument global function (Table 13.3) (Kishnani et al. 2016, 2017a; Tomazos et al. 2017). ERT resulted in improved mobility without use of the previously used walking assistive device was achieved in an adolescent after 6 months of AA treatment (Bowden and Adler 2018a), when striking improvements in bone radiographs was observed (Fig. 13.6g, h). In a 59 year-old adult with childhood-onset HPP, treatment with HPP significantly increased quality of life in terms of increased mobility, reduction in pain medication, and improved bone mineralization with healing of non-union fractures and no occurrence of new fractures (Remde et al. 2017). In two adult HPP patients with longstanding non-healing fractures, AA ERT dramatically improved healing and resolution of the fractures (Fig. 13.6i–l) (Klidaras et al. 2018).

13.7.4 Effects of ERT on Specific Clinical Features of HPP

As diverse and multi-systemic as HPP clinical manifestations are, so are therapeutic outcomes of AA ERT, and these deserve further attention and investigation. While AA ERT has produced dramatic improvements in skeletal manifestations of HPP, as described above, effects on other manifestations are not entirely uniform. Effects of ERT on biochemistry and various clinical aspects, including respiratory function, bone health, physical function, mobility, nephrocalcinosis, craniosynostosis, neurological manifestations, chronic pain, scoliosis, and dental tissues are outlined below.

13.7.4.1 Biochemistry

Low serum ALP levels with elevated serum PLP are diagnostic for HPP and these levels change dramatically after AA treatment and can be used to monitor the effect of, and patient compliance with ERT. As a result of three or six times weekly AA, levels of ALP have been reported as high as ~24,000 IU/L within 4 weeks of initiating ERT (Kitaoka et al. 2017), remaining at elevated levels of 3000–6000 IU/L even after 5 years of therapy (Whyte et al. 2016a). To date, these markedly elevated ALP levels do not appear to have harmful effects. In contrast, when AA therapy is discontinued completely, serum ALP levels may decrease to undetectable levels (similar to pretreatment level), as demonstrated in an adolescent boy who voluntarily discontinued AA ERT due to non-adherence (Bowden and Adler 2018b). This indicates that serum ALP level is a good marker for treatment adherence monitoring. Elevated serum PP_i and serum PLP decrease dra-

Fig. 13.6 (continued) skeletal deformities. (**d–f**) Chest radiographs of an infant with HPP at baseline and 6 and 12 months after AA ERT. Dramatic improvements in rib mineralization, chest structure, and thoracic volume were noted. (**g** and **h**) Hand radiographs of a 15-year-old patient with severe childhood HPP show marked metaphyseal fraying and characteristic "tongues" of radiolucency in the distal radius and ulna (yellow arrow) that are substantially resolved after 6 months of AA ERT. (**i–l**) Series of radiographs of a nonhealing subtrochanteric fracture in a 61-year-old male HPP patient showing lack of resolution 12 years (**i**) and 17 years (**j**) after fracture, with marked improvements after 11 months (**k**) and 14 months (**l**) of AA ERT. (Images in **a–c** reproduced with permission from Okazaki et al. 2016. Images in **d–f** reproduced with permission from Whyte et al. 2016b. Images in **g, h** reproduced with permission from Bowden and Adler 2018a, b. Images in **i–l** reproduced from Klidaras et al. 2018 and used under the terms of the Creative Commons CC BY license)

matically in response to treatment (Akiyama et al. 2018; Bowden and Adler 2018a; Kitaoka et al. 2017; Whyte et al. 2012, 2016a). Additional PLP-related metabolites, pyridoxal (a product of the enzymatic reaction of ALP) and 4-pyridoxic acid (a metabolite of pyridoxal) have been investigated as a diagnostic marker of HPP and an indicator of the AA treatment effect (Akiyama et al. 2018). The PLP-to-pyridoxal ratio (PLP/PL) is considered to reflect the activity of ALP (converting PLP to PL) and has been shown to be useful to evaluate the early treatment effect of AA ERT before the skeletal improvement occurs (Akiyama et al. 2018).

While calcium abnormalities are rare in adults with HPP, hypercalcemia is common in infants and children and often identified at the time of diagnosis. Some infants have continued to present hypercalcemia and hyperphosphatemia after AA ERT, requiring continued use of low-calcium and/or low phosphorus formula (Kitaoka et al. 2017). Hypocalcemic seizures were described in a neonate with perinatal HPP and serum calcium levels of 4.7 mg/dL, following 3 weeks of ERT initiated on the first day of life (Kitaoka et al. 2017). The seizures and hypocalcemia were resolved by increasing calcium supplementation. Hypocalcemia after AA treatment suggests "hungry bone" syndrome, or rapid formation and mineralization of previously deficient and hypomineralized bone that requires a significant influx of calcium. Dietary calcium restriction in children with HPP should be liberalized after AA ERT is initiated and hypercalcemia is no longer present, or when serum PTH levels increase, to prevent hypocalcemia and "hungry bones" (Whyte et al. 2012). Serum PTH increased during treatment, coinciding with the skeletal remineralization as evidenced by the radiographic improvements.

Falsely low serum testosterone level obtained by competitive radioimmunoassay has been reported in a patient taking asfotase alfa ERT, thought to be due to assay interference by exogenous TNSALP (Sofronescu et al. 2018).

13.7.4.2 Respiratory Function

In perinatal or infantile HPP, respiratory failure is often the cause of mortality. The etiology of respiratory insufficiency in HPP is multi-factorial and complex, and usually stems from thoracic deformity and fractures, as a result of the skeletal manifestations. Respiratory failure can also be due to pulmonary hypoplasia, muscle weakness, tracheomalacia, central nervous system dysfunction associated with craniosynostosis, episodic seizures, and increased susceptibility to infection due to low TNSALP levels in leukocytes (Whyte 2012). However, once skeletal mineralization improves, with stabilization of the chest wall and/or improved muscle strength, patients have improved respiratory function and can discontinue mechanical ventilation (Whyte et al. 2016b). Crying vital capacity has been utilized to evaluate respiratory status and is reported to be a good indicator for weaning mechanical ventilation (Shimada et al. 1979). Crying vital capacity of 15 ml/kg was used as a criterion for extubation and greater than 20 ml/kg as a criterion for discharge from the hospital in a patient with perinatal HPP who was discharged home at 6 months of life (Oyachi et al. 2018). Some children with perinatal or infantile HPP may not be fully weaned off the mechanical ventilation and may require tracheostomy with oxygen (Okazaki et al. 2016), or ventilation via tracheostomy only during sleep (8 h a day) (Hacihamdioglu et al. 2018).

13.7.4.3 Bone Health, Physical Function, and Mobility

As AA ERT is designed to bind to HAP in mineralized bone and tooth tissues, the effects on skeletal mineralization are dramatic and precede improved respiratory and motor function. At 6 months of AA treatment, improvements in skeletal health include diffusely increased bone mineralization, corrected or improved endochondral and membranous bone formation, reduced deformity, healing of fractures, and extensive modeling and remodeling of bone, with resolution of sclerosis (Whyte et al. 2012). While skeletal radiographs showing remarkable improvement after AA ERT have been consistently reported in clinical trials and case reports, reports on bone mineral density using dual-energy X-ray absorptiometry (DXA) after AA ERT have been limited.

In a 15 year-old boy with severe childhood HPP, the lumbar spine bone mineral density Z score was markedly low at −5.9 (height-adjusted Z score of −2.7). After 12 months of AA therapy, the absolute values of his lumbar spine BMD and total body bone mineral content and increased by 19% and 23%, respectively, but his height-adjusted Z scores for lumbar BMD decreased from baseline of −2.7 to −3.1 (Bowden and Adler 2018a). In an adult with late diagnosis of severe childhood HPP who suffered from multiple fractures and impaired mobility, treatment with AA for 12 months increased lumbar spine bone mineral density by 10% and improved mobility (Freitas et al. 2018). This patient also had improvement in bone microarchitecture parameters assessed by high-resolution peripheral quantitative computed tomography (HR-pQCT) (the first reported use of such imaging for bone assessment after AA ERT) in the distal tibia, with stabilization of bone parameters at the distal radius. Increased proteinaceous components of hypomineralized bones may affect bone density readings by DXA (Kishnani et al. 2017b). Further research is warranted to evaluate and characterize overall bone health and the therapeutic outcome of AA on bone density and bone microarchitecture by different bone imaging tools such as DXA or HR-pQCT.

Along with skeletal improvement, profound delays in growth and gross motor function in pediatric patients with severe HPP improved substantially from AA ERT, across all studies described above. Adolescents and adults patients with years of ambulatory disability, or even complete immobility, began to walk independently after 6–12 months of AA therapy (Bowden and Adler 2018a; Freitas et al. 2018; Klidaras et al. 2018; Remde et al. 2017), with one adult patient improving walking distance up to 4 miles (Klidaras et al. 2018). It is important to inform patients that continuation of AA therapy without interruption is critical to maintain optimal ALP and mineralization activity, and to prevent recurrence of clinical deterioration. Reappearance of hypomineralization of metaphyses, similar to the pre-treatment appearance, has been reported in an adolescent with severe childhood HPP who stopped AA therapy for 1 year due to non-adherence (Bowden and Adler 2018b).

Improvement in physical function, mobility and musculoskeletal function after AA therapy has been documented in clinical trials (Table 13.3) using objective, validated, and age-appropriate assessment tools: the Alberta Infant Motor Scale and Gross Motor Function Measure, the Bayley Scales of Infant and Toddler Development, and the Peabody Developmental Motor Scales in young children; Bruininks-Oseretsky Test of Motor Proficiency, Second Edition [BOT-2] (for age 4–21 years); and the 6-min walk test (6MWT) (for ambulatory children aged ≥5 years and adults). Pain (see pain section below) and disability assessment obtained by the parent reported Child Health Assessment Questionnaire (CHAQ) and Pediatric Outcomes Data Collection Instrument (PODCI) also showed substantial improvement from AA ERT (Whyte et al. 2016a). The 6MWT can be administered in the clinical setting and may be recorded on video to allow comparison of gait and mobility over time during AA therapy. Modified Performance-Oriented Mobility Assessment-Gait (mPOMA-G) (observational gait analysis from video footage during 6MWT) has been validated specifically in children with HPP and is strongly correlated with CHAQ, PODCI, and 6MWT scores (Phillips et al. 2018).

13.7.4.4 Nephrocalcinosis and Nephrolithiasis

Nephrocalcinosis is sometimes identified at the time of diagnosis in infants and children with HPP, and is thought to be secondary to hypercalcemia with hypercalciuria. During clinical trials to date, ERT seemed to cause no progression of nephrocalcinosis in children with infantile HPP (Kitaoka et al. 2017; Whyte et al. 2012), with some improvement even noted in some patients (Whyte et al. 2012). Monitoring of nephrocalcinosis by renal ultrasound is recommended at baseline and every 3 months in cases of perinatal and infantile HPP, and at baseline, 6 months, and then annually in childhood and adult HPP (Kishnani et al. 2017b).

In addition to nephrocalcinosis, nephrolithiasis with parietal calcifications can occur in adults with HPP (Freitas et al. 2018), likely from long-standing HPP disease with hypercalciuria, and/or long-term therapy with calcium and vitamin D for an 'osteoporosis' diagnosis. Nephrocalcinosis and nephrolithiasis can result in impaired renal function or chronic kidney failure in adults. AA ERT was safe and efficacious in an adult dialysis patient with HPP and stage 4 chronic kidney disease (Remde et al. 2017). AA was given immediately after hemodialysis three times a week at the recommended dose without the need for dose adjustment.

13.7.4.5 Craniosynostosis

Premature fusion of cranial sutures occurs in perinatal, infantile, and childhood forms of HPP. Clinical evidence of the abnormal skull shape (scaphocephaly or oxycephaly) secondary to the loss of one or several sutures, and/or the absence of head circumference growth, are clues to the diagnosis of craniosynostosis. Craniosynostosis is sometimes detected before HPP is diagnosed, and should alert clinicians to evaluate serum ALP. The underlying mechanism for craniosynostosis remains poorly understood. Clinical sequelae of craniosynostosis include papilledema, optic nerve atrophy, increased intracranial pressure with a copper-beaten skull on a skull x-ray (Poryo et al. 2016), and secondary ectopia of the cerebellar tonsils that can lead to hydrosyringomyelia (Collmann et al. 2009). To date, AA ERT has not prevented craniosynostosis, and the condition progressed during the treatment, requiring neurosurgical intervention (Whyte et al. 2012). The timing of ERT initiation may be critical for prevention of craniosynostosis, as a neonate with perinatal HPP receiving AA treatment from day 1 did not develop suture fusion (Okazaki et al. 2016), and similarly, early postnatal treatment in the mouse model of infantile HPP was also shown to prevent craniosynostosis (Liu et al. 2015).

13.7.4.6 Neurological Manifestations

In severe perinatal HPP, seizures occur as a result of defective PLP metabolism, and are an indicator of HPP severity and lethal prognosis, associated with 100% mortality rate (Whyte et al. 2016b). PLP is the active form of vitamin B6 that is unable to cross the cell's plasma membrane or the blood brain barrier. TNSALP dephosphorylates PLP into the pyridoxal form of vitamin B_6, able to cross the cell membrane and blood brain barrier and that is subsequently phosphorylated back into PLP, serving as a key co-factor for many enzymes. PLP in the brain is required for biosynthesis of many neurotransmitters in the brain (e.g. dopamine, norepinephrine, gamma-aminobutyric acid [GABA]). Reduced TNSALP enzymatic capability in HPP therefore results in elevated serum PLP and reduction of PLP in the brain, making severely affected neonates and infants with HPP prone to seizures. Provision of high doses of pyridoxine hydrochloride (vitamin B_6) can temporarily correct this deficiency and prevent seizures. AA ERT resulted in markedly improved survival rate of 77% in infants experiencing seizures (Whyte et al. 2016b) and prevented further seizures so that pyridoxine hydrochloride was discontinued (Oyachi et al. 2018). Pyridoxine-responsive seizures have been reported as the first symptom of infantile HPP in a neonate without bony abnormalities (Baumgartner-Sigl et al. 2007), further illustrating clinical heterogeneity in HPP. HPP was not diagnosed in this patient until age 7 months when clinical HPP became apparent with skeletal manifestations. Unfortunately, this child developed rib fractures and died from respiratory failure at age 9 months. Assessment of any neonate with pyridoxine-responsive seizures should include measurement of serum ALP to facilitate early detection of HPP, which can be life-saving due to the availability of ERT.

While seizures are not observed in older children or adults with HPP, other neurological symptoms such as fatigue, headaches, sleep disturbances, vertigo, depression, anxiety, neuropathy, and hearing loss occurred commonly (33–66%) in individuals with HPP, at a greater prevalence than the US general population (Colazo et al. 2018). Use of psychiatric medications for mental illness was observed in 65% of patients and preceded the diagnosis of HPP. Interestingly, patients with vestibulocochlear symptoms (hearing loss, tinnitus, and vertigo) had lower ALP levels than those without. The mental health issues and neurological symptoms described above should be included in the assessment and evaluation of HPP in clinical practice. Resolution of any of these symptoms after AA treatment should be documented to provide insights into this under-reported and under-studied aspect of HPP, and future prospective studies should be conducted to evaluate the effects of AA treatment on these neurological symptoms.

13.7.4.7 Chronic Pain

Musculoskeletal pain is a significant feature of HPP, experienced by many affected individuals across their lifespan. Infants and young children with HPP may not be able to communicate pain sensation, therefore, signs of irritability through vocalization, facial expression, or color changes, along with validated and age-appropriate pain scales such as the Neonatal Pain, Agitation and Sedation Scale (N-PASS), or the Face, Leg, Activity, Cry, Consolability Scale (FLACC) should be used to document pain and response to ERT (Phillips et al. 2016). In older children or adults, Wong-Baker FACES pain rating scale can be used. Pain is thought to be due to chronic inflammation with elevated prostaglandin levels (Girschick et al. 2006) caused by calcium pyrophosphate dihydrate (CPPD) crystal depositions (Beck et al. 2009). This can result in incapacitating bone and joint pain. Resolution of chronic musculoskeletal pain following AA therapy has been reported in all clinical trials and case reports. One adolescent male with severe childhood HPP discontinued daily analgesic use within 3 months of AA ERT (Bowden and Adler 2018a).

13.7.4.8 Scoliosis

Scoliosis has been described in four children with HPP in the literature (Arun et al. 2005; Bowden and Adler 2018a; Whyte et al. 2003). A child with infantile HPP had a normal spinal radiograph at age 2.5 months, but developed scoliosis by age 7 months (Whyte et al. 2003). Another two patients also exhibited early onset of severe scoliosis (Arun et al. 2005). The first child with severe infantile HPP was noted to have scoliosis of 62° at age 3 years, which progressed to 94° by age 7 years. The second child had odonto-HPP with no skeletal demineralization or deformity except for early onset of scoliosis of 74° at age 3.5 years, which progressed to 90° by age 10 years (Arun et al. 2005). These two patients subsequently had corrective spinal surgeries with good outcomes. The fourth reported individual with severe childhood HPP developed scoliosis during childhood and began AA ERT therapy at age 15 years when scoliosis was noted to be 60°. The patient had improved growth during first 9 months of AA therapy. However, after 12 months of AA treatment, scoliosis progressed to 110°, with a decrease in height, necessitating spinal fusion surgery (Bowden and Adler 2018a). It is important to screen for and monitor scoliosis during AA therapy in HPP patients, as untreated progressive scoliosis associated with metabolic bone disease can lead to pulmonary compromise and death (Collins 2006).

13.7.4.9 Dental Defects

To date, effects of AA ERT on dental manifestations of HPP have been limited to reports on three young children with lethal perinatal HPP. One child who started AA ERT at age 7.5 months showed better mineralization of teeth on lateral radiographs of the skull after 1 year of AA ERT; the other starting AA at 36 months showed improved stability of loose teeth (Whyte et al. 2012). Another child in Japan who started AA on the first day of life exhibited no tooth loss at age 3 years 5 months, but featured dentin and enamel defects and delayed formation of permanent teeth (Okawa et al. 2017b). These three patients represent the extent of dental findings reported for severe lethal HPP, while the potential for AA ERT to make improvements in dentoalveolar tissues in other forms of HPP remains unknown. Clinical trials to date have not incorporated dental examinations into data collection plans, to provide information on how the timing of intervention affects development and retention of primary and secondary teeth, and to determine whether AA is equally effective at improving cementum, dentin, enamel, and alveolar bone or periodontal manifestations of HPP. The opportunity to collect quantitative and qualitative dental findings in treated HPP subjects remains feasible, is a high priority for clinicians and researchers concerned with dental effects of HPP, and could prove important in justifying early administration of AA to prevent lifelong oral health problems and improve quality of life for individuals with severe dental manifestations of HPP. A registry has been initiated to collect qualitative and quantitative data on dental effects of HPP, allowing pre- and post-ERT comparisons to begin to determine efficacy of AA on ameliorating dental defects (http://u.osu.edu/hppstudy/).

13.7.5 Adverse Effects and Outcomes

Clinical trial results to date show very good safety profiles for AA ERT (Kitaoka et al. 2017; Whyte et al. 2012, 2016a). The most common adverse effects in patients treated with AA are injection-site reactions, occurring in about 75% of individuals treated (Kishnani et al. 2017b). Injection site reactions may include mild, localized, transient erythema, pain, induration, and lipohypertrophy or lipohypotrophy at injection sites. Localized lipohypertrophy persisted for more than 3 years in 6 out of 12 patients in the childhood HPP trial (Whyte et al. 2016a). In a patient undergoing AA ERT and not yet described in the literature, lipohypertrophy remained the same size with no regression even after discontinuing using the site for more than a year (personal observation) (Fig. 13.7). A rarer but more severe adverse effect reported has included anaphylactoid reaction (Kishnani et al. 2016). All patients enrolled in clinical trials developed anti-AA antibodies, but no tachyphylaxis has been reported to date (Kishnani et al. 2017b; Whyte et al. 2012).

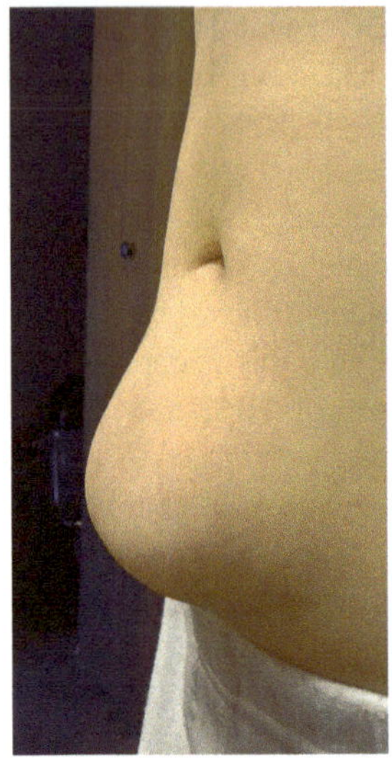

Fig. 13.7 Lipohypertrophy resulting from asfotase alfa injection. In an adolescent undergoing AA ERT, lipohypertrophy at an abdominal injection site remained the same size with no regression, even after discontinuing using the site for more than a year

Benefits of AA therapy are clearly evident in infants, children, and adults with HPP in clinical trials and case reports described above. Infants with life-threatening HPP survived; those who died were from sepsis, not attributed to AA therapy (Vidmar et al. 2017; Whyte et al. 2012), or from neurologic complications of craniosynostosis that occurred early in the treatment, possibly reflecting natural history and complications (Whyte et al. 2016b). Excellent clinical outcomes in all areas including physical function and quality of life continue after 5 years or more follow-up in those children and adults (Kishnani et al. 2017a). There has been one case of a term infant with perinatal HPP with unsuccessful treatment outcome due to irreversible pulmonary hypoplasia, despite improvement in rib mineralization after AA dose increased to 9 mg/kg/day (Costain et al. 2018). The poor outcome in this case may not be directly related to treatment failure, but possibly due to underlying genetic factor associated with failure of postnatal alveolar development.

13.7.6 Dosage and Administration of AA

AA is indicated for patients with pediatric-onset HPP. The recommended dosage of AA in perinatal- or infantile-onset HPP is 2 mg/kg three times weekly or 1 mg/kg six times weekly. The dose can be increased for lack of efficacy (e.g., no improvement in respiratory status, growth, or radiographic findings) up to 3 mg/kg three times weekly (maximum: 9 mg/kg per week). The dosage for childhood-onset HPP is also 2 mg/kg three times weekly or 1 mg/kg six times weekly. AA is administered subcutaneously only in the abdominal area, thigh, or deltoid. Rotation of the injection sites should be emphasized to reduce the risk of lipodystrophy (as described above and in Fig. 13.7). Patients or caregivers of pediatric patients should be instructed to administer AA within 1 h upon removal from refrigerator and to not administer injections in areas that are reddened, inflamed, or swollen. The 80 mg/0.8 mL concentration vial should not be used in pediatric patients. AA solution is clear, slightly opalescent or opalescent, colorless to slightly yellow; few small translucent or white particles may be present. Vial(s) not consistent with this appearance should be discarded. The maximum injection should not exceed 1 ml, using a 1 mL syringe with 1/2 inch needle (25–29 gauge). If more than 1 mL is required, the total volume should be split equally between two syringes, and two injections are administered at the same time using separate injection sites.

13.8 Future Directions for HPP Therapy

The preclinical studies described above in Sect. 5 employed repeated daily injections to achieve steady state concentrations of ALP activity with a recombinant protein that had a short half-life of 34 h in plasma. Novel therapeutic approaches are necessary to reduce adverse effects, improve efficacy, or serve as a secondary approach for those who do not respond, have adverse outcomes, or develop immune reactions to AA. Several of these concepts have been tested at the preclinical stage and these studies are summarized below.

13.8.1 Gene Therapy Approaches

Several gene therapy approaches employing single injections of TNSALP-D_{10} expressing viral vectors have been reported to show positive effects in correcting the HPP phenotype in *Alpl*$^{-/-}$ mice. In the first study, a lentiviral (LV) vector expressing TNSALP-D_{10} was injected IV into newborn pups at 1–3 dpn (Yamamoto et al. 2011). A high copy number of integrated vector was detected in liver (with lower copy numbers in other organs) and by 10 dpn, ALP was measured at 25-fold higher than untreated *Alpl*$^{-/-}$ mice and tenfold higher than WT control mice. While ALP in WT mice declined with age, treated *Alpl*$^{-/-}$ mice posted more than 70-fold increased ALP over WT by 60 dpn, indicating stable and even increasing enzyme production. LV expressed TNSALP-D_{10} normalized body size and eliminated seizures in *Alpl*$^{-/-}$ mice, extending lifespans of six of seven mice until 160 dpn, the termination of the study. Radiological and histological evaluation of long bones of treated *Alpl*$^{-/-}$ mice indicated dramatic improvements.

A second approach for viral vector mediated TNSALP-D_{10} gene therapy used adeno-associated virus (AAV) injected IV into newborn pups. This recombinant AAV expressed TNSALP-D_{10} under the tissue-nonspecific CAG promoter, a hybrid of the actin promoter and CMV-IE enhancer (Matsumoto et al. 2011). Recombinant AAV-mediated delivery also promoted phenotypic correction of *Alpl*$^{-/-}$ mice as indicated by radically increased ALP, lack of seizures, extended lifespan, and improved bone mineralization. As perinatal and infantile forms of HPP are the most severe and the most critical cases can sometimes be detected in utero, AAV-expressed TNSALP-D_{10} administration was also tested in utero by transuterine IP injection in a pregnant dame carrying *Alpl*$^{-/-}$ mouse fetuses on embryonic day 15 (E15), about 3–4 days before birth (Sugano et al. 2012). In total, seven of nine treated *Alpl*$^{-/-}$ mice treated in utero showed normalized weight and growth rate and absence of seizures during the study period of 2 months. Fetal gene therapy effectively transduced tissues including bone, muscle, heart, and liver, stably increasing ALP levels to tenfold greater than WT control mice. Elevated ALP activity was visualized at bone surfaces and in chondrocytes in treated mice, and skeletal phenotype was largely corrected.

An additional study examined the potential to employ a muscle-directed AAV-based therapy by using a novel AAV8 construct that expressed

TNSALP-D_{10} under the muscle creatine kinase (MCK) promoter (Nakamura-Takahashi et al. 2016). Muscle provides a large, easily accessible, and vascular target for AAV transduction. As with the previous gene therapy experiments, mice were injected at early postnatal age, though here injections were made intra-muscular (IM) bilaterally into the quadriceps femoris muscles. Using a control vector that expressed enhanced green fluorescent protein (EGFP), investigators demonstrated that this approach promoted high expression in muscle and heart, but not in other organs. Muscle-driven production of TNSALP-D_{10} proved capable of increasing ALP levels in *Alpl*$^{-/-}$ mice by more than tenfold WT levels and extending the lifespans of nine out of ten treated mice to 90 dpn. As with previous gene therapy approaches, muscle expression of TNSALP-D_{10} was accompanied by normalization of mobility, increased body size, and improved mineralization. However, abnormal chondrocyte arrangement in treated mice was associated with altered cortical and trabecular bone parameters, hypomineralization, and reduced length of long bones, indicating incomplete rescue of the HPP phenotype in *Alpl*$^{-/-}$ mice. Of the dentoalveolar tissues of treated *Alpl*$^{-/-}$ mice, only the jawbones showed clear signs of improvement, whereas teeth exhibited thin and hypomineralized root dentin, and detachment at the root surface indicative of defective cementum (Ikeue et al. 2018).

While concerns about the safety of gene therapy remain, the frequency of administration (three or six times weekly, as described in more detail above), injection site reactions in many subjects, production of costly recombinant enzyme, and potentially lifetime requirement for AA, all support investigation into gene or cell therapies in human subjects. Concerns regarding gene therapy in human subjects include potential for germline gene transfer or immune response to viral vector or gene product. As an alternative gene therapy approach to address some of these issues, Shimada and colleagues attempted ex vivo LV expressed TNSALP-D_{10} transduction of bone marrow cells (BMC) (Iijima et al. 2015). BMC harvested from 8 to 12 week-old WT mice were transduced with TNSALP-D_{10} LV or EGFP expressing control LV. Neonatal *Alpl*$^{-/-}$ mouse pups were irradiated on day 2 after birth to ablate the recipient BMC component, and donor BMC were injected IV. Engraftment of donor BMC was measured at approximately 30% over the 90-day study. Untransduced and EGFP expressing control BMC were unable to extend the lifespans or provide phenotypic correction in *Alpl*$^{-/-}$ mice. In contrast, TNSALP-D_{10} transduced BMC prolonged survival, significantly improved but did not totally normalize growth curves, and boosted ALP activity tenfold higher than WT mice and 400-fold higher than untreated or mock treated *Alpl*$^{-/-}$ mice. By histology, ALP activity was found at bone surfaces of TNSALP-D_{10} treated *Alpl*$^{-/-}$ mice, supporting the concept that enzyme produced by engrafted BMC made its way to mineralizing sites as predicted. Numerous cortical and trabecular bone parameters were improved in treated mice, however growth plates and bone length were not completely rescued, as seen in previous gene therapy approaches described above. While dental tissues were examined in mice receiving transduced BMC, there was little evidence for improvement in dentin, cementum, or periodontal attachment and function (Okawa et al. 2017a). This is not surprising because even though intervention was early, dental precursors arise from cranial neural crest derived ectomesenchyme, which is an embryological tissue distinct from BMC precursors. Lack of improvement in dentoalveolar tissues reflected insufficient enzyme reaching these tissues during critical periods of their development and mineralization.

13.8.2 Soluble TNSALP Enzyme Replacement Therapy

All of the AAV and LV expressed enzyme replacement therapies summarized so far in this section focused on use of the mineral-targeted TNSALP-D_{10} construct. An alternative approach to the concept of targeted treatment is to attempt to achieve sustained high levels of soluble (non-targeted) TNSALP. As summarized above, a recombinant anchorless TNSALP extended the lifespan of *Alpl*$^{-/-}$ mice, but was unable to correct skeletal and dental mineralization defects (Oikawa et al. 2014). AAV-mediated expression of soluble TNSALP showed far superior efficacy to injection of recombinant enzyme, with not only increased ALP levels and lifespan, but significant correction of bone mineralization by radiography (Matsumoto et al. 2011). Based on these successes with soluble forms of TNSALP, an additional strategy was attempted that employed a soluble intestinal-like chimeric alkaline phosphatase (ChimAP). ChimAP was engineered by substituting the crown domain of human intestinal phosphatase (IAP) with that of the placental isozyme (PLAP), making a chimeric form with IAP-like protein conformation, increased stability, Zn^{2+} binding in the enzyme active site, and narrowed substrate specificity (Kiffer-Moreira et al. 2014). Pharmacokinetic studies using bolus administration of 1, 8, or 16 mg/kg ChimAP indicated peak activity at 4 h and a half-life of 6 h, where residual ALP remained higher than untreated mice at 24 h after injection (Gasque et al. 2015). In a dose-response experimental design parallel to testing of TNSALP-D_{10} (Millan et al. 2008), ChimAP was administered 1, 8, or 16 mg/kg/day by daily SC injection. Median survival was increased to 44 dpn by 8 mg/kg/day, more than doubling lifespans of untreated or 1 mg/kg/day treated *Alpl*$^{-/-}$ mice, while mice treated with the highest dose survived to the termination of the experiment at 53 days. The highest ChimAP dose of 16 mg/kg/day normalized body weight and PP_i levels and significantly improved growth plate appearance, cortical bone parameters, and craniofacial shape. However, ChimAP was unable to completely normalize trabecular bone and all treated *Alpl*$^{-/-}$ mice featured increased osteoid accumulation compared to WT. While the high dose of ChimAP did seem to improve enamel and dentin in *Alpl*$^{-/-}$ mice, measurable dentin defects remained, substantial osteoid was observed on alveolar bone, and lack of cementum and periodontal breakdown was evident by histology, indicating a less robust response of dentoalveolar tissues. As of this writing, ChimAP (also known as recAP) has completed Phase I clinical trials and is being tested in Phase II trials (ClinicalTrials.gov identifier NCT02182440) in the U.S. and Europe sponsored by AM-Pharma (Bunnik, the Netherlands) for acute kidney injury (AKI), whereas ulcerative colitis (UC) and HPP studies remain in the preclinical stage. An oral formulation is in development for UC treatment. The rationale for application in AKI and UC lies in the potential for ALP to dephosphorylate lipopolysaccharide (LPS) that contributes to sepsis-associated AKI, and convert adenosine triphosphate (ATP) providing cell-protective and anti-inflammatory effects (Peters et al. 2013, 2014a, b, 2016a, b; Pickkers et al. 2012).

13.9 Application of Asfotase Alfa for Other Conditions

To date, AA has been studied almost entirely within the context of treating HPP. However, other metabolic conditions also feature altered or reduced ALP levels. Neurofibromatosis type 1 (NF1; OMIM# 162200) is caused by mutations in *NF1*, a cytoplasmic protein involved in a number of cell signaling cascades. NF1 clinical features include skin, ophthalmologic, and skeletal manifestations, with the latter reportedly including bone abnormalities, altered ALP and vitamin D levels, and osteopenia or osteoporosis in some affected individuals (Armstrong et al. 2013; Duman et al. 2008; Elefteriou et al. 2009; Lodish et al. 2012; Poyrazoglu et al. 2017; Rodari et al. 2018; Schnabel et al. 2013). A conditional knockout mouse model of NF1 featuring increased PP_i and reduced serum ALP exhibited skeletal hypomineralization (de la Croix Ndong et al. 2014; Wang et al. 2011). Treatment with TNSALP-D_{10} successfully increased bone growth and mineral density in these mice, providing preclinical data that use of ERT may improve skeletal effects of NF1 in affected individuals (de la Croix Ndong et al. 2014).

As TNSALP is perhaps the most critical enzyme for skeletal and dental mineralization, there is potential for AA to be applied to other hereditary mineral metabolism disorders, or even possibly used to promote more rapid fracture healing in healthy individuals, however no preclinical studies have yet been performed to support this.

13.10 Summary and Conclusions

AA ERT has been transformative as the first treatment for HPP, the last form of rickets to receive a medical therapy (Whyte 2017). Based on the clinical trials conducted to date, AA is approved for use in individuals of any age with pediatric-onset HPP. Clinical trials and case reports evaluated to date have targeted individuals with a high disease burden caused by perinatal, infantile, childhood, or adult HPP, specifically those with significant skeletal manifestations of HPP. Improved skeletal mineralization in response to AA ERT in turn leads to improved respiratory status and increased survival in severely affected infants, revolutionizing the outcome of a once fatal form of the disease. Improvement of skeletal manifestations by AA ERT also alleviates additional complications stemming from bone abnormalities, improving growth, mobility, physical function, and quality of life. Currently, evidence-based therapeutic recommendations for children and adults with less symptomatic HPP are lacking and unclear. A long-term follow-up care with at least annual assessment for progression of disease is needed to monitor for late manifestations of HPP (Mori et al. 2016). Judicious use of this high-cost ERT is imperative; risk-benefit ratio, feasibility and safety of treatment need to be considered. In childhood HPP, five key manifestations including mobility, pain, rickets, growth, and fracture, have been proposed as guidance for decision to treat (Rush 2018). Conservative management should be the first line of management: physical therapy for patients with hypotonia, mobility limitations, or gait abnormalities; analgesics (acetaminophen or nonsteroidal anti-inflammatory drugs) for those with musculoskeletal pain. If symptoms improve with the conservative treatment, then ERT is not indicated. AA treatment should be considered in patients with childhood HPP who have limited mobility that impair quality of life or debilitating pain unresponsive to conservative treatment. In those patients with symptomatic and disabling HPP, with or without fractures, for whom AA is considered, clinicians should establish treatment goals in order to monitor the patient's response and determine what clinical outcomes are to be achieved (Kishnani et al. 2017b). Given the wide clinical variability in disease manifestations of adult HPP, recently Shapiro and Lewiecki (2017) have suggested that AA treatment should be considered in adult HPP if one or more of the following is present and attributable to HPP: musculoskeletal pain requiring prescription pain medications, disabling polyarthropathy, disabling functional impairment assessed by validated measures, low-trauma fracture, delayed or incomplete fracture healing, repeated orthopedic surgeries for HPP bone disease, low bone mineral density T-score ≤ -2.5 in postmenopausal women and men age 50 years and older, or Z-score ≤ -2.0 in younger adult women and men in patients with fractures, and nephrocalcinosis.

Ultimately, it remains to be seen how AA will be employed in the long-term management of severely and mildly affected HPP patients, whether dose titration is efficacious for different disease severity, or any additional beneficial outcomes or adverse effects are discovered, and whether alternative approaches for delivery or achieving ERT are successful and transform HPP therapy yet again.

Acknowledgments We thank the subjects and their families for their cooperation and assistance with this research. This research was supported by research grants from Soft Bones, Inc. to BLF and grant AR 066110 to BLF from the National Institute of Arthritis and Musculoskeletal and Skin Diseases (NIAMS). We thank Dr. Jose Luis Millán for providing images for reproduction. We thank Mr. Michael B. Chavez for adaptation of the TNSALP 2D model in Fig. 13.2m. For more information on hypophosphatasia for affected individuals, researchers, or health care professionals, see the Soft Bones, Inc. website at https://www.softbones.org/.

Conflict of Interest BLF has served as a consultant and speaker for Alexion Pharmaceuticals, Inc., and received two research grants from Soft Bones, Inc., a nonprofit patient advocacy, support, and education group for families with hypophosphatsia. The authors report no other conflicts of interest in this work.

References

Akiyama T, Kubota T, Ozono K, Michigami T, Kobayashi D, Takeyari S, Sugiyama Y, Noda M, Harada D, Namba N et al (2018) Pyridoxal 5′-phosphate and related metabolites in hypophosphatasia: effects of enzyme replacement therapy. Mol Genet Metab 125(1–2):174–180

Albeggiani A, Cataldo F (1982) Infantile hypophosphatasia diagnosed at 4 months and surviving at 2 years. Helv Paediatr Acta 37(1):49–58

Anderson HC, Sipe JB, Hessle L, Dhanyamraju R, Atti E, Camacho NP, Millan JL, Dhamyamraju R (2004) Impaired calcification around matrix vesicles of growth plate and bone in alkaline phosphatase-deficient mice. Am J Pathol 164(3):841–847

Anderson HC, Harmey D, Camacho NP, Garimella R, Sipe JB, Tague S, Bi X, Johnson K, Terkeltaub R, Millan JL (2005) Sustained osteomalacia of long bones despite major improvement in other hypophosphatasia-related mineral deficits in tissue nonspecific alkaline phosphatase/nucleotide pyrophosphatase phosphodiesterase 1 double-deficient mice. Am J Pathol 166(6):1711–1720

Armstrong L, Jett K, Birch P, Kendler DL, McKay H, Tsang E, Stevenson DA, Hanley DA, Egeli D, Burrows M et al (2013) The generalized bone phenotype in children with neurofibromatosis 1: a sibling matched case-control study. Am J Med Genet A 161A(7):1654–1661

Arun R, Khazim R, Webb JK, Burn J (2005) Scoliosis in association with infantile hypophosphatasia: a case study in two siblings. Spine 30(16):E471–E476

Baumgartner-Sigl S, Haberlandt E, Mumm S, Scholl-Burgi S, Sergi C, Ryan L, Ericson KL, Whyte MP, Hogler W (2007) Pyridoxine-responsive seizures as the first symptom of infantile hypophosphatasia caused by two novel missense mutations (c.677t>c, p.M226t; c.1112c>t, p.T371i) of the tissue-nonspecific alkaline phosphatase gene. Bone 40(6):1655–1661

Beck C, Morbach H, Richl P, Stenzel M, Girschick H (2009) How can calcium pyrophosphate crystals induce inflammation in hypophosphatasia or chronic inflammatory joint diseases? Rheumatol Int 29(3):229–238

Beertsen W, VandenBos T, Everts V (1999) Root development in mice lacking functional tissue non-specific alkaline phosphatase gene: inhibition of acellular cementum formation. J Dent Res 78(6):1221–1229

Berkseth KE, Tebben PJ, Drake MT, Hefferan TE, Jewison DE, Wermers RA (2013) Clinical spectrum of hypophosphatasia diagnosed in adults. Bone 54(1):21–27

Biosse Duplan M, Coyac BR, Bardet C, Zadikian C, Rothenbuhler A, Kamenicky P, Briot K, Linglart A, Chaussain C (2017) Phosphate and vitamin d prevent periodontitis in x-linked hypophosphatemia. J Dent Res 96(4):388–395

Bisaz S, Russell RG, Fleisch H (1968) Isolation of inorganic pyrophosphate from bovine and human teeth. Arch Oral Biol 13(6):683–696

Bloch-Zupan A (2016) Hypophosphatasia: diagnosis and clinical signs – a dental surgeon perspective. Int J Paediatr Dent 26(6):426–438

Boileau G, Tenenhouse HS, Desgroseillers L, Crine P (2001) Characterization of phex endopeptidase catalytic activity: identification of parathyroid-hormone-related peptide107–139 as a substrate and osteocalcin, ppi and phosphate as inhibitors. Biochem J 355(Pt 3):707–713

Bowden SA, Adler BH (2018a) Asfotase alfa treatment for 1 year in a 16 year-old male with severe childhood hypophosphatasia. Osteoporos Int 29(2):511–515

Bowden SA, Adler BH (2018b) Reappearance of hypomineralized bone after discontinuation of asfotase alfa treatment for severe childhood hypophosphatasia. Osteoporos Int 29(9):2155–2156

Bruckner R, Rickles N, Porter D (1962) Hypophosphatasia with premature shedding of teeth and aplasia of cementum. Oral Surg Oral Med Oral Pathol 15:1351–1369

Cahill RA, Wenkert D, Perlman SA, Steele A, Coburn SP, McAlister WH, Mumm S, Whyte MP (2007) Infantile hypophosphatasia: transplantation therapy trial using bone fragments and cultured osteoblasts. J Clin Endocrinol Metab 92(8):2923–2930

Campos M, Couture C, Hirata IY, Juliano MA, Loisel TP, Crine P, Juliano L, Boileau G, Carmona AK (2003) Human recombinant endopeptidase phex has a strict s1' specificity for acidic residues and cleaves peptides derived from fibroblast growth factor-23 and matrix extracellular phosphoglycoprotein. Biochem J 373(Pt 1):271–279

Castells L, Cassanello P, Muñiz F, de Castro M-J, Couce ML (2018) Neonatal lethal hypophosphatasia: a case report and review of literature. Medicine 97(48):e13269

Clubb JS, Neale FC, Posen S (1965) The behavior of infused human placental alkaline phosphatase in human subjects. J Lab Clin Med 66(3):493–507

Colazo JM, Hu JR, Dahir KM, Simmons JH (2018) Correction to: neurological symptoms in hypophosphatasia. Osteoporos Int: J Established Result Cooperation Between Eur Found Osteoporos Natl Osteoporos Found USA 30:535

Collins MT (2006) Spectrum and natural history of fibrous dysplasia of bone. J Bone Miner Res 21(S2):P99–P104

Collmann H, Mornet E, Gattenlohner S, Beck C, Girschick H (2009) Neurosurgical aspects of childhood hypophosphatasia. Childs Nerv Syst 25(2):217–223

Costain G, Moore AM, Munroe L, Williams A, Shaul RZ, Rockman-Greenberg C, Offringa M, Kannu P (2018) Enzyme replacement therapy in perinatal hypophosphatasia: case report of a negative outcome and lessons for clinical practice. Mol Genet Metab Rep 14:22–26

Cruz T, Gleizes M, Balayssac S, Mornet E, Marsal G, Millan JL, Malet-Martino M, Nowak LG, Gilard V, Fonta C (2017) Identification of altered brain metabolites associated with tnap activity in a mouse model of hypophosphatasia using untargeted nmr-based metabolomics analysis. J Neurochem 140(6):919–940

Cundy T, Michigami T, Tachikawa K, Dray M, Collins JF, Paschalis EP, Gamsjaeger S, Roschger A, Fratzl-Zelman N, Roschger P et al (2015) Reversible deterioration in hypophosphatasia caused by renal failure with bisphosphonate treatment. J Bone Miner Res 30(9):1726–1737

Davit-Beal T, Gabay J, Antoniolli P, Masle-Farquhar J, Wolikow M (2014) Dental complications of rickets in early childhood: case report on 2 young girls. Pediatrics 133(4):e1077–e1081

de la Croix Ndong J, Makowski AJ, Uppuganti S, Vignaux G, Ono K, Perrien DS, Joubert S, Baglio SR, Granchi D, Stevenson DA et al (2014) Asfotase-alpha improves bone growth, mineralization and strength in mouse models of neurofibromatosis type-1. Nat Med 20(8):904–910

Deeb AA, Bruce SN, Morris AA, Cheetham TD (2000) Infantile hypophosphatasia: disappointing results of treatment. Acta Paediatr (Oslo, Norway: 1992) 89(6):730–733

Doetschman T, Gregg RG, Maeda N, Hooper ML, Melton DW, Thompson S, Smithies O (1987) Targetted correction of a mutant hprt gene in mouse embryonic stem cells. Nature 330(6148):576–578

Doshi KB, Hamrahian AH, Licata AA (2009) Teriparatide treatment in adult hypophosphatasia in a patient exposed to bisphosphonate: a case report. Clin Cases Miner Bone Metab 6(3):266–269

Duman O, Ozdem S, Turkkahraman D, Olgac ND, Gungor F, Haspolat S (2008) Bone metabolism markers and bone mineral density in children with neurofibromatosis type-1. Brain Dev 30(9):584–588

Durussel J, Liu J, Campbell C, Nam HK, Hatch NE (2016) Bone mineralization-dependent craniosynostosis and craniofacial shape abnormalities in the mouse model of infantile hypophosphatasia. Dev Dyn 245(2):175–182

Elefteriou F, Kolanczyk M, Schindeler A, Viskochil DH, Hock JM, Schorry EK, Crawford AH, Friedman JM, Little D, Peltonen J et al (2009) Skeletal abnormalities in neurofibromatosis type 1: approaches to therapeutic options. Am J Med Genet A 149A(10):2327–2338

El-Gharbawy AH, Peeden JN Jr, Lachman RS, Graham JM Jr, Moore SR, Rimoin DL (2010) Severe cleidocranial dysplasia and hypophosphatasia in a child with microdeletion of the c-terminal region of runx2. Am J Med Genet A 152a(1):169–174

Evans MJ, Kaufman MH (1981) Establishment in culture of pluripotential cells from mouse embryos. Nature 292(5819):154–156

Fedde KN, Blair L, Silverstein J, Coburn SP, Ryan LM, Weinstein RS, Waymire K, Narisawa S, Millan JL, MacGregor GR et al (1999) Alkaline phosphatase knock-out mice recapitulate the metabolic and skeletal defects of infantile hypophosphatasia. J Bone Miner Res 14(12):2015–2026

Feeney C, Stanford N, Lee S, Barry S (2018) Hypophosphatasia and the importance of the general dental practitioner – a case series and discussion of upcoming treatments. Br Dent J 224(12):937–943

Fisher LW, Fedarko NS (2003) Six genes expressed in bones and teeth encode the current members of the sibling family of proteins. Connect Tissue Res 44(Suppl 1):33–40

Fleisch H, Bisaz S (1962a) Isolation from urine of pyrophosphate, a calcification inhibitor. Am J Physiol 203:671–675

Fleisch H, Bisaz S (1962b) Mechanism of calcification: inhibitory role of pyrophosphate. Nature 195:911

Fleisch H, Schibler D, Maerki J, Frossard I (1965) Inhibition of aortic calcification by means of pyrophosphate and polyphosphates. Nature 207(5003):1300–1301

Fleisch H, Russell RG, Straumann F (1966) Effect of pyrophosphate on hydroxyapatite and its implications in calcium homeostasis. Nature 212(5065):901–903

Foster BL, Hujoel PP (2018) Vitamin d in dentoalveolar and oral health. In: Feldman D, Pike JW, Bouillon R (eds) Vitamin d, 4th edn. Academic, London, pp 497–520

Foster BL, Nagatomo KJ, Nociti FH Jr, Fong H, Dunn D, Tran AB, Wang W, Narisawa S, Millan JL, Somerman MJ (2012) Central role of pyrophosphate in acellular cementum formation. PLoS One 7(6):e38393

Foster BL, Nagatomo KJ, Tso HW, Tran AB, Nociti FH Jr, Narisawa S, Yadav MC, McKee MD, Millan JI, Somerman MJ (2013) Tooth root dentin mineralization defects in a mouse model of hypophosphatasia. J Bone Miner Res 28(2):271–282

Foster BL, Sheen CR, Hatch NE, Liu J, Cory E, Narisawa S, Kiffer-Moreira T, Sah RL, Whyte MP, Somerman MJ et al (2015) Periodontal defects in the a116t knock-in murine model of odontohypophosphatasia. J Dent Res 94(5):706–714

Foster BL, Kuss P, Yadav MC, Kolli TN, Narisawa S, Lukashova L, Cory E, Sah RL, Somerman MJ, Millan JL (2017) Conditional alpl ablation phenocopies dental defects of hypophosphatasia. J Dent Res 96(1):81–91

Freitas TQ, Franco AS, Pereira RMR (2018) Improvement of bone microarchitecture parameters after 12 months of treatment with asfotase alfa in adult patient with hypophosphatasia: case report. Medicine 97(48):e13210

Gasque KC, Foster BL, Kuss P, Yadav MC, Liu J, Kiffer-Moreira T, van Elsas A, Hatch N, Somerman MJ, Millan JL (2015) Improvement of the skeletal and dental hypophosphatasia phenotype in alpl−/− mice by administration of soluble (non-targeted) chimeric alkaline phosphatase. Bone 72:137–147

Genest F, Seefried L (2018) Subtrochanteric and diaphyseal femoral fractures in hypophosphatasia—not atypical at all. Osteoporos Int 1–11

Girschick H, Schneider P, Haubitz I, Hiort O, Collmann H, Beer M, Shin Y, Seyberth H (2006) Effective nsaid treatment indicates that hyperprostaglandinism is affecting the clinical severity of childhood hypophosphatasia. Orphanet J Rare Dis 1(1):24

Girschick HJ, Mornet E, Beer M, Warmuth-Metz M, Schneider P (2007) Chronic multifocal non-bacterial osteomyelitis in hypophosphatasia mimicking malignancy. BMC Pediatr 7(1):3

Goodwin HW, Robison R (1924) The possible significance of hexosephosphoric esters in ossification. Part iv. The phosphoric esters of the blood: preliminary communication. Biochem J 18(5):1161–1162

Greenberg CR, Taylor CL, Haworth JC, Seargeant LE, Philipps S, Triggs-Raine B, Chodirker BN (1993) A homoallelic gly317 – >asp mutation in alpl causes the perinatal (lethal) form of hypophosphatasia in Canadian mennonites. Genomics 17(1):215–217

Gurley K, Chen H, Guenther C, Nguyen E, Rountree R, Schoor M, Kingsley D (2006) Mineral formation in joints caused by complete or joint-specific loss of ank function. J Bone Miner Res 21(8):1238–1247

Gutman AB, Gutman EB (1938) An "acid" phosphatase occurring in the serum of patients with metastasizing carcinoma of the prostate gland. J Clin Invest 17(4):473–478

Hacihamdioglu B, Ozgurhan G, Pereira C, Tepeli E, Acar G, Comert S (2018) Perinatal form hypophosphatasia caused by a novel large duplication of alpl gene and one year follow-up under enzyme replacement therapy; a case report. J Clin Res Pediatr Endocrinol 10:175–178

Hall B, Limaye A, Kulkarni AB (2009) Overview: generation of gene knockout mice. Curr Protoct Cell Biol. Chapter 19:Unit 19 12 19 12 11–17

Harmey D, Hessle L, Narisawa S, Johnson KA, Terkeltaub R, Millan JL (2004) Concerted regulation of inorganic pyrophosphate and osteopontin by akp2, enpp1, and ank: an integrated model of the pathogenesis of mineralization disorders. Am J Pathol 164(4):1199–1209

Harmey D, Johnson KA, Zelken J, Camacho NP, Hoylaerts MF, Noda M, Terkeltaub R, Millan JL (2006) Elevated skeletal osteopontin levels contribute to the hypophosphatasia phenotype in akp2(−/−) mice. J Bone Miner Res 21(9):1377–1386

Hessle L, Johnson KA, Anderson HC, Narisawa S, Sali A, Goding JW, Terkeltaub R, Millan JL (2002) Tissue-nonspecific alkaline phosphatase and plasma cell membrane glycoprotein-1 are central antagonistic regulators of bone mineralization. Proc Natl Acad Sci U S A 99(14):9445–9449

Ho A, Johnson M, Kingsley D (2000) Role of the mouse ank gene in control of tissue calcification and arthritis. Science 289(5477):265–270

Hu JC, Plaetke R, Mornet E, Zhang C, Sun X, Thomas HF, Simmer JP (2000) Characterization of a family with dominant hypophosphatasia. Eur J Oral Sci 108(3):189–194

Iijima O, Miyake K, Watanabe A, Miyake N, Igarashi T, Kanokoda C, Nakamura-Takahashi A, Kinoshita H, Noguchi T, Abe S et al (2015) Prevention of lethal murine hypophosphatasia by neonatal ex vivo gene therapy using lentivirally transduced bone marrow cells. Hum Gene Ther 26(12):801–812

Ikeue R, Nakamura-Takahashi A, Nitahara-Kasahara Y, Watanabe A, Muramatsu T, Sato T, Okada T (2018) Bone-targeted alkaline phosphatase treatment of mandibular bone and teeth in lethal hypophosphatasia via an scaav8 vector. Mol Ther Methods Clin Dev 10:361–370

Johnson KA, Hessle L, Vaingankar S, Wennberg C, Mauro S, Narisawa S, Goding JW, Sano K, Millan JL, Terkeltaub R (2000) Osteoblast tissue-nonspecific alkaline phosphatase antagonizes and regulates pc-1. Am J Physiol Regul Integr Comp Physiol 279(4):R1365–R1377

Jung A, Russel RG, Bisaz S, Morgan DB, Fleisch H (1970) Fate of intravenously injected pyrophosphate-32p in dogs. Am J Physiol 218(6):1757–1764

Kay HD, Robison R (1924) The possible significance of hexosephosphoric esters in ossification. Part iii: the action of the bone enzyme on the organic phosphorus compounds in blood. Biochem J 18(3–4):755–764

Kiffer-Moreira T, Sheen CR, Gasque KC, Bolean M, Ciancaglini P, van Elsas A, Hoylaerts MF, Millan JL (2014) Catalytic signature of a heat-stable, chimeric human alkaline phosphatase with therapeutic potential. PLoS One 9(2):e89374

Kishnani PS, Madson KL, Whyte MP, Gayron M, Fujita K, Rockman-Greenberg C (2016) Biochemical and physical function outcomes in adolescents and adults with hypophosphatasia treated with asfotase alfa for up to 4 years: interim results from a phase ii study. Metabolic and genetic bone disorders. Endocrine Society, pp OR26-23-OR26-23

Kishnani PS, Rockman-Greenberg C, Denker AE, Moseley S, Whyte MP (2017a) Biochemical and physical function outcomes after 5 years of treatment with asfotase alfa in adolescents and adults with hypophosphatasia: phase 2 study results. Paper presented at: International conference on Children's Bone Health (ICCBH). Wurzberg

Kishnani PS, Rush ET, Arundel P, Bishop N, Dahir K, Fraser W, Harmatz P, Linglart A, Munns CF, Nunes ME (2017b) Monitoring guidance for patients with hypophosphatasia treated with asfotase alfa. Mol Genet Metab 122(1):4–17

Kitaoka T, Tajima T, Nagasaki K, Kikuchi T, Yamamoto K, Michigami T, Okada S, Fujiwara I, Kokaji M, Mochizuki H et al (2017) Safety and efficacy of treatment with asfotase alfa in patients with hypophosphatasia: results from a japanese clinical trial. Clin Endocrinol 87(1):10–19

Klidaras P, Severt J, Aggers D, Payne J, Miller PD, Ing SW (2018) Fracture healing in two adult patients with hypophosphatasia after asfotase alfa therapy. JBMR Plus 2(5):304–307

Lawrence JE, Saeed D, Bartlett J, Carrothers AD (2017) Adult-onset hypophosphatasia diagnosed following

bilateral atypical femoral fractures in a 55-year-old woman. Clin Cases Miner Bone Metab 14(3):347–353
Libby Kosnik-Infinger M, Gendron C, Gordon CB (2015) Enzyme replacement therapy for congenital hypophosphatasia allows for surgical treatment of related complex craniosynostosis: a case series. Neurosurg Focus 38(5):E10
Liu J, Nam HK, Campbell C, Gasque KC, Millan JL, Hatch NE (2014) Tissue-nonspecific alkaline phosphatase deficiency causes abnormal craniofacial bone development in the alpl(−/−) mouse model of infantile hypophosphatasia. Bone 67:81–94
Liu J, Campbell C, Nam HK, Caron A, Yadav MC, Millan JL, Hatch NE (2015) Enzyme replacement for craniofacial skeletal defects and craniosynostosis in murine hypophosphatasia. Bone 78:203–211
Lodish MB, Dagalakis U, Sinaii N, Bornstein E, Kim A, Lokie KB, Baldwin AM, Reynolds JC, Dombi E, Stratakis CA et al (2012) Bone mineral density in children and young adults with neurofibromatosis type 1. Endocr Relat Cancer 19(6):817–825
Lopez-Delgado L, Riancho-Zarrabeitia L, Garcia-Unzueta MT, Tenorio JA, Garcia-Hoyos M, Lapunzina P, Valero C, Riancho JA (2018) Abnormal bone turnover in individuals with low serum alkaline phosphatase. Osteoporos Int: J Established Result Cooperation Between Eur Found Osteoporos Natl Osteoporos Found USA 29(9):2147–2150
Mak TW (2007) Gene targeting in embryonic stem cells scores a knockout in Stockholm. Cell 131(6):1027–1031
Martland M, Robison R (1924) The possible significance of hexosephosphoric esters in ossification: part v. the enzyme in the early stages of bone development. Biochem J 18(6):1354–1357
Martland M, Robison R (1926) Possible significance of hexosephosphoric esters in ossification: part vi. Phosphoric esters in blood-plasma. Biochem J 20(4):847–855
Martland M, Robison R (1927) The possible significance of hexosephosphoric esters in ossification: part vii. The bone phosphatase. Biochem J 21(3):665–674
Martland M, Robison R (1929) The preparation and use of the bone phosphatase. Biochem J 23(2):237–242
Matsumoto T, Miyake K, Yamamoto S, Orimo H, Miyake N, Odagaki Y, Adachi K, Iijima O, Narisawa S, Millan JL et al (2011) Rescue of severe infantile hypophosphatasia mice by aav-mediated sustained expression of soluble alkaline phosphatase. Hum Gene Ther 22(11):1355–1364
McKee MD, Nakano Y, Masica DL, Gray JJ, Lemire I, Heft R, Whyte MP, Crine P, Millan JL (2011) Enzyme replacement therapy prevents dental defects in a model of hypophosphatasia. J Dent Res 90(4):470–476
Meyer JL (1984) Can biological calcification occur in the presence of pyrophosphate? Arch Biochem Biophys 231(1):1–8
Meyer JL, Fleisch H (1984) Calcification inhibitors in rat and human serum and plasma. Biochim Biophys Acta 799(2):115–121
Millán JL (2006) Mammalian alkaline phosphatases: from biology to applications in medicine and biotechnology. Wiley-VCH, Weinheim
Millan JL (2013) The role of phosphatases in the initiation of skeletal mineralization. Calcif Tissue Int 93(4):299–306
Millan JL, Whyte MP (2016) Alkaline phosphatase and hypophosphatasia. Calcif Tissue Int 98(4):398–416
Millan JL, Narisawa S, Lemire I, Loisel TP, Boileau G, Leonard P, Gramatikova S, Terkeltaub R, Camacho NP, McKee MD et al (2008) Enzyme replacement therapy for murine hypophosphatasia. J Bone Miner Res 23(6):777–787
Mori M, DeArmey SL, Weber TJ, Kishnani PS (2016) Case series: odontohypophosphatasia or missed diagnosis of childhood/adult-onset hypophosphatasia?–call for a long-term follow-up of premature loss of primary teeth. Bone Rep 5:228–232
Mornet E (2017) Genetics of hypophosphatasia. Arch Pediatr 24(5S2):5S51–55S56
Mornet E, Yvard A, Taillandier A, Fauvert D, Simon-Bouy B (2011) A molecular-based estimation of the prevalence of hypophosphatasia in the european population. Ann Hum Genet 75(3):439–445
Murshed M, Harmey D, Millan JL, McKee MD, Karsenty G (2005) Unique coexpression in osteoblasts of broadly expressed genes accounts for the spatial restriction of ecm mineralization to bone. Genes Dev 19(9):1093–1104
Nakamura-Takahashi A, Miyake K, Watanabe A, Hirai Y, Iijima O, Miyake N, Adachi K, Nitahara-Kasahara Y, Kinoshita H, Noguchi T et al (2016) Treatment of hypophosphatasia by muscle-directed expression of bone-targeted alkaline phosphatase via self-complementary aav8 vector. Mol Ther Methods Clin Dev 3:15059
Nam HK, Sharma M, Liu J, Hatch NE (2017) Tissue nonspecific alkaline phosphatase (tnap) regulates cranial base growth and synchondrosis maturation. Front Physiol 8:161
Narisawa S, Frohlander N, Millan JL (1997) Inactivation of two mouse alkaline phosphatase genes and establishment of a model of infantile hypophosphatasia. Dev Dyn 208(3):432–446
Narisawa S, Wennberg C, Millan JL (2001) Abnormal vitamin b6 metabolism in alkaline phosphatase knock-out mice causes multiple abnormalities, but not the impaired bone mineralization. J Pathol 193(1):125–133
Narisawa S, Huang L, Iwasaki A, Hasegawa H, Alpers D, Millán J (2003) Accelerated fat absorption in intestinal alkaline phosphatase knockout mice. Mol Cell Biol 23(21):7525–7530
Nishioka T, Tomatsu S, Gutierrez MA, Miyamoto K, Trandafirescu GG, Lopez PL, Grubb JH, Kanai R, Kobayashi H, Yamaguchi S et al (2006) Enhancement of drug delivery to bone: characterization of human tissue-nonspecific alkaline phosphatase tagged with an acidic oligopeptide. Mol Genet Metab 88(3): 244–255

Nociti FH Jr, Berry JE, Foster BL, Gurley KA, Kingsley DM, Takata T, Miyauchi M, Somerman MJ (2002) Cementum: a phosphate-sensitive tissue. J Dent Res 81(12):817–821

Offiah AC, Vockley J, Munns CF, Murotsuki J. 2018. Differential diagnosis of perinatal hypophosphatasia: radiologic perspectives. Pediatr Radiol

Oikawa H, Tomatsu S, Haupt B, Montano AM, Shimada T, Sly WS (2014) Enzyme replacement therapy on hypophosphatasia mouse model. J Inherit Metab Dis 37(2):309–317

Okawa R, Iijima O, Kishino M, Okawa H, Toyosawa S, Sugano-Tajima H, Shimada T, Okada T, Ozono K, Ooshima T et al (2017a) Gene therapy improves dental manifestations in hypophosphatasia model mice. J Periodontal Res 52(3):471–478

Okawa R, Kokomoto K, Yamamura-Miyazaki N, Michigami T, Nakano K (2017b) Oral findings in patient with lethal hypophosphatasia treated with enzyme replacement therapy. Pediatr Dent J 27(3):153–156

Okazaki Y, Kitajima H, Mochizuki N, Kitaoka T, Michigami T, Ozono K (2016) Lethal hypophosphatasia successfully treated with enzyme replacement from day 1 after birth. Eur J Pediatr 175(3):433–437

Oyachi M, Harada D, Sakamoto N, Ueyama K, Kondo K, Kishimoto K, Izui M, Nagamatsu Y, Kashiwagi H, Yamamuro M et al (2018) A case of perinatal hypophosphatasia with a novel mutation in the alpl gene: clinical course and review of the literature. Clin Pediatr Endocrinol 27(3):179–186

Peters E, van Elsas A, Heemskerk S, Jonk L, van der Hoeven J, Arend J, Masereeuw R, Pickkers P (2013) Alkaline phosphatase as a treatment of sepsis-associated acute kidney injury. J Pharmacol Exp Ther 344(1):2–7

Peters E, Heemskerk S, Masereeuw R, Pickkers P (2014a) Alkaline phosphatase: a possible treatment for sepsis-associated acute kidney injury in critically ill patients. Am J Kidney Dis 63(6):1038–1048

Peters E, Masereeuw R, Pickkers P (2014b) The potential of alkaline phosphatase as a treatment for sepsis-associated acute kidney injury. Nephron Clin Pract 127(1–4):144–148

Peters E, Ergin B, Kandil A, Gurel-Gurevin E, van Elsas A, Masereeuw R, Pickkers P, Ince C (2016a) Effects of a human recombinant alkaline phosphatase on renal hemodynamics, oxygenation and inflammation in two models of acute kidney injury. Toxicol Appl Pharmacol 313:88–96

Peters E, Mehta RL, Murray PT, Hummel J, Joannidis M, Kellum JA, Arend J, Pickkers P (2016b) Study protocol for a multicentre randomised controlled trial: safety, tolerability, efficacy and quality of life of a human recombinant alkaline phosphatase in patients with sepsis-associated acute kidney injury (stop-aki). BMJ Open 6(9):e012371

Phillips D, Case LE, Griffin D, Hamilton K, Lara SL, Leiro B, Monfreda J, Westlake E, Kishnani PS (2016) Physical therapy management of infants and children with hypophosphatasia. Mol Genet Metab 119(1):14–19

Phillips D, Griffin D, Przybylski T, Morrison E, Reeves AL, Vallee M, Fujita KP, Madson KL (2018) Development and validation of a modified performance-oriented mobility assessment tool for assessing mobility in children with hypophosphatasia. J Pediatr Rehabil Med 11(3):187–192

Pickkers P, Heemskerk S, Schouten J, Laterre PF, Vincent JL, Beishuizen A, Jorens PG, Spapen H, Bulitta M, Peters WH et al (2012) Alkaline phosphatase for treatment of sepsis-induced acute kidney injury: a prospective randomized double-blind placebo-controlled trial. Crit Care 16(1):R14

Poryo M, Meyer S, Eymann R, Yilmaz U, Nemat S, Rohrer T (2016) Clinical images: a cloudy skull-hypophosphatasia as reason for copper-beaten skull. Neuropediatrics 47(6):410

Poyrazoglu HG, Bas VN, Arslan A, Bastug F, Canpolat M, Per H, Gumus H, Kumandas S (2017) Bone mineral density and bone metabolic markers' status in children with neurofibromatosis type 1. J Pediatr Endocrinol Metab: JPEM 30(2):175–180

Reibel A, Maniere MC, Clauss F, Droz D, Alembik Y, Mornet E, Bloch-Zupan A (2009) Orodental phenotype and genotype findings in all subtypes of hypophosphatasia. Orphanet J Rare Dis 4:6

Remde H, Cooper MS, Quinkler M (2017) Successful asfotase alfa treatment in an adult dialysis patient with childhood-onset hypophosphatasia. J Endocr Soc 1(9):1188–1193

Righetti M, Wach J, Desmarchelier R, Coury F (2018) Teriparatide treatment in an adult patient with hypophosphatasia exposed to bisphosphonate and revealed by bilateral atypical fractures. Joint Bone Spine 85(3):365–367

Robison R (1923) The possible significance of hexosephosphoric esters in ossification. Biochem J 17(2):286–293

Robison R, Soames KM (1924) The possible significance of hexosephosphoric esters in ossification: part ii. The phosphoric esterase of ossifying cartilage. Biochem J 18(3–4):740–754

Robison R, Soames KM (1925) A chemical study of defective ossification in rachitic animals. Biochem J 19(1):153–161

Robison R, Macleod M, Rosenheim AH (1930) The possible significance of hexosephosphoric esters in ossification: calcification in vitro. Biochem J 24(6):1927–1941

Rodari G, Scuvera G, Ulivieri FM, Profka E, Menni F, Saletti V, Esposito S, Bergamaschi S, Ferrante E, Eller-Vainicher C et al (2018) Progressive bone impairment with age and pubertal development in neurofibromatosis type i. Arch Osteoporos 13(1):93

Rodrigues TL, Nagatomo KJ, Foster BL, Nociti FH, Somerman MJ (2011) Modulation of phosphate/pyrophosphate metabolism to regenerate the periodontium: a novel in vivo approach. J Periodontol 82(12):1757–1766

Rodrigues TL, Georgetti AP, Martins L, Pereira Neto JS, Foster BL, Nociti FH Jr (2012, June) Clinical correlate: Cementum and periodontal defects resulting from odontohypophosphatasia predispose for premature tooth loss. In: Somerman MJ, LK MC (eds) Mineralized tissues in Oral and craniofacial science: biological principles and clinical correlates

Rougier H, Desrumaux A, Bouchon N, Wroblewski I, Pin I, Nugues F, Mornet E, Baujat G (2018) Enzyme-replacement therapy in perinatal hypophosphatasia: case report and review of the literature. Arch Pediatr 25(7):442–447

Royce PM, Blumberg A, Zurbrugg RP, Zimmermann A, Colombo JP, Steinmann B (1988) Lethal osteogenesis imperfecta: abnormal collagen metabolism and biochemical characteristics of hypophosphatasia. Eur J Pediatr 147(6):626–631

Rush ET (2018) Childhood hypophosphatasia: to treat or not to treat. Orphanet J Rare Dis 13(1):116

Rutsch F, Vaingankar S, Johnson K, Goldfine I, Maddux B, Schauerte P, Kalhoff H, Sano K, Boisvert W, Superti-Furga A et al (2001) Pc-1 nucleoside triphosphate pyrophosphohydrolase deficiency in idiopathic infantile arterial calcification. Am J Pathol 158(2):543–554

Rutsch F, Ruf N, Vaingankar S, Toliat MR, Suk A, Hohne W, Schauer G, Lehmann M, Roscioli T, Schnabel D et al (2003) Mutations in enpp1 are associated with 'idiopathic' infantile arterial calcification. Nat Genet 34(4):379–381

Saraff V, Narayanan VK, Lawson AJ, Shaw NJ, Preece MA, Hogler W (2016) A diagnostic algorithm for children with low alkaline phosphatase activities: lessons learned from laboratory screening for hypophosphatasia. J Pediatr 172:181–186.e181

Schmidt T, Mussawy H, Rolvien T, Hawellek T, Hubert J, Rüther W, Amling M, Barvencik F (2017) Clinical, radiographic and biochemical characteristics of adult hypophosphatasia. Osteoporos Int 28(9):2653–2662

Schnabel C, Jett K, Friedman JM, Frieling I, Kruse HP, Mautner V (2013) Effect of vitamin d3 treatment on bone density in neurofibromatosis 1 patients: a retrospective clinical study. Joint Bone Spine 80(3):315–319

Sebastian-Serrano A, Engel T, de Diego-Garcia L, Olivos-Ore LA, Arribas-Blazquez M, Martinez-Frailes C, Perez-Diaz C, Millan JL, Artalejo AR, Miras-Portugal MT et al (2016) Neurodevelopmental alterations and seizures developed by mouse model of infantile hypophosphatasia are associated with purinergic signalling deregulation. Hum Mol Genet 25(19):4143–4156

Seshia S, Derbyshire G, Haworth J, Hoogstraten J (1990) Myopathy with hypophosphatasia. Arch Dis Child 65(1):130–131

Shao J, Engle M, Xie Q, Schmidt R, Narisawa S, Millan J, Alpers D (2000) Effect of tissue non-specific alkaline phosphatase in maintenance of structure of murine colon and stomach. Microsc Res Tech 51(2):121–128

Shapiro JR, Lewiecki EM (2017) Hypophosphatasia in adults: clinical assessment and treatment considerations. J Bone Miner Res 32:1977–1980

Shimada Y, Yoshiya I, Tanaka K, Yamazaki T, Kumon K (1979) Crying vital capacity and maximal inspiratory pressure as clinical indicators of readiness for weaning of infants less than a year of age. Anesthesiology 51(5):456–459

Siller AF, Whyte MP (2018) Alkaline phosphatase: discovery and naming of our favorite enzyme. J Bone Miner Res 33(2):362–364

Skoultchi AI, Nandi A, Roginski RS, Gregg RG, Smithies O (1987) Expression of genes inserted at the human beta-globin locus by homologous recombination. Prog Clin Biol Res 251:581–594

Sofronescu AG, Ross M, Rush E, Goldner W (2018) Spurious testosterone laboratory results in a patient taking synthetic alkaline phosphatase (asfotase alfa). Clin Biochem 58:118–121

Staines KA, MacRae VE, Farquharson C (2012) The importance of the sibling family of proteins on skeletal mineralisation and bone remodelling. J Endocrinol 214(3):241–255

Street SE, Kramer NJ, Walsh PL, Taylor-Blake B, Yadav MC, King IF, Vihko P, Wightman RM, Millan JL, Zylka MJ (2013) Tissue-nonspecific alkaline phosphatase acts redundantly with pap and nt5e to generate adenosine in the dorsal spinal cord. J Neurosci 33(27):11314–11322

Sugano H, Matsumoto T, Miyake K, Watanabe A, Iijima O, Migita M, Narisawa S, Millan JL, Fukunaga Y, Shimada T (2012) Successful gene therapy in utero for lethal murine hypophosphatasia. Hum Gene Ther 23(4):399–406

Sutton RA, Mumm S, Coburn SP, Ericson KL, Whyte MP (2012) "Atypical femoral fractures" during bisphosphonate exposure in adult hypophosphatasia. J Bone Miner Res 27(5):987–994

Terkeltaub R (2001) Inorganic pyrophosphate generation and disposition in pathophysiology. Am J Phys Cell Physiol 281(1):C1–C11

Thakker RV, Whyte MP, Eisman J, Igarashi T (2017) Genetics of bone biology and skeletal disease. Academic, San Diego

Thomas KR, Capecchi MR (1987) Site-directed mutagenesis by gene targeting in mouse embryo-derived stem cells. Cell 51(3):503–512

Tomazos IC, Moseley S, L'Italien G, Da Silva HG, Phillips D (2017) Improvements in the 6-minute walk test and correlation with quality-of-life measures in children and adults with hypophosphatasia treated with asfotase alfa. Paper presented at: ENDO 2017. Orlando

Ucakturk SA, Elmaogullari S, Unal S, Gonulal D, Mengen E (2018) Enzyme replacement therapy in hypophosphatasia. J Coll Physicians Surg Pak 28(9):S198–S200

Unger S, Mornet E, Mundlos S, Blaser S, Cole DE (2002) Severe cleidocranial dysplasia can mimic hypophosphatasia. Eur J Pediatr 161(11):623–626

Vallet M, Ralston SH (2016) Biology and treatment of paget's disease of bone. J Cell Biochem 117(2):289–299

van den Bos T, Handoko G, Niehof A, Ryan LM, Coburn SP, Whyte MP, Beertsen W (2005) Cementum and dentin in hypophosphatasia. J Dent Res 84(11):1021–1025

Vidmar AP, Ng C, Ganster A, Pitukcheewanont P (2017) Asfotase alfa treatment of an african-american infant with perinatal hypophosphatasia and homozygous hemoglobin sc disease. IBMS BoneKey 14

Wang W, Nyman JS, Ono K, Stevenson DA, Yang X, Elefteriou F (2011) Mice lacking nf1 in osteochondroprogenitor cells display skeletal dysplasia similar to patients with neurofibromatosis type i. Hum Mol Genet 20(20):3910–3924

Waymire KG, Mahuren JD, Jaje JM, Guilarte TR, Coburn SP, MacGregor GR (1995) Mice lacking tissue nonspecific alkaline phosphatase die from seizures due to defective metabolism of vitamin b-6. Nat Genet 11(1):45–51

Weber TJ, Sawyer EK, Moseley S, Odrljin T, Kishnani PS (2016) Burden of disease in adult patients with hypophosphatasia: results from two patient-reported surveys. Metabolism 65(10):1522–1530

Weiss MJ, Cole DE, Ray K, Whyte MP, Lafferty MA, Mulivor RA, Harris H (1988) A missense mutation in the human liver/bone/kidney alkaline phosphatase gene causing a lethal form of hypophosphatasia. Proc Natl Acad Sci U S A 85(20):7666–7669

Weiss MJ, Cole DE, Ray K, Whyte MP, Lafferty MA, Mulivor R, Harris H (1989a) First identification of a gene defect for hypophosphatasia: evidence that alkaline phosphatase acts in skeletal mineralization. Connect Tissue Res 21(1–4):99–104. discussion 104–106

Weiss MJ, Ray K, Fallon MD, Whyte MP, Fedde KN, Lafferty MA, Mulivor RA, Harris H (1989b) Analysis of liver/bone/kidney alkaline phosphatase mrna, DNA, and enzymatic activity in cultured skin fibroblasts from 14 unrelated patients with severe hypophosphatasia. Am J Hum Genet 44(5):686–694

Wenkert D, McAlister WH, Coburn SP, Zerega JA, Ryan LM, Ericson KL, Hersh JH, Mumm S, Whyte MP (2011) Hypophosphatasia: nonlethal disease despite skeletal presentation in utero (17 new cases and literature review). J Bone Miner Res 26(10): 2389–2398

Wennberg C, Hessle L, Lundberg P, Mauro S, Narisawa S, Lerner UH, Millan JL (2000) Functional characterization of osteoblasts and osteoclasts from alkaline phosphatase knockout mice. J Bone Miner Res 15(10):1879–1888

Whyte MP (2012) Hypophosphatasia. In: Glorieux FH, Pettifor JM, Juppner H (eds) Pediatric bone: biology & diseases, 2nd edn. Academic, London, pp 771–794

Whyte MP (2016) Hypophosphatasia – aetiology, nosology, pathogenesis, diagnosis and treatment. Nat Rev Endocrinol 12(4):233

Whyte MP (2017) Hypophosphatasia: enzyme replacement therapy brings new opportunities and new challenges. J Bone Miner Res 32(4):667–675

Whyte MP, Valdes R Jr, Ryan LM, McAlister WH (1982) Infantile hypophosphatasia: enzyme replacement therapy by intravenous infusion of alkaline phosphatase-rich plasma from patients with paget bone disease. J Pediatr 101(3):379–386

Whyte MP, McAlister WH, Patton LS, Magill HL, Fallon MD, Lorentz WB Jr, Herrod HG (1984) Enzyme replacement therapy for infantile hypophosphatasia attempted by intravenous infusions of alkaline phosphatase-rich paget plasma: results in three additional patients. J Pediatr 105(6):926–933

Whyte MP, Magill HL, Fallon MD, Herrod HG (1986) Infantile hypophosphatasia: normalization of circulating bone alkaline phosphatase activity followed by skeletal remineralization. Evidence for an intact structural gene for tissue nonspecific alkaline phosphatase. J Pediatr 108(1):82–88

Whyte MP, Kurtzberg J, McAlister WH, Mumm S, Podgornik MN, Coburn SP, Ryan LM, Miller CR, Gottesman GS, Smith AK et al (2003) Marrow cell transplantation for infantile hypophosphatasia. J Bone Miner Res 18(4):624–636

Whyte MP, Mumm S, Deal C (2007) Adult hypophosphatasia treated with teriparatide. J Clin Endocrinol Metab 92(4):1203–1208

Whyte MP, Wenkert D, McAlister WH, Mughal MZ, Freemont AJ, Whitehouse R, Baildam EM, Coburn SP, Ryan LM, Mumm S (2009) Chronic recurrent multifocal osteomyelitis mimicked in childhood hypophosphatasia. J Bone Miner Res 24(8):1493–1505

Whyte MP, Greenberg CR, Salman NJ, Bober MB, McAlister WH, Wenkert D, Van Sickle BJ, Simmons JH, Edgar TS, Bauer ML (2012) Enzyme-replacement therapy in life-threatening hypophosphatasia. N Engl J Med 366(10):904–913

Whyte MP, Zhang F, Wenkert D, McAlister WH, Mack KE, Benigno MC, Coburn SP, Wagy S, Griffin DM, Ericson KL (2015a) Hypophosphatasia: validation and expansion of the clinical nosology for children from 25years experience with 173 pediatric patients. Bone 75:229–239

Whyte MP, Zhang F, Wenkert D, McAlister WH, Mack KE, Benigno MC, Coburn SP, Wagy S, Griffin DM, Ericson KL et al (2015b) Hypophosphatasia: validation and expansion of the clinical nosology for children from 25 years experience with 173 pediatric patients. Bone 75:229–239

Whyte MP, Madson KL, Phillips D, Reeves AL, McAlister WH, Yakimoski A, Mack KE, Hamilton K, Kagan K, Fujita KP (2016a) Asfotase alfa therapy for children with hypophosphatasia. JCI Insight 1(9):e85971

Whyte MP, Rockman-Greenberg C, Ozono K, Riese R, Moseley S, Melian A, Thompson DD, Bishop N, Hofmann C (2016b) Asfotase alfa treatment improves survival for perinatal and infantile hypophosphatasia. J Clin Endocrinol Metab 101(1):334–342

Whyte MP, Coburn SP, Ryan LM, Ericson KL, Zhang F (2018) Hypophosphatasia: biochemical hallmarks validate the expanded pediatric clinical nosology. Bone 110:96–106

Williams DK, Pinzon C, Huggins S, Pryor JH, Falck A, Herman F, Oldeschulte J, Chavez MB, Foster BL, White SH et al (2018) Genetic engineering a large animal model of human hypophosphatasia in sheep. Sci Rep 8(1):16945

Wyckoff MH, El-Turk C, Laptook A, Timmons C, Gannon FH, Zhang X, Mumm S, Whyte MP (2005) Neonatal lethal osteochondrodysplasia with low serum levels of alkaline phosphatase and osteocalcin. J Clin Endocrinol Metab 90(2):1233–1240

Yadav MC, Lemire I, Leonard P, Boileau G, Blond L, Beliveau M, Cory E, Sah RL, Whyte MP, Crine P et al (2011) Dose response of bone-targeted enzyme replacement for murine hypophosphatasia. Bone 49(2):250–256

Yadav MC, de Oliveira RC, Foster BL, Fong H, Cory E, Narisawa S, Sah RL, Somerman M, Whyte MP, Millan JL (2012) Enzyme replacement prevents enamel defects in hypophosphatasia mice. J Bone Miner Res 27(8):1722–1734

Yamamoto S, Orimo H, Matsumoto T, Iijima O, Narisawa S, Maeda T, Millan JL, Shimada T (2011) Prolonged survival and phenotypic correction of akp2(−/−) hypophosphatasia mice by lentiviral gene therapy. J Bone Miner Res 26(1):135–142

Zankl A, Mornet E, Wong S (2008) Specific ultrasonographic features of perinatal lethal hypophosphatasia. Am J Med Genet A 146A(9):1200–1204

Zweifler LE, Patel MK, Nociti FH Jr, Wimer HF, Millan JL, Somerman MJ, Foster BL (2015) Counter-regulatory phosphatases tnap and npp1 temporally regulate tooth root cementogenesis. Int J Oral Sci 7(1):27–41

Rational Use of Pancreatic Enzymes for Pancreatic Insufficiency and Pancreatic Pain

14

Gyanprakash A. Ketwaroo and David Y. Graham

Abstract

We describe the rational use of enteric coated and unprotected replacement pancreatic enzymes for treatment of malabsorption due to pancreatic insufficiency and for pancreatic pain. Enteric coated formulations mix poorly with food allowing separation of enzymes and nutrients when emptying from the stomach. The site of dissolution of the enteric coating in the intestine is also unpredictable and enzymes may not be released until the distal intestine. Together, these barriers result in the lack of dose-response such that the strategy of increasing the dosage following a suboptimal effect is often ineffective. The ability to maintain the intragastric pH ≥4 with the combination of proton pump inhibitors and antacids suggests that it should be possible to reliably obtain a good response with uncoated enzymes. We also discuss the recognition, treatment and prevention of nutritional deficiencies associated with pancreatic insufficiency and recommend a test and treat strategy to identify and resolve nutritional deficits. Finally, we focus on mechanisms causing pain that may be amenable to therapy with pancreatic enzymes. Pain due to malabsorbed digestive contents can be prevented by successful therapy of malabsorption. Feedback inhibition of endogenous pancreatic secretion can prevent pain associated with pancreatic secretion but requires use of non-enteric coated formulations.

Keywords

Pancreatic enzymes · Pancreatic insufficiency · Chronic pancreatitis

Abbreviations

CCK	Cholecystokinin
CFA	Coefficient of fat absorption
FDA	Food and Drug Administration
INR	International normalized ratio
RBP	Retinol binding protein

Author Contributions: Each of the authors have been involved equally in study design, data acquisition, analysis, interpretation and drafting the article. All have read and approved the final manuscript. Each meets the criteria for authorship established by the International Committee of Medical Journal Editors and verify the validity of the results reported.

G. A. Ketwaroo · D. Y. Graham (✉)
Department of Medicine, Michael E. DeBakey VA Medical Center, Baylor College of Medicine, Houston, TX, USA
e-mail: dgraham@bcm.edu

N. Labrou (ed.), *Therapeutic Enzymes: Function and Clinical Implications*, Advances in Experimental Medicine and Biology 1148, https://doi.org/10.1007/978-981-13-7709-9_14

14.1 Introduction

Most food is ingested in the form of macromolecules that can only be absorbed after being reduced to smaller molecules. The pancreas is the primary source of enzymes involved in the digestion of carbohydrates, proteins, and fats. In addition to supplying enzymes, the pancreas produces bicarbonate to neutralize the gastric acid and provide the proper milieu for the enzymes to function. The absorptive and digestive capacity of the intestinal tract is large and has great redundancy such that the majority of the small intestine must be bypassed for successful bariatric surgery.

Lipid digestion and absorption is the most complicated requiring four distinct steps. The process begins by synthesis of lipases by pancreatic acinar cells which are then secreted through the pancreatic ducts into the duodenum in response to food entering the duodenum. The gastric contents entering the duodenum are acidic and the acidity must be neutralized by secretion of duodenal and pancreatic bicarbonate in order for the enzymes and bile acids to function properly. Lipase is irreversible inactivated if the pH falls to pH 4 or below. When there is insufficient duodenal pancreatic enzyme activity, exocrine pancreatic insufficiency occurs (Table 14.1). This condition can occur due to causes directly related to the pancreas such as loss of pancreatic acini, blockage of the pancreatic ducts preventing secretion of enzymes, or acidic duodenal contents which inactivate pancreatic lipase. Other causes are related to failure to stimulate pancreatic secretion and inability of the secreted enzymes to properly mix with duodenal contents (Table 14.1) (Singh et al. 2017).

Table 14.1 Causes of exocrine pancreatic insufficiency (insufficiency (insufficient intraluminal pancreatic enzyme activity)

Loss of pancreatic acini
Pancreatic inflammation, pancreatic resection, cystic fibrosis, pancreatic malignancy, cystic fibrosis,
Inability of secreted enzymes to enter the duodenum
Pancreatic ductular obstruction (fibrosis, stricture, stones, malignancy)
Altered anatomy (e.g., Roux-en-y gastric bypass
Acidic duodenal pH
Zollinger Ellison syndrome, defective pancreatic bicarbonate secretion
Insufficient stimulation of enzyme secretion (celiac disease)

14.2 Exocrine Pancreatic Insufficiency

14.2.1 Diagnosis

The diagnosis of pancreatic insufficiency requiring adjuvant enzyme replacement is typically based on clinical suspicion followed by laboratory confirmation or by confirmation of improvement of weight and nutritional deficiencies following enzyme replacement therapy. Fat malabsorption (steatorrhea) clinically presents as weight loss with large, foul smelling, pale, pasty stools. The stools may appear greasy and an oily sheen reflecting undigested triglycerides may be visible on the water in the toilet bowl. The presence of watery diarrhea and floating stools are often mentioned by students as important diagnostic features, but watery diarrhea is an uncommon presentation and stools float because of trapped air rather than the presence of fat in stools.

Pancreatic insufficiency can be confirmed by pancreatic function testing directly via the secretin pancreatic function test where duodenal juice is collected endoscopically or using a special "Dreiling" tube. The pancreatic fluid bicarbonate concentration is then measured, with normal being >80 mEq/L (Diamond et al. 1940; Dreiling and Hollander 1948; Ketwaroo et al. 2013; Pelley et al. 2012). This approach is highly sensitive and is able to stratify pancreatic dysfunction as mild, moderate or severe. However, the test is invasive, expensive and labor intensive (Diamond et al. 1940). Non-invasive tests are available and the gold standard non-invasive test is quantification of fat malabsorption by measuring 72-h fecal fat excretion. An abnormal result is excretion of more than 7% of ingested fat and is best expressed as a coefficient of fat absorption (e.g., >7 g while receiving a 100 g fat diet). Fecal fat measurement is often not offered because it requires collecting and handling of stools. This problem continues despite improved methodology that obviate the need for homogenization of stools, such as measuring fat content using near infrared spectrometry (Benini et al. 1989). Alternate indirect methods of assessing pancreatic exocrine function include measuring fecal concentrations of pancreatic enzymes, such as elastase 1 or chymotrypsin. Where available, breath tests are preferred. This approach assesses fat absorption directly follow administration of labeled triglycerides such as the carbon 13 triglyceride breath test (Afghani et al. 2014; Dominguez-Munoz et al. 2007). The most widely available test is measuring fecal elastase I.

14.2.2 Fecal Elastase 1

Elastase 1 is an enzyme produced by the pancreas. It is resistant to digestion and passes largely intact through the intestinal tract where its concentration is measured in the stool. The most common test format is as an enzyme-linked immunosorbent assay which uses a monoclonal antibody specific to human elastase I (Stein et al. 1996; Struyvenberg et al. 2017). The test result is therefore not influenced by the presence of the antigenically distinct exogenous porcine pancreatic enzymes so that enzyme therapy need not be withheld. The main caveat regarding interpretation is that the test is only accurate when done using formed stools (Struyvenberg et al. 2017). The cut-off value for a normal result is >200 μg/g feces. Values between 100 and 200 μg/g feces are considered indeterminate and values below 100 μg/g feces are highly suggestive of pancreatic insufficiency. However, as with any test, interpretation depends on the pretest probability and, in our experience, fecal elastase I testing is often ordered in the evaluation of patients with diarrhea where false positive test results are common. A recent review and meta-analysis of the role of fecal elastase testing in the diagnosis of exocrine pancreatic insufficiency concluded that a normal value was highly indicative of absence of pancreatic insufficiency (pooled sensitivity of 0.96 and specificity of 0.88). The false negative rate was 1.1% and the false positive rate, 11%. It followed that in high pretest probability conditions, only about 10% of cases of chronic exocrine insufficiency would be false negatives (Vanga et al. 2018).

14.2.3 Treatment of Exocrine Insufficiency

Pancreatic replacement enzymes have been available clinically since at least the late 1800s (Engesser 1879) and the most common source of pancreatic enzymes remains desiccated hog pancreas. Bovine pancreas preparations are also available but are used much less frequently and microbial lipases are just beginning to be used (Heubi et al. 2016; Lowe and Whitcomb 2015). It is expected that use of microbial derived enzymes will likely grow. Commercial products are described clinically in terms of lipase content (e.g., 20,000 USP lipase units).

Although it seems obvious that replacing the missing enzymes should be a successful strategy there are many myths (Table 14.2) and numerous impediments preventing normalization of digestion and correction of pancreatic exocrine insufficiency. Long ago, our forefathers discovered that simply feeding pancreatic enzymes did not reliably produce the desired effect and that gastric acid rapidly inactivated ingested pancreatic lipase (Chase 1905). Various methods have been attempted to overcome this acid barrier including administration of enzymes with antacids with or without anti-secretory agents and protecting the enzymes with enteric coating. As discussed below, none has proved reliably successful. Here, we discuss the limitations of replacement therapy as well as the weaknesses and misconceptions related to current practices. Current guidelines

Table 14.2 Myths related to treatment of exocrine pancreatic insufficiency

60,000 units of lipase should be administered with each meal
Increasing the dose (i.e., number of capsules or lipase units) reliably increases the effect
Increasing the enzyme dosage is a safe and effective strategy
Non-enteric coated preparations are almost always ineffective and should not be used
Proton pump co-therapy (e.g., 20 mg omeprazole) reliably improves treatment outcome
Enteric coated preparations are useful for treatment of pancreatic pain
The most reliably way to confirm therapy is effective is by symptom response

often seem to represent urban myths rather than recommendations based on scientifically sound principles. The treatment outcome is assessed based on the ability to normalize absorption of fats, which requires coordination of lipid hydrolysis, solubilization of the digestive products by bile, and absorption by the small intestine.

14.3 Pancreatic Enzymes

14.3.1 Dosing of Pancreatic Enzymes

The FDA approved package insert for a typical commercial product (e.g., Creon®) states that "the initial starting dose [of pancreatic enzymes] and increases in the dose per meal should be individualized based on clinical symptoms, the degree of steatorrhea present, and the fat content of the diet" (https://www.accessdata.fda.gov/drugsatfda_docs/label/2009/020725s000lbl.pdf). They refer to a clinical trial where patients received 72,000 lipase units per meal while consuming at least 100 g of fat per day and cite the Cystic Fibrosis Foundation Consensus Conferences Guidelines of 500 lipase units/kg of body weight per meal as the lowest starting dose (Stallings et al. 2008). They further note that "there is great inter-individual variation in response to enzymes; thus, a range of doses is recommended" and that "if doses exceed 2,500 lipase units/kg of body weight per meal, further investigation is warranted". Doses greater than 2500 lipase units/kg of body weight per meal (or greater than 10,000 lipase units/kg of body weight per day) should be used with caution" and that "patients currently receiving higher doses than 6000 lipase units/kg of body weight per meal should be examined and the dosage either immediately decreased or titrated downward to a lower range".

In contrast, Forsmark in Sleisenger and Fordtran's Gastrointestinal and Liver disease textbook suggests that 90,000 USP units of lipase are needed with each meal (Forsmark 2016). Broad recommendations such as therapy being individualized based on clinical symptoms, the degree of steatorrhea present, and the fat content of the diet do not identify which of the variables is best or whether all three are equivalent. Because clinicians rarely have access to the patients' degree of steatorrhea, the recommendation forces clinicians to rely on symptoms and fat content of the diet. However, no guidance is provided to advise the patient what characteristics they should use to judge the fat content of the diet or what adjustments they should make. In actual practice, following the advice of the package insert or the textbook will not reliably achieve the goal of resolving malabsorption or the nutritional consequences of pancreatic insufficiency.

Here, we attempt to provide a practical approach to assist patients and clinicians. First, we address the evidence regarding the quantity of lipase required to correct steatorrhea. Recommendations are given in terms of amount of lipase but this is confusing as lipase is described in different units in different countries. In the United States, FDA-approved products are described in USP lipase units (1 IU = 3 USP units). We will describe results of different studies in USP units. Current FDA-approved pancreatic enzyme products range from 3000 USP lipase units to 36,000 USP units per pill (Table 14.3). Outside of the United States a wide variety of preparations are available (Ianiro et al. 2016)

Under normal physiologic circumstances post prandial lipase secretion has been estimated at 9000–18,000 USP units/min (Keller et al. 1997; Keller and Layer 2005) totalling between 120,000 and 2,196,000 USP units in the 3 h post prandial period (DiMagno et al. 1977). Based on intubation studies in humans it has also been suggested that only 5–10% of normal pancreatic output is required for normal fat absorption (DiMagno et al. 1973; Kalser et al. 1968; Regan et al. 1979).

For steatorrhea to be abolished following oral administration of pancreatic enzymes requirements include that (a) the enzyme remain active and (b) mix and (c) empty with the meal which is (d) coordinated with the entry of bile into the duodenum and normal small intestinal motility and absorptive function. Studies have shown that administration of approximately 30,000 USP lipase units/meal of unprotected pancreatic enzymes can eliminate steatorrhea in those with absent or low acid secretion (Fig. 14.1) (Graham

Table 14.3 FDA approved pancreatic enzyme preparations

Drug	Lipase (USP units)	Preparation	Diameter (mm)
CREON®			
Creon 3000	3000	Capsule with enteric coated minimicrospheres	0.71–1.6
Creon 6000	6000	Capsule with enteric coated minimicrospheres	0.71–1.6
Creon 12000	12,000	Capsule with enteric coated minimicrospheres	0.71–1.6
Creon 24000	24,000	Capsule with enteric coated minimicrospheres	0.71–1.6
Creon 36000	36,000	Capsule with enteric coated minimicrospheres	0.71–1.6
Pancreaze®			
Pancreaze 4200	4200	Capsule with enteric coated microtablets	2
Pancreaze 10500	10,500	Capsule with enteric coated microtablets	2
Pancreaze 16800	16,800	Capsule with enteric coated microtablets	2
Pancreaze 21000	21,000	Capsule with enteric coated microtablets	2
Zenpep®			
Zenpep 3000	3000	Capsule with enteric coated beads	1.8–1.9
Zenpep 5000	5000	Capsule with enteric coated beads	1.8–1.9
Zenpep 10000	10,000	Capsule with enteric coated beads	2.2–2.5
Zenpep 15000	15,000	Capsule with enteric coated beads	2.2–2.5
Zenpep 20000	20,000	Capsule with enteric coated beads	2.2–2.5
Zenpep 25000	25,000	Capsule with enteric coated beads	2.2–2.5
Ultresa®			
Ultresa 13800	13,800	Capsule with enteric coated minitablet	2
Ultresa 20700	20,700	Capsule with enteric coated minitablet	2
Ultresa 23000	23,000	Capsule with enteric coated minitablet	2
Pertyze®			
Pertyze 8000	8000	Capsule with bicarbonate buffered enteric coated microsphere	0.8–2.2
Pertyze 16000	16,000	Capsule with bicarbonate buffered enteric coated microsphere	0.8–2.2
Viokace®			
Viokace 10440	10,440	Non-enteric coated	
Viokace 20800	20,880	Non-enteric coated	

(continued)

Table 14.3 (continued)

pH at or above which enzyme is designed to release most of the enzyme based on the package insert

1977). The duration of the postprandial gastric pH ~4 and the average duodenal pH was also shown to correlate with the percentage reduction in steatorrhea (i.e., the longer the gastric pH remained ~4, the higher the average duodenal pH, and the more reduction in steatorrhea achieved). Another study also showed complete resolution of steatorrhea in two of six patients with 18,000 USP lipase units/meal of enteric-coated microspheres given throughout the meal (Fig. 14.2) (Graham 1979). One can therefore conclude that in adults with pancreatic steatorrhea between 18,000 and 30,000 USP units of lipase per meal is sufficient to eliminate steatorrhea. The difficulty is how to "deliver a sufficient amount of active lipase at the right place, i.e., duodenum and proximal jejunum, and at the right time, i.e., in parallel with gastric emptying of nutrients," (Table 14.3) (Trang et al. 2014).

14.3.2 Barriers to Delivery of Sufficient Active Lipase to the Duodenum and Proximal Jejunum in Parallel with Gastric Emptying of Nutrients

There are remarkably few data available showing how to reliably achieve resolution of malabsorption with orally administered pancreatic enzymes. The normal digestive process provides integration of gastric emptying with pancreatobiliary

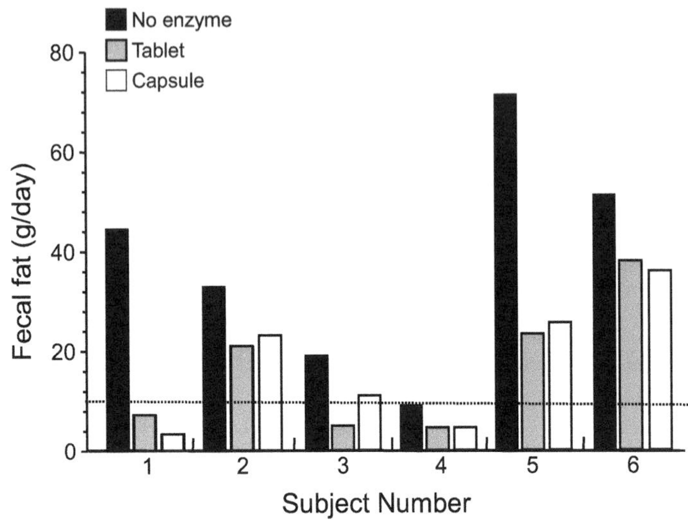

Fig. 14.1 Comparison of baseline and therapy with enzymes formulated as tablets or capsules in adults with exocrine pancreatic insufficiency. Approximately 30,000 USP units of lipase were given with meals. Correction of steatorrhea correlated with time the gastric pH was 4 or greater. (Adapted from Graham 1977. Copyright © 1977 Massachusetts Medical Society)

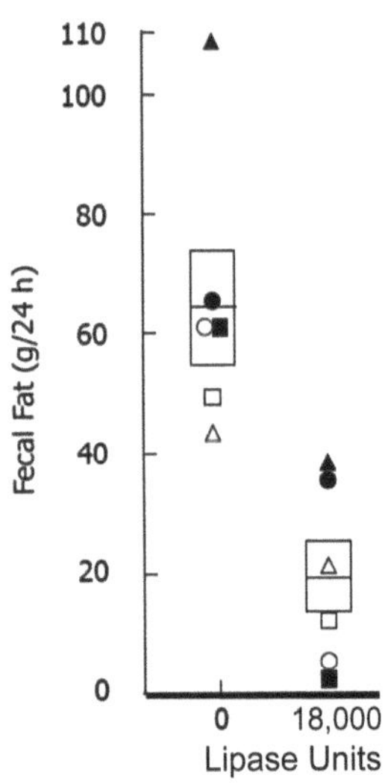

Fig. 14.2 Effect of increasing the enzyme dosage on fecal fat excretion while receiving a 100 g fat diet. Enzymes were given three times per day with meals providing 18,000 USP lipase units as enteric coated microspheres (i.e., three microsphere capsules with each meal). Each rectangle encloses the mean ± the standard deviation of the mean. The normal fecal fat is <6 g/24 h. (Adapted from Graham 1979)

secretion to provide ideal conditions in terms of pH and enzyme concentration to promote digestion and absorption of the nutritional elements despite marked differences in the composition and quantity of meals.

When pancreatic enzymes are administered orally they can either mix with or separate from the meal. They can also either survive or be destroyed by acid-pepsin. By 1905 it was noted that pancreatin "was rendered inert" by gastric juice (Chase 1905) and that enteric coating of the enzymes either failed to protect the enzymes or failed to dissolve rapidly enough "to allow the pancreatin to be of any service in digestion" (Chase 1905). The stomach is only one of the barriers to successful therapy as altered gastrointestinal motility and reduced pancreatic bicarbonate secretion also result in unpredictable destruction, transit or dissolution of administered enzymes (DiMagno et al. 1977; Layer et al. 1986). One common strategy has been to administer large quantities of pancreatic enzymes in an attempt to overpower the gastric barrier. Experience has shown that this rarely restores normal fat absorption (Beazell et al. 1941; DiMagno et al. 1977; Harris et al. 1955; Jordan and Grossman 1959; Littman and Hanscom 1969). The option of using enteric coating to protect the enzymes has also had limited success as it has been plagued both by separation of the enteric coated enzymes from the meal and the fact that the proximal intestine often remains acidic which delays dissolution of the coating and release of the enzymes to the distal small intestine and colon (Aloulou et al. 2008; Delchier et al. 1991).

14.3.3 Gastric Emptying Barrier

Gastric emptying is normally highly regulated by receptors in the duodenum that respond to the pH, osmolarity, and nutrient content of the contents entering from the stomach (Hunt 1983; Hunt and Knox 1968; Smith et al. 1984). The stomach acts as a reservoir which acidifies, grinds, and sieves the gastric contents such that small particles (e.g., <1 mm) suspended in liquid are the major form of the meal that exits into the duodenum (Meyer 1980; Meyer et al. 1988; Meyer and Lake 1997). The addition of nutrients to the stomach also results in robust acid secretion such that the pH is typically above four for only a short period after eating. The lack of pancreatic bicarbonate and enzymes results in more rapid emptying and inability of the duodenum to control the pH and maintain the ideal milieu for digestion (DiMagno et al. 1977; Layer et al. 1986).

14.3.4 Overcoming the pH Barrier

As noted previously, lipase is irreversibly inactivated at pH 4 or below. The gastric pH barrier often extends into the duodenum. Attempts to overcome the pH barrier include enteric coating of enzymes and/or the use of antacids or antisecretory drugs to increase the intragastric/duodenal pH.

14.3.5 Coating of Enzymes

Most pancreatic enzyme preparations available in the United States are packaged as enteric coated microspheres. The only exception is Viokace®. Pertyze® is enteric coated but also contains a small amount of bicarbonate in the outer layer. The amount of bicarbonate present is too small to be functionally important. While the enteric coated enzyme products are available in different dosages (Table 14.3), the amount of lipase is increased by packaging identical microspheres in larger capsules which containing more beads.

In 2004 the FDA mandated that all pancreatic enzymes be reformulated to meet new specifications including minimum and maximum amounts of enzyme and dissolution characteristics under defined conditions (Trang et al. 2014). The regulation was prompted by the wide variability of products including generic enteric coated products that often failed to protect the enzymes in transit through the stomach (Kuhn et al. 2007). The bar for clinical approval was very low as they only had to prove to be superior to placebo (Trang et al. 2014). The outcome was a reduced number of products and a large increase in price. Most FDA- mandated post-approval studies to better understand why the results were relatively poor have been completed but the results have not been revealed (Trang et al. 2014). Since 2010 only the newly approved products are available in the U.S., although over-the-counter products remain available at health food stores. These are typically not enteric coated and lipase activity is measured in different units such that interpretation required translation (Table 14.4.) (Scharpé et al. 1997).

Currently available enteric coated microbead enzymes are effective in protecting the acid-sensitive lipases from inactivation in the stomach and have proven more effective than placebo in reducing steatorrhea (Trang et al. 2014). However, they frequently fail to entirely correct malabsorption. Importantly, the strategy of administration of more microbeads (i.e., increasing the dosage) generally fails to provide a further reduction in steatorrhea (i.e., there is absence of a dose response) (da la Iglesia-García et al. 2017; Trang et al. 2014). The lack of a dose

Table 14.4 Conversion of relative potency of enzymes based on different units of measurement

Amylase: 1 Ph.Eur. Unit = 1 BP Unit = 1 FIP Unit ~ 4.15 USP Units
Lipase: 1 Ph.Eur. Unit = 1 BP Unit = 1 FIP Unit ~ 1 USP Unit = 1/3 IU
Protease: 1 Ph.Eur. Unit = 1 BP Unit = 1 FIP Unit ~ 62.5 USP Units

Ph.Eur European Pharmacopoeia, *BP* British Pharmacopoeia, *FIP* International Pharmaceutical Federation, *USP* United States Pharmacopoeia, *IU* International Units

response prevents dose escalation as an effective treatment strategy to achieve the desired clinical response (Trang et al. 2014). Overall, many patients do well by relying only on enteric coated microbeads despite only partial relief of steatorrhea but a proportion continues to experience nutritional deficiencies (da la Iglesia-García et al. 2017; Dominguez-Munoz et al. 2007; Lindkvist et al. 2015; Trang et al. 2014).

The lack of dose-response and the relatively poor treatment response is often related to the fact that enteric coated microspheres rapidly separate from bulk food/nutrients. They thus are neither uniformly distributed within the meal nor reliably emptied along with the nutrients (Trang et al.). This results in dietary fat being emptied into the duodenum without the accompanying lipase needed for lipid digestion. The enteric coating used is slow to dissolve even in highly buffered alkaline media in vitro (Trang et al.). Impaired bicarbonate secretion in the duodenum of patients with pancreatic insufficiency produces an acid milieu such that the microbeads may not dissolve and release the contents until in the distal jejunum, ileum, or colon (DiMagno et al. 1977; Layer et al. 1986; Trang et al. 2014). Attempts have been made to compensate for this incoordination by giving some enzymes immediately before, throughout, or after the meal (da la Iglesia-García et al. 2017; Dominguez-Munoz et al. 2005; Trang et al. 2014). The effects of this strategy have been studied in a number of FDA mandated studies. The fact that as of April 2018 the results have not been published or reported suggests that the issues with incoordination of the process have not been solved (e.g., Pancrease https://clinicaltrials.gov/ct2/show/study/NCT00676702?term=NCT00676702rank=1 –Completed with 13 participants but no results posted (NCT00676702). Pancrecarb https://clinicaltrials.gov/ct2/show/NCT00744250?term=NCT00744250rank=1 – And https://clinicaltrials.gov/ct2/show/NCT00749099?term=NCT00749099rank=1. Both terminated as no longer required by FDA, 3 enrolled. NCT00744250; 11 enrolled NCT00749099. Viokase – https://clinicaltrials.gov/ct2/show/NCT00559052?term=NCT00559052rank=1 completed with 22 participants, no results posted. NCT00559052).

Although increasing amounts of enzyme microbeads often fails to produce a meaningful reduction in steatorrhea, there is also a risk that dumping a high concentration of pancreatic enzymes or, more importantly, of the highly acidic enteric coating into the colon can result in development of colonic strictures. This is particularly a problem in children (Bakowski and Prescott 1997; Franzen et al. 2008; Gaia et al. 2001; Prescott and Bakowski 1999; Prieto et al. 2009; van Velzen et al. 1996). Colonic strictures were initially attributed to the high concentration of pancreatic enzymes but as other drugs using the same coating have caused colonic stricture the evidence suggests that the highly acid coating may actually be the agent responsible for colonic damage (Prescott and Bakowski 1999; van Ball et al. 1996).

14.3.6 Use of Adjuvant Antacids and Anti-secretory Agents

The recognition that unprotected pancreatic enzymes could be inactivated during transit through the stomach led early investigators to try antacids to prevent enzyme inactivation. The early studies used arbitrary amounts of antacids but showed that co-administration of sodium bicarbonate or aluminum hydroxide with enzymes was partially effective (Durie et al. 1980; Gow et al. 1981; Kalser et al. 1968; Kattwinkel et al. 1972; Veeger et al. 1962; Weber et al. 1976). Fordtran et al., provided a more scientific basis for effective use of antacids for healing of peptic ulcers disease based on timing and dosages of antacid administration designed to enhance and extend the buffering capacity of meals (Fordtran et al. 1973). However, the goal of antacids to heal peptic ulcers differs from what is required of antacids when used as adjuvants to protect pancreatic enzymes. The critical difference between the two objectives is the need to prevent the intragastric pH from falling to pH 4 or below while the enzymes are in the stomach.

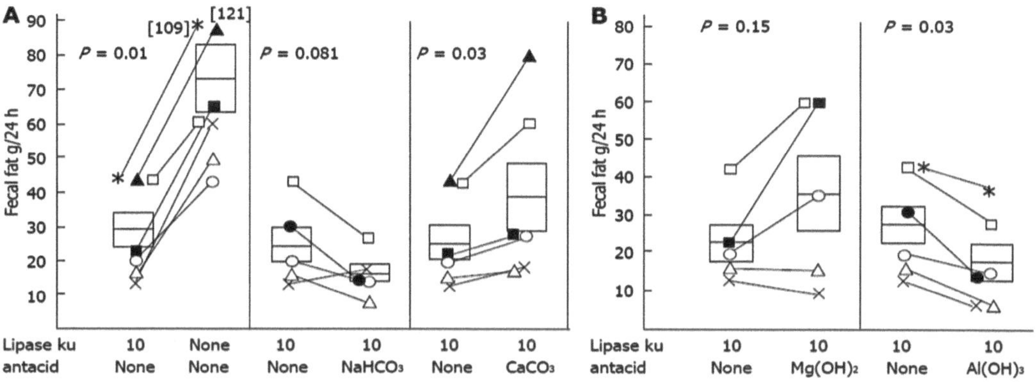

Fig. 14.3 Effect of antacids and enzymes on the effectiveness of 30,000 USP units of lipase per meal for the treatment of pancreatic steatorrhea. Each symbol represents a different patient. Box represents the mean ±SEM for the group. Number in [ ] = weight of stool. Sodium bicarbonate, magnesium aluminum hydroxide, aluminum hydroxide, or calcium carbonate were administered at the beginning and the termination of each meal. (Adapted from Graham 1982)

In a randomized study, we compared the effectiveness of sodium bicarbonate (1.3 g or 12 mEq), aluminum hydroxide (30 mL or 57 mEq), magnesium-aluminum hydroxide 30 mL or 72 mEq), or calcium carbonate (1 g or 21 mEq) administered before and immediately after each meal in improving steatorrhea in subjects receiving a low dose of lipase per meal while receiving 100 g fat/day (Graham 1982). The dose of lipase was expected to, on average, reduce steatorrhea by 50%. Those receiving adjuvant therapy with sodium bicarbonate or aluminum hydroxide experienced a reduction in steatorrhea (Fig. 14.3) (Graham 1982). Although all of the antacids lengthened the time the intragastric pH was >6 and increased duodenal pH and increased lipolysis, adjuvant therapy with calcium carbonate or magnesium-aluminum hydroxide resulted in worsening of steatorrhea and partially negated the benefits of enzyme therapy (Fig. 14.3) (Graham 1982; Graham and Sackman 1982). It was shown that the antacids did not impair lipase function and while calcium and magnesium-containing antacid therapy improved lipolysis, the released fatty acids combined with calcium or magnesium to produce calcium or magnesium soaps which were poorly absorbed (Graham 1982; Graham and Sackman 1982, 1983) introducing a new barrier to absorption.

H_2-receptor antagonists are generally incapable of maintaining the intragastric pH >4 which is required to prevent lipase inactivation (Graham 1982; Hunt et al. 1995; Jones et al. 1987). In contrast, while once daily administration of a proton pump inhibitor is able to increase the intragastric pH to ≥4, the duration is typically short (Bell et al. 1992; Graham and Tansel 2018). Studies with omeprazole and enteric coated pancreatic enzymes in cystic fibrosis patients with persistent steatorrhea despite use of enteric coated enzymes, confirmed that enzyme dose escalation failed to reduce steatorrhea whereas the strategy of increasing the enzymes along with adjuvant omeprazole was beneficial (Fig. 14.4) (Heijerman et al. 1991). Most subsequent studies with currently available enteric coated enzyme preparations have not demonstrated consistent benefits with adjuvant proton pump inhibitor therapy with the possible exception of those whose poor response was due to high gastric acid secretion (Bruno et al. 1994; Dominguez-Munoz et al. 2005; Marotta et al. 1989; Sander-Struckmeier et al. 2013).

Because of the general inability of adjuvant proton pump inhibitor therapy, as currently prescribed, to provide meaningful benefits this approach is not recommended for all patients (Dominguez-Munoz 2007). This admonition should now be reconsidered based on better understanding of how to use proton pump inhibitors to maintain the intragastric pH ≥4 (Graham and Tansel 2018).

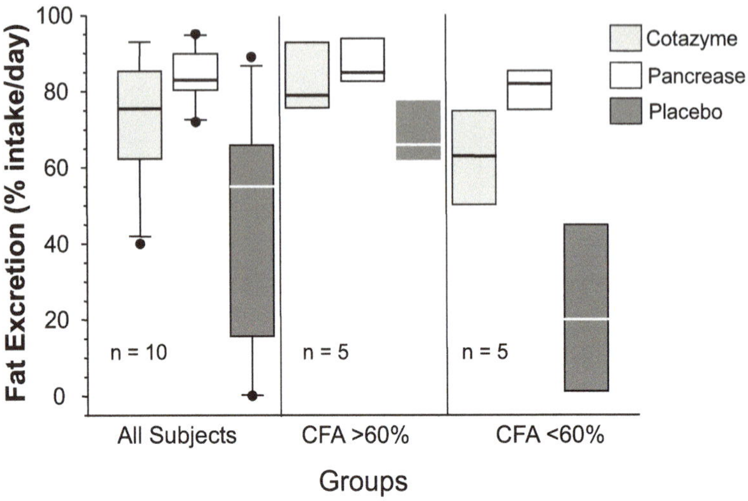

Fig. 14.4 Randomized cross-over comparison of similar amounts of lipase administered as unprotected enzyme capsules (Cotazyme®) or enteric-coated microspheres (Pancrease®) on coefficient of fat absorption (CFA) in cystic fibrosis patients with pancreatic insufficiency. Although the enteric coated preparation was better in those with the greatest degree of malabsorption (CFA <60%), neither formulation resulted in resolution of steatorrhea. (From Trang et al. 2014, with permission)

14.3.7 Combined Proton Pump Inhibitor and Antacid Adjuvant Therapy

The comparative effectiveness of different proton pump inhibitors in maintaining the intragastric pH above a desired pH (here, pH >4) for the entire 24 h day (pH4time) can be expressed in terms of omeprazole equivalents (Graham and Tansel 2018; Kirchheiner et al. 2009). Studies of pH4time are typically done after 5 days of therapy to ensure that steady state has been achieved. When different PPIs are given once daily the median pH 4 time increases linearly from approximately 30% (~7 h) following administration of about 2.5 mg omeprazole equivalents (equal to 10 mg of pantoprazole) to approximately 60% (~14 h) with about 70 mg omeprazole equivalents (equal to 40 mg of esomeprazole or rabeprazole). Most published studies of adjuvant proton pump therapy with pancreatic enzymes have used 20 mg of omeprazole once daily which produces a median pH 4 time of approximately 45% (10.8 h) (Fig. 14.5) (Graham and Tansel 2018). With 20 mg of omeprazole given twice daily the pH 4 time is approximately 70% (~17 h) and increases linearly to approximately 85% (~20 h) following administration of approximately 70 mg omeprazole equivalents twice daily. These results suggest that (a) the dose of omeprazole typically used in prior studies was insufficient to protect the pancreatic enzymes from inactivation in the stomach, (b) most of the beneficial effects would

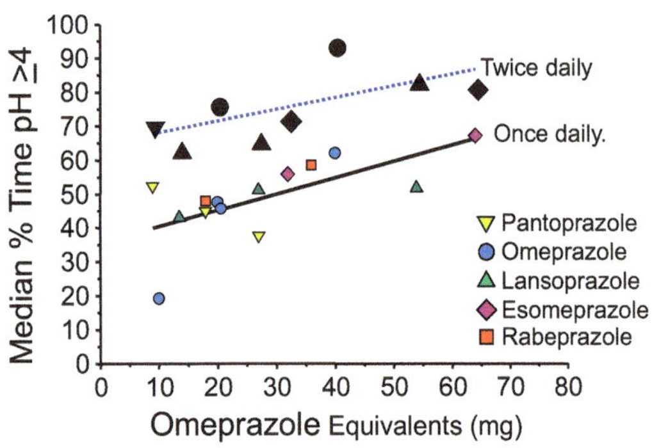

Fig. 14.5 Comparison of the effects of once-daily and twice-daily proton pump inhibitor administration as omeprazole equivalents on the proportion of the day the time the median intragastric pH remained at 4 or higher. Once-a-day proton pump inhibitor therapy ranging from 9 to 64 mg omeprazole equivalents. Twice-daily proton pump inhibitor administration ranged from 18 to 64 mg omeprazole equivalents. For both, the linear regression line is shown. For twice-daily administration the 95% CI is also shown. All data are after at least 5 days of therapy in Western populations. (From Graham and Tansel 2018)

likely have been at least partially related to improvement in duodenal pH which prevent inactivation of normally secreted enzymes (i.e., allow residual function to become active) and allow emptied enzyme to function properly, (c) the resulting change in the gastric contents might reduce the separation of the enteric coated beads from the meal in the stomach (which appears very unlikely unless it allowed some of the beads to dissolve in the stomach and release their enzymes) or, (d) improve bead dissolution in the proximal intestine.

As noted previously, if the intragastric pH remains high, unprotected enzymes are highly effective in reducing both steatorrhea and creatorrhea (Graham 1977). This degree of pH control is possible but to reliably achieve this probably requires twice daily proton pump inhibitor therapy as well as adjuvant antacids to neutralize the small amount of acid still being produced (Graham and Tansel 2018; Julapalli and Graham 2005) (Fig. 14.5). While the optimum doses of PPI and antacid for this indication have yet to be determined, we recommend that proof of principle experiments administer 60 or more mg omeprazole equivalents (e.g., 40 mg of esomeprazole or rabeprazole twice daily) which would be expected to provide a median pH4time of approximately 85%. This high dose of proton pump inhibitors can significantly inhibit acid secretion allowing very small amounts of antacid to have a profound and long lasting effect. Based on the data from prior studies, we would start with sodium bicarbonate (1.3 g; 12 mEq) or aluminum hydroxide (5 or 10 mL; 10 or 20 mEq) at the beginning and end of the meal, and possibly 1 and 3 h after the meal, for initial experiments with unprotected enzymes (Graham 1982). Subsequent experiments designed to identify the optimum proton pump inhibitor and antacid dosages and frequencies of administration and should also include measurements of fecal fat and intragastric pH. Possibly, the new and more potent and long acting competitive potassium blocker, vonoprazan alone would suffice without adjuvant antacids (Graham and Dore 2018).

14.3.8 Summary and Recommendations for Use of Enzymes

Our recommended approach to management of pancreatic insufficiency is illustrated in Fig. 14.6. Although it is recommended that one take into account the patient's diet and level of pancreatic insufficiency, these are hard to estimate and there is no evidence that they are actually important factors. We suggest starting with 18,000–30,000 USP units of lipase per meal with an enteric coated microbead product given in divided doses (e.g., before, at the beginning and mid-meal). Multiple administrations of enzymes to achieve the total dose are designed to achieve better coordination of enzyme and meal delivery to the duodenum (discussed in detail in reference (Trang et al.)). The most common approach to assessing effectiveness has been by patients' reported response and symptoms (Dominguez-Munoz 2011; Dominguez-Munoz and Iglesias-Garcia 2010). This is highly unreliable but repeated fecal fat or ^{13}C-mixed triglyceride breath testing are generally unavailable often making symptomatic assessment the only currently available practical approach for many clinicians. If there is an unsatisfactory response, increasing the enzyme (e.g., doubling the amount to a total of 50,000 or 60,000 units) is typically the next step but, as discussed above, one can expect little or no dose-response effect, such that the strategy is unlikely to be successful and likely only increases costs and side effects (da la Iglesia-García et al. 2017; Trang et al. 2014). An inadequate response to the initial dose of enzymes should prompt reconsideration of the presence of more than one diagnosis (Fig. 14.6). An alternate approach to increasing the dosage above 50,000–60,000 lipase units/meal is to instead add or substitute a non-enteric coated enzyme product just before or at the beginning of the meal (e.g., Viokace®). Probably a better alternative is to switch entirely to non-enteric coated enzymes along with reliable suppression of gastric acidity (Fig. 14.6) as described above (e.g., approximately 60 mg omeprazole equivalents BID and adjuvant antacids such as sodium bicarbonate or aluminum hydroxide).

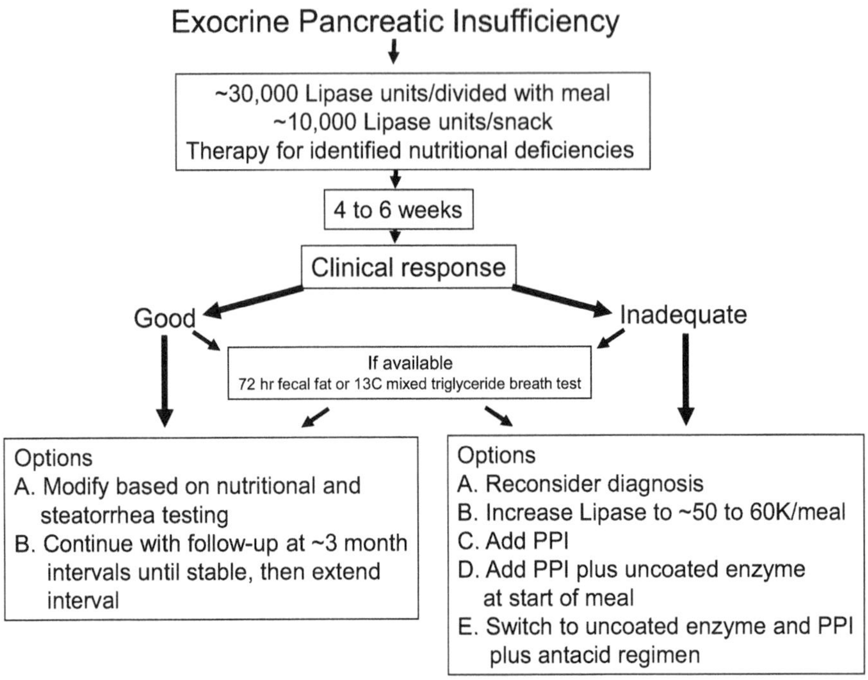

Fig. 14.6 Algorithm describing the recommended clinical approach to using replacement enzymes for the treatment of exocrine pancreatic insufficiency

This approach has not been tested with current formulations but treatment of those with low to absent gastric secretion with uncoated enzymes has proven highly successful in the past. With this approach enzymes should be taken immediately before and throughout the meal to ensure their mixing and emptying with the meal.

14.4 Chronic Pancreatitis

14.4.1 Nutritional Assessment of Patients with Chronic Pancreatitis

Chronic pancreatitis is associated with both endocrine and exocrine insufficiency. Patients with both endocrine and exocrine insufficiencies are particularly difficult to manage as they have difficulty absorbing ingested nutrients and in addition to malabsorption also experience calorie loss via urinary excretion of sugar. In these patients controlling blood sugar is often very difficult until malabsorption is controlled. Weight loss, symptoms associated with maldigestion and difficulty in controlling sugar are common presentations of pancreatic insufficiency. The focus on improving overall nutrition often does not receive the same attention in the literature or in practice as details regarding pancreatic enzymes replacement. Recent longitudinal cohort studies of patients with pancreatic insufficiency followed long term have also confirmed that there is an increased risk of mortality associated with chronic pancreatitis and that the mortality risk and poorest quality of life is greatest among those with low body mass index (da la Iglesia-García et al. 2018; Duggan et al. 2014). In recent studies, many patients with pancreatic insufficiency are either overweight or obese yet they demonstrate reduced functional capacity such as assessed by hand grip strength and muscle mass (Duggan et al. 2014).

Fecal elastase I levels do not relate to the presence or absence of micronutrient deficiencies and should not be used to guide whether deficiencies are present or whether one should evaluate micronutrient status. Nutritional deficiencies are common in patients with pancreatic insufficiency and we recommend that micronutrient status should be routinely and regularly assessed (Duggan et al. 2014). In the past, vitamin deficiencies were very common in this group of patients. Recent studies have confirmed that the problem remains although the prevalence of vitamin deficiencies is lower (Duggan et al. 2014). For example, a prospective study of 40 patients with chronic pancreatitis, many on treatment, found deficiencies of vitamin K (63%), vitamin D (53%), vitamin E (10%) and vitamin A (3%) as well as osteopenia (45%) and osteoporosis (10%) (Sikkens et al. 2013). Another study of those on long term treatment found vitamin A and D deficiencies in 14.5% and 24.5%, respectively (Duggan et al. 2014). However, some of these patients had excess vitamin A levels in the toxic range confirming the need for testing. Another recent study confirmed low levels of magnesium, hemoglobin, albumin, prealbumin, and retinol binding protein in patients with pancreatic insufficiency (Lindkvist et al. 2012). In that study a low serum magnesium (<2.05 mg/dL) highly correlated with the presence of pancreatic exocrine insufficiency. As noted earlier, both calcium and magnesium bind with fatty acids to form poorly soluble calcium or magnesium soaps and are malabsorbed resulting in hypomagnesemia and reduced bone density (Graham and Sackman 1982, 1983). This interaction requires calcium and magnesium replacement be separated from meals where the presence of calcium and magnesium could also interfere with fat absorption.

14.4.2 Recommended Testing for Vitamin Deficiencies and Nutritional Status

Evaluation of patients with pancreatic insufficiency should include anthropomorphic measurements and regular testing for specific nutritional deficiencies (Lindkvist et al. 2012). Hand strength testing is simple and is recommended. Initial testing will serve to identify if and which specific deficiencies are present and allow a patient-specific replacement strategy to be developed. Further testing is then required to ensure the deficiencies are corrected and hypervitaminoses do not occur. Regular assistance of a trained dietitian is extremely useful but not a guarantee of success (Sikkens et al. 2012). There are no recent high quality evidenced-based guideline defining which tests should be done or how often. Routine follow-up measurement of serum vitamin E, magnesium, and plasma proteins, notably retinol binding protein, albumin, and prealbumin levels has been recommended (Lindkvist et al. 2015).

The blood tests often used to assess nutritional status in pancreatic insufficiency are shown in Table 14.5. Initial nutritional status screening should be conducted at the time of diagnosis. We recommend that levels be rechecked after 3 months of starting enzyme replacement therapy and, if normal, subsequent testing of nutritional status should be done annually. More frequent laboratory testing should be individualized based on tolerance of oral feeding, whether high dosage vitamin supplementation (vitamin replacement therapy) has been instituted, and in the presence of continuing steatorrhea, nausea, vomiting or weight loss. The fat soluble vitamins (A, D, E, and K) are especially prone to being deficient. Among those with deficiency, vitamin D and E are most likely to be deficient.

The most dramatic manifestation of vitamin A deficiency is night blindness (i.e., Do you have difficulty driving at night?), however vitamin A deficiency is currently uncommonly seen in pancreatic insufficient patients (Duggan et al. 2014). Biochemical assessment of vitamin A involves measuring retinol binding protein and prealbumin (transthretin). Retinyl esters normally bind to retinol binding protein and prealbumin and are transported from the liver to the tissues. Retinol binding protein is a negative acute phase protein and thus levels fall during infection and inflammation. It has been suggested that rather than rely entirely on measurement of retinol binding protein, a better measure of vitamin A status is to assess the retinol binding protein:prealbumin ratio: a ratio of ≤ 0.36 is indicative of vitamin A deficiency (Rogers 2013). Zinc is required for synthesis of retinol binding protein such that failure to respond to supplemen-

Table 14.5 Recommended laboratory test for nutritional assessment, daily allowances and replacement recommendations

Test	Normal	Daily	Replacement
Vitamin A	32.5–78 µg/dL	10,000 IU	20,000 IU
Retinol binding protein (RBP)	1.6–6 mg/dL		
Prealbumin	16–30 mg/dL		
RBP:prealbumin ratio	≤0.36		
Vitamin D		800–2000 IU	1600–10,000 IU
25-hydroxy vitamin D	20–60 ng/mL		
Vitamin E	5.5–17.0 mg/L	200–400 IU	800–12,000 IU
Serum α-tocopherol	>0.7 ml/dL		
α-tocopherol:cholesterol ratio	<2.47 mg/g		
Vitamin K		300–500 µg	5–10 mg/week
INR (international normalized ratio)	<1.1		
Vitamin B12	200–800 pg/mL		
Magnesium	1.6–2.6 mEq/L		
Zinc	75–140 µg/dL		
Albumin	3.5–5.5 g/dL		
Cholesterol	<200 mg/dL		

ABIM Laboratory Test Reference Ranges January 2018

tal vitamin A suggests zinc deficiency. During replacement it is important that one avoid hypervitaminosis A which can manifest as nausea, vomiting, anorexia, and bone pain. Generally, one should recheck levels more often during high dose vitamin A replacement therapy. As noted above, in one study of patients on enzyme therapy, hypervitaminosis A was more common than deficiency (Duggan et al. 2014).

Vitamin D deficiency is especially common in chronic pancreatic insufficiency but it is also common in the general population. The recommendation is to provide 1500–2000 IU daily for those over 18 years of age with a low 25 hydroxy vitamin D level ≥20 but ≤30 ng/mL and increase the daily dose by 1600–6000 IU of vitamin D3 (Borowitz et al. 2002; Rogers 2013). Those with levels <20 ng/mL should receive 10,000 IU/day vitamin D3 for 3 months before rechecking levels and modifying treatment as required (Borowitz and Gelfond 2013). Because metabolic bone disease is a common problem in patients with pancreatic insufficiency periodic bone densitometry is recommended (Bernstein et al. 2003).

Vitamin E is an antioxidant and is assessed as serum α-tocopherol levels (normal >0.7 ml/dL). Vitamin E deficiency is one of the most common fat soluble vitamin deficiencies seen in pancreatic insufficiency. Serum levels correlate with plasma lipid levels such that an α-tocopherol:cholesterol ratio of <2.47 mg/g is considered indicative of deficiency. Vitamin K level is reflected by the prothrombin level usually assessed as the International Normalized Ratio (INR).

Vitamin B-12 deficiency may also be seen as pancreatic trypsin is required to dissociate intrinsic factor from R protein and make vitamin B12 available for absorption. Folate is usually normal but if folate and vitamin C levels are available, we recommend they also be checked initially.

14.4.3 Water-Miscible Replacement Vitamins

Vitamin dosing levels for adults are show in Table 14.5. While in children with cystic fibrosis many recommend water-miscible vitamins, water miscible vitamins are not necessary for adult patients on pancreatic enzyme replacement therapy. For those interested in acquiring water miscible vitamin preparations, data on individual preparations and their composition is available from the Cystic Fibrosis Foundation (https://www.cff.org/Life-With-CF/Daily-Life/Fitness-

and-Nutrition/Nutrition/Getting-Your-Nutrients/Vitamin-Comparison-Chart-for-CF-Specific-Multivitamins.pdf). Water-miscible fat-soluble vitamins are available from Aptalis (http://store.foundcare.com/aptalis/product/aquadeks-chewable-tablets/) as SourceCF or AquaADEK, or from Shear/Kershman laboratories as VITAMAX.

14.4.4 Enzymes for Treatment or Prevention of Pancreatic Pain

The management of pain in chronic pancreatitis is clinically challenging in part because the etiology of pain in this setting is poorly understood (Hobbs et al. 2016). A heterogeneous collection of theories of pancreatic pain abound, including pancreatic ductal obstruction/hypertension secondary to stones and strictures, fibrosis-induced increased interstitial pancreatic pressure, pancreatic ischemia, and pancreatic neuritis (Table 14.6). There are a number of excellent reviews that one can consult for specific details of pathogenesis and therapy (Hobbs et al. 2016; Poulsen et al. 2013). A variety of strategies to treat and prevent pain, some of which address these theories, have been tried (Hobbs et al. 2016). Administration of pancreatic enzymes remain a viable option in specific cases.

Chronic pancreatitis is characterized by ongoing pancreatic inflammation leading to disordered pancreatic structure and function. The characteristic pain of chronic pancreatitis is epigastric, precipitated by food and radiating to the back. One potential cause of pancreatic pain is related to pancreatic ductal hypertension attributed to inflammatory strictures or obstructing stones. Many studies have shown improvement of pain with decompression of a dilated pancreatic duct (Hobbs et al. 2016). However, ductal hypertension causing interstitial hypertension is not present in many with painful chronic pancreatitis and the pain is thought instead to be related to pancreatic ischemia and neuritis (Hobbs et al. 2016).

Table 14.6 Mechanisms of pain in chronic pancreatitis

Increased intraductal pressure
Ductal obstruction from strictures/stones
Increased intrapancreatic pressure (compartment-like syndrome)
Fibrosis causing lack of distensibility
Neuropathic
Entrapment of nerves
Damage of nerves by enzymes
Increased nerve tissue
Pancreatic ischemia
Worsened during increased enzyme secretion

Some patients have chronic pancreatitis pain responsive to pancreatic enzyme therapy. Pain associated with malabsorption can also arise from the presence of digestive products in the gastrointestinal tract and correction of malabsorption will reduce or eliminate this pain (Hobbs et al. 2016). A second mechanism for pain reduction is to prevent increased pancreatic pressure by feedback inhibition of pancreatic secretion.

14.4.5 Pancreatic Enzymes and Negative Feedback

Observational studies have noted reduction in pancreatic pain with pancreatic enzyme therapy in some patients with chronic pancreatitis (Hobbs et al. 2016). This has been attributed to exogenous pancreatic enzymes reducing endogenous secretion of enzymes in response to meals which reduces the increase in ductal and parenchymal pressure associated with secretion of pancreatic juice and prevents or reduces pain (Hobbs et al. 2016).

The normal human pancreas secretes continuously which increases in the post-prandial period. Entry of food and fatty acids into the duodenum triggers secretion of cholecystokinin (CCK) and secretin which stimulate pancreatic enzyme and bicarbonate secretion (Layer and Keller 1999). Negative pancreatic feedback inhibition has been demonstrated in rats, chickens and pigs (Chernick et al. 1948; Corring 1973; Green and Lyman 1972; Ihse et al. 1979; Louie et al. 1986; Rausch et al. 1987; Shiratori et al. 1986). In healthy humans, pancreatic enzyme output suppression is dose-dependent occurring with the intraduodenal infusion of proteases: the minimum dose is 0.5 mg/mL of trypsin and maximal suppression occurred with 1.0 mg/mL (Owyang et al. 1986a).

Suppression also correlated with the decline in blood CCK levels (Owyang et al. 1986a). It remains unclear how mg of trypsin/mL relate to USP units of protease activity used to describe pancreatic enzymes. The Worthington catalog suggests the conversion is about 3000 USP units/mg bovine trypsin (http://www.worthington-biochem.com/try/cat.html).

Pancreatic outputs have been compared in patients with differing severity of chronic pancreatitis and healthy controls (Slaff et al. 1984) and the infusion of 10 mg/mL of trypsin was found to reduce pancreatic secretion by approximately 32% in patients with reduced pancreatic output vs. 74% in those with normal pancreatic secretion. No inhibition was noted in patients with low pancreatic bicarbonate secretion and steatorrhea (Slaff et al. 1984). However, chronic pancreatic enzyme therapy resulted in a 27% decrease in basal pancreatic secretion compared to a 46% decrease with amino acid stimulated secretion. In that study, the minimum trypsin concentration required to inhibit pancreatic exocrine secretion was 0.9 mg/mL with maximum suppression at 2.5 mg/mL. Chymotrypsin (10 mg/mL) also decreased amino acid-stimulated trypsin output whereas protease-free lipase and amylase have no effect. Overall, the data are consistent with the notions that (a) intraduodenal trypsin and chymotrypsin both suppress human pancreatic secretion, (b) that suppression is minimal in advanced pancreatic insufficiency and (c) patients who fail to suppress pancreatic secretion often do not experience pain relief with enzyme supplementation (Slaff et al. 1984). The data regarding control of pancreatic secretion in human are consistent with several distinct feedback pathways, one mediated by proteases (e.g., trypsin/chymotrypsin) (Adler et al. 1988a, b; Ebbehoj et al. 1990; Liener et al. 1988) and another by acetylcholine (Owyang et al. 1986b).

14.4.6 How Well Does Enzyme Therapy Reduce Pancreatic Pain?

There have been numerous studies and several large meta-analyses of the use of pancreatic enzymes in the treatment of abdominal pain in chronic pancreatitis (Hobbs et al. 2016). The available studies are heterogenous in relation to severity of exocrine insufficiency, etiology of pancreatitis, clinical presentation, presence or absence of narcotic use, and importantly, to enzyme formulation and dosage and relation to meals. Together, these caveats greatly inhibit one's ability to evaluate the effect of enzyme therapy on pain relief. Individual studies have, however, shown reduced pancreatic pain with both enteric and non-enteric coated enzymes compared with placebo and have reported improved quality of life with pancreatic enzymes (Czako et al. 2003; Ramesh et al. 2013).

Overall, the data confirm that some patients with pancreatic pain will respond to enzyme therapy, however, studies showing excellent or good effects are in the minority (Hobbs et al. 2016; Mossner 1991). One issue is that inhibition of pancreatic secretion is protease-specific and requires a threshold concentration of trypsin/chymotrypsin. Most studies have used enteric-coated enzymes which are unlikely to provide sufficient intraduodenal trypsin activity to provide feedback effective inhibition. In addition, most of the patients involved have severe insufficiency and are thus were the least likely group to respond.

While the data regarding use of pancreas enzymes to treat pain in chronic pancreatitis is poor, long term studies have shown improved outcome in terms of absorption and pain relief associated with the use of pancreatic enzymes. This is consistent with pain associated with malabsorption of nutrients being an important and treatable factor (Czako et al. 2003; Gubergrits et al. 2011; Hobbs et al. 2016; Ramesh et al. 2013). Studies with non-enteric coated enzyme preparations given while preventing gastric inactivation are needed to adequately test the role of the negative feedback loop and to rest the pancreas and also provide pain relief.

14.5 Conclusions

The most common uses for pancreatic enzymes are as replacement therapy for treatment of exocrine pancreatic sufficiency and for pain associated with chronic pancreatitis. Exocrine pancreatic insufficiency is one of the most common causes of malabsorption. The most common etiologies are chronic pancreatitis, cystic fibrosis, and surgical resection. We discuss the details of use of pancreatic enzymes to replace those needed for normal digestion as well as the barriers that must be successfully dealt with to achieve that goal. We also discuss the use of pancreatic enzymes in pancreatic pain and the various mechanisms that may produce pain in chronic pancreatic disease. Finally, we discuss the nutritional deficiencies common in patients with pancreatic insufficiency and the approach to identifying, monitoring, and treating these deficiencies.

The major hurdle to providing successful therapy has been destruction of pancreatic enzymes during transit through the stomach. The introduction of enteric coated enzymes packages as microspheres helped overcome this barrier but also proved to have significant limitations in that the microspheres tend to separate from the meal and empty separately, introducing a new barrier. In the natural process, the enzymes and meal are mixed along with bile salts at the proper pH to maximize digestion and absorption. Separation of the microspheres from the meal and their slow dissolution results in a new barrier made worse by the fact that microspheres may not release their contents until deep within the small intestine. Nonetheless they are partially effective and were more reliable than uncoated enzymes. However, there is no dose response, as increasing the microsphere dosage has minimal or no further effect on efficacy and generally only results in increased costs and side effects. We discuss how to maximize the benefits with microspheres but for most questions there are no clinical trials to confirm improved efficacy such as whether adding non-coated enzymes at the beginning of the meal would improve efficacy.

Until recently it was unknown how to reliably overcome the pH barrier caused by lipase being irreversibly inactivated at pH 4 or below. Recent understandings of relative PPI potency and how best to administer PPIs to maximize the time the intragastric pH remains above four suggests that it should now be possible to utilize non-enteric coated enzymes effectively. For example, administration of 60–70 mg of omeprazole or its equivalent twice-a-day (e.g., 75 mg of lansoprazole, or 40 mg of esomeprazole or rabeprazole twice-a-day) possibly with a small amount of an appropriate antacid (e.g., aluminum hydroxide or sodium bicarbonate) at the beginning and end of the meal or 1 h after the meal should provide a milieu to protect the enzymes, allow mixing and emptying along with the meal, and provide maximum benefit. This hypothesis remains to be tested. In 2004, the FDA mandated that all pancreatic enzymes must prove efficacy and the research has been company-sponsored studies to prove that the new products were superior to placebo. There have been a few company-sponsored studies looking at some important variables, such as microsphere emptying and separation from the meals, but none of those data have been published or made available on request and support for addressing the many clinically important questions noted above remains lacking.

Acknowledgement Dr. Graham is in part by the Research Service Department of Veterans Affairs and by Public Health Service grant DK56338 which funds the Texas Medical Center Digestive Diseases Center. The authors acknowledge the excellent advice received from Temitope Osinaiya, DHSc, MS, RD, LDN regarding the nutritional assessment of patients with pancreatic insufficiency.

Competing Interests Dr. Graham is a consultant for RedHill Biopharma regarding novel *H. pylori* therapies and has received research support for culture of *Helicobacter pylori* and is the PI of an international study of the use of antimycobacterial therapy for Crohn's disease. He is also a consultant for BioGaia in relation to probiotic therapy for *H. pylori* infection and for Takeda in relation to *H. pylori* therapies. Dr. Ketwaroo does not have any relevant disclosures.

References

Adler G, Mullenhoff A, Bozkurt T, Goke B, Koop I, Arnold R (1988a) Comparison of the effect of single and repeated administrations of a protease inhibitor (Camostate) on pancreatic secretion in man. Scand J Gastroenterol 23(2):158–162

Adler G, Mullenhoff A, Koop I, Bozkurt T, Goke B, Beglinger C, Arnold R (1988b) Stimulation of pancreatic secretion in man by a protease inhibitor (camostate). Eur J Clin Investig 18(1):98–104

Afghani E, Sinha A, Singh VK (2014) An overview of the diagnosis and management of nutrition in chronic pancreatitis. Nutr Clin Pract 29(3):295–311

Aloulou A, Puccinelli D, Sarles J, Laugier R, Leblond Y, Carriere F (2008) In vitro comparative study of three pancreatic enzyme preparations: dissolution profiles, active enzyme release and acid stability. Aliment Pharmacol Ther 27(3):283–292

Bakowski MT, Prescott P (1997) Patterns of use of pancreatic enzyme supplements in fibrosing colonopathy: implications for pathogenesis. Pharmacoepidemiol Drug Saf 6(5):347–358

Beazell JM, Schmidt CR, Ivy AC (1941) The diagnosis and treatment of achylia pancreatica. JAMA 116(25):2735–2739

Bell NJ, Burget D, Howden CW, Wilkinson J, Hunt RH (1992) Appropriate acid suppression for the management of gastro-oesophageal reflux disease. Digestion 51(Suppl 1):59–67

Benini L, Caliari S, Guidi GC, Vaona B, Talamini G, Vantini I, Scuro LA (1989) Near infrared spectrometry for faecal fat measurement: comparison with conventional gravimetric and titrimetric methods. Gut 30(10):1344–1347

Bernstein CN, Leslie WD, Leboff MS (2003) AGA technical review on osteoporosis in gastrointestinal diseases. Gastroenterology 124(3):795–841

Borowitz D, Gelfond D (2013) Intestinal complications of cystic fibrosis. Curr Opin Pulm Med 19(6):676-680. Available from: PM:24060981

Borowitz D, Baker RD, Stallings V (2002) Consensus report on nutrition for pediatric patients with cystic fibrosis. J Pediatr Gastroenterol Nutr 35(3):246–259

Bruno MJ, Rauws EA, Hoek FJ, Tytgat GN (1994) Comparative effects of adjuvant cimetidine and omeprazole during pancreatic enzyme replacement therapy. Dig Dis Sci 39(5):988–992

Chase RF (1905) The therapeutic value of some digestive preparations, and the indications for use of pepsin, in diseases of the stomach. Boston Med Surg J 152:572–574

Chernick SS, Lepkovsky S, Chaikoff IL (1948) A dietary factor regulating the enzyme content of the pancreas; changes induced in size and proteolytic activity of the chick pancreas by the ingestion of raw soy-bean meal. Am J Phys 155(1):33–41

Corring T (1973) Mechanism of the exocrine pancreatic secretion in the pig feed-back regulation. Ann Biol Amin Biochim Biophys 13:755–756

Czako L, Takacs T, Hegyi P, Pronai L, Tulassay Z, Lakner L, Dobronte Z, Boda K, Lonovics J (2003) Quality of life assessment after pancreatic enzyme replacement therapy in chronic pancreatitis. Can J Gastroenterol 17(10):597–603

Delchier JC, Vidon N, Saint-Marc Girardin MF, Soule JC, Moulin C, Huchet B, Zylberberg P (1991) Fate of orally ingested enzymes in pancreatic insufficiency: comparison of two pancreatic enzyme preparations. Aliment Pharmacol Ther 5(4):365–378

Diamond JS, Siegel SA, Kantor JL, Jour DD (1940) The secretin test in the diagnosis of pancreatic diseases with a report of one hundred thirty tests. Am J Dig Dis 7(10):435–442

DiMagno EP, Go VL, Summerskill WH (1973) Relations between pancreatic enzyme outputs and malabsorption in severe pancreatic insufficiency. N Engl J Med 288(16):813–815

DiMagno EP, Malagelada JR, Go VL, Moertel CG (1977) Fate of orally ingested enzymes in pancreatic insufficiency comparison of two dosage schedules. N Engl J Med 296(23):1318–1322

Dominguez-Munoz JE (2007) Pancreatic enzyme therapy for pancreatic exocrine insufficiency. Curr Gastroenterol Rep 9(2):116–122

Dominguez-Munoz JE (2011) Pancreatic enzyme replacement therapy for pancreatic exocrine insufficiency: when is it indicated, what is the goal and how to do it? Adv Med Sci 56(1):1–5

Dominguez-Munoz JE, Iglesias-Garcia J (2010) Oral pancreatic enzyme substitution therapy in chronic pancreatitis: is clinical response an appropriate marker for evaluation of therapeutic efficacy? JOP 11(2):158–162

Dominguez-Munoz JE, Iglesias-Garcia J, Iglesias-Rey M, Figueiras A, Vilarino-Insua M (2005) Effect of the administration schedule on the therapeutic efficacy of oral pancreatic enzyme supplements in patients with exocrine pancreatic insufficiency: a randomized, three-way crossover study. Aliment Pharmacol Ther 21(8):993–1000

Dominguez-Munoz JE, Iglesias-Garcia J, Vilarino-Insua M, Iglesias-Rey M (2007) 13C-mixed triglyceride breath test to assess oral enzyme substitution therapy in patients with chronic pancreatitis. Clin Gastroenterol Hepatol 5(4):484–488

Dreiling DA, Hollander F (1948) Studies in pancreatic function; preliminary series of clinical studies with the secretin test. Gastroenterology 11(5):714–729

Duggan SN, Smyth ND, O'Sullivan M, Feehan S, Ridgway PF, Conlon KC (2014) The prevalence of malnutrition and fat-soluble vitamin deficiencies in chronic pancreatitis. Nutr Clin Pract 29(3):348–354

Durie PR, Bell L, Linton W, Corey ML, Forstner GG (1980) Effect of cimetidine and sodium bicarbonate on pancreatic replacement therapy in cystic fibrosis. Gut 21(9):778–786

Ebbehoj N, Borly L, Bulow J, Rasmussen SG, Madsen P (1990) Evaluation of pancreatic tissue fluid pressure and pain in chronic pancreatitis a longitudinal study. Scand J Gastroenterol 25(5):462–466

Engesser H (1879) Beitrage zur therapeutischen Verwendung der Bauchspeicheldruse von Schlachtthieren und deren praparate. Dtsch Arch Klin Med 24:539–582

Fordtran JS, Morawski SG, Richardson CT (1973) In vivo and in vitro evaluation of liquid antacids. N Engl J Med 288(18):923–928

Forsmark CE (2016) Chronic pancreatitis. In: Feldman M, Friedman LS, Brandt LJ (eds) Sleisenger and

Fordtran's gastrointestinal and liver disease, 10th edn. Elsevier Saunders, Philadelphia, p 1020

Franzen D, Went P, Buhlmann U (2008) Fibrosing colonopathy in absence of pancreatic enzyme supplementation in one adult patient with cystic fibrosis. Indian J Gastroenterol 27(3):133–134

Gaia E, Sambatoro A, De Giuli P, Angeli A (2001) Adult fibrosing colonopathy associated with mesalazine treatment. Am J Gastroenterol 96(8):2508–2509

Gow R, Bradbear R, Francis P, Shepherd R (1981) Comparative study of varying regimens to improve steatorrhoea and creatorrhoea in cystic fibrosis: effectiveness of an enteric-coated preparation with and without antacids and cimetidine. Lancet 2(8255): 1071–1074

Graham DY (1977) Enzyme replacement therapy of exocrine pancreatic insufficiency in man relations between in vitro enzyme activities and in vivo potency in commercial pancreatic extracts. N Engl J Med 296(23):1314–1317

Graham DY (1979) An enteric-coated pancreatic enzyme preparation that works. Dig Dis Sci 24(12):906–909

Graham DY (1982) Pancreatic enzyme replacement: the effect of antacids or cimetidine. Dig Dis Sci 27(6):485–490

Graham DY, Dore MP (2018) Update on the use of vonoprazan: a competitive acid blocker. Gastroenterology 154(154):462–466

Graham DY, Sackman JW (1982) Mechanism of increase in steatorrhea with calcium and magnesium in exocrine pancreatic insufficiency: an animal model. Gastroenterology 83(3):638–644

Graham DY, Sackman JW (1983) Solubility of calcium soaps of long-chain fatty acids in simulated intestinal environment. Dig Dis Sci 28(8):733–736

Graham DY, Tansel A (2018) Interchangeable use of proton pump inhibitors based on relative potency. Clin Gastroenterol Hepatol 16(6):800–808

Green GM, Lyman RL (1972) Feedback regulation of pancreatic enzyme secretion as a mechanism for trypsin inhibitor-induced hypersecretion in rats. Proc Soc Exp Biol Med 140(1):6–12

Gubergrits N, Malecka-Panas E, Lehman GA, Vasileva G, Shen Y, Sander-Struckmeier S, Caras S, Whitcomb DC (2011) A 6-month, open-label clinical trial of pancrelipase delayed-release capsules (Creon) in patients with exocrine pancreatic insufficiency due to chronic pancreatitis or pancreatic surgery. Aliment Pharmacol Ther 33(10):1152–1161

Harris R, Norman AP, Payne WW (1955) The effect of pancreatin therapy on fat absorption and nitrogen retention in children with fibrocystic disease of the pancreas. Arch Dis Child 30(153):424–427

Heijerman HG, Lamers CB, Bakker W (1991) Omeprazole enhances the efficacy of pancreatin (pancrease) in cystic fibrosis. Ann Intern Med 114(3):200–201

Heubi JE, Schaeffer D, Ahrens RC, Sollo N, Strausbaugh S, Graff G, Jain R, Witte S, Forssmann K (2016) Safety and efficacy of a novel microbial lipase in patients with exocrine pancreatic insufficiency due to cystic fibrosis: a randomized controlled clinical trial. J Pediatr 176:156–161e1

Hobbs PM, Johnson WG, Graham DY (2016) Management of pain in chronic pancreatitis with emphasis on exogenous pancreatic enzymes. World J Gastrointest Pharmacol Ther 7(3):370–386

Hunt JN (1983) Mechanisms and disorders of gastric emptying. Annu Rev Med 34:219–229

Hunt JN, Knox MT (1968) Control of gastric emptying. Am J Dig Dis 13(4):372–375

Hunt RH, Cederberg C, Dent J, Halter F, Howden C, Marks IN, Rune S, Walt RP (1995) Optimizing acid suppression for treatment of acid-related diseases. Dig Dis Sci 40(2 Suppl):24S–49S

Ianiro G, Pecere S, Giorgio V, Gasbarrini A, Cammarota G (2016) Digestive enzyme supplementation in gastrointestinal diseases. Curr Drug Metab 17(2):187–193

Ihse I, Lilja P, Lundquist I (1979) Trypsin as a regulator of pancreatic secretion in the rat. Scand J Gastroenterol 14(7):873–880

Jones DB, Howden CW, Burget DW, Kerr GD, Hunt RH (1987) Acid suppression in duodenal ulcer: a meta-analysis to define optimal dosing with antisecretory drugs. Gut 28(9):1120–1127

Jordan PH Jr, Grossman MI (1959) Effect of dosage schedule on the efficacy of substitution therapy in pancreatic insufficiency. Gastroenterology 36(4):447–451

Julapall VR, Graham DY (2005) Appropriate use of intravenous proton pump inhibitors in the management of bleeding peptic ulcer. Dig Dis Sci 50(7):1185–1193

Kalser MH, Leite CA, Warren WD (1968) Fat assimilation after massive distal pancreatectomy. N Engl J Med 279(11):570–576

Kattwinkel J, Agus SG, Taussig LM, di Sant'Agnese PA, Laster L (1972) The use of L-arginine and sodium bicarbonate in the treatment of malabsorption due to cystic fibrosis. Pediatrics 50(1):133–137

Keller J, Layer P (2005) Human pancreatic exocrine response to nutrients in health and disease. Gut 54(Suppl 6):vi1–v28

Keller J, Runzi M, Goebell H, Layer P (1997) Duodenal and ileal nutrient deliveries regulate human intestinal motor and pancreatic responses to a meal. Am J Phys 272(3 Pt 1):G632–G637

Ketwaroo G, Brown A, Young B, Kheraj R, Sawhney M, Mortele KJ, Najarian R, Tewani S, Dasilva D, Freedman S, Sheth S (2013) Defining the accuracy of secretin pancreatic function testing in patients with suspected early chronic pancreatitis. Am J Gastroenterol 108(8):1360–1366

Kirchheiner J, Glatt S, Fuhr U, Klotz U, Meineke I, Seufferlein T, Brockmoller J (2009) Relative potency of proton-pump inhibitors-comparison of effects on intragastric pH. Eur J Clin Pharmacol 65(1):19–31

Kuhn RJ, Eyting S, Henniges F, Potthoff A (2007) In vitro comparison of physical parameters, enzyme activity, acid resistance, and pH dissolution characteristics of enteric-coated pancreatic enzyme preparations: implications for clinical variability and pharmacy substitution. J Pediatr Pharmacol Ther 12(2):115–128 da la Iglesia-García D, Huang W, Szatmary P, Baston-Rey I, Gonzalez-Lopez J, Prada-Ramallal G, Mukherjee R, Nunes QM, Dominguez-Munoz JE, Sutton R (2017) Efficacy of pancreatic enzyme replacement therapy in chronic pancreatitis: systematic review and meta-analysis. Gut 66(8):1354–1355

da la Iglesia-García D, Vallejo-Senra N, Iglesias-Garcia J, Lopez-Lopez A, Nieto L, Dominguez-Munoz JE (2018) Increased risk of mortality associated with pancreatic exocrine insufficiency in patients with chronic pancreatitis. J Clin Gastroenterol 52(8):e63–e72

Layer P, Keller J (1999) Pancreatic enzymes: secretion and luminal nutrient digestion in health and disease. J Clin Gastroenterol 28(1):3–10

Layer P, Go VL, DiMagno EP (1986) Fate of pancreatic enzymes during small intestinal aboral transit in humans. Am J Phys 251(4 Pt 1):G475–G480

Liener IE, Goodale RL, Deshmukh A, Satterberg TL, Ward G, DiPietro CM, Bankey PE, Borner JW (1988) Effect of a trypsin inhibitor from soybeans (Bowman-Birk) on the secretory activity of the human pancreas. Gastroenterology 94(2):419–427

Lindkvist B, Dominguez-Munoz JE, Luaces-Regueira M, Castineiras-Alvarino M, Nieto-Garcia L, Iglesias-Garcia J (2012) Serum nutritional markers for prediction of pancreatic exocrine insufficiency in chronic pancreatitis. Pancreatology 12(4):305–310

Lindkvist B, Phillips ME, Dominguez-Munoz JE (2015) Clinical, anthropometric and laboratory nutritional markers of pancreatic exocrine insufficiency: prevalence and diagnostic use. Pancreatology 15(6):589–559

Littman A, Hanscom DH (1969) Current concepts: pancreatic extracts. N Engl J Med 28(4):201–204

Louie DS, May D, Miller P, Owyang C (1986) Cholecystokinin mediates feedback regulation of pancreatic enzyme secretion in rats. Am J Physiol 250(2 Pt 1):G252–G259

Lowe ME, Whitcomb DC (2015) Next generation of pancreatic enzyme replacement therapy: recombinant microbial enzymes and finding the perfect lipase. Gastroenterology 149(7):1678–1681

Marotta F, O'Keefe SJ, Marks IN, Girdwood A, Young G (1989) Pancreatic enzyme replacement therapy: importance of gastric acid secretion, H2-antagonists, and enteric coating. Dig Dis Sci 34(3):456–461

Meyer JH (1980) Gastric emptying of ordinary food: effect of antrum on particle size. Am J Phys 239(3):G133–G135

Meyer JH, Lake R (1997) Mismatch of duodenal deliveries of dietary fat and pancreatin from enterically coated microspheres. Pancreas 15(3):226–235

Meyer JH, Elashoff J, Porter-Fink V, Dressman J, Amidon GL (1988) Human postprandial gastric emptying of 1-3-millimeter spheres. Gastroenterology 94(6):1315–1325

Mossner J (1991) Treatment of pain in chronic pancreatitis with pancreatic enzymes another point of view. In: Lankisch PG (ed) Pancreatic enzymes in health and disease. Springer-Verlag, Berlin, pp 103–112

Mullady DK, Yadav D, Amann ST, O'Connell MR, Barmada MM, Elta GH, Scheiman JM, Wamsteker EJ, Chey WD, Korneffel ML, Weinman BM, Slivka A, Sherman S, Hawes RH, Brand RE, Burton FR, Lewis MD, Gardner TB, Gelrud A, DiSario J, Baillie J, Banks PA, Whitcomb DC, Anderson MA (2011) Type of pain, pain-associated complications, quality of life, disability and resource utilisation in chronic pancreatitis: a prospective cohort study. Gut 60(1):77–84

Owyang C, Louie DS, Tatum D (1986a) Feedback regulation of pancreatic enzyme secretion suppression of cholecystokinin release by trypsin. J Clin Invest 77(6):2042–2047

Owyang C, May D, Louie DS (1986b) Trypsin suppression of pancreatic enzyme secretion differential effect on cholecystokinin release and the enteropancreatic reflex. Gastroenterology 91(3):637–643

Pelley JR, Gordon SR, Gardner TB (2012) Abnormal duodenal [HCO3-] following secretin stimulation develops sooner than endocrine insufficiency in minimal change chronic pancreatitis. Pancreas 41(3):481–484

Poulsen JL, Olesen SS, Malver LP, Frokjaer JB, Drewes AM (2013) Pain and chronic pancreatitis: a complex interplay of multiple mechanisms. World J Gastroenterol 19(42):7282–7291

Prescott P, Bakowski MT (1999) Pathogenesis of fibrosing colonopathy: the role of methacrylic acid copolymer. Pharmacoepidemiol Drug Saf 8(6):377–384

Prieto G, Perez-Moneo B, Molina M, Ramos E, Sarria J, Larrauri J, Tovar JA (2009) Fibrosing colonopathy associated with treatment with enteric-coated mesalazine pills. Inflamm Bowel Dis 15(10):1452–1453

Ramesh H, Reddy N, Bhatia S, Rajkumar JS, Bapaye A, Kini D, Kalla M, Thorat V (2013) A 51-week, open-label clinical trial in India to assess the efficacy and safety of pancreatin 40000 enteric-coated minimicrospheres in patients with pancreatic exocrine insufficiency due to chronic pancreatitis. Pancreatology 13(2):133–139

Rausch U, Weidenbach H, Adler G, Kern HF (1987) Stimulation of pancreatic secretory process in the rat by low-molecular weight proteinase inhibitor II regulation of total protein and individual enzyme biosynthesis. Cell Tissue Res 249(1):63–67

Regan PT, Malagelada JR, DiMagno EP, Go VL (1979) Reduced intraluminal bile acid concentrations and fat maldigestion in pancreatic insufficiency: correction by treatment. Gastroenterol 77(2):285–289

Rogers CL (2013) Nutritional management of the adult with cystic fibrosis – part I. Pract Gastroenterol 113(1):10–24

Sander-Struckmeier S, Beckmann K, Janssen-van SG, Pollack P (2013) Retrospective analysis to investigate the effect of concomitant use of gastric acid-suppressing drugs on the efficacy and safety of pancrelipase/pancreatin (CREON(R)) in patients with pancreatic exocrine insufficiency. Pancreas 42(6):983–989

Scharpé S, Uyttenbroeck W, Samyn N (1997) In: Lauwers A, Scharpé S (eds) Pancreatic enzyme replacement. Taylor Francis, London, pp 187–221

Shiratori K, Chen YF, Chey WY, Lee KY, Chang TM (1986) Mechanism of increased exocrine pancreatic secretion in pancreatic juice-diverted rats. Gastroenterology 91(5):1171–1178

Sikkens EC, Cahen DL, Eijck C, Kuipers EJ, Bruno MJ (2012) Patients with exocrine insufficiency due to chronic pancreatitis are undertreated: a Dutch national survey. Pancreatology 12(1):71–73

Sikkens EC, Cahen DL, Koch AD, Braat H, Poley JW, Kuipers EJ, Bruno MJ (2013) The prevalence of fat-soluble vitamin deficiencies and a decreased bone mass in patients with chronic pancreatitis. Pancreatology 13(3):238–242

Singh VK, Haupt ME, Geller DE, Hall JA, Quintana Diez PM (2017) Less common etiologies of exocrine pancreatic insufficiency. World J Gastroenterol 23(39):7059–7076

Slaff J, Jacobson D, Tillman CR, Curington C, Toskes P (1984) Protease-specific suppression of pancreatic exocrine secretion. Gastroenterology 87(1):44–52

Smith JL, Jiang CL, Hunt JN (1984) Intrinsic emptying pattern of the human stomach. Am J Phys 246(6 Pt 2):R959–R962

Stallings VA, Stark LJ, Robinson KA, Feranchak AP, Quinton H (2008) Evidence-based practice recommendations for nutrition-related management of children and adults with cystic fibrosis and pancreatic insufficiency: results of a systematic review. J Am Diet Assoc 108(5):832–839

Stein J, Jung M, Sziegoleit A, Zeuzem S, Caspary WF, Lembcke B (1996) Immunoreactive elastase I: clinical evaluation of a new noninvasive test of pancreatic function. Clin Chem 42(2):222–226

Struyvenberg MR, Martin CR, Freedman SD (2017) Practical guide to exocrine pancreatic insufficiency – breaking the myths. BMC Med 15(1):29

Trang T, Chan J, Graham DY (2014) Pancreatic enzyme replacement therapy for pancreatic exocrine insufficiency in the 21(st) century. World J Gastroenterol 20(33):11467–11485

Vanga RR, Tansel A, Sidiq S, El-Serag HB, Othman M (2018) Diagnostic performance of measurement of fecal elastase-1 in detection of exocrine pancreatic insufficiency – systematic review and meta-analysis. Clin Gastroenterol Hepatol 16(8):1220–1228.e4

Veeger RW, AbelsS J, Hellemans N, Nieweg HO (1962) Effect of sodium bicarbonate and pancreatin on the absorption of vitamin B12 and fat in pancreatic insufficiency. N Engl J Med 267:1341–1344

van Velzen D, Ball LM, Dezfulian AR, Southgate A, Howard CV (1996) Comparative and experimental pathology of fibrosing colonopathy. Postgrad Med J 72(Suppl 2):S39–S48

Weber AM, Roy CC, Chartrand L, Lepage G, Dufour OL, Morin CL, Lasalle R (1976) Relationship between bile acid malabsorption and pancreatic insufficiency in cystic fibrosis. Gut 17(4):295–299

White JV, Guenter P, Jensen G, Malone A, Schofield M (2012) Consensus statement: Academy of Nutrition and Dietetics and American Society for Parenteral and Enteral Nutrition: characteristics recommended for the identification and documentation of adult malnutrition (undernutrition). JPEN J Parenter Enter Nutr 36(3):275–283

Fibrinolytic Enzymes for Thrombolytic Therapy

15

Swaroop S. Kumar and Abdulhameed Sabu

Abstract

Cardiovascular diseases are a group of disorders consisting importantly of coronary heart disease, peripheral arterial disease, cerebrovascular disease, rheumatic heart disease, congenital heart disease, deep vein thrombosis and pulmonary embolism. Severe cardiovascular disease conditions lead to acute myocardial infarction and stroke. One of the reasons for this is formation of blood clots inside the vessel. Anticoagulants and antiplatelet drugs are used for managing cardiovascular diseases for a long time. However, they were unable to dissolve an existing thrombus. Fibrinolytic enzymes have become more substantial for treating cardiovascular diseases since they could lyse the fibrin clot within the blood vessel. Inability of plasma fibrinolytic system demands better thrombolytic drugs. Major thrombolytic enzymes belonging to plasminogen activators and plasmin like enzymes. Currently used fibrinolytic enzymes and their limitations are revisited in the present chapter. Reported enzymes from various sources with potential to be used as cardiovascular therapeutic is also discussed here.

Keywords

Fibrinolytic enzymes · Plasminogen activators · Microbial thrombolytic enzymes · Nattokinase · Serrapeptase

Abbrevations

°C	Degree celsius
ADP	Adenosine diphosphate
AIS	Acute ischemic stroke
Ala	Alanine
AMI	Acute myocardial infarction
APAN	p-amidinophenyl p-anisate hydrochloride
APSAC	Anisoylated plasminogen streptokinase activator complex
Arg	Arginine
Asn	Asparagine
Asp	Aspartic acid
ASSENT	Assessment of Safety and Efficacy of a New Thrombolytic agent
AT	Antithrombin III
BK	Bacillokinase
bpDJ-2	Bacillopeptidase DJ-2
CCU	Coronary care units
CDP	*Codium diaricatum* protease
CHO	Chinese Hasmster Ovary
CIP	*Codium intricatum* protease
CLP	*Codium latum* protease
COMPASS	Comparative Trial of Saruplase Versus Streptokinase

S. S. Kumar · A. Sabu (✉)
Department of Biotechnology & Microbiology, School of Life Sciences, Kannur University, Palayad, Kerala, India

N. Labrou (ed.), *Therapeutic Enzymes: Function and Clinical Implications*, Advances in Experimental Medicine and Biology 1148, https://doi.org/10.1007/978-981-13-7709-9_15

COX	Cyclooxygenase
Cys	Cysteine
DIAS	Desmoteplase in Acute Ischemic Stroke
DNA	Deoxyribo nucleic acid
DSPAS	Desmodus salivary plasminogen activators
ECLT	Euglobulin clot lysis time
EDTA	Eethylene diamine tetraacetic acid
EFE	*Eisenia fetida* enzymes
EGF	Epidermal growth factor
ESPRIT	European/Australasian Stroke Prevention in Reversible Ischaemia Trial
FDA	Food and Drug Administration
FFP	*Fomitella fraxinea* proteases
GISSI	Gruppo Italiano per la Sperimentazione della Streptochinasi nell'Infarto Miocardico
Gln	Glutamine
Glu	Glutamic acid
Gly	Glycine
GP IIb-IIIa	Glycoproteins (IIb-IIIa)
GUSTO	Global Utilization of Streptokinase and Tissue Plasminogen Activator for Occluded Coronary Arteries
His	Histidine
HIT	Heparin induced thrombocytopenia
Ile	Isoleucine
InTIME-II	Intravenous NPA for the treatment of infarcting myocardium early
ISIS	International Study of Infarct Survival Collaborative Group
kDa	Kilo daltons
Leu	Leucine
LK	Lumbrokinase
LMWHs	Low molecular weight heparins
Lys	Lysine
Met	Methionine
MI	Myocardial infarction
NK	Nattokinase
NO	Nitric oxide
n-PA	Novel plasminogen activator
P2Y	Puronogenic ADP receptors
PAGE	Polyacrylamide gel electrophoresis
PAIs	Plasminogen activator inhibitors
PAs	Plasminogen activators
PATT	Partial thromboplastin time
PERSIST	PERsistence Study of Ibandronate verSus alendronaTe
pH	Hydrogen ion
PH_2	Prostaglnadin H_2
Phe	Phenylalanine
Pro	Proline
PROCAT	Prolyse in Acute Cerebral Thromboembolism
Pro-UK	Pro-urokinase
r-PA	Recombinant plasminogen activator
r-proUK	Recombinant prourokinase
r-SAK	Recombinant staphylokinase
rSK	Recombinant streptokinase
rt-PA	Recombinant tissue plasminogen activator
SAK	Staphylokinase
Scu-PA	Single chain urokinase plasminogen activator
SDS	Sodium dodecyl sulphate
Ser	Serine
SERPINS	Serine protease inhibitors
SESAM	Study in Europe with Saruplase and Alteplase in Myocardial Infarction
SK	Streptokinase
SP	Serrapeptase
TAFI	Thrombin-activatable fibrinolysis inhibitor
Thr	Threonine
TIMI	Thrombolysis in Myocardial Infarction
TNK	Tenecteplase
t-PA	Tissue-type plasminogen activator
Trp	Tryptophan
TxA2	Thromboxane A2
Tyr	Tyrosine
u-PA	Urokinase plasminogen activator
USA	United States of America
USFDA	United States Food and Drug Administration
Val	Valine
vWF	vonWillebrand factor
WHO	World Health Organization
wt-SAK	Wild type staphylokinase
α1-PI	α1-Antitrypsin

15.1 Introduction

Coronary thrombosis was identified as fatal in the early nineteenth century itself, but little was known about its cause and risk factors. It was rather regarded as a medical curiosity and for many years it was considered causing sudden death every time. Until the dawn of twentieth century pathophysiology of acute myocardial infarction (AMI) was unclear. Krehl in 1901 proposed that thrombosis doesn't always cause immediate death (Krehl 1901). Later it was found that recovery from AMI is possible and advocated bed rest for post infarction recovery (Herrick 1912). During that period morphine was recommended to relieve pain after AMI (Parkinson and Bedford 1928). Thus management of AMI was palliative than curative during the earlier times. Aspirin was reported to prevent myocardial infarction (MI) during 1950s and used widely for preventing primary and secondary MIs though it has not been proved useful for acute phases of infarction (Craven 1950, 1953). Coronary care units (CCU) were introduced in early 1960s for having continuous cardiac monitoring and treating fatal arrhythmias, thereby reducing the mortality rate due to MI (Julian 1961).

Whenever an injury or trauma occurs, the blood flow through the vessels should be immediately arrested to prevent the loss. This is achieved by hemostasis, characterized by equilibrium between blood coagulation and fibrinolysis. The hemostatic system includes primary hemostasis (platelet plug formation), secondary hemostasis (coagulation) and tertiary hemostasis (fibrinolysis). Primary hemostasis initiates through vascular spasm or local vaso-constriction occurring immediately after injury and thereby reducing the blood flow through the vessel followed by platelet aggregation to form platelet plug. Coagulation factors normally exist as pro-enzymes and circulate through blood stream along with pro-coagulant and anticoagulant factors such as von Willebrand factor, tissue factor, platelet activating factors, tissue factor pathway inhibitor, endothelium-derived relaxing factor or nitric oxide (NO), tissue plasminogen activator (t-PA) and prostacyclin (Lane et al. 2005; Pearson 1994). Once trauma occurs the above factors get activated. Coagulation or secondary hemostasis usually referred as coagulation cascade initiates a series of enzyme reactions which consists of intrinsic and extrinsic pathways (Macfarlane 1964). Chief coagulating enzyme thrombin, formed from Factor Xa initiates fibrin polymerization and thereby blood clotting (Kovalenko et al. 2017). However, removal of clot after wound healing is also important in order to ensure the proper blood circulation. Tertiary hemostasis is characterized by dissolution of clot or fibrinolysis controlled by fibrinolytic system. Plasmin is the enzyme that degrades clot, and it is formed form plasminogen with the aid of plasminogen activators. However, this is regulated by plasminogen activator inhibitor 1 (PAI-1), thrombin-activatable fibrinolysis inhibitor (TAFI), α2-antiplasmin and α2-macroglobulin etc. (Stassen et al. 2004).

Normal functioning of the circulatory system is achieved through hemostatic balance between fibrin formation and degradation (Fig. 15.1). Improper accumulation of blood clot inside the vessel is called thrombus which is usually removed by plasmin. Nevertheless, when the balance shifts towards the inappropriate accumulation of fibrin, normal oxygen supply towards cells will be disturbed causing tissue damage, leading to thrombolytic diseases such as acute myocardial infarction and stroke (Arnout et al. 2006). According to WHO report 17.7 million deaths (below 70 years old) are due to heart diseases representing 31% of total mortality occurred globally during the year 2015 (WHO 2017). Among these 7.4 million were due to coronary heart disease while 6.7 million were due to stroke. Cardiovascular conditions such as acute myocardial infarction, ischemic heart disease and stroke often occur due to blockage of blood flow and many factors play a role in these obstructions. Removal of fibrin clot is essential for proper management of thrombolytic diseases. Cardiovascular disease management often involves use of anticoagulants such as warfarin and heparin, antiplatelets such as dipyridamol and aspirin, removal of blood clot either by surgery or by using thrombolytic enzymes.

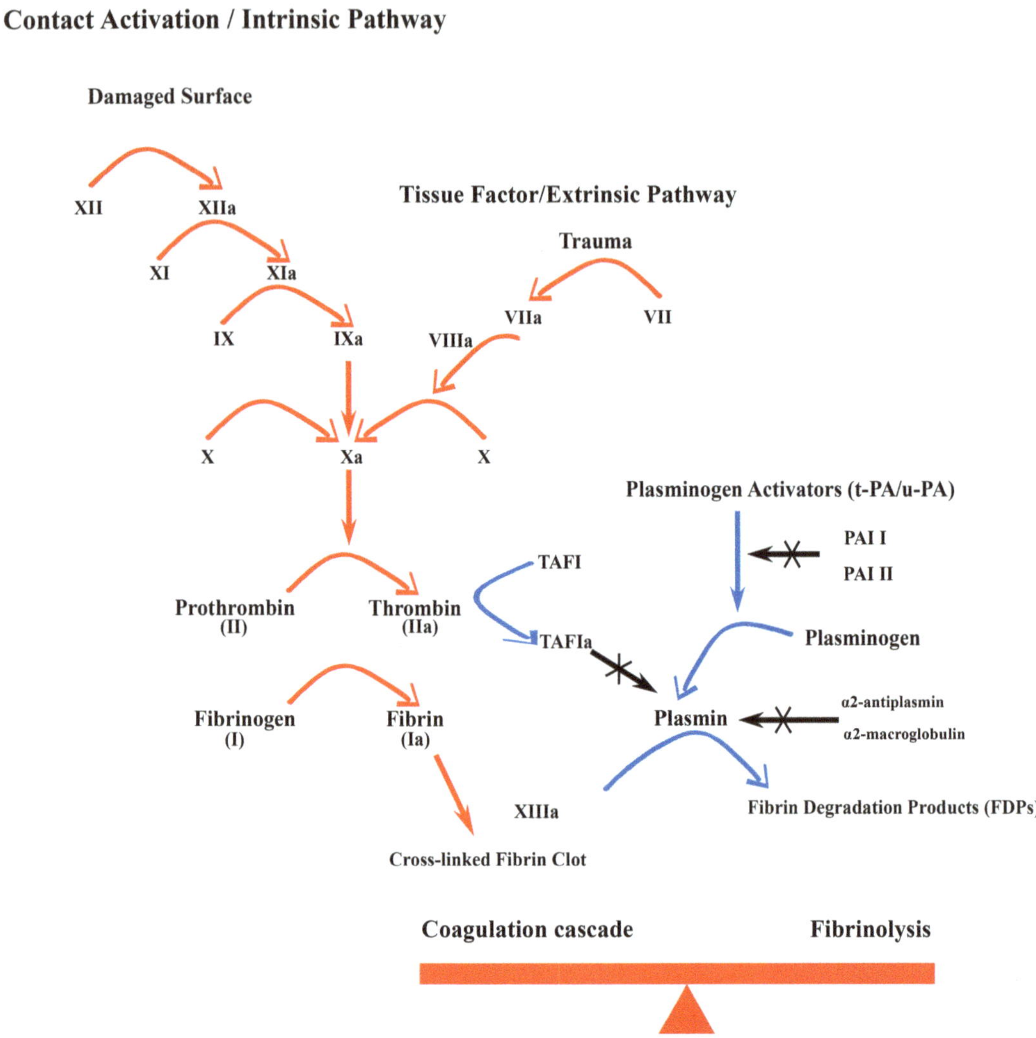

Fig. 15.1 Normal hemostasis- equilibrium between coagulation and fibrinolysis (*TAFI* Thrombin Activatable Fibrinolysis Inhibitor, *PAI* Plasminogen Activator Inhibitor, *FDPs* Fibrin degradation products)

15.2 Status of Anticoagulants and Antiplatelets

15.2.1 Heparin and Derivatives

Anticoagulants are known to inhibit blood clotting and were used in preventing MI from early twentieth century. A fat soluble anticoagulant was discovered while searching for procoagulants from dog liver in 1915 by McLean under the guidance of Howell. Later, Howell and Holt isolated fat soluble anticoagulant heparin which was claimed to be distinct from the anticoagulant molecule obtained by McLean. The term heparin was coined from the Greek word 'hepar' meaning liver by Howell indicating the source of isolation (Howell and Holt 1918). However the credit of discovery was disputed since McLean claimed himself that he discovered heparin (Marcum 2000). Heparin is a glycosaminoglycan inhibiting blood clotting. Though it was discovered a century ago it took long time for its clinical application. However, earlier studies conducted by Mayo Clinic, Minnesota, USA showed that heparin causes side effects such as headache, fever and nausea (Mason 1924). Later, researchers from various parts of the world worked on purification and improvement of heparin preparation and thereby eliminating the side effects to a great extent.

Murray et al. (1937) showed that heparin could prevent thrombus formation in traumatized veins of dogs. Later it was used as anticoagulant in humans in the year 1937. It enhanced the clotting time during 2 h infusion (Murray et al. 1937). Several reports expressed successful use of heparin (McLean and Johnson 1946; Lange et al. 1945), but clinical practice of the molecule reflected certain shortcomings such as heparin induced thrombocytopenia (HIT), bleeding complications etc. which have to be constantly monitored as well as multiple and regular administration of heparin on a daily basis is essential (Cohen 1999). Despite these limitations, heparin still remains as one of the widely used cardiovascular drugs, though it is the oldest therapeutic used as anticoagulant. Apart from heparin, low molecular weight heparins (LMWHs) derived from heparin by its depolymerization also find their application as anticoagulants. They unveil better therapeutic index with lower frequency of HIT in comparison with heparin. But, LMWHs are not appropriate for management of already established HIT, since the antibodies produced against heparin react with them. Each LMWH is distinct in its molecular weight and therapeutic properties (Bick and Fareed 1999; Cohen 1999). On the other hand, LMWHs also need to be injected and the major route of clearance is renal. Hence, the half-life of the molecule is higher in patients with renal failure leading to its accumulation (Hirsh et al. 2001). A few of these LMWHs include enoxaparin, dalteparin, nadroparin, tinzaparin, certoparin, reviparin, ardeparin, parnaparin and bemiparin.

15.2.2 Coumarin Derivatives

Anticoagulants like coumarin derivatives were discovered in 1939 by the identification of dicumarol from spoiled sweet clover hay which is a vitamin K antagonist by Link and Campbell (Stahmann et al. 1941). Further studies on coumarin derivatives led to the discovery of warfarin which was initially used as rodenticide and later approved for clinical use as anticoagulant. It was the first oral thrombin inhibitor and found useful in preventing embolic strokes (Aguilar and Hart 2005). Limitations of these vitamin K antagonists were bleeding complications, interaction with food, necrosis and hair loss (Dantas et al. 2004; Ansell et al. 2008; Pirmohamed 2006). Even though other coumarin derivatives such as phenprocoumon and acenocoumarol also were used as anticoagulants, warfarine is the most common vitamin K antagonist in practice and remains as affordable in cardiovascular disease management.

15.2.3 Thrombin Inhibitors

Several thrombin inhibitors have been developed during the past couple of decades and proved their efficacy as anticoagulants. Hirudin is one of the most important, naturally occurring thrombin inhibitor molecule isolated from leach. Administration of hirudin has been associated with increased risk of bleeding as well as formation of non-neutralizing antibodies in patients (Hoppensteadt et al. 2008). Later, many thrombin inhibitors with better efficacy and therapeutic potential have been developed such as argatroban, bivalirudin, lepirudin and dabigatran etexilate. The first used thrombin inhibitor is Argatroban and it is now widely used in Japan. It was recommended as an alternate anticoagulant for patients suffering from HIT and its clinical use significantly reduced bleeding complications in comparison to heparin (Lewis et al. 2001, 2003). Bivalirudin is a bivalent reversible inhibitor and when compared to heparin and LMWHs, they declined the bleeding complications by almost 50% whereas the efficacy remained same for all of them (Carswell and Plosker 2002; Ahrens et al. 2007). Another thrombin inhibitor lepirudin was found to be marginally superior to heparin and more suitable for patients with previously reported HIT though continuous monitoring is required here also (Lubenow et al. 2004). Dabigatran etexilate is an oral prodrug that would get converted into active dabigatran, a direct thrombin inhibitor, upon intestinal absorption (Lee and Ansell 2011). All those thrombin inhibitors described here are FDA approved for preventing various cardiovascular diseases.

15.2.4 Factor Xa Inhibitors

Factor Xa is a serine protease and since it is part of prothrombinase complex, it carries crucial role in blood coagulation. Factor Xa inhibitors are another class of anticoagulant drugs reducing thrombin formation. These small molecules bind with active site of factor X (Xa) which in turn inhibits the formation of thrombin. They demonstrate varying structural and functional diversity and can be either naturally occurring or synthetic molecule and chemically it can be peptide, proteins or heparin saccharidic sequences (Fareed et al. 1999; Leone et al. 2004; Bauer 2006; Samama et al. 1997). Their mode of action could be either direct by binding to factor Xa or indirect through binding with antithrombin III (AT), leading to Factor Xa inactivation. The interaction of Xa inhibitors could be reversible or irreversible (Harenberg and Fenyvesi 2004; Walenga et al. 2002). Fondaparinux is a heparin pentasaccharide, FDA approved this for its use after surgery and also for initial management of thrombotic disorders such as pulmonary embolism and deep vein thrombosis trials (Hoppensteadt et al. 2008). Another indirect acting FXa inhibitor is Idraparinux (SANORG-34006), which is a hyper methylated variant of fondaparinux and showed more affinity towards antithrombin III (AT) (PERSIST investigators 2004). Rivaroxaban was an orally active, direct thrombin inhibitor and extensively studied for its use in preventing venous thromboembolism during knee and hip replacement (Turpie et al. 2005; Eriksson et al. 2006). A few other direct Factor Xa inhibitors include otamixaban, apixaban, edoxaban etc. Although most of the Factor Xa inhibitors are either under clinical trial or in different developmental stages, still they are considered safer than thrombin inhibitors (Hoppensteadt et al. 2008).

15.2.5 Antiplatelet Drugs

Platelet activation is initiated by vonWillebrand factor (vWF) interacting with collagen and platelet surface glycoprotein receptors resulting in its adherence to vascular sub-endothelium. Activated platelets further release thromboxane A2 (TxA2), thrombin and adenosine diphosphate (ADP). These series of events eventually leads to the expression of platelet surface glycoproteins (GP IIb-IIIa) subsequently leading to platelet aggregation and thrombus formation (Oprea and Popescu 2013). Antiplatelet drugs are used for prevention and management of thrombotic disorders. First antiplatelet drug used was aspirin, though its mechanism of action was unresolved at the time of use. Later it was found that platelet-connective tissue reaction can be considerably damaged by aspirin uptake. This inhibits the platelet plug formation and thereby prolongs the clotting time (Weiss and Aledort 1967). Further studies revealed that antiplatelet effect of aspirin is largely due to its irreversible inactivation of cyclooxygenase enzyme, primarily cyclooxygenase 1 (COX1). Inhibition of prostaglandin synthase/cyclooxygenase enzyme impedes the prostaglandin produced by platelets. Prostaglnadin (PH_2) is the substrate for synthesis of the TxA2 by thromboxane-A synthase (Vane 1971; Smith and Willis 1971; Oprea and Popescu 2013) and thus, COX1 inhibition by aspirin is accountable for its use as antiplatelet drug. Another class of antiplatelet therapeutics is ADP receptor inhibitor which blocks binding of ADP to its specific receptor on blood platelet surface. Ticlopidine was the earliest ADP receptor antagonist, later discontinued due to its adverse effects such as thrombocytopenia. Successively a series of ADP receptor inhibitors are developed like clopidogrel, prasugrel, and ticagrelor. Clopidogrel is a $P2Y_{12}$ receptor inhibitor which irreversibly blocks binding of ADP. Prasugrel is another irreversible inhibitor of $P2Y_{12}$ receptor. Ticagrelor on the other hand mainly blocks $P2Y_1$ receptor and it partially inhibits $P2Y_{12}$ receptor too. The inhibitory mechanism is instead reversible here. A third class of antiplatelet drugs belongs to glycoproteins (GP IIb-IIIa) antagonists. GP IIb-IIIa are chief integrins responsible for platelet aggregation and subsequent thrombus formation. Abxicimab, tirofiban and eptifibatide are a few GP IIb-IIIa inhibitors that are FDA approved and antagonism exhibited by these molecules are reversible (Oprea and Popescu 2013).

15.3 Thrombolytic Drugs

Despite all the advancements in the area, anticoagulants and antiplatelet drugs are with significant short comings such as bleeding complications, but are in practice for a longer period of time. However, major drawback of this therapy is that these drugs cannot act on an existing clot. Demand for clot buster drugs was there for a longer time. Enzyme therapies are becoming more significant during the current scenario. Fibrinolytic enzyme therapy is one among the many potential applications of therapeutic enzymes. Accidental discovery of streptokinase paved the way for development of thrombolytic therapy. Streptokinase was regarded as a wonder drug since its discovery and still a drug of choice in most of the developing countries.

Major advantage of treatment using fibrinolytic enzymes over anticoagulants and antiplatelets is that they could act upon an existing clot. Primary focus of these enzymatic actions is on the protein (fibrin) either directly or indirectly. They are often called as clot bustering enzymes and categorized into two types based on the mechanism of action. First category is plasminogen activators; converting plasminogen to active plasmin that cleaves the fibrin clot formed. eg tissue-type plasminogen activator (t-PA) and urokinase. They are widely used for treatment of cardiovascular diseases. t-PA was approved in 1996 by Food and Drug A administration (FDA) for intravenous injection against thrombosis. Recombinant tissue plasminogen activator alteplase (Activase®) against cardiovascular diseases is the first genetically engineered enzyme to get approved by United States Food and Drug Administration (USFDA). Many plasminogen activators such as alteplase, reteplase, tenecteplase, urokinase, streptokinase and anistreplase are available for clinical use and have been commonly used in treating cardiovascular diseases for the past five to six decades (Kotb 2014). Second category of thrombolytic enzyme is plasmin like enzymes that are direct acting fibrinolytic enzymes which do not require activation of plasminogen. Instead, they can perform clot dissolution by acting upon it, eg. plasmin, nattokinase and lumbrokinase.

15.3.1 Plasminogen Activators as Thrombolytic Drugs

Plasminogen activators, also called indirect thrombolytic drugs are serine proteases that stimulate formation of plasmin from plasminogen. They are classified into three categories namely first, second and third generation plasminogen activators. First generation plasminogen activators (streptokinase and urokinase) were less fibrin specific and caused bleeding complications. Second generation plasminogen activators were developed to increase fibrin specificity and thereby reducing the bleeding complications. Though it reduced the complications associated with first generation plasminogen activators, they still possess risk of bleeding. Third generation plasminogen activators are mostly genetically engineered t-PA variants (except staphylokinase and desmoteplase) to achieve superior properties such as fibrin selectivity, improved half-life and prolonged renal clearance time. Development of plasminogen activators as thrombolytic drugs are discussed below.

15.3.1.1 First Generation Plasminogen Activators as Thrombolytic Drugs

With the discovery of streptokinase, a new line of treatment for cardiovascular diseases were started i.e., dissolution of clot using thrombolytic enzymes. First generation plasminogen activators converts circulating plasminogen to active plasmin which in turn cleaves fibrin as well as other proteins such as clotting factors, fibrinogen etc. The first generation thrombolytics thus are not fibrin specific and lead to systemic bleeding complications, but offered a significant hope in the new therapeutic approach towards cardiovascular diseases (Flemmig and Melzig 2012).

Streptokinase

First generation fibrinolytic enzyme therapy began with the discovery of streptokinase during 1930s. Streptococci a beta hemolytic bacterium was known to produce a fibrinolytic substance which was isolated from wound exudates (Tillett and Garner 1933). This substance was earlier named streptococcal fibrinolysin and later named as streptokinase by Christensen. Plasminogen in human plasma gets activated by these kinases to active plasmin which is responsible for fibrinolysis (Christensen 1945). Further studies proved that plasmin could digest not only fibrin but fibrinogen too. Plasmin inhibitors prevented fibrinolysis in normal course. Partially purified streptokinase was used to study the therapeutic potential in earlier studies (Christensen 1947). However, the prospective of using streptokinase (SK) as thrombolytic drug was acknowledged only with the experiment of Johnson and Tillett (1952) when they successfully dissolved intravascular clots in the marginal ear veins of rabbit. Later thrombolysis in femoral artery by streptokinase was reported (Sherry 1954).

Large scale purification of human plasminogen was achieved by Kline (1953). This led to direct use of plasmin as clot buster. Intravenous administration of streptokinase-activated human plasmin dissolved clots formed in canine vessels (Cliffton et al. 1953). The major concern was whether direct use of plasmin or use of plasminogen activators was beneficial. Mechanism of action by streptokinase turned out to be a two-step process. Initially it formed a complex with human plasma factor and converted the plasma factor to plasminogen activator and finally this plasminogen activator converted plasminogen to plasmin (Davies et al. 1964; Sherry 1954). Direct use of plasmin not only dissolved clot but also other proteins such as clotting factors and fibrinogen. Thus, it was assumed that plasminogen activators are better to be used as therapeutic agent rather than circulating plasmin (Alkjaersig et al. 1959; Fletcher et al. 1959b). It was established that streptokinase infusion via the intravenous route after AMI significantly reduced mortality (Fletcher et al. 1958, 1959a). Streptokinase as cardiovascular therapeutic was found promising since it was the only drug that could improve the condition after AMI at that time. Dissolution of intracoronary clots by streptokinase was proved for first time by Ruegsegger and colleagues (1959).

Correspondingly, the adverse effects of SK were reported earlier by Tillett and Sherry (1949). Usage of SK leads to pyrogenic reaction associated with symptoms like headache, malaise, arthralgia and occasionally nausea (Tillett and Sherry 1949). However, it was impossible to produce streptokinase with reduced pyrogenic reaction (Sherry 1981). During those days the purity of streptokinase attained was only 10% (Sikri and Bardia 2007). In order to reduce the side effects, further purification of SK was carried out. Purification of SK was achieved and the molecular weight was determined to be 47 kDa (De Renzo *et al.* 1967). It is a single polypeptide composed of 415 amino acids (Jackson and Tang 1982). Continuous evolution of SK has been achieved through various attempts. With advances in recombinant DNA technology, cloning and expression of streptokinase was made possible to yield better preparations (Malke and Ferretti 1984; Muller et al. 1989; Wong et al. 1994). Recombinant SK (rSK) thus produced was superior with lesser antigenicity compared to wild type SK (Collen et al. 1996).

A large number of clinical trials were performed involving both animals and humans by various groups to determine the safety and therapeutic potential of SK. Some of these studies were representing only a smaller population (Koren et al. 1985). Extensive studies involving large number of populations were conducted during the early 1980s. GISSI (Gruppo Italiano per la Sperimentazione della Streptochinasi nell'Infarto Miocardico) was the first among them. In this study of 11,806 patients in 176 coronary care units were monitored for a period of 17 months. Survival rate was more for patients who received SK during early onset of AMI (GISSI 1986).

The scenario changed after the introduction of t-PA for treating AMI, and finding the best thrombolytic agent among SK and t-PA was the new challenge. t-PA become popular in developed countries for its superior action and t-PA as thrombolytic agent will be discussed later. But SK is still the drug of choice in the case of developing nations and poor countries due to its reduced cost. Streptase®, Kabikinase®, Heberkinasa® etc. are a few of the commercially available streptokinase formulations of which Heberkinasa® is a recombinant streptokinase (rSK) (Kumar and Sabu 2017).

Urokinase Plasminogen Activator (u-PA)
Macfarlane and Pilling (1947) reported the fibrinolytic activity of normal urine for the first time. Urine dilutions of 1/2000–1/4000 dissolved blood clots. Urokinase was later purified from urine (Ploug and Kjeldgaard 1957). Studies of adult human kidney have revealed that primary culture of adult kidney cells produced plasminogen activator. Apart from hepatic cells, primary cultures of other tissues (both prenatal and post natal) such as ureter, renal blood vessels, bladder and lung too produced plasminogen activators, which were antigenically similar to Urokinase (Bernik and Kwaan 1967, 1969). Urokinase is released as an inactive single-chain precursor molecule of 55 kDa and proteolytic enzymes like plasmin cleaves the precursor giving rise to active form of urokinase (Kasai et al. 1985; Petersen et al. 1988; Bernik 1973). The proteolytic cleavage at Lys157 gives the active urokinase, which is a disulfide linked heterodimer of A chain with a molecular weight of 20 kDa and B chain with molecular weight 34 of kDa. The A chain consists of an epidermal growth factor-like domain and a kringle domain whereas B chain is the catalytic unit (Kasai et al. 1985; Appella et al. 1987; Patthy et al. 1984). Plasma is devoid of urokinase inhibitors and hence it is freely available at the site of action.

15.3.1.2 Second Generation Plasminogen Activators

First generation thrombolytics were nonspecific towards fibrin (Flemmig and Melzig 2012). Side effects such as bleeding complication demanded more specific thrombolytic drugs. Second generation thrombolytics were developed to overcome the shortcomings of first generation plasminogen activators.

Development of t-PA as a Thrombolytic Agent
Earlier it was reported that fibrinokinase, a plasminogen activator is present in animal tissues and isolation of the enzyme with chaotropic agents were described by Stage and Astrup (1952). Later it was named as tissue type plasminogen activator (t-PA). Purification of these plasminogen activators was attempted from many sources. Extraction of human t-PA was achieved with high purity, though the yield was as low as 1 mg of t-PA from 5 kg uterine tissue. Later it was found that t-PA is formed in vascular endothelial cells and then released in to circulatory system (Rijken et al. 1979, 1980). However, cloning and expression of human t-PA was achieved which resolved the difficulties with respect to low yield and cost of production. t-PA is a serine protease of 527 amino acids with a molecular weight of 70 kDa (Ranby et al. 1982; Pennica et al. 1983). It is secreted as single polypeptide which is cleaved by plasmin into a two chain form, linked together by disulfide bonds. The C-terminal light chain carries the catalytic serine protease domain while the N-terminal heavy chain comprises an epidermal growth factor domain, a finger domain and two kringle domains (Rijken and Groeneveld 1986). Plasminogen interacts with kringle 2 domain alone, while fibrin binds to kringle 2 domain and fibronectin type III finger domain of t-PA, pointing towards the stimulatory effect of fibrin on t-PA (van Zonneveld et al. 1986). From the kinetic studies of t-PA also, it was found that the activation of plasminogen by the enzyme is remarkably enhanced in the presence of fibrin. Thus it's specificity towards fibrin is substantiated (Hoylaerts et al. 1982). Alteplase is a recombinant t-PA approved in1996 by FDA, and it is the first recombinant therapeutic enzyme to get FDA approval. Since then it is in clinical practice. Alteplase is available in market under the trade name Activase® (Kumar and Sabu 2017). It

is used for treating acute ischemic stroke, pulmonary embolism and acute myocardial infarction and is regulated to be used within 3 h after ischaemic stroke. Also, it is having a short half-life of 4–8 min and unsuitable for bolus injections (Ali et al. 2014). t-PA is clinically used as therapeutic for acute ischemic stroke even though it is often associated with the risk of cerebral hemorrhage and edema (Kaur et al. 2004).

Studies such as GUSTO (Global Utilization of Streptokinase and Tissue Plasminogen Activator for Occluded Coronary Arteries) (GUSTO Angiographic Investigators 1993), Third International Study of Infarct Survival Collaborative Group, ISIS-3 (ISIS-3 1992) and GISSI-2 (Gruppo Italiano per so Studio della Sopravvienza nell'Infarto Miocardico) (GISSI-2 1990) were used to evaluate the therapeutic potential of SK to t-PA. All these studies didn't show any significant difference in death rate for those received either streptokinase or t-PA. However, follow-up of 1 year showed significant reduction in mortality for those treated with t-PA according to GUSTO. t-PA was having a slight edge over streptokinase in the case of patients either having chance to receive drug at the earliest, or at younger age, or earlier treated with streptokinase for AMI (Gershlick and More 1998).

Prourokinase (scu-PA)

Prourokinase or single-chain urokinase-type plasminogen activator (scu-PA) is an inactive precursor form of urokinase and in a surface-controlled process, plasmin associated with fibrin network cleaves the single chain prourokinase into two chain urokinase which in turn converts plasminogen to plasmin. Major advantage of scu-PA is that it shows specificity towards fibrin unlike native urokinase (Bernik 1973; Fleury and Angles-Cano 1991). Pro-UK showed increased half-life of more than 24 h in circulation compared to urokinase and complete lysis of fibrin was attained within 1.5 h whereas urokinase required 3 h for the fibrinolysis (Zamarron et al. 1984). Saruplase or prourokinase (scu-PA) is a naturally occurring glycoprotein carrying 411 amino acids. Recombinant production of prourokinase (r-proUK) can be achieved in either bacterial system (*E.coli*) or mammalian cell line. Prolyse® (Abbott Laboratories) is a recombinant prourokinase (r-proUK) yet to be approved by FDA and production is achieved in murine hybridoma cell line. It was found effective in stroke and myocardial infarctions (Sun et al. 1998; Ouriel 2002). These recombinant prourokinases have undergone a large number of clinical studies. SESAM (Study in Europe with Saruplase and Alteplase in Myocardial Infarction) revealed that saruplase was safe and effective as altplase in acute myocardial infarction (Bar et al. 1997). Another study, PROCAT (Prolyse in Acute Cerebral Thromboembolism) showed the recanalization efficacy in acute thromboembolic strokes (del Zoppo et al. 1998). COMPASS (Comparative Trial of Saruplase Versus Streptokinase) study showed comparable 30-day mortality rate for both saruplase and streptokinase (Tebbe et al. 1998). According to PROCAT II intra-arterial administration of pro-urokinase within 6 h after the acute ischemic stroke considerably enhanced clinical outcome at 90 days (Furlan et al. 1999).

Anistreplase (APSAC)

Anistreplase (anisoylated plasminogen streptokinase activator complex or APSAC) is an equimolar mixture of streptokinase and plasminogen. The active site of plasminogen is protected from plasmin inhibitors by reversible acylation with p-amidinophenyl p-anisate hydrochloride (APAN). Upon injection, deacylation starts by hydrolysis of p-anisate group leaving active plasminogen-streptokinase complex. APSAC is more stable with improved half-life and fibrin binding capacity (Sherry 1990). APSAC is a readily available plasminogen activator and shows more specificity towards plasminogen bound to clot (Smith et al. 1981). ISIS-3 reported that APSAC and streptokinase showed similar results in mortality. Also, it is comparable to streptokinase in terms of antigenicity and safety (Sherry 1990; Munger and Forrence 1990; ISIS-3 1992). Eminase® is a commercially available anistreplase formulation. However, its usage is discontinued in USA recently (Ali et al. 2014).

15.3.1.3 Third Generation Plasminogen Activators (by Protein Engineering)

Third generation plasminogen activators with improved properties were developed for thrombolytic therapy. Most of these classes of drugs are genetically engineered t-PA variants with improved resistance to inhibitors, more safe upon administration, prolonged half-life etc. Their increased clearance rate makes them suitable for bolus administration. A few non t-PA variants such as staphylokinase and desmoteplase also belong to this category.

Reteplase (Recombinant Plasminogen Activator, r-PA, BM 06.022)

Reteplase is an unglycosylated shortened single chain deletion variant of t-PA, recombinatly produced in *E.coli* and having a molecular weight of 39 kDa. It consists of 355 of 527 amino acids from native t-PA; precisely 1–3 and 176–527 amino acids of t-PA. Thus reteplase includes only catalytic protease domains along with kringle 2 domain while it lacks epidermal growth factor domain, finger domain and kringle 1 domains of t-PA (Kohnert et al. 1992; Sturzebecher et al. 1992). The molecules attained enhanced half-life from 4 to 15 min as it lacks kringle 1 domain which is responsible for quick renal clearance (Martin et al. 1991b; Flemmig and Melzig, 2012). Absence of fibrin binding finger domain decreases the binding affinity of the molecule, although kringle 2 domain interacts with fibrin (Kohnert et al. 1992; Martin et al. 1991a). The mortality rate and hemorrhagic stroke for 30 days are similar for reteplase and alteplase (GUSTO III investigators 1997). Reteplase is commercially available as Retavase® (Centocor, Inc.) and Rapilysin® (Roche).

Tenecteplase (TNK-tPA)

Tenecteplase is a recombinant mutant variant of t-PA. It is a 65 kDa glycoprotein with 527 amino acids. The mutations include Asp103Thr, Glu117Asn and 4 alanine substitutions (Lys296Ala, His297Ala, Arg298Ala, and Arg299Ala). These point mutations increase the fibrin specificity by 14-fold and resistance to plasminogen activator inhibitor 1 (PAI-1) by 80-fold (Smalling 1996). Mutations in kringle 1 domain prolongs the renal clearance time and thus increases the half-life to about 10–24 min (Cannon et al. 1997). This makes it suitable for bolus administration. Clinical trials such as TIMI 10B (Thrombolysis in Myocardial Infarction) trial, ASSENT-1 and ASSENT-2 (Assessment of Safety and Efficacy of a New Thrombolytic agent) trials reported similar efficacy and safety while used for treatment of AMI. However, bleeding rate was lesser compared to altepase (Zeymer and Neuhaus 1999; Cannon et al. 1998; Van de Werf et al. 1999a, 1999b). Trade name of tenectplase is TNKase™ (Genentech, Inc.) and was approved by USFDA in 2000 for treatment of myocardial infarction.

Lanoteplase/Novel Plasminogen Activator (n-PA)

Lanoteplase is a novel truncated mutant variant of t-PA created by removing finger domain and epidermal growth factor domain. Apart from that, it has point mutation at kringle1 domain Asn117Gln. This in turn results in deletion of glycosylation site. These modifications increase the renal clearance time and the plasma half-life of the molecule turns out to be 37–45 min. It is also suitable for bolus administration. The enzyme is a single polypeptide with a molecular weight of 53.6 kDa and recombinant production is achieved in Chinese Hasmster Ovary (CHO) cell lines (Larsen et al. 1989, 1991; Nordt et al. 1999; Collen et al. 1988). In comparison with alteplase higher patency rate was shown by lanetoplase for 90 min while similar mortality rate was shown at 30 days (den Heijer et al. 1998; InTIME-II investigators 2000). However mild bleeding complications with haemorrhagic stroke were observed more than that found for alteplase (Verstraete 2000).

Pamiteplase/Solinase

Pamiteplase is another shortened mutant of t-PA, devoid of kringle 1 domain (residues 92–173) along with point mutation Arg274Glu. This variant is resistant to cleavage by plasmin and having a half-life of about 30–47 min (Flemmig and

Melzig 2012). In comparison with t-PA, much higher fibrinolysis was shown by pamiteplase under an in vivo study using canine coronary thrombosis model (Kawasaki et al. 1993). Another trial reported comparable results for t-PA and pamiteplase, where TIMI-3 flow rate was 25% at 30 min and 50% after 60 min for pamiteplase while it was respectively 16% and 48% for t-PA (Flemmig and Melzig 2012). Recombinant production is achieved in CHO cell lines like lanoteplase (Yokota and Tanji 2008).

Duteplase (met-t-PA)

Duteplase in another variant of t-PA with a single point mutation Val245Met and recombinant production is achieved in CHO cell lines. Duteplase is a nonglycosylated, two-chain form as opposed to single chain form of alteplase. According to ESPRIT study, incidence of patency in infarct-related coronary artery was achieved in 70% patients at 90 min with weight-adjusted duteplase infusion along with oral aspirin and intravenous heparin. Safety of the molecule was similar to that of alteplase (Kalbfleisch et al. 1992; Malcolm et al. 1996).

Monteplase (E6010)

Another mutated variant of t-PA is monteplase where a point mutation has been achieved in epidermal growth factor domain (Cys84Ser). However, extended half-life of about 23 min makes the mutant suitable for bolus administration (Suzuki et al. 1991; Kawai et al. 1997). TIMI-3 flow rate was 69% at 60 min post monteplase treatment for AMI (Flemmig and Melzig 2012). Higher rate of early recanalization was reported for the molecule in comparison to tisokinase, a native t-PA variant in a randomized double-blind study conducted in Japan. Also, fatal bleeding wasn't observed during the trial (Kawai et al. 1997).

Amediplase (MEN 9036)

Amediplase is a chimeric enzyme of 365 amino acids with kringle 2 domain of t-PA (1–3 and 176–275 residues) and the protease domain of single-chain urokinase, scu-PA (159–411). Molecular weight of the protein is 39.9 kDa and recombinant production is achieved in CHO cell lines (Agnelli et al. 1993; Muller et al. 1994). Amediplase showed ten times lesser binding affinity towards fibrin in comparison with alteplase under internal and external clot lysis methods. However clot penetration ability of the molecule was superior to t-PA (Rijken et al. 2004). However, amediplase has shown slightly better activity than tenecteplase and scu-PA in another external clot lysis method (Guimaraes et al. 2006).

Desmoteplase (Bat-PA, DSPAα1, v-PA)

Desmoteplase is third generation plasminogen activator, a non t-PA variant. Desmodus salivary plasminogen activators (DSPAS) is a group of four enzymes isolated from common vampire bat, *Desmodus rotundus* and desmoteplase is one among them, called DSPAα1 or Bat-PA (Kratzschmar et al. 1991). It is a single chain 50 kDa protein which could be produced in both CHO cell lines and insect cells by recombinant techniques. It has an extended terminal half-life when compared with other thrombolytic enzymes such as t-PA (Hildebrand et al. 1996). DSPAα1 contains a single kringle domain quite similar to the kringle 1 domain of t-PA but it is devoid of any lysine binding site. Apart from kringle domain, it also contains a signal peptide, an epidermal growth factor (EGF) domain, a serine protease domain, and a finger domain. Also the protease domain lacks plasmin cleavage site. The bimolecular rate constant of plasminogen activation was 13,000 for DSPAα1 and 72 for t-PA in the presence of fibrin versus fibrinogen (fibrin selectivity), proving it to be highly fibrin specific (Bringmann et al. 1995). Tranexamic acid is a synthetic lysine analog that can inhibit desmoteplase, therefore possible for managing any bleeding complications, occuring during the usage of the enzyme (Niego et al. 2008). Canine model of arterial thrombosis showed desmoteplase to be better in thrombolysis compared to t-PA (Mellott et al. 1992). Intravenous administration directly into brain causes excitotoxic injury leading to brain damage in case of t-PA, whereas desmoteplase does not encourage excitotoxic injury and thus proves to be a better therapeutic for ischemic stroke (Reddrop et al. 2005). DSPAα1 definitely has therapeutic advantages over t-PA and preclinical studies have already showed better specificity and selectivity towards fibrin. Most thrombolytic drugs have a time window of less than 3–4.5 h. However, desmoteplase is supposed to demonstrate better clinical advantages by extending the therapeutic window of patients with ischemic stroke. Desmoteplase in Acute Ischemic Stroke (DIAS) trial; a phase II trial showed that administration of this drug within 3–9 h after the onset of the symptoms of stroke has got a clinical benefit with use of DSPAα1 (Hacke et al. 2005). However the DIAS-2 failed to prove any significant advantage of the drug within 3–9 h time window after onset of stroke (Hacke et al. 2009). According to DIAS 3 and DIAS 4 trials, acceptable safety profile was observed for DSPAα1, but no evidence to extend therapeutic window up to 9 h for ischemic stroke was found (Albers et al. 2015; Kummer et al. 2016). Satisfactory reperfusion efficacy was shown by desmoteplase after 3 h since onset of AIS; however more studies with well-designed clinical trials are needed to be evaluated for the effectiveness of the drug to improvise the therapeutic window in treating AIS patients (Li et al. 2017). Although intracranial hemorrhages are low for desmoteplase, it did not showed any neurological improvement or survival rate improvement (Hacke et al. 2009; Albers et al. 2015; Kummer et al. 2016).

Staphylokinase (SAK)

Staphylokinase is another non t-PA variant belonging to the third generation plasminogen activators. It is an extracellular protein produced by *Staphylococcus aureus*, which forms 1:1 stoichiometric complex with plasmin or plasminogen and activates more plasminogen molecules. SAK is a monomer of 136 amino acids without any disulfide bond and molecular weight is about 15.5 KDa. SAK complexes with plasminogen molecules that are bound to partially degraded fibrin (Sakharov et al. 1996). α2-antiplasmin inhibits plasminogen-SAK complex in the absence of fibrin, however, in presence of fibrin the lysine binding domain of the complex is occupied, preventing the inhibition by α2-antiplasmin and there is a fourfold increment in activity with presence of fibrin, thus more specificity towards fibrin (Lijnen et al. 1991).

Recombinant production of SAK is achieved in *Escherichia coli*, *Bacillus subtilis* and yeast *Hansenula polymorpha* (Sako 1985; Gerlach et al. 1988; Moussa et al. 2012). Recombinant wild type SAK (wt-SAK) showed more fibrinolysis and less fibrinogenolysis compared to that of streptokinase (Collen and Van de Werf 1993). Yet, therapeutic application of wt-SAK is associated with antigenicity and hence its repetitive use as a drug is ineffective (Vanderschueren et al. 1997). Antigenic epitopes can be removed without affecting the activity, by site directed mutagenesis (Laroche et al. 2000). Since then many site directed mutants were produced with reduced immunogenicity. Many of the in vivo studies and clinical trials showed the therapeutic efficacy and safety of recombinant SAK over t-PA. It did not show any systemic fibrinogenolysis or allergic reactions during the earlier studies (Vanderschueren et al. 1995a, b, 1996, 1997; Armstrong et al. 2000, 2003).

Two hybrid proteins, SAK-RGD-K2-Hir and SAK-RGD-K2-Hirul, carrying fibrinolytic C-terminal part of recombinant staphylokinase (r-SAK), fibrin binding kringle 2 domain of tissue-type plasminogen activator (t-PA), antiplatelet aggregating tripeptide sequence Arg-Gly-Asp and anticoagulant hirudin (65 amino acids) or hirulog (hirudin analogue of 20 amino acids), were produced by cloning and expression in pESP-3 expression vector. In vitro studies proposed these fusion proteins to be better thrombolytic agents than recombinant wt-SAK. However SAK-RGD-K2-Hirul didn't show any clinical benefit over SAK-RGD-K2-Hir and it also caused bleeding complications (Szemraj et al. 2005, 2011).

The major disadvantage of plasminogen activators (PAs) is high cost and side effects such as bleeding, nausea and hemorrhage. Upon systemic administration, plasminogen activators (PAs) exceed the plasminogen activator inhibitors (PAIs) and start dissolving the clot. Apart from clot dissolution, they also disrupt the vascular plug formed at the site of injury. Thus bleeding complications are inevitable in treatment with plasminogen activators that could lead to serious situations such as intracranial hemorrhage (Gore et al. 1991; Marder 1979; Marder and Sherry 1988). Their disadvantages necessitated the discovery of better thrombolytic agents.

15.3.2 Direct Acting Plasmin Like Thrombolytic Drugs

Research on plasminogen activators resulted in development of recombinant and chimeric protein production. Significant improvement in fibrin specificity, safety, extended plasma life etc. has been achieved so far. But still these thrombolytic agents possess risk of bleeding complication, allergic reaction and repercussion. Hence search for new and better drugs continues. Catheter based local administration of plasminogen activators fastens the process of thrombolysis, yet systemic bleeding cannot be avoided completely. Direct acting fibrinolytic enzymes were reported from various sources which could by-pass plasminogen activation and disintegrate fibrin clot. They were isolated from non-microbial as well as microbial sources.

15.3.2.1 Non-microbial Thrombolytic Enzymes

Non microbial thrombolytic enzymes include plasmin, variants of plasmin, fibrinolytic proteases from earth worm, snake venom, plant and algae. Few among them offered hope as cardiovascular therapeutic, however further clinical studies are required to determine their safety and efficacy. Even though reported with thrombolytic potential, further therapeutic usage of plant and algal proteases were not reported so far.

Plasmin

Plasmin is the fibrin degrading enzyme which can be administered without any bleeding complication and allergic reactions. Systematically administered plasmin often gets inhibited by α2-antiplasmin, thereby losing the therapeutic benefit (Marder 2011). But when given locally plasmin doesn't get blocked by its inhibitors, and will lead to fibrinolysis. Later it gets inhibited by α2-antiplasmin when it is released to circulation after thrombolysis (Marder and Novokhatny 2010; Marder 2011). In vivo animal models and clinical trials suggested the clinical benefit and safety of plasmin as a cardiovascular drug that could accomplish fibrinolysis (Hoefer et al. 2009; Sadeghi et al. 2003; Jahan et al. 2008; Shlansky-Goldberg et al. 2008). Plasmin is a heterodimer with an N-terminal heavy chain of molecular weight 65 kDa and a C-terminal light chain of molecular weight 25 kDa linked together by two disulfide bonds. Heavy chain contains 5 kringle domains and a 77 residue N-terminal peptide, whereas light chain carries a serine protease domain. It binds to fibrin through lysine binding sites at kringle domain, chiefly at kringle 1 domain (Wiman et al. 1979; Robbins et al. 1967; Sottrup-Jensen et al. 1978; Violand and Castellino 1976; Wiman 1977, Wiman and Wallen 1977; Lucas et al. 1983; Wu et al. 1994).

Miniplasmin

Numerous variants of plasmin have been constructed for therapeutic applications. Miniplasmin is one among them and it is a truncated mutant obtained from plasminogen by proteolysis using porcine pancreatic elastase. It contains kringle 5 domain and serine protease domain of plasmin (Sottrup-Jensen et al. 1978). Kringle 5 domain mediates binding of fibrin to miniplasmin and inhibition of this protein is lower compared to that of plasmin (Wiman et al. 1978). Recombinant production of miniplasminogen is achieved in *E. coli* using T7 expression system and activated to form miniplasmin by urokinase (Moroz 1981). However, the kinetic data for thrombolysis by miniplasmin is contradictory, which is showing higher and lower efficiency in two separate studies (Moroz 1981; Kolev et al. 1996). Using canine model of femoral artery thrombosis it was observed that occurrence of reperfusion was much higher upon miniplasmin administration, when compared to rt-PA. Also, miniplasmin treated group left without any reocclusion, whereas 20% reocculsion was showed by rt-PA treated group. This points towards the lower incidence of hemorrhage while administering miniplasmin over rt-PA (Fu et al. 2008).

Microplasmin

Microplasmin is another plasmin variant generated by autolysis of plasmin at high alkaline pH (pH 11) and is a 29 kDa protein. It is a heterodimer made up of 'A' chain with 31 amino acids and 'B' chain with 230 amino acids (Wu et al. 1987). Micro-plasminogen can be recombinantly produced in *Pichia pastoris* and *E. coli.* (Nagai et al. 2003; Ma et al. 2007). Bioengineering significantly reduces the overall production cost. On the other hand it doesn't bind with fibrin since it is devoid of all the kringle domains of plasmin. Also, rate of inhibition by α2-antiplasmin is very low with respect to native plasmin (Nagai et al. 2003). In vivo animal models and clinical trials proved its safety to be used as a therapeutic since it was not causing any hemorrhage (Suzuki et al. 2004; Chen et al. 2007; Thijs et al. 2009). Microplasmin (ocriplasmin) also finds application as therapeutic for vitreomacular adhesion. It causes vitreomacular detachment of fibronectin and laminin (Chen et al. 2008; de Smet et al. 2009). Ocriplasmin received FDA approval for treating vitreomacular adhesion and is commercially available under trade name JETREA®. However, its use as a cardiovascular drug is yet to be validated by further clinical studies.

Delta Plasmin

Third and most important plasmin variant is Delta plasmin, which is a truncated mutant formed by removing krinlgle 2–5 domains leaving only kringle 1 domain attached to the serine protease domain of plasmin. Recombinant production of the Delta plasminogen [Delta(K2-K5) Pg] was carried out in *E. coli* T7 expression system and found functionally similar to human plasminogen. It was converted to active Delta plasmin [Delta (K2-K5) Pm] by plasminogen activators such as t-PA, urokinase and streptokinase. Even though it showed fibrinolytic efficacy equivalent to plasmin (Hunt et al. 2008) further validation is required.

Plasmin and its variants are serine proteases that directly dissolves clot, but gets inhibited by serine protease inhibitors (SERPINS) in plasma. Most common serpin in circulation is α2-antiplasmin, which immediately inhibits plasmin upon systemic administration. Plasmin however doesn't show bleeding complications since it get inhibited rapidly in circulation. However catheter based local administration of plasmin is more effective in dissolving thrombus (Marder et al. 2001). Delivery of thrombolytics through sophisticated catheter based system limited their use (Marder and Novokhatny 2010).

Common serine proteases such as trypsin, chymotrypsin and elastase are capable of dissolving thrombi (Sarkar 1960; Sherry 1954; Komorowicz et al. 1998). Both plasmin and plasminogen activators (PAs) are also chymotrypsin like serine proteases. However, the broad substrate specificity of these common proteases makes them less suitable for therapeutic applications (Madison et al. 1995). The SERPINS found in circulation is another important factor affecting the efficacy of serine proteases while performing administration. One among them is α1-antitrypsin (α1-PI) which is found in abundance and characterized by its ability to inhibit serine proteases such as plasmin and thrombin. They also could directly impede trypsin, chymotrypsin and elastases. Even though reactivation of these proteases by α2-macroglobulin is possible, the slow rate of reactivation is also a limiting factor (Aubry and Bieth 1977). Thus, systemic administration of these proteases is not recommendable for treating cardiovascular disorders. Limitations of enzymes involved in plasma fibrinolytic system (plasmin, t-PA) and common proteases (trypsin, chymotrypsin and elastase) demand administration of fibrinolytic proteases from external sources. Few of the external non microbial sources include earthworms *Eisenia fetida*, *Lumbricus rubellus* (Mihara et al. 1991; Roch 1979), snake venoms (Markland 1998), *Catharsius molossus* (Ahn et al. 2003), *Spirodela polyrhiza* (Choi and Sa 2001) etc.

Snake Venom Thrombolytic Enzymes

Another important source for fibrinolytic enzymes is venom of snakes. Fibrolase, is a fibrinolytic enzyme isolated from snake *Agkistrodon contortrix contortrix*. It is a zinc metalloprotease of 203 amino acids having molecular weight of approximately 29 kDa. A truncated mutant of 201 amino acids is produced from fibrolase called alfimeprase with improved fibrinolytic properties (sixfold higher activity than plasminogen activators such as t-PA and urokinase) without bleeding complications upon administration. Two deletions and one substitution changes EQRFPQR of fibrolase to SFPQR in alfimeprase. Animal models as well as Phase I and II clinical studies proved efficacy and safety of the drug upon administration (Randolph et al. 1992; Deitcher et al. 2006). Lebetase is another direct acing fibrinolytic enzyme isolated from *Vipera lebetina*. It is a single chain polypeptide of 214 amino acids constituting 23.7 kDa and is a metalloprotease (inhibited by EDTA). Lebetase achieves fibrinogenolysis by rapid dissolution of Aα-chain together with slow lysis of Bβ-chain. It appears to be safer since it showed only lower hemorrhagic activity (Siigur and Siigur 1991).

Earthworm Thrombolytic Enzymes

Earthworms are rich source of proteases with thrombolytic potential. Initial characterization of earthworm fibrinolytic enzyme was performed from *Lumbricus rubellus* and named lumbrokinases (LK) representing a group of protease enzymes (Mihara et al. 1991). *Eisenia fetida, E. andrei L. bimastus* etc. are a few other earthworm species identified as fibrinolytic enzyme producers. They produce multiple thrombolytic proteases. *L. rubellus* produced around six lumbrokinase variants and *E. fetida* yields seven variants of fibrinolytic enzymes (EFE a-g) (Cho et al. 2004; Wang et al. 2003). These enzymes show more specificity towards fibrin in comparison with plasminogen activators such as t-PA and urokinases. Lumbrokinases are direct acting fibrinolytic enzymes, also showing functions like activation of plasminogen, reduction of platelet aggregation, blood thinning etc. Molecular mass of these enzymes ranges from 20 to 35 kDa and they retains their activity over a wide range of pH (pH 1–11) (Pan et al. 2010). In vivo animal models revealed their potential to be used for oral administration (Lee et al. 2007; Yan et al. 2010). Recombinant production of this enzyme can be achieved in *pichia pastoris* (Ge et al. 2005). LK has its main advantage as a safer therapeutic enzyme with lower risk of bleeding disorders (Tjandrawinata et al. 2016); but yet to get approval from FDA. However, Boluoke® is a commercially available LK approved by China Food and Drug Administration. *E. fetida* enzymes belongs to serine proteases with molecular mass ranges from 24.5 to 29.5 kDa and few of them were quite similar to lumbrokinases (Wang et al. 2003).

Plant Based Thrombolytic Enzymes

Herbal formulations are known for therapeutic usage since it is widely used in traditional medicines. Plants are rich source of proteolyitc enzymes, especially plant latex. Crude preparations of plants are reported to show in vitro clot dissolution and they show considerable stability over a wide range of temperature and pH. *Fagonia Arabica, Bacopa monnieri, Leea indica, Leucas aspera, Clausena suffruticosa, Urena sinuate and Trema orientalis* are few among the plants producing thrombolytic enzymes (Prasad et al. 2007; Rahman et al. 2013; Emran et al. 2015). Latex enzymes from various plants such as *Calotropis gigantea, Wrightia tinctoria and Synadenium grantii, Cynanchum puciflorum, Asclepias curassavica and Pergularia extensa* showed procoagulant activity (clot inducing) as well as clot dissolving activity. These enzymes are either serine proteases or cysteine proteases. Though, protease from *C. gigantea* was found to be most potent thrombolytic agent, it showed high risk of hemorrhage also; whereas enzymes from *S. grantii* and *W. tinctoria* showed no hemorrhage (Rajesh et al. 2007; Shivaprasad et al. 2009). Cysteine protease enzymes from fruits of *Bromelia hieronymi, B. balansae* and *Pseudananas macrodontes* was evaluated for its thrombolytic potential and found that at low concentration they showed procoagulant activity and as concentration increases it showed fibrinolytic activity (Errasti et al. 2016). Extracts from *Litchi chinensis* was shown to have antiplatelet, thrombolytic and anticoagulant potential. The extract showed increase in fibrinolytic activity in a dose dependent manner (Sung et al. 2012). Fibrinolytic proteases are also reported from latex of *Cryptostegia grandiflora* and *Plumeria rubra* (Viana et al. 2013). A fibrinogenolytic enzyme Eumiliin was purified and characterized from the latex of *Euphorbia milii* var. *hislopii*. Eumiliin is a 30 kDa cysteine protease and showed stability over a wide range of temperature and pH. It lost complete activity at 80 °C and predominantly degraded Aα chain of fibrinogen (Fonseca et al. 2010). A serine protease, EuP-82, with fibrinogenolytic potential was isolated form *Euphorbia* cf. *lactea*. It is a homodimer and retains its stability over a wide range of pH and temperature (Siritapetawee et al. 2015). Hirtin is a 34 kDa serine protease isolated from the latex of *Euphorbia hirta* with both fibrinolytic as well as fibrinogenolytic potential. It was stable between pH 4–8 and temperature from 30 to 70 °C. Hirtin also showed more affinity towards Aα chain of fibrinogen during fibrinogenolysis (Patel et al. 2012). Another fibrinolytic protease, AMP-48, was isolated from the latex of *Artocarpus heterophyllus*. It is a serine protease of molecular weight 48 kDa hydrolyzing fibrinogen and readily dissolving α subunit followed by partial hydrolysis of β and γ subunits just like other fibrinolytics of plant origin (Siritapetawee et al. 2012). Some of the plant fibrinolytic enzymes along with their properties are summarized in Table 15.1.

Algal Thrombolytic Enzymes

Thrombolytic enzymes are isolated from algal sources too. Seaweeds from the genus *codium* such as *Codium intricatum,* C. *latum and* C. *diaricatum* are reported to produce multiple fibrinolytic enzymes. CIP I (*C. intricatum* protease I) and CIP II (*C. intricatum* protease II) are two direct acting fibrinolytic enzymes isolated from *C. intricatum.* Molecular mass of CIP I and CIP II were approximately 20 kDa under denaturing conditions. Protease isolated from *C. latum* designated as CLP (*C. latum* protease) also showed fibrinolytic potential. Molecular weight of the enzyme was 23 KDa and N terminal sequencing confirms it to be distinct from CIP. A 31 KDa thrombolytic enzyme CDP (*C. diaricatum* protease) was isolated from C. *diaricatum.* Although the fibrinolytic enzymes from these seaweeds are diverse they showed almost similar properties. They are serine proteases and predominantly hydrolyzed Aα chain of fibrinogen while prolonged incubation cleaved all subunits (Matsubara et al. 1998, 1999, 2000) Fibrinolytic enzymes from algae with their properties are listed in Table 15.2.

15.3.2.2 Microbial Thrombolytic Enzymes

Though fibrinolytic enzymes were reported from various sources, most of the fibrinolytic enzymes reported were from microbial sources. Few of the microbial producers includes fungi such as *Aspergillus ochraceus* (Batomunkueva and Egorov 2001), *Fusarium pallidoroseum* (El-Aassar 1995), *Pleurotus ostreatus* (Choi and Shin 1998), algae such as *Codium* sp. (Matsubara et al. 1998) and bacteria such as *Streptomyces* sp. (Wang et al. 1999), *Pseudomonas* sp. (Wang

Table 15.1 Fibrinolytic enzymes produced by plants and their characteristics

Organism	Enzyme	Type	Molecular weight, optimum temperature and pH	Mode of action	References
Euphorbia milii var. hislopii	Eumiliin	Cysteine protease	30 kDa pH 8.0	Preferentially cleaved Aα chain of fibrinogen	Fonseca et al. (2010)
Euphorbia cf. lactea	EuP-82	Serine protease	35 kDa (monomeric form) 35 °C, pH 11	Cleaved all subunits of human fibrinogen	Siritapetawee et al. (2015)
Euphorbia hirta	Hirtin	Serine protease	34 kDa 50 °C, pH 7.2	Preferentially cleaved Aα chain of fibrinogen followed by Bβ and γ chains	Patel et al. (2012)
Artocarpus heterophyllus	AMP-48	Serine protease	48 kDa 55–60 °C, pH 8	Preferentially cleaved α followed by partial hydrolysis of β and γ subunits	Siritapetawee et al. (2012)

Table 15.2 Fibrinolytic enzymes produced by algae and their characteristics

Organism	Enzyme	Type	Molecular weight, and pH	Mode of action	References
Codium intricatum	CIP	Serine protease	20 kDa pH 8–9	Preferentially cleaved Aα chain followed Bβ and γ chains	Matsubara et al. (1998)
Codium latum	CLP	Serine protease	23 kDa pH 10	Preferentially cleaved Aα chain followed Bβchain and γ chain to end	Matsubara et al. (1999)
Codium diaricatum	CDP	Serine protease	31 kDa pH 9	Cleaved Aα chain, very little affinity towards Bβchain and γ chain	Matsubara et al. (2000)

et al. 2009b), *Staphylococcus* (Gerheim 1948) and *Bacillus* sp. (Sumi et al. 1987; Peng et al. 2003; Ko et al. 2004). Major advantages of microbial enzymes are its ease of production, lesser bleeding complications, oral administration etc. (Goldhaber and Bounameaux 2001; Tough 2005). Past couple of decades witnessed the discovery of thrombolytic agents from numerous microorganisms and this elicited the medical interest in development of better therapeutic agents from them with lesser or no side effects. Direct acting plasmin like fibrinolytic enzymes from microbial sources are discussed below.

Fungal Thrombolytic Enzymes

Fungi are rich source of proteolytic enzymes and some of them are having fibrinolytic potential too. Here we categorize fungus under microbial producers though micro and macro fungi are discussed. The ability of fungi to grow on solid substrates such as agro-industrial waste residues made them more suitable for economic production of fibinolytic enzymes. *Aspergillus* species are most widely occurring fungal group and few of them turn to be thrombolytic enzyme producers. Protease enzyme produced by *Aspergillus ochraceus* 513 was reported to have fibrinolytic potential and subsequent purification steps yielded a 36.5 kDa protease with both fibrinolytic as well as anticoagulant potential (Batomunkueva and Egorov 2001). *Aspergillus oryzae* KSK-3 produced a 30 kDa serine protease with thrombolytic potential. It remained stable upto temperatures of up to 50 °C and pH ranging from 4 to 9 (Shirasaka et al. 2012). There are numerous reports of fungal fibrinolytic enzymes of *fusarium* species (Tao et al. 1997, 1998). A thrombolytic enzyme producing fungus *Fusarium oxysporum* was successfully grown on a solid substrate rice chaff (Tao et al. 1997, 1998). *Fusarium pallidoroseum* produced a metalloprotease with plasmin like activities (El-Aassar 1995). *Fusarium BLB* isolated from hibiscus leaf produced an alkaline serine protease with fibrinolytic potential. This 27 kDa protein exhibited pH tolerance and retained its activity between pH 2.5 and 11.5 (Ueda et al. 2007). Fu-P is a fibrinolytic enzyme isolated from *Fusarium* sp. CPCC 480097. It is a chymotrypsin like enzyme, with a molecular weight of 28 kDa and has high affinity towards α-chain but showing only low affinity towards β and γ chain during fibrinogenolysis (Wu et al. 2009). Fungal strains such as *Cochliobolus lunatus* and *Penicillium chrysogenum* H9 were also reported to produce fibrinolytic enzymes (Abdel-Fattah and Ismail 1984; El-Aassar et al. 1990). *Rhizopus chinensis*-12 is another fungal producer, of a hydrosulfuryl-metalloprotease with fibrinolytic activity. The enzyme is a single polypeptide of 18 kDa according to SDS-PAGE results and showed similar affinity towards α, β and γ chains of fibrinogen (Xiao-lan et al. 2005). Purification of a fibrinolytic metalloprotease was achieved from the mycelium of *Perenniporia fraxinea*. The 42 kDa protein thus obtained showed affinity towards α-chain of fibrin (Kim et al. 2008).

Thermophilic fungus *Oidiodendron flavum* produced a 22 kDa fibrinolytic enzyme which showed stability over a pH range of 6–11 and completely lost its activity at 70 °C (Tharwat 2006). PTEFP is another fungal fibrinolytic enzyme extracted from *Paecilomyces tenuipes*. It is a serine protease with a molecular mass of 14 kDa. PTEFP showed affinity towards Aα-chain only during fibrinogenolysis (Kim et al. 2011). A

52 kDa metalloprotease with fibrionolytic potential as well as anticoagulant activity was isolated from the mycelium of *Ganoderma lucidum* (Choi and Sa 2000). Properties of thrombolytics from fungi excluding those from mushrooms are given in Table 15.3.

Mushrooms are also known producers of thrombolytic enzymes including some edible mushrooms. Mushrooms belonging to the genus *Cordyceps* such as C. *sinensis* and C. *militaris* yield direct acting plasmin like fibrinolytic enzymes. C. *sinensis* produce a 31 kDa serine protease called CSP capable of hydrolyzing Aα chain of fibrinogen. C. *militaris* produced another fibrinolytic enzyme, CMase is with a molecular mass of 27.3 kDa and it is a metalloprotease (Li et al. 2007; Cui et al. 2008). *Flammulina velutipes* proteases were reported having thrombolytic potential. The organism produced FVP-I, a 37 kDa metalloprotease showing higher degree of affinity for α-chain during fibrinolysis (Park et al. 2007). Zn metalloprotease isolated from *Pleurotus ostreatus* dissolves fibrin as well as Bβ and γ chains of fibrinogen (Choi and Shin 1998) whereas *Pleurotus eryngii* produce a 14 kDa serine protease with specificity towards Aα-chain of fibrinogen during hydrolysis (Cha et al. 2010). Medicinal mushroom *Schizophyllum commune* is also reported to have fibrinolytic activity. The enzyme isolated was a metalloprotease with a molecular mass of 18.2 kDa (Lu et al. 2010a). Another metalloprotease enzyme, TSMEP1with fibrinolytic activity was isolated from fruiting bodies of *Tricholoma saponaceum* (Kim and Kim 2001). AMMP isolated from mycelium of *Armillaria mellea* is also a fibrinolytic metallpro-

Table 15.3 Fibrinolytic enzymes produced by fungi (excluding mushrooms) and their characteristics

Microorganism	Enzyme	Type	Molecular weight, optimum temperature and pH	Mode of action	References
Aspergillus ochraceus 513	–	–	36.5 kDa	–	Batomunkueva and Egorov (2001)
Aspergillus oryzae KSK-3	–	Serine protease	30 kDa 50 °C, pH 6.0	–	Shirasaka et al. (2012)
Fusarium pallidoroseum	–	Metalloprotease	40 °C, pH 7.0	–	El-Aassar (1995)
Fusarium sp. BLB	–	Serine protease	27 kDa 50 °C, pH 9.5	–	Ueda et al. (2007)
Fusarium sp. CPCC 480097	Fu-P	Serine metalloprotease	28 kDa 45 °C, pH 8.5	Cleaves α-chain fibrinogen, low affinity towards β and γ chain	Wu et al. (2009)
Cochliobolus lunatus	–	–	pH 6.98	–	Abdel-Fattah and Ismail (1984)
Rhizopus chinensis-12	–	Hydrosulfuryl-metalloprotease	18 kDa 45 °C, pH 10.5	Similar affinity towards α, β and γ chains of fibrinogen	Xiao-lan et al. (2005)
Perenniporia fraxinea	–	Metalloprotease	42 kDa 35–40 °C, pH 6.0	Cleaves α-chain of fibrinogen followed by β and γ chains	Kim et al. (2008)
Oidiodendron flavum	–	–	22 kDa 45–55 °C, pH 8.0		Tharwat 2006
Paecilomyces tenuipes	PTEFP	Serine protease	14 kDa 35 °C, pH 5.0	Quickly dissolved Aα chain of fibrinogen, no affinity towards β and γ chains	Kim et al. (2011)
Ganoderma lucidum	–	Metalloprotease	52 kDa pH 7.0–7.5	Cleaves Aα and Bβ chain of fibrinogen	Choi and Sa (2000)

tease of 21 kDa (Lee et al. 2005). *Fomitella fraxinea* is another fibrinolytic enzyme producer contributing two potential enzymes, a 32 kDa serine protease called FFP1 (*F. fraxenia* proteases 1) and 42 kDa metalloprotease, FFP2 (*F. fraxenia* proteases 2) (Lee et al. 2006). *Lyophyllum shimeji*, an edible mushroom, also has thrombolytic potential and the enzyme derived from it is 21 kDa serine protease which performs rapid Aα fibrinogenolysis, followed by Bβ and γ chain hydrolysis (Moon et al. 2014). The enzymes with thrombolytic potential from mushrooms are shown in Table 15.4.

Bacterial Thrombolytic Enzymes

Bacteria are the major fibrinolytic enzyme producers and they are the most preferred sources also. It is anticipated that bacterial proteins are suitable for oral administration (Sumi et al. 1990; Omura et al. 2004). Another advantage of bacterial enzymes is ease for large scale production. Plasminogen activators such as streptokinase and staphylokinase of bacterial origin were previously discussed while a large number of direct acting plasmin like enzymes also were reported from bacterial sources.

From *Bacillus* sp.

Most of the bacterial producers belong to the genus *Bacillus* and nattokinase isolated from *Bacillus* sp. turned to be the most prospective fibrinolytic enzyme among them. Nattokinase was initially isolated from *Bacillus subtilis natto* (Sumi et al. 1987). Various other bacterial strains were also reported to produce nattokinase enzyme and it will be discussed separately later.

Table 15.4 Fibrinolytic enzymes produced by fungi (mushrooms) and their characteristics

Microorganism	Enzyme	Type	Molecular weight, optimum temperature and pH	Mode of action	References
Cordyceps sinensis	CSP	Serine protease	31 kDa 40 °C pH 7.0	Hydrolyses Aα chain of fibrinogen	Li et al. (2007)
Cordyceps militaris	CMase	Metalloprotease	27.3 kDa 25 °C, pH 6.0	–	Cui et al. (2008)
Flammulina velutipes	FVP-I	Metalloprotease	37 kDa 20–30 °C, pH 6.0	Higher degree of affinity for α-chain followed by β and γ chain	Park et al. (2007)
Pleurotus ostreatus	–	Zn metalloprotease	12 kDa pH 7.5–8.0	Cleaves aα, Bβ and γ chains of fibrinogen	Choi and Shin (1998)
Pleurotus eryngii	–	Serine protease	14 kDa 40 °C, pH 5.0	Specificity towards Aα-chain of fibrinogen followed by Bβ chain	Cha et al. (2010)
Schizophyllum commune	–	Metalloprotease	21.32 kDa 45 °C, pH 7.4		Lu et al. (2010a)
Tricholoma saponaceum	TSMEP1	Metalloprotease	18 kDa 55 °C, pH 7.5	Cleaves Aα and Bβ chain of fibrinogen, no affinity for γ chain	Kim and Kim (2001)
Armillaria mellea	AMMP	Metallprotease	21 kDa 33 °C, pH 6.0	Preferentially cleaved Aα chain	Lee et al. (2005)
Fomitella fraxinea	FFP1	Serine protease	32 kDa 40 °C, pH 10.0	Aα and Bβ chain of fibrinogen, no affinity for γ chain	Lee et al. (2006)
Fomitella fraxinea	FFP2	Metalloprotease	42 kDa 40 °C, pH 5.0	Aα chain and Bβ chain of fibrinogen	Lee et al. (2006)
Lyophyllum shimeji	–	Serine protease	21 kDa 37 °C pH 8.0	Preferentially cleaves Aα chain of fibrinogen, followed by Bβ and γ chain	Moon et al. (2014)

Apart from Nattokinase several other fibrinolytic enzymes were isolated from *Bacillus* sp. Subtilisin QK1 and subtilisin QK2 are two serine fibrinoytic proteases having a molecular weight of 42 kDa and 28 kDa respectively, isolated from B. *subtilis* QK02 (Ko et al. 2004). Bafibrinase, a thrombolytic enzyme of 32.3 kDa is produced by *Bacillus* sp. strain AS-S20-I (Mukherjee et al. 2012) and a 31 kDa thrombolytic enzyme is produced by *B. subtilis* BK-17 (Jeong et al. 2001). Subtilisin BSF1 from *Bacillus subtilis* A26 and BAF1 produced by *Bacillus amyloliquefaciens* An6 are two serine proteases with molecular weight of 28 and 30 kDa respectively, both having fibrinolytic potential (Agrebi et al. 2009, 2010). Thrombolytic activity along with plasminogen activation was exhibited by two proteolytic enzymes having molecular weight 29 kDa and 29.5 kDa, produced by *Bacillus halodurans* IND18 and *Bacillus cereus* IND1 respectively (Vijayaraghavan et al. 2016b; Vijayaraghavan and Vincent 2014). A 20.5 kDa metalloprotease was purified from *Bacillus subtilis* K42 which also showed fibrinolytic as well as anticoagulant potential (Hassanein et al. 2011). URAK enzyme produced by *Bacillus cereus* NK1 (Deepak et al. 2010) and bacillopeptidase DJ-2 (bpDJ-2) from *Bacillus* sp. DJ-2 are two other thrombolytic enzymes from the *Bacillus* species (Choi et al. 2005). A few fibrinolytic enzymes with their properties are summarized in Table 15.5.

From *Streptomyces* sp.

Second largest genus of bacterial fibrinolytic enzyme producers is *Streptomyces.* Thermophilic *Streptomyces megasporus* strain SD5 produced a chymotrypsin-like serine peptidase of molecular weight 35 KDa with fibrinolytic potential (Chitte and Dey 2000). Another member of the genus, *Streptomyces omiyaensis* yielded a thrombolytic serine protease, SOT (Uesugi et al. 2011). Low molecular weight serine metalloprotease of 18 kDa has been isolated from *Streptomyces* sp. CS624 and named as FES624. It quickly dissolved Aα, Bβ and γ chain of fibrinogen (Mander et al. 2011), whereas a 20 kDa fibrinolytic serine protease SFE1 of *Streptomyces* sp. XZNUM 00004 showed preference towards Aα chain (Ju et al. 2012). FP84 is another *Streptomyces* derived fibrinolytic enzyme with a molecular weight of 35 kDa contributed by *Streptomyces* sp. CS684. It is a serine metalloprotease with affinity towards Bβ chain during fibrinogenolysis (Simkhada et al. 2010). SW-1 is a fibrinolytic enzyme isolated from *Streptomyces* sp. Y405 and it was found to be a 30 kDa serine metallo protease with a lysine binding site (Wang et al. 1999). FSP3, a 44 kDa serine protease was extracted from *Streptomyces* sp. P3. It is direct acting thrombolytic serine protease which also acts as plasminogen activator (Cheng et al. 2015). *Streptomyces* sp. DPUA 1576 produced a fibrinolytic serine protease of 39 kDa that could digest Aα-chain, Bβ-chain, and γ-chain during fibrinogenolysis (Silva et al. 2016). A few fibrinolytic enzymes from actinomycetes with their properties are summarized in Table 15.6.

Other Bacterial Fibrinolytic Enzyme Producers

Various bacterial species excluding *Bacillus* sp. and *Streptomyces* sp. are also reported to produce fibrinolytic enzymes although those two genera contribute most of the fibrinolytic enzymes discovered so far. A 63.3 kDa serine protease was isolated from *Paenibacillus polymyxa* EJS-3 named PPFE-I. It was found dissolving Aα-chain of fibrinogen quickly and subsequently Bβ-chain and followed by γ-chain during fibrinogenolysis (Lu et al. 2010b). *Paenibacillus* sp. IND8 is also known to produce fibrinolytic enzyme (Vijayaraghavan et al. 2016a). A 50 kDa metalloprotease with thrombolytic effect was reported from *Serratia* sp. RSPB11 and *Serratia* sp. KG-2-1 (Lakshmi and Prakasham 2013; Taneja et al. 2017). Another fibrinolytic enzyme producer is *Shewanella* sp. IND20 which contributed a 55.5 kDa thrombolytic enzyme (Vijayaraghavan and Vincent 2015). *Psuedoalteromonas* sp. IND11 yielded a 64 kDa protease which acted as plasminogen activator and disintegrated blood clot directly (Vijayaraghavan et al. 2015). *Treponema denticola* also reported with thrombolytic enzyme production (Rosen et al. 1994). Fibrinolytic enzyme from *Proteus penneri* showed higher affinity towards α-chain followed by β-chain, showing

Table 15.5 Fibrinolytic enzymes produced by Bacteria (*Bacillus* sp.) and their characteristics

Microorganism	Enzyme	Type	Molecular weight, optimum temperature and pH	Mode of action	References
Bacillus subtilis HK176	AprE 176	Serine metalloprotease	27 kDa 40 °C pH 8.0	Rapid dissolution of Aα chain followed by Bβ of fibrinogen	Jeong et al. (2015)
B. subtilis CH3-5	Apr E2	Serine protease	29 kDa 40 °C, pH 7.0	Rapid dissolution of Aα and Bβ chains of fibrinogen only	Jeong et al. (2014)
Bacillus subtilis natto	Nattokinase (NK)	Serine protease	27.7 kDa		Sumi et al. (1987)
Bacillus subtilis QK02	Subtilisin QK1	Serine protease	42 kDa	Readily hydrolyzed Aα and Bβ and γ chains of fibrin	Ko et al. (2004)
Bacillus subtilis QK02	Subtilisin QK2	Serine protease	28 kDa 55 °C pH 8.5	Readily hydrolyzed Aα and Bβ and γ chains of fibrin	Ko et al. (2004)
Bacillus sp. strain AS-S20-I	Bafibrinase	Serine protease	32.3 kDa 37 °C pH 7.4	Preferentially cleaved Aα and Bβ chains of fibrinogen	Mukherjee et al. (2012)
Bacillus subtilis BK-17	Bacillokinase (BK)	Serine protease	31 kDa 50 °C pH 7.0–8.0		Jeong et al. (2001) Bek et al. (2005)
Bacillus subtilis A26	Subtilisin BSF1	Serine protease	28 kDa 60 °C pH 9.0		Agrebi et al. (2009)
Bacillus amyloliquefaciens An6	BAF1	Serine protease	30 kDa 60 °C pH 9.0		Agrebi et al. (2010)
Bacillus halodurans IND18	–	–	29.5 kDa 60 °C pH 9.0		Vijayaraghavan et al. (2016b)
Bacillus cereus IND1	–	–	29.5 kDa 60 °C pH 8.0		Vijayaraghavan and Vincent (2014)
Bacillus subtilis K42	K42	Metalloprotease	20.5 kDa 40 °C pH 9.4		Hassanein et al. (2011)
Bacillus cereus NK1	URAK	–	46 kDa		Deepak et al. (2010)
Bacillus sp. DJ-2	Bacillopeptidase DJ-2 (bpDJ-2)	Serine protease	42 kDa 60 °C pH 9		Choi et al. (2005)

Table 15.6 Fibrinolytic enzymes produced by *Streptomyces* sp. and their characteristics

Microorganism	Enzyme	Type	Molecular weight, optimum temperature and pH	Mode of action	References
Streptomyces sp. Y405	SW-1	Serine metalloprotease	30 kDa pH 8.0	–	Wang et al. (1999)
Streptomyces megasporus strain SD5	–	Serine protease	35 kDa		Chitte and Dey (2000)
Streptomyces sp. CS684	FP84	Serine metalloprotease	35 kDa 45 °C, pH 7.0–8.0	Preference towards Bβ chain of fibrinogen	Simkhada et al. (2010)
Streptomyces omiyaensis	SOT	Serine protease	23 kDa	–	Uesugi et al. (2011)
Streptomyces sp. CS624	FES624	Serine metalloprotease	18 kDa 60 °C, pH 7.0	Rapid dissolution of Aα, Bβ and γ chain of fibrinogen	Mander et al. (2011)
Streptomyces sp. XZNUM 00004	SFE1	Serine protease	20 kDa 35 °C, pH 7.8	Preferentially cleaved Aα chain of fibrinogen followed Bβ-chain and γ-chain at the end	Ju et al. (2012)
Streptomyces sp. P3	FSP3	Serine protease	44 kDa 50 °C, pH 6.5	–	Cheng et al. (2015)
Streptomyces sp. DPUA 1576	–	Serine protease	39 kDa 45 °C, pH 7.5	Cleaved Aα, Bβ, and γ-chains of fibrinogen	Silva et al. (2016)

lower affinity towards γ-chain. It leaves partially hydrolyzed γ-chain after fibrinolysis (Jhample et al. 2015). Nattokinase enzyme production was reported from *Pseudomonas* sp. TKU015 with molecular weight of 24 kDa as determined by gel filtration chromatography (Wang et al. 2009b). Bacterial fibrinolytic enzymes other than those from *Bacillus* sp. and actinomycetes are shown in Table 15.7.

Nattokinase (NK) Subtilisin NAT

Fibrinolytic activity of Japanese fermented food natto was first described by Sumi and team. The enzyme extracted was termed nattokinase (Sumi et al. 1987). Subsequent years witnessed studies leading to the characterization and functional analysis of the enzyme (Sumi et al. 1990). Later, nattokinase production was achieved from various bacterial strains. A few among them are *Bacillus subtilis* isolated from Thua nao a fermented food of Thailand (Inatsu et al. 2006), *Bacillus subtilis* ICTF-1 isolated from marine sources (Mahajan et al. 2012), *Bacillus subtilis* 1A752 (Deepak et al. 2008) and *Pseudomonas sp.* TKU015 (Wang et al. 2009b). Nattokinase (EC 3.4.21.62) belongs to subtilisin family and also known as subtilisin NAT. It was derived from *Bacillus subtilis natto* encoded by *apr N* gene (Nakamura et al. 1992). It is a 27 kDa serine protease devoid of cysteine and hence lacking disulfide bonds (Fujita et al. 1993; Wang et al. 2009a). Fibrinolysis by nattokinase is achieved by different ways. In vivo studies involving oral administration of nattokinase or natto showed slight enhancement in plasma fibrinolysis, thus it turned to be direct acting fibrinolyic enzyme. Apart from this, another important factor aiding fibrinolysis is that the enzyme degrades plasminogen activator inhibitor-1(PAI-1). Also it increases the production of plasminogen activators and there by aiding plasmin formation (Sumi et al. 1990; Urano et al. 2001). Hence, it leads to combined fibrinolysis, by NK and plasmin. NK shows more specificity towards cross-linked fibrin and less affinity towards fibrinogen (Fujita et al. 1995c). NK shares immense homology with other subtilisins such as subtilisin E (99.5%) and subtilisin J (99.3%) (Kurosawa et al. 2015), but it exhibits higher substrate specificity towards fibrin in comparison with other subtilisins, making it more

Table 15.7 Fibrinolytic enzymes produced by other bacteria (excluding *Bacillus* sp. and *Streptomyces* sp.) and their characteristics

Microorganism	Enzyme	Type	Molecular weight, optimum temperature and pH	Mode of action	References
Paenibacillus polymyxa EJS-3	PPFE-I	Serine protease	63.3 kDa 37 °C, pH 7.5	Preferentially cleaved Aα-chain of fibrinogen followed Bβ-chain and γ-chain at the end	Lu et al. (2010b)
Paenibacillus sp. IND8	–	–	44 kDa 40 °C, pH 7.0	–	Vijayaraghavan et al. (2016a)
Serratia sp. RSPB11	–	Metalloprotease	50 kDa 37 °C, pH 9.0	Preferentially cleaved Aα chain of fibrinogen followed Bβ-chain and γ-chain at the end	Lakshmi and Prakasham (2013)
Serratia sp. KG-2-1	–	Metalloprotease	50 kDa 40 °C, pH 8.0	–	Taneja et al. (2017)
Shewanella sp. IND20	–	–	55.5 kDa 50 °C, pH 8.0	–	Vijayaraghavan and Vincent (2015)
Psuedoalteromonas sp. IND11	–	–	64 kDa pH 8.0, 40 °C	–	Vijayaraghavan et al. (2015)

suitable for therapeutic applications (Peng et al. 2005). In vitro studies substantiated that NK is a superior thrombolytic enzyme and six times more active than plasmin (Fujita et al. 1995c). NK is the only microbial fibrinolytic therapeutic enzyme whose in vivo studies have been conducted in the best possible manner. Early in vivo studies involving dogs showed that after oral administration, complete clot dissolution was achieved within 5 h and blood circulation was restored (Sumi et al. 1990). Another study proved that after intraduodenal administration, NK was transported across intestinal tract of rat and plasma fibrinogen was hydrolyzed thereafter (Fujita et al. 1995b). On a molar basis, NK was found to be more effective in clot lysis and restoration of arterial blood flow in comparison with plasmin and elastase (Fujita et al. 1995a). NK prevented platelet aggregation, along with reduction in euglobulin clot lysis time (ECLT) and extension of partial thromboplastin time (PATT) upon its oral administration in rats (Park et al. 2012). Clinical trials were performed on three groups viz. healthy group, cardiovascular group (patients with cardiovascular risk factors), and dialysis group (patients undergoing dialysis) and showed that NK reduced fibrinogen, factor VII, and factor VIII in plasma without affecting the blood lipids (Hsia et al. 2009). Trials involving healthy volunteers in Japan proved that fibrinolysis and anti-coagulation was considerably enhanced by single-dose of NK administration (Kurosawa et al. 2015). Effect of nattokinase on North American hypertensive population showed that its consumption reduced blood pressure and von Willebrand factor (Jensen et al. 2016). NK is under Phase II clinical trial study in the USA for its competence in inhibiting atherothrombosis (Weng et al. 2017). Improvement of NK as a fibrinolytic as well as antithrombotic drug that could significantly recover blood circulation is highly anticipated.

Serrapeptase/Serralysin

Serrapeptase (SP) also known as serralysin is a proteolytic enzyme derived from enterobacterium *Serratia* E-15, isolated from silkworm *Bombyx mori* (Miyata et al. 1970). It comprises of 470 amino acids with a molecular weight of approximately 50 kDa (Nakahama et al. 1986) and is a metalloprotease with zinc ion, essential for its activity (Miyata et al. 1971). Later, it was identified from various species of *Serratia, Xenorhabdus, Pseudomonas* etc. (Massaoud et al. 2011; Wu et al. 2016; Lakshmi and Prakasham 2013; Louis et al. 1998). It has been used as an anti-inflammatory agent from 1960s. Because of the anti-inflammatory properties along with anti-edemic and analgesic properties, SP frequently finds its application in surgery, gynaecology, orthopedics etc. Also, the enzyme shows fibrinolytic potential making it suitable for treatment of atherosclerosis (Bhagat et al. 2013). It can dissolve only dead and damaged tissue without damaging the living tissue. Thus, removal of deposits from the arteries such as cholesterol, fatty substances, cellular waste products etc. can be achieved by SP. Elimination of atherosclerotic plaques and blood thinning along with clot dissolution also can be achieved by the enzyme (Kotb 2013). An in vitro study indicated serralysin from *Serratia* sp. RSPB11 as a direct acting thrombolytic agent and shown to be more active than plasmin (Lakshmi and Prakasham 2013). In vivo studies proved intestinal absorption of serrapeptase upon oral administration and it remained in blood stream in active form (Moriya et al. 1994). Rare cases of pneumonitis have been reported as side effects associated with use of SP (Hirahara et al. 1989; Sasaki et al. 2000). Apart from that, not much information is available on its side effects ever since the literatures are available as monographs from pharmaceutical companies. SP has been studied extensively for its anti-inflammatory properties. However, regarding its thrombolytic potential in vivo studies or trials have not been conducted.

15.4 Conclusion

During their development thrombolytic enzymes have gone through many phases since the discovery streptokinase. Both plasminogen activators and plasmin like enzymes were reported with its advantages and disadvantages. More and more thrombolytic enzymes have been discovered from microbial sources in recent times. In spite of all these developments a thrombolytic enzyme without any serious adverse effects such as bleeding complications, suitable for bolus administration and also has lower production cost is not in use currently. For the past few decades recombinant DNA technology has been emerged as a powerful tool in genetic manipulation and can be used for the development of thrombolytic enzymes without systemic bleeding and improved circulation time. Chimeric enzymes (fusion proteins) with improved properties can be achieved through cloning and expression. With cutting edge technologies, therapeutic enzymes for cardiovascular disease management are expecting new avenues.

References

Abdel-Fattah AF, Ismail AM (1984) Purification and some properties of pure *Cochliobolus lunatus* fibrinolytic enzyme. Biotechnol Bioeng 26:407–411

Agnelli G, Pascucci C, Nenci GG, Mele A, Burgi R, Heim J (1993) Thrombolytic and haemorrhagic effects of bolus doses of tissue-type plasminogen activator and a hybrid plasminogen activator with prolonged plasma half-life (K2tu-PA: CGP 42935). Thromb Haemost 70:294–300

Agrebi R, Haddar A, Hmidet N, Jellouli K, Manni L, Nasri M (2009) BSF1 fibrinolytic enzyme from a marine bacterium *Bacillus subtilis* A26: purification, biochemical and molecular characterization. Process Biochem 44:1252–1259

Agrebi R, Hmidet N, Hajji M, Ktari N, Haddar A, Fakhfakh-Zouari N, Nasri M (2010) Fibrinolytic serine protease isolation from *Bacillus amyloliquefaciens* An6 grown on *Mirabilis jalapa* tuber powders. Appl Biochem Biotechnol 162:75–88

Aguilar MI, Hart R (2005) Oral anticoagulants for preventing stroke in patients with non-valvular atrial fibrillation and no previous history of stroke or transient ischemic attacks. Cochrane Database Syst Rev 3:CD001927

Ahn MY, Hahn BS, Ryu KS, Kim JW, Kim I, Kim YS (2003) Purification and characterization of a serine protease with fibrinolytic activity from the dung beetles, *Catharsius molossus*. Thromb Res 112:339–347

Ahrens I, Smith BK, Bode C, Peter K (2007) Direct thrombin inhibition with bivalirudin as an antithrombotic strategy in general and interventional cardiology. Expert Opin Drug Metab Toxicol 3:609–620

Albers GW, von Kummer R, Truelsen T, Jensen JKS, Ravn GM, Gronning BA, Chabriat H, Chang KC, Davalos AE, Ford GA, Grotta J, Kaste M, Schwamm LH, Shuaib A, DIAS-3 Investigators (2015) Safety and efficacy of desmoteplase given 3-9 h after ischaemic stroke in patients with occlusion or high-grade stenosis in major cerebral arteries (DIAS-3): a double-blind, randomised, placebo-controlled phase 3 trial. Lancet Neurol 14:575–584

Ali MR, Salim Hossain M, Islam MA, Saiful Islam Arman M, Sarwar Raju G, Dasgupta P, Noshin TF (2014) Aspect of thrombolytic therapy: a review. Sci World J 2014:586510

Alkjaersig N, Fletcher AP, Sherry S (1959) The mechanism of clot dissolution by plasmin. J Clin Invest 38:1086–1095

Ansell J, Hirsh J, Hylek E, Jacobson A, Crowther M, Palareti G (2008) Pharmacology and management of the vitamin K antagonists: American College of Chest Physicians Evidence-Based Clinical Practice Guidelines (8th edition). Chest 133:160S–198S

Appella E, Robinson EA, Ullrich SJ, Stoppelli MP, Corti A, Cassani G, Blasi F (1987) The receptor-binding sequence of urokinase. A biological function for the growth-factor module of proteases. J Biol Chem 262:4437–4440

Armstrong PW, Burton JR, Palisaitis D, Thompson CR, Van de Werf F, Rose B, Collen D, Teo KK (2000) Collaborative angiographic patency trial of recombinant staphylokinase (CAPTORS). Am Heart J 139:820–823

Armstrong PW, Burton J, Pakola S, Molhoek PG, Betriu A, Tendera M, Bode C, Adgey AA, Bar F, Vahanian A, Van de Werf F, CAPTORS II Investigators (2003) Collaborative angiographic patency trial of recombinant staphylokinase (CAPTORS II). Am Heart J 146:484–488

Arnout J, Hoylaerts MF, Lijnen HR (2006) Haemostasis. In: Moncada S, Higgs A (eds) The vascular endothelium II. Handbook of experimental pharmacology, vol 176/II. Springer, Berlin, pp 1–41

Aubry M, Bieth J (1977) Kinetics of the inactivation of human and bovine trypsins and chymotrypsins by α1-proteinase inhibitor and of their reactivation by α2-macroglobulin. Clin Chim Acta 78:371–380

Bar FW, Meyer J, Vermeer F, Michels R, Charbonnier B, Haerten K, Spiecker M, Macaya C, Hanssen M, Heras M, Boland JP, Morice MC, Dunn FG, Uebis R, Hamm C, Ayzenberg O, Strupp G, Withagen AJ, Klein W, Windeler J, Hopkins G, Barth H, von Fisenne MJ, SESAM Study Group (1997) Comparison of saruplase and alteplase in acute myocardial infarction. The Study in Europe with Saruplase and Alteplase in Myocardial Infarction. Am J Cardiol 79:727–732

Batomunkueva BP, Egorov NS (2001) Isolation, purification, and resolution of the extracellular proteinase complex of *Aspergillus ochraceus* 513 with fibrinolytic and anticoagulant activities. Microbiology 70:519–523

Bauer KA (2006) New anticoagulants. Hematology 1:450–456

Bek H, Hak-Seob L, Kyung Kae C, Choi Y, Byung Tae C, Min-Jeong S, Ji-Eun K, Eun-Ju R, Man Kyu H, Joo WH, Young Kee J (2005) Characterization of a novel fibrinolytic enzyme produced from *Bacillus subtilis* BK-17. J Life Sci 15:987–993

Bernik MB (1973) Increased plasminogen activator (urokinase) in tissue culture after fibrin deposition. J Clin Invest 52:823–834

Bernik MB, Kwaan HC (1967) Origin of fibrinolytic activity in cultures of the human kidney. J Lab Clin Med 70:650–661

Bernik MB, Kwaan HC (1969) Plasminogen activator activity in cultures from human tissues. An immunological and histochemical study. J Clin Invest 48:1740–1753

Bhagat S, Agarwal M, Roy V (2013) Serratiopeptidase: a systematic review of the existing evidence. Int J Surg 11:209–217

Bick RL, Fareed J (1999) Low molecular weight heparins: differences and similarities in approved preparations in the United States. Clin Appl Thromb Hemost 5:S63–S66

Bringmann P, Gruber D, Liese A, Toschi L, Kratzschmar J, Schleuning WD, Donner P (1995) Structural features mediating fibrin selectivity of vampire bat plasminogen activators. J Biol Chem 270:25596–25603

Cannon CP, McCabe CH, Gibson CM, Ghali M, Sequeira RF, McKendall GR, Breed J, Modi NB, Fox NL, Tracy RP, Love TW, Braunwald E (1997) TNK-tissue plasminogen activator in acute myocardial infarction. Results of the thrombolysis in myocardial infarction (TIMI) 10A dose-ranging trial. Circulation 95:351–356

Cannon CP, Gibson CM, McCabe CH, Adgey AA, Schweiger MJ, Sequeira RF, Grollier G, Giugliano RP, Frey M, Mueller HS, Steingart RM, Weaver WD, Van de Werf F, Braunwald E (1998) Thrombolysis in Myocardial Infarction (TIMI) 10B Investigators. TNK-tissue plasminogen activator compared with front-loaded alteplase in acute myocardial infarction: results of the TIMI 10B trial. Circulation 98:2805–2814

Carswell CI, Plosker GL (2002) Bivalirudin: a review of its potential place in the management of acute coronary syndromes. Drugs 62:841–870

Cha WS, Park SS, Kim SJ, Choi D (2010) Biochemical and enzymatic properties of a fibrinolytic enzyme from *Pleurotus eryngii* cultivated under solid-state conditions using corn cob. Bioresour Technol 101:6475–6481

Chen F, Suzuki Y, Nagai N, Sun X, Wang H, Yu J, Marchal G, Ni Y (2007) Microplasmin and tissue plasminogen activator: comparison of therapeutic effects in rat stroke model at multiparametric MR imaging. Radiology 244:429–438

Chen W, Huang X, Ma XW, Mo W, Wang WJ, Song HY (2008) Enzymatic vitreolysis with recombinant microplasminogen and tissue plasminogen activator. Eye 22:300–307

Cheng G, He L, Sun Z, Cui Z, Du Y, Kong Y (2015) Purification and biochemical characterization of a novel fibrinolytic enzyme from *Streptomyces* sp. P3. J Microbiol Biotechnol 25:1449–1459

Chitte RR, Dey S (2000) Potent fibrinolytic enzyme from a thermophilic *Streptomyces megasporus* strain SD5. Lett Appl Microbiol 31:405–410

Cho IH, Choi ES, Lim HG, Lee HH (2004) Purification and characterization of six fibrinolytic serine-proteases from earthworm *Lumbricus rubellus*. J Biochem Mol Biol 37:199–205

Choi HS, Sa YS (2000) Fibrinolytic and antithrombotic protease from *Ganoderma lucidum*. Mycologia 92:545–552

Choi HS, Sa YS (2001) Fibrinolytic and antithrombotic protease from *Spirodela polyrhiza*. Biosci Biotechnol Biochem 65:781–786

Choi HS, Shin HH (1998) Purification and partial characterization of a fibrinolytic protease in *Pleurotus ostreatus*. Mycologia 90:674–679

Choi NS, Yoo KH, Hahm JH, Yoon KS, Hyun BH, Maeng PJ, Kim SH (2005) Purification and characterization of a new peptidase, bacillopeptidase DJ-2, having fibrinolytic activity: produced by *Bacillus* sp. DJ-2 from Doen-Jang. J Microbiol Biotechnol 15:72–79

Christensen LR (1945) Streptococcal fibrinolysis: a proteolytic reaction due to a serum enzyme activated by streptococcal fibrinolysin. J Gen Physiol 28:363–383

Christensen LR (1947) Protamine purification of streptokinase and effect of pH and temperature on reversible inactivation. J Gen Physiol 30:465–473

Cliffton EE, Cannamela DA, Grossi C (1953) In vivo studies of human plasmin. Intravenous injection in dogs and rabbits. J Appl Physiol 6:143–150

Cohen M (1999) Heparin-induced thrombocytopenia and the clinical use of low molecular weight heparins in acute coronary syndromes. Semin Hematol 36:33–36

Collen D, Stassen JM, Larsen G (1988) Pharmacokinetics and thrombolytic properties of deletion mutants of human tissue-type plasminogen activator in rabbits. Blood 71:216–219

Collen D, Van de Werf F (1993) Coronary thrombolysis with recombinant staphylokinase in patients with evolving myocardial infarction. Circulation 87:1850–1853

Collen D, Moreau H, Stockx L, Vanderschueren S (1996) Recombinant staphylokinase variants with altered immunoreactivity: II: thrombolytic properties and antibody induction. Circulation 94:207–216

Craven LL (1950) Acetylsalicylic acid, possible preventive of coronary thrombosis. Ann West Med Surg 4:95

Craven LL (1953) Experiences with aspirin (acetylsalicylic acid) in the nonspecific prophylaxis of coronary thrombosis. Miss Valley Med J 75:38–44

Cui L, Dong MS, Chen XH, Jiang M, Lv X, Yan G (2008) A novel fibrinolytic enzyme from *Cordyceps militaris*, a Chinese traditional medicinal mushroom. World J Microbiol Biotechnol 24:483–489

Dantas GC, Thompson BV, Manson JA, Tracy CS, Upshur REG (2004) Patients' perspectives on taking warfarin:

qualitative study in family practice. BMC Fam Pract 5:15
Davies MC, Englert ME, Derenzo EC (1964) Interaction of streptokinase and human plasminogen. I. Combining of streptokinase and plasminogen observed in the ultracentrifuge under a variety of experimental conditions. J Biol Chem 239:2651–2656
De Renzo EC, Siiteri PK, Hutchings BL, Bell PH (1967) Preparation and certain properties of highly purified streptokinase. J Biol Chem 242:533–542
de Smet MD, Valmaggia C, Zarranz-Ventura J, Willekens B (2009) Microplasmin: ex vivo characterization of its activity in porcine vitreous. Invest Ophthalmol Vis Sci 50:814–819
Deepak V, Kalishwaralal K, Ramkumarpandian S, Babu SV, Senthilkumar SR, Sangiliyandi G (2008) Optimization of media composition for Nattokinase production by *Bacillus subtilis* using response surface methodology. Bioresour Technol 99:8170–8174
Deepak V, Ilangovan S, Sampathkumar MV, Victoria MJ, Pasha SPBS, Pandian SBRK, Gurunathan S (2010) Medium optimization and immobilization of purified fibrinolytic URAK from *Bacillus cereus* NK1 on PHB nanoparticles. Enzym Microb Technol 47:297–304
Deitcher SR, Funk WD, Buchanan J, Liu S, Levy MD, Toombs CF (2006) Alfimeprase: a novel recombinant direct-acting fibrinolytic. Expert Opin Biol Ther 6:1361–1369
del Zoppo GJ, Higashida RT, Furlan AJ, Pessin MS, Rowley HA, Gent M (1998) PROACT: a phase II randomized trial of recombinant pro-urokinase by direct arterial delivery in acute middle cerebral artery stroke. Stroke 29:4–11
den Heijer P, Vermeer F, Ambrosioni E, Sadowski Z, López Sendón JL, von Essen R, Beaufils P, Thadani U, Adgey J, Pierard L, Brinker J, Davies RF, Smalling RW, Wallentin L, Caspi A, Pangerl A, Trickett L, Hauck C, Henry D, Chew P (1998) Evaluation of a weight-adjusted single-bolus plasminogen activator in patients with myocardial infarction: a double-blind, randomized angiographic trial of lanoteplase versus alteplase. Circulation 98:2117–2125
El-Aassar SA (1995) Production and properties of fibrinolytic enzyme in solid state cultures of *Fusarium pallidoroseum*. Biotechnol Lett 17:943–948
El-Aassar SA, El-Badry HM, Abdel-Fattah AF (1990) The biosynthesis of proteases with fibrinolytic activity in immobilized cultures of *Penicillium chrysogenum* H9. Appl Microbiol Biotechnol 33:26–30
Emran TB, Rahman MA, Uddin MMN, Rahman MM, Uddin MZ, Dash R, Layzu C (2015) Effects of organic extracts and their different fractions of five Bangladeshi plants on in vitro thrombolysis. BMC Complement Altern Med 15:128
Eriksson BI, Borris L, Dahl OE, Haas S, Huisman MV, Kakkar AK, Misselwitz F, Kälebo P, ODIXa-HIP Study Investigators (2006) Oral, direct factor Xa inhibition with BAY 59-7939 for the prevention of venous thromboembolism after total hip replacement. J Thromb Haemost 4:121–128
Errasti ME, Prospitti A, Viana CA, Gonzalez MM, Ramos MV, Rotelli AE, Caffini NO (2016) Effects on fibrinogen, fibrin, and blood coagulation of proteolytic extracts from fruits of *Pseudananas macrodontes*, *Bromelia balansae*, and *B. hieronymi* (Bromeliaceae) in comparison with bromelain. Blood Coagul Fibrinolysis 27:441–449
Fareed J, Walenga JM, Hoppensteadt D, Kaiser B, Jeske W (1999) Factor Xa inhibitors in the control of thrombogenesis. Hamostaseologie 19:55–62
Flemmig M, Melzig MF (2012) Serine-proteases as plasminogen activators in terms of fibrinolysis. J Pharm Pharmacol 64:1025–1039
Fletcher AP, Alkjaersig N, Smyrniotis FE, Sherry S (1958) The treatment of patients suffering from early myocardial infarction with massive and prolonged streptokinase therapy. Trans Assoc Am Phys 71:287–296
Fletcher AP, Alkjaersig N, Sherry S (1959a) The maintenance of a sustained thrombolytic state in man. I. Induction and effects. J Clin Invest 38:1096–1110
Fletcher AP, Sherry S, Alkjaersig N, Smyrniotis FE, Jick S (1959b) The maintenance of a sustained thrombolytic state in man. II. Clinical observations on patients with myocardial infarction and other thromboembolic disorders. J Clin Invest 38:1111–1119
Fleury V, Angles-Cano E (1991) Characterization of the binding of plasminogen to fibrin surfaces: the role of carboxy-terminal lysines. Biochemistry 30:7630–7638
Fonseca KC, Morais NCG, Queiroz MR, Silva MC, Gomes MS, Costa JO, Mamede CCN, Torres FS, Penha-Silva N, Beletti ME, Canabrava HAN, Oliveira F (2010) Purification and biochemical characterization of Eumiliin from *Euphorbia milii* var. *hislopii* latex. Phytochemistry 71:708–715
Fu J, Ren J, Zou L, Bian G, Li R, Lu Q (2008) The thrombolytic effect of miniplasmin in a canine model of femoral artery thrombosis. Thromb Res 122:683–690
Fujita M, Nomura K, Hong K, Ito Y, Asada A, Nishimuro S (1993) Purification and characterization of a strong fibrinolytic enzyme (nattokinase) in the vegetable cheese natto, a popular soybean fermented food in Japan. Biochem Biophys Res Commun 197:1340–1347
Fujita M, Hong K, Ito Y, Fujii R, Kariya K, Nishimuro S (1995a) Thrombolytic effect of nattokinase on a chemically induced thrombosis model in rat. Biol Pharm Bull 18:1387–1391
Fujita M, Hong K, Ito Y, Misawa S, Takeuchi N, Kariya K, Nishimuro S (1995b) Transport of nattokinase across the rat intestinal tract. Biol Pharm Bull 18:1194–1196
Fujita M, Ito Y, Hong K, Nishimuro S (1995c) Characterization of nattokinase-degraded products from human fibrinogen or cross-linked fibrin. Fibrinolysis 9:157–164
Furlan A, Higashida R, Wechsler L, Gent M, Rowley H, Kase C, Pessin M, Ahuja A, Callahan F, Clark WM, Silver F, Rivera F (1999) Intra-arterial prourokinase for acute ischemic stroke. The PROACT II study: a randomized controlled trial. Prolyse in Acute Cerebral Thromboembolism. JAMA 282:2003–2011

Ge T, Sun ZJ, Fu SH, Liang GD (2005) Cloning of thrombolytic enzyme (lumbrokinase) from earthworm and its expression in the yeast *Pichia pastoris*. Protein Expr Purif 42:20–28

Gerheim EB (1948) Staphylococcal coagulation and fibrinolysis. Nature 162:732

Gerlach D, Kraft R, Behnke D (1988) Purification and characterization of the bacterial plasminogen activator staphylokinase secreted by a recombinant *Bacillus subtilis*. Zentralbl Bakteriol Mikrobiol Hyg A 269:314–322

Gershlick AH, More RS (1998) Treatment of myocardial infarction. BMJ 316:280–284

GISSI (1986) Effectiveness of intravenous thrombolytic treatment in acute myocardial infarction. Gruppo Italiano per la Sperimentazione della Streptochinasi nell'Infarto Miocardico (GISSI). Lancet 327:397–402

GISSI-2 (1990) A factorial randomised trial of alteplase versus streptokinase and heparin versus no heparin among 12,490 patients with acute myocardial infarction. Gruppo Italiano per lo studio della Sopravvivenza nell'Infarto Miocardico. Lancet 336:65–71

Goldhaber SZ, Bounameaux H (2001) Thrombolytic therapy in pulmonary embolism. Semin Vasc Med 01:213–220

Gore JM, Sloan M, Price TR, Randall AM, Bovill E, Collen D, Forman S, Knatterud GL, Sopko G, Terrin ML (1991) Intracerebral hemorrhage, cerebral infarction, and subdural hematoma after acute myocardial infarction and thrombolytic therapy in the Thrombolysis in Myocardial Infarction study. Thrombolysis in Myocardial Infarction, phase II, pilot and clinical trial. Circulation 83:448–459

Guimaraes AHC, Barrett-Bergshoeff MM, Criscuoli M, Evangelista S, Rijken DC (2006) Fibrinolytic efficacy of Amediplase, Tenecteplase and scu-PA in different external plasma clot lysis models: sensitivity to the inhibitory action of thrombin activatable fibrinolysis inhibitor (TAFI). Thromb Haemost 96:325–330

GUSTO Angiographic Investigators (1993) The effects of tissue plasminogen activator, streptokinase, or both on coronary-artery patency, ventricular function, and survival after acute myocardial infarction. N Engl J Med 329:1615–1622

GUSTO III Investigators (1997) Global Use of Strategies to Open Occluded Coronary Arteries (GUSTO III). A comparison of reteplase with alteplase for acute myocardial infarction. N Engl J Med 337:1118–1123

Hacke W, Albers G, Al-Rawi Y, Bogousslavsky J, Davalos A, Eliasziw M, Fischer M, Furlan A, Kaste M, Lees KR, Soehngen M, Warach S, DIAS Study Group (2005) The Desmoteplase in Acute Ischemic Stroke Trial (DIAS): a phase II MRI-based 9-hour window acute stroke thrombolysis trial with intravenous desmoteplase. Stroke 36:66–73

Hacke W, Furlan AJ, Al-Rawi Y, Davalos A, Fiebach JB, Gruber F, Kaste M, Lipka LJ, Pedraza S, Ringleb PA, Rowley HA, Schneider D, Schwamm LH, Leal JS, Söhngen M, Teal PA, Wilhelm-Ogunbiyi K, Wintermark M, Warach S (2009) Intravenous desmoteplase in patients with acute ischaemic stroke selected by MRI perfusion–diffusion weighted imaging or perfusion CT (DIAS-2): a prospective, randomised, double-blind, placebo-controlled study. Lancet Neurol 8:141–150

Harenberg J, Fenyvesi T (2004) Heparin, thrombin and factor Xa inhibitors. Hamostaseologie 24:261–278

Hassanein WA, Kotb E, Awny NM, El-Zawahry YA (2011) Fibrinolysis and anticoagulant potential of a metallo protease produced by *Bacillus subtilis* K42. J Biosci 36:773–779

Herrick JB (1912) Certain clinical features of sudden obstruction of the coronary arteries. JAMA 59:2012

Hildebrand M, Bhargava AS, Bringmann P, Schutt A, Verhallen P (1996) Pharmacokinetics of the novel plasminogen activator *Desmodus rotundus* plasminogen activator in animals and extrapolation to man. Fibrinolysis 10:269–276

Hirahara K, Saitoh T, Terada I, Uno K, Nagai A, Kioi S, Arakawa M (1989) A case of pneumonitis due to serrapeptase. Nihon Kyobu Shikkan Gakkai Zasshi 27:1231–1236

Hirsh J, Anand SS, Halperin JL, Fuster V (2001) Guide to anticoagulant therapy: heparin: a statement for healthcare professionals from the American Heart Association. Circulation 103:2994–3018

Hoefer IE, Stroes ESG, Pasterkamp G, Levi MM, Reekers JA, Verhagen HJM, Meijers JC, Humphries JE, Rotmans JI (2009) Locally applied recombinant plasmin results in effective thrombolysis in a porcine model of arteriovenous graft thrombosis. J Vasc Interv Radiol 20:951–958

Hoppensteadt DA, Jeske W, Walenga J, Fareed J (2008) The future of anticoagulation. Semin Respir Crit Care Med 29:90–99

Howell W, Holt E (1918) Two new factors in blood coagulation-heparin and pro-antithrombin. Am J Phys 47:328–341

Hoylaerts M, Rijken DC, Lijnen HR, Collen D (1982) Kinetics of the activation of plasminogen by human tissue plasminogen activator. Role of fibrin. J Biol Chem 257:2912–2919

Hsia CH, Shen MC, Lin JS, Wen YK, Hwang KL, Cham TM, Yang NC (2009) Nattokinase decreases plasma levels of fibrinogen, factor VII, and factor VIII in human subjects. Nutr Res 29:190–196

Hunt JA, Petteway SR, Scuderi P, Novokhatny V (2008) Simplified recombinant plasmin: production and functional comparison of a novel thrombolytic molecule with plasma-derived plasmin. Thromb Haemost 100:413–419

Inatsu Y, Nakamura N, Yuriko Y, Fushimi T, Watanasiritum L, Kawamoto S (2006) Characterization of *Bacillus subtilis* strains in Thua nao, a traditional fermented soybean food in northern Thailand. Lett Appl Microbiol 43:237–242

InTIME-II Investigators (2000) Intravenous NPA for the treatment of infarcting myocardium early; InTIME-II, a double-blind comparison of single-bolus lanoteplase vs accelerated alteplase for the treatment of

patients with acute myocardial infarction. Eur Heart J 21:2005–2013
ISIS-3 (1992) A randomised comparison of streptokinase vs tissue plasminogen activator vs anistreplase and of aspirin plus heparin vs aspirin alone among 41,299 cases of suspected acute myocardial infarction. ISIS-3 (Third International Study of Infarct Survival) Collaborative Group. Lancet 339:753–770
Jackson KW, Tang J (1982) Complete amino acid sequence of streptokinase and its homology with serine proteases. Biochemistry 21:6620–6625
Jahan R, Stewart D, Vinters HV, Yong W, Vinuela F, Vandeberg P, Marder VJ (2008) Middle cerebral artery occlusion in the rabbit using selective angiography: application for assessment of thrombolysis. Stroke 39:1613–1615
Jensen GS, Lenninger M, Ero MP, Benson KF (2016) Consumption of nattokinase is associated with reduced blood pressure and von Willebrand factor, a cardiovascular risk marker: results from a randomized, double-blind, placebo-controlled, multicenter North American clinical trial. Integr Blood Press Control 9:95–104
Jeong YK, Park JU, Baek H, Park SH, Kong IS, Kim DW, Joo WH (2001) Purification and biochemical characterization of a fibrinolytic enzyme from *Bacillus subtilis* BK-17. World J Microbiol Biotechnol 17:89–92
Jeong SJ, Cho KM, Lee CK, Kim GM, Shin JH, Kim JS, Kim JH (2014) Overexpression of aprE2, a fibrinolytic enzyme gene from *Bacillus subtilis* CH3-5, in Escherichia coli and the properties of AprE2. J Microbiol Biotechnol 24:969–978
Jeong SJ, Heo K, Park JY, Lee KW, Park JY, Joo SH, Kim JH (2015) Characterization of AprE176, a fibrinolytic enzyme from *Bacillus subtilis* HK176. J Microbiol Biotechnol 25:89–97
Jhample SB, Bhagwat PK, Dandge PB (2015) Statistical media optimization for enhanced production of fibrinolytic enzyme from newly isolated *Proteus penneri* SP-20. Biocatal Agric Biotechnol 4:370–379
Johnson AJ, Tillett WS (1952) The lysis in rabbits of intravascular blood clots by the streptococcal fibrinolytic system (streptokinase). J Exp Med 95:449–464
Ju X, Cao X, Sun Y, Wang Z, Cao C, Liu J, Jiang J (2012) Purification and characterization of a fibrinolytic enzyme from *Streptomyces* sp. XZNUM 00004. World J Microbiol Biotechnol 28:2479–2486
Julian DG (1961) Treatment of cardiac arrest in acute myocardial ischæmia and infarction. Lancet 278:840–844
Kalbfleisch J, Thadani U, LittleJohn JK, Brown G, Magorien R, Kutcher M, Taylor G, Maddox WT, Campbell WB, Perry J, Spann JF, Vetrovec G, Kent R, Armstrong PW (1992) Evaluation of a prolonged infusion of recombinant tissue-type plasminogen activator (Duteplase) in preventing reclusion following successful thrombolysis in acute myocardial infarction. Am J Cardiol 69:1120–1127
Kasai S, Arimura H, Nishida M, Suyama T (1985) Primary structure of single-chain pro-urokinase. J Biol Chem 260:12382–12389
Kaur J, Zhao Z, Klein GM, Lo EH, Buchan AM (2004) The neurotoxicity of tissue plasminogen activator. J Cereb Blood Flow Metab 24:945–963
Kawai C, Yui Y, Hosoda S, Nobuyoshi M, Suzuki S, Sato H, Takatsu F, Motomiya T, Kanmatsuse K, Kodama K, Yabe Y, Minamino T, Kimata S, Nakashima M (1997) A prospective, randomized, double-blind multicenter trial of a single bolus injection of the novel modified t-PA E6010 in the treatment of acute myocardial infarction: comparison with native t-PA. E6010 Study Group. J Am Coll Cardiol 29:1447–1453
Kawasaki T, Katoh M, Kaku S, Gushima H, Takenaka T, Yui Y, Kawai C (1993) Thrombolytic activity of a novel modified tissue-type plasminogen activator, YM866, in a canine model of coronary artery thrombosis. Jpn J Pharmacol 63:9–16
Kim JH, Kim YS (2001) Characterization of a metalloenzyme from a wild mushroom, *Tricholoma saponaceum*. Biosci Biotechnol Biochem 65:356–362
Kim JS, Kim JE, Choi BS, Park SE, Sapkota K, Kim S, Lee HH, Kim CS, Park Y, Kim MK, Kim YS, Kim SJ (2008) Purification and characterization of fibrinolytic metalloprotease from *Perenniporia fraxinea* mycelia. Mycol Res 112:990–998
Kim HC, Choi BS, Sapkota K, Kim S, Lee HJ, Yoo JC, Kim SJ (2011) Purification and characterization of a novel, highly potent fibrinolytic enzyme from *Paecilomyces tenuipes*. Process Biochem 46:1545–1553
Kline DL (1953) The purification and crystallization of plasminogen (profibrinolysin). J Biol Chem 204:949–955
Ko JH, Yan JP, Zhu L, Qi YP (2004) Identification of two novel fibrinolytic enzymes from *Bacillus subtilis* QK02. Comp Biochem Physiol C Toxicol Pharmacol 137:65–74
Kohnert U, Rudolph R, Verheijen JH, Weening-Verhoeff EJ, Stern A, Opitz U, Martin U, Lill H, Prinz H, Lechner M (1992) Biochemical properties of the kringle 2 and protease domains are maintained in the refolded t-PA deletion variant BM 06.022. Protein Eng 5:93–100
Kolev K, Komorowicz E, Owen WG, Machovich R (1996) Quantitative comparison of fibrin degradation with plasmin, miniplasmin, neurophil leukocyte elastase and cathepsin G. Thromb Haemost 75:140–146
Komorowicz E, Kolev K, Lerant I, Machovich R (1998) Flow rate-modulated dissolution of fibrin with clot-embedded and circulating proteases. Circ Res 82:1102–1108
Koren G, Weiss AT, Hasin Y, Appelbaum D, Welber S, Rozenman Y, Lotan C, Mosseri M, Sapoznikov D, Luria MH (1985) Prevention of myocardial damage in acute myocardial ischemia by early treatment with intravenous streptokinase. N Engl J Med 313:1384–1389
Kotb E (2013) Activity assessment of microbial fibrinolytic enzymes. Appl Microbiol Biotechnol 97:6647–6665

Kotb E (2014) The biotechnological potential of fibrinolytic enzymes in the dissolution of endogenous blood thrombi. Biotechnol Prog 30:656–672

Kovalenko TA, Panteleev MA, Sveshnikova AN (2017) The mechanisms and kinetics of initiation of blood coagulation by the extrinsic tenase complex. Biophysics 62:291–300

Kratzschmar J, Haendler B, Langer G, Boidol W, Bringmann P, Alagon A, Donner P, Schleuning WD (1991) The plasminogen activator family from the salivary gland of the vampire bat *Desmodus rotundus*: cloning and expression. Gene 105:229–237

Krehl L (1901) Diseases of the heart muscle and nervous heart disease. Alfred Holder, Vienna

Kumar SS, Sabu A (2017) Therapeutic enzymes. In: Sugathan S, Pradee NS, Sabu A (eds) Bioresources and bioprocess in biotechnology: volume 2: exploring potential biomolecules. Springer Singapore, Singapore, pp 45–73

Kurosawa Y, Nirengi S, Homma T, Esaki K, Ohta M, Clark JF, Hamaoka T (2015) A single-dose of oral nattokinase potentiates thrombolysis and anti-coagulation profiles. Sci Rep 5:11601

Lakshmi BP, Prakasham RS (2013) A fibrinolytic, alkaline and thermostable metalloprotease from the newly isolated *Serratia* sp RSPB11. Int J Biol Macromol 61:479–486

Lane DA, Philippou H, Huntington JA (2005) Directing thrombin. Blood 106:2605–2612

Lange K, Boyd LJ, Loewe L (1945) The functional pathology of frostbite and the prevention of gangrene in experimental animals and humans. Science 102:151–152

Laroche Y, Heymans S, Capaert S, De Cock F, Demarsin E, Collen D (2000) Recombinant staphylokinase variants with reduced antigenicity due to elimination of B-lymphocyte epitopes. Blood 96:1425–1432

Larsen GR, Metzger M, Henson K, Blue Y, Horgan P (1989) Pharmacokinetic and distribution analysis of variant forms of tissue-type plasminogen activator with prolonged clearance in rat. Blood 73:1842–1850

Larsen GR, Timony GA, Horgan PG, Barone KM, Henson KS, Angus LB, Stoudemire JB (1991) Protein engineering of novel plasminogen activators with increased thrombolytic potency in rabbits relative to activase. J Biol Chem 266:8156–8161

Lee CJ, Ansell JE (2011) Direct thrombin inhibitors. Br J Clin Pharmacol 72:581–592

Lee SY, Kim JS, Kim JE, Sapkota K, Shen MH, Kim S, Chun HS, Yoo JC, Choi HS, Kim MK, Kim SJ (2005) Purification and characterization of fibrinolytic enzyme from cultured mycelia of *Armillaria mellea*. Protein Expr Purif 43:10–17

Lee JS, Hyung-Suk B, Sang-Shin P (2006) Purification and characterization of two novel fibrinolytic proteases from mushroom, *Fomitella fraxinea*. J Microbiol Biotechnol 16:264–271

Lee CK, Shin JS, Kim BS, Cho IH, Kim YS, Lee EB (2007) Antithrombotic effects by oral administration of novel proteinase fraction from earthworm *Eisenia andrei* on venous thrombosis model in rats. Arch Pharm Res 30:475–480

Leone G, Rossi E, Leone AM, De Stefano V (2004) Novel antithrombotic agents: indirect synthetic inhibitors of factor Xa and direct thrombin inhibitors. Evidences from clinical studies. Curr Med Chem Cardiovasc Hematol Agents 2:311–326

Lewis BE, Wallis DE, Berkowitz SD, Matthai WH, Fareed J, Walenga JM, Bartholomew J, Sham R, Lerner RG, Zeigler ZR, Rustagi PK, Jang IK, Rifkin SD, Moran J, Hursting MJ, Kelton JG (2001) Argatroban anticoagulant therapy in patients with heparin-induced thrombocytopenia. Circulation 103:1838–1843

Lewis BE, Wallis DE, Leya F, Hursting MJ, Kelton JG, Argatroban-915 Investigators (2003) Argatroban anticoagulation in patients with heparin-induced thrombocytopenia. Arch Intern Med 163:1849–1856

Li H, Hu Z, Yuan J, Fan H, Chen W, Wang S, Zheng S, Zheng Z, Zou G (2007) A novel extracellular protease with fibrinolytic activity from the culture supernatant of *Cordyceps sinensis*: purification and characterization. Phytother Res 21:1234–1241

Li X, Ling L, Li C, Ma Q (2017) Efficacy and safety of desmoteplase in acute ischemic stroke patients: a systematic review and meta-analysis. Medicine 96:e6667

Lijnen HR, Van Hoef B, De Cock F, Okada K, Ueshima S, Matsuo O, Collen D (1991) On the mechanism of fibrin-specific plasminogen activation by staphylokinase. J Biol Chem 266:11826–11832

Louis D, Bernillon J, Wallach JM (1998) Specificity of *Pseudomonas aeruginosa* serralysin revisited, using biologically active peptides as substrates. Biochim Biophys Acta Protein Struct Mol Enzymol 1387:378–386

Lu CL, Chen S, Chen SN (2010a) Purification and characterization of a novel fibrinolytic protease from *Schizophyllum commune*. J Food Drug Anal 18:69–76

Lu F, Lu Z, Bie X, Yao Z, Wang Y, Lu Y, Guo Y (2010b) Purification and characterization of a novel anticoagulant and fibrinolytic enzyme produced by endophytic bacterium *Paenibacillus polymyxa* EJS-3. Thromb Res 126:349–355

Lubenow N, Eichler P, Lietz T, Farner B, Greinacher A (2004) Lepirudin for prophylaxis of thrombosis in patients with acute isolated heparin-induced thrombocytopenia: an analysis of 3 prospective studies. Blood 104:3072–3077

Lucas MA, Fretto LJ, McKee PA (1983) The binding of human plasminogen to fibrin and fibrinogen. J Biol Chem 258:4249–4256

Ma Z, Lu W, Wu S, Chen J, Sun Z, Liu JN (2007) Expression and characterization of recombinant human micro-plasminogen. Biotechnol Lett 29:517–523

Macfarlane RG (1964) An enzyme cascade in the blood clotting mechanism, and its function as a biochemical amplifier. Nature 202:498–499

Macfarlane RG, Pilling J (1947) Fibrinolytic activity of normal urine. Nature 159:779

Madison EL, Coombs GS, Corey DR (1995) Substrate specificity of tissue type plasminogen activator char-

acterization of the fibrin independent specificity of t-PA for plasminogen. J Biol Chem 270:7558–7562
Mahajan PM, Nayak S, Lele SS (2012) Fibrinolytic enzyme from newly isolated marine bacterium *Bacillus subtilis* ICTF-1: media optimization, purification and characterization. J Biosci Bioeng 113:307–314
Malcolm A, Keltai M, Walsh M (1996) ESPRIT: a european study of the prevention of reocclusion after initial thrombolysis with duteplase in acute myocardial infarction. Eur Heart J 17:1522–1531
Malke H, Ferretti JJ (1984) Streptokinase: cloning, expression, and excretion by *Escherichia coli*. Proc Natl Acad Sci U S A 81:3557–3561
Mander P, Cho SS, Simkhada JR, Choi YH, Yoo JC (2011) A low molecular weight chymotrypsin-like novel fibrinolytic enzyme from *Streptomyces* sp. CS624. Process Biochem 46:1449–1455
Marcum JA (2000) The origin of the dispute over the discovery of heparin. J Hist Med Allied Sci 55:37–66
Marder VJ (1979) The use of thrombolytic agents: choice of patient, drug administration, laboratory monitoring. Ann Intern Med 90:802–808
Marder VJ (2011) Historical perspective and future direction of thrombolysis research: the re-discovery of plasmin. J Thromb Haemost 1:364–373
Marder VJ, Novokhatny V (2010) Direct fibrinolytic agents: biochemical attributes, preclinical foundation and clinical potential. J Thromb Haemost 8:433–444
Marder VJ, Sherry S (1988) Thrombolytic therapy: current status (1). N Engl J Med 318:1512–1520
Marder VJ, Landskroner K, Novokhatny V, Zimmerman TP, Kong M, Kanouse JJ, Jesmok G (2001) Plasmin induces local thrombolysis without causing hemorrhage: a comparison with tissue plasminogen activator in the rabbit. Thromb Haemost 86:739–745
Markland FS (1998) Snake venoms and the hemostatic system. Toxicon 36:1749–1800
Martin U, von Mollendorff E, Akpan W, Kientsch-Engel R, Kaufmann B, Neugebauer G (1991a) Dose-ranging study of the novel recombinant plasminogen activator BM 06.022 in healthy volunteers. Clin Pharmacol Ther 50:429–436
Martin U, von Mollendorff E, Akpan W, Kientsch-Engel R, Kaufmann B, Neugebauer G (1991b) Pharmacokinetic and hemostatic properties of the recombinant plasminogen activator BM 06.022 in healthy volunteers. Thromb Haemost 66:569–574
Mason EC (1924) A note on the use of heparin in blood transfusion. J Lab Clin Med 10:203–206
Massaoud MK, Marokházi J, Venekei I (2011) Enzymatic characterization of a serralysin-like metalloprotease from the entomopathogen bacterium, *Xenorhabdus*. Biochim Biophys Acta, Proteins Proteomics 1814:1333–1339
Matsubara K, Sumi H, Hori K, Miyazawa K (1998) Purification and characterization of two fibrinolytic enzymes from a marine green alga, *Codium intricatum*. Comp Biochem Physiol B Biochem Mol Biol 119:177–181
Matsubara K, Hori K, Matsuura Y, Miyazawa K (1999) A fibrinolytic enzyme from a marine green alga, *Codium latum*. Phytochemistry 52:993–999
Matsubara K, Hori K, Matsuura Y, Miyazawa K (2000) Purification and characterization of a fibrinolytic enzyme and identification of fibrinogen clotting enzyme in a marine green alga, *Codium divaricatum*. Comp Biochem Physiol B Biochem Mol Biol 125:137–143
McLean J, Johnson AB (1946) Gangrene following fracture treated with heparin; papaverine, and intermittent venous occlusion; report of a case; reasons for using heparin. Surgery 20:324–336
Mellott MJ, Stabilito II, Holahan MA, Cuca GC, Wang S, Li P, Barrett JS, Lynch JJ, Gardell SJ (1992) Vampire bat salivary plasminogen activator promotes rapid and sustained reperfusion without concomitant systemic plasminogen activation in a canine model of arterial thrombosis. Arterioscler Thromb 12:212–221
Mihara H, Sumi H, Yoneta T, Mizumoto H, Ikeda R, Seiki M, Maruyama M (1991) A novel fibrinolytic enzyme extracted from the earthworm, *Lumbricus rubellus*. Jpn J Physiol 41:461–472
Miyata K, Maejima K, Tomoda K, Isono M (1970) Serratia protease. Part I. Purification and general properties of the enzyme. Agric Biol Chem 34:310–318
Miyata K, Tomoda K, Isono M (1971) Serratia protease. Part III. Characteristics of the enzyme as metalloenzyme. Agric Biol Chem 35:460–467
Moon SM, Kim JS, Kim HJ, Choi MS, Park BR, Kim SG, Ahn H, Chun HS, Shin YK, Kim JJ, Kim DK, Lee SY, Seo YW, Kim YH, Kim CS (2014) Purification and characterization of a novel fibrinolytic α chymotrypsin like serine metalloprotease from the edible mushroom, *Lyophyllum shimeji*. J Biosci Bioeng 117:544–550
Moriya N, Nakata M, Nakamura M, Takaoka M, Iwasa S, Kato K, Kakinuma A (1994) Intestinal absorption of serrapeptase (TSP) in rats. Biotechnol Appl Biochem 20:101–108
Moroz LA (1981) Mini-plasminogen: a mechanism for leukocyte modulation of plasminogen activation by urokinase. Blood 58:97–104
Moussa M, Ibrahim M, El Ghazaly M, Rohde J, Gnoth S, Anton A, Kensy F, Mueller F (2012) Expression of recombinant staphylokinase in the methylotrophic yeast *Hansenula polymorpha*. BMC Biotechnol 12:96
Mukherjee AK, Rai SK, Thakur R, Chattopadhyay P, Kar SK (2012) Bafibrinase: a non-toxic, non-hemorrhagic, direct-acting fibrinolytic serine protease from *Bacillus* sp. strain AS-S20-I exhibits in vivo anticoagulant activity and thrombolytic potency. Biochimie 94:1300–1308
Muller J, Reinert H, Malke H (1989) Streptokinase mutations relieving *Escherichia coli* K-12 (prlA4) of detriments caused by the wild-type skc gene. J Bacteriol 171:2202–2208
Muller D, Domon B, Karas M, van Oostrum J, Richter WJ (1994) Characterization and direct glycoform profiling of a hybrid plasminogen activator by matrix-assisted laser desorption and electrospray mass spectrometry:

correlation with high-performance liquid chromatographic and nuclear magnetic resonance analyses of the released glycans. Biol Mass Spectrom 23:330–338
Munger MA, Forrence EA (1990) Anistreplase: a new thrombolytic for the treatment of acute myocardial infarction. Clin Pharm 9:530–540
Murray DWG, Jaques LB, Perrett TS, Best CH (1937) Heparin and the thrombosis of veins following injury. Surgery 2:163–187
Nagai N, Demarsin E, Van Hoef B, Wouters S, Cingolani D, Laroche Y, Collen D (2003) Recombinant human microplasmin: production and potential therapeutic properties. J Thromb Haemost 1:307–313
Nakahama K, Yoshimura K, Marumoto R, Kikuchi M, Lee IS, Hase T, Matsubara H (1986) Cloning and sequencing of serratia protease gene. Nucleic Acids Res 14:5843–5855
Nakamura T, Yamagata Y, Ichishima E (1992) Nucleotide sequence of the subtilisin NAT gene, aprN, of *Bacillus subtilis* (natto). Biosci Biotechnol Biochem 56:1869–1871
Niego B, Horvath A, Coughlin PB, Pugsley MK, Medcalf RL (2008) Desmoteplase-mediated plasminogen activation and clot lysis are inhibited by the lysine analogue tranexamic acid. Blood Coagul Fibrinolysis 19:322–324
Nordt TK, Moser M, Kohler B, Kubler W, Bode C (1999) Pharmacokinetics and pharmacodynamics of lanoteplase (n-PA). Thromb Haemost 82:121–123
Omura K, Hitosugi M, Kaketani K, Zhu X, Maeda H, Tokudome S (2004) Fibrinolytic and anti-thrombotic effect of NKCP, the protein layer from *Bacillus subtilis* (natto). Biofactors 22:185–187
Oprea AD, Popescu WM (2013) Perioperative management of antiplatelet therapy. Br J Anaesth 111:i3–i17
Ouriel K (2002) Current status of thrombolysis for peripheral arterial occlusive disease. Ann Vasc Surg 16:797–804
Pan R, Zhang ZJ, He RQ (2010) Earthworm protease. Appl Environ Soil Sci 2010:294258
Park SE, Li MH, Kim JS, Sapkota K, Kim JE, Choi BS, Yoon YH, Lee JC, Lee HH, Kim CS, Kim SJ (2007) Purification and characterization of a fibrinolytic protease from a culture supernatant of *Flammulina velutipes* mycelia. Biosci Biotechnol Biochem 71:2214–2222
Park KJ, Kang JI, Kim TS, Yeo IH (2012) The antithrombotic and fibrinolytic effect of natto in hypercholesterolemia rats. Prev Nutr Food Sci 17:78–82
Parkinson J, Bedford DE (1928) Cardiac infarction and coronary thrombosis. Lancet 211:4–11
Patel GK, Kawale AA, Sharma AK (2012) Purification and physicochemical characterization of a serine protease with fibrinolytic activity from latex of a medicinal herb *Euphorbia hirta*. Plant Physiol Biochem 52:104–111
Patthy L, Trexler M, Vali Z, Banyai L, Varadi A (1984) Kringles: modules specialized for protein binding. Homology of the gelatin-binding region of fibronectin with the kringle structures of proteases. FEBS Lett 171:131–136
Pearson JD (1994) Endothelial cell function and thrombosis. Baillieres Clin Haematol 7:441–452
Peng Y, Huang Q, Zhang R, Zhang Y (2003) Purification and characterization of a fibrinolytic enzyme produced by *Bacillus amyloliquefaciens* DC-4 screened from douchi, a traditional Chinese soybean food. Comp Biochem Physiol B Biochem Mol Biol 134:45–52
Peng Y, Yang X, Zhang Y (2005) Microbial fibrinolytic enzymes: an overview of source, production, properties, and thrombolytic activity in vivo. Appl Microbiol Biotechnol 69:126–132
Pennica D, Holmes WE, Kohr WJ, Harkins RN, Vehar GA, Ward CA, Bennett WF, Yelverton E, Seeburg PH, Heyneker HL, Goeddel DV, Collen D (1983) Cloning and expression of human tissue-type plasminogen activator cDNA in *E. coli*. Nature 301:214–221
PERSIST investigators (2004) A novel long-acting synthetic factor Xa inhibitor (SanOrg34006) to replace warfarin for secondary prevention in deep vein thrombosis. A phase II evaluation. J Thromb Haemost 2:47–53
Petersen LC, Lund LR, Nielsen LS, Dano K, Skriver L (1988) One-chain urokinase-type plasminogen activator from human sarcoma cells is a proenzyme with little or no intrinsic activity. J Biol Chem 263:11189–11195
Pirmohamed M (2006) Warfarin: almost 60 years old and still causing problems. Br J Clin Pharmacol 62:509–511
Ploug J, Kjeldgaard NO (1957) Urokinase an activator of plasminogen from human urine I. Isolation and properties. Biochim Biophys Acta 24:278–282
Prasad S, Kashyap RS, Deopujari JY, Purohit HJ, Taori GM, Daginawala HF (2007) Effect of *Fagonia arabica* (Dhamasa) on in vitro thrombolysis. BMC Complement Altern Med 7:36
Rahman MA, Sultana R, Bin Emran T, Islam MS, Rahman MA, Chakma JS, Rashid H, Hasan CMM (2013) Effects of organic extracts of six Bangladeshi plants on in vitro thrombolysis and cytotoxicity. BMC Complement Altern Med 13:25
Rajesh R, Shivaprasad HV, Gowda CD, Nataraju A, Dhananjaya BL, Vishwanath BS (2007) Comparative study on plant latex proteases and their involvement in hemostasis: a special emphasis on clot inducing and dissolving properties. Planta Med 73:1061–1067
Ranby M, Bergsdorf N, Pohl G, Wallen P (1982) Isolation of two variants of native one-chain tissue plasminogen activator. FEBS Lett 146:289–292
Randolph A, Chamberlain SH, Chu HL, Retzios AD, Markland FS, Masiarz FR (1992) Amino acid sequence of fibrolase, a direct-acting fibrinolytic enzyme from *Agkistrodon contortrix contortrix* venom. Protein Sci 1:590–600
Reddrop C, Moldrich RX, Beart PM, Farso M, Liberatore GT, Howells DW, Petersen KU, Schleuning WD, Medcalf RL (2005) Vampire bat salivary plasminogen activator (desmoteplase) inhibits tissue-type plasminogen activator-induced potentiation of excitotoxic injury. Stroke 36:1241–1246
Rijken DC, Groeneveld E (1986) Isolation and functional characterization of the heavy and light chains of

human tissue-type plasminogen activator. J Biol Chem 261:3098–3102
Rijken DC, Wijngaards G, Zaal-de Jong M, Welbergen J (1979) Purification and partial characterization of plasminogen activator from human uterine tissue. Biochim Biophys Acta 580:140–153
Rijken DC, Wijngaards G, Welbergen J (1980) Relationship between tissue plasminogen activator and the activators in blood and vascular wall. Thromb Res 18:815–830
Rijken DC, Barrett-Bergshoeff MM, Jie AFH, Criscuoli M, Sakharov DV (2004) Clot penetration and fibrin binding of amediplase, a chimeric plasminogen activator (K2 tu-PA). Thromb Haemost 91:52–60
Robbins KC, Summaria L, Hsieh B, Shah RJ (1967) The peptide chains of human plasmin. Mechanism of activation of human plasminogen to plasmin. J Biol Chem 242:2333–2342
Roch P (1979) Protein analysis of earthworm coelomic fluid: 1. polymorphic system of the natural hemolysin of *Eisenia fetida andrei*. Dev Comp Immunol 3:599–608
Rosen G, Naor R, Kutner S, Sela MN (1994) Characterization of fibrinolytic activities of *Treponema denticola*. Infect Immun 62:1749–1754
Ruegsegger P, Nydick I, Hutter RC, Freiman AH, Bang NU, Cliffton EE, Ladue JS (1959) Fibrinolytic (plasmin) therapy of experimental coronary thrombi with alteration of the evolution of myocardial infarction. Circulation 19:7–13
Sadeghi S, Marder VJ, Stewart D, Kong M, Humphries J, Baumbach GA, Jesmok G (2003) Safety of plasmin in the setting of concomitant aspirin and heparin administration in an animal model of bleeding. J Thromb Haemost 1:2621–2625
Sakharov DV, Lijnen HR, Rijken DC (1996) Interactions between staphylokinase, plasmin (ogen), and fibrin. Staphylokinase discriminates between free plasminogen and plasminogen bound to partially degraded fibrin. J Biol Chem 271:27912–27918
Sako T (1985) Overproduction of staphylokinase in *Escherichia coli* and its characterization. Eur J Biochem 149:557–563
Samama M, Walenga J, Kaiser B, Fareed J (1997) Specific factor Xa inhibitors. In: Verstraete M, Fuster V, Topol E (eds) Cardiovascular thrombosis: thrombocardiology. Lippincott-Raven, Brussels, pp 173–188
Sarkar NK (1960) Mechanism of clot lysis. Nature 185:624–625
Sasaki S, Kawanami R, Motizuki Y, Nakahara Y, Kawamura T, Tanaka A, Watanabe S (2000) Serrapeptase-induced lung injury manifesting as acute eosiniphilic pneumonia. Nihon Kokyuki Gakkai Zasshi 38:540–544
Sherry S (1954) The fibrinolytic activity of streptokinase activated human plasmin. J Clin Invest 33:1054–1063
Sherry S (1981) Personal reflections on the development of thrombolytic therapy and its application to acute coronary thrombosis. Am Heart J 102:1134–1138
Sherry S (1990) Pharmacology of anistreplase. Clin Cardiol 13:3–10
Shirasaka N, Naitou M, Okamura K, Fukuta Y, Terashita T, Kusuda M (2012) Purification and characterization of a fibrinolytic protease from *Aspergillus oryzae* KSK-3. Mycoscience 53:354–364
Shivaprasad HV, Riyaz M, Venkatesh Kumar R, Dharmappa KK, Tarannum S, Siddesha JM, Rajesh R, Vishwanath BS (2009) Cysteine proteases from the *Asclepiadaceae* plants latex exhibited thrombin and plasmin like activities. J Thromb Thrombolysis 28:304–308
Shlansky-Goldberg RD, Matsumoto AH, Baumbach GA, Siegel JB, Raabe RD, Murphy TP, Deng C, Ray Dawkins J, Marder VJ (2008) A first-in-human phase I trial of locally delivered human plasmin for hemodialysis graft occlusion. J Thromb Haemost 6:944–950
Siigur E, Siigur J (1991) Purification and characterization of lebetase, a fibrinolytic enzyme from *Vipera lebetina* (snake) venom. Biochim Biophys Acta 1074:223–229
Sikri N, Bardia A (2007) A history of streptokinase use in acute myocardial infarction. Tex Heart Inst J 34:318–327
Silva GMM, Bezerra RP, Teixeira JA, Silva FO, Correia JM, Porto TS, Lima-Filho JL, Porto ALF (2016) Screening, production and biochemical characterization of a new fibrinolytic enzyme produced by *Streptomyces* sp. (Streptomycetaceae) isolated from Amazonian lichens. Acta Amazon 46:323–332
Simkhada JR, Mander P, Cho SS, Yoo JC (2010) A novel fibrinolytic protease from *Streptomyces* sp. CS684. Process Biochem 45:88–93
Siritapetawee J, Thumanu K, Sojikul P, Thammasirirak S (2012) A novel serine protease with human fibrino(geno)lytic activities from *Artocarpus heterophyllus* latex. Biochim Biophys Acta 1824:907–912
Siritapetawee J, Sojikul P, Klaynongsruang S (2015) Biochemical characterization of a new glycosylated protease from *Euphorbia* cf. *lactea* latex. Plant Physiol. Biochemist 92:30–38
Smalling RW (1996) Molecular biology of plasminogen activators: what are the clinical implications of drug design? The am. J Cardiol 78:2–7
Smith JB, Willis AL (1971) Aspirin selectively inhibits prostaglandin production in human platelets. Nat New Biol 231:235
Smith RA, Dupe RJ, English PD, Green J (1981) Fibrinolysis with acyl-enzymes: a new approach to thrombolytic therapy. Nature 290:505–508
Sottrup-Jensen L, Claeys H, Zajdel M, Petersen T, Magnusson S (1978) The primary structure of human plasminogen: isolation of two lysine binding fragments and one mini plasminogen (MW 38000) by elastase catalyzed specific limited proteolysis. In: Davidson JF, Rowan RM, Samama MM, Desnoyers PC (eds) Progress in chemical fibrinolysis and thrombolysis. Raven Press, New York, pp 191–209
Stage A, Astrup T (1952) Isolation of a soluble fibrinolytic activator from animal tissue. Nature 170:929
Stahmann MA, Huebner CF, Link KP (1941) Studies on the hemorrhagic sweet clover disease: v. Identification and synthesis of the hemorrhagic agent. J Biol Chem 138:513–527

Stassen JM, Arnout J, Deckmyn H (2004) The hemostatic system. Curr Med Chem 11:2245–2260

Sturzebecher J, Neumann U, Kohnert U, Kresse GB, Fischer S (1992) Mapping of the catalytic site of CHO-t-PA and the t-PA variant BM 06.022 by synthetic inhibitors and substrates. Protein Sci 1:1007–1013

Sumi H, Hamada H, Tsushima H, Mihara H, Muraki H (1987) A novel fibrinolytic enzyme (nattokinase) in the vegetable cheese Natto; a typical and popular soybean food in the Japanese diet. Experientia 43:1110–1111

Sumi H, Hamada H, Nakanishi K, Hiratani H (1990) Enhancement of the fibrinolytic activity in plasma by oral administration of nattokinase. Acta Haematol 84:139–143

Sun AL, Hua ZC, Yao J, Yang YH, Yin DQ (1998) Fusion expression of human pro-urokinase with *E. coli* thioredoxin. Biochem Mol Biol Int 46:479–486

Sung YY, Yang WK, Kim HK (2012) Antiplatelet, anticoagulant and fibrinolytic effects of *Litchi chinensis* Sonn. extract. Mol Med Rep 5:721–724

Suzuki S, Saito M, Suzuki N, Kato H, Nagaoka N, Yoshitake S, Mizuo H, Yuzuriha T, Yui Y, Kawai C (1991) Thrombolytic properties of a novel modified human tissue-type plasminogen activator (E6010): a bolus injection of E6010 has equivalent potency of lysing young and aged canine coronary thrombi. J Cardiovasc Pharmacol 17:738–746

Suzuki Y, Nagai N, Collen D (2004) Comparative effects of microplasmin and tissue-type plasminogen activator (tPA) on cerebral hemorrhage in a middle cerebral artery occlusion model in mice. J Thromb Haemost 2:1617–1621

Szemraj J, Walkowiak B, Kawecka I, Janiszewska G, Buczko W, Bartkowiak J, Chabielska E (2005) A new recombinant thrombolytic and antithrombotic agent with higher fibrin affinity-a staphylokinase variant. I. In vitro study. J Thromb Haemost 3:2156–2165

Szemraj J, Zakrzeska A, Brown G, Stankiewicz A, Gromotowicz A, Grędzinski T, Chabielska E (2011) New derivative of staphylokinase SAK-RGD-K2-Hirul exerts thrombolytic effects in the arterial thrombosis model in rats. Pharmacol Rep 63:1169–1179

Taneja K, Bajaj BK, Kumar S, Dilbaghi N (2017) Production, purification and characterization of fibrinolytic enzyme from *Serratia* sp. KG-2-1 using optimized media. 3 Biotech 7:184

Tao S, Peng L, Beihui L, Deming L, Zuohu L (1997) Solid state fermentation of rice chaff for fibrinolytic enzyme production by *Fusarium oxysporum*. Biotechnol Lett 19:465–467

Tao S, Peng L, Beihui L, Deming L, Zuohu L (1998) Successive cultivation of *Fusarium oxysporum* on rice chaff for economic production of fibrinolytic enzyme. Bioprocess Eng 18:379–381

Tebbe U, Michels R, Adgey J, Boland J, Caspi A, Charbonnier B, Windeler J, Barth H, Groves R, Hopkins GR, Fennell W, Betriu A, Ruda M, Mlczoch J (1998) Randomized, double-blind study comparing saruplase with streptokinase therapy in acute myocardial infarction: the COMPASS equivalence trial. Comparison Trial of Saruplase and Streptokinase (COMASS) Investigators. J Am Coll Cardiol 31:487–493

Tharwat N (2006) Purification and biochemical characterization of fibrinolytic enzyme produced by thermophilic fungus *Oidiodendron flavum*. Biotechnology 5:160–165

Thijs VNS, Peeters A, Vosko M, Aichner F, Schellinger PD, Schneider D, Neumann-Haefelin T, Rother J, Davalos A, Wahlgren N, Verhamme P (2009) Randomized, placebo-controlled, dose-ranging clinical trial of intravenous microplasmin in patients with acute ischemic stroke. Stroke 40:3789–3795

Tillett WS, Garner RL (1933) The fibrinolytic activity of hemolytic *Streptococci*. J Exp Med 58:485–502

Tillett WS, Sherry S (1949) The effect in patients of streptococcal fibrinolysin (streptokinase) and streptococcal desoxyribonuclease on fibrinous, purulent, and sanguinous pleural exudations. J Clin Invest 28:173–190

Tjandrawinata RR, Yunaidi DA, Susanto LW (2016) The safety and tolerability of lumbrokinase DLBS1033 in healthy adult subjects. Drug Res (Stuttg) 66:293–299

Tough J (2005) Thrombolytic therapy in acute myocardial infarction. Nurs Stand 19:55–64

Turpie AGG, Fisher WD, Bauer KA, Kwong LM, Irwin MW, Kalebo P, Misselwitz F, Gent M, OdiXa-Knee Study Group (2005) BAY 59-7939: an oral, direct factor Xa inhibitor for the prevention of venous thromboembolism in patients after total knee replacement. A phase II dose-ranging study. J Thromb Haemost 3:2479–2486

Ueda M, Kubo T, Miyatake K, Nakamura T (2007) Purification and characterization of fibrinolytic alkaline protease from *Fusarium sp*. BLB Appl Microbiol Biotechnol 74:331

Uesugi Y, Usuki H, Iwabuchi M, Hatanaka T (2011) Highly potent fibrinolytic serine protease from *Streptomyces*. Enzym Microb Technol 48:7–12

Urano T, Ihara H, Umemura K, Suzuki Y, Oike M, Akita S, Tsukamoto Y, Suzuki I, Takada A (2001) The profibrinolytic enzyme subtilisin NAT purified from *Bacillus subtilis* cleaves and inactivates plasminogen activator inhibitor type 1. J Biol Chem 276:24690–24696

Van De Werf F, Adgey J, Ardissino D, Armstrong PW, Aylward P, Barbash G, Betriu A, Binbrek AS, Califf R, Diaz R, Fanebust R, Fox K, Granger C, Heikkila J, Husted S, Jansky P, Langer A, Lupi E, Maseri A, Meyer J, Mlczoch J, Mocceti D, Myburgh D, Oto A, Paolasso E, Pehrsson K, Seabra-Gomes R, Soares-Piegas L, Sùgrue D, Tendera M, Topol E, Toutouzas P, Vahanian A, Verheugt F, Wallentin L, White H (1999a) Single-bolus tenecteplase compared with front-loaded alteplase in acute myocardial infarction: the ASSENT-2 double-blind randomised trial. Lancet 354:716–722

Van de Werf F, Cannon CP, Luyten A, Houbracken K, McCabe CH, Berioli S, Bluhmki E, Sarelin H, Wang-Clow F, Fox NL, Braunwald E (1999b) Safety assessment of single-bolus administration of TNK tissue-plasminogen activator in acute myocardial infarction: the ASSENT-1 trial. The ASSENT-1 Investigators. Am Heart J 137:786–791

van Zonneveld AJ, Veerman H, MacDonald ME, van Mourik JA, Pannekoek H (1986) Structure and function of human tissue-type plasminogen activator (t-PA). J Cell Biochem 32:169–178

Vanderschueren S, Barrios L, Kerdsinchai P, Van den Heuvel P, Hermans L, Vrolix M, De Man F, Benit E, Muyldermans L, Collen D (1995a) A randomized trial of recombinant staphylokinase versus alteplase for coronary artery patency in acute myocardial infarction. The STAR Trial Group. Circulation 92:2044–2049

Vanderschueren S, Stockx L, Wilms G, Lacroix H, Verhaeghe R, Vermylen J, Collen D (1995b) Thrombolytic therapy of peripheral arterial occlusion with recombinant staphylokinase. Circulation 92:2050–2057

Vanderschueren S, Collen D, van de Werf F (1996) A pilot study on bolus administration of recombinant staphylokinase for coronary artery thrombolysis. Thromb Haemost 76:541–544

Vanderschueren S, Dens J, Kerdsinchai P, Desmet W, Vrolix M, De Man F, Van den Heuvel P, Hermans L, Collen D, Van de Werf F (1997) Randomized coronary patency trial of double-bolus recombinant staphylokinase versus front-loaded alteplase in acute myocardial infarction. Am Heart J 134:213–219

Vane JR (1971) Inhibition of prostaglandin synthesis as a mechanism of action for aspirin-like drugs. Nat New Biol 231:232–235

Verstraete M (2000) Third-generation thrombolytic drugs. Am J Med 109:52–58

Viana CA, Oliveira JS, Freitas CDT, Alencar NMN, Carvalho CPS, Nishi BC, Ramos MV (2013) Thrombin and plasmin-like activities in the latices of *Cryptostegia grandiflora* and *Plumeria rubra*. Blood Coagul Fibrinolysis 24:386–392

Vijayaraghavan P, Vincent SGP (2014) Statistical optimization of fibrinolytic enzyme production using agroresidues by *Bacillus cereus* IND1 and its thrombolytic activity in vitro. Biomed Res Int 2014:725064

Vijayaraghavan P, Vincent SGP (2015) A low cost fermentation medium for potential fibrinolytic enzyme production by a newly isolated marine bacterium, *Shewanella* sp. IND20. Biotechnol Rep 7:135–142

Vijayaraghavan P, Raj F, Vincent SGP (2015) Purification and characterization of fibrinolytic enzyme from *Pseudoalteromonas* sp. IND11 and its in vitro activity on blood clot. Int J Biol Chem 9:11–20

Vijayaraghavan P, Vincent SGP, Arasu MV (2016a) Purification, characterization of a novel fibrinolytic enzyme from *Paenibacillus* sp. IND8, and its in vitro thrombolytic activity. SIJBS 2:434–444

Vijayaraghavan P, Vincent SGP, Arasu MV, Al-Dhabi NA (2016b) Bioconversion of agro-industrial wastes for the production of fibrinolytic enzyme from *Bacillus halodurans* IND18: purification and biochemical characterization. Electron J Biotechnol 20:1–8

Violand BN, Castellino FJ (1976) Mechanism of the urokinase-catalyzed activation of human plasminogen. J Biol Chem 251:3906–3912

von Kummer R, Mori E, Truelsen T, Jensen JKS, Gronning BA, Fiebach JB, Lovblad KO, Pedraza S, Romero JM, Chabriat H, Chang KC, Davalos A, Ford GA, Grotta J, Kaste M, Schwamm LH, Shuaib A, Albers GW (2016) Desmoteplase 3 to 9 hours after major artery occlusion stroke: the DIAS-4 trial (efficacy and safety study of desmoteplase to treat acute ischemic stroke). Stroke 47:2880–2887

Walenga JM, Jeske WP, Samama MM, Frapaise FX, Bick RL, Fareed J (2002) Fondaparinux: a synthetic heparin pentasaccharide as a new antithrombotic agent. Expert Opin Investig Drugs 11:397–407

Wang J, Wang M, Wang Y (1999) Purification and characterization of a novel fibrinolytic enzyme from *Streptomyces* spp. Chin J Biotechnol 15:83–89

Wang F, Wang C, Li M, Gui L, Zhang J, Chang W (2003) Purification, characterization and crystallization of a group of earthworm fibrinolytic enzymes from *Eisenia fetida*. Biotechnol Lett 25:1105–1109

Wang C, Du M, Zheng D, Kong F, Zu G, Feng Y (2009a) Purification and characterization of nattokinase from *Bacillus subtilis* natto B-12. J Agric Food Chem 57:9722–9729

Wang SL, Chen HJ, Liang TW, Lin YD (2009b) A novel nattokinase produced by *Pseudomonas* sp. TKU015 using shrimp shells as substrate. Process Biochem 44:70–76

Weiss HJ, Aledort LM (1967) Impaired platelet-connective-tissue reaction in man after aspirin ingestion. Lancet 2:495–497

Weng Y, Yao J, Sparks S, Wang KY (2017) Nattokinase: an oral antithrombotic agent for the prevention of cardiovascular disease. Int J Mol Sci 18:523

WHO (2017) World Health Organization fact sheet fs317 cardiovascular diseases (CVDs) updated May 2017

Wiman B (1977) Primary structure of the B-chain of human plasmin. Eur J Biochem 76:129–137

Wiman B, Wallen P (1977) The specific interaction between plasminogen and fibrin. A physiological role of the lysine binding site in plasminogen. Thromb Res 10:213–222

Wiman B, Boman L, Collen D (1978) On the kinetics of the reaction between human antiplasmin and a low-molecular-weight form of plasmin. Eur J Biochem 87:143–146

Wiman B, Lijnen HR, Collen D (1979) On the specific interaction between the lysine-binding sites in plasmin and complementary sites in alpha2-antiplasmin and in fibrinogen. Biochim Biophys Acta 579:142–154

Wong SL, Ye R, Nathoo S (1994) Engineering and production of streptokinase in a *Bacillus subtilis* expression-secretion system. Appl Environ Microbiol 60:517–523

Wu HL, Shi GY, Wohl RC, Bender ML (1987) Structure and formation of microplasmin. Proc Natl Acad Sci U S A 84:8793–8795

Wu TP, Padmanabhan KP, Tulinsky A (1994) The structure of recombinant plasminogen kringle 1 and the fibrin binding site. Blood Coagul Fibrinolysis 5:157–166

Wu B, Wu L, Chen D, Yang Z, Luo M (2009) Purification and characterization of a novel fibrinolytic protease from *Fusarium* sp. CPCC 480097. J Ind Microbiol Biotechnol 36:451–459

Wu D, Li P, Zhou J, Gao M, Lou X, Ran T, Wu S, Wang W, Xu D (2016) Identification of a toxic serralysin family protease with unique thermostable property from *S. marcescens* FS14. Int J Biol Macromol 93:98–106

Xiao-lan L, Lian-xiang D, Fu-ping L, Xi-qun Z, Jing X (2005) Purification and characterization of a novel fibrinolytic enzyme from *Rhizopus chinensis* 12. Appl Microbiol Biotechnol 67:209–214

Yan XM, Kim CH, Lee CK, Shin JS, Cho IH, Sohn UD (2010) Intestinal absorption of fibrinolytic and proteolytic lumbrokinase extracted from earthworm, *Eisenia andrei*. Korean J Physiol Pharmacol 14:71–75

Yokota M, Tanji Y (2008) Analysis of cell-cycle-dependent production of tissue plasminogen activator analogue (pamiteplase) by CHO cells. Biochem Eng J 39:297–304

Zamarron C, Lijnen HR, Van Hoef B, Collen D (1984) Biological and thrombolytic properties of proenzyme and active forms of human urokinase-I. Fibrinolytic and fibrinogenolytic properties in human plasma in vitro of urokinases obtained from human urine or by recombinant DNA technology. Thromb Haemost 52:19–23

Zeymer U, Neuhaus KL (1999) Clinical trials in acute myocardial infarction. Curr Opin Cardiol 14:392

Index

N. Labrou (ed.), *Therapeutic Enzymes: Function and Clinical Implications*, Advances in Experimental Medicine and Biology 1148, https://doi.org/10.1007/978-981-13-7709-9

O

P

Q

R

S

T

FSC
www.fsc.org
MIX
Papier | Fördert
gute Waldnutzung
FSC® C083411

Kolibri 360 GmbH
Ferdinand-Jühlke-Straße 7
99095 Erfurt, Deutschland
produktsicherheit@kolibri360.de